Handbook of Mathematical Models and Algorithms in Computer Vision and Imaging

Ke Chen · Carola-Bibiane Schönlieb ·
Xue-Cheng Tai · Laurent Younes
Editors

Handbook of Mathematical Models and Algorithms in Computer Vision and Imaging

Mathematical Imaging and Vision

Volume 1

With 553 Figures and 72 Tables

Springer

Editors
Ke Chen
Department of Mathematical Sciences
The University of Liverpool
Liverpool, UK

Xue-Cheng Tai
Hong Kong Center for
Cerebrocardiovascular Health
Engineering (COCHE)
Shatin, Hong Kong, China

Carola-Bibiane Schönlieb
Department of Applied Mathematics and
Theoretical Physics
University of Cambridge
Cambridge, UK

Laurent Younes
Department of Applied Mathematics and Statistics
Johns Hopkins University
Baltimore, MD, USA

ISBN 978-3-030-98660-5   ISBN 978-3-030-98661-2 (eBook)
https://doi.org/10.1007/978-3-030-98661-2

© Springer Nature Switzerland AG 2023

This work is subject to copyright. All rights are reserved by the Publisher, whether the whole or part of the material is concerned, specifically the rights of translation, reprinting, reuse of illustrations, recitation, broadcasting, reproduction on microfilms or in any other physical way, and transmission or information storage and retrieval, electronic adaptation, computer software, or by similar or dissimilar methodology now known or hereafter developed.

The use of general descriptive names, registered names, trademarks, service marks, etc. in this publication does not imply, even in the absence of a specific statement, that such names are exempt from the relevant protective laws and regulations and therefore free for general use.

The publisher, the authors, and the editors are safe to assume that the advice and information in this book are believed to be true and accurate at the date of publication. Neither the publisher nor the authors or the editors give a warranty, expressed or implied, with respect to the material contained herein or for any errors or omissions that may have been made. The publisher remains neutral with regard to jurisdictional claims in published maps and institutional affiliations.

This Springer imprint is published by the registered company Springer Nature Switzerland AG.
The registered company address is: Gewerbestrasse 11, 6330 Cham, Switzerland.

Preface

The rapid development of new imaging hardware, the advance in medical imaging, the advent of multi-sensor data fusion and multimodal imaging, as well as the advances in computer vision have sparked numerous research endeavours leading to highly sophisticated and rigorous mathematical models and theories. Motivated by the increasing use of variational models, shapes and flows, differential geometry, optimisation theory, numerical analysis, statistical/Bayesian graphical models, machine learning, and deep learning, we have invited contributions from leading researchers and publish this handbook to review and capture the state of the art of research in Computer Vision and Imaging.

This constantly improving technology that generates new demands not readily met by existing mathematical concepts and algorithms provides a compelling justification for such a book to meet the ever-growing challenges in applications and to drive future development. As a consequence, new mathematical models have to be found, analysed and realised in practice. Knowing the precise state-of-the-art developments is key, and hence this book will serve the large community of mathematics, imaging, computer vision, computer sciences, statistics, and, in general, imaging and vision research. Our primary audience are

- Graduate students
- Researchers
- Imaging and vision practitioners
- Applied mathematicians
- Medical imagers
- Engineers
- Computer scientists

Viewing discrete images as data sampled from functional surfaces enables the use of advanced tools from calculus, functions and calculus of variations, and optimisation and provides the basis of high-resolution imaging through variational models. No other framework can provide the comparable accuracy and precision to imaging and vision.

Although our initial emphasis is on the variational methods, which represent the optimal solutions to class of imaging and vision problems, and on effective algorithms, which are necessary for the methods to be translated to practical use in various applications, the editors recognise that the range of effective and efficient methods for solving problems from computer vision and imaging go beyond variational methods and have enlarged our coverage to include mathematical models and algorithms. So, the book title reflects this viewpoint and a big vision for the reference book.

All chapters will have introductions so that the book is readily accessible to graduate students. We have divided the 53 chapters of this book into 3 sections, namely

(a) Convex and Non-convex Large-Scale Optimisation in Imaging
(b) Model- and Data-Driven Variational Imaging Approaches
(c) Shape Spaces and Geometric Flows

to facilitate browsing the content list. However, such a division is artificial because, these days, research becomes increasingly intra-disciplinary as well as inter-disciplinary, and ideas from one topic often directly or indirectly inspire or transpire another topic. This is very exciting.

For newcomers to the field, the book provides a comprehensive and fast track introduction to the core research problems, to save time and get on with tackling new and emerging challenges, rather than running the risk of reproducing/comparing to some old works already done or reinventing same results. For researchers, exposure to the state of the art of research works leads to an overall view of the entire field so as to guide new research directions and avoid pitfalls in moving the field forward and looking into the next 25 years of imaging and information sciences.

The dreadful Covid-19 pandemic starting from 2020 has affected lives of everyone, of course including all researchers. We are still not out of the woods. The editors are very much grateful to the book authors who have endured much hardship during the last 3 years and overcome many difficulties to have completed their chapters on time. We are also indebted to many anonymous reviewers who provided valuable reviews and helpful criticism to improve presentations of our chapters.

The original gathering of all editors was in 2017 when the first three editors co-organised the prestigious Isaac Newton Institute programme titled "*Variational methods and effective algorithms for imaging and vision*" (https://www.newton.ac.uk/event/vmv/), partially supported by UK EPSRC GR/EP F005431 and Isaac Newton Institute for Mathematical Sciences. During the programme, Mr Jan Holland from Springer-Nature kindly suggested the idea of a book. We are grateful to his suggestion which sparked the editors' fruitful collaboration in the last few years. The large team of publishers who have offered immense help to us include Michael Hermann (Springer), Allan Cohen (Palgrave) and Salmanul Faris Nedum Palli (Springer). We thank them all.

Finally, we wish all readers a happy reading.

The editorial team:

Liverpool, UK	Ke Chen (Lead)
Cambridge, UK	Carola-Bibiane Schönlieb
Shatin, Hong Kong	Xue-Cheng Tai
Baltimore, USA	Laurent Younes
February 2023	

Contents

Volume 1

Part I Convex and Non-convex Large-Scale Optimization in Imaging 1

1. **Convex Non-convex Variational Models** 3
   Alessandro Lanza, Serena Morigi, Ivan W. Selesnick, and Fiorella Sgallari

2. **Subsampled First-Order Optimization Methods with Applications in Imaging** 61
   Stefania Bellavia, Tommaso Bianconcini, Nataša Krejić, and Benedetta Morini

3. **Bregman Methods for Large-Scale Optimization with Applications in Imaging** 97
   Martin Benning and Erlend Skaldehaug Riis

4. **Fast Iterative Algorithms for Blind Phase Retrieval: A Survey** 139
   Huibin Chang, Li Yang, and Stefano Marchesini

5. **Modular ADMM-Based Strategies for Optimized Compression, Restoration, and Distributed Representations of Visual Data** 175
   Yehuda Dar and Alfred M. Bruckstein

6. **Connecting Hamilton-Jacobi Partial Differential Equations with Maximum a Posteriori and Posterior Mean Estimators for Some Non-convex Priors** 209
   Jérôme Darbon, Gabriel P. Langlois, and Tingwei Meng

7. **Multi-modality Imaging with Structure-Promoting Regularizers** 235
   Matthias J. Ehrhardt

8 Diffraction Tomography, Fourier Reconstruction, and Full Waveform Inversion 273
Florian Faucher, Clemens Kirisits, Michael Quellmalz, Otmar Scherzer, and Eric Setterqvist

9 Models for Multiplicative Noise Removal 313
Xiangchu Feng and Xiaolong Zhu

10 Recent Approaches to Metal Artifact Reduction in X-Ray CT Imaging 347
Soomin Jeon and Chang-Ock Lee

11 Domain Decomposition for Non-smooth (in Particular TV) Minimization 379
Andreas Langer

12 Fast Numerical Methods for Image Segmentation Models 427
Noor Badshah

13 On Variable Splitting and Augmented Lagrangian Method for Total Variation-Related Image Restoration Models 503
Zhifang Liu, Yuping Duan, Chunlin Wu, and Xue-Cheng Tai

14 Sparse Regularized CT Reconstruction: An Optimization Perspective 551
Elena Morotti and Elena Loli Piccolomini

15 Recent Approaches for Image Colorization 585
Fabien Pierre and Jean-François Aujol

16 Numerical Solution for Sparse PDE Constrained Optimization ... 623
Xiaoliang Song and Bo Yu

17 Game Theory and Its Applications in Imaging and Vision 677
Anis Theljani, Abderrahmane Habbal, Moez Kallel, and Ke Chen

18 First-Order Primal–Dual Methods for Nonsmooth Non-convex Optimization 707
Tuomo Valkonen

Volume 2

Part II Model- and Data-Driven Variational Imaging Approaches 749

19 Learned Iterative Reconstruction 751
Jonas Adler

20	An Analysis of Generative Methods for Multiple Image Inpainting	773
	Coloma Ballester, Aurélie Bugeau, Samuel Hurault, Simone Parisotto, and Patricia Vitoria	
21	Analysis of Different Losses for Deep Learning Image Colorization	821
	Coloma Ballester, Hernan Carrillo, Michaël Clément, and Patricia Vitoria	
22	Influence of Color Spaces for Deep Learning Image Colorization	847
	Aurélie Bugeau, Rémi Giraud, and Lara Raad	
23	Variational Model-Based Deep Neural Networks for Image Reconstruction	879
	Yunmei Chen, Xiaojing Ye, and Qingchao Zhang	
24	Bilevel Optimization Methods in Imaging	909
	Juan Carlos De los Reyes and David Villacís	
25	Multi-parameter Approaches in Image Processing	943
	Markus Grasmair and Valeriya Naumova	
26	Generative Adversarial Networks for Robust Cryo-EM Image Denoising..	969
	Hanlin Gu, Yin Xian, Ilona Christy Unarta, and Yuan Yao	
27	Variational Models and Their Combinations with Deep Learning in Medical Image Segmentation: A Survey	1001
	Luying Gui, Jun Ma and Xiaoping Yang	
28	Bidirectional Texture Function Modeling	1023
	Michal Haindl	
29	Regularization of Inverse Problems by Neural Networks	1065
	Markus Haltmeier and Linh Nguyen	
30	Shearlets: From Theory to Deep Learning	1095
	Gitta Kutyniok	
31	Learned Regularizers for Inverse Problems	1133
	Sebastian Lunz	
32	Filter Design for Image Decomposition and Applications to Forensics	1155
	Robin Richter, Duy H. Thai, Carsten Gottschlich, and Stephan F. Huckemann	

| 33 | Deep Learning Methods for Limited Data Problems in X-Ray Tomography 1183
Johannes Schwab | |
|---|---|---|
| 34 | MRI Bias Field Estimation and Tissue Segmentation Using Multiplicative Intrinsic Component Optimization and Its Extensions 1203
Samad Wali, Chunming Li, and Lingyan Zhang | |
| 35 | Data-Informed Regularization for Inverse and Imaging Problems 1235
Jonathan Wittmer and Tan Bui-Thanh | |
| 36 | Randomized Kaczmarz Method for Single Particle X-Ray Image Phase Retrieval 1273
Yin Xian, Haiguang Liu, Xuecheng Tai, and Yang Wang | |
| 37 | A Survey on Deep Learning-Based Diffeomorphic Mapping 1289
Huilin Yang, Junyan Lyu, Roger Tam, and Xiaoying Tang | |

Volume 3

Part III Shape Spaces and Geometric Flows 1323

| 38 | Stochastic Shape Analysis 1325
Alexis Arnaudon, Darryl Holm, and Stefan Sommer | |
|---|---|---|
| 39 | Intrinsic Riemannian Metrics on Spaces of Curves: Theory and Computation 1349
Martin Bauer, Nicolas Charon, Eric Klassen, and Alice Le Brigant | |
| 40 | An Overview of SaT Segmentation Methodology and Its Applications in Image Processing 1385
Xiaohao Cai, Raymond Chan, and Tieyong Zeng | |
| 41 | Recent Development of Medical Shape Analysis via Computational Quasi-conformal Geometry 1413
Hei-Long Chan and Lok-Ming Lui | |
| 42 | A Survey of Topology and Geometry-Constrained Segmentation Methods in Weakly Supervised Settings 1437
Ke Chen, Noémie Debroux, and Carole Le Guyader | |

43	**Recent Developments of Surface Parameterization Methods Using Quasi-conformal Geometry**........................... Gary P. T. Choi and Lok Ming Lui	1483
44	**Recent Geometric Flows in Multi-orientation Image Processing via a Cartan Connection**..................................... R. Duits, B. M. N. Smets, A. J. Wemmenhove, J. W. Portegies, and E. J. Bekkers	1525
45	**PDE-Constrained Shape Optimization: Toward Product Shape Spaces and Stochastic Models**............................... Caroline Geiersbach, Estefania Loayza-Romero, and Kathrin Welker	1585
46	**Iterative Methods for Computing Eigenvectors of Nonlinear Operators**... Guy Gilboa	1631
47	**Optimal Transport for Generative Models**..................... Xianfeng Gu, Na Lei, and Shing-Tung Yau	1659
48	**Image Reconstruction in Dynamic Inverse Problems with Temporal Models**.. Andreas Hauptmann, Ozan Öktem, and Carola Schönlieb	1707
49	**Computational Conformal Geometric Methods for Vision**........ Na Lei, Feng Luo, Shing-Tung Yau, and Xianfeng Gu	1739
50	**From Optimal Transport to Discrepancy**...................... Sebastian Neumayer and Gabriele Steidl	1791
51	**Compensated Convex-Based Transforms for Image Processing and Shape Interrogation**..................................... Antonio Orlando, Elaine Crooks, and Kewei Zhang	1827
52	**The Potts Model with Different Piecewise Constant Representations and Fast Algorithms: A Survey**................ Xuecheng Tai, Lingfeng Li, and Egil Bae	1887
53	**Shape Spaces: From Geometry to Biological Plausibility**......... Nicolas Charon and Laurent Younes	1929

Index 1959

About the Editors

Prof. Ke Chen, PhD received his BSc, MSc and PhD degrees in applied mathematics, respectively, from the Dalian University of Technology (China), the University of Manchester (UK) and the University of Plymouth (UK). He is a computational mathematician specialised in developing novel and fast numerical algorithms for various scientific computing (especially imaging) applications. He has been the director of multidisciplinary research at the Centre for Mathematical Imaging Techniques (CMIT) since 2007, and the director of the EPSRC Liverpool Centre of Mathematics in Healthcare (LCMH) since 2015. He heads a large group of computational imagers, tackling novel analysis of real-life images. His group's imaging work in variational modelling and algorithmic development is mostly interdisciplinary, strongly motivated by emerging real-life problems and their challenges: image restoration, image inpainting, tomography, image segmentation and registration.

Carola-Bibiane Schönlieb is Professor of Applied Mathematics at the University of Cambridge. There, she is head of the Cambridge Image Analysis group and co-director of the EPSRC Cambridge Mathematics of Information in Healthcare Hub. Since 2011, she is a fellow of Jesus College Cambridge and since 2016 a fellow of the Alan Turing Institute, London. She also holds the chair of the Committee for Applications and Interdisciplinary Relations (CAIR) of the EMS. Her current research interests focus on variational methods, partial differential equations and machine learning for image analysis, image processing and inverse imaging problems. She has active interdisciplinary collaborations with clinicians, biologists and physicists on biomedical imaging topics, chemical engineers and plant scientists on image sensing, as well as collaborations with artists and art conservators on digital art restoration.

Her research has been acknowledged by scientific prizes, among them the LMS Whitehead Prize 2016, the Philip Leverhulme Prize in 2017, the Calderon Prize 2019, a Royal Society Wolfson fellowship in 2020 and a doctorate honoris causa from the University of Klagenfurt in 2022, and by invitations to give plenary lectures at several renowned applied mathematics conferences, among them the SIAM conference on Imaging Science in 2014, the SIAM conference on Partial Differential Equations in 2015, the SIAM annual meeting in 2017, the Applied Inverse Problems Conference in 2019, the FOCM 2020 and the GAMM 2021.

Carola graduated from the Institute for Mathematics, University of Salzburg (Austria), in 2004. From 2004 to 2005, she held a teaching position in Salzburg. She received her PhD degree from the University of Cambridge (UK) in 2009. After 1 year of postdoctoral activity at the University of Göttingen (Germany), she became a lecturer at Cambridge in 2010, promoted to reader in 2015 and promoted to professor in 2018.

About the Editors

Prof. Xue-Cheng Tai is a chief research scientist and executive programme director at Hong Kong Center for Cerebro-cardiovascular Health Engineering (COCHE), Hong Kong Science Park. He is a professor and head of the Department of Mathematics at Hong Kong Baptist University (China) since 2017. Before 2017, he served as a professor in the Department of Mathematics at Bergen University (Norway). His research interests include numerical PDEs, optimisation techniques, inverse problems and image processing. He has done significant research work in his research areas and published more than 250 top-quality international conference and journal papers. He is the winner of the 8th Feng Kang Prize for scientific computing. Prof Tai has served as organising and programme committee member for a number of international conferences and has been often invited at international conferences. He has served as referee and reviewer for many premier conferences and journals.

Laurent Younes is a professor in the Department Applied Mathematics and Statistics, Johns Hopkins University (USA), which he joined in 2003, after 10 years as a researcher for the CNRS in France. He is a former student of Ecole Normale Supérieure (Paris) and of the University of Paris 11 from which he received his PhD in 1988. His work includes contributions to applied probability, statistics, graphical models, shape analysis and computational medicine. He is a fellow of the IMS and of the AMS.

Contributors

Jonas Adler Department of Mathematics, KTH – Royal Institute of Technology, Stockholm, Sweden

Alexis Arnaudon Department of Mathematics, Imperial College, London, UK

Blue Brain Project, École polytechnique fédéral de Lausanne (EPFL), Geneva, Switzerland

Jean-François Aujol Univ. Bordeaux, Bordeaux INP, CNRS, IMB, UMR 5251, Talence, France

Noor Badshah Department of Basic Sciences, University of Engineering and Technology, Peshawar, Pakistan

Egil Bae Norwegian Defence Research Establishment (FFI), Kjeller, Norway

Coloma Ballester IPCV, DTIC, University Pompeu Fabra, Barcelona, Spain

Martin Bauer Department of Mathematics, Florida State University, Tallahassee, FL, USA

E. J. Bekkers Amsterdam Machine Learning Lab, University of Amsterdam, Amsterdam, The Netherlands

Stefania Bellavia Dipartimento di Ingegneria Industriale, Università degli Studi di Firenze (INdAM-GNCS members), Firenze, Italia

Martin Benning The School of Mathematical Sciences, Queen Mary University of London, London, UK

Tommaso Bianconcini Verizon Connect, Firenze, Italia

Alfred M. Bruckstein Computer Science Department, Technion – Israel Institute of Technology, Haifa, Israel

Aurélie Bugeau LaBRI, CNRS, Université de Bordeaux, Talence, France

Tan Bui-Thanh Department of Aerospace Engineering and Engineering Mechanics, The Oden Institute for Computational Engineering and Sciences, UT Austin, Austin, TX, USA

Xiaohao Cai School of Electronics and Computer Science, University of Southampton, Southampton, UK

Hernan Carrillo LaBRI, CNRS, Bordeaux INP, Université de Bordeaux, Bordeaux, France

Hei-Long Chan Chinese University of Hong Kong, Hong Kong, China

Raymond Chan Department of Mathematics, College of Science, City University of Hong Kong, Kowloon Tong, Hong Kong, China

Huibin Chang School of Mathematical Sciences, Tianjin Normal University, Tianjin, China

Nicolas Charon Department of Applied Mathematics and Statistics, Johns Hopkins University, Baltimore, MD, USA

Center for Imaging Science, Johns Hopkins University, Baltimore, MD, USA

Ke Chen Department of Mathematical Sciences, Centre for Mathematical Imaging Techniques, University of Liverpool, Liverpool, UK

Yunmei Chen Department of Mathematics, University of Florida, Gainesville, FL, USA

Gary P. T. Choi Department of Mathematics, Massachusetts Institute of Technology, Cambridge, MA, USA

Michaël Clément LaBRI, CNRS, Bordeaux INP, Université de Bordeaux, Bordeaux, France

Elaine Crooks Department of Mathematics, Swansea University, Swansea, UK

Yehuda Dar Electrical and Computer Engineering Department, Rice University, Houston, TX, USA

Jérôme Darbon Division of Applied Mathematics, Brown University, Providence, RI, USA

Noémie Debroux Pascal Institute, University of Clermont Auvergne, Clermont-Ferrand, France

Juan Carlos De los Reyes Research Center for Mathematical Modelling (MODEMAT), Escuela Politécnica Nacional, Quito, Ecuador

Yuping Duan Center for Applied Mathematics, Tianjin University, Tianjin, China

R. Duits Applied Differential Geometry, Department of Mathematics and Computer Science, Eindhoven University of Technology, Eindhoven, The Netherlands

Matthias J. Ehrhardt Institute for Mathematical Innovation, University of Bath, Bath, UK

Contributors

Florian Faucher Faculty of Mathematics, University of Vienna, Vienna, Austria

Xiangchu Feng School of Mathematics and Statistics, Xidian University, Xi'an, China

Caroline Geiersbach Weierstrass Institute, Berlin, Germany

Guy Gilboa Technion – IIT, Haifa, Israel

Rémi Giraud Univ. Bordeaux, CNRS, IMS UMR5251, Bordeaux INP, Talence, France

Carsten Gottschlich Institute for Mathematical Stochastics, University of Göttingen, Göttingen, Germany

Markus Grasmair NTNU, Trondheim, Norway

Hanlin Gu Hong Kong University of Science and Technology, Hong Kong, China

Xianfeng Gu Stony Brook University, Stony Brook, NY, USA

Luying Gui Department of Mathematics, Nanjing University of Science and Technology, Nanjing, China

Abderrahmane Habbal Modeling and Data Science, Mohammed VI Polytechnic University Benguerir, Morocco

Université Côte d'Azur, Inria, Sophia Antipolis, France

Michal Haindl Institute of Information Theory and Automation, Czech Academy of Sciences, Prague, Czechia

Markus Haltmeier Department of Mathematics, University of Innsbruck, Innsbruck, Austria

Andreas Hauptmann Research Unit of Mathematical Sciences, University of Oulu, Oulu, Finland

Darryl Holm Department of Mathematics, Imperial College, London, UK

Stephan F. Huckemann Felix-Bernstein-Institute for Mathematical Statistics in the Biosciences, University of Göttingen, Göttingen, Germany

Samuel Hurault Bordeaux INP, CNRS, IMB, Université de Bordeaux, Talence, France

Soomin Jeon Department of Radiology, Massachusetts General Hospital and Harvard Medical School, Boston, MA, USA

Moez Kallel Laboratory for Mathematical and Numerical Modeling in Engineering Science (LAMSIN), University of Tunis El Manar, National Engineering School of Tunis, Tunis-Belvédère, Tunisia

Clemens Kirisits Faculty of Mathematics, University of Vienna, Vienna, Austria

Eric Klassen Department of Mathematics, Florida State University, Tallahassee, FL, USA

Nataša Krejić Department of Mathematics and Informatics, Faculty of Sciences, University of Novi Sad, Novi Sad, Serbia

Gitta Kutyniok Ludwig-Maximilians-Universität München, Mathematisches Institut, München, Germany

Andreas Langer Centre for Mathematical Sciences, Lund University, Lund, Sweden

Gabriel P. Langlois Division of Applied Mathematics, Brown University, Providence, RI, USA

Alessandro Lanza Department of Mathematics, University of Bologna, Bologna, Italy

Alice Le Brigant Department of Applied Mathematics, University Paris, Paris, France

Carole Le Guyader INSA Rouen Normandie, Laboratory of Mathematics, Normandie University, Rouen, France

Chang-Ock Lee Department of Mathematical Sciences, KAIST, Daejeon, Republic of Korea

Na Lei Dalian University of Technology, Dalian, China

Chunming Li School of Information and Communication Engineering, University of Electronic Science and Technology of China, Chengdu, China

Lingfeng Li Department of Mathematics, Hong Kong Baptist University, Kowloon Tong, Hong Kong, China

Department of Mathematics, Southern University of Science and Technology, Shenzhen, China

Haiguang Liu Microsoft Research-Asian, Beijing, China

Zhifang Liu School of Mathematical Sciences, Tianjin Normal University, Tianjin, China

Estefania Loayza-Romero Institute for Analysis and Numerics, University of Münster, Münster, Germany

Lok Ming Lui Department of Mathematics, The Chinese University of Hong Kong, Hong Kong, China

Sebastian Lunz Department of Applied Mathematics and Theoretical Physics, University of Cambridge, Cambridge, UK

Feng Luo Rutgers University, Piscataway, NJ, USA

Junyan Lyu Department of Electronic and Electrical Engineering, Southern University of Science and Technology, Shenzhen, Guangdong, China

Jun Ma Department of Mathematics, Nanjing University of Science and Technology, Nanjing, China

Stefano Marchesini SLAC National Laboratory, Menlo Park, CA, USA

Tingwei Meng Division of Applied Mathematics, Brown University, Providence, RI, USA

Serena Morigi Department of Mathematics, University of Bologna, Bologna, Italy

Benedetta Morini Dipartimento di Ingegneria Industriale, Università degli Studi di Firenze (INdAM-GNCS members), Firenze, Italia

Elena Morotti Department of Political and Social Sciences, University of Bologna, Bologna, Italy

Valeriya Naumova Machine Intelligence Department, Simula Consulting and SimulaMet, Oslo, Norway

Sebastian Neumayer Institute of Mathematics, TU Berlin, Berlin, Germany

Linh Nguyen Department of Mathematics, University of Idaho, Moscow, ID, USA

Ozan Öktem Department of Information Technology, Division of Scientific Computing, Uppsala University, Uppsala, Sweden

Antonio Orlando CONICET, Departamento de Bioingeniería, Universidad Nacional de Tucumán, Tucumán, Argentina

Simone Parisotto DAMTP, University of Cambridge, Cambridge, UK

Elena Loli Piccolomini Department of Computer Science and Engineering, University of Bologna, Bologna, Italy

Fabien Pierre LORIA, UMR CNRS 7503, Université de Lorraine, INRIA projet Tangram, Nancy, France

J. W. Portegies Center for Analysis, Scientific Computing and Applications, Department of Mathematics and Computer Science, Eindhoven University of Technology, Eindhoven, The Netherlands

Michael Quellmalz Institute of Mathematics, Technical University Berlin, Berlin, Germany

Lara Raad LIGM, CNRS, Univ Gustave Eiffel, Marne-la-Vallée, France

Robin Richter Felix-Bernstein-Institute for Mathematical Statistics in the Biosciences, University of Göttingen, Göttingen, Germany

Erlend Skaldehaug Riis The Department of Applied Mathematics and Theoretical Physics, Cambridge, UK

Otmar Scherzer Faculty of Mathematics, University of Vienna, Vienna, Austria

Carola Schönlieb Department of Applied Mathematics and Theoretical Physics, University of Cambridge, Cambridge, UK

Johannes Schwab Department of Mathematics, University of Innsbruck, Innsbruck, Austria

Ivan W. Selesnick Department of Electrical and Computer Engineering, New York University, New York, NY, USA

Eric Setterqvist Johann Radon Institute for Computational and Applied Mathematics (RICAM), Linz, Austria

Fiorella Sgallari Department of Mathematics, University of Bologna, Bologna, Italy

B. M. N. Smets Applied Differential Geometry, Department of Mathematics and Computer Science, Eindhoven University of Technology, Eindhoven, The Netherlands

Stefan Sommer Department of Computer Science (DIKU), University of Copenhagen, Copenhagen E, Denmark

Xiaoliang Song School of Mathematical Sciences, Dalian University of Technology, Dalian, Liaoning, China

Gabriele Steidl Institute of Mathematics, TU Berlin, Berlin, Germany

Xuecheng Tai Hong Kong Center for Cerebro-cardiovascular Health Engineering (COCHE), Shatin, Hong Kong, China

Roger Tam School of Biomedical Engineering, The University of British Columbia, Vancouver, BC, Canada

Xiaoying Tang Department of Electronic and Electrical Engineering, Southern University of Science and Technology, Shenzhen, Guangdong, China

Duy H. Thai Department of Mathematics, Colorado State University, Fort Collins, CO, USA

Anis Theljani Department of Mathematical Sciences, University of Liverpool Mathematical Sciences Building, Liverpool, UK

Ilona Christy Unarta Hong Kong University of Science and Technology, Hong Kong, China

Tuomo Valkonen Center for Mathematical Modeling, Escuela Politécnica Nacional, Quito, Ecuador

Department of Mathematics and Statistics, University of Helsinki, Helsinki, Finland

David Villacís Research Center for Mathematical Modelling (MODEMAT), Escuela Politécnica Nacional, Quito, Ecuador

Patricia Vitoria IPCV, DTIC, University Pompeu Fabra, Barcelona, Spain

Samad Wali School of Information and Communication Engineering, University of Electronic Science and Technology of China, Chengdu, China

Yang Wang Hong Kong University of Science and Technology, Hong Kong, SAR, China

Kathrin Welker Faculty of Mechanical Engineering and Civil Engineering, Helmut-Schmidt-University/University of the Federal Armed Forces Hamburg, Hamburg, Germany

A. J. Wemmenhove Applied Differential Geometry, Department of Mathematics and Computer Science, Eindhoven University of Technology, Eindhoven, The Netherlands

Jonathan Wittmer Department of Aerospace Engineering and Engineering Mechanics, UT Austin, Austin, TX, USA

Chunlin Wu School of Mathematical Sciences, Nankai University, Tianjin, China

Yin Xian Hong Kong Applied Science and Technology Research Institute (ASTRI), Hong Kong, China

TCL Research Hong Kong, Hong Kong, SAR, China

Huilin Yang Department of Electronic and Electrical Engineering, Southern University of Science and Technology, Shenzhen, Guangdong, China

Li Yang School of Mathematical Sciences, Tianjin Normal University, Tianjin, China

Xiaoping Yang Department of Mathematics, Nanjing University, Nanjing, China

Yuan Yao Hong Kong University of Science and Technology, Hong Kong, China

Shing-Tung Yau Harvard University, Cambridge, MA, USA

Xiaojing Ye Department of Mathematics and Statistics, Georgia State University, Atlanta, GA, USA

Laurent Younes Center for Imaging Science, Johns Hopkins University, Baltimore, MD, USA

Bo Yu School of Mathematical Sciences, Dalian University of Technology, Dalian, Liaoning, China

Tieyong Zeng Department of Mathematics, The Chinese University of Hong Kong, Shatin, Hong Kong, China

Qingchao Zhang Department of Mathematics, University of Florida, Gainesville, FL, USA

Lingyan Zhang School of Information and Communication Engineering, University of Electronic Science and Technology of China, Chengdu, China

Kewei Zhang School of Mathematical Sciences, University of Nottingham, Nottingham, UK

Xiaolong Zhu School of Mathematics and Statistics, Xidian University, Xi'an, China

Part I
Convex and Non-convex Large-Scale Optimization in Imaging

Convex Non-convex Variational Models

Alessandro Lanza, Serena Morigi, Ivan W. Selesnick, and Fiorella Sgallari

Contents

Introduction..	4
Convex or Non-convex: Main Idea and Related Works.............................	10
Sparsity-Inducing Separable Regularizers...	11
CNC Models with Sparsity-Inducing *Separable* Regularizers..........................	16
Sparsity-Inducing Non-separable Regularizers......................................	22
CNC Models with Sparsity-Inducing *Non-separable* Regularizers......................	24
Construction of Matrix B..	25
A Simple CNC Example..	27
Path of Solution Components..	29
Forward-Backward Minimization Algorithms.......................................	30
FB Strategy for Separable CNC Models.......................................	32
FB Strategy for Non-separable CNC Models...................................	35
Efficient Solution of the Backward Steps by ADMM...........................	36
Numerical Examples...	41
Examples Using CNC Separable Models......................................	46
Examples Using CNC Non-separable Models..................................	49
Conclusion...	57
References...	57

Abstract

An important class of computational techniques to solve inverse problems in image processing relies on a variational approach: the optimal output is obtained by finding a minimizer of an energy function or "model" composed of two terms,

---

A. Lanza · S. Morigi · F. Sgallari (✉)
Department of Mathematics, University of Bologna, Bologna, Italy
e-mail: alessandro.lanza2@unibo.it; serena.morigi@unibo.it; fiorella.sgallari@unibo.it

I. W. Selesnick
Department of Electrical and Computer Engineering, New York University, New York, NY, USA
e-mail: selesi@nyu.edu

© Springer Nature Switzerland AG 2023
K. Chen et al. (eds.), *Handbook of Mathematical Models and Algorithms in Computer Vision and Imaging*, https://doi.org/10.1007/978-3-030-98661-2_61

the data-fidelity term, and the regularization term. Much research has focused on models where both terms are convex, which leads to convex optimization problems. However, there is evidence that non-convex regularization can improve significantly the output quality for images characterized by some sparsity property. This fostered recent research toward the investigation of optimization problems with non-convex terms. Non-convex models are notoriously difficult to handle as classical optimization algorithms can get trapped at unwanted local minimizers. To avoid the intrinsic difficulties related to non-convex optimization, the convex non-convex (CNC) strategy has been proposed, which allows the use of non-convex regularization while maintaining convexity of the total cost function. This work focuses on a general class of parameterized non-convex sparsity-inducing separable and non-separable regularizers and their associated CNC variational models. Convexity conditions for the total cost functions and related theoretical properties are discussed, together with suitable algorithms for their minimization based on a general forward-backward (FB) splitting strategy. Experiments on the two classes of considered separable and non-separable CNC variational models show their superior performance than the purely convex counterparts when applied to the discrete inverse problem of restoring sparsity-characterized images corrupted by blur and noise.

Keywords

Convex non-convex optimization · Sparsity regularization · Image restoration · Alternating direction method of multipliers · Forward backward algorithm

Introduction

A wide class of linear systems derived from the discretization of linear ill-posed inverse problems in data processing is characterized by high dimensionality, ill-conditioned matrices, and noise-corrupted data. In this class of discrete inverse problems, a noisy indirect observation $b \in \mathbb{R}^m$ of an original unknown image $x \in \mathbb{R}^n$ is modeled as

$$b = Ax, \qquad (1)$$

where $A \in \mathbb{R}^{m \times n}$ accounts for the data-acquisition system. For instance, A can be a convolution matrix modeling optical blurring, a wavelet or Fourier transform matrix in image synthesis, a radon transform matrix in X-ray computerized tomography, a sampling matrix in compressed sensing, a binary selection matrix in image inpainting, or the identity matrix in image denoising and segmentation.

When $m < n$, the linear system (1) is underdetermined and among the infinity of solutions, it is common to seek an approximate solution with minimal norm, that is, one solves the constrained optimization problem

$$\min_{x \in \mathbb{R}^n} \|x\|_2^2 \quad \text{subject to} \quad b = Ax, \tag{2}$$

where $\|v\|_2$ denotes the ℓ_2 norm of vector v.

On the other hand, when $m > n$, the linear system (1) is overdetermined; in general there is no solution, and it is common to seek for the least squares solution, that is, the solution which minimizes the residual norm; in formula,

$$\min_{x \in \mathbb{R}^n} \|b - Ax\|_2^2. \tag{3}$$

Even in the most favorable case that $m = n$, so that the linear system (1) can admit a unique solution, ill-conditioning of matrix A typically makes the problem very difficult from a numerical point of view.

Indeed, for many image processing applications of practical interest, problems in form (1) are *ill-posed* linear inverse problems. The term *ill-posed* was coined in the early twentieth century by Hadamard who defined a linear problem to be well-posed if it satisfies the following three requirements:

- Existence: The problem must have a solution.
- Uniqueness: The problem must have only one solution.
- Stability: The solution must depend continuously on the data.

If the problem violates one or more of these requirements, it is said to be ill-posed (Hansen 1997).

A violation of the stability condition implies that arbitrarily small perturbations of the data can produce arbitrarily large perturbations in the solution. Noise is a typical unavoidable perturbation component in the digital data acquisition process which, coupled with ill-conditioning of matrix A, makes inverse problems in imaging typically ill-posed.

In this work, we assume that the noise is additive white Gaussian (AWG), so that the observed noisy image $b \in \mathbb{R}^m$ is related to the underlying true image $x \in \mathbb{R}^n$ by means of the following degradation model

$$b = Ax + \eta, \tag{4}$$

with $\eta \in \mathbb{R}^m$ the realization of an m-dimensional random vector having Gaussian distribution with zero mean and scalar covariance matrix. In many practical cases, the matrix A is so ill-conditioned (if not numerically singular) that recovering x given b and A by means of a naive (not regularized) least-squares procedure leads to meaningless results. Some sort of regularization is required. The key aspect is to reformulate the problem such that the solution to the new problem is less sensitive to the perturbations. We say that we stabilize or regularize the problem.

Regularization strategies in traditional variational methods are usually problem-dependent and take advantage of a priori information specific to any particular imaging application. In this paper, we focus on those applications which involve sparsity in the solution, or in its representation, or in a function of the solution. For instance, images of stars from a telescope are sparse themselves, while images of humans are sparse under the wavelet transform. Sparsity plays an important role in image processing and machine learning. How to build appropriate sparse-based models, how to numerically find solutions of the sparse-based models, and how to derive theoretical guarantees of the correctness of the solutions are essential for the success of sparsity in a wide range of applications (Bruckstein et al. 2009).

We focus on regularized variational methods where an approximate solution $x^* \in \mathbb{R}^n$ of the inverse problem (4) is sought among the (global) minimizers of a cost function $\mathcal{J}: \mathbb{R}^n \to \mathbb{R}$ which takes the following form

$$x^* \in \arg\min_{x \in \mathbb{R}^n} \mathcal{J}(x), \qquad \mathcal{J}(x) = \frac{1}{2}\|Ax - b\|_2^2 + \mu \Psi(x). \qquad (5)$$

The quadratic term in (5) is the so-called L_2 *fidelity term*, which forces closeness of solution(s) x^* to data b according to the linear acquisition model (4) and to the assumed noise Gaussian distribution. The term $\Psi(x)$ in (5) represents the sparsity-inducing *regularization term* and encodes some sparsity priors on the unknown sought image. Finally, the positive scalar μ, referred to as the *regularization parameter* of variational model (5), is a free parameter which allows to control the trade-off between data fidelity and regularization.

In this work, we are particularly interested in sparsity-promoting regularization terms $\Psi: \mathbb{R}^n \to \mathbb{R}$ having the following general form

$$\Psi(x) := \Phi(x, y), \qquad y := G(z) \quad z := Lx, \qquad (6)$$

with

- $L \in \mathbb{R}^{r \times n}$ the regularization matrix
- $G: \mathbb{R}^r \to \mathbb{R}^s$ a possibly nonlinear vector-valued function with $g_i: \mathbb{R}^r \to \mathbb{R}$, $i = 1, \ldots, s$, representing its scalar-valued components
- $y \in \mathbb{R}^s$ the features vector to be sparsified
- $\Phi: \mathbb{R}^n \times \mathbb{R}^s \to \mathbb{R}$ a sparsity-promoting penalty function (Selesnick and Bayram 2014; Selesnick et al. 2015; Lanza et al. 2016a)

It is important for the purposes of this work to introduce a partition of the class of sparsity-promoting regularizers Ψ defined in (6) into two sub-classes based on *separable* and *non-separable* penalty functions Φ.

Definition 1 (Separable and non-separable sparsity-promoting regularizers). A sparsity-inducing regularizer Ψ of the form in (6) is referred to as *separable* (with respect to the feature vector y to be sparsified) if the penalty function Φ only depends on y and is additively separable with respect to the y components; in formula,

1 Convex Non-convex Variational Models

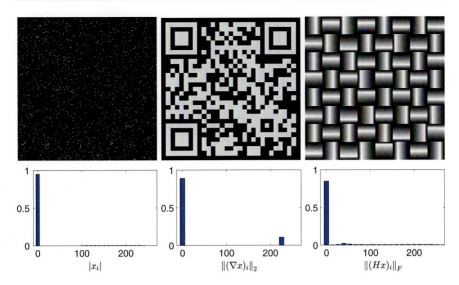

Fig. 1 Prototypical example images characterized by different sparse feature vectors (first row) and their associated normalized histograms (second row)

$$\Phi(y) = \sum_{i=1}^{s} \phi_i(y_i), \quad \text{with } \phi_i : \mathbb{R} \to \mathbb{R}, \tag{7}$$

otherwise, it is named *non-separable*.

Examples of image feature vectors $y = G(Lx)$ which can be characterized by a sparsity property in typical application scenarios are, e.g., the vectorized image itself (for predominantly zero images), the vector of image gradient magnitudes (for piecewise constant images), the vector of image Hessian Frobenious norms (for piecewise affine images), and the vector of coefficients of the image in a transformed domain (e.g., Fourier, wavelet,...).

Examples of predominantly zero, piecewise constant, and piecewise affine images are depicted in the first row of Fig. 1. They are characterized, from left to right, by a sparse vector y of components $y_i = |x_i|$, $y_i = \left\|(\nabla x)_i\right\|_2$ and $y_i = \|(Hx)_i\|_F$, $i = 1, \ldots, n$, respectively, where $(\nabla x)_i \in \mathbb{R}^2$ and $(Hx)_i \in \mathbb{R}^{2 \times 2}$ represent the gradient and the Hessian matrix of image x at pixel i, respectively. In the second row of Fig. 1, the reported normalized histograms of the corresponding y vector values clearly highlight their sparsity.

Although the three images above represent almost ideal prototypes, also many images from real-life applications commonly exhibit sparsity features. In Fig. 2, we show three realistic images characterized by increasing level of sparsity of the gradient magnitudes, together with their associated histograms. This indicates the practical importance of sparse-regularized variational models which, in many application scenarios, hold the potential for very high quality results.

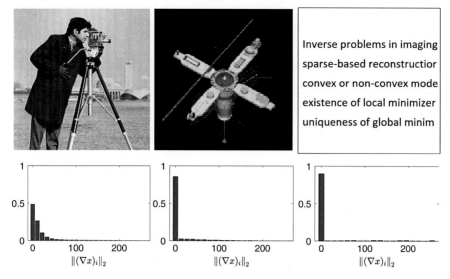

Fig. 2 Realistic images characterized, from left to right, by increasing level of sparsity of the gradient magnitudes (first row) and their associated normalized histograms (second row)

Some interesting models of the form (5)–(6) are characterized by the following well-known matrices A and L:

- **TV-L_2 Restoration**: In image restoration, the popular Total Variation (TV)-L_2 (Rudin et al. 1992) calls for a matrix A characterizing the image blur, or $A = I_n$ for image denoising. For what concerns the linear operator L, it is defined as $L := (D_h^T, D_v^T)^T \in \mathbb{R}^{2n \times n}$ with $D_h, D_v \in \mathbb{R}^n$ finite difference matrices discretizing the first-order horizontal and vertical partial derivatives, respectively, $g_i(z) := \|(z_i, z_{i+n})\|_2$ or $g_i(z) - \|(z_i, z_{i+n})\|_1$, $i = 1, \ldots, n$, for isotropic and anisotropic TV regularization, respectively, and Φ the ℓ_1 norm function; in formulas

$$\mathrm{TV}(x) = \|G(Lx)\|_1 = \sum_{i=1}^{n} |g_i(Lx)| = \begin{cases} \sum_{i=1}^{n} \sqrt{(D_h x)_i^2 + (D_v x)_i^2} & \text{(isotropic)} \\ \sum_{i=1}^{n} \left(|(D_h x)_i| + |(D_v x)_i|\right) & \text{(anisotropic)} \end{cases} \tag{8}$$

- **Sparse Reconstruction (Analysis)**: A full rank, $L := W$ with W an orthogonal basis or an overcomplete dictionary, which satisfies the tight frame condition, i.e., $L^T L = \delta I_n$, $\delta > 0$, Parekh and Selesnick (2015).

- **Sparse Reconstruction (Synthesis):** $A := W^{-1}$, $L = I_n$, and G the identity operator.

The main difficulties in solving variational models of the considered form in (5)–(6) stems from the facts that the involved optimization domain is typically of very high dimension (the number of pixels in the image), the linear operator A can be ill-conditioned or even singular, and, more importantly, the regularization term Ψ is preferably a non-convex non-smooth function in order to effectively promote sparsity of vector y. Summarizing, (5)–(6) is a very challenging large-scale optimization problem. The real challenge comes from possible non-convexity of the problem, which yields all the well-known associated intricacies, namely, the existence of local minimizers and the problematic convergence of minimization algorithms.

A very interesting approach proposed in literature to address this issue is the so-called CNC strategy. It consists in constructing and then minimizing convex cost functions containing non-convex (sparsity-promoting) regularization terms. This can be obtained by using regularizers parameterized such that their degree of non-convexity can be tuned. By suitably setting the parameters of the regularizer, one can thus obtain a convex variational model containing a non-convex regularizer which holds the potential to induce sparsity of the solution more effectively than any convex regularizer. As it will be shown in this work, suitably parameterized non-separable regularizers of the form in (6) allow to apply the CNC strategy to the solution of any linear inverse problem in imaging, thus overcoming the intrinsic limitations of separable regularizers.

The chapter contents will be organized as follows. In section "Convex or Non-convex: Main Idea and Related Works," we outline the main ideas at the basis of the CNC strategy and shortly review the most related approaches. In section "Sparsity-Inducing Separable Regularizers," we present separable non-convex parameterized regularizers, and then in section "CNC Models with Sparsity-Inducing *Separable* Regularizers," we illustrate the associated CNC models and the related convexity condition results. In section "Sparsity-Inducing Non-separable Regularizers," we present non-separable non-convex parameterized regularizers, while their integration into suitable CNC models is described in section "CNC Models with Sparsity-Inducing *Non-separable* Regularizers," together with the construction of the related matrix B which leads to convexity of the total cost function. An illustrative example of CNC separable and non-separable models is given in section "A Simple CNC Example." In section "Forward-Backward Minimization Algorithms," we outline the optimization algorithms for solving the illustrated classes of CNC variational models, based on the FB splitting strategy and the Alternating Direction Method of Multipliers (ADMM) for the related subproblems. Finally, in section "Numerical Examples," we evaluate experimentally the performance of the two CNC classes when applied to the linear ill-posed inverse problem of restoring images corrupted by blur and noise.

Convex or Non-convex: Main Idea and Related Works

Convexity is a sufficient condition for all local minima to be global minima. If $\mathcal{J}$ is non-convex, it may have many local minima which are not global minima. This means that classical convex optimization algorithms applied to a non-convex cost function $\mathcal{J}$ will almost certainly get trapped at a local minimimum that is of higher cost than the global minimum. Moreover, which local minimum is reached will depend strongly on the starting point of the algorithm.

However, non-convex non-smooth optimization problems arise more and more frequently in image processing, neural network training, and machine learning, where suitable non-convex regularizers have shown superior performance with respect to their convex counterparts (Nikolova 2011; Bruckstein et al. 2009). In the literature, for example, the most natural sparsity-inducing penalty is the ℓ_0 pseudo-norm, which, however, leads to NP-hard and non-convex optimization problems.

Literature on non-convex optimization dates back to the 1950s. An important class of non-convex optimization problems that has been extensively studied in the past is related to the specific set of non-convex cost functions that can be defined as the difference of convex functions, or DC functions for short; we refer to the seminal papers Tuy (1995) and Hartman (1959) and the more recent work Yuille and Rangarajan (2003) for more details on DC functions and optimization. Other important approaches to optimization in the non-convex regime are represented, e.g., by simulated annealing, see Geman and Geman (1984); genetic algorithms, see Jensen and Nielsen (1992); the Mean Field Annealing by Geiger and Girosi, which provides a deterministic version of simulated annealing (Geiger and Girosi 1991); and the Graduated Non-Convexity (GNC) strategy introduced in Blake and Zisserman (1987) by Blake and Zisserman.

The basic idea of the popular GNC algorithmic strategy is to construct a modified, parameterized cost function $\mathcal{J}_\lambda$, governed by a control parameter $\lambda \in [0, 1]$, chosen so that $\mathcal{J}_0 = \mathcal{J}$, the true cost function, and $\mathcal{J}_1 = \mathcal{J}_c$, a convex approximation to $\mathcal{J}$. Then GNC computes a solution to the non-convex problem by starting from its convex approximation $\mathcal{J}_c$, which must have a global minimum, and gradually changing λ (i.e., gradually increasing the amount of non-convexity) until the original non-convex function $\mathcal{J}$ is recovered. The solution obtained at each iteration is used as initial guess for the subsequent iteration. In the construction of a suitable convex surrogate function $\mathcal{J}_c$, the authors in Blake and Zisserman (1987) introduced the concept of "balancing" the positive second derivatives in the first term (fidelity) against the negative second derivatives in the regularization term. This represents the seminal idea behind the CNC strategy, namely, designing non-convex parameterized penalty terms which allow to maintain convexity of the total cost function.

This simple concept, later called the CNC strategy (Lanza et al. 2015), has been applied by Nikolova (1998) in the context of denoising of binary images and then extended to many other sparse-regularized variational problems (Bayram

2016; Selesnick and Bayram 2014; Lanza et al. 2017), including 1D and 2D total variation denoising (Lanza et al. 2016a; Malek-Mohammadi et al. 2016; Zou et al. 2019; Du and Liu 2018), transform-based denoising (Parekh and Selesnick 2015; Ding and Selesnick 2015), low-rank matrix estimation (Parekh and Selesnick 2016), decomposition and segmentation of images and scalar fields over surfaces (Chan et al. 2017; Huska et al. 2019a,b), as well as machine fault detection (Cai et al. 2018; Wang et al. 2019).

The flexibility and effectiveness of the CNC approach depends on the construction of non-trivial separable and non-separable convex functions. It turns out that Moreau envelopes and infimal convolutions are useful for this purpose (Selesnick 2017a,b; Carlsson 2016; Soubies et al. 2015). Based on convex analysis, families of non-convex non-separable penalty functions have been proposed in Selesnick (2017a) that do maintain convexity of the cost functional $\mathcal{J}$ for any matrix A, but only in the special case where both G and L in (6) are identity operators. More recently, a convex approach was applied in Lanza et al. (2019) where a general CNC framework is proposed for constructing non-separable non-convex regularizers starting from any convex regularizer, any matrix A and L, and quite general functions G. In particular, an infimal convolution is subtracted from a convex regularizer, such as the ℓ_1-norm, leading to a resulting non-convex regularizer.

Non-convex penalties of various functional forms have been proposed too for overcoming limitations of the ℓ_1 norm by using penalties that promote sparsity more strongly (Castella and Pesquet 2015; Candés et al. 2008; Nikolova 2011; Nikolova et al. 2010; Chartrand 2014; Chouzenoux et al. 2013; Portilla and Mancera 2007; Shen et al. 2016). However, these methods do not aim to maintain convexity of the cost function to be minimized. Moreover, for what concerns non-separable sparsity-inducing penalties in (6), pioneering work has been conducted in Tipping (2001) and Wipf et al. (2011); however, also such penalties were not designed to maintain cost function convexity.

We finally note that infimal convolution (related to the Moreau envelope) has been used to define generalized TV regularizers (Setzer et al. 2011; Chambolle and Lions 1997; Burger et al. 2016; Becker and Combettes 2014). However, the aims and methodologies of these past works are quite different from those considered here. In fact, in these works, the ℓ_1 norm is replaced by an infimal convolution; the resulting regularizer is convex.

Sparsity-Inducing Separable Regularizers

In this section, we first recall some definitions which will be useful for the rest of the work, and, in particular, we report some results from convex analysis. We then review some popular sparsity-inducing separable regularizers and discuss their properties.

In this work, we denote by $\mathbb{R}_+$ and $\mathbb{R}_{++}$ the sets of nonnegative and positive real numbers, respectively, by I_n the identity matrix of order n, by 0_n the n-dimensional null vector, by null(M) the null space of matrix M, and by $\Gamma_0(\mathbb{R}^n)$ the set of proper lower semicontinuous convex functions from $\mathbb{R}^n$ to $\overline{\mathbb{R}} := \mathbb{R} \cup \{+\infty\}$.

Definition 2 (infimal convolution). Let $f, g : \mathbb{R}^n \to \overline{\mathbb{R}}$. The infimal convolution of f and g is defined by

$$(f \square g)(x) = \inf_{v \in \mathbb{R}^n} \{f(v) + g(x - v)\}. \tag{9}$$

and it is said to be exact and denoted by $f \boxdot g$ if the infimum above is attained for any $x \in \mathbb{R}^n$, namely, $(f \boxdot g)(x) = \min_{v \in \mathbb{R}^n} \{f(v) + g(x - v)\}$, for any $x \in \mathbb{R}^n$.

Definition 3 (Moreau envelope). Let $f \in \Gamma_0(\mathbb{R}^n)$ and let $a \in \mathbb{R}_{++}$. The Moreau envelope of f with parameter a is defined by

$$\mathrm{env}_f^a(x) = \left(f \boxdot \frac{a}{2} \|\cdot\|_2^2 \right)(x) = \min_{v \in \mathbb{R}^n} \left\{ f(v) + \frac{a}{2} \|x - v\|_2^2 \right\}. \tag{10}$$

Definition 4 (proximity operator). Let $f \in \Gamma_0(\mathbb{R}^n)$ and let $a \in \mathbb{R}_{++}$. The proximity operator of f with parameter a is defined by

$$\mathrm{prox}_f^a(x) = \arg\min_{v \in \mathbb{R}^n} \left\{ f(v) + \frac{a}{2} \|x - v\|_2^2 \right\}. \tag{11}$$

We notice that, for any $f \in \Gamma_0(\mathbb{R}^n), a \in \mathbb{R}_{++}$, the cost function $f(v) + \frac{a}{2} \|x - v\|_2^2$ in (10)–(11) is strongly convex in v; hence it admits a unique (global) minimizer.

Definition 5 (Huber function). The Huber function $h_a : \mathbb{R} \to \mathbb{R}_+$ with parameter $a \in \mathbb{R}_{++}$ is defined by

$$h_a(t) = \mathrm{env}_{|\cdot|}^a(t) = \min_{v \in \mathbb{R}} \left\{ |v| + \frac{a}{2}(t-v)^2 \right\} = \begin{cases} \dfrac{a}{2} t^2 & \text{for } |t| \in [0, 1/a], \\ |t| - \dfrac{1}{2a} & \text{for } |t| \in]1/a, +\infty[. \end{cases} \tag{12}$$

Definition 6 (minimax concave penalty function). The minimax concave (MC) penalty function $\phi_{\mathrm{MC}} : \mathbb{R} \to \mathbb{R}_+$ with parameter $a \in \mathbb{R}_{++}$ is defined by

$$\phi_{\mathrm{MC}}(t; a) = |t| - h_a(t) = \begin{cases} -\dfrac{a}{2} t^2 + |t| & \text{for } |t| \in [0, 1/a], \\ \dfrac{1}{2a} & \text{for } |t| \in]1/a, +\infty[. \end{cases} \tag{13}$$

Proposition 1 (Moreau envelope gradient). *Let $f \in \Gamma_0(\mathbb{R}^n)$ and let $a \in \mathbb{R}_{++}$. Then, the Moreau envelope of f with parameter a is a differentiable function with gradient given by*

$$\nabla \left(\text{env}_f^a \right)(x) = a \left(x - \text{prox}_f^a(x) \right). \tag{14}$$

Proposition 2. *Let $h_a : \mathbb{R} \to \mathbb{R}$ be the Huber function defined in (12). Then, for any value of the parameter $a \in \mathbb{R}_{++}$ the function*

$$f_a(z) := h_a\left(\|x\|_2\right), \quad x \in \mathbb{R}^n, \tag{15}$$

is continuously differentiable and its gradient is given by

$$\nabla f_a(x) = \min\left\{ a, \frac{1}{\|x\|_2} \right\} x. \tag{16}$$

Proof. Recalling the Huber function definition in (12), the function f_a in (15) takes the explicit form

$$f_a(x) = \begin{cases} \dfrac{a}{2} \sum_{i=1}^n x_i^2 & \text{for } \|x\|_2 \in [0, 1/a], \\ \sqrt{\sum_{i=1}^n x_i^2} - \dfrac{1}{2a} & \text{for } \|x\|_2 \in \,]1/a, +\infty[. \end{cases} \tag{17}$$

The two pieces of function f_a in (17) are clearly both continuously differentiable on their domain with gradients given by

$$\nabla f_a(x) = \begin{cases} a\, x & \text{for } \|x\|_2 \in [0, 1/a], \\ \dfrac{1}{\|x\|_2} x & \text{for } \|x\|_2 \in \,]1/a, +\infty[. \end{cases} \tag{18}$$

It follows easily from (18) that, for any $a \in \mathbb{R}_{++}$, the gradient function $\nabla f_a(x)$ is continuous also at points x on the spherical surface $\|x\|_2 = 1/a$ separating its two pieces. Finally, the compact form of ∇f_a given in (16) comes straightforwardly from (18). □

Among separable sparsity-promoting regularizers (see Definition 1), the most natural choice is represented by the ℓ_0 pseudo-norm of the features vector y to sparsify, namely, $\Phi(y) = \|y\|_0 = \#\{i : y_i \neq 0\}$, as it directly measures the sparsity of y by counting the number of non-zero elements in it (see the dashed magenta line in Fig. 3a). However, ℓ_0 regularization leads to non-convex NP-hard optimization problems. Intrinsic difficulties involved in using the ℓ_0 pseudo-norm can be overcome by using the ℓ_1 norm, namely, $\Phi(y) = \|y\|_1 = \sum_{i=1}^s |y_i|$ (see

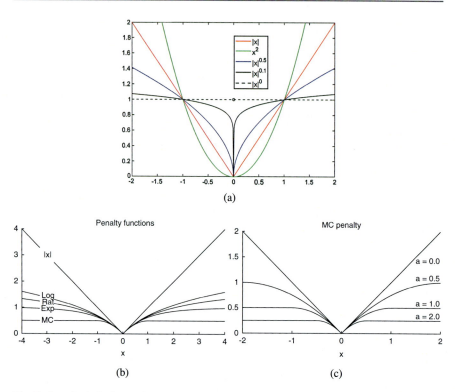

Fig. 3 Sparsity-inducing scalar penalties: (**a**) ℓ_p penalty for some different p values, (**b**) some parameterized non-convex penalties satisfying assumptions 1–5 (see Table 1) and the MC penalty (see definition in (13)) all with concavity parameter $a = 1$, and (**c**) MC penalty for some different values of the concavity parameter a

the solid red curve in Fig. 3a). In fact, this choice very likely leads to a convex sparsity-inducing regularizer and, hence, to a convex variational model which can be solved numerically by standard convex optimization algorithms. However, it is well known that the ℓ_1 norm penalty function tends to underestimate high-amplitude components of the vector to which it is applied, in our case $y = G(Lx)$. More generally, it is well known that non-convex penalty functions hold the potential for inducing sparsity more effectively than convex penalty functions. A natural non-convex separable alternative to the ℓ_1 norm is the ℓ_p quasi-norm penalty (Sidky et al. 2014), $\Phi(y) = \frac{1}{p}\|y\|_p^p = \frac{1}{p}\sum_{i=1}^{s}|y_i|^p$, $0 < p < 1$; see the solid blue and black curves in Fig. 3a, corresponding to $p = 0.5$ and $p = 0.1$, respectively. However, such a non-convex family of penalties can not be used to the purpose of constructing CNC variational models. In fact, since the infimum of the second-order derivative of the ℓ_p penalty is equal to $-\infty$ for any $p \in\,]0, 1[$, it is not possible to obtain a total convex model even when coupling the regularizer with a strongly convex quadratic fidelity term.

1 Convex Non-convex Variational Models

To the aim of constructing CNC models with separable sparsity-promoting regularizers characterized by tunable degree of non-convexity, one can usefully consider the class of parameterized scalar penalty functions $\phi(t; a) : \mathbb{R} \to \mathbb{R}$ which, for any value of the parameter $a \in \mathbb{R}_+$, satisfy the following assumptions:

1. $\phi(t; a) \in C^0(\mathbb{R}) \cap C^2(\mathbb{R} \setminus \{0\})$
2. $\phi(t; a) = \phi(-t; a) \quad \forall t \in \mathbb{R}_{++}$
3. $\phi'(t; a) \geq 0 \quad \forall t \in \mathbb{R}_{++}$
4. $\phi''(t; a) \leq 0 \quad \forall t \in \mathbb{R}_{++}$
5. $\phi(0; a) = 0, \quad \inf_{t \in \mathbb{R}_{++}} \phi''(t; a) = -a$

We denoted by $\phi'(t; a)$ and $\phi''(t; a)$ the first-order and second-order derivatives of ϕ with respect to the variable t, respectively. Assumptions 1–5 above are quite standard and encompass a wide class of continuous but non-smooth non-convex sparsity-promoting penalty functions (Geman and Geman 1984). The parameter a, referred to as the penalty concavity parameter, is directly related to the degree of non-convexity of the penalty function, as defined in assumption 5.

In Table 1, we report the definitions of four widely used sparsity-promoting parameterized scalar penalty functions, referred to as $\phi_{\log}$, ϕ_{rat}, ϕ_{atan}, and $\phi_{\exp}$, which satisfy all the assumptions 1–5 and have been considered, e.g., in Selesnick and Bayram (2014), Chen and Selesnick (2014), and Lanza et al. (2015, 2016a). In particular, the penalty ϕ_{atan} has been proposed in Selesnick and Bayram (2014) as the maximally sparsity-inducing function among those characterized by a first-order derivative of inverse quadratic polynomial type.

In order to mimic in a more faithful manner not only the asymptotically constant behavior of the ℓ_0 pseudo-norm, a class of piecewise defined truncated penalties has been introduced in literature. One of the most popular and effective representatives of this class is the so-called minimax concave (MC) penalty function, formally defined in (13) and also reported in the last row of Table 1. In the rest of this work, we will use the MC penalty within all the illustrated separable CNC variational models.

Table 1 Four popular non-convex, sparsity-promoting, parameterized scalar penalty functions $\phi(t; a): \mathbb{R} \to \mathbb{R}_+$ satisfying assumptions 1–5 and, in the last row, the MC penalty function

$\phi_{\log}(t; a)$	=	$\dfrac{\log(1 + at)}{a}$
$\phi_{\text{rat}}(t; a)$	=	$\dfrac{t}{1 + at/2}$
$\phi_{\text{atan}}(t; a)$	=	$\dfrac{\operatorname{atan}\left(\frac{1+2at}{\sqrt{3}}\right) - \frac{\pi}{6}}{a\sqrt{3}/2}$
$\phi_{\exp}(t; a)$	=	$\dfrac{1 - e^{-at}}{a}$
$\phi_{\text{MC}}(t; a)$	=	$\begin{cases} -\dfrac{a}{2}t^2 + \lvert t \rvert & \text{for } \lvert t \rvert \in [0, 1/a] \\ \dfrac{1}{2a} & \text{for } \lvert t \rvert \in \,]1/a, +\infty[\end{cases}$

To give a visual insight of the considered parameterized penalty functions, in Fig. 3b we depict the graphs of some of the penalties in Table 1, all with concavity parameter $a = 1$, whereas in Fig. 3c we illustrate the MC penalty for some different values of the concavity parameter a.

CNC Models with Sparsity-Inducing *Separable* Regularizers

This section is concerned with the formulation of CNC variational models with *separable* sparsity-promoting regularization terms; see Definition 1. The general form of such models reads

$$x^* \in \arg\min_{x \in \mathbb{R}^n} \mathcal{J}_S(x; a), \tag{19}$$

$$\mathcal{J}_S(x; a) = \frac{1}{2}\|Ax - b\|_2^2 + \mu \Psi_S(x; a), \quad \Psi_S(x; a) = \sum_{i=1}^{s} \phi_{\mathrm{MC}}\left(g_i(Lx); a_i\right), \tag{20}$$

where, we recall, $A \in \mathbb{R}^{m \times n}$ and $L \in \mathbb{R}^{r \times n}$ are the coefficient matrices of two bounded linear operators, $g_i : \mathbb{R}^r \to \mathbb{R}$, $i = 1, \ldots, s$ are the components of a possibly nonlinear vector-valued function $G : \mathbb{R}^r \to \mathbb{R}^s$, $\mu \in \mathbb{R}_{++}$ is the regularization parameter, $\phi_{\mathrm{MC}} : \mathbb{R} \to \mathbb{R}_+$ is the non-convex MC penalty function defined in (13), and where we introduced the vector $a := (a_1, \ldots, a_s)^T \in \mathbb{R}_{++}^s$ containing the concavity parameters of all the s instances of the MC penalty in the regularizer Ψ_S. We refer to (19)–(20) as the class of CNC separable (least-squares) models, abbreviated CNC-S-L$_2$ models.

In order to refer to models (19)–(20) as CNC, we clearly need to derive and then impose convexity conditions for the objective function $\mathcal{J}_S$. More precisely, we seek sufficient conditions on the operators A, L, and G and on the parameters μ and a_i, $i = 1, \ldots, s$, to ensure that the function $\mathcal{J}_S$ in (20) is convex (strongly convex) on its entire domain $x \in \mathbb{R}^n$. It is worth noting that, in practice, the operators A, L, and G are commonly prescribed by the specific application at hand. In fact, operator A typically comes from a (more or less accurate) modeling of the image acquisition process, whereas operators L and G are related to the expected properties of sparsity of the sought solution. This implies that the derived convexity conditions can be regarded as constraints on the free parameters μ and a_i of model (19)–(20).

In Lemma 1, we give some useful reformulations of the separable regularizer Ψ_S defined in (20); then in Theorem 1, we derive conditions for convexity of $\mathcal{J}_S$.

Lemma 1. *The separable regularizer Ψ_S in (20) can be rewritten as*

$$\Psi_S(x; a) = \|G(Lx)\|_1 - \mathcal{H}_S(x; a), \tag{21}$$

where the function $\mathcal{H}_S$ in (21) takes the following equivalent forms:

1 Convex Non-convex Variational Models

$$\mathcal{H}_S(x;a) = \sum_{i=1}^{s} h_{a_i}\left(g_i(Lx)\right) \tag{22}$$

$$= \left(\|\cdot\|_1 \,\square\, \frac{1}{2}\|W\cdot\|_2^2\right)(G(Lx)) \tag{23}$$

$$= \text{env}^1_{\|W^{-1}\cdot\|_1}(WG(Lx)), \tag{24}$$

with h_{a_i} the Huber function defined in (12) and $W \in \mathbb{R}^{s \times s}$ the matrix defined by

$$W := \text{diag}\left(\sqrt{a_1}, \ldots, \sqrt{a_s}\right). \tag{25}$$

In the special case that $a_i = \bar{a} \;\; \forall i = 1, \ldots, s$, $\bar{a} \in \mathbb{R}_{++}$, then (23) and (24) reduce to

$$\mathcal{H}_S(x;a) = \text{env}^{\bar{a}}_{\|\cdot\|_1}(G(Lx)). \tag{26}$$

Proof. First, recalling the MC penalty definition in (13), Ψ_S in (20) can be rewritten as

$$\Psi_S(x;a) = \sum_{i=1}^{s}\left(\left|g_i(Lx)\right| - h_{a_i}\left(g_i(Lx)\right)\right) = \|G(Lx)\|_1 - \underbrace{\sum_{i=1}^{s} h_{a_i}\left(g_i(Lx)\right)}_{\mathcal{H}_S(x;a)}, \tag{27}$$

which proves (21)–(22). Then, based on the Huber function definition in (12), the function $\mathcal{H}_S(x;a)$ in (27) can be manipulated as follows:

$$\mathcal{H}_S(x;a) = \sum_{i=1}^{s} \text{env}^{a_i}_{|\cdot|}(g_i(Lx))$$

$$= \sum_{i=1}^{s} \min_{v_i \in \mathbb{R}}\left\{|v_i| + \frac{a_i}{2}\left(g_i(Lx) - v_i\right)^2\right\}$$

$$= \min_{v \in \mathbb{R}^s} \sum_{i=1}^{s}\left(|v_i| + \frac{a_i}{2}\left(g_i(Lx) - v_i\right)^2\right)$$

$$= \min_{v \in \mathbb{R}^s}\left\{\sum_{i=1}^{s}|v_i| + \frac{1}{2}\sum_{i=1}^{s}\left(\sqrt{a_i}\left(g_i(Lx) - v_i\right)\right)^2\right\}$$

$$= \min_{v \in \mathbb{R}^s}\left\{\|v\|_1 + \frac{1}{2}\|W(G(Lx) - v)\|_2^2\right\} \tag{28}$$

$$= \left(\|\cdot\|_1 \,\square\, \frac{1}{2}\|W\cdot\|_2^2 \right)(G(Lx)), \tag{29}$$

with matrix W defined in (25). The last equality (29), which proves (23), comes straightforwardly from the definition of infimal convolution in (9).

Starting from (28), and noting that by assumption the square diagonal matrix W in (25) is invertible (in fact, $a_i \in \mathbb{R}_{++} \,\forall i = 1, \ldots, s$), we can write

$$\mathcal{H}_S(x; a) = \min_{v \in \mathbb{R}^s} \left\{ \|v\|_1 + \frac{1}{2} \|WG(Lx) - Wv\|_2^2 \right\}$$

$$= \min_{z \in \mathbb{R}^s} \left\{ \|W^{-1}z\|_1 + \frac{1}{2} \|WG(Lx) - z\|_2^2 \right\}$$

$$= \mathrm{env}^1_{\|W^{-1}\cdot\|_1}(WG(Lx)),$$

which completes the proof of (24). Statement (26) follows easily. □

In the following result, we define the set of sub-vectors $\{z^{(i)}\}_{i=1}^s$, $z^{(i)} \in \mathbb{R}^{r_i}$, as a *partition* of vector $z \in \mathbb{R}^r$ if $z^{(i)} = P^{(i)}z$, with $P^{(i)} \in \mathbb{R}^{r_i \times n}$ binary selection matrices satisfying $\left((P^{(1)})^T, (P^{(2)})^T, \ldots, (P^{(s)})^T \right)^T = P$, with $P \in \mathbb{R}^{r \times r}$ a permutation matrix, so that $\left((z^{(1)})^T, (z^{(2)})^T, \ldots, (z^{(s)})^T \right)^T = Pz$, a permuted version of z.

Theorem 1. *If the components $g_i : \mathbb{R}^r \to \mathbb{R}$ of function G are all lower semicontinuous functions, then for any matrices A, L and any value of parameters $\mu \in \mathbb{R}_{++}$, $a \in \mathbb{R}_{++}^s$, the objective function $\mathcal{J}_S : \mathbb{R}^n \to \mathbb{R}$ defined in (20) is lower semicontinuous and bounded from below by zero.*

Moreover, if any g_i is either linear or a lower semicontinuous convex and nonnegative function, then a sufficient condition for $\mathcal{J}_S$ to be convex (strongly convex) is that the function

$$\mathcal{J}_1(x) := \|Ax\|_2^2 - \mu \|WG(Lx)\|_2^2 \quad \text{is convex (strongly convex)}, \tag{30}$$

with matrix W defined in (25).

In particular, in the special cases that G is the identity operator or a function defined by

$$G(z) = \left(\|z^{(1)}\|_2, \ldots, \|z^{(s)}\|_2 \right)^T, \quad \text{with } \{z^{(i)}\}_{i=1}^s \text{ partition of } z \in \mathbb{R}^r, \tag{31}$$

then it follows from (30) that $\mathcal{J}_S$ is convex (strongly convex) if

1 Convex Non-convex Variational Models

$$Q := A^T A - \mu L^T W^2 L \succeq 0 \; (\succ 0). \tag{32}$$

Finally, in case that $a_i = \tilde{a} \; \forall i = 1, \ldots, s$, (32) reduces to

$$Q = A^T A - \mu \tilde{a} L^T L \succeq 0 \; (\succ 0), \tag{33}$$

that is,

$$\tilde{a} = \tau_c \frac{\rho_{A,L}}{\mu}, \quad \tau_c \in [0, 1] \; \left(\tau_c \in [0, 1[\right), \quad \rho_{A,L} := \frac{\sigma_{A,\min}^2}{\sigma_{L,\max}^2}, \tag{34}$$

with $\sigma_{A,\min}$ and $\sigma_{L,\max}$ denoting the minimum singular value of matrix A and the maximum singular value of matrix L, respectively.

Proof. Since the MC penalty function defined in (13) is continuous and bounded from below by zero, if functions g_i are all lower semicontinuous, then the regularizer Ψ_S and, hence, the total objective function $\mathcal{J}_S$ in (20) are both lower semicontinuous and bounded from below by zero.

In order to derive convexity conditions for $\mathcal{J}_S$, we first introduce the function $q_a : \mathbb{R} \to \mathbb{R}_+$ defined by

$$q_a(t) := \frac{a}{2} t^2 + |t| - h_a(t) = \begin{cases} |t| & \text{for } |t| \in [0, 1/a], \\ \frac{a}{2} t^2 + \frac{1}{2a} & \text{for } |t| \in]1/a, +\infty[, \end{cases} \tag{35}$$

where the second equality in (35) comes from the Huber function definition in (12). It is easy to prove that, for any value of the parameter $a \in \mathbb{R}_{++}$, the function q_a in (35) is convex on $\mathbb{R}$, continuously differentiable on $\mathbb{R} \setminus \{0\}$, and monotonically increasing on $\mathbb{R}_+$.

Based on results in Lemma 1, in particular (21)–(22), and on definition of the Huber function in (12), the expression of function $\mathcal{J}_S$ in (20) can be manipulated and equivalently rewritten as follows:

$$\begin{aligned}
\mathcal{J}_S(x; a) &= \frac{1}{2} \|Ax - b\|_2^2 + \mu \left(\|G(Lx)\|_1 - \sum_{i=1}^{s} h_{a_i} \left(g_i(Lx) \right) \right) \\
&= \frac{1}{2} \|Ax - b\|_2^2 + \mu \sum_{i=1}^{s} \left[|g_i(Lx)| - h_{a_i} \left(g_i(Lx) \right) \right] \\
&= \frac{1}{2} \|Ax - b\|_2^2 + \mu \sum_{i=1}^{s} \left[|g_i(Lx)| - h_{a_i} \left(g_i(Lx) \right) \right]
\end{aligned}$$

$$+\frac{a_i}{2}\left(g_i\left(Lx\right)\right)^2 - \frac{a_i}{2}\left(g_i\left(Lx\right)\right)^2\bigg]$$

$$=\frac{1}{2}\|Ax-b\|_2^2 - \frac{\mu}{2}\sum_{i=1}^{s} a_i\left(g_i\left(Lx\right)\right)^2 + \mu \sum_{i=1}^{s} q_{a_i}\left(g_i(Lx)\right)$$

$$=\frac{1}{2}\left(\|Ax-b\|_2^2 - \mu\,\|WG(Lx)\|_2^2\right) + \mu \sum_{i=1}^{s} q_{a_i}\left(g_i(Lx)\right)$$

$$=\frac{1}{2}\mathcal{J}_1(x) + \underbrace{(1/2)\|b\|_2^2 - b^{\mathrm{T}}Ax}_{\mathcal{J}_2(x)} + \underbrace{\mu \sum_{i=1}^{s} q_{a_i}\left(g_i(Lx)\right)}_{\mathcal{J}_3(x)}, \tag{36}$$

with function $\mathcal{J}_1(x)$ defined in (30). Function $\mathcal{J}_2(x)$ in (36) is affine; hence it clearly does not affect convexity of the total objective function $\mathcal{J}_S$. Recalling that, given two convex functions $f_1 : \mathbb{R}^n \to \mathbb{R}$ and $f_2 : \mathbb{R} \to \mathbb{R}$, if f_1 is linear or f_2 is monotonically increasing, then the composite function $f_2 \circ f_1 : \mathbb{R}_n \to \mathbb{R}$ is convex, function $\mathcal{J}_3(x)$ in (36) is convex. In fact, since the functions q_{a_i} are all convex on $\mathbb{R}$ and monotonically increasing on $\mathbb{R}_+$ and, by assumption in the theorem statement, all functions g_i are either linear or lower semicontinuous, convex, and nonnegative, each term of the summation defining $\mathcal{J}_3$ in (36) is a convex function of x. Finally, since $\mu \in \mathbb{R}_{++}$, it follows that a sufficient condition for $\mathcal{J}_S$ to be convex (strongly convex) is that the term $\mathcal{J}_1$ in (30) is convex (strongly convex). This proves (30).

If G is the identity operator or G has the form in (31), then we have

$$\mathcal{J}_1(x) = x^{\mathrm{T}}\left(A^{\mathrm{T}}A - \mu\, L^{\mathrm{T}}W^2 L\right)x, \tag{37}$$

from which convexity condition (32) follows easily.

Finally, condition (33) comes straightforwardly from (32) after recalling the definition of matrix W in (25) and the equivalent condition (34) on $\tilde{a}$ has been proved in Lanza et al. (2017). □

In order to apply in practice the CNC strategy with separable regularizers, one has to compute the value of the scalar $\rho_{A,L}$ defined in (34), depending on the minimum singular value of the measurement matrix A, $\sigma_{A,\min}$, and on the maximum singular value of the regularization matrix L, $\sigma_{L,\max}$. In many important imaging applications, the values of $\sigma_{A,\min}$ and $\sigma_{L,\max}$ can be obtained by explicit formulas. In a general case where no explicit expressions for $\sigma_{A,\min}$ and $\sigma_{L,\max}$ are available, efficient numerical procedures can be used for their accurate estimation.

The parameter τ_c in (34) is referred to as the *convexity coefficient* of the separable CNC variational model in (19)–(20), as it allows to tune the degree of convexity of the model cost function $\mathcal{J}_S$. In particular, we notice that for τ_c approaching 0 from above, the separable regularizer Ψ_S tends toward the standard convex ℓ_1 norm-based sparsity-promoting regularizer $\|G(Lx)\|_1$, whereas for τ_c approaching 1 from below, the regularizer Ψ_S tends to be maximally non-convex (hence, potentially, maximally sparsity-promoting) under the CNC constraint that the total cost function $\mathcal{J}_S$ must be convex.

In Corollary 1 below, we highlight some important properties of the introduced class of separable CNC variational models which hold when the null spaces of the measurement matrix A and the regularization matrix L have trivial intersection. In fact, this is an important case, as it almost always occurs in practical applications.

Corollary 1. *Under the same settings of Theorem 1 with G the identity operator or a function of the form in* (31), *in case that* null(A) $\cap$ null(L) = $\{0_n\}$ *we have:*

C1. *Convexity condition* (32) *can be satisfied (with strict or weak inequality) only if matrix A has full column rank.*
C2. *If A has full column rank, and condition* (32) *is satisfied with strict inequality, then the function $\mathcal{J}_S$ in* (20) *is strongly convex; hence it admits a unique global minimizer.*
C3. *If A has full column rank, and condition* (32) *is satisfied with weak inequality, then the function $\mathcal{J}_S$ in* (20) *is convex and coercive; hence it admits a compact convex set of global minimizers.*

Proof. We prove C1 by contradiction. Let us assume that A has not full column rank, such that $A^T A$ has at least one null eigenvalue. Let v be an eigenvector associated with a null eigenvalue of $A^T A$, and let us consider the restriction $Z(t)$ of the quadratic function $x^T Q x$ – with Q the matrix defined in (32) – to the line tv, $t \in \mathbb{R}$:

$$Z(t) = tv^T Q tv = \cancel{tv^T A^T A tv} - \mu tv^T L^T W^T W L tv = -t^2 \mu \|W L v\|_2^2. \tag{38}$$

Under the considered assumption that null$(A) \cap$ null$(L) = \{0_n\}$, Lv is different from the null vector. Then, recalling that W is a positive definite diagonal matrix and that $\mu \in \mathbb{R}_{++}$, we have $\mu \|WLv\|_2^2 > 0$; hence $Z(t)$ is a quadratic concave function. This proves C1. C2 does not need a proof. For what concerns C3, first we notice that when A has full column rank, the quadratic fidelity term in (20) is coercive. Moreover, since the MC penalty defined in (13) is bounded below (by zero) for any $a \in \mathbb{R}_{++}$, then the regularizer Ψ_S in (20) is also bounded below (by zero). This implies that the total function $\mathcal{J}_S$ in (20) is coercive and C3 follows easily. □

It is an important consequence of statement C1 in Corollary 1 that if the measurement matrix $A \in \mathbb{R}^{m \times n}$ in the considered imaging application is wide, namely, $m < n$ (this is the case of many important applications, ranging from image inpainting to compressed sensing), then the CNC strategy with separable sparsity-inducing regularizers can not be used. This strongly motivated the introduction of

CNC models with non-separable regularizers, which will be illustrated in the next two sections.

Sparsity-Inducing Non-separable Regularizers

As pointed out in previous section, when the measurement matrix A is not full column rank, then a CNC formulation is not possible using a separable sparsity-promoting regularizer. However, in Lanza et al. (2019) and Selesnick et al. (2020), a general strategy to construct parameterized sparsity-promoting non-convex non-separable regularizers has been proposed which allows to tackle also the case of A not being full column rank. This is of great importance, as it enables us to apply the CNC approach to practically any linear inverse problem in imaging.

In accordance with Lanza et al. (2019) and Selesnick et al. (2020), we present a general strategy for constructing non-separable sparsity-promoting regularizers Ψ_{NS} starting from any convex sparsity-promoting regularizer $\mathcal{R}$ and then subtracting its generalized Moreau envelope. In particular, we consider regularizers $\mathcal{R}$ of the form

$$\mathcal{R}(x) := \Theta(y), \quad y = G(Lx), \tag{39}$$

where, coherently with the definitions given in previous sections, $L \in \mathbb{R}^{r \times n}$, $G : \mathbb{R}^r \to \mathbb{R}^s$ is a possibly nonlinear function, $y \in \mathbb{R}^s$ represents the image features vector to be sparsified, and $\Theta : \mathbb{R}^s \to \mathbb{R}$ is some function promoting sparsity of its argument. Following Lanza et al. (2019), the introduced regularizer and the matrix $B \in \mathbb{R}^{q \times n}$ – the meaning of which will be clarified later – must satisfy the following assumptions:

B1. $\mathcal{R} \in \Gamma_0(\mathbb{R}^n)$, bounded below by 0 with $\mathcal{R}(0) = 0$.
B2. $\Theta(G(\cdot))$ is proper, lower semicontinuous, and coercive.
B3. B has full row rank and satisfies $\mathrm{null}(B) \cap \mathrm{null}(L) = \{0_n\}$.

The non-separable sparsity-promoting regularizer Ψ_{NS} is defined as follows:

$$\Psi_{NS}(x; B) := \mathcal{R}(x) - \mathcal{H}_{NS}(x; B), \tag{40}$$

with

$$\mathcal{H}_{NS}(x; B) := \left(\mathcal{R} \square \tfrac{1}{2}\|B \cdot \|_2^2\right)(x; B) = \min_{v \in \mathcal{R}^n} \left\{\mathcal{R}(v) + \tfrac{1}{2}\|B(x - v)\|_2^2\right\}, \tag{41}$$

where B is a matrix-valued parameter which plays the same role of the parameter vector a in the class of separable regularizers illustrated in section "CNC Models with Sparsity-Inducing *Separable* Regularizers". Indeed, the introduced class of non-separable regularizers in (40)–(41) can be regarded as a sort of generalization of the class of separable regularizers defined in (20) and equivalently reformulated in (21), (22), (23), (24). The square diagonal matrix W in (25), containing the square root of the parameter vector a on the main diagonal, is replaced in (40)–(41) by a more general (not necessarily square and diagonal) parameter matrix B, and the term $\|G(Lx)\|_1$ in (21) is substituted by a more general convex function $\mathcal{R}(x) = \Theta(G(Lx))$, according to definitions (39)–(40).

We notice that the introduced regularizer in (40)–(41) can not be written as a function of only the vector to be sparsified $y = G(Lx)$, hence, coherently with Definition 1, is non-separable and takes the general form $\Psi_{\mathrm{NS}}(x; B) = \Phi(x, y)$.

We also note that if $C^\mathrm{T} C = B^\mathrm{T} B$, then $\mathcal{H}_{\mathrm{NS}}(x; B) = \mathcal{H}_{\mathrm{NS}}(x; C)$ for all $x \in \mathbb{R}^n$. That is, the function $\mathcal{H}_{\mathrm{NS}}(x; B)$ depends only on $B^\mathrm{T} B$ and not B itself. Therefore, without loss of generality, we may assume B has full row rank. In fact, if a given matrix B does not have full row rank, then there is another matrix C with full row rank such that $C^\mathrm{T} C = B^\mathrm{T} B$ which yields the same function $\mathcal{H}_{\mathrm{NS}}(x; B)$.

In the sequel, we outline some properties of function $\mathcal{H}_{\mathrm{NS}}(x; B)$, proved in Lanza et al. (2019).

Proposition 3. *The function $\mathcal{H}_{\mathrm{NS}}(x; B)$ in (41) exhibits the following properties:*

1. *For any matrix B, $\mathcal{H}_{\mathrm{NS}}(x; B)$ is proper, continuous, and convex and satisfies*

$$0 \leq \mathcal{H}_{\mathrm{NS}}(x; B) \leq \mathcal{R}(x), \quad \forall x \in \mathbb{R}^n, \tag{42}$$

$$\mathcal{H}_{\mathrm{NS}}(x; B) \leq \mathcal{H}_{\mathrm{NS}}(x; \alpha I_n), \quad \forall x \in \mathbb{R}^n, \forall \alpha \geq \|B\|_2. \tag{43}$$

2. *For any full row rank matrix B, $\mathcal{H}_{\mathrm{NS}}(x; B)$ is a differentiable function, with gradient given by*

$$\nabla \mathcal{H}_{\mathrm{NS}}(x; B) = B^\mathrm{T} B \left(x - \arg\min_{v \in \mathbb{R}^n} \left\{ \frac{1}{2} \|B(x-v)\|_2^2 + \mathcal{R}(v) \right\} \right). \tag{44}$$

Moreover, $\mathcal{H}_{\mathrm{NS}}(x; B)$ can be expressed in terms of a Moreau envelope as

$$\mathcal{H}_{\mathrm{NS}}(x; B) = \left(\mathrm{env}^1_{d \circ B^+} \circ B \right)(x), \tag{45}$$

where $d \colon \mathbb{R}^n \to \mathbb{R}$ is the convex function

$$d(x) = \min_{w \in \mathrm{null}(B)} \mathcal{R}(x - w). \tag{46}$$

By the way of illustration, in Fig. 4 we show a simple example of non-separable non-convex regularizer $\Psi_{\mathrm{NS}}(x; B)$ (third column) obtained – in accordance with the definition in (40)–(41) – by subtracting from the convex regularizer $\mathcal{R}(x) = \|x\|_1$ (first column) its generalized Moreau envelope $\mathcal{H}_{\mathrm{NS}}(x; B)$ (second column), for a

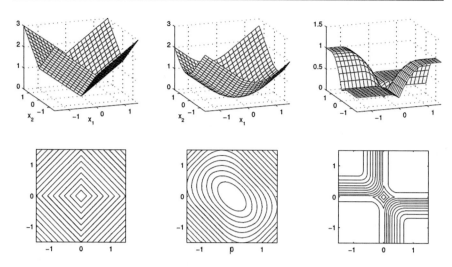

Fig. 4 Example of construction of a non-separable regularizer of the form in (40)–(41) with parameter matrix B defined in (47): $\mathcal{R}(x) = \|x\|_1$ (first column), $\mathcal{H}_{NS}(x; B)$ (second column), and $\Psi_{NS}(x; B) = \mathcal{R}(x) - \mathcal{H}_{NS}(x; B)$ (third column); the associated contour plots are shown in the bottom row

vector $x \in \mathbb{R}^2$, and a (rectangular) parameter matrix $B \in \mathbb{R}^{3 \times 2}$ defined as

$$B = \begin{bmatrix} 1 & 0 \\ 1 & 1 \\ 0 & 1 \end{bmatrix}. \tag{47}$$

CNC Models with Sparsity-Inducing *Non-separable* Regularizers

This section is concerned with the formulation of CNC variational models containing *non-separable* sparsity-promoting regularizers (see Definition 1) having the form introduced in (40)–(41). The considered non-separable CNC models thus read

$$x^* \in \arg\min_{x \in \mathbb{R}^n} \mathcal{J}_{NS}(x; B), \tag{48}$$

$$\mathcal{J}_{NS}(x; B) := \frac{1}{2}\|Ax - b\|_2^2 + \mu \Psi_{NS}(x; B), \quad \Psi_{NS}(x; B) := \mathcal{R}(x) - \mathcal{H}_{NS}(x; B), \tag{49}$$

with function $\mathcal{H}_{NS}$ defined in (41) and the matrix B and the regularizer $\mathcal{R}$ satisfying assumptions B1–B3 outlined in the previous section. We refer to (48)–(49) as the class of CNC non-separable (least-squares) models, abbreviated CNC-NS-L$_2$.

1 Convex Non-convex Variational Models

In Theorem 2, we give conditions on the parameter matrix B of the regularizer Ψ_{NS} in order to guarantee convexity (strong convexity) of the total cost function $\mathcal{J}_{NS}$ in (48)–(49); then in Corollary 2 we discuss existence and uniqueness of its minimizer(s), that is, of the solution(s) x^* of the introduced class of CNC-NS-L$_2$ variational models.

Theorem 2. *Let $\mathcal{R}$ and B satisfy assumptions B1–B3, and let Ψ_{NS} be the function defined in (49) with $\mathcal{H}_{NS}$ given in (41). Then, the function $\mathcal{J}_{NS}$ in (49) is proper, lower semicontinuous, and bounded below by zero. Moreover, a sufficient condition for $\mathcal{J}_{NS}$ to be convex (strongly convex) is that the matrix of parameters B satisfies*

$$Q := A^\mathrm{T} A - \mu\, B^\mathrm{T} B \succeq 0 \ (\succ 0)\,. \tag{50}$$

Corollary 2. *Under the same assumptions of Theorem 2, if function $\mathcal{J}_{NS}$ in (49) is strongly convex, then it admits a unique global minimizer. If, instead, $\mathcal{J}_{NS}$ is only convex, with Q weakly satisfying (50), and $\mathrm{null}(A) \cap \mathrm{null}(L) = \{0_n\}$, then $\mathcal{J}_{NS}$ is coercive; hence it admits compact convex set of global minimizers.*

The proofs of Theorem 2 and Corollary 2 are reported in Lanza et al. (2019).

Remark 1. All the previous derivations are valid for any function $\Theta : \mathbb{R}^r \to \mathbb{R}$ in the definition of the convex regularizer $\mathcal{R}$ in (39), provided that assumptions B1–B3 are satisfied. However, since $\mathcal{R} = \Theta(G(L\,\cdot\,))$ must be a convex regularizer inducing (as effectively as possible) sparsity of the features vector $y = G(Lx)$, then it is very reasonable to consider convex, sparsity-promoting, additively separable functions Θ of the form

$$\Theta(y) = \sum_{i=1}^{s} \theta(y_i), \tag{51}$$

with $\theta : \mathbb{R} \to \mathbb{R}_+$ even, continuous, convex, monotonically increasing on $\mathbb{R}_+$ and such that $\theta(0) = 0$. In particular, one of the best (and most natural) choices is to consider $\Theta = \|\cdot\|_1$, corresponding to $\theta = |\cdot|$. If one aims at avoiding non-differentiability (which is not the case in this work), a good alternative is to consider as θ the Huber function in place of the absolute value function.

Construction of Matrix B

Convexity condition (50) for the cost function $\mathcal{J}_{NS}$ in (49) sets an inequality constraint on $B^\mathrm{T} B$, hence on the matrix B of free parameters in the non-separable regularizer Ψ_{NS}. In the sequel, we illustrate a few simple strategies for choosing B.

The first and simplest strategy consists in setting $B = \sqrt{\gamma/\mu}\, A$, that is,

$$B^\mathrm{T} B = \frac{\gamma}{\mu} A^\mathrm{T} A, \quad \gamma \in [0, 1], \tag{52}$$

which clearly fulfills condition (50). We notice that, analogously to τ_c in (34) for the CNC separable models, the scalar parameter γ in (52) controls the degree of non-convexity of the non-separable regularization term Ψ_NS, hence the degree of convexity of the total objective $\mathcal{J}_\mathrm{NS}$: the greater the γ, the more non-convex the Ψ_NS and, hence, the less convex the $\mathcal{J}_\mathrm{NS}$. In particular, for γ approaching 0 from above, B tends to the null matrix, and hence, the non-separable regularizer Ψ_NS tends to the convex regularizer $\mathcal{R}$. On the other side, for γ approaching 1 from below, the regularizer Ψ_NS tends to be maximally non-convex (hence, potentially, maximally sparsity-promoting) under the CNC constraint that the total cost function $\mathcal{J}_\mathrm{NS}$ must be convex.

A more sophisticated and flexible strategy for constructing a matrix $B^\mathrm{T} B$ satisfying convexity condition (50) can be derived by considering the eigenvalue decomposition of the symmetric, positive semidefinite matrix $A^\mathrm{T} A$,

$$A^\mathrm{T} A = V E V^\mathrm{T}, \quad E, V \in \mathbb{R}^{n \times n}, \quad E = \mathrm{diag}(e_1, \ldots, e_n), \quad V^\mathrm{T} V = V V^\mathrm{T} = I_n, \tag{53}$$

with e_i, $i = 1, \ldots, n$, indicating the real non-negative eigenvalues of $A^\mathrm{T} A$. We set

$$B^\mathrm{T} B = \frac{1}{\mu} V \Gamma E V^\mathrm{T}, \quad \Gamma := \mathrm{diag}(\gamma_1, \ldots, \gamma_n), \quad \gamma_i \in [0, 1] \; \forall i \in \{1, \ldots, n\}, \tag{54}$$

so that, replacing (54) into convexity condition (50), we have

$$Q = V(E - \Gamma E) V^\mathrm{T} \succeq 0 \; (\succ 0) \iff E(I_n - \Gamma) \succeq 0 \; (\succ 0), \tag{55}$$

which is clearly satisfied given the definition of matrix Γ in (54). We notice that when one chooses $\gamma_1 = \gamma_2 = \cdots = \gamma_n = \gamma \in [0, 1]$, then (54) reduces to (52), that is, strategy (52) is included in the more general strategy (54).

Finally, in Park and Burrus (1987) another method for prescribing the matrix $B^\mathrm{T} B$, hence B, for the specific purpose of image processing with TV regularization has been proposed. In particular, the diagonal matrix Γ in (54) is set to represent a two-dimensional dc-notch filter (a type of band stop filter) defined by $\Gamma := I - H$, where H is a two-dimensional low-pass filter with a dc-gain of unity and $H \preceq I$. A simple choice for H is $H = H_0^\mathrm{T} H_0$ with H_0 a moving-average filter having square support. Hence, H is a row-column separable two-dimensional filter given by convolution with a triangle sequence (Park and Burrus 1987).

A Simple CNC Example

In this section, we provide some visual insights on the properties of the considered non-convex separable and non-separable sparsity-promoting regularizers, Ψ_S and Ψ_{NS}, respectively defined in (20) and (40). To this aim, we consider the three two-dimensional variational models defined by minimizing the cost functions

$$\mathcal{J}_R(x) := \frac{1}{2} \|Ax - b\|_2^2 + \mu \mathcal{R}(x), \qquad \mathcal{R}(x) = \|Lx\|_1, \tag{56}$$

$$\mathcal{J}_S(x; a) := \frac{1}{2} \|Ax - b\|_2^2 + \mu \Psi_S(x; a), \quad \Psi_S(x; a) = \mathcal{R}(x) - \mathcal{H}_S(x; a), \tag{57}$$

$$\mathcal{J}_{NS}(x; B) := \frac{1}{2} \|Ax - b\|_2^2 + \mu \Psi_{NS}(x; B), \quad \Psi_{NS}(x; B) = \mathcal{R}(x) - \mathcal{H}_{NS}(x; B), \tag{58}$$

where (56) represents the model containing the baseline convex ℓ_1 norm-based sparsity-inducing regularizer, the functions $\mathcal{H}_S$ in (57) and $\mathcal{H}_{NS}$ in (58) are defined in (24) and (41), respectively, and we set

$$\mu = 1.5, \quad b = \begin{bmatrix} 0 \\ 0 \end{bmatrix}, \quad A = \begin{bmatrix} 0.4 & 1.5 \\ -1.0 & 0.8 \end{bmatrix}, \quad L = \begin{bmatrix} -2.0 & -1.0 \\ 0.5 & -2.5 \end{bmatrix}. \tag{59}$$

Moreover, according to the convexity conditions in (34) and (52), for the CNC separable and non-separable models in (57) and (58), we choose

$$a_1 = a_2 = \bar{a} = \tau_c \frac{\rho_{A,L}}{\mu}, \quad \tau_c = 0.99, \tag{60}$$

$$B = \sqrt{\frac{\gamma}{\mu}} A, \quad \gamma = 0.99, \tag{61}$$

respectively, so that both the CNC models are pushed toward their convexity limit.

In Fig. 5, we show the regularizer $\mathcal{R}$ and total cost function $\mathcal{J}_R$ of the baseline convex model (56), in Fig. 6 the regularizer Ψ_S and total cost function $\mathcal{J}_S$ of the separable CNC model (57), and in Fig. 7 the regularizer Ψ_{NS} and total cost function $\mathcal{J}_{NS}$ of the non-separable CNC model (58). All function graphs are accompanied, in the bottom row, by their associated contour plots. The solid red and blue lines in the contour plot figures represent the hyperplanes Y_1 and Y_2, respectively, with $Y_i := \{x \in \mathbb{R}^2 : L_i x = 0\}$, $i \in \{1, 2\}$, and L_i the i-th row of matrix L.

From the left columns of Figs. 5, 6, and 7, it can be noticed that the baseline regularizer $\mathcal{R}(x)$ is clearly convex, but not strictly convex, whereas the separable and non-separable regularizers $\Psi_S(x; a)$ and $\Psi_{NS}(x; B)$ are non-convex. In fact, according to their definitions in (57) and (58), they are obtained by subtracting

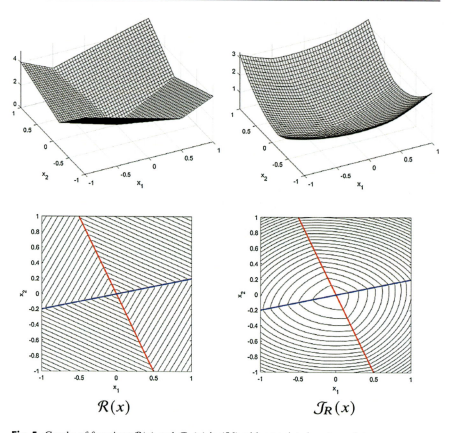

Fig. 5 Graphs of functions $\mathcal{R}(x)$ and $\mathcal{J}_R(x)$ in (56) with associated contour plots

from $\mathcal{R}(x)$ the convex terms $\mathcal{H}_S(x; a)$ and $\mathcal{H}_{NS}(x; B)$, respectively. The non-convex regularizers $\Psi_S(x; a)$ and $\Psi_{NS}(x; B)$ thus hold the potential for promoting sparsity of the vector $Lx = (L_1 x, L_2 x)^T$ more effectively than the convex regularizer $\mathcal{R}(x)$.

The plots in the right columns of Figs. 5, 6, and 7 confirm, first, that the total cost function $\mathcal{J}_R(x)$ is clearly convex and then, more interestingly, that the cost functions $\mathcal{J}_S(x; a)$ and $\mathcal{J}_{NS}(x; B)$ of the separable and non-separable CNC models in (57) and (58) are also both convex, as prescribed by the CNC rationale and as expected due to our settings $\tau_c = \gamma = 0.99 < 1$.

As a final interesting experiment, we push both the separable and non-separable CNC models in (57), (58) outside their guaranteed convexity regimes, as defined by sufficient conditions (34), (52), respectively. More precisely, we set $\tau_c, \gamma > 1$ in (60), (61), thus obtaining the total cost functions $\mathcal{J}_S(x; a)$, $\mathcal{J}_{NS}(x; B)$ depicted in Fig. 8. It can be noticed from the graphs in the top row and, more clearly, from the associated contour plots in the bottom row that both the cost functions are non-convex, as expected from theory.

1 Convex Non-convex Variational Models

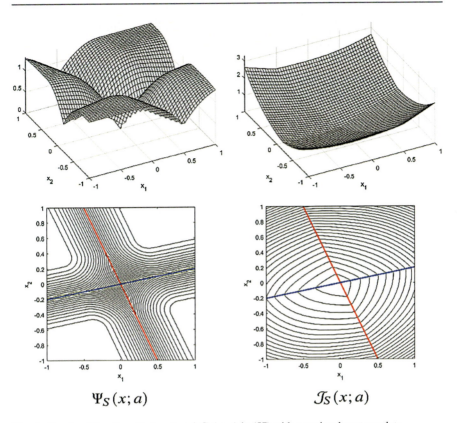

Fig. 6 Graphs of functions $\Psi_S(x;a)$ and $\mathcal{J}_S(x;a)$ in (57) with associated contour plots

Path of Solution Components

The different behavior of standard ℓ_1 norm convex regularization versus its associated non-convex non-separable regularization can be illustrated by observing the solution path as the regularization parameter μ varies. In particular, we denote by x_{L_1} the solution of the minimization problem (56) with $L = I$ and by x_{NS} the solution of its associated non-separable CNC model (58). When μ is sufficiently large, both the solutions x_{L_1} and x_{NS} will be the all-zero vector. When μ is sufficiently close to zero, the solution using either regularizations will approximate the unconstrained least-squares solution. However, as μ varies between these two extremes, the solutions obtained using the two regularization methods will sweep different paths. This is illustrated in Fig. 2.1 in Hastie et al. (2015) which concerns an example of least-squares problem with ℓ_1 norm regularization where matrix A is of size 50×5. This example is reproduced in Fig. 9. As in Hastie et al. (2015), the solution path is shown as a function of the fraction: the ℓ_1 norm of x_{L_1} divided by the ℓ_1 norm of the unconstrained (unregularized) least-squares solution x_{LS}; this fraction varies between zero and one. Repeating the same example using non-

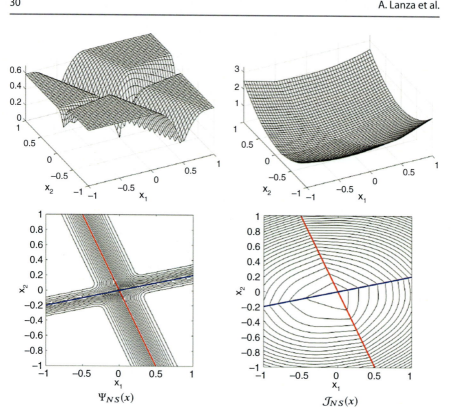

Fig. 7 Graphs of functions $\Psi_{NS}(x;B)$ and $\mathcal{J}_{NS}(x;B)$ in (58) with associated contour plots

separable non-convex regularization in (58) instead of ℓ_1 norm regularization, we obtain a different solution path for x_{NS}, as shown in Fig. 9. It can be seen that the x_{NS} solution is more sparse than the ℓ_1 norm solution x_{L_1} for most of the solution component path. The x_{NS} solution starts to have two non-zero components when the x_{L_1} solution already has three non-zero components. It can also be seen that along most of the solution path, non-zero components of the x_{NS} solution are greater in absolute value than those of the x_{L_1} solution. The solution paths show that components of the x_{NS} solution become non-zero later (along this axis) than components of the x_{L_1} solution.

Forward-Backward Minimization Algorithms

In this section, we introduce optimization algorithms for the numerical solution of the illustrated separable and non-separable CNC variational models, based on the iterative FB strategy within the general framework of splitting, commonly used when the objective function is the sum of two convex but not necessarily differentiable functions. This iterative method, proposed in Beck and Teboulle

1 Convex Non-convex Variational Models

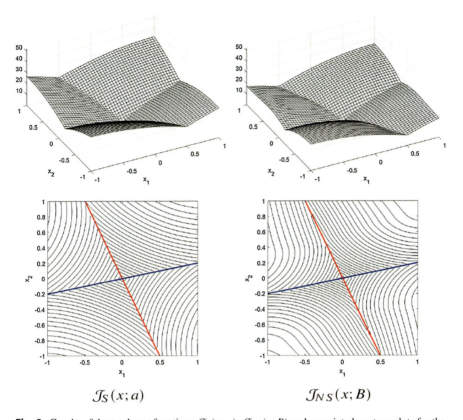

Fig. 8 Graphs of the total cost functions $\mathcal{J}_S(x;a)$, $\mathcal{J}_{NS}(x;B)$ and associated contour plots for the separable and non-separable variational models in (57), (58) pushed beyond their convexity limit, that is, for $\tau_c, \gamma > 1$

(2009), has attracted extensive interests due to its simplicity and several important advantages. It is well-known that this method uses little storage, readily exploits the separable structure of the minimization problem, and is easily implemented to practical applications. It relies on a forward gradient step (an explicit step) followed by a backward proximal step (an implicit step).

In the separable case (section "FB Strategy for Separable CNC Models"), it reduces to a standard proximal gradient or subgradient splitting minimization algorithm. In the non-separable case (section "FB Strategy for Non-separable CNC Models"), a more general form of the FB algorithm aimed to solve monotone inclusion problems is used. The solution of the minimization problems in the backward steps of the FB applied to both the separable and non-separable cases relies on a very efficient ADMM strategy (section "Efficient Solution of the Backward Steps by ADMM").

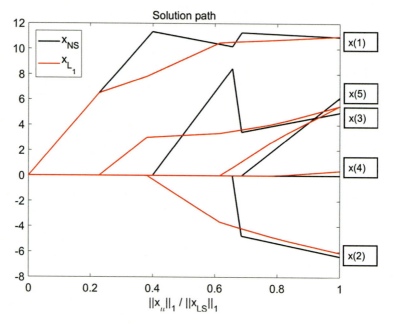

Fig. 9 Path of the five solution components for the regularized least squares example in Hastie et al. (2015); $\|x_\mu\|/\|x_{LS}\|$ is the red colored path for $x_\mu = x_{L_1}$ and the black colored path for $x_\mu = x_{\text{NS}}$, for increasing values of the regularization parameter μ

FB Strategy for Separable CNC Models

Based on Lemma 1, in particular expression (21) for the separable sparsity-promoting regularizer Ψ_S, with function $\mathcal{H}_S$ in the forms (22) and (24), the class of considered separable CNC variational models defined in (19)–(20) can be equivalently rewritten in the following equivalent form:

$$x^* \in \arg\min_{x \in \mathbb{R}^n} \mathcal{J}_S(x; a), \tag{62}$$

$$\mathcal{J}_S(x; a) = \frac{1}{2}\|Ax - b\|_2^2 - \mu \sum_{i=1}^{s} h_{a_i}\left(g_i(Lx)\right) + \mu \|G(Lx)\|_1 \tag{63}$$

$$= \underbrace{\frac{1}{2}\|Ax - b\|_2^2 - \mu\,\text{env}_{\|W^{-1}\cdot\|_1}^1\left(WG(Lx)\right)}_{\mathcal{J}_1(x; a)} + \underbrace{\mu\|G(Lx)\|_1}_{\mathcal{J}_2(x)}. \tag{64}$$

Based on results in Theorem 1, first we notice that if convexity condition (30) is satisfied – which is the case of interest for us – both the total objective $\mathcal{J}_S$ and the two terms $\mathcal{J}_1$ and $\mathcal{J}_2$ in (64) are proper, lower semicontinuous, and convex functions. Then, the term $\mathcal{J}_2$ is in general – i.e., for the great majority of reasonable functions G – a non-differentiable function, whereas $\mathcal{J}_1$ can be differentiable or non-differentiable depending on G. Indeed, some popular regularizers are defined in terms of G functions yielding differentiability of $\mathcal{J}_1$, as it will be illustrated in Proposition 4.

Hence, we propose to compute approximate solutions x^* of the CNC separable model in (62), (63), and (64) by means of the FB iterative scheme outlined in Proposition 5. The forward step consists of a subgradient – or gradient, depending on G – descent step of the term $\mathcal{J}_1$, whereas the backward step is a proximal step of $\mathcal{J}_2$. In Proposition 4, we preliminarily derive the expression of the subgradient – or gradient – of the function $\mathcal{J}_1$.

Proposition 4. *Let* $\mathcal{J}_1 : \mathbb{R}^n \to \mathbb{R}$ *be the function defined in (64), and let the convexity conditions (30) for $\mathcal{J}_S$ be satisfied. Then, in the general case of a possibly non-differentiable function G, the subdifferential* $\partial \mathcal{J}_1 : \mathbb{R}^n \rightrightarrows \mathbb{R}^n$ *takes the form*

$$\partial \mathcal{J}_1(x;a) = A^T(Ax - b)$$
$$- \mu L^T \partial G(Lx) W \left(WG(Lx) - \mathrm{prox}_{\|W^{-1}\cdot\|_1}\left(WG(Lx) \right) \right), \quad (65)$$

with $\partial \mathcal{J}_1$ and ∂G replaced by $\nabla \mathcal{J}_1$ and ∇G if G is differentiable.

In the special case that G is a non-differentiable function of the form in (31) with the partition of vector $z = Lx$ defined by a permutation matrix $P = \left((P^{(1)})^T, \ldots, (P^{(s)})^T \right)^T \in \mathbb{R}^{r \times r}$, $P^{(i)} \in \mathbb{R}^{r_i \times r}$, $i = 1, \ldots, s$, then the function $\mathcal{J}_1$ in (64) is differentiable with gradient $\nabla \mathcal{J}_1 : \mathbb{R}^n \to \mathbb{R}^n$ given by

$$\nabla \mathcal{J}_1(x;a) = A^T(Ax - b) - \mu L^T P^T C(Lx) P L x, \quad C = \mathrm{diag}\left(C^{(1)}, \ldots, C^{(s)} \right), \quad (66)$$

where $C : \mathbb{R}^r \to \mathbb{R}^{r \times r}$ is a block-diagonal matrix-valued function with scalar diagonal blocks defined by

$$C^{(i)}(z) = \min\left\{ a_i, \frac{1}{\|P^{(i)}z\|_2} \right\} I_{r_i}, \quad i = 1, \ldots, s. \quad (67)$$

Proof. The quadratic term in $\mathcal{J}_1$ – namely, the data fidelity term – is clearly differentiable with gradient given by $A^T(Ax - b)$. Recalling that the Moreau envelope is a differentiable function (see Proposition 1), the second term in $\mathcal{J}_1$ is differentiable if the function G is differentiable. In fact, in this case the term is composition of differentiable functions. If G is non-differentiable, then the term can be non-differentiable or, for some special G, also differentiable.

In the general case of a possibly non-differentiable function G, expression (65) for the subdifferential of $\mathcal{J}_1$ comes from applying the chain rule of differentiation to the calculus of the subdifferential of function $\mathcal{J}_1$ in the form (64) and from recalling the expression of the gradient of the Moreau envelope function given in (14).

To demonstrate (66)–(67), first we notice that if G has the form in (31), we can write:

$$\mathcal{H}_S(x;a) = \sum_{i=1}^{s} h_{a_i}(g_i(Lx)) = \sum_{i=1}^{s} h_{a_i}(\|z^{(i)}\|_2) = \sum_{i=1}^{s} f_{a_i}(P^{(i)}z), \quad z = Lx,$$

with f_a the function defined in (17). Hence, we have

$$\mathcal{H}_S(x;a) = H(Lx;a), \quad \text{with} \quad H(z;a) := \sum_{i=1}^{s} f_{a_i}(P^{(i)}z). \tag{68}$$

It follows from Proposition 2 that the function $H(z;a)$ above is differentiable (sum of differentiable functions) with gradient given by

$$\nabla_z H(z) = \sum_{i=1}^{s} \left[(P^{(i)})^T \nabla_z f_{a_i}(P^{(i)}z) \right]$$

$$= \sum_{i=1}^{s} \left((P^{(i)})^T \min\left\{ a_i, 1/\|P^{(i)}z\|_2 \right\} P^{(i)} z \right)$$

$$= \left(\sum_{i=1}^{s} \left((P^{(i)})^T \min\left\{ a_i, 1/\|P^{(i)}z\|_2 \right\} P^{(i)} \right) \right) z$$

$$= P^T C(z) P z,$$

with C the diagonal matrix-valued function defined in (66)–(67). The function $\mathcal{H}_S(x;a)$ in (68) is thus differentiable with gradient given by

$$\nabla_x \mathcal{H}_S(x;a) = L^T \nabla_z H(Lx;a) = L^T P^T C(Lx) P L x.$$

Recalling the definition of function $\mathcal{J}_1$ in (63), it is thus clear that it is a differentiable function with gradient given in (66)–(67). □

Proposition 5. *Let $\mathcal{J}_S(x;a) : \mathbb{R}^n \to \mathbb{R}$ be the function defined in (62), (63), and (64), with parameters $a \in \mathbb{R}_{++}^s$ satisfying convexity condition in (30). Then, a global minimizer x^* of $\mathcal{J}_S$ can be obtained as the limit point of the sequence of iterates $\{x^{(k)}\}_{k=1}^{\infty}$ generated by the following FB iterative scheme:*

for $k = 0, 1, 2, \ldots$

$$\omega^{(k)} \in \partial \mathcal{J}_1\left(x^{(k)}\right)$$

$$w^{(k)} = x^{(k)} - \lambda^{(k)} \omega^{(k)}$$

1 Convex Non-convex Variational Models

$$x^{(k+1)} = \text{prox}_{\mathcal{J}_2}^{1/\lambda^{(k)}}\left(w^{(k)}\right) = \arg\min_{x \in \mathbb{R}^n}\left\{\|G(Lx)\|_1 + \frac{1}{2\lambda^{(k)}\mu}\|x - w^{(k)}\|_2^2\right\}$$

end

where the variable stepsizes $\lambda^{(k)}$ are chosen according to the strategy in Bello Cruz (2017) if $\mathcal{J}_1$ is non-differentiable, or $\lambda^{(k)} = \lambda \in\,]0, 2/\rho[$ with ρ the Lipschitz constant of the gradient of $\mathcal{J}_1$, if $\mathcal{J}_1$ is differentiable.

For a generic non-differentiable G function, (62), (63), and (64) is a non-smooth convex optimization problem with an objective function which is the sum of two non-differentiable convex functions, $\mathcal{J}_1$ and $\mathcal{J}_2$. In this case, the proximal FB splitting iteration in Proposition 5 – in particular, the computation of $\omega^{(k)}$ in the forward step – relies on the subdifferential (65). For the convergence of this particular FB case, we refer the reader to Bello Cruz (2017).

In case that G is a differentiable function (e.g., G is the identity function) or a non-differentiable function of the special form in (31), the proximal FB splitting iteration in Proposition 5 uses the gradient given in (66). Therefore, the convergence follows the classical results in Beck and Teboulle (2009).

FB Strategy for Non-separable CNC Models

Even though the proposed class of non-separable regularization functions Ψ_{NS} in (49) does not have a simple explicit formula, a global minimizer of the total sparse-regularized cost function $\mathcal{J}_{NS}$ in (49) can be readily calculated using proximal algorithms.

As described in Lanza et al. (2019), in order to compute the solution x^* of the minimization problem in (48)–(49) by using proximal algorithms, it is useful to rewrite it as an equivalent saddle-point problem:

$$\{x^*, v^*\} = \arg\min_{x \in \mathbb{R}^n}\max_{v \in \mathbb{R}^n} \mathcal{F}(x, v; B), \tag{69}$$

$$\mathcal{F}(x, v; B) = \frac{1}{2}\|Ax - b\|_2^2 + \mu \mathcal{R}(x) - \mu \mathcal{R}(v) - \frac{\mu}{2}\|B(x - v)\|_2^2, \tag{70}$$

where, we recall, the regularization function $\mathcal{R}(x) = \Theta(G(Lx))$ and the parameter matrix B satisfy assumptions B1–B3 outlined at the beginning of section "Sparsity-Inducing Non-separable Regularizers".

The solution of the saddle-point problem above can be calculated using a general form of the FB algorithm (Theorem 25.8 in Bauschke and Combettes 2011). This form of the FB algorithm is formulated to solve the general class of monotone inclusion problems, of which the saddle-point problem (69)–(70) is a special case.

The algorithm, which we will refer to as Primal-Dual FB (PDFB) (as in Lanza et al. (2019)), is outlined in Proposition 6. It involves operators A, A^T, B, and B^T and the proximity operator of the regularization term $\mathcal{R}$.

Proposition 6. *Let $\mathcal{F}(x, v; B) : \mathbb{R}^{2n} \to \mathbb{R}$ be the function defined in (70) with the parameters matrix B set as in (53)–(54). Then, a saddle-point $\{x^*, v^*\}$ of $\mathcal{F}$ can be obtained as the limit point of the sequence of iterates $\{x^{(k)}, v^{(k)}\}_{k=1}^{\infty}$ generated by the following PDFB iterative scheme:*

$$\text{set} \quad \rho = \max_{i} \left\{ \frac{1 - 2\gamma_i + 2\gamma_i^2}{1 - \gamma_i} e_i \right\}$$

$$\text{set} \quad \lambda \in \,]0, 2/\rho\,[$$

for $k = 0, 1, 2, \ldots$

$$w^{(k)} = x^{(k)} - \lambda \left[A^T \left(A x^{(k)} - b \right) + \mu \, B^T B \left(v^{(k)} - x^{(k)} \right) \right]$$

$$u^{(k)} = v^{(k)} - \lambda \mu \, B^T B (v^{(k)} - x^{(k)})$$

$$x^{(k+1)} = \arg\min_{x \in \mathbb{R}^n} \left\{ \mathcal{R}(x) + \frac{1}{2\lambda\mu} \|x - w^{(k)}\|_2^2 \right\}$$

$$v^{(k+1)} = \arg\min_{v \in \mathbb{R}^n} \left\{ \mathcal{R}(v) + \frac{1}{2\lambda\mu} \|v - u^{(k)}\|_2^2 \right\}$$

end

where e_i and γ_i are defined in (53)–(54) and k is the iteration counter.

Efficient Solution of the Backward Steps by ADMM

The backward steps in the FB and PDFB algorithms outlined in Propositions 5 and 6 for the numerical solution of the separable and non-separable CNC variational models illustrated in sections "CNC Models with Sparsity-Inducing *Separable* Regularizers" and "CNC Models with Sparsity-Inducing *Non-separable* Regularizers", respectively, all consist of solving the same class of minimization problems, which, in the general case, does not admit a closed-form solution. More precisely, the computations of $x^{(k+1)}$ in the FB algorithm in Proposition 5 and of $x^{(k+1)}$ and $v^{(k+1)}$ in the PDFB algorithm in Proposition 6 all correspond to calculating the proximal operator of a regularization function $\mathcal{R} : \mathbb{R}^n \to \mathbb{R}$ of the form $\mathcal{R} = \Upsilon(G(L \cdot))$ with proximity parameter $\alpha := 1/(\lambda\mu) \in \mathbb{R}_{++}$ at a point $p \in \mathbb{R}^n$ (equal to $w^{(k)}$ for $x^{(k+1)}$ and to $u^{(k)}$ for $v^{(k+1)}$). We have thus to solve the following minimization problem:

$$t^* = \text{prox}_{\mathcal{R}}^{\alpha}(p) = \arg\min_{t \in \mathbb{R}^n} \left\{ \mathcal{R}(t) + \frac{\alpha}{2} \|t - p\|_2^2 \right\}$$

$$= \arg\min_{t \in \mathbb{R}^n} \left\{ \Upsilon(G(L\,t)) + \frac{\alpha}{2} \|t - p\|_2^2 \right\}. \tag{71}$$

For both the FB and PDFB cases, the matrix L and the function G – hence, the image features vector $y = G(L \cdot)$ to be sparsified – are defined as in section "Introduction", whereas the function $\Upsilon : \mathbb{R}^s \to \mathbb{R}$ is to the ℓ_1 norm function $\|\cdot\|_1$ for FB and the function Θ for PDFB. In both cases, it follows from the considered convexity assumptions/conditions that the regularizer $\mathcal{R} = \Upsilon(G(L \cdot))$ is convex; hence the cost function in (71) is strongly convex and admits a unique (global) minimizer t^*.

As it will be later discussed, in most cases of practical interest the function $\Upsilon(G(\cdot))$ is easily proximable, that is, its proximity operator admits a closed form expression or can be calculated very efficiently. Hence, we suggest to solve the minimization problem in (71) by means of the following ADMM-based approach.

First, we rewrite (71) in the equivalent linearly constrained form:

$$\{t^*, z^*\} = \arg\min_{t,z} \left\{ \Upsilon(G(z)) + \frac{\alpha}{2} \|t - p\|_2^2 \right\} \quad \text{s.t.} : \quad z = L\,t, \tag{72}$$

where $z \in \mathbb{R}^r$ is an auxiliary variable (the notation has been chosen for coherence with definition in (6)). Then, we introduce the augmented Lagrangian function,

$$\mathcal{L}(t, z, \rho) = \Upsilon(G(z)) + \frac{\alpha}{2} \|t - p\|_2^2 - \langle \rho, z - L\,t \rangle + \frac{\beta}{2} \|z - L\,t\|_2^2, \tag{73}$$

where $\beta > 0$ is a scalar penalty parameter and $\rho \in \mathbb{R}^r$ is the dual variable, i.e., the vector of Lagrange multipliers associated with the set of r linear constraints in (72). Solving (72) is tantamount to seek for the saddle point of the augmented Lagrangian function in (73). The saddle-point problem reads as follows:

$$\{t^*, z^*\} = \arg\min_{t,z} \max_{\rho} \mathcal{L}(t, z, \rho). \tag{74}$$

Upon suitable initialization, and for any $j = 0, 1, 2, \ldots$, the j-th iteration of the ADMM applied to solving the saddle-point problem (74) with the augmented Lagrangian function $\mathcal{L}$ defined in (73) reads as follows:

$$t^{(j+1)} = \arg\min_{t \in \mathbb{R}^n} \mathcal{L}(t, z^{(j)}, \rho^{(j)})$$

$$= \left(\epsilon I_n + L^T L \right)^{-1} \left(\epsilon p + L^T \left(z^{(j)} - \frac{1}{\beta} \rho^{(j)} \right) \right), \quad \epsilon = \frac{\alpha}{\beta}, \tag{75}$$

$$z^{(j+1)} = \arg\min_{z \in \mathbb{R}^r} \mathcal{L}(t^{(j+1)}, z, \rho^{(j)}) = \arg\min_{z \in \mathbb{R}^r} \left\{ \Upsilon(G(z)) + \frac{\beta}{2} \left\| z - q^{(j)} \right\|_2^2 \right\}$$

$$= \operatorname{prox}^{\beta}_{\Upsilon(G(\cdot))} \left(q^{(j)} \right), \quad q^{(j)} = L t^{(j+1)} + \frac{1}{\beta} \rho^{(j)}, \tag{76}$$

$$\rho^{(j+1)} = \rho^{(j)} - \beta \left(z^{(j+1)} - L t^{(j+1)} \right). \tag{77}$$

The ADMM scheme outlined above has guaranteed convergence and, in most cases of practical interest, allows to compute very efficiently the solution t^* of (71).

In the general case, the computational cost of the ADMM iteration (75), (76), and (77) is dominated by the solution of the two subproblems for the primal variables t and z, as the cost for updating the dual variable $\rho \in \mathbb{R}^r$ by (77) is linear in r, hence in the number of pixels n (we do not consider the cost of multiplication by matrix L since the term $L t^{(j+1)}$ in (77) must have been previously computed for solving (76)).

The subproblem for t in (75) consists in solving an $n \times n$ linear system with symmetric positive definite (hence, invertible) coefficient matrix $\epsilon I_n + L^T L$. For ADMM implementations with iteration-independent penalty parameter β, the matrix is constant along the ADMM iterations, and for FB (or PDFB) implementations with iteration-independent stepsize λ, it is also constant along the (outer) FB (or PDFB) iterations. The linear system can thus be solved by direct methods, namely, Cholesky factorization carried out once for all before starting iterations and solution of (75) by forward and backward substitution, or by iterative methods. In particular, when L is a sparse matrix, the (suitably preconditioned) conjugate gradient method equipped with some variable stopping tolerance strategy represents a good (i.e., efficient) choice. If L is a diagonal matrix, or some unitary matrix (e.g., the 2D discrete Fourier or cosine transform matrix, so as to sparsify the sought image coefficients in the Fourier or cosine basis), or the matrix of some overcomplete dictionary satisfying the tight frame condition $L^T L = \delta I_n$, $\delta \in \mathbb{R}_{++}$, then the coefficient matrix is diagonal and (75) can be solved very efficiently. Finally, in the special but practically very important case where $L = \left(L_1^T, \ldots, L_c^T \right)^T$ with $L_i \in \mathbb{R}^{n \times n}$ convolution matrices, $i = 1, \ldots, c$, the linear system can also be solved very efficiently by fast 2D discrete transforms. In particular, by assuming periodic, symmetric, or anti-symmetric boundary conditions for the unknown image t, the linear system in (75) can be solved by using the fast 2D discrete Fourier, cosine, or sine transforms, respectively, all characterized by $O(n \log_2(n))$ computational complexity. This is the case of the TV regularizer (isotropic and anisotropic), the Hessian-based regularizers and, more in general, of the whole important class of widely used regularizers aimed to sparsify some (discretized) differential quantity of the sought image.

Based on Remark 1 in section "CNC Models with Sparsity-Inducing *Non-separable* Regularizers", for both the FB and PDFB cases the subproblem for z in (76) can be written as

1 Convex Non-convex Variational Models

$$\hat{z} = \arg\min_{z \in \mathbb{R}^r} \left\{ \sum_{i=1}^{s} \upsilon\left(g_i(z)\right) + \frac{\beta}{2} \|z - q\|_2^2 \right\}, \tag{78}$$

where, to simplify notations, we dropped the iteration index superscripts (namely, $\hat{z} = z^{(j+1)}$ and $q = q^{(j)}$) and where the function $\upsilon : \mathbb{R} \to \mathbb{R}_+$ is defined by $\upsilon = |\cdot|$ for FB and by $\upsilon = \theta$ for PDFB. Then, for the important case of sparsified image feature vectors $y = G(Lx)$ characterized by the function G being the identity operator or a function of the form in (31), the r-dimensional minimization problem in (78) is separable into the following s independent (and lower-dimensional) sub-problems:

$$\hat{z}^{(i)} = \arg\min_{z^{(i)} \in \mathbb{R}^{r_i}} \left\{ \upsilon\left(\|z^{(i)}\|_2\right) + \frac{\beta}{2} \|z^{(i)} - q^{(i)}\|_2^2 \right\} \tag{79}$$

$$= \operatorname{prox}_{\upsilon(\|\cdot\|_2)}^{\beta}\left(q^{(i)}\right), \quad i = 1, \ldots, s, \tag{80}$$

where, in accordance with (31), the set of (sub-)vectors $\{z^{(i)}\}_{i=1}^{s}$, $z^{(i)} \in \mathbb{R}^{r_i}$, $\sum_{i=1}^{s} r_i = r$, represents a partition of vector $z \in \mathbb{R}^r$, i.e., $\left(\left(z^{(1)}\right)^T, \ldots, \left(z^{(s)}\right)^T\right)^T = Pz$, with $P \in \mathbb{R}^{r \times r}$ a permutation matrix. Clearly, the (sub-)vectors $\hat{z}^{(i)}, q^{(i)} \in \mathbb{R}^{r_i}$ in (79)–(80) are defined according to an analogous partition of vectors $\hat{z}, q \in \mathbb{R}^r$ in (78). The s minimization problems in (79) may have different dimensionality – in fact, in the considered general case, the integers r_i are not assumed to be equal – but they all have the same structure corresponding to the proximal map of the composite function $\upsilon(\|\cdot\|_2)$, as outlined in (80). Based on results in Proposition 7 below, under quite general and very reasonable, i.e., very likely to be satisfied in practice, assumptions on function υ, each sub-problem in (79)–(80) reduces to a 1-d strongly convex box-constrained (well-posed) minimization problem which, for most popular υ functions, admits a closed-form solution. In particular, based on (83)–(84), if υ is the absolute value function, then (79)–(80) reduces to

$$z^{(i)} = \begin{cases} 0_{r_i} & \text{if } \|q^{(i)}\|_2 = 0, \\ \max\left\{1 - \frac{1}{\beta \|q^{(i)}\|_2}, 0\right\} q & \text{if } \|q^{(i)}\|_2 > 0, \end{cases} \quad i = 1, 2, \ldots, s. \tag{81}$$

Recalling that, based on definition (31), the vectors $q^{(i)}$ form a partition of $q \in \mathbb{R}^r$ and, hence, $s \leq r$, the computation in (81) – including calculation of all the ℓ_2 norm terms $\|q^{(i)}\|_2$ – has linear complexity in the dimension r of the codomain of matrix $L \in \mathbb{R}^{r \times n}$, hence in the number of pixels n.

Proposition 7. *For any proper, lower semicontinuous, convex, and monotonically increasing function $\upsilon : \mathbb{R}_+ \to \overline{\mathbb{R}}$, the composite function $\upsilon(\|\cdot\|_2) : \mathbb{R}^r \to \overline{\mathbb{R}}$*

is proper, lower semicontinuous, and convex and its proximal map with proximity parameter $\beta \in \mathbb{R}_{++}$ evaluated at point $q \in \mathbb{R}^r$ is given by

$$\operatorname{prox}^{\beta}_{\upsilon(\|\cdot\|_2)}(q) = \arg\min_{z \in \mathbb{R}^r} \left\{ \upsilon\left(\|z\|_2\right) + \frac{\beta}{2} \|z - q\|_2^2 \right\} \tag{82}$$

$$= \begin{cases} 0_r & \text{if } \|q\|_2 = 0, \\ \hat{\xi} \dfrac{q}{\|q\|_2} & \text{if } \|q\|_2 > 0, \end{cases} \quad \hat{\xi} = \arg\min_{\xi \in [0, \|q\|_2]} \left\{ \upsilon(\xi) + \frac{\beta}{2} \left(\xi - \|q\|_2\right)^2 \right\}. \tag{83}$$

In particular, if υ is the identity function, then $\hat{\xi}$ in (83) is given by

$$\hat{\xi} = \max\left\{ \|q\|_2 - \frac{1}{\beta}, 0 \right\}. \tag{84}$$

Proof. First, all the stated properties of composite function $\upsilon(\|\cdot\|_2)$ come easily from the assumed properties of functions υ and from the ℓ_2 norm function $\|\cdot\|_2$ being continuous and convex on the entire domain $\mathbb{R}^r$.

Then, convexity of $\upsilon(\|\cdot\|_2)$ yields strong convexity of the cost function in (82) which, hence, admits a unique (global) minimizer $\hat{z} \in \mathbb{R}^r$. If $\|q\|_2 = 0$ or, equivalently, q is the null vector, then the cost function in (82) reduces to $\upsilon(\|z\|_2) + (\beta/2)\|z\|_2^2$, which is a monotonically increasing function of $\|z\|_2$. The solution of (82) in this case is thus $\hat{z} = 0_r$. If q is not the null vector, i.e., $\|q\|_2 > 0$, then it is easy to prove (see the initial part of the proof of Proposition 1 in Sidky et al. 2014) that, under the considered assumptions on function υ, the solution of (82) must belong to the closed segment of extremes 0_r and q. By thus considering the restriction of the cost function in (82) to that segment, parameterized by $z = \xi q/\|q\|_2, \xi \in [0, \|q\|_2]$, one easily obtains the 1-D constrained minimization problem in (83). Finally, the closed-form solution in (84) obtained when υ is the identity function represents the quite popular multidimensional soft-thresholding operator. Its derivation can be found, e.g., in the proof of Proposition 1 in Sidky et al. (2014). □

It is worth noticing that in the special case where the regularization matrix L is the identity matrix (e.g., when one wants to sparsify the image itself as it is expected to be predominantly zero-valued, or in general in the synthesis-based sparse reconstruction framework), then the backward step in (71) reduces to

$$t^* = \arg\min_{t \in \mathbb{R}^n} \left\{ \varUpsilon(G(t)) + \frac{\alpha}{2} \|t - p\|_2^2 \right\} = \operatorname{prox}^{\alpha}_{\varUpsilon(G(\cdot))}(p). \tag{85}$$

Hence, ADMM is not required since problem (85) consists in computing only one proximal map of the same type as in the ADMM sub-problem for z in (76), which in its turn reduces to solving the s lower-dimensional problems in (79)–(80) by, e.g., (81).

1 Convex Non-convex Variational Models

Finally, we notice that a suitable warm-start strategy can be used in both the FB and PDFB approaches in order to further speedup the backward step computation by ADMM. More precisely, at each (outer) iteration of the FB and PDFB algorithms, the (inner) iterative ADMM scheme in (75), (76), and (77) is initialized with the results of previous (outer) iteration, in terms of both the primal variables t, z and the dual variable ρ. This allows to significantly decrease the number of ADMM iterations.

Numerical Examples

In this section, we test the non-convex separable and non-separable sparsity-promoting regularization terms introduced in sections "Sparsity-Inducing Separable Regularizers" and "Sparsity-Inducing Non-separable Regularizers", respectively. More precisely, we are interested in evaluating experimentally the performance of the two classes of separable CNC-S-L$_2$ and non-separable CNC-NS-L$_2$ variational models illustrated in sections "CNC Models with Sparsity-Inducing *Separable* Regularizers" and "CNC Models with Sparsity-Inducing *Non-separable* Regularizers", respectively, when applied to the linear discrete inverse problem of restoring images corrupted by blur and AWG noise. More broadly, the goal of this numerical session is to investigate experimentally if and how convex variational models containing non-convex sparsity-inducing regularizers, i.e., the class of CNC models, can improve upon standard convex models containing convex sparsity-promoting regularizers in case the sought solution really exhibits some sparsity property.

At this aim, we consider the three gray-scale test images SPD0, SPD1, SPD2 shown in Fig. 1 and reported again in the first row of Fig. 10. They all have resolution 256×256 pixels and, we recall, they are characterized, from left to right, by strong sparsity of the three feature vectors

$$y^{(j)} \in \mathbb{R}^n, \quad \text{with} \quad y_i^{(0)} = |x_i|, \quad y_i^{(1)} = \left\|(\nabla x)_i\right\|_2, \quad y_i^{(2)} = \left\|(Hx)_i\right\|_F, \tag{86}$$

$i = 1, \ldots, n$, respectively, where $(\nabla x)_i \in \mathbb{R}^2$ and $(Hx)_i \in \mathbb{R}^{2\times 2}$ denote the discrete gradient and Hessian matrix of image x at pixel i. In a nutshell, the SPD0, SPD1, and SPD2 images are representatives of the three important classes of predominantly zero, piecewise constant, and piecewise affine images, respectively. For each test image, in the second, third, and fourth row of Fig. 11 we also show the three associated *binary sparsity masks* $M^{(0)}, M^{(1)}, M^{(2)}$, respectively, with 0-value pixels in black and 1-value pixels in yellow. Such masks, defined by

$$M_i^{(j)} = \begin{cases} 0 & \text{if } y_i^{(j)} = 0 \\ 1 & \text{if } y_i^{(j)} \neq 0 \end{cases}, \quad j = 0, 1, 2, \quad i = 1, \ldots, n, \tag{87}$$

provide an immediate idea of the level of sparsity of each image in terms of the three feature vectors considered in (86). In Table 2, we report, for each image, the

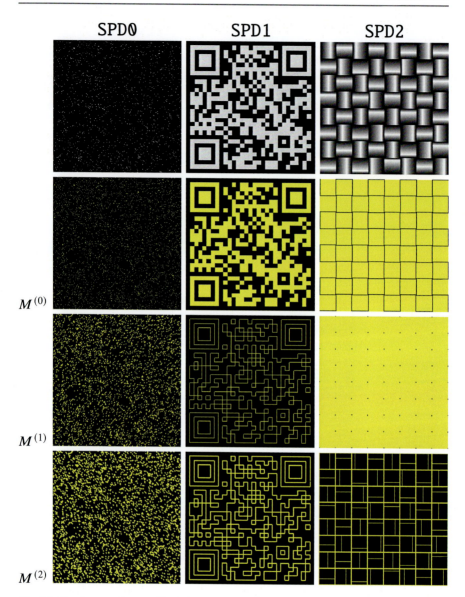

Fig. 10 The three test images SPD0, SPD1, SPD2 (first row) and their associated binary sparsity masks $M^{(j)}$, $j = 0, 1, 2$ (second-fourth rows) defined in (87) in terms of the feature vectors $y^{(j)}$, $j = 0, 1, 2$, given in (86)

total number of pixels n and the three total numbers of 0-value pixels of the binary sparsity masks defined by $\zeta^{(j)} := n - \sum_{i=1}^{n} M_i^{(j)}$, $j = 0, 1, 2$. As expected, the SPD0, SPD1, and SPD2 images exhibit the highest level of sparsity, i.e., the largest number of 0-value pixels, for the features vectors $y^{(0)}$, $y^{(1)}$, and $y^{(2)}$, respectively.

1 Convex Non-convex Variational Models

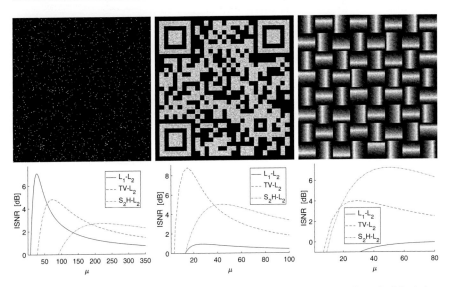

Fig. 11 The three test images SPD0, SPD1, SPD2 corrupted by AWG noise of standard deviation σ yielding $\mathrm{BSNR}(b,\bar{x}) = 15$ (first row) and the associated ISNR graphs for the three purely convex baseline models L_1-L_2, TV-L_2, $\mathcal{S}_2 H$-L_2 defined in (89), (90), and (91) (second row)

Table 2 Sparsity levels of the three test images SPD0, SPD1, SPD2 shown in the first row of Fig. 10 in terms of the features vectors $y^{(0)}, y^{(1)}, y^{(2)}$ defined in (86)

	SPD0	SPD1	SPD2
n	65536	65536	65536
$\zeta^{(0)}$	62255	35178	4096
$\zeta^{(1)}$	56172	**58367**	128
$\zeta^{(2)}$	48144	51463	**55680**

In accordance with the sparsity properties of the three considered test images, to evaluate the performance of the proposed CNC separable and non-separable models, we will compare them with the corresponding purely convex (i.e., with convex regularizers) models, namely, the minimal L_1 norm model (89), referred to as L_1-L_2 model, the isotropic TV-L_2 model (90), and the $\mathcal{S}_2 H$-L_2 model (91) containing the $\mathcal{S}_2 H$ regularizer which induces sparsity of the image Hessian Shatten 2-norm (Lefkimmiatis et al. 2013). More precisely, we consider the following three variational models:

$$x^* = \arg\min_{x \in \mathbb{R}^n} \mathcal{J}^{(j)}(x), \quad j = 0, 1, 2, \tag{88}$$

with cost functions defined by

$$\mathbf{L_1 - L_2}: \quad \mathcal{J}^{(0)}(x) = \frac{1}{2} \|Ax - b\|_2^2 + \mu \underbrace{\sum_{i=1}^{n} |x_i|}_{L_1(x)}, \tag{89}$$

$$\mathbf{TV-L_2}: \quad \mathcal{J}^{(1)}(x) = \frac{1}{2}\|Ax-b\|_2^2 + \mu \underbrace{\sum_{i=1}^{n}\|(\nabla x)_i\|_2}_{TV(x)}, \quad (90)$$

$$\mathbf{S_2H-L_2}: \quad \mathcal{J}^{(2)}(x) = \frac{1}{2}\|Ax-b\|_2^2 + \mu \underbrace{\sum_{i=1}^{n}\|(Hx)_i\|_F}_{S_2H(x)}. \quad (91)$$

We thus assume that the above three models are representative of the class of purely convex models, and we compare their performance with those of the proposed associated separable CNC-S-L$_2$ and non-separable CNC-NS-L$_2$ models which, we recall, are also convex but contain non-convex regularizers.

It is worth to point out that the three models in (89), (90), and (91) can be represented in a unified form according to definition (6) of the considered class of sparsity-promoting regularizers:

$$\mathcal{J}^{(j)}(x) = \frac{1}{2}\|Ax-b\|_2^2 + \mu \|y^{(j)}\|_1, \quad y^{(j)} = G^{(j)}\left(L^{(j)}x\right), \quad j=0,1,2. \quad (92)$$

In particular, the nonlinear vector-valued functions $G^{(j)}: \mathbb{R}^{r_j} \to \mathbb{R}^n$ read

$$G^{(j)}(z) = \left(g_1^{(j)}(z),\ldots,g_n^{(j)}(z)\right)^T, \quad z \in \mathbb{R}^{r_j}, \quad r_j = (j+1)n, \quad j=0,1,2, \quad (93)$$

with components defined by

$$g_i^{(0)}(z) = |z_i|, \quad g_i^{(1)}(z) = \left\|(z_i, z_{i+n})\right\|_2, \quad g_i^{(2)}(z) = \left\|(z_i, z_{i+n}, z_{i+2n})\right\|_2, \quad (94)$$

$i=1,\ldots,n$, whereas the linear operators $L^{(j)} \in \mathbb{R}^{r_j \times n}$ are

$$L^{(0)} = I_n, \quad L^{(1)} = \left(D_h^T, D_v^T\right)^T, \quad L^{(2)} = \left(D_{hh}^T, D_{vv}^T, \sqrt{2}D_{hv}^T\right)^T, \quad (95)$$

with $D_h, D_v, D_{hh}, D_{vv}, D_{hv} \in \mathbb{R}^{n \times n}$ finite difference operators discretizing the first-order horizontal and vertical and the second-order horizontal, vertical, and mixed horizontal-vertical partial derivatives, respectively. The discrete gradient and Hessian operators in (90) and (91) are thus defined in terms of these matrices as follows:

1 Convex Non-convex Variational Models

$$(\nabla x)_i = \begin{bmatrix} (D_h x)_i \\ (D_v x)_i \end{bmatrix}, \quad (Hx)_i = \begin{bmatrix} (D_{hh} x)_i & (D_{hv} x)_i \\ (D_{hv} x)_i & (D_{vv} x)_i \end{bmatrix}, \quad i = 1, \ldots, n. \tag{96}$$

Finally, for what concerns the actual discretization of the gradient and Hessian operators, in all the experiments matrices $D_h, D_v, D_{hh}, D_{vv}, D_{hv}$ are the 2-D convolution matrices (with periodic boundary conditions) associated with the following point-spread functions:

$$D_h \to (+1, -\mathbf{1}), \quad D_v \to \begin{pmatrix} +1 \\ -\mathbf{1} \end{pmatrix},$$

$$D_{hh} \to (+1, -\mathbf{2}, +1), \quad D_{vv} \to \begin{pmatrix} +1 \\ -\mathbf{2} \\ +1 \end{pmatrix}, \quad D_{hv} \to \begin{pmatrix} +1 & -1 \\ -1 & +\mathbf{1} \end{pmatrix},$$

with boldface cells indicating the center of application of the PSF.

For all numerical examples, the experimental setting is as follows. The original test image $\bar{x}$ is synthetically degraded according to the measurement model (4). First, $\bar{x}$ is corrupted by space-invariant Gaussian blur under the assumption of periodic boundary conditions. The acquisition matrix $A \in \mathbb{R}^{n \times n}$, referred to as blurring matrix in this case, is thus block-circulant with circulant blocks and is constructed starting from the Gaussian convolution kernel, or point-spread function, generated by the Matlab command `fspecial('gaussian',band,sigma)`. The parameters `band` and `sigma` determine the bandwidth and the values of each circulant block in A, respectively. In particular, `band` represents the side length (in pixels) of the square support of the kernel, whereas `sigma` is the standard deviation of the circular, zero-mean, bivariate Gaussian probability density function representing the Gaussian point-spread function in the continuous setting. The blurred image $A\bar{x} \in \mathbb{R}^n$ is then corrupted by AWG noise with standard deviation σ to obtain the observed image $b \in \mathbb{R}^n$. Given A and b, the goal is to determine as accurately as possible estimates x^* of the original uncorrupted image $\bar{x}$ by using variational models containing sparsity-promoting separable and non-separable regularization terms.

Regarding the optimization algorithms, the considered models are numerically solved by using the FB and PDFB splitting algorithms described in section "Forward-Backward Minimization Algorithms" and applying the illustrated ADMM strategy for the efficient computation of the backward steps. In all the experiments and for all the models, we use the observed corrupted image as the initial iterate, i.e., $x^{(0)} = b$, and we terminate the iterations as soon as two successive iterates satisfy

$$\delta_k^{(x)} := \frac{\left\| x^{(k)} - x^{(k-1)} \right\|_2}{\left\| x^{(k-1)} \right\|_2} < 10^{-5}. \tag{97}$$

The quality of the observed degraded images b and of the restored images x^* (in comparison with the original uncorrupted image $\bar{x}$) are measured by means of the Blurred Signal-to-Noise Ratio (BSNR) and the Improved Signal-to-Noise Ratio (ISNR), respectively. They are defined by

$$\text{BSNR}(b, \bar{x}) = \text{SNR}(b, A\bar{x}), \quad \text{ISNR}(x^*, b, \bar{x}) = \text{SNR}(x^*, \bar{x}) - \text{SNR}(b, \bar{x}), \tag{98}$$

with the Signal-to-Noise Ratio (SNR) quality measure of an image I versus a reference image $\bar{I}$ given by

$$\text{SNR}(I, \bar{I}) := 10 \log_{10} \left(\frac{\|\bar{I} - E[\bar{I}]\|_2^2}{\|\bar{I} - I\|_2^2} \right) \; [dB], \tag{99}$$

where $E[\bar{I}]$ denotes the image with constant intensity equal to the mean value of $\bar{I}$. The larger the BSNR value, the lower is the intensity, i.e., the standard deviation σ, of the AWG noise corrupting the observation b (hence, the easier is the image restoration problem); the larger the ISNR value, the higher the quality of the restored image x^* obtained by the considered variational model. In all the experiments, after choosing the blurring operator A and computing the blurred image $A\bar{x}$, we set the desired BSNR value of the observation b and then exploit the BSNR definition in (98)–(99) in order to determine the (unique) value of the AWG noise standard deviation σ yielding the selected BSNR value:

$$\sigma = \frac{\|A\bar{x} - E[A\bar{x}]\|_2}{\sqrt{n} \, 10^{\frac{\text{BSNR}}{20}}}. \tag{100}$$

Examples Using CNC Separable Models

We consider the problem of denoising the three considered test images SPD0, SPD1, SPD2 corrupted only by AWG noise (no blur, i.e. $A = I_n$ in the acquisition model (4) as well as in the baseline convex variational models (89), (90), and (91)) with standard deviation σ yielding $\text{BSNR}(b, \bar{x}) = 15$, as shown in the first row of Fig. 11. The three separable CNC variational models, referred to as CNC-S-L_1-L_2, CNC-S-TV-L_2, and CNC-S-S_2H-L_2, to be compared with the baseline purely convex models L_1-L_2, TV-L_2 and S_2H-L_2 defined in (89), (90), and (91), read as follows:

$$x^* = \arg \min_{x \in \mathbb{R}^n} \mathcal{J}_S^{(j)}(x; a), \quad j = 0, 1, 2, \tag{101}$$

with cost functions defined by

CNC – S – L$_1$ – L$_2$:

$$\mathcal{J}_S^{(0)}(x;a) = \frac{1}{2} \|Ax - b\|_2^2 + \mu \underbrace{\sum_{i=1}^{n} \phi_{MC}(|x_i|;a)}_{S-L_1(x;a)}, \qquad (102)$$

CNC – S – TV – L$_2$:

$$\mathcal{J}_S^{(1)}(x;a) = \frac{1}{2} \|Ax - b\|_2^2 + \mu \underbrace{\sum_{i=1}^{n} \phi_{MC}\left(\|(\nabla x)_i\|_2;a\right)}_{S-TV(x;a)}, \qquad (103)$$

CNC – S – S_2H – L$_2$:

$$\mathcal{J}_S^{(2)}(x;a) = \frac{1}{2} \|Ax - b\|_2^2 + \mu \underbrace{\sum_{i=1}^{n} \phi_{MC}\left(\|(Hx)_i\|_F;a\right)}_{S-S_2H(x;a)}, \qquad (104)$$

where ϕ_{MC} is the scalar MC penalty function defined in (13) and where we are assuming a space-invariant, i.e., constant for all pixel locations, concavity parameter $a \in \mathbb{R}_{++}$ for ϕ_{MC}. It follows from Theorem 1, in particular, condition (33), that sufficient conditions for the three cost functions above to be convex (strongly convex) are the following:

$$Q^{(j)} = I_n - \mu a \left(L^{(j)}\right)^T L^{(j)} \succeq 0 \; (\succ 0), \quad j = 0, 1, 2, \qquad (105)$$

where we used the fact that $A = I_n$ for the considered case of image denoising and where the regularization matrices $L^{(j)}$ are defined in (95). According to the statement of Theorem 1, the sufficient conditions in (105) can be equivalently and usefully rewritten as follows:

$$a = \tau_c \frac{1}{\mu \kappa^{(j)}}, \quad \tau_c \in [0,1] \; (\tau_c \in [0,1[\,), \quad j = 0, 1, 2, \qquad (106)$$

with the scalar coefficients $\kappa^{(j)}$ given by

$$k^{(j)} = \sigma_{max}^2\left(L^{(j)}\right), \; j = 0, 1, 2 \; \implies \; k^{(0)} = 1, \; k^{(1)} = 8, \; k^{(2)} = 64. \qquad (107)$$

As a preliminary experiment, we evaluate the performance of the three baseline convex models L$_1$-L$_2$, TV-L$_2$, and S_2H-L$_2$ defined in (89), (90), and (91) when applied to the three corrupted images illustrated in the first row of Fig. 11. The plots

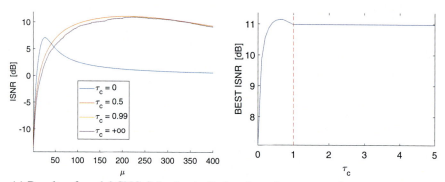

(a) Results of model CNC-S-L_1-L_2 applied to the noise-corrupted test image SPD0

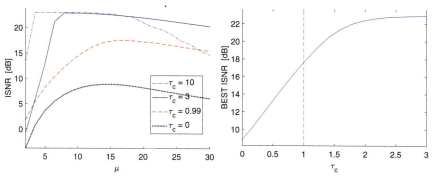

(b) Results of model CNC-S-TV-L_2 applied to the noise-corrupted test image SPD1

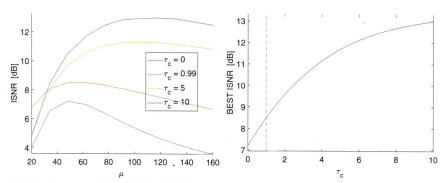

(c) Results of model CNC-S-S_2H-L_2 applied to the noise-corrupted test image SPD2

Fig. 12 ISNR results of separable CNC models CNC-S-L_1-L_2, CNC-S-TV-L_2, and CNC-S-S_2H-L_2 defined in (102), (103), and (104) when applied to the noise-corrupted images SPD0, SPD1, and SPD2, respectively. First column: ISNR values as a function of the regularization parameter μ for some different τ_c values. Second column: highest achieved ISNR values as a function of the convexity coefficient τ_c. The dashed vertical red lines, corresponding to $\tau_c = 1$, separate, for each model, the pure convex and CNC regimes ($\tau_c \in [0, 1]$) from the pure non-convex regime ($\tau_c \in]1, +\infty[$).

in Fig. 11 (second row) represent the ISNR values achieved by the three models as a function of the regularization parameter μ for the three corresponding noise-corrupted test images SPD0, SPD1, and SPD2 illustrated in the first row. From a visual inspection, column by column, of Fig. 11 (second row), we observe that, as expected, the best ISNR values are obtained by models L_1-L_2, TV-L_2, and S_2H-L_2 on images SPD0, SPD1 and SPD2, respectively. This is completely in accordance with the sparsity properties of the three images. The regularizers of models L_1-L_2, TV-L_2, and S_2H-L_2 are in fact suitable for predominantly zero, piecewise constant, and piecewise affine images, respectively, as they promote sparsity of the intensities and of the first- and second-order intensity derivatives of the restored image.

In the next experiment, we compare the best assessed regularization models in the three convexity regimes: pure convex ($\tau_c = 0$), CNC ($\tau_c \in (0, 1]$), and pure non-convex regime ($\tau_c > 1$). In other words, we now test the three separable CNC models CNC-S-L_1-L_2, CNC-S-TV-L_2, and CNC-S-S_2H-L_2 defined in (102), (103), and (104) on the corresponding test images for different τ_c values. In Fig. 12, for each test image SPD0 (first row), SPD1 (second row), and SPD2 (third row), we report some interesting ISNR curves for the associated best-performing models CNC-S-L_1-L_2, CNC-S-TV-L_2, and CNC-S-S_2H-L_2, respectively. In particular, the plots in the first column represent, for some different τ_c values, the achieved ISNR values as a function of the regularization parameter μ. The curves in the second column depict, for a fine grid of τ_c values, the highest ISNR values achieved by letting μ vary in its entire domain.

In Figs. 13, 14, and 15, we report the best (i.e., with highest associated ISNR value) denoising results obtained by applying models CNC-S-L_1-L_2, CNC-S-TV-L_2, and CNC-S-S_2H-L_2 to the noise-corrupted test images SPD0, SPD1, and SPD2, respectively, with different τ_c values. In particular, in the first column of Figs. 13, 14, and 15, we show the denoised images, whereas in the second column we report the associated absolute error images.

From ISNR plots reported in the second column of Fig. 12, we can first observe that usefulness of using high τ_c values becomes larger as the order of image derivatives sparsified by the regularizer increases. For model CNC-S-L_1-L_2, the best results are obtained in the CNC regime, i.e., for $\tau_c \in]0, 1]$. We recall that in this case the upper limit of the CNC regime ($\tau_c = 1$) corresponds to using $\|x\|_0$ as the regularizer, such that the solution is obtained by a pixel-wise hard thresholding of the noisy observation b. For the CNC-S-TV-L_2 model, the ISNR gain obtained by the CNC regime is remarkable, whereas for the CNC-S-S_2H-L_2 model, such gain is smaller. In other words, pushing the model in pure non-convex regime ($\tau_c > 1$) is much more appealing for CNC-S-S_2H-L_2 than for CNC-S-TV-L_2.

Examples Using CNC Non-separable Models

In this section, we test the performance of the proposed non-separable CNC variational models when applied to image denoising and deblurring problems. In fact, unlike the separable CNC strategy, the non-separable CNC approach can be

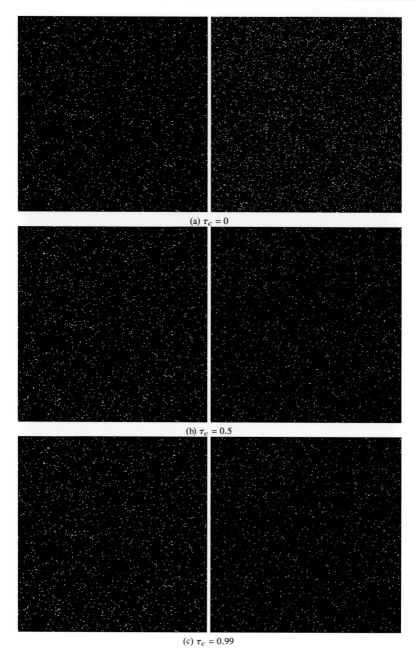

Fig. 13 *Separable CNC models*. Best denoising results obtained by CNC-S-L_1-L_2 on image SPD0 for different τ_c values (left column) and associated absolute error images (right column)

1 Convex Non-convex Variational Models

Fig. 14 *Separable CNC models.* Best denoising results obtained by CNC-S-TV-L_2 on image SPD1 for different τ_c values (left column) and associated absolute error images (right column)

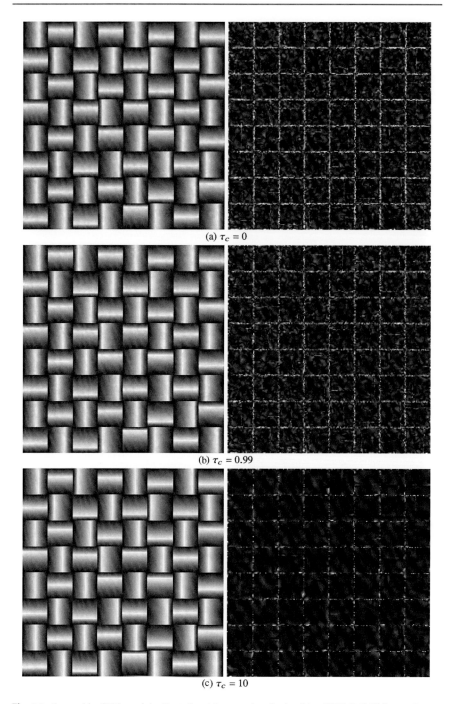

Fig. 15 *Separable CNC models.* Best denoising results obtained by CNC-S-S_2H-L_2 on image SPD2 for different τ_c values (left column) and associated absolute error images (right column)

usefully applied for any acquisition matrix A, also when A is very ill-conditioned or even numerically singular like it is often the case in deblurring problems. More precisely, we consider the restoration of the piecewise constant image SPD1 and the piecewise affine image SPD2 depicted in the first row of Fig. 10 which, we recall, are characterized by sparse first- and second-order derivatives, respectively.

In accordance with the considered degradation model in (4), the two test images SPD1 and SPD2 have been synthetically corrupted by space-invariant Gaussian blur and AWG noise, as described at the beginning of section "Numerical Examples". In particular, for the denoising experiment, clearly A is the identity operator, and no synthetic blur is applied, whereas for the deblurring experiment, the Gaussian point-spread function is generated with parameters band = 7, sigma = 1.5. We then add AWG noise corruptions of standard deviations σ yielding $\text{BSNR}(b, \bar{x}) = 15$ for the denoising case and $\text{BSNR}(b, \bar{x}) = 7.6$ for the deblurring case.

For the restoration, i.e., denoising and/or deblurring, of the degraded SPD1 and SPD2 test images, we consider the non-separable CNC versions, referred to as CNC-NS-TV-L_2 and CNC-NS-S_2H-L_2, of the two separable CNC models CNC-S-TV-L_2 and CNC-S-S_2H-L_2 defined in (103) and (104), respectively. We also consider a slightly different but interesting version of the CNC-NS-S_2H-L_2 model, referred to as CNC-NS-S_1H-L_2, where the Shatten 2-norm (Frobenious norm) has been replaced by the Shatten 1-norm (nuclear norm).

The three considered non-separable CNC models thus read

$$x^* = \arg\min_{x \in \mathbb{R}^n} \mathcal{J}_{\text{NS}}^{(j)}(x; B), \quad j = 1, 2, 3, \tag{108}$$

with cost functions defined by

CNC $-$ NS $-$ TV $-$ L$_2$:

$$\mathcal{J}_{\text{NS}}^{(1)}(x; B) = \frac{1}{2} \|Ax - b\|_2^2 + \mu \underbrace{\left(\text{TV}(x) - \left(\text{TV} \square \tfrac{1}{2}\|B \cdot \|_2^2 \right)(x) \right)}_{\text{NS-TV}(x; B)}, \tag{109}$$

CNC $-$ NS $-$ S_2H $-$ L$_2$:

$$\mathcal{J}_{\text{NS}}^{(2)}(x; B) = \frac{1}{2} \|Ax - b\|_2^2 + \mu \underbrace{\left(S_2\text{H}(x) - \left(S_2\text{H} \square \tfrac{1}{2}\|B \cdot \|_2^2 \right)(x) \right)}_{\text{NS-}S_2\text{H}(x; B)}, \tag{110}$$

CNC $-$ NS $-$ S_1H $-$ L$_2$:

$$\mathcal{J}_{\text{NS}}^{(3)}(x; B) = \frac{1}{2} \|Ax - b\|_2^2 + \mu \underbrace{\left(S_1\text{H}(x) - \left(S_1\text{H} \square \tfrac{1}{2}\|B \cdot \|_2^2 \right)(x) \right)}_{\text{NS-}S_1\text{H}(x; B)}. \tag{111}$$

Table 3 ISNR values obtained by restoring the test images SPD1 and SPD2 corrupted by zero-mean AWG noise (Denoise) and space-invariant Gaussian blur (Deblur)

Image	Model	Denoise	Deblur	Image	Model	Denoise	Deblur
SPD1	TV-L_2	**8.84**	**6.64**	SPD2	TV-L_2	7.21	**3.00**
	S_1H-L_2	5.56	2.00		S_1H-L_2	**7.67**	2.50
	S_2H-L_2	4.54	1.90		S_2H-L_2	6.65	2.73
	CNC-NS-TV-L_2	**20.35**	**6.72**		CNC-NS-TV-L_2	4.11	**3.20**
	CNC-NS-S_1H-L_2	11.34	2.11		CNC-NS-S_1H-L_2	**12.33**	2.83
	CNC-NS-S_2H-L_2	9.13	2.00		CNC-NS-S_2H-L_2	10.57	2.73

The parameter matrix B has been constructed using dc-notch filters as described at the end of section "Construction of Matrix B", so that the three total cost functions above are all convex, and hence, the three models are CNC.

Quantitative and qualitative (visual) results have been produced. In Table 3, we report the ISNR values obtained by the three considered non-separable CNC models on the two test images for both the denoising and deblurring experiments. For comparison, we also report the ISNR values achieved by using the associated purely convex baseline models. For each experiment, the best ISNR results within each class of models are marked in boldface. Figures 16 and 17 show the corrupted images (top rows) and the best restored images computed by the two classes of purely convex models (center rows) and non-separable CNC models (bottom rows), in case of denoising and deblurring, respectively, see the associated ISNR values marked in boldface in Table 3.

From the ISNR values in Table 3 and the visual inspection of the restored images in Figs. 16 and 17, the improvement in accuracy provided by the considered non-convex non-separable regularizers versus the corresponding convex separable baseline regularizers is evident, particularly for the denoising case, and in agreement with the sparsity characteristics of the two images. It is worth remarking that such improvement is obtained without renouncing any of the well-known advantages of (strongly) convex optimization, namely, the existence of a unique (global) minimizer and of numerical algorithms with proved convergence toward such minimizer.

Furthermore, for the denoising results we could also extend the comparison to the CNC models with separable regularizers, which were demonstrated in section "Examples Using CNC Separable Models" to outperform the baseline purely convex models in inducing sparsity of the gradient magnitudes or the Hessian Shatten 2-norms in the denoised images.

To conclude, we notice that for both the separable and non-separable CNC considered models, the regularization parameter μ has been set manually so as to achieve the best accuracy results in terms of ISNR. In practical applications, clearly this procedure can not be used (the true image $\bar{x}$ is unknown), and also manually tuning μ by visually inspecting the attained results is not practical. Hence, some sort of automatic parameter selection strategy is always highly desirable. Actually,

1 Convex Non-convex Variational Models

Fig. 16 *Non-separable CNC models*. Denoising results on SPD1 (left column) and SPD2 (right column) corrupted by AWG noise. First row: degraded images (BSNR = 15). Second row: restorations by TV-L$_2$ (ISNR=8.84), left, and by S_1H-L$_2$ (ISNR=7.67), right. Third row: restorations by CNC-NS-TV-L$_2$ (ISNR=20.35), left, and by CNC-NS-S_1H-L$_2$ (ISNR=12.33), right

Fig. 17 *Non-separable CNC models*. Deblurring results on SPD1 (left column) and SPD2 (right column) corrupted by blur and AWG noise. First row: degraded images (BSNR = 7.6). Second row: restorations by TV-L_2 (ISNR=6.64), left, and by S_1H-L_2 (ISNR=3.00), right. Third row: restorations by CNC-NS-TV-L_2 (ISNR=6.72), left, CNC-NS-S_1H-L_2 (ISNR=3.20), right

the proposed FB and PDFB numerical solution algorithms can be quite easily equipped with such an automatic strategy. In particular, if one wants to select μ according to the very popular *discrepancy principle* or to the less popular but very effective *residual whiteness principle*, the ADMM approach proposed for solving the backward denoising step can benefit from the adaptive strategies proposed in Lanza et al. Lanza et al. (2016b, 2021, 2020) for the more general class of deblurring problems.

Conclusion

We discussed a CNC strategy for sparsity-inducing regularization of linear least-squares inverse problems. To avoid the intrinsic difficulties related to non-convex optimization, the CNC strategy allows the use of non-convex regularization while maintaining convexity of the total cost function. In this work we analyzed a general class of parameterized non-convex sparsity-promoting separable and non-separable regularizers and their associated CNC variational models. We derived convexity conditions for the total cost functions and we discussed related theoretical properties. A general forward-backward splitting strategy has been presented and applied for the numerical solution of the CNC models considered and a theoretical proof of convergence has been given. A series of numerical experiments related to image denoising and deblurring have been carried out, and the reported results strongly indicate that the considered non-convex regularizers hold the potential for achieving high quality results while remaining in a convex, safe regime.

Acknowledgments This research was supported in part by the National Group for Scientific Computation (GNCS-INDAM), Research Projects 2019/2020.

References

Bauschke, H.H., Combettes, P.L.: Convex Analysis and Monotone Operator Theory in Hilbert Spaces. Springer, New York (2011)

Bayram, I.: On the convergence of the iterative shrinkage/thresholding algorithm with a weakly convex penalty. IEEE Trans. Signal Process. **64**(6), 1597–1608 (2016)

Beck, A., Teboulle, M.: A fast iterative shrinkage-thresholding algorithm for linear inverse problems. SIAM J. Imag. Sci. **2**(1), 183–202 (2009)

Becker, S., Combettes, P.L.: An algorithm for splitting parallel sums of linearly composed monotone operators, with applications to signal recovery. J. Nonlinear Convex Anal. **15**(1), 137–159 (2014)

Bello Cruz, J.Y.: On proximal subgradient splitting method for minimizing the sum of two nonsmooth convex functions. Set-Valued Var. Anal **25**, 245–263 (2017)

Blake, A., Zisserman, A.: Visual Reconstruction. MIT Press, Cambridge, MA (1987)

Bruckstein, A., Donoho, D., Elad, M.: From sparse solutions of systems of equations to sparse modeling of signals and images. SIAM Rev. **51**(1), 34–81 (2009)

Burger, M., Papafitsoros, K., Papoutsellis, E., Schönlieb, C.B.: Infimal convolution regularisation functionals of BV and Lp spaces. J. Math. Imaging Vis. **55**(3), 343–369 (2016)

Cai, G., Selesnick, I.W., Wang, S., Dai, W., Zhu, Z.: Sparsity enhanced signal decomposition via generalized minimax-concave penalty for gearbox fault diagnosis. J. Sound Vib. **432**, 213–234 (2018)

Candés, E.J., Wakin, M.B., Boyd, S.: Enhancing sparsity by reweighted l1 minimization. J. Fourier Anal. Appl.**14**(5), 877–905 (2008)

Carlsson, M.: On convexification/optimization of functionals including an l2-misfit term. arXiv preprint arXiv:1609.09378 (2016)

Castella, M., Pesquet, J.C.: Optimization of a Geman-McClure like criterion for sparse signal deconvolution. In: IEEE International Workshop on Computational Advances Multi-sensor Adaptive Processing, pp. 309–312 (2015)

Chambolle, A., Lions, P.L.: Image recovery via total variation minimization and related problems. Numerische Mathematik **76**, 167–188 (1997)

Chan, R., Lanza, A., Morigi, S., Sgallari, F.: Convex non-convex image segmentation. Numerische Mathematik **138**(3), 635–680 (2017)

Chartrand, R.: Shrinkage mappings and their induced penalty functions. In: International Conference on Acoustics, Speech and Signal Processing (ICASSP), pp. 1026–1029 (2014)

Chen, P.Y., Selesnick, I.W.: Group-sparse signal denoising: non-convex regularization, convex optimization. IEEE Trans. Signal Proc. **62**, 3464–3478 (2014)

Chouzenoux, E., Jezierska, A., Pesquet, J., Talbot, H.: A majorize-minimize subspace approach for l2–l0 image regularization. SIAM J. Imag. Sci. **6**(1), 563–591 (2013)

Ding, Y., Selesnick, I.W.: Artifact-free wavelet denoising: nonconvex sparse regularization, convex optimization. IEEE Signal Process. Lett. **22**(9), 1364–1368 (2015)

Du, H., Liu, Y.: Minmax-concave total variation denoising. Signal Image Video Process. **12**(6), 1027–1034 (2018)

Geiger, D., Girosi, F.: Parallel and deterministic algorithms from MRF's: surface reconstruction. IEEE Trans. Pattern Anal. Mach. Intell. **13**(5), 410–412 (1991)

Geman, S., Geman, D.: Stochastic relaxation, Gibbs distribution, and the Bayesian restoration of images. IEEE PAMI **6**(6), 721–741 (1984)

Hansen, P.C.: Rank-Deficient and Discrete Ill-Posed Problems. SIAM, Philadelphia (1997)

Hartman, P.: On functions representable as a difference of convex functions. Pac. J. Math. **9**(3), 707–713 (1959)

Hastie, T., Tibshirani, R., Wainwright, M.: Statistical Learning with Sparsity: The Lasso and Generalizations. CRC Press, Boca Raton (2015)

Huska, M., Lanza, A., Morigi, S., Sgallari, F.: Convex non-convex segmentation of scalar fields over arbitrary triangulated surfaces. J. Comput. Appl. Math. **349**, 438–451 (2019a)

Huska, M., Lanza, A., Morigi, S., Selesnick, I.W.: A convex-nonconvex variational method for the additive decomposition of functions on surfaces. Inverse Problems **35**, 124008–124041 (2019b)

Jensen, J.B., Nielsen, M.: A simple genetic algorithm applied to discontinuous regularization. In: Proceedings IEEE workshop on NNSP, Copenhagen (1992)

Lanza, A., Morigi, S., Sgallari, F.: Convex image denoising via non-convex regularization. Scale Space Variat. Methods Comput. Vis. **9087**, 666–677 (2015)

Lanza, A., Morigi, S., Sgallari, F.: Convex image denoising via nonconvex regularization with parameter selection. J. Math. Imaging Vis. **56**(2), 195–220 (2016a)

Lanza, A., Morigi, S., Sgallari, F.: Constrained TV$_p$-ℓ_2 model for image restoration. J. Sci. Comput. **68**, 64–91 (2016b)

Lanza, A., Morigi, S., Selesnick, I.W., Sgallari, F.: Nonconvex nonsmooth optimization via convex-nonconvex majorization minimization. Numerische Mathematik **136**(2), 343–381 (2017)

Lanza, A., Morigi, S., Sgallari, F.: Automatic parameter selection based on residual whiteness for convex non-convex variational restoration. In: Mathematical Methods in Image Processing and and Inverse Problems (eds) Tai XC, Wei S, Liu H. Springer, Singapore, **360**, (2021). https://doi.org/10.1007/978-981-16-2701-9

Lanza, A., Morigi, S., Selesnick, I.W., Sgallari, F.: Sparsity-inducing nonconvex nonseparable regularization for convex image processing. SIAM J. Imag. Sci. **12**(2), 1099–1134 (2019)

Lanza, A., Pragliola, M., Sgallari, F.: Residual whiteness principle for parameter-free image restoration. Electron. Trans. Numer. Anal. **53**, 329–351 (2020)

Lefkimmiatis, S., Ward, J., Unser, M.: Hessian Schatten-Norm regularization for linear inverse problems. IEEE Trans. Image Process. **22**, 1873–1888 (2013)

Malek-Mohammadi, M., Rojas, C.R., Wahlberg, B.: A class of nonconvex penalties preserving overall convexity in optimization based mean filtering. IEEE Trans. Signal Process. **64**(24), 6650–6664 (2016)

Nikolova, M.: Estimation of binary images by minimizing convex criteria. Proc. IEEE Int. Conf. Image Process. **2**, 108–112 (1998)

Nikolova, M.: Energy minimization methods. In: Scherzer, O. (ed.) Handbook of Mathematical Methods in Imaging, Chapter 5, pp. 138–186. Springer, Berlin (2011)

Nikolova, M., Ng, M.K., Tam, C.P.: Fast nonconvex nonsmooth minimization methods for image restoration and reconstruction. IEEE Trans. Image Process. **19**(12), 3073–3088 (2010)

Parekh, A., Selesnick, I.W.: Convex denoising using non-convex tight frame regularization. IEEE Signal Process. Lett. **22**(10), 1786–1790 (2015)

Parekh, A., Selesnick, I.W.: Enhanced low-rank matrix approximation. IEEE Signal Process. Lett. **23**(4), 493–497 (2016)

Park, T.W., Burrus, C.S.: Digital Filter Design. Wiley, New York (1987)

Portilla, J., Mancera, L.: L0-based sparse approximation: two alternative methods and some applications. In: Proceedings of SPIE, San Diego, vol. 6701 (Wavelets XII) (2007)

Rudin, L.I., Osher, S., Fatemi, E.: Nonlinear total variation based noise removal algorithms. Physics D **60**(1–4), 259–268 (1992)

Selesnick, I.W.: Sparse regularization via convex analysis. IEEE Trans. Signal Process. **65**(17), 4481–4494 (2017a)

Selesnick, I.W.: Total variation denoising via the Moreau envelope. IEEE Signal Process. Lett. **24**(2), 216–220 (2017b)

Selesnick, I.W., Bayram, I.: Sparse signal estimation by maximally sparse convex optimization. IEEE Trans. Signal Proc. **62**(5), 1078–1092 (2014)

Selesnick, I.W., Parekh, A., Bayram, I.: Convex 1-D total variation denoising with non-convex regularization. IEEE Signal Process. Lett. **22**, 141–144 (2015)

Selesnick, I.W., Lanza, A., Morigi, S., Sgallari, F.: Non-convex total variation regularization for convex denoising of signals. J. Math. Imag. Vis. **62**, 825–841 (2020)

Setzer, S., Steidl, G., Teuber, T.: Infimal convolution regularizations with discrete l1-type functionals. Commun. Math. Sci. **9**(3), 797–827 (2011)

Shen, L., Xu, Y., Zeng, X.: Wavelet inpainting with the l0 sparse regularization. J. Appl. Comp. Harm. Anal. **41**(1), 26–53 (2016)

Sidky, E.Y., Chartrand, R., Boone, J.M., Pan, X.: Constrained TpV–minimization for enhanced exploitation of gradient sparsity: application to CT image reconstruction. IEEE J. Trans. Eng. Health Med. **2**, 1–18 (2014)

Soubies, E., Blanc-Féraud, L., Aubert, G.: A continuous exact L0 penalty (CEL0) for least squares regularized problem. SIAM J. Imag. Sci.**8**(3), 1607–1639 (2015)

Tipping, M.E.: Sparse Bayesian learning and the relevance vector machine. J. Mach. Learn. Res. **1**, 211–244 (2001)

Tuy, H.: DC optimization: theory, methods and algorithms. In: Handbook of Global Optimization, pp. 149–216. Springer, Boston, (1995)

Wang, S., Selesnick, I.W., Cai, G., Ding, B., Chen, X.: Synthesis versus analysis priors via generalized minimax-concave penalty for sparsity-assisted machinery fault diagnosis. Mech. Syst. Signal Process. **127**, 202–233 (2019)

Wipf, D.P., Rao, B.D., Nagarajan, S.: "Latent variable Bayesian models for promoting sparsity. In: IEEE Trans. Inf. Theory **57**(9), 6236–6255 (2011)

Yuille, A.L., Rangarajan, A.: The concave-convex procedure. Neural Comput. **15**(4), 915–936 (2003)

Zhang, C.-H.: Nearly unbiased variable selection under minimax concave penalty. Ann. Stat. **38**(2), 894–942 (2010)

Zou, J., Shen, M., Zhang, Y., Li, H., Liu, G., Ding, S.: Total variation denoising with non-convex regularizers. IEEE Access **7**, 4422–4431 (2019)

Subsampled First-Order Optimization Methods with Applications in Imaging

2

Stefania Bellavia, Tommaso Bianconcini, Nataša Krejić, and Benedetta Morini

Contents

Introduction	62
Convolutional Neural Networks	64
Convolutional Layer	67
Max Pooling Layer	68
Stochastic Gradient and Variance Reduction Methods	69
Gradient Methods with Adaptive Steplength Selection Based on Globalization Strategies	77
Accuracy Requirements	81
Stochastic Line Search	82
Adaptive Regularization and Trust-Region	84
Numerical Experiments	85
The Neural Network in Action	86
Training the Neural Network	88
Implementation Details	90
Results	90
Conclusion	91
References	92

S. Bellavia · B. Morini (✉)
Dipartimento di Ingegneria Industriale, Università degli Studi di Firenze (INdAM-GNCS members), Firenze, Italia
e-mail: stefania.bellavia@unifi.it; benedetta.morini@unifi.it

T. Bianconcini
Verizon Connect, Firenze, Italia
e-mail: tommaso.bianconcini@verizonconnect.com

N. Krejić
Department of Mathematics and Informatics, Faculty of Sciences, University of Novi Sad, Novi Sad, Serbia
e-mail: natasak@uns.ac.rs

© Springer Nature Switzerland AG 2023
K. Chen et al. (eds.), *Handbook of Mathematical Models and Algorithms in Computer Vision and Imaging*, https://doi.org/10.1007/978-3-030-98661-2_78

Abstract

This work presents and discusses optimization methods for solving finite-sum minimization problems which are pervasive in applications, including image processing. The procedures analyzed employ first-order models for the objective function and stochastic gradient approximations based on subsampling. Among the variety of methods in the literature, the focus is on selected algorithms which can be cast into two groups: algorithms using gradient estimates evaluated on samples of very small size and algorithms relying on gradient estimates and machinery from standard globally convergent optimization procedures. Neural networks and convolutional neural networks widely used for image processing tasks are considered, and a classification problem of images is solved with some of the methods presented.

Keywords

Finite-sum minimization · First-order methods · Stochastic gradient · Neural networks · Convolutional neural networks · Image classification

Introduction

The focus of this paper is on finite-sum minimization

$$\min_{x \in \mathbb{R}^n} f(x), \qquad (1)$$

where $f : \mathbb{R}^n \to \mathbb{R}$ is a Lipschitz smooth function of the form

$$f(x) = \frac{1}{N} \sum_{i=1}^{N} f_i(x), \qquad (2)$$

and each f_i is such that $f_i : \mathbb{R}^n \to \mathbb{R}$. We assume that f is bounded from below in $\mathbb{R}^n$.

The case of interest here is when problem dimension n and N are large numbers. Such finite-sum minimization comprises a variety of applications including problems from machine learning Bottou et al. (2018) and plays an important role in image processing, e.g., in tasks such as image classification, object detection, and image segmentation (Aggarwal 2018; Chollet 2017; Forsyth et al. 2002; Goodfellow et al. 2016; Patterson et al. 2017; Shanmugamani 2018).

In a large-scale regime, working with the objective function f and its gradient in first-order methods, or even Hessian in the second-order methods, may be prohibitively expensive. In order to reduce the computational cost, typically f and its derivatives are approximated using a subset of the summation terms. In particular, such approximation is carried out by *subsampling*, i.e., considering summation terms corresponding to a random sample of indices $\mathcal{S} \subseteq \{1, \ldots, N\}$. The random sample set $\mathcal{S}$ is also called mini-batch if it is a small subset of $\{1, \ldots, N\}$.

Considering first-order methods, let k be the iteration index and f_k^0 and g_k be subsampled approximation of $f(x_k)$ and $\nabla f(x_k)$, respectively, i.e.,

$$f_k^0 = \frac{1}{|\mathcal{S}_{k,f}|} \sum_{i \in \mathcal{S}_{k,f}} f_i(x_k), \qquad (3)$$

$$g_k = \frac{1}{|\mathcal{S}_{k,g}|} \sum_{i \in \mathcal{S}_{k,g}} \nabla f_i(x_k), \qquad (4)$$

where $\mathcal{S}_{k,f}$ and $\mathcal{S}_{k,g}$ are random subsets of $\{1, \ldots, N\}$ and $|\mathcal{S}_{k,f}|$, $|\mathcal{S}_{k,g}|$ denote their cardinality. Then, the kth iteration of the stochastic gradient procedures we are dealing with has the form

$$x_{k+1} = x_k - \alpha_k g_k, \qquad (5)$$

where α_k is a positive steplength. By construction, $\{x_k\}$ is a stochastic process whose behavior depends on the randomly selected samples.

Choosing the size of the sample set and the steplength along the iterations clearly represents the main issue in the realization of subsampled first-order methods and characterizes the procedures. Since there is a large variety of approaches, classifying the large number of methods in the literature on the basis of their features is not a trivial task. In this work, we cast renowned stochastic first-order procedures into two groups along the following arguments. Methods in the first group employ subsampled gradient estimates on very small batch sizes (in some approaches full gradient evaluations are occasionally performed) and do not perform checks for acceptance of the new iterate x_{k+1}, i.e., the computed step is accepted in every iteration. Consequently, the computational cost per iteration is low, and their implementation is simple. The original idea can be traced back to Robbins and Monro (Robbins et al. 1951), who proposed the famous Stochastic Approximation method. With careful and problem-dependent choices of the steplength sequence $\{\alpha_k\}$, theoretical results establish the behavior of the expected function values and gradient norm values. Methods (Andradottir 1996; Delyon and Juditsky 1993; Kesten 1958; Kiefer 1952; Krejić et al. 2013, 2015; Nemirovski et al. 2009; Robbins et al. 1951; Spall 2003; Tan et al. 2016; Yousefian et al. 2012; Xu et al. 2012) belong to such class. The performance of these methods is sensitive to the steplength selection and to stochastic variance reduction techniques (Defazio et al. 2014; Johnson et al. 2013; Kingma and Ba 2015; Nguyen et al. 2017; Schmidt et al. 2017).

Methods in the second class rely on machinery from standard globally convergent optimization procedures such as line search, trust-region, or adaptive overestimation strategies (Bellavia et al. 2019, 2020c; Birgin et al. 2018; Blanchet et al. 2019; Cartis et al. 2018; Chen et al. 2018; Curtis et al. 2019; Krejić et al. 2016; Krejić N et al. 2013; Krejić et al. 2015; Paquette et al. 2020; Tripuraneni et al. 2018) and have been proposed with the aim of overcoming the need of problem-dependent steplengths. In fact, by using subsampled function and gradient estimates, steplength selection is adaptive and made on the basis of some globalization strategy and knowable quantities. The choice of the sample size can vary from simple heuristics to sophisticated schemes that take into account the progress made by the optimization process itself. A further relevant distinction from the methods in the first group is that, except for Curtis et al. (2019), the accuracy of the function and gradient estimates is controlled adaptively along the iterations and plays a central role in the convergence analysis. Assuming that the variance of random functions and gradients is bounded, specific accuracy requirements can be fulfilled by means of a sufficiently large sample size estimated using probabilistic arguments (Bellavia et al. 2019; Tripuraneni et al. 2018; Tropp 2015). Some approaches Bellavia et al. (2020c), Birgin et al. (2018), Krejić N et al. (2013); Krejić et al. (2015); Krejić et al. (2016) reach eventually full precision functions and gradients, and thus the convergence results are deterministic; in the remaining methods, convergence is stated in terms of probability statements, either in mean square or almost sure.

The work is organized as follows. In section "Convolutional Neural Networks", we briefly introduce neural networks and convolutional neural networks which are widely used for image processing tasks. In section "Stochastic Gradient and Variance Reduction Methods", we describe subsampled first-order methods in the first group, while in section "Gradient Methods with Adaptive Steplength Selection Based on Globalization Strategies" we present methods belonging to the second group. Finally, in section "Numerical Experiments", we solve a classification problem of images, discussing the neural network used, implementation issues, and results obtained with some of the methods presented. All norms in the paper are Euclidean $\|\cdot\| \stackrel{\text{def}}{=} \|\cdot\|_2$ and given a random variable A; the symbols $Pr(A)$ and $E[A]$ denote the probability and expected value of A, respectively.

Convolutional Neural Networks

Neural networks (NNs) have become a state-of-the-art methodology for classification and regression tasks in artificial intelligence field (Bishop 2006; Hastie et al. 2001). NNs are used to approximate functions $\phi : \mathbb{R}^s \to \mathbb{R}^t$ whose value is known only at a given set of points $\mathbf{d}_i \in \mathbb{R}^s$, $i = 1, \ldots, N$. Letting $\hat{\mathbf{y}}_i = \phi(\mathbf{d}_i)$ for $i = 1, \ldots, N$, the pairs $\{(\mathbf{d}_i, \hat{\mathbf{y}}_i)\}_{i=1,\ldots,N} \in \mathbb{R}^s \times \mathbb{R}^t$, are available and can be used to train the neural network that is supposed to approximate values of $\phi(\mathbf{d})$ for $\mathbf{d} \neq \mathbf{d}_i$, $i = 1, \ldots, N$.

A neural network is a model which is typically represented by a network diagram as the one in Fig. 1. It consists of layers $L_1, \ldots, L_m$, $m \geq 2$; L_1 is called *input layer*, L_m is the *output layer*, and, when $m > 2$, $L_2, \ldots, L_{m-1}$ are called *hidden layers*. Each layer L_i contains a finite number n_i of neurons, subject to the constraints $n_1 = s$, $n_m = t$. Given an input data $\mathbf{d} \in \mathbb{R}^s$, the neural network returns an output vector in $\mathbb{R}^t$.

Given an input data $\mathbf{d} \in \mathbb{R}^s$, a neuron of a NN is modeled as shown in Fig. 2. Let $\mathbf{v}_i = (v_{i,1}, \ldots, v_{i,n_i})^T \in \mathbb{R}^{n_i}$ be the output of layer L_i and $\boldsymbol{\sigma}_i = (\sigma_{i,1}, \ldots, \sigma_{i,n_i})^T \in \mathbb{R}^{n_i}$ contain the *activation* functions $\sigma_{i,j} : \mathbb{R} \to \mathbb{R}$. Thus, the output of the jth neuron of the layer L_i, for $i = 2, \ldots, m$ is the scalar

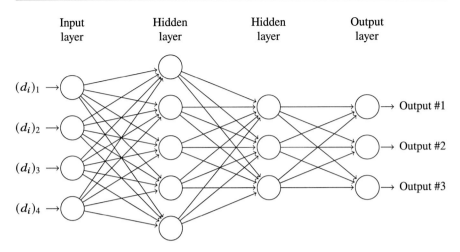

Fig. 1 An example of neural network with two hidden layers, s=4, t=3

$$v_{i,j} = \sigma_{i,j}\left(\sum_{k=1}^{n_{i-1}} x_{i,j,k} v_{i-1,k} + b_{i,j}\right), \tag{6}$$

where $b_{i,j} \in \mathbb{R}$ is called *bias* and the parameters $x_{i,j,k}$ are called *weights*. Vector $\mathbf{v}_1$ coincides with the input data $\mathbf{d}$. Letting $\mathbf{X}_i \in \mathbb{R}^{n_i} \times \mathbb{R}^{n_{i-1}}$ be the matrix with (j,k)-entry given by $x_{i,j,k}$, for $1 \leq j \leq n_i$, $1 \leq k \leq n_{i-1}$ and $\mathbf{b}_i = \left(b_{i,1}, \ldots, b_{i,n_i}\right)^T \in \mathbb{R}^{n_i}$, the output of the whole layer L_i is

$$\mathbf{v}_i = \sigma_i\left(\mathbf{X}_i \mathbf{v}_{i-1} + \mathbf{b}_i\right). \tag{7}$$

In fact, the output of each layer is defined recursively by (7) and depends on the output of the previous layer.

Common examples of activation functions are (Bishop 2006; Goodfellow et al. 2016):

- *Linear*: $\sigma(z) = z$
- *Sigmoid* or *logistic*: $\sigma(z) = 1/(1 + e^{-z})$
- *Tanh*: $\sigma(z) = \tanh(z)$
- *Relu*: $\sigma(z) = \max(0, z)$
- *Elu*: $\sigma(z) = z \cdot \mathbb{X}_{[x \geq 0]} + (e^z - 1) \cdot \mathbb{X}_{[x < 0]}$

where $\mathbb{X}_I : \mathbb{R} \to \mathbb{R}$ is the indicator function, defined by

$$\mathbb{X}_I(x) = \begin{cases} 1 & x \in I \subseteq \mathbb{R} \\ 0 & \text{otherwise} \end{cases}. \tag{8}$$

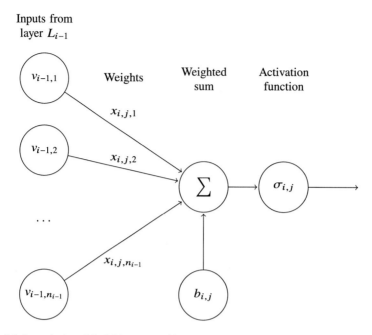

Fig. 2 Mathematical model of jth neuron of layer L_i

The procedure for choosing the parameters $\left\{ (\mathbf{X}_i, \mathbf{b}_i) \right\}_{i=1,\ldots,m}$ is referred to as *training phase*. Let x be the vectorization of $\left\{ (\mathbf{X}_i, \mathbf{b}_i) \right\}_{i=1,\ldots,m}$. Given the set of known data $\left\{ (\mathbf{d}_i, \hat{\mathbf{y}}_i) \right\}_{i=1,\ldots,N}$ (*training set*), the aim is to choose the parameters so that the output $\mathbf{v}_m(x; \mathbf{d}_i)$ of the neural network corresponding to the input $\mathbf{d}_i$ is as close as possible to the value $\hat{\mathbf{y}}_i$ for every $i = 1, \ldots, N$.

In order to do that, it is necessary to select a function $\mathbb{E} : \mathbb{R}^t \times \mathbb{R}^t \to \mathbb{R}$ for measuring the error made by the network on the prediction of each given data and minimize the so-called *loss* function:

$$\frac{1}{N} \sum_{i=1}^{N} \mathbb{E}(\mathbf{v}_m(x; \mathbf{d}_i), \hat{\mathbf{y}}_i). \qquad (9)$$

Since $\mathbf{d}_i$ and $\hat{\mathbf{y}}_i$ are known, the loss function is a special case of (2) where

$$f_i(x) = \mathbb{E}(\mathbf{v}_m(x; \mathbf{d}_i), \hat{\mathbf{y}}_i), \quad \text{for } i = 1, \ldots, N.$$

We underline that the minimization of suitable loss functions gives rise to prediction functions that *generalize* information from the available data and avoid *overfitting* of the training set (Bottou et al. 2018, §2).

Convolutional neural networks (CNNs) are a specialized kind of neural network for processing data with a grid-like topology, such as images represented as a two- or three-dimensional grid of pixels. CNNs extract features from the input image which are in some way representative of local neighboring portions of the image. This choice is motivated by the fact that important connections in an image are local (Strang 2019) and that reducing the dimension of weight matrices speeds up the process. This task is achieved exploiting filters commonly used in the computer vision context, such as convolution filter, which are able to extract low level features such as edges, color, and gradient orientation (Forsyth et al. 2002, Chap. 4). These filters are combined with standard neural network layers, so that all the low-level features are combined together. In the following, we give an overview on the main layers used in CNNs and refer the interested reader to Goodfellow et al. (2016, Chap. 9) and (Chollet 2017; Patterson et al. 2017) for additional details.

Convolutional Layer

We consider an image I as a three-dimensional $w \times h \times c$ array, where w is the image width, h is the image height, and c is the number of channels.

Discrete convolution aims to reduce the noise of a signal by applying a weighted average of each entry of the signal and its neighbors. Given an image I sized $w \times h \times c$, an integer $k \geq 1$, a three-dimensional $(2k + 1) \times (2k + 1) \times c$ array W called *kernel*, and a scalar b called *bias*, the discrete convolution between I and W, denoted by I * W, is the two-dimensional array defined by

$$(I * W)(i, j) = \sum_{s} \sum_{t} \sum_{u=1}^{c} I(s, t, u) \cdot W(s - i + k + 1, t - j + k + 1, u) + b, \quad (10)$$

for $i = 1, \ldots, w - 2k$ and $j = 1, \ldots, h - 2k$, where s and t range over all allowed subscripts for I and W, namely, $s = \max\{1, i - k\}, \ldots, \min\{i + k, w\}$, $t = \max\{1, j - k\}, \ldots, \min\{j + k, h\}$.

The application of a filter to the input yields a two-dimensional array instead of a three-dimensional; see index u in (10). Typically, convolutional layers apply m different filters of the same dimension to the input. Consider m kernels $\{W_\ell\}_{\ell=1,\ldots,m}$, each one sized $(2k + 1) \times (2k + 1) \times c$. The output of the convolutional layer is the 3D array defined by

$$(I * *W)(i, j, \ell) = (I * W_\ell)(i, j),$$

where $i = 1, \ldots, w - 2k$, $j = 1, \ldots, h - 2k$ and $\ell = 1, \ldots, m$; thus, every filter adds a channel to the output array. Hence, the output of convolutional layers with m kernels is given by an array of width and length $w - 2k$ and $h - 2k$, respectively, while the new number of channels is equal to the number of filters which have been applied.

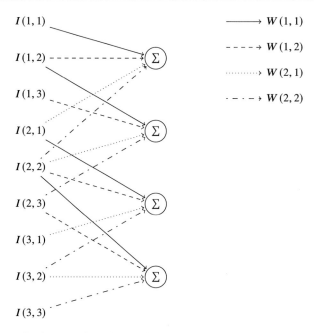

Fig. 3 An example of convolutional layer acting on a $3 \times 3 \times 1$ array I. The kernel W dimension is $2 \times 2 \times 1$. Weights W are shared among different neurons. In this example, the output of the convolutional layer consists of four neurons. Biases and activation functions have been omitted

CNNs are networks composed by at least one convolutional layer and standard layers. In convolutional layers, the entries of the filters are the parameters which are updated during the training. Hence, a convolutional layer consists of $m \cdot ((2k+1) \cdot (2k+1) \cdot c + 1)$ trainable parameters, $(2k+1) \cdot (2k+1) \cdot c + 1$ for each filter, bias term included. Each element of the array resulting from a convolution can be viewed as a neuron of the type shown in Fig. 2, where some of the connections, corresponding to the indices falling outside the ranges defined in (10), have been dropped (i.e., the corresponding weights are set to 0). In contrast with standard NN layers, convolutional layers share weights among different neurons. The kernel weights are in fact the same in each output neuron, as shown in Fig. 3.

Max Pooling Layer

In order to speed up the training phase by reducing the dimension of the object involved, the max pooling strategy is commonly used in CNN architectures for imaging (Strang 2019). It consists in replacing, for every channel, a square neighborhood with its maximum. More formally, given an image I, max pooling process MP acts as follows:

$$MP(I)(i,j,k) = \max_{(s,t) \in S(i,j)} I(s,t,k), \qquad (11)$$

where $S(i,j)$ is a neighborhood of (i,j).

The square neighborhood is defined by mean of two hyperparameters $\chi_{sl} \geq \chi_{st}$, which are the *spatial extent*, the length of the square edge, and the *stride*, the step which is used to move the square around the image, respectively. When $\chi_{sl} > \chi_{st}$, we talk about *overlapping* pooling. By construction, the max pooling layer does not call for parameters to be trained, and the dimension of the output of MP is smaller than that of the input and given by

$$((w - \chi_{sl})/\chi_{st} + 1) \times ((h - \chi_{sl})/\chi_{st} + 1) \times c,$$

where $w \times h \times c$ is the input dimension. This strategy can be viewed also as a *downsampling* in order to mitigate overfitting during the training.

Stochastic Gradient and Variance Reduction Methods

In this section, we present the widely used stochastic gradient descent (SGD) method (Robbins et al. 1951) and incremental gradient algorithms based on variance reduction such as stochastic variance reduction gradient (SVRG) method (Johnson et al. 2013), SVRG method with Barzilai-Borwein steplengths (SVRG - BB) (Tan et al. 2016), StochAstic Recursive grAdient algoritHm (SARAH) method (Nguyen et al. 2017), stochastic average gradient (SAG) method (Schmidt et al. 2017), and SAGA (Defazio et al. 2014). In the presentation of the convergence properties of these methods, we will make use of the specific form (2) of the problem and of following assumptions.

Assumption 1. Each function $f_i : \mathbb{R}^n \to \mathbb{R}$ has Lipschitz continuous gradient, i.e., there exists a constant $L \geq 0$ such that

$$\|\nabla f_i(x) - \nabla f_i(y)\| \leq L\|x - y\| \quad x,y \in \mathbb{R}^n.$$

This assumption clearly implies that the gradient of objective function is also L-Lipschitz continuous:

$$\|\nabla f(x) - \nabla f(y)\| \leq L\|x - y\| \quad x,y \in \mathbb{R}^n.$$

Assumption 2. The function $f : \mathbb{R}^n \to \mathbb{R}$ is μ strongly convex, i.e., there exists a constant $\mu > 0$ such that

$$f(x) \geq f(y) + (\nabla f(y))^T(x-y) + \frac{\mu}{2}\|x-y\|^2 \quad \text{for all} \quad (x,y) \in \mathbb{R}^n \times \mathbb{R}^n. \qquad (12)$$

In case of convex (strongly convex) problems, we denote x_* an (the unique) optimal solution.

The standard gradient descent GD method employing the full (true) gradient (FG) is defined by the following iterative formula:

$$x_{k+1} = x_k - \alpha_k \nabla f(x_k).$$

The steplength α_k can be fixed in a number of ways, for example, one can apply a line search procedure based on specific requirements on f or take a constant value, $\alpha_k = \alpha, \forall k \geq 0$. If f is convex and Assumption 1 holds, method FG with fixed steplength α converges sublinearly and satisfies the following error bound:

$$f(x_k) - f(x_*) = \mathcal{O}(1/k),$$

provided that $0 < \alpha < 2/L$ (Nesterov 1998, Th. 2.1.13). If additionally f is strongly convex and $0 < \alpha < 2/(\mu + L)$, then FG achieves linear convergence:

$$f(x_k) - f(x_*) = \mathcal{O}(\rho^k),$$

with ρ depending on the condition number L/μ (Nesterov 1998, Th. 2.1.14).

In the case where the number of component functions f_i is large, such as in machine learning applications, the computation of the full gradient is very expensive, and SGD (stochastic gradient descent) appears as an appealing alternative. The method was first proposed in the seminal paper of Robbins and Monro as SA (stochastic approximation) method (Robbins et al. 1951). The main idea of SGD is to replace the expensive gradient $\nabla f(x_k)$ with a significantly cheaper stochastic vector g_k. Here we focus on the case where g_k is an unbiased approximation to $\nabla f(x_k)$, i.e., $E[g_k] = \nabla f(x_k)$, built via (4) with $\mathcal{S}_{k,g}$ chosen uniformly at random from $\{1, \ldots, N\}$.

Intuition for using subsampled functions evaluated on random small size sample sets comes from the fact that the training set is often highly redundant, see, e.g., (Bottou et al. 2018). Sample sets $\mathcal{S}_{k,g}$ with small cardinality $|\mathcal{S}_{k,g}|$, in the limit equal to one, are generally used. Whenever $|\mathcal{S}_{k,g}| > 1$, the stochastic approximation of the full gradient is denoted as *mini-batch*; on the other hand, if the sample set reduces to a single element, the stochastic approximation is called *simple* or *basic*. In the following algorithm, without loss of generality, we present SGD referring to the latter case.

ALGORITHM SGD

Step 0: Initialization. Choose an initial point x_0 and a sequence of strictly positive steplengths $\{\alpha_k\}$. Set $k = 0$.

Step 1. Stochastic gradient computation. Choose randomly and uniformly $i_k \in \{1, \ldots, N\}$. Set $g_k = \nabla f_{i_k}(x_k)$.

(continued)

Step 2. Iterate computation. Set $x_{k+1} = x_k - \alpha_k g_k$. Increment k by one and go to Step 1.

Since $\{x_k\}$ is a stochastic process whose behavior depends on the random variables $\{i_k\}$, convergence analysis has to be carried out in expectation. Given that one iteration of SGD requires a single gradient $\nabla f_{i_k}(x_k)$, each iteration of the SGD method is significantly cheaper than FG method. Due to the variance introduced by the approximations g_k, in case of fixed steplength, it is not possible to prove convergence of the method to the solution even in the strongly convex case. On the other hand, it can be proved that if there exist positive scalars M_1 and M_2 such that at each iteration of SGD

$$E[\|g_k\|^2] \leq M_1 + M_2 \|\nabla f(x_k)\|^2, \tag{13}$$

and if $\alpha \leq \mu/(LM_2)$, then the expected optimality gap $f(x_k) - f(x_*)$ falls below a problem-dependent value (Bottou et al. 2018, Th. 4.6).

Convergence in expectation can be proved assuming to employ diminishing steplengths, i.e., the sequence $\{\alpha_k\}$ satisfies $\sum_{k=1}^{\infty} \alpha_k = \infty$, $\sum_{k=1}^{\infty} \alpha_k^2 < \infty$. It can be shown (see Nemirovski et al. (2009, p. 1578)) that for strongly convex functions, properly chosen steplengths such as $\alpha_k = \theta/k$ with $\theta > 1/(2\mu)$, and random gradient approximations having bounded variance, one can get

$$E[\|x_k - x_*\|] = \mathcal{O}(1/\sqrt{k}).$$

A further result on expected optimality gap for strongly convex functions is given below.

Theorem 1 (Bottou et al. 2018, Th. 4.7). *Suppose that Assumptions* 1 *and* 2 *hold and let x_* be the minimizer of f. Assume that* (13) *holds at each iteration. Then, if SGD is run with $\alpha_k = \frac{\beta}{\gamma+k}$, $\beta > \frac{1}{\mu}$ and $\gamma > 0$ such that $\alpha_1 \leq \frac{1}{LM_2}$, there exists a scalar $\nu > 0$ such that*

$$E[f(x_k)] - f(x_*) \leq \frac{\nu}{\gamma + k}. \tag{14}$$

The theorem above shows that, in the case of strongly convex problems, SGD converges slower (sublinearly) than FG method and this depends on the variance of the random sampling. Note that the larger M_2 is, the smaller the steplength is, and this implies slow convergence.

Theoretical results for SGD applied to nonconvex optimization problems are available (Bottou et al. 2018, §4.3). In particular, if f is bounded, in expectation $-g_k$ is a direction of sufficient descent for f at x_k and SGD is applied with diminishing steplengths $\{\alpha_k\}$ satisfying $\sum_{k=1}^{\infty} \alpha_k = \infty$, $\sum_{k=1}^{\infty} \alpha_k^2 < \infty$, then it can be shown that the expected gradient norms cannot stay bounded away from zero (Bottou et al. 2018, Th. 4.9).

If the approximate gradient g_k has a large variance, SGD may show slow convergence and bad performance. Taking a larger sample size for $\mathcal{S}_{k,g}$ could help to reduce gradient variance, but large sample may deteriorate the overall computational efficiency of stochastic gradient optimization. In order to improve convergence with respect to SGD, stochastic variance reduction methods have been proposed, see, e.g., Defazio et al. (2014), Johnson et al. (2013), Nguyen et al. (2017), Tan et al. (2016), Schmidt et al. (2017), Wang et al. (2013). In particular, in Wang et al. (2013), a variance reduction technique is proposed by making use of control variates (Ross 2006) to augment the gradient approximation and consequently reduce its variance.

Variance reduction is the core of SVRG (stochastic variance reduction gradient) method presented in Johnson et al. (2013); the algorithm is given below.

ALGORITHM SVRG

Step 0: Initialization. Choose an initial point $x_0 \in \mathbb{R}^n$, an inner loop size $m > 0$, a steplength $\alpha > 0$, and the option for the iterate update. Set $k = 1$.
Step 1: Outer iteration, full gradient evaluation.
Set $\tilde{x}_0 = x_{k-1}$. Compute $\nabla f(\tilde{x}_0)$.
Step 2: Inner iterations
For $t = 0, \ldots, m-1$
    Uniformly and randomly choose $i_t \in \{1, \ldots, N\}$.
    Set $\tilde{x}_{t+1} = \tilde{x}_t - \alpha(\nabla f_{i_t}(\tilde{x}_t) - \nabla f_{i_t}(\tilde{x}_0) + \nabla f(\tilde{x}_0))$.
Step 3: Outer iteration, iterate update.
Set $x_k = \tilde{x}_m$ (Option I). Increment k by one and go to Step 1.
Set $x_k = \tilde{x}_t$ for randomly chosen $t \in \{0, \ldots, m-1\}$ (Option II). Increment k by one and go to Step 1.

SVRG consists of outer and inner iterations. At each outer iteration k, the full gradient at x_k is computed. Then a prefixed number m of inner iterations is performed using stochastic gradients and fixed steplength α; the internal iterates are $\tilde{x}_0, \tilde{x}_1, \ldots, \tilde{x}_m$. At the tth inner iteration, the stochastic gradient used has the form

$$\nabla f_{i_t}(\tilde{x}_t) - \nabla f_{i_t}(\tilde{x}_0) + \nabla f(\tilde{x}_0),$$

with i_t chosen uniformly and randomly in $\{1, \ldots, N\}$. This quantity is an unbiased estimation of the gradient. Finally, the new iterate is either the last computed iterate $\tilde{x}_m$ (Option I) or one of the vectors $\tilde{x}_0, \ldots, \tilde{x}_{m-1}$ (Option II). Although Option I, taking the new iterate as the last outcome of inner loop, is intuitively more appealing, the convergence results from Johnson et al. (2013) are valid for Option II only. The results presented in Johnson et al. (2013) cover both the convex and nonconvex cases. For the sake of simplicity, here, we consider the strongly convex case.

Theorem 2 (Johnson et al. 2013, Th 1). *Suppose that Assumptions* 1 *and* 2 *hold and that all f_i are convex, and let x_* be the minimizer of f. If m and α satisfy*

$$\theta = \frac{1}{\mu\alpha(1-2L\alpha)m} + \frac{2L\alpha}{1-2L\alpha} < 1, \tag{15}$$

then Algorithm SVRG with Option II generates a sequence such that

$$E[f(x_k)] - f(x_*) \leq \theta^k (f(x_0) - f(x_*)).$$

The above statement clearly demonstrates that convergence in expectation depends on m and α and it is guaranteed taking both a sufficiently large loop size m and a sufficiently small steplength α. Note that θ in (15) depends on the scalars L and μ and condition (15) imposes the following restrictions to α and m: $\alpha < 1/(4L)$ and $m > 2/(\mu\alpha)$.

The linear convergence in expectation of the sequence of the iterates generated by the same algorithm with Option I has been proved later in Tan et al. (2016), and it is given below.

Theorem 3 (Tan et al. 2016, Corollary 1). *Suppose that Assumptions* 1 *and* 2 *hold and let x_* be the minimizer of f. If m and α satisfy*

$$\theta = (1 - 2\alpha\mu(1-\alpha L)^m) + \frac{4\alpha L^2}{\mu(1-\alpha L)} < 1,$$

then Algorithm SVRG with Option I generates a sequence which converges linearly in expectation

$$E[\|x_k - x_*\|^2] \leq \theta^k \|x_0 - x_*\|^2.$$

The value of m is most often of order $\mathcal{O}(n)$; in Johnson et al. (2013), it is suggested to take $m = 2n$ for convex problems and $m = 5n$ for nonconvex problems. Numerical studies that concentrate on the influence of m and α are available in Tan et al. (2016) as well as the comparison with the method of SVRG type employing adaptive steplengths. Further, in practical applications, it can be convenient to replace the full gradient at outer iterations with a mini-batch stochastic gradient. Application of SVRG to nonconvex problems is briefly discussed in Johnson et al. (2013, §3). Notice that SVRG requires the full gradient which is stored in memory during the whole inner loop execution. Instead of storing all gradients $\nabla f_i(\tilde{x}_0)$ separately, at each inner iteration $\nabla f_{i_t}(\tilde{x}_0)$ is evaluated along with $\nabla f_{i_t}(\tilde{x}_t)$; this increases the computational cost but reduces the memory requirement drastically. In applications where gradient evaluation is very expensive, the full gradient is typically replaced with a mini-batch stochastic gradient (Lei et al. 2017). Further, we mention a limited memory approach which gives rise to k-SVRG (Raj et al. 2018).

A variant of SVRG borrows ideas from the spectral gradient method (Barzilai et al. 1988; Raydan et al. 1997) which is very popular modification of the classical FG. The spectral gradient method is based on the idea of approximating the Hessian matrix in each iteration with a multiple of the identity matrix which minimizes the discrepancy from the secant equation and yields an adaptive steplength in each iteration of the gradient method. This steplength is known as Barzilai-Borwein steplength or the spectral coefficient. The adaptive steplengths overcome hand-tuning and do not need to be small, i.e., of order $1/L$ when the Lipschitz constant is large. Therefore, it is reasonable to expect that some advantages of similar type might be expected in the framework of SGD and SVRG methods. The following algorithm is developed in Tan et al. (2016), introducing the Barzilai-Borwein steplengths in the SVRG framework.

ALGORITHM SVRG - BB
Step 0: Initialization. Choose an initial point $x_0 \in \mathbb{R}^n$, an inner loop size $m > 0$, an initial steplength $\alpha_0 > 0$. Set $k = 1$.
Step 1: Outer iteration, full gradient evaluation.
Set $\tilde{x}_0 = x_{k-1}$. Compute $\nabla f(\tilde{x}_0)$.
If $k > 0$, then set $\alpha_k = \dfrac{1}{m} \dfrac{\|x_k - x_{k-1}\|^2}{(x_k - x_{k-1})^T (\nabla f(x_k) - \nabla f(x_{k-1}))}$
Step 2: Inner iterations
For $t = 0, \ldots, m-1$
    Uniformly and randomly choose $i_t \in \{1, \ldots, N\}$.
    Set $\tilde{x}_{t+1} = \tilde{x}_t - \alpha_k (\nabla f_{i_t}(\tilde{x}_t) - \nabla f_{i_t}(\tilde{x}_0) + \nabla f(\tilde{x}_0))$
Step 3: Outer iteration, iterate update. Set $x_k = \tilde{x}_m$. Increment k by one and go to Step 1.

Note that at the first outer iteration, the steplength is the input data α_0, while at the successive outer iterations, the steplengths α_k are adaptively chosen and used within inner iterations. The following results are established for strongly convex functions.

Theorem 4 (Tan et al. 2016, Th. 3.8). *Suppose that Assumptions 1 and 2 hold and let x_* be the minimizer of f. Define $\theta = (1 - e^{-2\mu/L})/2$. If m is chosen such that*

$$m > \max\left\{\frac{2}{\log(1 - 2\theta) + 2\mu/L}, \frac{4L^2}{\theta \mu^2} + \frac{L}{\mu}\right\},$$

then SVRG-BB converges linearly in expectation

$$E[\|x_k - x_*\|^2] < (1-\theta)^k \|\tilde{x}_0 - x_*\|^2.$$

A number of practical issues regarding the application of variance reduction gradient methods is considered in the literature. All of these methods compute the full gradient at each outer iteration, and this represents the main cost of these algorithms. Results presented in Babanezhad et al. (2015) show that it is possible to perform the outer iterations with increasing batch size for the gradient approximation without compromising the linear convergence rate. Mini-batch methods in inner loop iterations are also considered in Babanezhad et al. (2015).

SAG (Schmidt et al. 2017) method is based on average gradient approximation, which represent an alternative to the gradient estimators previously described. The main idea is to accumulate previously computed stochastic gradient values. The basic version of SAG method Schmidt et al. (2017) is presented in the algorithm below.

ALGORITHM SAG
Step 0: Initialization. Choose an initial point $x_0 \in \mathbb{R}^n$, positive steplengths $\{\alpha_k\}$, $y_i = 0$, for $i = 1, \ldots, N$. Set $k = 0$.
Step 1: Stochastic gradient update. Uniformly and randomly choose $i_k \in \{1, \ldots, N\}$. Set $y_{i_k} = \nabla f_{i_k}(x_k)$.
Step 2: Iterate update. Set $x_{k+1} = x_k - \frac{\alpha_k}{N} \sum_{i=1}^{N} y_i$. Increment k by one and go to Step 1.

SAG method uses a gradient estimation for $\nabla f(x_k)$ composed of the sum along all terms in the gradient, in the spirit of FG, but the cost of each iteration is the same as SDG. Remarkably, at the price of keeping track of a $N \times n$ matrix containing the gradient values computed through the iterations, SAG achieves almost the same convergence rate than FG. In fact, unlike SDG, convergence of SAG can be achieved taking constant steplength $\alpha_k = 1/(16L)$, $\forall k \geq 0$ and the optimality gap on average iterates achieve the same error bound $\mathcal{O}(1/k)$ as FG for convex function and linear convergence for strongly convex functions (Schmidt et al. 2017, Th. 1). If the Lipschitz constant is not available, a strategy for its estimation is given in Schmidt et al. (2017, §4.6). The following result concerns strongly convex problems.

Theorem 5 (Schmidt et al. 2017, Th. 1). *Suppose that Assumptions 1 and 2 hold. Let x_* be the minimizer of f. If $\alpha = 1/(16L)$, then*

$$E[f(x_k)] - f(x_*) \leq \left(1 - \min\left\{\frac{\mu}{16L}, \frac{1}{8N}\right\}\right)^k C_0,$$

where $C_0 > 0$ depends on x_*, x_0, f, L, N.

Note that for ill-conditioned problems where $N < (2L)/\mu$, N does not play a role in the convergence rate, and the SAG algorithm has nearly the same convergence rate as the FG method with a step size of $1/(16L)$, even though it uses iterations which are N times cheaper. This indicates that in case of ill-conditioned problems, the convergence rate is not affected by the use of out-of-date gradient values. A SAG extension, called SAGA, has been also proposed in Defazio et al. (2014). SAGA exploits SVRG-like unbiased approximations of the gradient and combines ideas of SAG and SVRG algorithms; a fixed steplength is employed. The interested reader can find additional details about SAGA in Defazio et al. (2014).

SARAH method Nguyen et al. (2017) is a further variant of SGD based on accumulated stochastic information. Unlike SAGA, SARAH is based on the idea of variance reduction and biased estimations of the gradient; the algorithm is sketched below.

ALGORITHM SARAH

Step 0: Initialization. Choose an initial point $x_0 \in \mathbb{R}^n$, an inner loop size $m > 0$, a steplength $\alpha > 0$. Set $k = 1$.

Step 1: Outer iteration, full gradient evaluation.
Set $\tilde{x}_0 = x_{k-1}$. Compute $y_0 = \nabla f(\tilde{x}_0)$. Set $\tilde{x}_1 = \tilde{x}_0 - \alpha y_0$.

Step 2: Inner iterations.
For $t = 1, \ldots, m-1$
    Uniformly and randomly choose $i_t \in \{1, \ldots, N\}$.
    Compute $y_t = \nabla f_{i_t}(\tilde{x}_t) - \nabla f_{i_t}(\tilde{x}_{t-1}) + y_{t-1}$.
    Set $\tilde{x}_{t+1} = \tilde{x}_t - \alpha y_t$.

Step 3: Outer iteration, iterate update. Set $x_k = \tilde{x}_t$ for randomly chosen $t \in \{0, \ldots, m\}$. Increment k by one and go to Step 1.

As already mentioned, y_t is a biased estimator of the gradient as

$$E[y_t] = \nabla f(\tilde{x}_t) - \nabla f(\tilde{x}_{t-1}) + y_{t-1} \neq \nabla f(\tilde{x}_t).$$

The convergence results presented in Nguyen et al. (2017) cover both the convex and strongly convex cases, as well as address complexity analysis; the result for the strongly convex case is given below.

Theorem 6 (Nguyen et al. 2017, Th. 4). *Suppose that Assumptions* 1 *and* 2 *hold and that each function* f_i, $1 \leq i \leq N$ *is convex. If* α *and* m *are such that*

$$\sigma = \frac{1}{\mu \alpha (m+1)} + \frac{\alpha L}{2 - \alpha L} < 1, \tag{16}$$

then the sequence $\{\|\nabla f(x_k)\|\}$ *generated by Algorithm SARAH satisfies*

$$E[\|\nabla f(x_k)\|^2] \leq \sigma^k \|\nabla f(x_0)\|^2.$$

We observe that condition (16) imposes the upper bound $1/L$ on the steplength α, while the analogous condition (15) governing the convergence of SVRG imposes the tighter bound $\alpha < 1/(4L)$; further, for any α and m, it holds $\sigma < \theta$. An additional advantage of SARAH is that if α is small enough, then the stochastic steps computed converge linearly in the inner loop in expectation.

Theorem 7 (Nguyen et al. 2017, Th. 1b). *Suppose that Assumption* 1 *holds and each function* f_i, $1 \leq i \leq N$ *is* μ*-strongly convex with* $\mu > 0$. *If* $\alpha \leq 2/(\mu + L)$, *then for any* $t \geq 1$

$$E[\|y_t\|^2] \leq \left(1 - \frac{2\mu L\alpha}{\mu + L}\right) E[\|y_{t-1}\|^2] \leq \left(1 - \frac{2\mu L\alpha}{\mu + L}\right)^t E[\|\nabla f(\tilde{x}_0)\|^2].$$

Gradient Methods with Adaptive Steplength Selection Based on Globalization Strategies

Gradient methods discussed in the previous section employ stochastic (possibly and occasionally full) gradient estimates and do not rely on any machinery from standard globally convergent optimization procedures such as line search, trust-region, or adaptive overestimation strategies. On the other hand, a few and recent papers (Bellavia et al. 2019, 2020c; Blanchet et al. 2019; Cartis et al. 2018; Chen et al. 2018; Curtis et al. 2019; Paquette et al. 2020) rely on such strategies for selecting the steplength and part of them mimic traditional step acceptance rules using stochastic estimates of functions and gradients. The purpose of these methods is to partially overcome the dependence of the steplengths from the Lipschitz constant of the gradient, i.e., lack of natural scaling, which appears in the convergence results of SGD and its variants given in section "Stochastic Gradient and Variance Reduction Methods"; see Curtis et al. (2019, §1).

One relevant proposal in the field of stochastic trust-region methods is TRish (Trust-Region-*ish*) algorithm (Curtis et al. 2019). TRish uses a stochastic gradient estimate g_k of $\nabla f(x_k)$ and a careful steplength selection which, to a certain extent, mimics a trust-region strategy. TRish algorithm is sketched below.

ALGORITHM TRISH
Step 0: Initialization. Choose an initial point $x_0 \in \mathbb{R}^n$, positive steplengths $\{\alpha_k\}$, positive $\{\gamma_{1,k}\}$ and $\{\gamma_{2,k}\}$ such that $\gamma_{1,k} > \gamma_{2,k}$, $\forall k \geq 0$. Set $k = 0$.
Step 1: Step computation. Compute a gradient estimate $g_k \in \mathbb{R}^n$.

(continued)

Step 2: Steplength selection. Set

$$s_k = \begin{cases} -\gamma_{1,k}\alpha_k g_k & \text{if } \|g_k\| \in \left[0, \frac{1}{\gamma_{1,k}}\right) \\ -\alpha_k \frac{g_k}{\|g_k\|} & \text{if } \|g_k\| \in \left[\frac{1}{\gamma_{1,k}}, \frac{1}{\gamma_{2,k}}\right] \\ -\gamma_{2,k}\alpha_k g_k & \text{if } \|g_k\| \in \left(\frac{1}{\gamma_{2,k}}, \infty\right). \end{cases} \quad (17)$$

Set $x_{k+1} = x_k + s_k$, increment k by one, and go to Step 1.

The relationship between the norms of $s_k = x_{k+1} - x_k$ and g_k is shown in Fig. 4. The norm of the step, as function of the norm of the stochastic gradient, is continuous. When $\|g_k\| \in \left[\frac{1}{\gamma_{1,k}}, \frac{1}{\gamma_{2,k}}\right]$, the step s_k can be viewed as a trust-region step since it solves the trust-region problem:

$$\min_{\|s\| \le \alpha_k} f(x_k) + g_k^T s. \quad (18)$$

If the norm of the stochastic gradient is below $1/\gamma_{1,k}$, then the steplength is $\gamma_{1,k}\alpha_k$, while if the norm is larger than $1/\gamma_{2,k}$, then the steplength is $\gamma_{2,k}\alpha_k$ with $\gamma_{2,k} < \gamma_{1,k}$. Note that the trust-region machinery is used for building the step, but unlike standard trust-region strategies, it does not employ step acceptance conditions and therefore it does not affect the choice of the steplengths $\{\alpha_k\}$. Examples in Curtis et al. (2019, §2) show that a pure trust-region algorithm, taking steps from (18) independently of the norm of the stochastic gradient, is not guaranteed to converge; this would be the case if $\gamma_{1,k} \gg 0$ and $\gamma_{2,k} \approx 0$. Hence, the convergence theory of TRish is based on an appropriate upper bound for $\gamma_{1,k}/\gamma_{2,k}$. The theoretical results for TRish are similar to those of SGD since both methods take steps along the stochastic gradient; on the other hand, SGD possesses no natural scaling, while

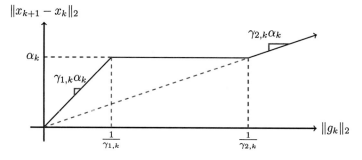

Fig. 4 Relationship between $\|x_{k+1} - x_k\|$ and $\|g_k\|$

`TRish` exploits normalized steps whenever $\|g_k\| \in \left[\frac{1}{\gamma_{1,k}}, \frac{1}{\gamma_{2,k}}\right]$. This issue can be interpreted as an adaptive choice of the steplength which is $\alpha_k/\|g_k\|$ instead of α_k itself; it is expected to improve numerical performance upon traditional SGD, and this is confirmed by the numerical results provided in Curtis et al. (2019, §2) and in the subsequent section "Numerical Experiments".

We summarize some results from the convergence analysis presented in Curtis et al. (2019). Let us assume that Assumption 1 holds, g_k is an unbiased estimator of $\nabla f(x_k)$ satisfying inequality (13) for any $k \geq 0$, f is bounded below by $f_* = \inf_{x \in \mathbb{R}^n} f(x) \in \mathbb{R}$, and the Polyak-Lojasiewicz condition holds at any $x \in \mathbb{R}^n$ with $\mu > 0$, i.e.,

$$2\mu(f(x) - f_*) \leq \|\nabla f(x)\|^2, \quad \forall x \in \mathbb{R}^n. \tag{19}$$

Note that (19) holds if f is μ-strongly convex.

The first convergence result of `TRish` deals with constant choices for the parameters $\gamma_{1,k} = \gamma_1$, $\gamma_{2,k} = \gamma_2$, and $\alpha_k = \alpha$ for all $k \geq 0$ (Curtis et al. 2019, Theorem 1). Provided that γ_1/γ_2 and α are bounded from above by quantities involving μ, L, and M_1, M_2 in (13), then `TRish` has expected optimality gap:

$$\mathbb{E}\left[f(x_k)\right] - f_* \leq c_1 + c_2^{k-2}(f(x_0) - f_* - c_1),$$

where $c_1 > 0$ and $c_2 \in (0, 1)$ are scalars depending on α, γ_1, γ_2. In fact, using a constant steplength depending on the Lipschitz constant L, the expected optimality gap is guaranteed to be reduced below a given threshold as in SGD. A comparison of the steplength bound in `TRish` with that in the classical SGD method can be found in Curtis et al. (2019, p.207).

Convergence can be proved to be linear if the variance of the stochastic gradient decreases linearly (Curtis et al. 2019, Theorem 4). Specifically, if additionally the stochastic gradient satisfies

$$\mathbb{E}\left[\|g_k\|^2\right] \leq c\zeta^{k-1} + \|\nabla f(x_k)\|^2, \tag{20}$$

for all $k \geq 0$ and some $c > 0$, $\zeta \in (0, 1)$, then

$$\mathbb{E}\left[f(x_k)\right] - f_* \leq \omega \rho^{k-1},$$

where $\omega > 0$ and $\rho \in (0, 1)$. Assumption (20) on gradients can be satisfied if g_k is computed by subsampling with increasing sample size.

A further convergence result covers the cases of sublinearly diminishing steplengths Curtis et al. (2019, Theorem 2) and resembles the corresponding result for SGD method. If the steplengths α_k are sublinearly diminishing, i.e., $\alpha_k = \beta/(\nu + k)$ for some positive β and ν properly chosen, $\gamma_{1,k} = \gamma_1 > 0$, $\gamma_1 - \gamma_{2,k} = \eta\alpha_k$, $\forall k$ and some $\eta \in (0, 1)$, then

$$E\left[f(x_k)\right] - f_* \leq \frac{\phi}{\nu + k},$$

for all k, with ϕ positive. We refer to Curtis et al. (2019) for more convergence results, including the case where the Polyak-Lojasiewicz condition is not satisfied.

Other approaches exploit globalization procedures more closely than TRish, with the aim of computing the steplength adaptively and testing, at each iteration, some verifiable criterion on progress toward optimality. To establish such control, they need stochastic estimates of functions, in addition to gradient estimates required in all the approaches described so far, and impose dynamic accuracy in stochastic function and gradient approximations. The general scheme for such procedures is given below. We will say that iteration k is successful whenever the acceptance criterion tested in Step 2 is fulfilled, unsuccessful otherwise. Acceptance criteria employed in literature will be presented in the sections "Stochastic Line Search" and "Adaptive Regularization and Trust-Region".

ALGORITHM LSANDTR

Step 0: Initialization. Choose an initial point $x_0 \in \mathbb{R}^n$, $\alpha_0 > 0$, parameters governing the steplength selection, and the accuracy requirement in gradient and function. Set $k = 0$.

Step 1: Step computation. Compute a gradient estimate $g_k \in \mathbb{R}^n$ and form a step $s_k = -\alpha_k g_k$.

Step 2: Step acceptance. Compute estimates f_k^0 and f_k^s of $f(x_k)$ and $f(x_k + s_k)$ and test for acceptance of $x_k + s_k$. If the iteration is successful, set $x_{k+1} = x_k + s_k$; otherwise, set $x_{k+1} = x_k$.

Step 3: Parameters' update. Compute α_{k+1} and update parameters governing the accuracy requirements in the computation of functions and gradients. Increment k by one and go to Step 1.

The above scheme includes the stochastic line search method proposed in Paquette et al. (2020), the stochastic trust-region method proposed in Blanchet et al. (2019) and Chen et al. (2018), and the adaptive overestimation method proposed in Bellavia et al. (2019). Accuracy in function and gradient approximations is controlled acknowledging that f has a central role since it is the quantity we ultimately wish to decrease. Specifically, it is assumed that f_k^0, f_k^s, and g_k are sufficiently accurate in probability, conditioned on the past, and an adaptive absolute accuracy for the objective function and an adaptive relative accuracy for the gradient are imposed. These requirements are supposed to be satisfied probabilistically. The method given in Cartis et al. (2018) belongs to the previous framework but uses the exact function in Step 2. Thus, it only imposes adaptive relative accuracy on the gradient.

Accuracy Requirements

As a general setting, let g_k be an estimate of $\nabla f(x_k)$, $\epsilon_g > 0$ be the accuracy requirement, and I_k be the event defined as

$$I_k = \{\|g_k - \nabla f(x_k)\| \leq \epsilon_g\}, \quad \epsilon_g > 0. \tag{21}$$

A gradient estimate g_k is said to be p_g-probabilistically sufficiently accurate whenever

$$Pr(\mathbb{1}_{I_k} = 1) \geq p_g \text{ with } p_g \in (0,1), \tag{22}$$

with $\mathbb{1}_{I_k} = 1$ if g_k is such that the event I_k holds, $\mathbb{1}_{I_k} = 0$ otherwise. In a similar way, let f_k^0 and f_k^s be estimates of $f(x_k)$ and $f(x_k + s_k)$, $\epsilon_f > 0$ be the accuracy requirement, and J_k be the event defined as

$$J_k = \{|f_k^0 - f(x_k)| \leq \epsilon_f \text{ and } |f_k^s - f(x_k + s_k)| \leq \epsilon_f\}, \quad \epsilon_f > 0. \tag{23}$$

Estimates f_k^0 and f_k^s are said to be p_f-probabilistically sufficiently accurate whenever the event J_k in (23) satisfies the condition

$$Pr(\mathbb{1}_{J_k} = 1) \geq p_f, \text{ with } p_f \in (0,1). \tag{24}$$

As for problem (2), the computation of f_k^0, f_k^s and g_k can be performed by averaging functions f_i and gradients ∇f_i in uniformly and randomly selected subsamples of the set $\{1,\ldots,N\}$. In order to satisfy (22) and (24) probabilistically, the size of uniform sampling $|\mathcal{S}_{k,f}|$ and $|\mathcal{S}_{k,g}|$ can be bounded below via the Bernstein inequality (Tropp 2015). In particular, in Bellavia et al. (2019, Theorem 6.2) it is shown that given $\epsilon_g > 0$, g_k is p_g-probabilistically sufficiently accurate if the cardinality $|\mathcal{S}_{k,g}|$ of the set $\mathcal{S}_{k,g}$ in (4) satisfies

$$|\mathcal{S}_{k,g}| \geq \min\left\{N, \left\lceil \frac{2}{\epsilon_g}\left(\frac{V_g}{\epsilon_g} + \frac{2\omega_g(x_k)}{3}\right) \log\left(\frac{n+1}{1-p_g}\right) \right\rceil\right\}, \tag{25}$$

where $E(\|\nabla f_i(x) - \nabla f(x)\|^2) \leq V_g$ and $\max_{i \in \{1,\ldots,N\}} |\nabla f_i(x)| \leq \omega_g(x)$, or

$$|\mathcal{S}_{k,g}| \geq \min\left\{N, \left\lceil \frac{4\omega_g(x_k)}{\epsilon_g}\left(\frac{2\omega_g(x_k)}{\epsilon_g} + \frac{1}{3}\right) \log\left(\frac{n+1}{1-p_g}\right) \right\rceil\right\}. \tag{26}$$

Similarly, given $\epsilon_f > 0$, f_k^0 is p_f-probabilistically sufficiently accurate if the cardinality $|\mathcal{S}_{k,f}|$ of the set $\mathcal{S}_{k,f}$ in (3) satisfies

$$|\mathcal{S}_{k,f}| \geq \min\left\{N, \left\lceil \frac{2}{\epsilon_f}\left(\frac{V_f}{\epsilon_f} + \frac{2\omega_f(x_k)}{3}\right)\log\left(\frac{2}{1-p_f}\right)\right\rceil\right\}, \quad (27)$$

where $E(|f_i(x) - f(x)|^2) \leq V_f$ and $\max_{i \in \{1,\ldots,N\}} |f_i(x)| \leq \omega_f(x)$, or

$$|\mathcal{S}_{k,f}| \geq \min\left\{N, \left\lceil \frac{4\omega_f(x_k)}{\epsilon_f}\left(\frac{2\omega_f(x_k)}{\epsilon_f} + \frac{1}{3}\right)\log\left(\frac{n}{1-p_f}\right)\right\rceil\right\}. \quad (28)$$

It is worth noting that in (25)–(28) failure probabilities $1 - p_f$, $1 - p_g$ appear in the logarithmic terms and therefore their contribution is damped even if they are very small. Specific accuracy requirements made will be specialized in the following subsections.

Stochastic Line Search

A stochastic line search method, which falls into the general scheme LSandTR, is given in Paquette et al. (2020). At iteration k, the computation of the step s_k and the stochastic line search are performed using a constant $\theta \in (0, 1)$ and a positive parameter δ_k. Given α_k, a probability $p_g \in (0, 1)$, a constant $\kappa > 0$, and letting $\epsilon_g = \kappa \alpha_k \|g_k\|$, the gradient estimate g_k formed in Step 1 is supposed to be p_g-probabilistically sufficiently accurate, i.e., to satisfy (22) with $\epsilon_g = \kappa \alpha_k \|g_k\|$.

With g_k at hand, the step s_k in Step 1 takes the form $s_k = -\alpha_k g_k$, and in Step 2 the Armijo condition

$$f_k^s \leq f_k^0 - \theta \alpha_k \|g_k\|^2, \quad (29)$$

is tested for acceptance. This condition is a stochastic variant of Armijo condition (Armijo et al. 1966) as f_k^0 and f_k^s are stochastic estimates of $f(x_k)$ and $f(x_k + s_k)$. Values f_k^0 and f_k^s are supposed to meet two requirements. First, given a probability $p_f \in (0, 1)$ and letting $\epsilon_f = \kappa \alpha_k^2 \|g_k\|^2$, f_k^0 and f_k^s are required to satisfy (24), namely, to be p_f-probabilistically sufficiently accurate with $\epsilon_f = \kappa \alpha_k^2 \|g_k\|^2$. Second, given a constant $\kappa_f > 0$, the sequence of estimates $\{f_k^0, f_k^s\}$ is supposed to satisfy the following variance conditions for all $k \geq 0$:

$$E[|f_k^0 - f(x_k)|^2] \leq \max\{\kappa_f \alpha_k^2 \|\nabla f(x_k)\|^4, \theta^2 \delta_k^4\},$$
$$E[|f_k^s - f(x_k + s_k)|^2] \leq \max\{\kappa_f \alpha_k^2 \|\nabla f(x_k)\|^4, \theta^2 \delta_k^4\}.$$

Note that both accuracy requirements on functions and gradients are adaptive and the function has to be approximated with higher accuracy than the gradient.

Moreover, observe that the variance condition depends on the parameter δ_k, the steplength α_k, and the norm of the true gradient.

The kth iteration is successful if (29) is met, unsuccessful otherwise. Whenever the iteration is successful, parameters are updated in Step 3 as follows:

$$\alpha_{k+1} = \max\{\gamma \alpha_k, \alpha_{\max}\}$$

$$\delta_{k+1}^2 = \begin{cases} \gamma \delta_k^2 & \text{if } \alpha_k \|g_k\|^2 \geq \delta_k^2 \\ \gamma^{-1} \delta_k^2 & \text{otherwise} \end{cases}$$

for some fixed $\gamma > 1$ and $\alpha_{\max} > 0$. On the other hand, when the iteration is unsuccessful, Step 3 consists in updating

$$\alpha_{k+1} = \gamma^{-1} \alpha_k, \qquad \delta_{k+1}^2 = \gamma^{-1} \delta_k^2.$$

The rules for choosing α_k and δ_k either enlarge or reduce accuracy in stochastic estimates based on fulfillment of the decrease condition (29) and the magnitude of the expected improvement of f_k^s over f_k^0. In fact, the parameter α_k affects the accuracy of gradient and function estimates and is enlarged when the iteration is successful, diminished otherwise. On the other hand, the parameter δ_k affects the variance of function estimates and is intended to guess how much the true function decreases. In fact, the decrease obtained in (29) does not guarantee a similar reduction in the true function as well. Hence, δ_k^2 is enlarged only in the case where the iteration is successful, and $\alpha_k \|g_k\|^2$ is not smaller than δ_k^2, that is, when the variance of function values is not larger than the square of the decrease in the approximate function. Interestingly, $\alpha_k \|g_k\|$ may not diminish as $\|g_k\|$ decreases and consequently accuracy requirements do not necessarily become more stringent along iterations.

In Paquette et al. (2020) stochastic complexity results have been established for convex, strongly convex, and general nonconvex, smooth problems; they imply convergence results. In case of μ-strongly convex problems, under suitable assumptions on the stochastic process, Paquette et al. (2020, Th. 4.18) shows that there exist probabilities p_g, p_f sufficiently close to one and satisfying $p_g p_f > \frac{1}{2}$ and a constant $\nu \in (0, 1)$ such that the expected number T_ϵ of iterations needed to satisfy

$$f(x_k) - f(x_*) \leq \epsilon$$

is such that

$$E[T_\epsilon] \leq \mathcal{O}(1) \frac{p_g p_f}{2 p_g p_f - 1} \frac{(L \kappa \alpha_{\max})^3}{\mu} (\log(\Phi_0) + \log(\epsilon^{-1}))$$

where x_* is the minimizer of f and Φ_0 is a problem-dependent positive scalar. We refer to Paquette et al. (2020) for the complete set of results. As a final comment, the implementation of the above stochastic line search method encounters the problem that $\epsilon_g = \kappa \alpha_k \|g_k\|$ depends on the norm of the vector g_k that has to be computed. Following Cartis et al. (2018), the computation of the approximated gradient g_k by subsampling can be performed via an inner iterative process. The approximated gradient g_k is computed via (25) or (26) using a predicted sample size. Then, if the predicted accuracy is larger than the required accuracy, the sample size is progressively increased until the accuracy requirement is satisfied.

Adaptive Regularization and Trust-Region

Trust-region and adaptive regularization methods are classes of optimization methods based on a nonlinear steplength control and can be cast into a unifying framework as shown in Toint (2013). Variants of these methods based on estimates for functions and derivatives are proposed in Bellavia et al. (2019), Blanchet et al. (2019), Chen et al. (2018), Wang and Yuan (2019). Here we focus on the case where first-order models are used at each iterations and discuss the adaptive regularization method named AR1DA (Adaptive Regularization with Dynamic Accuracy and first-order model) developed in Bellavia et al. (2019). It shares similarities with STORM, and we refer to Blanchet et al. (2019, §3) for details on this latter algorithm and its stochastic properties. The AR1DA method employs first-order random models with adaptive regularization of order two. The regularization parameter $\sigma_k > 0$ controls the steplength, and a parameter $\omega_k \in (0, 1)$ controls the level of accuracy required in the estimate f_k^0, f_k^s, and g_k. In fact, the gradient estimate g_k formed in Step 1 is supposed to be p_g-probabilistically sufficiently accurate, with $\epsilon_g = \omega_k \|g_k\|$. Once g_k has been computed, the step s_k in Step 1 is found by minimizing a regularized first-order random model model $m_k(s)$ for $f(x_k + s)$ around x_k:

$$\min_{s \in \mathbb{R}^n} m_k(s) = f_k^0 + g_k^T s + \frac{1}{2}\|s\|^2,$$

with f_k^0 being an approximation to $f(x_k)$. Trivially the step takes the form $s_k = -\frac{1}{\sigma_k} g_k$, i.e., $\alpha_k = \frac{1}{\sigma_k}$ in Step 1 of the general scheme LSandTR.

Acceptance of the step is tested using the rules employed in trust-region and regularization methods, but different from the standard approaches, here the function values and the gradient involved are approximated. Using function estimates f_k^0 and f_k^s for $f(x_k)$ and $f(x_k + s_k)$, the test for acceptance is

$$\rho_k = \frac{f_k^0 - f_k^s}{f_k^0 - (f_k^0 + g_k^T s_k)} \geq \eta_1, \quad \eta_1 \in (0, 1). \tag{30}$$

Values f_k^0 and f_k^s are supposed to be p_f-probabilistically sufficiently accurate with $\epsilon_f = \omega_k (g_k^T s_k)$.

Summarizing, the iteration is successful, i.e., the trial point $x_k + s_k$ is accepted as the new iterate, if $\rho_k \geq \eta_1$, unsuccessful otherwise. The updating rule for σ_k and ω_k is

$$\sigma_{k+1} = \begin{cases} \max\{\gamma^{-1}\sigma_k, \sigma_{\min}\} & \text{if } \rho_k \geq \eta_1 \\ \gamma \sigma_k & \text{otherwise} \end{cases},$$

and

$$\omega_{k+1} = \min\left(\kappa_\omega, \frac{1}{\sigma_{k+1}}\right),$$

for some fixed $\gamma > 1$, $\sigma_{\min} > 0$, and $\kappa_\omega \in (0, 1/(2\eta_1))$. Specifically, in case of successful iterations, the regularization parameter is decreased, and the parameter that rules the accuracy requirements is increased. On the other hand, in case of unsuccessful iterations, σ_k is increased and tighter accuracy requirements are imposed on function and gradient approximations.

In Bellavia et al. (2019), complexity analysis in high probability for AR1DA is carried out. Assume for sake of simplicity $p_g = p_f$ and let $\bar{p} \in (0, 1)$ be a prescribed probability for meeting the approximate first-order optimality condition:

$$\|\nabla f(x_k)\| \leq \epsilon, \tag{31}$$

with $\epsilon > 0$. In Bellavia et al. (2019, Th. 7.1), it is shown that if $1 - p_g = \mathcal{O}\left((1-\bar{p})\epsilon^2/3\right)$, then AR1DA needs at most $\mathcal{O}(\epsilon^{-2})$ iterations and approximate evaluations of the objective function to satisfy (31) with probability at least $\bar{p}$.

From a practical point of view, the approximated gradient g_k is computed via (25) or (26) using a predicted accuracy requirement, say, ϵ_p. Then, with g_k at hand, if $\epsilon_p > \omega_k\|g_k\|$, then ϵ_p is progressively decreased and g_k recomputed until $\epsilon_p \leq \omega_k\|g_k\|$ or $\epsilon_p < \epsilon$. We finally mention that the algorithm is stopped whenever the condition

$$\|g_k\| \leq \frac{\epsilon}{1+\omega_k}$$

holds. Remarkably, the accuracy requirement $\epsilon_g = \omega_k\|g_k\|$ guarantees that (31) holds with probability at least p_g.

Numerical Experiments

In this section, we show the performance of three methods previously discussed: SG, SVRG, and TRish applied in the training phase of a CNN. We train a neural network on cifar-10 (Krizhevsky 2009), a classical image recognition dataset. This dataset contains 60000 colored images with a resolution of 32 × 32 pixels divided

Fig. 5 Some random images from each class of cifar-10 dataset (Image taken from https://www.cs.toronto.edu/~kriz/cifar.html)

into a training set (5/6 of the images) and a testing set (1/6 of the images). The images are classified into ten homogeneously distributed classes: airplanes, cars, birds, cats, deer, dogs, frogs, horses, ships, and trucks. In Fig. 5, we show some images from the dataset. The color model of cifar-10 images is RGB, i.e., each pixel of an image is represented by three numbers (typically integers) which vary between 0 and 255 and represent the intensity of each channel; hence, the image can be viewed as a $32 \times 32 \times 3$ matrix. It is common to normalize the intensity of each channel between 0 and 1.

The training set is constituted by $N = 50000$ data $\left\{ \left(\mathbf{d}_i, \hat{\mathbf{y}}_i \right) \right\}_{i=1,\ldots,N}$, where $\mathbf{d}_i \in \mathbb{R}^{3072}$ is the vector containing the ith image stacked by columns and $\hat{\mathbf{y}}_i \in \mathbb{R}^{10}$ contains value 1 for the actual category of the ith image and 0 for any other category.

The Neural Network in Action

We describe the NN used in our experiments which consists of 14 layers and is displayed in Fig. 6.

The first layer of our network is convolutional (see section "Convolutional Layer") with 32 filters and a 3×3 kernel; the activation function is *elu*. The number of filters reshapes the tensor so that the number of channels becomes equal to the number of filters in the convolutional layer. The width and the height of the image

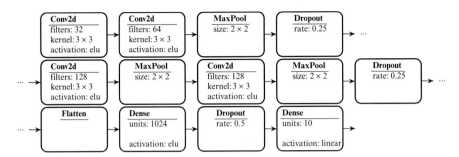

Fig. 6 Architecture of the neural network used for cifar-10. Four convolutional layers mixed with max pooling layers are followed by two dense layers

are changed too, accordingly to section "Convolutional Layer", and become both equal to 30. Summarizing, the output of the first layer has size $30 \times 30 \times 32$ and is received by the second layer which is again a convolutional layer with 64 filters, a 3×3 kernel, and *elu* as the activation function. After the second layer, the tensor shape becomes $28 \times 28 \times 64$. The third layer is a max pooling layer (see section "Max Pooling Layer"), which applies a 2×2 max filter on every channel; this halves the dimension of every slice of the tensor. The fourth layer is a Dropout layer with rate 0.25 which does not alter the shape of the tensor but randomly selects 25% of the values of the tensor and sets them to 0; this phase is commonly performed to avoid overfitting. Next, we apply two times a convolutional layer with 128 filters and a 3×3 kernel followed by a max pooling. After such four layers, a further Dropout layer with rate 0.25 is used; the resulting tensor shape is $2 \times 2 \times 128$. At this stage, the process for transforming the tensor into an array of probabilities is started. First, a Flatten layer vectorizes the $2 \times 2 \times 128$ tensor and returns a one-dimensional array with 512 values. Second, a Dense layer with 1024 neurons is used; the input array with 512 entries is transformed using the *elu* activation function. Third, a Dropout layer with rate 0.5 is used, and, finally, a Dense layer with ten neurons returns an array with 10 entries. Since the network output is expected to be a vector $\mathbf{v_m} = (v_{m,1}, \ldots, v_{m,10})^T$ such that $v_{m,j}$ represents the probability of an input image of being part of the jth category for $j = 1, \ldots, 10$, in the last layer, we use the *softmax* function defined as

$$SM(\mathbf{z}) = \frac{e^{\mathbf{z}}}{\sum_{j=1}^{t} e^{z_j}}, \qquad (32)$$

where $\mathbf{z} \in \mathbb{R}^t$. This function resembles all the outputs of the neurons within the very last layer and produces positive estimates that sum up to 1.

Every layer of the network, except the last, can be viewed as a step forward in generating information to be used for classification. The vector of dimension 1024 built at the penultimate layer is essentially a set of features which have been extracted from the original image. More insight into the outputs of intermediate

Fig. 7 An image of a frog from cifar-10 dataset

layers, after training out network, we fed it with the image of the frog in Fig. 7 and analyzed the output of the four convolutional layers. These outputs are displayed in Fig. 8; the channels are plotted side by side for a total of 16 channels per row. In the first plot, we display the 32 channels of the tensor built at the first convolutional layer; the shape of the frog is pretty recognizable in all channels. After the second and the third layer, the image of the frog is no longer recognizable. Even if, after the fourth convolutional layer, the 4×4 pixels of each channel have not apparent connection with the original image, they still contain enough information. The dimension of the input has been reduced, and the condensed information contained in the array is used to generate the 1024 entries which provide the features needed for the final classification. As we will see in the numerical results subsection, the information spread by the network allows, after network training, to correctly classify new entries with satisfactory accuracy.

Training the Neural Network

In the training phase, in order to measure the error made by the network on the prediction of each data, we used the loss function (9) where $\mathbb{E}$ is *categorical cross-entropy function* defined as

$$\mathbb{E}(\mathbf{v}_m(x; \mathbf{d}_i), \hat{\mathbf{y}}_i) = -\sum_{j=1}^{10} \hat{y}_{ij} log\left(v_{m,j}(x; \mathbf{d}_i)\right).$$

In the training phase, the weights of each layer of the network are updated via the minimization of the loss function; any of the methods previously described can be applied.

The training procedure consists in shuffling the training dataset and splitting it into mini-batches. The neural network is fed with each of such mini-batches in order to compute the approximated value of the gradient and to update the network

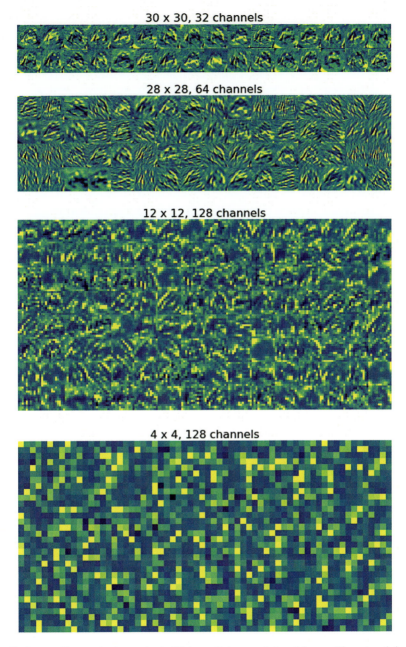

Fig. 8 *Intermediate activation*: output of intermediate convolutional layers. The network is fed with the image of a frog in Fig. 7. The color gradient we used for the intensity spans from yellow (lowest intensity) to blue (highest)

weights using any of the methods described in previous sections. Once the whole dataset has been used, the procedure is repeated. In machine learning terminology, the number of iterations needed to the neural network to handle each entry of the dataset is called an *epoch* of the training.

Implementation Details

We implemented the neural network and the training routine using the Python library Keras (https://keras.io/) and Tensorflow (https://www.tensorflow.org/) for handling the backend on the GPU, a NVIDIA Quadro M1000M. Keras comes with an utility to get the cifar-10 dataset split in training and test. We adapted one of the examples contained into Keras library (https://www.tensorflow.org/tutorials/images/cnn) to develop the network architecture previously described.

The SGD optimizer, presented in section "Stochastic Gradient and Variance Reduction Methods", is included in Keras. After fine-tuning, we ran it using steplength $\alpha_k = 10^{-2}$, $\forall k \geq 0$. SVRG, presented in section "Stochastic Gradient and Variance Reduction Methods", was run using an available implementation (https://github.com/idiap/importance-sampling); in such implementation, the SVRG gradient update rules are wrapped around the Keras framework. The full gradient on the outer iteration of SVRG was replaced by a SG computed on a mini-batch of 1000 training samples; the outer iteration was scheduled to be performed 32 times per epoch. The steplength for the inner iteration was set to 10^{-2}. TRish optimizer presented in section "Gradient Methods with Adaptive Steplength Selection Based on Globalization Strategies" has been implemented from scratch. After fine-tuning, the hyperparameters were set as follows: $\alpha_k = 10^{-1}$, $\forall k \geq 0$, $\gamma_{1,k} = 1$, $\forall k \geq 0$, and $\gamma_{2,k} = 10^{-3}$, $\forall k \geq 0$.

All the three methods have been implemented in a mini-batch manner as described at the end of the previous section. The batch size used for all training runs is 32, i.e., g_k was computed through (4) with $|\mathcal{S}_{k,g}| = 32$. The methods under comparison do not use the objective function at all; then its approximation is not needed.

Results

SGD, SVRG, and TRish were run imposing a number of 25 epochs. At the end of each epoch, the accuracy on both training and testing sets was measured. The accuracy is defined as the percentage of samples for which the classifier assigned the highest probability to the actual class. In Fig. 9, we report the accuracy achieved by each method both on the training and on the testing set during the training. The accuracy is evaluated at the end of each epoch.

TRish method appears to be the most effective in classification. Our experience showed that in the large majority of TRish iterations, the normalized step arising from the minimization of the trust-region subproblem (18) is selected. We recall that the key difference in the gradient methods under investigation is that TRish

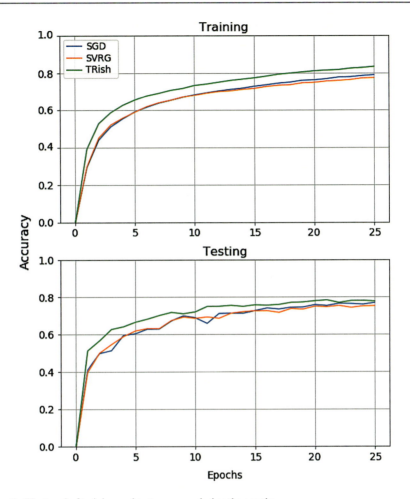

Fig. 9 The trend of training and test accuracy during the epochs

can take normalized steps and this can be viewed as an adaptive steplength selection as the step taken is $s_k = -\frac{\alpha_k}{\|g_k\|} g_k$ instead of $-\alpha_k g_k$. The adaptive approach used in TRish clearly improves classification on the testing set with respect to SGD and SVRG run with prefixed steplength. In fact, after only two epochs, TRish is already more accurate than SGD and SVRG and gives approximately 74% of accuracy on the test set after 12 epoch.

Conclusion

Optimization methods play a key role in machine learning applications. In this work, several subsampled first-order optimization methods suited for machine learning applications have been revised both from a theoretical and algorithmic point of view. Stochastic procedures for solving convex and nonconvex problems applicable to neural networks and convolutional neural networks have been discussed, and numerical experience on a convolutional neural network designed for classifying images has been presented. Our presentation aims to show how the specific features of the optimization problems arising in the training phase of neural networks give rise to stochastic procedures which can address the numerical solution of convex and nonconvex problems.

The presented procedures are recent and part of the state of the art in optimization for machine learning. The literature on this topic is immense and steadily increasing, and this presentation is not comprehensive of the variety of existing first-order methods. We focused on methods with well-assessed convergence analysis. However we are aware of widely adopted methods which are less theoretically well founded than the procedures presented but are successful in machine learning. At this regard, we would like to mention `SGD with momentum` (Rumelhart et al. 1986; Loizou 2017) and `ADAM` (Kingma and Ba 2015; Sashank 2018). Both methods aim to speed the convergence rate of `SGD` method in the solution of ill-conditioned problems where the surface in a neighborhood of local optima curves more steeply in one direction than in another. In fact, in such cases a common drawback of steepest descent methods is that iterates zigzag toward the solution (Nocedal et al. 1999; Sutton 1986). To avoid that, `SGD with momentum` makes use of a search direction which is a combination of the current gradient approximation and the step (first-order momentum of the stochastic gradient) used at the previous iteration. `ADAM` method computes individual adaptive steplengths for updating the iterate component-wise on the basis of the current first- and second-order momentum of the stochastic gradient.

We conclude underling a current growing interest in second-order methods for nonconvex finite-sum optimization problems; see, e.g., Aggarwal (2018), Bellavia et al. (2020, 2021, 2019, 2020a,b), Berahas et al. (2020), Bollapragada et al. (2019), Bottou et al. (2018), Byrd et al. (2016), Byrd et al. (2012), Erdogdu et al. (2015), Liu et al. (2018), Roosta-Khorasani et al. (2019), Strang (2019), Xu et al. (2016, 2019).

Acknowledgments The financial support of INdAM-GNCS Projects 2019 and 2020 is gratefully acknowledged by the first and by the fourth authors. Thanks are due the referee whose comments improved the presentation of this paper.

References

Aggarwal, C.C.: Neural Networks and Deep Learning. Springer (2018)
Andradottir, S.: A scaled stochastic approximation algorithm. Manag. Sci. **42**, 475–498 (1996)
Armijo, L.: Minimization of functions having lipschitz continuous first partial derivatives. Pac. J. Math. **16**, 1–3 (1966)
Babanezhad, R., Ahmed, M.O., Virani, A., Schmidt, M., Konečný, J., Sallinen S.: Stop wasting my gradients: Practical SVRG. In: Proceedings of the 28th International Conference on Neural Information Processing Systems, Vol. 2 pp. 2251–2259 (2015)

Barzilai, J., Borwein, J.: Two-point step size gradient. IMA J. Numer. Anal. **8**, 141–148 (1988)

Bellavia, S., Gurioli, G.: Complexity analysis of a Stochastic cubic regularisation method under inexact gradient evaluations and dynamic Hessian accuracy, (2020). arXiv:2001.10827

Bellavia, S., Gurioli, G., Morini, B., Adaptive cubic regularization methods with dynamic inexact Hessian information and applications to finite-sum minimization. IMA J. Numer. Anal. **41**, 764–799 (2021). https://doi.org/10.1093/imanum/drz076

Bellavia, S., Gurioli, G., Morini, B., Toint, P.L.: Adaptive regularization algorithms with inexact evaluations for nonconvex optimization. SIAM J. Optimiz. **29**, 2881–2915 (2019)

Bellavia, S., Gurioli, G., Morini, B., Toint, P.L.: High-order Evaluation Complexity of a Stochastic Adaptive Regularization Algorithm for Nonconvex Optimization Using Inexact Function Evaluations and Randomly Perturbed Derivatives (2020a) arXiv:2005.04639

Bellavia, S., Krejić, N., Krklec Jerinkić, N.: Subsampled Inexact Newton methods for minimizing large sums of convex function. IMA J. Numer. Anal. **40**, 2309–2341 (2020b)

Bellavia, S., Krejić, N., Morini, B.: Inexact restoration with subsampled trust-region methods for finite-sum minimization. Comput. Optim. Appl. **76**, 701–736 (2020c)

Berahas, A.S., Bollapragada, R., Nocedal, J.: An investigation of Newton-sketch and subsampled Newton methods. Optim. Method Softw. **35**, 661–680 (2020)

Bertsekas, D.P., Tsitsiklis, J.N.: Gradient convergence in gradient methods with errors. SIAM J. Optimiz. **10**, 627–642 (2000)

Bishop, C.M.: Pattern Recognition and Machine Learning (Information Science and Statistics). Springer (2006)

Birgin, G.E., Krejić, N., Martínez, J.M.: On the employment of Inexact Restoration for the minimization of functions whose evaluation is subject to programming errors. Math. Comput. **87**, 1307–1326 (2018)

Blanchet, J., Cartis, C., Menickelly, M., Scheinberg, K.: Convergence rate analysis of a stochastic trust region method via submartingales. INFORMS J. Optim. **1**, 92–119 (2019)

Bollapragada, R., Byrd, R., Nocedal, J.: Exact and inexact subsampled Newton methods for optimization. IMA J. Numer. Anal. **39**, 545–578 (2019)

Bottou, L., Curtis, F.C., Nocedal, J.: Optimization methods for large-scale machine learning. SIAM Rev. **60**, 223–311 (2018)

Byrd, R.H., Hansen, S.L., Nocedal, J., Singer Y.: A stochastic quasi-Newton method for large-scale optimization. SIAM J. Optimiz. **26**, 1008–1021 (2016)

Byrd, R.H., Chin, G.M., Nocedal, J., Wu, Y.: Sample size selection in optimization methods for machine learning. Math. Program. **134**, 127–155 (2012)

Cartis, C., Scheinberg, K.: Global convergence rate analysis of unconstrained optimization methods based on probablistic models. Math. Program. **169**, 337–375 (2018)

Chen, R., Menickelly, M., Scheinberg, K.: Stochastic optimization using a trust-region method and random models. Math. Program. **169**, 447–487 (2018)

Chollet, F.: Deep Learning with Python. Manning Publications Co. (2017)

Curtis, F.E., Scheinberg, K., Shi, R.: A stochastic trust region algorithm based on careful step normalization. INFORMS J. Optimiz. **1**, 200–220 (2019)

Defazio, A., Bach, F., Lacoste-Julien, S.: SAGA: A fast incremental gradient method with support for non-strongly convex composite objectives. Advances in Neural Information Processing Systems 27 (NIPS 2014)

Delyon, B., Juditsky, A.: Accelerated stochastic approximation. SIAM J. Optimiz. **3**, 868–881 (1993)

Erdogdu, M.A., Montanari, A.: Convergence rates of sub-sampled Newton methods, Advances in Neural Information Processing Systems 28 (NIPS 2015)

Forsyth, D.A., Ponce, J.: Computer Vision: A Modern Approach. Pearson (2002)

Friedlander, M.P., Schmidt, M.: Hybrid deterministic-stochastic methods for data fitting. SIAM J. Sci. Comput. **34**, 1380–1405 (2012)

Goodfellow, I., Bengio, Y., Courville, A.: Deep Learning. The MIT Press (2016)

Hastie, T., Tibshirani, R., Friedman, J.: The Elements of Statistical Learning. Springer New York Inc. (2001)

Johnson, R., Zhang, T.: Accelerating stochastic gradient descent using predictive variance reduction. In: Proceedings of the 26th International Conference on Neural Information Processing Systems 26 (NIPS 2013)

Lei, L., Jordan, M.I.: Less than a single pass: Stochastically controlled stochastic gradient method. In: Proceedings of the Twentieth Conference on Artificial Intelligence and Statistics (AISTATS) (2017)

Liu, L., Liu, X., Hsieh, C.-J., Tao, D.: Stochastic second-order methods for non-convex optimization with inexact Hessian and gradient (2018) arXiv:1809.09853

Loizou, N., Richtárik, P.: Momentum and stochastic momentum for stochastic gradient, Newton, proximal point and subspace descent methods (2017) arXiv:1712.09677

Kesten, H.: Accelerated stochastic approximation. Ann. Math. Statist. **29**, 41–59 (1958)

Kiefer, J., Wolfowitz, J.: Stochastic estimation of the maximum of a regression function. Ann. Math. Stat. **23**, 462–466 (1952)

Kingma, D.P., Ba, J.: Adam: A Method for Stochastic Optimization, 3rd International Conference on Learning Representations, ICLR 2015 (2015) arXiv: 1412.6980

Krejić, N., Lužanin, Z., Stojkovska, I.: A gradient method for unconstrained optimization in noisy environment. App. Numer. Math. **70**, 1–21 (2013)

Krejić, N., Lužanin, Z., Ovcin, Z., Stojkovska, I.: Descent direction method with line search for unconstrained optimization in noisy environment. Optim. Method. Soft. **30**, 1164–1184 (2015)

Nocedal, J., Wright, S.J.: Numerical Optimization. Springer Series in Operations Research. Springer (1999)

Krejić, N., Martínez, J.M.: Inexact restoration approach for minimization with inexact evaluation of the objective function. Math. Comput. **85**, 1775–1791 (2016)

Krejić, N., Krklec, N.: Line search methods with variable sample size for unconstrained optimization. J. Comput. Appl. Math. **245**, 213–231 (2013)

Krejić, N., Krklec Jerinkić N.: Nonmonotone line search methods with variable sample size. Numer. Algorithms **68**, 711–739 (2015)

Krizhevsky, A.: Learning multiple layers of features from tiny images. Technical Report, University of Toronto (2009)

Nemirovski, A., Juditsky, A., Lan, G., Shapiro, A.: Robust stochastic approximation approach to stochastic programming. SIAM J. Optimiz. **19**, 1574–1609 (2009)

Nesterov, Y.: Introductory lectures on convex programming, Volume I: Basic course. Lecture Notes (1998)

Nocedal, J., Sartenaer, A., Zhu, C.: On the behavior of the gradient norm in the steepest descent method. Comput. Optim. Appl. **22**, 5–35 (2002)

Nguyen, L.M., Liu, J., Scheinberg, K., Takač, M., SARAH: A novel method for machine learning problems using stochastic recursive gradient. In: Proceedings of the 34th International Conference on Machine Learning (2017) pp. 2613–2621

Paquette, C., Scheinberg, K.: A stochastic line search method with expected complexity analysis. SIAM J. Optim. **30**, 349–376 (2020)

Patterson, J., Gibson, A.: Deep Learning: A Practitioner's Approach. O'Reilly Media, Inc (2017)

Pilanci, M., Wainwright, M.J.: Newton sketch: A near linear-time optimization algorithm with linear-quadratic convergence. SIAM J. Optimiz. **27**, 205–245 (2017)

Polak, E., Royset, J.O.: Efficient sample sizes in stochastic nonlinear programing. J. Comput. Appl. Math. **217**, 301–310 (2008)

Raj, A., Stich, S.U.: k-SVRG: Variance Reduction for Large Scale Optimization (2018) arXiv:1805.00982

Raydan, M.: The Barzilai and Borwein gradient method for the large scale unconstrained minimization problem. SIAM J. Optim. **7**, 26–33 (1997)

Ross, S.: Simulation. Elsevier, 4th edn. (2006)

Roosta-Khorasani, F., Mahoney, M.W.: Sub-sampled Newton methods. Math. Progr. **174**, 293–326 (2019)

Robbins, H., Monro, S.: A stochastic approximation method. Ann. Math. Stat. **22**, 400–407 (1951)

Rumelhart, D.E., Hinton, G.E., Williams, R.J.: Learning representations by back-propagating errors. Nature **323**, 533–536 (1986)

Sashank, J.R., Satyen, K.A., Sanjiv, K.U.: On the convergence of Adam and beyond. In: 6th International Conference on Learning Representations (ICLR 2018)

Schmidt, M., Le Roux, N., Bach, F.: Minimizing finite sums with the stochastic average gradient. Math. Program. **162**, 83–112 (2017)

Shanmugamani, R.: Deep Learning for Computer Vision: Expert Techniques to Train Advanced Neural Networks Using TensorFlow and Keras. Packt Publishing (2018)

Spall J.C.: Introduction to Stochastic Search and Optimization. Wiley-Interscience series in discrete mathematics (2003)

Spall J.C.: Implementation of the simultaneous perturbation algorithm for stochastic optimization. IEEE Trans. Aerosp. Electron. Syst. **34**, 817–823 (1998)

Shanmugamani, R.: Deep Learning for Computer Vision: Expert Techniques to Train Advanced Neural Networks Using TensorFlow and Keras. Packt Publishing (2018)

Strang, G.: Linear Algebra and Learning from Data. Wellesley-Cambridge Press (2019)

Sutton, R.: Two problems with back propagation and other steepest descent learning procedures for networks. In: Proceedings of the Eighth Annual Conference of the Cognitive Science Society, pp. 823—832 (1986)

Tan, C., Ma, S., Dai, Y., Qian, Y.: Barzilai-Borwein step size for stochastic gradient descent. Advances in Neural Information Processing Systems **29**, 685–693 (2016)

Toint, P.L.: Nonlinear stepsize control, trust regions and regularizations for unconstrained optimization. Optimiz. Meth. Softw. **28**, 82–95 (2013)

Tripuraneni, N., Stern, M., Regier, M., Jordan, M.I.: Stochastic cubic regularization for fast nonconvex optimization. Adv. Neural Inf. Proces. Syst. **31**, 2899–2908 (2018)

Tropp, J.: An introduction to matrix concentration inequalities. Found. Trends Mach. Learn. **8**(1–2), 1–230 (2015)

Yousefian, F., Nedic, A., Shanbhag, U.V.: On stochastic gradient and subgradient methods with adaptive steplength sequences. Automatica **48**, 56–67 (2012)

Wang, C., Chen, X., Smola, A., Xing, E.: Variance reduction for stochastic gradient optimization. Adv. Neural Inf. Proces. Syst. **26**, 181–189 (2013)

Xu, Z., Dai, Y.H.: New stochastic approximation algorithms with adaptive step sizes. Optim. Lett. **6**, 1831–1846 (2012)

Xu, P., Yang, J., Roosta-Khorasani, F., Ré, C., Mahoney, M.W.: Sub-sampled Newton methods with non-uniform sampling. Adv. Neural Inf. Proces. Syst. **29**, 3008–3016 (2016)

Xu, P., Roosta-Khorasani, F., Mahoney, M.W.: Newton-type methods for non-convex optimization under inexact Hessian information. Math. Program. (2019). https://doi.org/10.1007/s10107-019-01405-z

Wang, X., Yuan, Y.X.: Stochastic Trust Region Methods with Trust Region Radius Depending on Probabilistic Models (2019). arXiv:1904.03342

Bregman Methods for Large-Scale Optimization with Applications in Imaging

3

Martin Benning and Erlend Skaldehaug Riis

Contents

Introduction	98
Bregman Proximal Methods	99
A Unified Framework for Implicit and Explicit Gradient Methods	101
Bregman Proximal Gradient Method	102
Bregman Iteration	104
Linearized Bregman Iteration as Gradient Descent	104
Bregman Iterations as Iterative Regularization Methods	106
Inverse Scale Space Flows	107
Accelerated Bregman Methods	108
Incremental and Stochastic Bregman Proximal Methods	110
Stochastic Mirror Descent	111
The Sparse Kaczmarz Method	111
Deep Neural Networks	113
Bregman Incremental Aggregated Gradient	114
Bregman Coordinate Descent Methods	116
The Bregman Itoh–Abe Method	117
Equivalencies of Certain Bregman Coordinate Descent Methods	119
Saddle-Point Methods	120
Alternating Direction Method of Multipliers	121
Primal-Dual Hybrid Gradient Method	122
Applications	124
Robust Principal Component Analysis	125
Deep Learning	127
Student-t Regularized Image Denoising	129
Conclusions and Outlook	131
References	132

M. Benning (✉)
The School of Mathematical Sciences, Queen Mary University of London, London, UK
e-mail: m.benning@qmul.ac.uk

E. S. Riis
The Department of Applied Mathematics and Theoretical Physics, Cambridge, UK

Abstract

In this chapter we review recent developments in the research of Bregman methods, with particular focus on their potential use for large-scale applications. We give an overview on several families of Bregman algorithms and discuss modifications such as accelerated Bregman methods, incremental and stochastic variants, and coordinate descent-type methods. We conclude this chapter with numerical examples in image and video decomposition, image denoising, and dimensionality reduction with auto-encoders.

Keywords

Optimization · Bregman proximal methods · Bregman iterations · Inverse problems · Nesterov acceleration · Mirror descent · Kaczmarz method · Coordinate descent · Itoh-Abe method · Alternating direction method of multipliers · Primal-dual hybrid gradient · Robust principal components analysis · Deep learning · Image denoising

Introduction

Bregman methods have a long history in mathematical research areas such as optimization, inverse and ill-posed problems, statistical learning theory, and machine learning. In this review, we mainly focus on the areas of optimization and inverse and ill-posed problems and the application of popular Bregman methods to potentially large-scale problems. Following Lev Bregman's seminal work in 1967 (Bregman 1967), it was not before the work of Censor and Lent (1981) in 1981 that the use of Bregman methods has slowly but steadily been popularized in the area of mathematical optimization, shortly followed by the advent of the mirror descent algorithm (Nemirovsky and Yudin 1983). Bregman proximal methods, which we discuss in greater detail in the following section, were first introduced by Censor and Zenios in their seminal work in 1992 (Censor and Zenios 1992), shortly followed by Teboulle (1992), Teboulle and Chen (1993), and Eckstein (1993). Bregman methods have been extensively studied since, see, for example, Bauschke et al. (2003) and references therein, and many notable extensions were developed, with one of the most popular ones in the context of inverse and ill-posed problems being the so-called Bregman iteration (Osher et al. 2005), which is based on a generalized Bregman distance notion (Kiwiel 1997b). Bregman iterations have been shown to possess favorable regularization properties over traditional linear iterative regularization methods, especially in the context of imaging and image processing applications, and therefore gained a lot of attention in those research fields. We refer to Osher et al. (2005), Burger (2016), and Benning and Burger (2018) for an overview on Bregman iterations.

The goal of this chapter is to provide a non-exhaustive overview over some recent developments in the adaptation of Bregman methods to handle potentially large-scale problems. These extensions range from simple linearizations to accelerated versions of Bregman methods, incremental and stochastic adaptations, and coordinate descent variants to Bregman extensions of popular primal-dual frameworks. The chapter is therefore structured as follows. In section "Bregman Proximal Methods" we give an overview over Bregman proximal methods and some notable extensions. In section "Accelerated Bregman Methods" we discuss accelerations of the linearized Bregman iteration, before we focus on incremental and stochastic variants in section "Incremental and Stochastic Bregman Proximal Methods." Subsequently, we discuss coordinate descent-type Bregman methods in section "Bregman Coordinate Descent Methods" and saddle-point formulations of Bregman algorithms in section "Saddle-Point Methods." We present several application examples in section "Applications" before concluding this chapter with section "Conclusions and Outlook."

Bregman Proximal Methods

The Bregman proximal method or Bregman proximal algorithm is defined as the following iterative procedure. Starting with an initial value $x^0 \in \mathbb{R}^n$, we compute

$$x^{k+1} = \arg\min_{x \in \mathbb{R}^n} \left\{ F(x) + D_R(x, x^k) \right\}, \tag{1}$$

for $k \in \mathbb{N}$. Here $F : \mathbb{R}^n \to \mathbb{R}$ is a function that we wish to minimize via (1). We assume that F is bounded from below and that both F and R satisfy conditions that guarantee existence and uniqueness of the solution of (1), without discussing them in greater detail. The term $D_R(x, y)$ denotes the Bregman distance w.r.t. a convex and continuously differentiable function $R : \mathbb{R}^n \to \mathbb{R}$, which is defined as

$$D_R(x, y) = R(x) - R(y) - \langle \nabla R(y), x - y \rangle, \tag{2}$$

for all $x, y \in \mathbb{R}^n$, see Bregman (1967) and Censor and Lent (1981). In the following example, we recall a few relevant examples of Bregman distances.

Example 1 (Bregman distances). For a symmetric, positive semi-definite matrix $Q \in \mathbb{R}^{n \times n}$ and the function $R(x) := \frac{1}{2}\langle Qx, x \rangle$, we observe

$$D_R(x, y) = \frac{1}{2}\langle Q(x - y), x - y \rangle.$$

Special cases include the squared Euclidean distance if Q is the identity matrix and the squared Mahalanobis distance (cf. Mahalanobis 1936) if Q is a covariance matrix.

The generalized Kullback-Leibler divergence, i.e.,

$$D_R(x, y) = \sum_{j=1}^{n} \left[x_j \log\left(\frac{x_j}{y_j}\right) + y_j - x_j \right],$$

can be obtained by choosing R as the (shifted, negative) Boltzmann-Shannon entropy, i.e., $R(x) := \sum_{j=1}^{n} \left[x_j \log(x_j) - x_j \right]$. Other notable examples include the Itakura–Saito distance (cf. Itakura 1968) and the Hellinger distance (cf. Hellinger 1909).

Note that $D_R(x, y) \geq 0$ is guaranteed for all $x, y \in \mathbb{R}^n$ due to the convexity of R. Before we are briefly going to discuss how this Bregman framework unifies implicit and explicit gradient methods in the following section, we want to recall some basic and well-known properties of (1).

Corollary 1. *Let $F : \mathbb{R}^n \to \mathbb{R}$ and $R : \mathbb{R}^n \to \mathbb{R}$ be continuously differentiable functions, where R is also convex, and suppose for some $\bar{x} \in \mathbb{R}^n$ that x^* is defined as*

$$x^* := \arg\min_{x \in \mathbb{R}^n} \left\{ F(x) + D_R(x, \bar{x}) \right\}. \qquad (3)$$

Then, the following identity holds:

$$F(x^*) + D_F(x, x^*) + D_R(x, x^*) + D_R(x^*, \bar{x}) = F(x) + D_R(x, \bar{x}). \qquad (4)$$

Corollary 1 can easily be verified by computing the optimality condition of (3), subsequent computation of the inner product of the optimality condition with $x^* - x$, and the use of the three-point identity for Bregman distances, first proven in Chen and Teboulle (1993, Lemma 3.1). Corollary 1 allows us to verify the following convergence result of the Bregman method with convergence rate $1/k$ for convex functions F.

Theorem 1. *Let $F : \mathbb{R}^n \to \mathbb{R}$ and $R : \mathbb{R}^n \to \mathbb{R}$ be continuously differentiable and convex functions. Suppose $\hat{x}$ is a global minimizer of F that exists. Then, for any x^0, the iterates (1) satisfy*

$$F(x^k) - F(\hat{x}) \leq \frac{D_R(\hat{x}, x^0) - D_R(\hat{x}, x^k)}{k},$$

for $k \in \mathbb{N}$.

Proof. Applying Corollary 1 for $x^* = x^{k+1}$, $\bar{x} = x^k$, and $x = \hat{x}$ yields

$$F(x^{k+1}) + D_F(\hat{x}, x^{k+1}) + D_R(\hat{x}, x^{k+1}) + D_R(x^{k+1}, x^k) = F(\hat{x}) + D_R(\hat{x}, x^k),$$

which implies

$$F(x^{k+1}) - F(\hat{x}) \leq D_R(\hat{x}, x^k) - D_R(\hat{x}, x^{k+1}),$$

due to the convexity of F and R. Summing up this inequality from $k = 0, \ldots, K-1$ leads to

$$\sum_{k=0}^{K-1} F(x^{k+1}) - K F(\hat{x}) \leq D_R(\hat{x}, x^0) - D_R(\hat{x}, x^K).$$

Applying Corollary 1 again – but this time for $x^* = x^{k+1}$, $\bar{x} = x^k$ and $x = x^k$ – leaves us with

$$F(x^{k+1}) + D_F(x^k, x^{k+1}) + D_R(x^k, x^{k+1}) + D_R(x^{k+1}, x^k) = F(x^k) + \underbrace{D_R(x^k, x^k)}_{=0},$$

which in return implies $F(x^{k+1}) \leq F(x^k)$ due to the convexity of F and R (which is also an immediate consequence of the variational formulation of the Bregman method). Hence, we observe $K F(x^K) \leq \sum_{k=0}^{K-1} F(x^{k+1})$, which concludes the proof.

Remark 1. Note that the conditions on F and R in Theorem 1 alone do not necessarily guarantee uniqueness or even existence of x^{k+1} in (1). However, if the solution exists and is unique and computable, then Theorem 1 applies.

Let us now turn our attention to implicit and explicit gradient methods and how they can both be formulated as special cases of (1).

A Unified Framework for Implicit and Explicit Gradient Methods

While it is common in numerical analysis to distinguish between implicit and explicit methods, a feature of the Bregman framework is that it covers both types of methods. This can be seen by considering (1), i.e.,

$$x^{k+1} = \arg\min_{x \in \mathbb{R}^n} \left\{ F(x) + D_J(x, x^k) \right\}, \tag{5}$$

for the special choice of $J : \mathbb{R}^n \to \mathbb{R}$ with

$$J(x) := \begin{cases} R(x) & \text{implicit} \\ \frac{1}{\tau} R(x) - F(x) & \text{explicit} \end{cases}. \tag{6}$$

Evaluating the Bregman distance w.r.t. J turns (5) into

$$x^{k+1} = \arg\min_{x \in \mathbb{R}^n} \begin{Bmatrix} F(x) + D_R(x, x^k) & \text{implicit} \\ F(x^k) + \langle \nabla F(x^k), x - x^k \rangle + \frac{1}{\tau} D_R(x, x^k) & \text{explicit} \end{Bmatrix};$$

Hence, we can construct Bregman methods that are either implicit or explicit w.r.t. ∇F. Whenever we use J as the notation of our function throughout this manuscript, we implicitly refer to J as defined in (6). Whenever we use R, we refer to a function R that is not of the form $\frac{1}{\tau} R - F$. Note that we rediscover the traditional gradient descent algorithm for the choice $R(x) = \frac{1}{2} \|x\|^2$ as a special case of the explicit formulation. Furthermore, note that the explicit formulation

$$x^{k+1} = \arg\min_{x \in \mathbb{R}^n} \left\{ F(x^k) + \langle \nabla F(x^k), x - x^k \rangle + \frac{1}{\tau} D_R(x, x^k) \right\} \quad (7)$$

is also known as mirror descent (Ben-Tal et al. 2001; Beck and Teboulle 2003; Juditsky et al. 2011), Bregman gradient method (Teboulle 2018), or recently also as NoLips (Bauschke et al. 2017). In order to guarantee convergence of (5), one usually has to guarantee convexity of J. In the explicit setting, this implies that τ and R have to be chosen to ensure convexity of $\frac{1}{\tau} R - F$ or equivalently that F is $1/\tau$-smooth if R is also a quadratic function. The latter condition has basically been proposed in Bauschke et al. (2017) and further discussed in Benning et al. (2017a,b) and Bolte et al. (2018). It has also been shown that if the step size τ is chosen such that $c R - F$ is convex, for a some constant $c > 0$ and a function F, the estimate $0 < \tau \leq \left((1 + \gamma(R)) - \delta\right)/c$ is sufficient to guarantee convergence under mild assumptions that are outlined in detail in Bauschke et al. (2017). Here $\gamma(R)$ denotes the symmetry coefficient defined as

$$\gamma(R) := \inf \left\{ D_R(x, y)/D_R(y, x) \,\middle|\, (x, y) \in (\text{int dom } R)^2 \setminus \{x, y \mid x = y\} \right\} \in [0, 1],$$

and δ is a constant that satisfies $\delta \in (0, 1 + \gamma(R))$. In the following section, we want to review the special case of Bregman gradient methods where F is the sum of two functions.

Bregman Proximal Gradient Method

An interesting, special case frequently considered in the literature is the case where F is a sum of two functions L and S, i.e., the Bregman method reads

$$x^{k+1} = \arg\min_{x \in \mathbb{R}^n} \left\{ L(x) + S(x) + D_J(x, x^k) \right\}, \quad (8)$$

where we assume that $L : \mathbb{R}^n \to \mathbb{R}$ is a continuously differentiable function. The function $S : \mathbb{R}^n \to \overline{\mathbb{R}}$ on the other hand is proper, lower semi-continuous (l.s.c.) and convex, for $\overline{\mathbb{R}} := \mathbb{R} \cup \{\infty\}$. If we choose $J(x) := \frac{1}{2\tau}\|x\|^2 - L(x)$ in the spirit of (6), then (8) reads

$$x^{k+1} = \arg\min_{x \in \mathbb{R}^n} \left\{ \frac{1}{2}\left\| x - \left(x^k - \tau \nabla L(x^k)\right) \right\|^2 + \tau S(x) \right\},$$

$$=: (I + \tau S)^{-1}\left(x^k - \tau \nabla L(x^k)\right),$$

where $(I + \tau S)^{-1} : \mathbb{R}^n \to \mathbb{R}^n$ is known as the proximal map or resolvent, see, for instance, (Parikh et al. 2014). This is the classical proximal gradient method, also known as forward backward splitting (Lions and Mercier 1979). More general proximal gradient methods can be derived for different choices of J and S, for example, the entropic mirror descent algorithm (Nemirovsky and Yudin 1983; Beck and Teboulle 2003; Beck 2017; Doan et al. 2018), i.e.,

$$x_j^{k+1} = \frac{x_j^k \exp\left(-\tau(\nabla L(x^k))_j\right)}{\sum_{j=1}^n x_j^k \exp\left(-\tau(\nabla L(x^k))_j\right)},$$

for $j \in \{1, \ldots, n\}$, the difference of the negative Boltzmann Shannon entropy as defined in Example 1 and the function L, i.e., $J(x) := \frac{1}{\tau}\sum_{j=1}^n \left[x_j \log(x_j) - x_j\right] - L(x)$ with the convention $0 \log(0) \equiv 0$, and the characteristic function

$$S(x) := \begin{cases} 0 & x \in \Sigma \\ +\infty & x \notin \Sigma \end{cases},$$

over the simplex constraint

$$\Sigma := \left\{ x \in \mathbb{R}^n \,\middle|\, x_j \geq 0,\ \forall j \in \{1, \ldots, n\},\ \sum_{j=1}^n x_j = 1 \right\}.$$

We also mention *variable metric proximal gradient methods*, an important class of algorithms which may be viewed as an instance of Bregman proximal gradient methods where the Bregman function J_k is iteration-dependent. Denoting by $(A_k)_{k \in \mathbb{N}}$ a sequence of symmetric positive definite matrices, which act as preconditioners, we define $J_k(x) := \frac{1}{2\tau_k}\langle x, A_k x\rangle - L(x)$. Note that if $S \equiv 0$, $A_k = \nabla^2 L(x^k)$, and $\tau_k = 1$, then one recovers the Newton method for L

$$x^{k+1} = x^k - (\nabla^2 L(x^k))^{-1} \nabla L(x^k).$$

More generally when $S \not\equiv 0$, one may choose A_k to be an approximation to the Hessian of L at x^k, so as to incorporate elements of quasi-Newton methods to the proximal gradient scheme. These schemes were studied by Bonnans et al. (1995) and later studied for non-convex objective functions (Chouzenoux et al. 2014; Frankel et al. 2015), Hilbert spaces (Combettes and Vũ 2014), and extensions to inertial methods (Bonettini et al. 2018), to mention a few examples.

In the next section, we focus on extensions of the Bregman proximal methods to convex but nonsmooth functions.

Bregman Iteration

A very important generalization of (1), first proposed in Osher et al. (2005), allows us to also use convex but nonsmooth functions J as defined in (6) instead of convex and continuously differentiable functions J. Suppose we are given a proper, l.s.c. and convex function $J : \mathbb{R}^n \to \overline{\mathbb{R}}$. Then its subdifferential, defined as

$$\partial J(y) := \left\{ p \in \mathbb{R}^n \mid J(x) - J(y) \geq \langle p, x - y \rangle, \forall x \in \mathbb{R}^n \right\},$$

is non-empty. It therefore makes sense to extend the definition (2) to a generalized Bregman distance (Kiwiel 1997a) for subdifferentiable functions, i.e.,

$$D_J^p(x, y) = J(x) - J(y) - \langle p, x - y \rangle,$$

for $p \in \partial J(y)$. A generalization of (1), commonly known as Bregman iteration, can then be defined as

$$x^{k+1} = \arg\min_{x \in \mathbb{R}^n} \left\{ F(x) + D_J^{p^k}(x, x^k) \right\}, \tag{9a}$$

$$p^{k+1} = p^k - \nabla F(x^{k+1}), \tag{9b}$$

for initial values $x^0 \in \mathbb{R}^n$ and $p^0 \in \partial J(x^0)$. Note that Corollary 1 and Theorem 1 also apply to Bregman iterations (cf. Benning and Burger 2018, Corollary 6.5), as those statements did not utilize any potential differentiability of J. Furthermore, note that the explicit variant of the Bregman iteration is known as the linearized Bregman iteration and has extensively been studied in Yin et al. (2008), Cai et al. (2009a,b,c), and Yin (2010).

Linearized Bregman Iteration as Gradient Descent

With the particular choice $J(x) = \frac{1}{2\tau}\|x\|^2 + \frac{1}{\tau}R(x) - F(x)$, the Bregman iteration (9) turns into the linearized Bregman iteration, which reads

$$x^{k+1} = \arg\min_{x \in \mathbb{R}^n} \left\{ F(x^k) + \langle \nabla F(x^k), x - x^k \rangle + \frac{1}{2\tau} \|x - x^k\|^2 + \frac{1}{\tau} D_R^{q^k}(x, x^k) \right\},$$

$$= (I + \partial R)^{-1} \left(x^k + q^k - \tau \nabla F(x^k) \right), \tag{10a}$$

$$q^{k+1} = q^k - \left(x^{k+1} - x^k + \tau \nabla F(x^k) \right), \tag{10b}$$

where $(I + \partial R)^{-1}$ denotes the proximal mapping w.r.t. the function R and $q^k \in \partial R(x^k)$ the subgradient of R at x^k that is iteratively defined via (10b) and some initial value $q^0 \in \partial R(x^0)$. Suppose we assume that $(x^k + q^k)/\tau - \nabla F(x^k)$ is in the range of some matrix $A \in \mathbb{R}^{m \times n}$ and that we therefore can substitute $\tau A^\top b^k := x^k + q^k - \tau \nabla F(x^k)$. Then (10) can be written as

$$x^{k+1} = (I + \partial R)^{-1}(\tau A^\top b^k), \tag{11a}$$

$$A^\top b^{k+1} = A^\top b^k - \nabla F(x^{k+1}). \tag{11b}$$

In the following, we want to focus on the special case $F(x) = \frac{1}{2}\|Ax - b^\delta\|^2$ with $\nabla F(x) = A^\top (Ax - b^\delta)$ for a matrix $A \in \mathbb{R}^{m \times n}$, for which (11) simplifies to

$$x^{k+1} = (I + \partial R)^{-1}(\tau A^\top b^k), \tag{12a}$$

$$b^{k+1} = b^k - \left(Ax^{k+1} - b^\delta \right), \tag{12b}$$

with initial value $b^0 = b^\delta$, given the assumption that the initial values of the original formulation were $x^0 = 0$ and $p^0 = 0$. Note that we can also write (12) as

$$b^{k+1} = b^k - \left(A(I + \partial R)^{-1} \left(\tau A^\top b^k \right) - b^\delta \right). \tag{13}$$

Hence, if we can identify an energy G_τ for which we can associate its gradient ∇G_τ with $A(I + \partial R)^{-1} \left(\tau A^\top \cdot \right) - b^\delta$, we can consider the linearized Bregman iteration a gradient descent method applied to this specific energy. In Yin (2010) and Huang et al. (2013), this energy has been identified as

$$G_\tau(b) := \frac{\tau}{2}\|A^\top b\|^2 - \langle b, b^\delta \rangle - \frac{1}{\tau}\tilde{R}(\tau A^\top b),$$

where $\tilde{R}$ denotes the Moreau-Yosida regularization of R (cf. Moreau 1965; Yosida 1964), i.e.,

$$\tilde{R}(z) := \inf_{x \in \mathbb{R}^n} \left\{ R(x) + \frac{1}{2}\|x - z\|^2 \right\}.$$

Since the gradient of the Moreau-Yosida regularization of R reads $\nabla \tilde{R}(z) = z - (I + \partial R)^{-1}(z)$ (see, for instance, Attouch et al. 2014, Proposition 17.2.1), we easily verify

$$\nabla G_\tau(b) = A(I + \partial R)^{-1}(\tau A^\top b) - b^\delta.$$

As a consequence, (13) is equivalent to

$$b^{k+1} = b^k - \nabla G_\tau(b^k),$$

and the linearized Bregman iteration for $F(x) = \frac{1}{2}\|Ax - b^\delta\|^2$ reduces to a gradient descent method. This equivalence will be useful when studying acceleration methods.

Bregman Iterations as Iterative Regularization Methods

Bregman iterations are not only useful for solving optimization problems but are also extremely important in the context of solving inverse and ill-posed problems. The reason for this is that Bregman iterations can be used as iterative regularization methods. If we consider the deterministic linear inverse problem

$$Ax^\dagger = b^\dagger, \tag{14}$$

for a given matrix $A \in \mathbb{R}^{m \times n}$, the aim of solving this inverse problem is to approximate $x^\dagger$ in (14), for given A and data b^δ with $\|b^\dagger - b^\delta\| \leq \delta$. Here, δ is a known, positive bound on the error of the measured data b^δ and the data $b^\dagger$ that satisfies (14).

Suppose we consider a convex function F that depends on A and b^δ, which we will denote as F_{b^δ}. It then can easily be shown that the iterates of (9) satisfy

$$D_J^{p^{k+1}}(x^\dagger, x^{k+1}) < D_J^{p^k}(x^\dagger, x^k),$$

for all indices $k \leq k^*(\delta)$ that satisfy Morozov's discrepancy principle (Morozov 1966), i.e.,

$$F_{b^\delta}(x^{k^*(\delta)}) \leq \eta \delta < F_{b^\delta}(x^k),$$

for a parameter $\eta \geq 1$, see Osher et al. (2005) and Burger et al. (2007). Note that for $\eta > 1$ it can be guaranteed that $k^*(\delta)$ is finite. With the additional regularity assumption that $x^\dagger$ satisfies the so-called range condition (Benning and Burger 2018, Definition 5.8), i.e.,

$$x^\dagger \in \arg\min_{x \in \mathbb{R}^n} \{F_g(x) + R(x)\},$$

for some data $g \in \mathbb{R}^m$, one can prove the error estimate

$$D_J^{p^k}(x^\dagger, x^k) \leq \frac{\|w\|^2}{2k} + \delta\|w\| + \delta^2 k,$$

for the special case $F_{b^\delta}(x) := \frac{1}{2}\|Ax - b^\delta\|^2$, see Burger et al. (2007, Theorem 4.3). Here, w is defined as $w := g - Ax^\dagger \in \mathbb{R}^m$, which satisfies the source condition $A^*w \in \partial J(x^\dagger)$, cf. (Chavent and Kunisch 1997; Burger and Osher 2004). If $k^*(\delta)$ is of order $1/\delta$, we therefore observe

$$D_J^{p^{k^*(\delta)}}(x^\dagger, x^{k^*(\delta)}) = \mathcal{O}(\delta);$$

Hence, $x^{k^*(\delta)}$ converges to $x^\dagger$ in terms of the Bregman distances if δ converges to zero.

For more details on how to use Bregman iterations in the context of (linear) inverse problems, we refer the reader to Osher et al. (2005), Resmerita and Scherzer (2006), Schuster et al. (2012), Burger (2016), and Benning and Burger (2018). For the remainder of this paper, we want to discuss modifications of Bregman iterations and Bregman proximal methods that are suitable to large-scale optimization and inverse problems.

Inverse Scale Space Flows

In what follows, we describe the *inverse scale space* (ISS) flow, a system of differential equations which can be derived as the continuous time limit of the Bregman iterations. For a Bregman function $J : \mathbb{R}^n \to \overline{\mathbb{R}}$ and objective function $F : \mathbb{R}^n \to \mathbb{R}$, this flow is given by

$$\dot{p}(t) = -\nabla F(x(t)), \quad p(t) \in \partial J(x(t)). \tag{15}$$

It is straightforward to verify that Bregman iterations (9b) and linearized Bregman iterations (10) can be derived, respectively, as the forward and backward Euler discretization of (15).

The term *inverse scale space flow* was coined by Scherzer and Groetsch (2001) in 2001. In addition to its connection to Bregman schemes, the ISS flow itself is an active topic of research. Initially studied by Burger et al. (2006, 2007, 2013), and Burger (2016), it has found applications in nonlinear spectral analysis by Burger et al. (2016), Gilboa et al. (2016), and Schmidt et al. (2018).

The ISS flow itself has largely been studied in the context of scale space methods and data filtering, where the objective functions generally take the more specific forms $\|x - b^\dagger\|^2/2$ or $\|Ax - b^\dagger\|^2/2$. We mention some papers that address questions regarding the existence and uniqueness results for solutions to (15). Burger et al. (2007) proved existence, uniqueness, and certain regularity properties of the solution to the flow when J is the total variation seminorm. These results were extended by Frick and Scherzer (2007) to all convex, proper, lower semicontinuous functions J, while in Burger et al. (2013), Burger et al. characterize the solution to the flow explicitly for the case $J = \|\cdot\|_1$. We note that while these studies do not assume strict convexity of J, strong convexity is ensured for F by the $\|\cdot\|^2$ term in F (restricted to the range of the linear operator A), so that the iterations (and flow) are still well-defined.

By supposing that J were twice continuously differentiable and μ-convex for some $\mu > 0$ (i.e., strongly convex with parameter μ, see Hiriart-Urruty and Lemaréchal 1993), we can provide an additional interpretation of the ISS flow, rewriting (15) as

$$\dot{x}(t) = -(\nabla^2 J(x(t)))^{-1} \nabla F(x(t)). \tag{16}$$

With this formulation, one can interpret the Hessian of $J(x(t))$ as a preconditioner for the flow. Furthermore, by using the chain rule, we derive an energy dissipation law for the system

$$\frac{d}{dt} F(x(t)) = \langle \dot{x}(t), \nabla F(x(t)) \rangle = -\langle \dot{x}(t), \nabla^2 J(x(t)) \dot{x}(t) \rangle \leq -\mu \|\dot{x}(t)\|^2,$$

where the final inequality follows from μ-convexity of J. Furthermore, observe that if $J = F$, (16) reduces to a continuous-time variant of Newton's method. One may tie this back to the variable metric proximal gradient methods, which were designed to incorporate quasi-Newton preconditioning to proximal gradient methods.

In section "The Bregman Itoh–Abe Method," we describe the Bregman Itoh–Abe (BIA) method (Benning et al. 2020), an iterative system derived by applying structure-preserving methods from numerical integration to the flow. Thus the ISS flow provides an alternative way to consider variational formulations for formulating Bregman schemes.

Accelerated Bregman Methods

Not only when dealing with large-scale problems, reducing the number of iterations is an important goal to achieve when designing an algorithm. In Theorem 1 we have seen that the Bregman proximal method (1) has a convergence rate of order $1/k$. In the wake of Nesterov (1983), many acceleration strategies have been developed for first-order optimization methods that aim at minimizing convex functions. As we focus on Bregman methods, we want to highlight the following adaptation of Nesterov (1983), first developed in Huang et al. (2013) for quadratic functions F. There, the authors consider the linearized Bregman iteration, i.e., (9) for the choice $J(x) = \frac{1}{2\tau}\|x\|^2 + \frac{1}{\tau}R(x) - F(x)$, as shown in (10). We have seen that (10) can be formulated as the gradient descent (13) for the special case $F(x) = \frac{1}{2}\|Ax - b^\delta\|^2$. The authors in Huang et al. (2013) have applied the idea of Nesterov acceleration to formulation (13), which reads

$$b^{k+1} = (1 + \beta_k)b^k - \beta_k b^{k-1} - \nabla G_\tau((1+\beta_k)b^k - \beta_k b^{k-1}), \tag{17}$$

where $\{\beta_k\}_{k \in \mathbb{N}}$ is a sequence of positive scalars. Applying $\tau A^\top$ to both sides of the equation and substituting $\tau A^\top b^k = x^k + q^k - \tau A^\top(Ax^k + b^\delta)$ then yields the equivalent formulation

$$x^{k+1} = \arg\min_{x \in \mathbb{R}^n}\left\{F(x) + (1+\beta_k)D_J^{p^k}(x, x^k) - \beta_k D_J^{p^{k-1}}(x, x^{k-1})\right\}, \tag{18a}$$

$$p^{k+1} = (1+\beta_k)p^k - \beta_k p^{k-1} - \nabla F(x^{k+1}), \tag{18b}$$

for $J(x) = \frac{1}{2\tau}\|x\|^2 + \frac{1}{\tau}R(x) - F(x)$, $F(x) = \frac{1}{2}\|Ax - b^\delta\|^2$, $p^k = \frac{1}{\tau}(x^k + q^k) - \nabla F(x^k) \in \partial J(x^k)$, and $q^k \in \partial R(x^k)$ for all $k \in \mathbb{N}$.

Remark 2. We want to emphasize that the equivalence between (17) and (18) does not hold for arbitrary functions F as we have exploited the linearity of ∇F by making use of $\nabla F((1+\beta_k)x^k - \beta_k x^{k-1}) = (1+\beta_k)\nabla F(x^k) - \beta_k \nabla F(x^{k-1})$.

Note that (17) can also be written in less compact form as

$$x^{k+1} = (I + \partial R)^{-1}(z^k), \tag{19a}$$

$$y^{k+1} = z^k - \tau \nabla F(x^{k+1}), \tag{19b}$$

$$z^{k+1} = (1 + \beta_{k+1})y^{k+1} - \beta_{k+1}y^k, \tag{19c}$$

if we substitute $y^k = \tau A^\top b^k$. Following the same approach as in Chambolle and Dossal (2015), (19) can also be written as

$$x^{k+1} = (I + \partial R)^{-1}(z^k), \tag{20a}$$

$$y^{k+1} = z^k - \tau \nabla F(x^{k+1}), \tag{20b}$$

$$z^{k+1} = \left(1 - \frac{1}{t_{k+1}}\right)y^{k+1} + \frac{1}{t_{k+1}}u^{k+1}, \tag{20c}$$

$$u^{k+1} = y^k + t_{k+1}(y^{k+1} - y^k). \tag{20d}$$

for $\beta_k := (t_k - 1)/t_{k+1}$ and a sequence $\{t_k\}_{k \in \mathbb{N}}$ of positive parameters.

An open problem which has attracted interest in recent years concerns whether accelerated versions of Bregman (proximal) gradient methods with generic, strongly convex Bregman distances are possible (Teboulle 2018). In a recent work by Dragomir et al. (2019), this question is partly answered in the negative, concluding that for Bregman distances, based on smooth functions R or functions R that satisfy that $\frac{1}{\tau}R - F$ is convex, the $\mathcal{O}(1/k)$ convergence rate is optimal for first-order methods that use previous gradient and Bregman proximal evaluations. However, for more restrictive function classes, faster convergence rates can be achieved, as has been shown in Hanzely et al. (2018) and Gutman and Peña (2018).

Acceleration strategies such as Nesterov acceleration have also been analyzed in the context of iterative regularization strategies (e.g., (9) combined with early stopping as described in section "Bregman Iterations as Iterative Regularization Methods"), see, for instance, Matet et al. (2017), Neubauer (2017), Garrigos et al. (2018), and Calatroni et al. (2019).

Incremental and Stochastic Bregman Proximal Methods

Many large-scale problems, in particular in machine learning, involve the minimization of functions of the form

$$F(x) := \frac{1}{m} \sum_{i=1}^{m} f_i(x). \tag{21}$$

In other words, the objective function is a sum of m individual functions. If m happens to be extremely large, computing the gradient of F can be computationally extremely expensive, rendering the application of traditional methods such as (1) or (18) computationally infeasible. Feasible alternatives are methods that make use of gradients that are only based on a subset $B \subset \{1, \ldots, m\}$ of all indices. Such methods include incremental gradient methods (Bertsekas et al. 2011a) and stochastic gradient methods (Robbins and Monro 1951). If we assume that F in (21) is of the form

$$F(x) = L(x) + S(x) = \frac{1}{m} \sum_{i=1}^{m} \ell_i(x) + \frac{1}{m} \sum_{i=1}^{m} s_i(x), \tag{22}$$

an incremental version of the Bregman proximal gradient as in (8) can be formulated as

$$x^k = \arg\min_{x \in \mathbb{R}^n} \left\{ \ell_{i(k)}(x) + s_{i(k)}(x) + D_{J_k}(x, x^{k-1}) \right\}. \tag{23}$$

Here $i : \mathbb{N} \to \{1, \ldots, m\}$ denotes the index function $i(x) := x$ modulo m, although other cycle orderings are certainly possible as well. A special case of (23) is the classical incremental proximal gradient method (Bertsekas et al. 2011b)

$$x^k = \left(I + \tau_k \partial s_{i(k)}\right)^{-1} \left(x^{k-1} - \tau_k \nabla \ell_{i(k)}(x^{k-1})\right)$$

for the choice of $J_k(x) = \frac{1}{2\tau_k}\|x\|^2 - \ell_{i(k)}(x)$. If we further pick $s_i \equiv 0$ for all i, we obtain the classical incremental gradient descent (Widrow and Hoff 1960; Bertsekas et al. 2011a), i.e.,

$$x^k = x^{k-1} - \tau_k \nabla \ell_{i(k)}(x^{k-1}),$$
$$= x^{k-1} - \tau_k \nabla f_{i(k)}(x^{k-1}), \tag{24}$$

as a special case.

In the following sections, we discuss extensions of stochastic gradient descent (SGD) and Kaczmarz methods in the Bregman framework, before highlighting the connection between single cycles of incremental Bregman proximal methods and deep neural network architectures.

Stochastic Mirror Descent

Stochastic gradient descent generalizes naturally to the Bregman proximal setting with the *stochastic mirror descent* (SMD) method (recall that mirror descent is equivalent to the Bregman gradient or linearized Bregman iteration). SMD is one of the most popular families of methods for stochastic optimization, and the method is defined as Nemirovski et al. (2009)

$$x^{k+1} = \arg\min_{x \in \mathbb{R}^n} \{\tau_k \langle \nabla f_{i(k)}(x^k), x \rangle + D_J^{p^k}(x, x^k)\}. \tag{25}$$

As in the setting of incremental descent methods, $i(k) \in \{1, \ldots, n\}$ represents a sequence of indices, which in the setting of SMD are typically randomized.

SMD was originally introduced by Nemirovsky and Yudin (1983), while subsequent, significant contributions include Nemirovski et al. (2009), Nesterov (2009), and Xiao (2010). The framework and its convergence analysis were further extended by Duchi et al. (2012) to cases where the samples from the distribution are not assumed to be independent.

Similar to SGD, the SMD algorithms are suitable for large-scale optimization and online learning settings, yet furthermore they come with the added benefits of Bregman iterations of exploiting structures in the data. Because of this, SMD is one of the most widely used family of methods for large-scale stochastic optimization (Azizan and Hassibi 2018; Zhou et al. 2017).

In the aforementioned works on SMD, the Bregman function J is assumed to be differentiable. In contrast, the use of nonsmooth Bregman functions, e.g., that invoke the ℓ^1-norm, is significant in the context of Bregman iterations and sparse signal processing. In the following section, we cover a Bregman method for sparse reconstruction of linear systems which can be seen as an instance of SMD, using the nonsmooth Bregman function $J(x) = \|x\|^2/2 + \lambda \|x\|_1$.

The Sparse Kaczmarz Method

The Kaczmarz method is a scheme for solving quadratic problems of the form $\min_x \langle x, Ax \rangle/2 - \langle b, x \rangle$. The method was originally introduced by Kaczmarz

(1937) and later by Gordon et al. (1970) under the name *algebraic reconstruction technique*. In this section, we review the extension of *Kaczmarz methods* to *sparse Kaczmarz methods* (Lorenz et al. 2014b) and their block variants. The motivation for sparse Kaczmarz methods is to find sparse solutions to linear problems $Ax = b$ via the problem formulation

$$\min_{x \in \mathbb{R}^n} \left\{ \frac{1}{2} \|x\|^2 + \lambda \|x\|_1 : Ax = b \right\}. \tag{26}$$

We first briefly review the original Kaczmarz method. For $x^0 = 0$, time steps $\tau_k > 0$, and a sequence of indices $(i(k))_{k \in \mathbb{N}}$, the (randomized) Kaczmarz method is given by

$$x^{k+1} = x^k - \tau_k(\langle a_{i(k)}, x^k \rangle - b_{i(k)})a_{i(k)}. \tag{27}$$

Here $a_{i(k)}$ denotes the ith row vector of A. If $i(k)$ comprise a subset of indices, then the block-variant of the Kaczmarz method is given by

$$x^{k+1} = x^k - \tau_k a_{i(k)}^\dagger (a_{i(k)} x^k - b_{i(k)}),$$

where $a_{i(k)}$ denotes the submatrix formed by the row vectors of A indexed by $i(k)$ and $a_{i(k)}^\dagger$ denotes the Moore-Penrose pseudo-inverse of $a_{i(k)}$. The iterates of the randomized Kaczmarz methods converge linearly to a solution of $Ax = b$ (Gower and Richtárik 2015).

Lorenz et al. (2014b) proposed a sparse Kaczmarz method as follows. Given starting points $x^0 = z^0 = 0$, the updates are given by

$$\begin{aligned} z^{k+1} &= z^k - \tau_k(\langle a_{i(k)}, x^k \rangle - b_{i(k)})a_{i(k)}, \\ x^{k+1} &= S_\lambda(z^{k+1}). \end{aligned} \tag{28}$$

Here S_λ denotes the soft-thresholding operator with threshold λ. The iterates $(x^k)_{k \in \mathbb{N}}$ converge linearly to a solution of (26) (Schöpfer and Lorenz 2019, Theorem 3.2).

A block variant of the sparse Kaczmarz method was proposed in Lorenz et al. (2014b). For blocks of rows of A denoted by sets of indices $i(k)$, it consists of the updates

$$\begin{aligned} z^{k+1} &= z^k - \tau_k a_{i(k)}^\top (a_{i(k)} x^k - b_{i(k)}), \\ x^{k+1} &= S_\lambda(z^{k+1}). \end{aligned} \tag{29}$$

Note that this uses the transpose $a_{i(k)}^\top$, unlike the standard block Kaczmarz method which uses the pseudo-inverse $a_{i(k)}^\dagger$. This too converges to a solution of (26) (Lorenz et al. 2014a, Corollary 2.9).

The sparse (block-)Kaczmarz method (29) has connections to two aforementioned Bregman schemes. First, one may verify that it corresponds to the SMD method (25) for $J(x) = \|x\|^2/2 + \lambda\|x\|_1$ and $F(x) = \sum_{i=1}^{n} |\langle a_i, x \rangle - b_i|^2$. Second, if one takes the entire matrix A as each block, then one recovers the linearized Bregman method for the same J (Lorenz et al. 2014b).

As with the general SMD method, the sparse Kaczmarz method is particularly suitable in online reconstruction settings, where the rows of the linear system A and/or data entries b are not all available instantly but successively are made available over time. We refer the reader to Lorenz et al. (2014b) for numerical examples which include the application of online compressed sensing.

Deep Neural Networks

We can generalize the incremental Bregman proximal gradient (23) by including an additional, potentially nonlinear projection $H_k : \mathbb{R}^{n_{k-1}} \to \mathbb{R}^{n_k}$, to obtain

$$x^k = \arg\min_{x \in \mathbb{R}^{n_k}} \left\{ \ell_k(x) + s_k(x) + D_{J_k}(x, H_k(x^{k-1})) \right\}, \tag{30}$$

for a sequence of dimensions $\{n_k\}_{k=1}^{l}$ with $n_k \in \mathbb{N}$ for all $k = 1, \ldots, l$. We are interested in a single cycle of this incremental Bregman proximal method only, which is why we have simplified the indexing notation from $i(k)$ to k throughout this subsection. In the following, we want to demonstrate how certain deep neural network architectures are special cases of (30). This connection was first investigated in the context of variational networks by Kobler et al. (2017), in the context of Bregman methods by Benning and Burger (2018), and in the context of proximal gradient methods by Frerix et al. (2017), Combettes and Pesquet (2018), and Bertocchi et al. (2019). Gradient-based learning with Bregman algorithms has also been studied in the context of image segmentation by Ochs et al. in (2015), and Bregman distances are used to analyze regularization strategies based on neural networks (Li et al. 2020). With the following example, we want to demonstrate how a class of feedforward neural networks coincides with (30).

Example 2 (Feedforward neural network with ReLU activation function). In this example we want to demonstrate how basic feedforward neural networks can be interpreted as variants of Algorithm (30). If we, for instance, choose $\{\ell_k\}_{k=1}^{l}$ to be of the form

$$\ell_k(x) := \frac{1}{2}\langle (I - M_k)x - 2b_k, x \rangle,$$

for quadratic matrices $\{M_k\}_{k=1}^{l}$ and vectors $\{b_k\}_{k=1}^{l}$ with $M_k \in \mathbb{R}^{n_k \times n_k}$ and $b_k \in \mathbb{R}^{n_k}$, which has the gradient

$$\nabla \ell_k(x) = \left(I - \frac{1}{2}\left(M_k + M_k^\top \right) \right) x - b_k,$$

and if we choose $\{s_k\}_{k=1}^l$ of the form

$$s_k(x) := \chi_{\geq 0}(x) = \begin{cases} 0 & \forall j : x_j \geq 0 \\ \infty & \exists j : x_j < 0 \end{cases}$$

for all $k \in \{1, \ldots, l\}$, then we easily verify that for the choice $J_k(x) = \|x\|^2/2 - \ell_k(x)$ the update

$$x^k = \max\left(0, A_k(x^{k-1}) + b_k \right),$$

with $A_k := \frac{1}{2}(M_k + M_k^T) \circ H_k$ is the unique solution of (30). Hence, we can consider this l-layer feedforward neural network with rectified linear units (ReLU) as activation functions (Nair and Hinton 2010) as a special case of the modified incremental Bregman gradient method (30) if we further guarantee that x^0 is chosen to be the input of the network.

Many other neural network architectures can be recovered in similar fashion to Example 2, where different activation functions can be recovered as proximal mappings for different choices of functions s_k, such as in Combettes and Pesquet (2018), and Bertocchi et al. (2019). For a recent overview of machine learning algorithms in the context of inverse problems, we refer to Arridge et al. (2019).

Bregman Incremental Aggregated Gradient

Two particularly interesting instances of incremental Bregman proximal methods are the *incremental aggregated gradient* (IAG) method (Blatt et al. 2007) and its stochastic counterpart *stochastic averaged gradient* (SAG) (Schmidt et al. 2017). For the sake of brevity, we focus on the incremental version in this paper. The IAG method reads

$$x^{k+1} = x^k - \frac{\tau_k}{m} g^k, \tag{31a}$$

$$g^{k+1} = g^k - \nabla f_{i(k+1)}(x^{k+1-m}) + \nabla f_{i(k+1)}(x^{k+1}). \tag{31b}$$

Here $\{\tau_k\}_{k \in \mathbb{N}}$ is a sequence of positive scalars and $i : \mathbb{N} \to \{1, \ldots, m\}$ is defined as in section "A Unified Framework for Implicit and Explicit Gradient Methods." Please also note that m arbitrary points $x^{1-m}, x^{2-m}, \ldots, x^0$ have to be chosen as initialization. It is easy to see and has also been pointed out in Blatt et al. (2007) that (31) can be rewritten as

$$x^{k+1} = x^k - \frac{\tau_k}{m} \sum_{l=0}^{m-1} \nabla f_{i(k-l)}(x^{k-l}), \tag{32}$$

for $k \geq m$. Note that this is equivalent to the following characterization in terms of Bregman distances, in analogy to the explicit gradient descent characterization in section "A Unified Framework for Implicit and Explicit Gradient Methods": if we rewrite (21) to $F(x) = \sum_{l=0}^{m-1} f_{i(k-l)}(x)$ for any $k \in \mathbb{N}$ and suppose we consider a Bregman method of the form

$$x^{k+1} = \arg\min_{x \in \mathbb{R}^n} \left\{ F(x) + \frac{1}{2\tau_k}\|x - x^k\|^2 - \frac{1}{m}\sum_{l=0}^{m-1} D_{f_{i(k-l)}}(x, x^{k-l}) \right\}, \tag{33a}$$

$$= \arg\min_{x \in \mathbb{R}^n} \left\{ \frac{1}{m}\sum_{l=0}^{m-1} \left[f_{i(k-l)}(x^{k-l}) + \langle \nabla f_{i(k-l)}(x^{k-l}), x - x^{k-l} \rangle \right] \right.$$

$$\left. + \frac{1}{2\tau_k}\|x - x^k\|^2 \right\}, \tag{33b}$$

then it becomes evident from computing the optimality condition of (33a) that the update (33b) is equivalent to (32) and hence (31) for $k \geq m$. Note that we can rewrite (33a) to

$$x^{k+1} = \arg\min_{x \in \mathbb{R}^n} \left\{ F(x) - \frac{1}{m}\sum_{l=1}^{m-1} D_{f_{i(k-l)}}(x, x^{k-l}) + D_{J_k}(x, x^k) \right\}, \tag{34}$$

for $J_k(x) := \frac{1}{2\tau_k}\|x\|^2 - \frac{1}{m}f_{i(k)}(x)$. The notable difference to the conventional IAG method is that we can replace the Bregman distance $D_{J_k}(x, x^k)$ in (34) with more generic Bregman distances. As in section "A Unified Framework for Implicit and Explicit Gradient Methods," we can for example choose $J_k(x) = \frac{1}{2\tau_k}\|x\|^2 + \frac{1}{\tau_k}R(x) - \frac{1}{m}f_{i(k)}(x)$ and therefore derive incremental Bregman iterations of the form

$$x^{k+1} = (I + \partial R)^{-1}\left(x^k + q^k - \frac{\tau_k}{m}g^k\right)$$

$$q^{k+1} = q^k - \left(x^{k+1} - x^k + \frac{\tau_k}{m}g^k\right),$$

$$g^{k+1} = g^k - \nabla f_{i(k+1)}(x^{k+1-m}) + \nabla f_{i(k+1)}(x^{k+1}),$$

where $q^k \in \partial R(x^k)$ for all k. Hence, substituting $y^k = x^k + q^k - \frac{\tau_k}{m}g^k$ yields the equivalent formulation

$$x^{k+1} = (I + \partial R)^{-1}\left(y^k\right),$$

$$g^{k+1} = g^k - \nabla f_{i(k+1)}(x^{k+1-m}) + \nabla f_{i(k+1)}(x^{k+1}),$$

$$y^{k+1} = y^k - \frac{\tau_{k+1}}{m} g^{k+1}.$$

If F is of the form (22), where $s_i = s$ for some (convex) function $s : \mathbb{R}^n \to \overline{\mathbb{R}}$ for all indices $i \in \{1, \ldots, m\}$ and if we choose $J_k(x) = \frac{1}{m\tau_k} R(x) - \frac{1}{m}\ell_{i(k)}(x)$ for continuously differentiable R, we recover the proximal-like incremental aggregated gradient (PLIAG) method, recently proposed in Zhang et al. (2017), which reads

$$x^{k+1} = \arg\min_{x \in \mathbb{R}^n} \left\{ s(x) + \sum_{l=0}^{m-1}\left[\ell_{i(k-l)}(x^{k-l}) + \langle \nabla \ell_{i(k-l)}(x^{k-l}), x - x^{k-l}\rangle\right] + \frac{1}{\tau_k} D_R(x, x^k) \right\}.$$

Needless to say, many different IAG or SAG methods can be derived for different choices of $\{J_k\}_{k=1}^m$. Choosing J_k such that convergence of the above algorithms is guaranteed is a delicate issue and involves carefully chosen assumptions, cf. Zhang et al. (2017, Section 2.3). Convergence guarantees for J_k as defined above with an arbitrary (proper, convex, and l.s.c.) function R which is an open problem. Having considered incremental variants of Bregman proximal algorithms, we now want to review coordinate descent adaptations of this algorithm in the following section.

Bregman Coordinate Descent Methods

In the previous section, we have reviewed Bregman adaptations of popular algorithms for minimizing objective functions that are sums of individual objective functions that occur in numerous large-scale applications, such as empirical risk minimization in machine learning.

In this section, we want to focus on Bregman adaptations of algorithms that aim to minimize multi-variable functions $F : \mathbb{R}^n \to \mathbb{R}$ by minimizing the objective with respect to one variable at a time. If we consider (1) for example, a simple coordinate descent adaption is

$$x_i^{k+1} = \arg\min_{x \in \mathbb{R}} \left\{ F(x_1^{k+1}, x_2^{k+1}, \ldots, x_{i-1}^{k+1}, x, x_{i+1}^k, \ldots, x_n^k) + D_{J_i}(x, x_i^k) \right\},$$

See, for example, Hua and Yamashita (2016), Corona et al. (2019a,b), Ahookhosh et al. (2019), Benning et al. (2020), and Gao et al. (2020). In the following, we want to give a brief overview on Bregman coordinate descent-type methods, with particular emphasis on an Itoh-Abe discrete gradient-based method, and also highlight their connections to traditional coordinate descent algorithms (and their Bregman adaptations) such as successive over-relaxation (SOR).

The Bregman Itoh–Abe Method

The Bregman Itoh–Abe (BIA) method (Benning et al. 2020) is a particular form for coordinate descent, derived by applying the discrete gradient method to the ISS flow (15). Discrete gradients are methods from geometric numerical integration for solving differential equations while preserving geometric structures – for details on geometric numerical integration, see, e.g., Hairer et al. (2006) and McLachlan and Quispel (2001) – and have found several applications to optimization, e.g., Benning et al. (2020), Grimm et al. (2017), Ehrhardt et al. (2018), Riis et al. (2018), and Ringholm et al. (2018) due to their ability to preserve energy dissipation laws.

A discrete gradient is an approximation to a gradient that must satisfy two properties as follows.

Definition (Discrete gradient). Let F be a continuously differentiable function. A *discrete gradient* is a continuous map $\overline{\nabla} F : \mathbb{R}^n \times \mathbb{R}^n \to \mathbb{R}^n$ such that for all $x, y \in \mathbb{R}^n$,

$$\langle \overline{\nabla} F(x, y), y - x \rangle = F(y) - F(x) \quad \text{(Mean value)}, \tag{35}$$

$$\lim_{y \to x} \overline{\nabla} F(x, y) = \nabla F(x) \quad \text{(Consistency)}. \tag{36}$$

Given a choice of $\overline{\nabla} F$, starting points x^0, $p^0 \in \partial J(x^0)$, and time steps $(\tau_k)_{k \in \mathbb{N}}$, the Bregman discrete gradient scheme is defined as

$$p^{k+1} = p^k - \tau_k \overline{\nabla} F(x^k, x^{k+1}), \qquad p^{k+1} \in \partial J(x^{k+1}). \tag{37}$$

As with the other Bregman schemes, this is a discretization of (15). Furthermore, the following dissipation property is an immediate consequence of the definition of discrete gradients.

Remark 3. When $J(x) = \|x\|^2/2$, then the ISS flow reduces to the Euclidean gradient flow, and we refer to the corresponding BIA method simply as the Itoh–Abe (IA) method.

Proposition. *Suppose J is μ-convex and that (x^{k+1}, p^{k+1}) solves the update (35) given (x^k, p^k) and time step $\tau_k > 0$. Then*

$$F(x^{k+1}) - F(x^k) = -\frac{1}{\tau_k} D_J^{symm}(x^k, x^{k+1}) \leq -\frac{\mu}{\tau_k} \|x^k - x^{k+1}\|^2, \tag{38}$$

where $D_J^{symm}(x, y)$ is the symmetrized Bregman distance defined as

$$D_J^{symm}(x, y) := D_J^p(x, y) + D_J^q(y, x) = \langle p-q, y-x \rangle \text{ for } p \in \partial J(y), q \in \partial J(x).$$

Proof. By (35) and (37) respectively, we have

$$F(x^{k+1}) - F(x^k) = \langle \overline{\nabla} F(x^k, x^{k+1}), x^{k+1} - x^k \rangle = -\frac{1}{\tau_k} \langle p^{k+1} - p^k, x^{k+1} - x^k \rangle.$$

The result then follows from monotonicity of convex functions, see, e.g., Hiriart-Urruty and Lemaréchal (1993, Theorem 6.1.2).

While there are various discrete gradients (see, e.g., McLachlan et al. 1999), the *Itoh–Abe discrete gradient* (Itoh and Abe 1988) (also known as the coordinate increment discrete gradient) is of particular interest in optimization as it is derivative-free and can be implemented for nonsmooth functions. It is defined as

$$\overline{\nabla} F(x, y) = \begin{pmatrix} \frac{F(y_1, x_2, \ldots, x_n) - F(x)}{y_1 - x_1} \\ \frac{F(y_1, y_2, x_3, \ldots, x_n) - F(y_1, x_2, \ldots, x_n)}{y_2 - x_2} \\ \vdots \\ \frac{F(y) - F(y_1, \ldots, y_{n-1}, x_n)}{y_n - x_n} \end{pmatrix}, \tag{39}$$

where $0/0$ is interpreted as $\partial_i F(x)$.

The BIA method is derived by plugging in the Itoh–Abe discrete gradient for $\overline{\nabla} F$ in (37). Provided that J is separable in the coordinates, i.e., $J(x) = \sum_{i=1}^n J_i(x_i)$, for $J_i : \mathbb{R} \to \mathbb{R}$, then this method reduces to sequential updates along the coordinates. Specifically, it can be written as

$$p_i^{k+1} = p_i^k - \tau_{k,i} \frac{F(y^{k,i}) - F(y^{k,i-1})}{x_i^{k+1} - x_i^k}, \quad p_i^{k+1} \in \partial J_i(y_i^{k,i}),$$
$$y^{k,i} = [x_1^{k+1}, \ldots, x_i^{k+1}, x_{i+1}^k, \ldots, x_n^k], \quad i = 1, \ldots, n. \tag{40}$$

In addition to having a derivative-free formulation, the BIA method has convergence guarantees for a large group of objective functions. In particular, if the Bregman function J is nonsmooth and strongly convex, and if F is locally Lipschitz continuous with a regularity assumption (see Benning et al. 2020 for details), the BIA scheme converges to a set of *Clarke stationary points* (Benning et al. 2020, Theorem 4.5). Clarke stationarity refers to the optimality criteria $0 \in \partial^C F(x)$, where $\partial^C F(x)$ denotes the *Clarke subdifferential* of F at x (Clarke 1990).

This scheme comes with the cost that the updates (40) are in general implicit. However, for the cases

$$J(x) = \frac{1}{2}\|x\|^2, \qquad J(x) = \frac{1}{2}\|x\|^2 + \lambda\|x\|_1,$$

$$F(x) = \frac{1}{2}\|Ax - b^\delta\|^2, \qquad F(x) = \frac{1}{2}\|Ax - b^\delta\|^2 + \gamma\|x\|_1,$$

the updates are explicit (Benning et al. 2020).

In section "Student-t Regularized Image Denoising," we present an example of a nonsmooth, nonconvex image denoising model, previously considered in Benning et al. (2020), for which one can significantly speed up convergence by exploiting sparsity in the residual $x^* - x^\delta$.

Equivalencies of Certain Bregman Coordinate Descent Methods

In what follows, we briefly discuss and draw connections between various approaches to coordinate descent methods using Bregman distances. This builds on the observation by Miyatake et al. (2018) that the Itoh–Abe method applied to quadratic functions $F(x) = \langle x, Ax \rangle/2 - \langle b, x \rangle$ is equivalent to the Gauss–Seidel and successive-over-relaxation (SOR) methods (Young 1971).

The explicit coordinate descent method (Beck and Tetruashvili 2013; Wright 2015) for minimizing F is given by

$$\begin{aligned} y^{k,0} &= x^k \\ y^{k,i} &= y^{k,i-1} - \overline{\tau}_i [\nabla F(y^{k,i-1})]_i e^i, \\ x^{k+1} &= y^{k,n}, \end{aligned} \qquad (41)$$

where $\overline{\tau}_i > 0$ is the time step and e^i denotes the ith basis vector. As mentioned in Wright (2015), the SOR method is also equivalent to the coordinate descent method with F as above and the time steps scaled coordinate-wise by $1/A_{i,i}$. Hence, in this setting, the Itoh–Abe discrete gradient method is equivalent not only to SOR methods but to explicit coordinate descent.

Furthermore, these equivalencies extend to discretizations of the inverse scale space flow for certain quadratic objective functions and certain forms of Bregman functions J. Consider a quadratic function $F(x) = \langle x, Ax \rangle/2 - \langle b, x \rangle$ where A is symmetric and positive definite, and denote by B the diagonal matrix for which $A_{i,i} = B_{i,i}$ for each i. Given a scaling parameter $\omega > 0$ and the Bregman function

$$J(x) = \frac{1}{2\omega}\langle x, Bx \rangle + \lambda\|x\|_1, \qquad (42)$$

The Itoh–Abe method yields a sparse SOR scheme as detailed in Benning et al. (2020). We may compare this to a *Bregman linearized coordinate descent* scheme

$$y^{k,0} = x^k, \quad p^k \in \partial J(x^k),$$

$$z_i = \arg\min_y [\nabla F(y^{k,i-1})]_i \cdot y + D_J^{p^k}(y^{k,i-1}, y^{k,i-1} + ye^i),$$

$$y^{k,i} = y^{k,i-1} + z_i e^i,$$

$$x^{k+1} = y^{k,n},$$

where J is given by (42) for some $\omega = \omega_E \in (0, 2)$. One can verify that these schemes are equivalent if one sets $\omega_E = \frac{1}{1/\omega + 1/2}$. We furthermore mention that these equivalencies also hold if we were to consider (implicit) Bregman iterations rather than linearized ones.

Remark 4. It is worth noting at this stage that while the Kaczmarz method (27) is closely related to SOR (Oswald and Zhou 2015), this connection does not carry over to the BIA method versus the sparse Kaczmarz method.

Saddle-Point Methods

Many problems in imaging (Chambolle and Pock 2016a) and machine learning (Goldstein et al. 2015; Adler and Öktem 2018) can be formulated as minimization problems of the form

$$\min_{x \in \mathbb{R}^n, z \in \mathbb{R}^m} G(x) + F(z) \quad \text{subject to} \quad K(x, z) = c. \tag{43}$$

Here $G : \mathbb{R}^n \to \overline{\mathbb{R}}$ and $F : \mathbb{R}^m \to \overline{\mathbb{R}}$ are proper and lower semi-continuous and usually also convex functions, the operator $K : \mathbb{R}^n \times \mathbb{R}^m \to \mathbb{R}^s$ is a bounded, and usually linear operator and $c \in \mathbb{R}^s$ are a vector. A classical linear example for K is

$$K(x, z) = Ax + Bz,$$

where $A \in \mathbb{R}^{s \times n}$ and $B \in \mathbb{R}^{s \times m}$ are matrices (Boyd et al. 2011).

In terms of optimization, the equality constraint can be incorporated with the help of a Lagrange multiplier $y \in \mathbb{R}^s$. We can then re-formulate (43) as finding a saddle point of an augmented Lagrange function, i.e., we solve

$$\min_{x \in \mathbb{R}^n, z \in \mathbb{R}^m} \max_{y \in \mathbb{R}^s} \mathcal{L}_\delta(x, z; y)$$

for the augmented Lagrangian

$$\mathcal{L}_\delta(x, z; y) := G(x) + F(z) + \langle y, K(x, z) - c \rangle + \frac{1}{2\delta} \|K(x, z) - c\|^2, \tag{44}$$

where $\delta > 0$ is a positive scalar. For the special case $K(x, z) = Ax - z$ and $c \equiv 0$, one can replace $F(Ax)$ with its convex conjugate and formulate the alternative saddle-point problem

$$\min_{x \in \mathbb{R}^n} \max_{y \in \mathbb{R}^m} G(x) + \langle Ax, y \rangle - F^*(y), \tag{45}$$

where the convex conjugate or Fenchel conjugate F^* of F is defined as

$$F^*(y) := \sup_{x \in \mathbb{R}^n} \langle x, y \rangle - F(x).$$

We want to emphasize that extensions for nonconvex functions (Li and Pong 2015; Moeller et al. 2015; Möllenhoff et al. 2015) and extensions for nonlinear operators A (Valkonen 2014; Benning et al. 2015; Clason and Valkonen 2017) or nonlinear replacements of the dual product (Clason et al. 2019) exist. In the following, we review Bregman algorithms for the numerical computation of solutions of those saddle-point formulations.

Alternating Direction Method of Multipliers

The alternating direction method of multipliers (ADMM), (Gabay 1983), is a coordinate descent method applied to the augmented Lagrangian functional (44). The augmented Lagrangian is furthermore modified to also include appropriate penalization terms, so that we compute

$$x^{k+1} = \arg\min_{x \in \mathbb{R}^n} \mathscr{L}_\delta(x, z^k; \mu^k) + D_{J_x}(x, x^k), \tag{46a}$$

$$z^{k+1} = \arg\min_{z \in \mathbb{R}^m} \mathscr{L}_\delta(x^{k+1}, z; y^k) + D_{J_z}(z, z^k), \tag{46b}$$

$$y^{k+1} = \arg\max_{y \in \mathbb{R}^m} \mathscr{L}_\delta(x^{k+1}, z^{k+1}; y) - D_{J_y}(y, y^k), \tag{46c}$$

in an alternating fashion. To our knowledge, the first adaptation of ADMM to more general Bregman functions was proposed in Wang and Banerjee (2014). In the setting discussed here, the functions J_x, J_z, and J_y are convex and continuously differentiable functions. In the most basic scenario, we choose $K(x, z) = Ax + Bz$, J_x, and J_y as the zero functions, i.e., $J_x(x) = 0$ and $J_z(z) = 0$ for all $x \in \mathbb{R}^n$ $z \in \mathbb{R}^m$, while J_y is chosen to be a positive multiple of the squared Euclidean norm $J_y(y) := \frac{1}{2\tau} \|y\|^2$. Then (46) reduces to the classical ADMM setting (cf. Boyd et al. 2011)

$$x^{k+1} = \left(A^\top A + \delta \partial G\right)^{-1} \left(A^\top \left(c - (Bz^k + \delta y^k)\right)\right),$$

$$z^{k+1} = \left(B^\top B + \delta \partial F\right)^{-1} \left(B^\top \left(c - (Ax^{k+1} + \delta y^k)\right)\right),$$

$$y^{k+1} = y^k + \tau \left(Ax^{k+1} + Bz^{k+1} - c\right).$$

Depending on the choices of J_x, J_z, and J_y, many other useful variants are possible, such as

$$x^{k+1} = (I + \tau_x \delta \partial G)^{-1} \left(x^k - \tau_x A^\top \left(Ax^k + Bz^k + \delta y^k - c \right) \right),$$

$$z^{k+1} = (I + \tau_z \delta \partial F)^{-1} \left(z^k - \tau_z B^\top \left(Ax^{k+1} + Bz^k + \delta y^k - c \right) \right),$$

$$y^{k+1} = y^k + \tau_y \left(Ax^{k+1} + Bz^{k+1} - c \right),$$

for the choices $J_x(x) = \frac{1}{2\delta \tau_x} \|x\|^2 - \frac{1}{2\delta} \|Ax\|^2$, $J_z(z) = \frac{1}{2\delta \tau_z} \|z\|^2 - \frac{1}{2\delta} \|Bz\|^2$, and $J_y(y) = \frac{1}{2\tau_y} \|y\|^2$, which is fully explicit with respect to the operators A and B. Moreover, J_x is convex for $0 < \tau_x < \|A\|^2$, while J_z is convex for $0 < \tau_z < \|B\|^2$. A unified Bregman framework for primal-dual algorithms is discussed in greater detail in Zhang et al. (2011).

Primal-Dual Hybrid Gradient Method

In this section we focus on the special saddle-point formulation (45). It is straightforward to verify that for convex G and F a saddle point $(\hat{x}, \hat{y})^\top$ is characterized by the optimality system

$$0 \in \partial G(\hat{x}) + A^\top \hat{y}, \tag{47a}$$

$$0 \in \partial F^*(\hat{y}) - A\hat{x}. \tag{47b}$$

It is sensible and has indeed been suggested in Chambolle and Pock (2016b), and Hohage and Homann (2014) to solve this nonlinear inclusion problem with a fixed point algorithm of the form

$$\begin{pmatrix} 0 \\ 0 \end{pmatrix} \in \begin{pmatrix} \partial G(x^{k+1}) + A^\top y^{k+1} \\ \partial F^*(y^{k+1}) - Ax^{k+1} \end{pmatrix} + \partial J(x^{k+1}, y^{k+1}) - \partial J(x^k, y^k). \tag{48}$$

Here ∂J denotes the subdifferential of some convex function $J : \mathbb{R}^n \times \mathbb{R}^m \to \mathbb{R}$. For the choice

$$J(x, y) := \frac{1}{2} \left\| \begin{pmatrix} x \\ y \end{pmatrix} \right\|_M^2 \quad \text{with} \quad \left\| \begin{pmatrix} x \\ y \end{pmatrix} \right\|_M := \sqrt{\left\langle M \begin{pmatrix} x \\ y \end{pmatrix}, \begin{pmatrix} x \\ y \end{pmatrix} \right\rangle}$$

$$\text{and} \quad M := \begin{pmatrix} \frac{1}{\tau} I & -A^\top \\ -A & \frac{1}{\sigma} I \end{pmatrix},$$

and $\tau\sigma \|A\|^2 < 1$, we obtain the conventional primal-dual hybrid gradient (PDHG) method (with relaxation parameter set to one) as proposed and discussed in Zhu and Chan (2008), Pock et al. (2009), Esser et al. (2010), and Chambolle and Pock (2011, 2016a), which reads

$$x^{k+1} = (I + \tau \partial G)^{-1} \left(x^k - \tau A^\top y^k \right), \tag{49a}$$

$$y^{k+1} = (I + \sigma \partial F^*)^{-1} \left(y^k + \sigma A(2x^{k+1} - x^k) \right). \tag{49b}$$

Note that we can reformulate (48) to

$$\begin{pmatrix} 0 \\ 0 \end{pmatrix} \in \begin{pmatrix} \partial G(x^{k+1}) - \partial G(\hat{x}) + A^\top (y^{k+1} - \hat{y}) \\ \partial F^*(y^{k+1}) - \partial F^*(\hat{y}) - A(x^{k+1} - \hat{x}) \end{pmatrix}$$

$$+ \partial J(x^{k+1}, y^{k+1}) - \partial J(\hat{x}, \hat{y}) - \left(\partial J(x^k, y^k) - \partial J(\hat{x}, \hat{y}) \right), \tag{50}$$

if we add the optimality system (47) to (48), for a saddle point $(\hat{x}, \hat{y})^\top$. Taking a dual product of

$$\begin{pmatrix} \partial G(x^{k+1}) - \partial G(\hat{x}) + A^\top (y^{k+1} - \hat{y}) \\ \partial F^*(y^{k+1}) - \partial F^*(\hat{y}) - A(x^{k+1} - \hat{x}) \end{pmatrix}$$

with $(x^{k+1} - \hat{x}, y^{k+1} - \hat{y})^\top$ therefore yields

$$\left\langle \begin{pmatrix} \partial G(x^{k+1}) - \partial G(\hat{x}) + A^\top (y^{k+1} - \hat{y}) \\ \partial F^*(y^{k+1}) - \partial F^*(\hat{y}) - A(x^{k+1} - \hat{x}) \end{pmatrix}, \begin{pmatrix} x^{k+1} - \hat{x} \\ y^{k+1} - \hat{y} \end{pmatrix} \right\rangle$$

$$= D_G^{\text{symm}}(x^{k+1}, \hat{x}) + D_{F^*}^{\text{symm}}(y^{k+1}, \hat{y}) \geq 0.$$

Here $D_J^{\text{symm}}(x, y)$ denotes the symmetric Bregman distance $D_J^{\text{symm}}(x, y) = D_J^q(x, y) + D_J^p(y, x) = \langle p - q, x - y \rangle$, for subgradients $p \in \partial J(x)$ and $q \in \partial J(y)$, which is also known as Jeffreys–Bregman divergence and closely related to other symmetrizations such as Jensen–Bregman divergences (Nielsen and Boltz 2011) and Burbea Rao distances (Burbea and Rao 1982a,b). As an immediate consequence, we observe

$$0 \geq \left\langle \partial J(x^{k+1}, y^{k+1}) - \partial J(x^k, y^k), \begin{pmatrix} x^{k+1} - \hat{x} \\ y^{k+1} - \hat{y} \end{pmatrix} \right\rangle$$

$$= D_J\left(\begin{pmatrix} \hat{x} \\ \hat{y} \end{pmatrix}, \begin{pmatrix} x^{k+1} \\ y^{k+1} \end{pmatrix} \right) - D_J\left(\begin{pmatrix} \hat{x} \\ \hat{y} \end{pmatrix}, \begin{pmatrix} x^k \\ y^k \end{pmatrix} \right) + D_J\left(\begin{pmatrix} x^{k+1} \\ y^{k+1} \end{pmatrix}, \begin{pmatrix} x^k \\ y^k \end{pmatrix} \right),$$

where we have made use of the three-point identity for Bregman distances (Chen and Teboulle 1993). Thus, we can conclude

$$D_J\left(\begin{pmatrix}\hat{x}\\\hat{y}\end{pmatrix},\begin{pmatrix}x^{k+1}\\y^{k+1}\end{pmatrix}\right) + D_J\left(\begin{pmatrix}x^{k+1}\\y^{k+1}\end{pmatrix},\begin{pmatrix}x^k\\y^k\end{pmatrix}\right) \le D_J\left(\begin{pmatrix}\hat{x}\\\hat{y}\end{pmatrix},\begin{pmatrix}x^k\\y^k\end{pmatrix}\right)$$

for all iterates. Consequently, the iterates are bounded in the Bregman distance setting with respect to J. Summing up the dual product of (48) with $(x^{k+1} - \hat{x}, y^{k+1} - \hat{y})^\top$ therefore yields

$$\sum_{k=0}^N \left[D_G^{\mathrm{symm}}(x^{k+1},\hat{x}) + D_{F^*}^{\mathrm{symm}}(y^{k+1},\hat{y}) \right] + \sum_{k=0}^N D_J\left(\begin{pmatrix}x^{k+1}\\y^{k+1}\end{pmatrix},\begin{pmatrix}x^k\\y^k\end{pmatrix}\right)$$

$$= \sum_{k=0}^N \left[D_J\left(\begin{pmatrix}\hat{x}\\\hat{y}\end{pmatrix},\begin{pmatrix}x^k\\y^k\end{pmatrix}\right) - D_J\left(\begin{pmatrix}\hat{x}\\\hat{y}\end{pmatrix},\begin{pmatrix}x^{k+1}\\y^{k+1}\end{pmatrix}\right) \right] \le D_J\left(\begin{pmatrix}\hat{x}\\\hat{y}\end{pmatrix},\begin{pmatrix}x^0\\y^0\end{pmatrix}\right)$$

$$< +\infty.$$

Hence, we can conclude $D_G^{\mathrm{symm}}(x^N,\hat{x}) \to 0$, $D_{F^*}^{\mathrm{symm}}(y^N,\hat{y}) \to 0$, and $D_J\left(\begin{pmatrix}x^N & y^N\end{pmatrix}^\top, \begin{pmatrix}x^k & y^k\end{pmatrix}^\top\right) \to 0$ for $N \to \infty$. If G and F^* are at least convex and if J is strongly convex with respect to some norm, one can further guarantee convergence of the corresponding iterates in norm to a saddle-point (x, y) solution of (45) with standard arguments. For more details, analysis, and extensions of PDHG methods, we refer the reader to Chambolle and Pock (2016a).

Applications

In the following we want to show applications for some of the Bregman algorithms discussed in this review chapter. We want to emphasize that none of the applications shown are really large-scale applications. The idea of this section is rather to demonstrate that the algorithms are applicable to a wide range of different problems, offering the potential to enhance actual large-scale problems. We focus on three combinations of applications and algorithms: robust principal component analysis via the accelerated linearized Bregman iteration, deep learning with an incremental proximal Bregman architecture, and image denoising via the Bregman Itoh–Abe method.

Robust Principal Component Analysis

Robust principal component analysis is an extension of principal component analysis first proposed in Candès et al. (2011). The key idea is to decompose a matrix $X \in \mathbb{R}^{m \times n}$ into a low-rank matrix $L \in \mathbb{R}^{m \times n}$ and a sparse matrix $S \in \mathbb{R}^{m \times n}$ by solving the optimization problem

$$\min_{L,S} \alpha_1 \|L\|_* + \alpha_2 \|S\|_1 \quad \text{subject to} \quad X = L + S. \tag{51}$$

Here $\|S\|_1$ is the one norm of the matrix S, i.e., $\|S\|_1 = \sum_{i=1}^{m} \sum_{j=1}^{n} |s_{ij}|$, while $\|L\|_*$ denotes the nuclear norm of L, which is the one norm of the singular values of L, i.e., $\|L\|_* = \sum_{j=1}^{\min(n,m)} \sigma_j$, for $L = U \Sigma V^*$ with $\Sigma_{ij} = \begin{cases} \sigma_j & i = j \\ 0 & i \neq j \end{cases}$ and U and V being orthogonal. There are numerous strategies for solving (51) numerically (Bouwmans et al. 2018); we focus on using the accelerated linearized Bregman iteration as discussed in section "Accelerated Bregman Methods." For this we use formulation (12) of the linearized Bregman iteration, respectively (19), in the accelerated case. We choose $A = \begin{pmatrix} I & I \end{pmatrix}^\top$, $b^\delta = X$, and $R = \alpha_1 \|\cdot\|_* + \alpha_2 \|\cdot\|_1$ and therefore obtain

$$L^{k+1} = (I + \alpha_1 \partial \|\cdot\|_*)^{-1} \left(\tau X^k\right),$$

$$S^{k+1} = (I + \alpha_2 \partial \|\cdot\|_1)^{-1} \left(\tau X^k\right),$$

$$X^{k+1} = X^k - \left(L^{k+1} + S^{k+1} - X\right),$$

in the case of (12), respectively

$$L^{k+1} = (I + \alpha_1 \partial \|\cdot\|_*)^{-1} \left(\tau Y^k\right),$$

$$S^{k+1} = (I + \alpha_2 \partial \|\cdot\|_1)^{-1} \left(\tau Y^k\right),$$

$$X^{k+1} = Y^k - \left(L^{k+1} + S^{k+1} - X\right),$$

$$Y^{k+1} = (1 + \beta_{k+1}) X^{k+1} - \beta_{k+1} X^k,$$

in the case of (17), for $X^0 := X$. We choose the parameters to be $\tau = 1/\|A\|^2 = 1/2$, $\alpha_1 = 10\sqrt{\max(m,n)}$, $\alpha_2 = 10$, and $\beta_k = (k-1)/(k+3)$ for $k \geq 1$. Note that the latter automatically implies $Y^0 = X$. We run the algorithm on two test datasets; inspired by Brunton and Kutz (2019), the first one is the Yale Faces B dataset (Lee et al. 2005), and the second one is a video sequence of a Cornell box with a moving

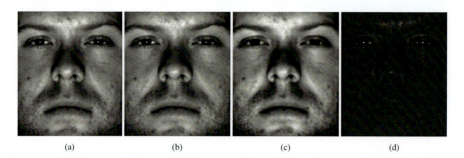

Fig. 1 From left to right: the first image of the Yale B faces database, its approximation which is the sum of a low-rank and a sparse matrix, the low-rank matrix, and the sparse matrix. (**a**) Original (**b**) Approximation (**c**) Low-rank part (**d**) Sparse part

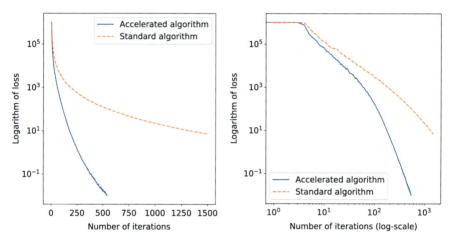

Fig. 2 This is an empirical validation of the different convergence rates of the linearized Bregman iteration and its accelerated counterpart (with regular scaling of the iterations on the left-hand side and a logarithmic scaling on the right-hand side)

shadow, from Benning et al. (2007). Figure 1 shows the first image of the Yale B faces database, its approximation, and its decomposition into a low-rank and a sparse part.

The more important aspect in terms of this review paper is certainly the comparison between the linearized Bregman iteration and its accelerated counterpart. A log-scale plot of the decrease of the loss function $\frac{1}{2}\|L + S - X\|_F^2$, where $\|\cdot\|_F$ denotes the Frobenius norm, over the course of the iterations of the two algorithms is visualized in Fig. 2. The plot is an empirical validation that (18) converges at rate $\mathcal{O}(1/k^2)$ as opposed to the $\mathcal{O}(1/k)$ rate of its non-accelerated counterpart.

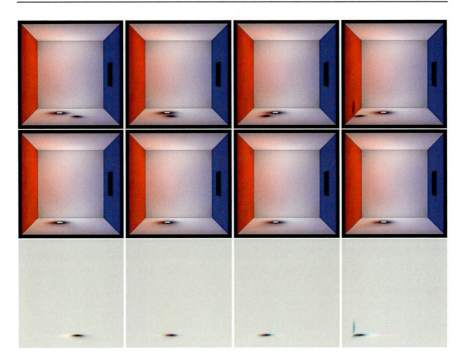

Fig. 3 First row: the 1st, 50th, 100th, and 150th frame of the original video sequence from Benning et al. (2007). Second row: the same frames of the computed low-rank part. Third row: the same frames of the computed sparse part

In Fig. 3 we see the 1st, 50th, 100th, and 150th frame of the original Cornell box video sequence from Benning et al. (2007), together with a low-rank approximation and a sparse component computed with the accelerated linearized Bregman iteration.

Deep Learning

Ever since Alexnet entered the scene in 2012 (Krizhevsky et al. 2012), thwarting then state-of-the-art image classification approaches in terms of accuracy in the process, deep neural networks (DNNs) have been central to research in computer vision and imaging. In this section, we merely want to support the analogy between incremental Bregman proximal methods and DNNs as shown in section "Deep Neural Networks" with a practical example, rather than engaging in a discussion of when and why DNNs based on (30) should be used or what advantages or shortcomings they possess compared to other neural network architectures. For a comprehensive overview over developments in deep learning, we refer the reader to Goodfellow et al. (2016).

In this example, we set up a DNN-based auto-encoder for dimensionality reduction and compare it to classical dimensionality reduction via singular value decomposition. The auto-encoder is of the form

$$x^k = \left(I + \partial \|\cdot\|_1\right)^{-1} \left(A_k x^{k-1} + b_k\right),$$
$$= S_1 \left(A_k x^{k-1} + b_k\right),$$

for $k \in \{1, 2, 3, 4\}$ and $x^0 = x$, where x denotes the input of the network, $A_k := \frac{1}{2}(M_k + M_k^\top) \circ H_k$ for matrices $M_k \in \mathbb{R}^{m_k \times m_k}$ dimensions $m_1 = 196$, $m_2 = 49$, $m_3 = 196$, and $m_4 = 784$, and where H_1 and H_2 are two-dimensional average pooling operators with window size 2×2 and H_3 and H_4 are nearest-neighbor interpolation operators that upscale by a factor of two. The vectors $\{b_k\}_{k=1}^4$ are bias vectors of dimensions $\{m_k\}_{k=1}^4$, and the operator S_1 is the soft-shrinkage operator as described in section "The Sparse Kaczmarz Method." Please note that this auto-encoder architecture is of the form (30) and represents a parametrized mapping Φ_Θ from $\mathbb{R}^{784}$ to $\mathbb{R}^{784}$, where $\Theta = (\{M_k\}_{k=1}^4, \{b_k\}_{k=1}^4)$ denotes the collection of parameters. We train the auto-encoder by minimizing the empirical risk based on the mean-squared error for a set of samples $\{x_i\}_{i=1}^s$, $s = 60000$, via stochastic gradient descent (which is the randomized version of (24)), i.e., we approximately estimate optimal parameters $\hat{\Theta}$ via

$$\hat{\Theta} = \arg\min_\Theta \frac{1}{2s} \sum_{i=1}^s \left(\Phi_\Theta(x_i) - x_i\right)^2.$$

We emphasize that the soft-thresholding activation function S_1 leaves Φ_Θ as not differentiable, which is why the application of (24) is technically a stochastic subgradient method. We train the auto-encoder with the help of PyTorch for a fixed number of epochs (500) and fixed step size $\tau = 2$ with batch size 100 on the MNIST training dataset (LeCun et al. 1998). In Fig. 4, we visualized several samples and the corresponding transformed outputs of the auto-encoder. In Fig. 5, we have visualized random images from the same dataset in comparison to their truncated singular value decomposition reconstructions where all but the first 49 singular values are cut off. As to be expected, nonlinear dimensionality reduction can outperform linear dimensionality reduction, achieving visually superior results for the same subspace dimensionality.

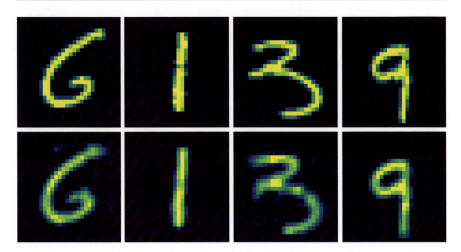

Fig. 4 Top row: random samples from the MNIST dataset. Bottom row: the corresponding approximations with the trained auto-encoder

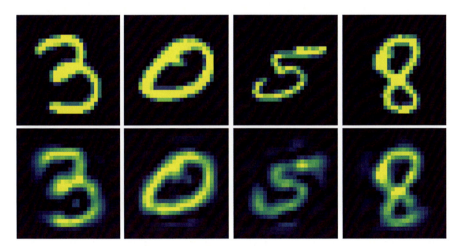

Fig. 5 Top row: random samples from the MNIST dataset. Bottom row: the corresponding approximations with the first 49 singular vectors

Student-t Regularized Image Denoising

In what follows, we apply BIA methods for solving a nonsmooth, nonconvex image denoising model, previously presented in Ochs et al. (2014). A priori knowledge of the noise distribution allows the use of Bregman functions $J(x)$ that exploit sparsity structures of the problem. As we will see, this yields significantly improved convergence rates in comparison with the default Itoh–Abe scheme (i.e., $J(x)' = \|x\|^2/2$). The application of the BIA method for this example was previously presented in Benning et al. (2020).

The objective function is given by

$$F : \mathbb{R}^n \to \mathbb{R}, \quad F(x) := \sum_{i=1}^{N} \varphi_i \Phi(K_i x) + \|x - x^\delta\|_1. \tag{52}$$

Here $\{K_i\}_{i=1}^{N}$ is a collection of linear filters, $(\varphi_i)_{i=1}^{N} \subset [0, \infty)$ are coefficients, $\Phi : \mathbb{R}^n \to \mathbb{R}$ is the nonconvex function based on the student-t distribution, defined as

$$\Phi(x) := \sum_{j=1}^{n} \psi(x_i), \quad \psi(x) := \log(1 + x^2),$$

and x^δ is an image corrupted by impulse noise (salt and pepper noise).

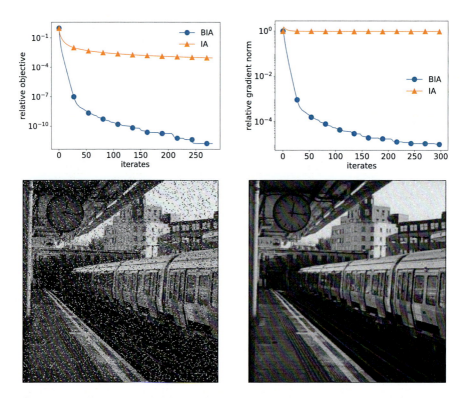

Fig. 6 Comparison of BIA and IA methods, for student-t regularized image denoising. First: convergence rate for relative objective. Second: convergence rate for relative gradient norm. Third: input data. Fourth: reconstruction

As impulse noise only affects a fraction of pixels, we use the data fidelity term $x \mapsto \|x - x^\delta\|_1$ to promote sparsity of $x^* - x^\delta$ for $x^* \in \arg\min F(x)$. As linear filters, we consider the simple case of finite difference approximations to first-order derivatives of x. We note that by applying a gradient flow to this regularization function, we observe a similarity to Perona–Malik diffusion (Perona and Malik 1990).

For the BIA method, we consider the Bregman function

$$J(x) := \frac{1}{2}\|x\|^2 + \gamma \|x - x^\delta\|_1,$$

to account for the sparsity of the residual $x^* - x^\delta$ and compare the method to the regular Itoh–Abe discrete gradient method (abbreviated to IA).

We set the starting point $x^0 = x^\delta$ and the parameters to $\tau_k = 1$ for all k, $\gamma = 0.5$, and $\varphi_i = 2$, $i = 1, 2$. For the impulse noise, we use a noise density of 10%. In the case where x_i^{k+1} is not set to x_i^δ, we use the scalar root solver *scipy.optimize.brenth* on Python. Otherwise, the updates are in closed form.

See Fig. 6 for numerical results. By gradient norm, we mean $\text{dist}(\partial^C F(x^k), 0)$.

Conclusions and Outlook

In this review paper, we gave a selective overview on a range of topics concerning adaptations of Bregman algorithms suited for large-scale problems in imaging. In particular, we discussed Nesterov accelerations of the Bregman (proximal) gradient or linearized Bregman iteration, incremental variants of Bregman methods, and coordinate descent-type Bregman algorithms with a particular focus on a Bregman Itoh–Abe scheme.

Despite the variety of numerous adaptations, a lot of research on Bregman algorithms is yet to be done. We conclude this chapter by discussing some open problems as well as ongoing directions of research.

Examples of open problems are adaptations for nonconvex objectives (following recent advances in papers such as Ahookhosh et al. 2019), extensions to nonlinear inverse problems (Bachmayr and Burger 2009) or inverse problems with non-quadratic data fidelity terms (Benning and Burger 2011) and the closer analysis and numerical realization of neural network architectures inspired by Bregman algorithms. We also want to emphasize that Bregman variants of incremental or stochastic variants of ADMM or the PDHG method in the spirit of Ouyang et al. (2013) and Chambolle et al. (2018) are still open problems.

Another important topic of ongoing research is to understand the scope for and limitations of accelerated Bregman methods, as stated by Teboulle (2018). Dragomir et al. (2019) point out the open problem of whether accelerated Bregman methods are possible if one makes further assumptions on the objective and Bregman functions or by allowing access to second-order information. Another interesting approach is to consider ODEs – see, e.g., Krichene et al. (2015) in which Krichene et al. investigate accelerating mirror descent via the ODE interpretation of Nesterov's acceleration (Su et al. 2016).

Going from optimization to sampling, some recent papers consider methods for sampling of distributions which incorporate elements of mirror descent in the underlying dynamics. Hsieh et al. (2018) propose a framework for sampling from constrained distributions, termed *mirrored Langevin dynamics*. In a similar vein, Zhang et al. (2020) propose a Mirror Langevin Monte Carlo algorithm, to improve the smoothness and convexity properties for the distribution.

Acknowledgments MB thanks Queen Mary University of London for their support. ESR acknowledges support from the London Mathematical Society.

References

Adler, J., Öktem, O.: Learned primal-dual reconstruction. IEEE Trans. Med. Imaging **37**(6), 1322–1332 (2018)

Ahookhosh, M., Hien, L.T.K., Gillis, N., Patrinos, P.: Multi-block Bregman proximal alternating linearized minimization and its application to sparse orthogonal nonnegative matrix factorization. arXiv preprint arXiv:1908.01402 (2019)

Arridge, S., Maass, P., Öktem, O., Schönlieb, C.-B.: Solving inverse problems using data-driven models. Acta Numerica **28**, 1–174 (2019)

Attouch, H., Buttazzo, G., Michaille, G.: Variational analysis in Sobolev and BV spaces: applications to PDEs and optimization. SIAM (2014)

Azizan, N., Hassibi, B.: Stochastic gradient/mirror descent: Minimax optimality and implicit regularization. arXiv preprint arXiv:1806.00952 (2018)

Bachmayr, M., Burger, M.: Iterative total variation schemes for nonlinear inverse problems. Inverse Prob. **25**(10), 105004 (2009)

Bauschke, H.H., Bolte, J., Teboulle, M.: A descent lemma beyond Lipschitz gradient continuity: first-order methods revisited and applications. Math. Oper. Res. **42**(2), 330–348 (2017)

Bauschke, H.H., Borwein, J.M., Combettes, P.L.: Bregman monotone optimization algorithms. SIAM J. Control. Optim. **42**(2), 596–636 (2003)

Beck, A.: First-Order Methods in Optimization, Vol. 25. SIAM (2017)

Beck, A., Teboulle, M.: Mirror descent and nonlinear projected subgradient methods for convex optimization. Oper. Res. Lett. **31**(3), 167–175 (2003)

Beck, A., Tetruashvili, L.: On the convergence of block coordinate descent type methods. SIAM J. Optim. **23**(4), 2037–2060 (2013)

Ben-Tal, A., Margalit, T., Nemirovski, A.: The ordered subsets mirror descent optimization method with applications to tomography. SIAM J. Optim. **12**(1), 79–108 (2001)

Benning, M., Betcke, M., Ehrhardt, M., Schönlieb, C.-B.: Gradient descent in a generalised Bregman distance framework. In: Geometric Numerical Integration and its Applications, Vol. 74, pp. 40–45. MI Lecture Notes series of Kyushu University (2017)

Benning, M., Betcke, M.M., Ehrhardt, M.J., Schönlieb, C.-B.: Choose your path wisely: gradient descent in a Bregman distance framework. SIAM Journal on Imaging Sciences (SIIMS). arXiv preprint arXiv:1712.04045 (2017)

Benning, M., Burger, M.: Error estimates for general fidelities. Electron. Trans. Numer. Anal. **38**(44–68), 77 (2011)

Benning, M., Burger, M.: Modern regularization methods for inverse problems. Acta Numerica **27**, 1–111 (2018)

Benning, M., Knoll, F., Schönlieb, C.-B., Valkonen, T.: Preconditioned ADMM with nonlinear operator constraint. In: IFIP Conference on System Modeling and Optimization, pp. 117–126. Springer (2015)

Benning, M., Lee, E., Pao, H., Yacoubou-Djima, K., Wittman, T., Anderson, J.: Statistical filtering of global illumination for computer graphics. IPAM Research in Industrial Projects for Students (RIPS) Report (2007)

Benning, M., Riis, E.S., Schönlieb, C.-B.: Bregman Itoh–Abe methods for sparse optimisation. In print: J. Math. Imaging Vision (2020)

Bertocchi, C., Chouzenoux, E., Corbineau, M.-C., Pesquet, J.-C., Prato, M.: Deep unfolding of a proximal interior point method for image restoration. Inverse Prob. **36**, 034005 (2019)

Bertsekas, D.P.: Incremental gradient, subgradient, and proximal methods for convex optimization: A survey. Optim. Mach. Learn. **2010**(1–38), 3 (2011)

Bertsekas, D.P.: Incremental proximal methods for large scale convex optimization. Math. Program. **129**(2), 163 (2011)

Blatt, D., Hero, A.O., Gauchman, H.: A convergent incremental gradient method with a constant step size. SIAM J. Optim. **18**(1), 29–51 (2007)

Bolte, J., Sabach, S., Teboulle, M., Vaisbourd, Y.: First order methods beyond convexity and Lipschitz gradient continuity with applications to quadratic inverse problems. SIAM J. Optim. **28**(3), 2131–2151 (2018)

Bonettini, S., Rebegoldi, S., Ruggiero, V.: Inertial variable metric techniques for the inexact forward–backward algorithm. SIAM J. Sci. Comput. **40**(5), A3180–A3210 (2018)

Bonnans, J.F., Gilbert, J.C., Lemaréchal, C., Sagastizábal, C.A.: A family of variable metric proximal methods. Math. Program. **68**(1–3), 15–47 (1995)

Bouwmans, T., Javed, S., Zhang, H., Lin, Z., Otazo, R.: On the applications of robust PCA in image and video processing. Proc. IEEE **106**(8), 1427–1457 (2018)

Boyd, S., Parikh, N., Chu, E., Peleato, B., Eckstein, J., et al.: Distributed optimization and statistical learning via the alternating direction method of multipliers. Found. Trends® Mach. Learn. **3**(1), 1–122 (2011)

Bregman, L.M.: The relaxation method of finding the common point of convex sets and its application to the solution of problems in convex programming. USSR Comput. Math. Math. Phys. **7**(3), 200–217 (1967)

Brunton, S.L., Kutz, J.N.: Data-Driven Science and Engineering: Machine Learning, Dynamical Systems, and Control. Cambridge University Press (2019)

Burbea, J., Rao, C.: On the convexity of higher order Jensen differences based on entropy functions (corresp.). IEEE Trans. Inf. Theory **28**(6), 961–963 (1982)

Burbea, J., Rao, C.: On the convexity of some divergence measures based on entropy functions. IEEE Trans. Inf. Theory **28**(3), 489–495 (1982)

Burger, M.: Bregman distances in inverse problems and partial differential equations. In: Advances in Mathematical Modeling, Optimization and Optimal Control, pp. 3–33. Springer (2016)

Burger, M., Frick, K., Osher, S., Scherzer, O.: Inverse total variation flow. Multiscale Model. Simul. **6**(2), 366–395 (2007)

Burger, M., Gilboa, G., Moeller, M., Eckardt, L., Cremers, D.: Spectral decompositions using one-homogeneous functionals. SIAM J. Imag. Sci. **9**(3), 1374–1408 (2016)

Burger, M., Gilboa, G., Osher, S., Xu, J.: Nonlinear inverse scale space methods. Commun. Math. Sci. **4**(1), 179–212 (2006)

Burger, M., Moeller, M., Benning, M., Osher, S.: An adaptive inverse scale space method for compressed sensing. Math. Comput. **82**(281), 269–299 (2013)

Burger, M., Osher, S.: Convergence rates of convex variational regularization. Inverse Prob. **20**(5), 1411 (2004)

Burger, M., Resmerita, E., He, L.: Error estimation for Bregman iterations and inverse scale space methods in image restoration. Computing **81**(2–3), 109–135 (2007)

Cai, J.-F., Osher, S., Shen, Z.: Convergence of the linearized Bregman iteration for ℓ^1-norm minimization. Math. Comput. **78**(268), 2127–2136 (2009)

Cai, J.-F., Osher, S., Shen, Z.: Linearized Bregman iterations for compressed sensing. Math. Comput. **78**(267), 1515–1536 (2009)

Cai, J.-F., Osher, S., Shen, Z.: Linearized Bregman iterations for frame-based image deblurring. SIAM J. Imag. Sci. **2**(1), 226–252 (2009)

Calatroni, L., Garrigos, G., Rosasco, L., Villa, S.: Accelerated iterative regularization via dual diagonal descent. arXiv preprint arXiv:1912.12153 (2019)

Candès, E.J., Li, X., Ma, Y., Wright, J.: Robust principal component analysis? J. ACM **58**(3), 11 (2011)

Censor, Y., Lent, A.: An iterative row-action method for interval convex programming. J. Optim. Theory Appl. **34**(3), 321–353 (1981)

Censor, Y., Stavros Zenios, A.: Proximal minimization algorithm with d-functions. J. Optim. Theory Appl. **73**(3), 451–464 (1992)

Chambolle, A., Dossal, C.: On the convergence of the iterates of the "fast iterative shrinkage/thresholding algorithm. J. Optim. Theory Appl. **166**(3), 968–982 (2015)

Chambolle, A., Ehrhardt, M.J., Richtárik, P., Carola-Schonlieb, B.: Stochastic primal-dual hybrid gradient algorithm with arbitrary sampling and imaging applications. SIAM J. Optim. **28**(4), 2783–2808 (2018)

Chambolle, A., Pock, T.: A first-order primal-dual algorithm for convex problems with applications to imaging. J. Math. Imaging Vision **40**(1), 120–145 (2011)

Chambolle, A., Pock, T.: An introduction to continuous optimization for imaging. Acta Numerica **25**, 161–319 (2016)

Chambolle, A., Pock, T.: On the ergodic convergence rates of a first-order primal–dual algorithm. Math. Prog. **159**(1–2), 253–287 (2016)

Chavent, G., Kunisch, K.: Regularization of linear least squares problems by total bounded variation. ESAIM Control Optim. Calc. Var. **2**, 359–376 (1997)

Chen, G., Teboulle, M.: Convergence analysis of a proximal-like minimization algorithm using Bregman functions. SIAM J. Optim. **3**(3), 538–543 (1993)

Chouzenoux, E., Pesquet, J.-C., Repetti, A.: Variable metric forward–backward algorithm for minimizing the sum of a differentiable function and a convex function. J. Optim. Theory Appl. **162**(1), 107–132 (2014)

Clarke, F.H.: Optimization and Nonsmooth Analysis. Classics in Applied Mathematics, 1st edn. SIAM, Philadelphia (1990)

Clason, C., Mazurenko, S., Valkonen, T.: Acceleration and global convergence of a first-order primal-dual method for nonconvex problems. SIAM J. Optim. **29**(1), 933–963 (2019)

Clason, C., Valkonen, T.: Primal-dual extragradient methods for nonlinear nonsmooth PDE-constrained optimization. SIAM J. Optim. **27**(3), 1314–1339 (2017)

Combettes, P.L., Pesquet, J.-C.: Deep neural network structures solving variational inequalities. arXiv preprint arXiv:1808.07526 (2018)

Combettes, P.L., Vũ, B.C.: Variable metric forward–backward splitting with applications to monotone inclusions in duality. Optimization **63**(9), 1289–1318 (2014)

Corona, V., Benning, M., Ehrhardt, M.J., Gladden, L.F., Mair, R., Reci, A., Sederman, A.J., Reichelt, S., Schönlieb, C.-B.: Enhancing joint reconstruction and segmentation with nonconvex Bregman iteration. Inverse Prob. **35**(5), 055001 (2019)

Corona, V., Benning, M., Gladden, L.F., Reci, A., Sederman, A.J., Schoenlieb, C.-B.: Joint phase reconstruction and magnitude segmentation from velocity-encoded MRI data. arXiv preprint arXiv:1908.05285 (2019)

Doan, T.T., Bose, S., Nguyen, D.H., Beck, C.L.: Convergence of the iterates in mirror descent methods. IEEE Control Syst. Lett. **3**(1), 114–119 (2018)

Dragomir, R.-A., Taylor, A., d'Aspremont, A., Bolte, J.: Optimal complexity and certification of Bregman first-order methods. arXiv preprint arXiv:1911.08510 (2019)

Duchi, J.C., Agarwal, A., Johansson, M., Jordan, M.I.: Ergodic mirror descent. SIAM J. Optim. **22**(4), 1549–1578 (2012)

Eckstein, J.: Nonlinear proximal point algorithms using Bregman functions, with applications to convex programming. Math. Oper. Res. **18**(1), 202–226 (1993)

Ehrhardt, M.J., Riis, E.S., Ringholm, T., Schönlieb, C.-B.: A geometric integration approach to smooth optimisation: Foundations of the discrete gradient method. ArXiv e-prints (2018)

Esser, E., Zhang, X., Chan, T.F.: A general framework for a class of first order primal-dual algorithms for convex optimization in imaging science. SIAM J. Imag. Sci. **3**(4), 1015–1046 (2010)

Frankel, P., Garrigos, G., Peypouquet, J.: Splitting methods with variable metric for Kurdyka–Łojasiewicz functions and general convergence rates. J. Optim. Theory Appl. **165**(3), 874–900 (2015)

Frerix, T., Möllenhoff, T., Moeller, M., Cremers, D.: Proximal backpropagation. arXiv preprint arXiv:1706.04638 (2017)

Frick, K., Scherzer, O.: Convex inverse scale spaces. In: International Conference on Scale Space and Variational Methods in Computer Vision, pp. 313–325. Springer (2007)

Gabay, D.: Chapter ix applications of the method of multipliers to variational inequalities. In: Studies in Mathematics and Its Applications, Vol. 15, pp. 299–331. Elsevier (1983)

Gao, T., Lu, S., Liu, J., Chu, C.: Randomized Bregman coordinate descent methods for non-Lipschitz optimization. arXiv preprint arXiv:2001.05202 (2020)

Garrigos, G., Rosasco, L., Villa, S.: Iterative regularization via dual diagonal descent. J. Math. Imaging Vision **60**(2), 189–215 (2018)

Gilboa, G., Moeller, M., Burger, M.: Nonlinear spectral analysis via one-homogeneous functionals: Overview and future prospects. J. Math. Imaging Vision **56**(2), 300–319 (2016)

Goldstein, T., Li, M., Yuan, X.: Adaptive primal-dual splitting methods for statistical learning and image processing. In: Advances in Neural Information Processing Systems, pp. 2089–2097 (2015)

Goodfellow, I., Bengio, Y., Courville, A.: Deep Learning. MIT Press (2016)

Gordon, R., Bender, R., Herman, G.T.: Algebraic reconstruction techniques (ART) for three-dimensional electron microscopy and X-ray photography. J. Theor. Biol. **29**(3), 471–481 (1970)

Gower, R.M., Richtárik, P.: Randomized iterative methods for linear systems. SIAM J. Matrix Anal. Appl. **36**(4), 1660–1690 (2015)

Grimm, V., McLachlan, R.I., McLaren, D.I., Quispel, G.R.W., Schönlieb, C.-B.: Discrete gradient methods for solving variational image regularisation models. J. Phys. A **50**(29), 295201 (2017)

Gutman, D.H., Peña, J.F.: A unified framework for Bregman proximal methods: subgradient, gradient, and accelerated gradient schemes. arXiv preprint arXiv:1812.10198 (2018)

Hairer, E., Lubich, C., Wanner, G.: Geometric Numerical Integration: Structure-Preserving Algorithms for Ordinary Differential Equations, Vol. 31, 2nd edn. Springer Science & Business Media, Berlin (2006)

Hanzely, F., Richtarik, P., Xiao, L.: Accelerated Bregman proximal gradient methods for relatively smooth convex optimization. arXiv preprint arXiv:1808.03045 (2018)

Hellinger, E.: Neue begründung der theorie quadratischer formen von unendlichvielen veränderlichen. Journal für die reine und angewandte Mathematik (Crelles Journal) **1909**(136), 210–271 (1909)

Hiriart-Urruty, J.-B., Lemaréchal, C.: Convex analysis and minimization algorithms I: Fundamentals, volume 305 of Grundlehren der Mathematischen Wissenschaften [Fundamental Principles of Mathemati- cal Sciences], 2nd edn. Springer, Berlin (1993)

Hohage, T., Homann, C.: A generalization of the chambolle-pock algorithm to Banach spaces with applications to inverse problems. arXiv preprint arXiv:1412.0126 (2014)

Hsieh, Y.-P., Kavis, A., Rolland, P., Cevher, V.: Mirrored Langevin dynamics. In: Advances in Neural Information Processing Systems, pp. 2878–2887 (2018)

Hua, X., Yamashita, N.: Block coordinate proximal gradient methods with variable Bregman functions for nonsmooth separable optimization. Math. Program. **160**(1–2), 1–32 (2016)

Huang, B., Ma, S., Goldfarb, D.: Accelerated linearized Bregman method. J. Sci. Comput. **54**(2–3), 428–453 (2013)

Itakura, F.: Analysis synthesis telephony based on the maximum likelihood method. In: The 6th International Congress on Acoustics, 1968, pp. 280–292 (1968)

Itoh, T., Abe, K.: Hamiltonian-conserving discrete canonical equations based on variational difference quotients. J. Comput. Phys. **76**(1), 85–102 (1988)

Juditsky, A., Nemirovski, A., et al.: First order methods for nonsmooth convex large-scale optimization, I: General purpose methods. Optim. Mach. Learn. 121–148 (2011). https://doi.org/10.7551/mitpress/8996.003.0007

Kaczmarz, M.S.: Angenäherte Auflösung von Systemen linearer Gleichungen. Bulletin International de l'Académie Polonaise des Sciences et des Lettres. Classe des Sciences Mathématiques et Naturelles. Série A, Sciences Mathématiques **35**, 355–357 (1937)

Kiwiel, K.C.: Free-steering relaxation methods for problems with strictly convex costs and linear constraints. Math. Oper. Res. **22**(2), 326–349 (1997)

Kiwiel, K.C.: Proximal minimization methods with generalized Bregman functions. SIAM J. Control. Optim. **35**(4), 1142–1168 (1997)

Kobler, E., Klatzer, T., Hammernik, K., Pock, T.: Variational networks: connecting variational methods and deep learning. In: German Conference on Pattern Recognition, pp. 281–293. Springer (2017)

Krichene, W., Bayen, A., Bartlett, P.L.: Accelerated mirror descent in continuous and discrete time. In: Advances in Neural Information Processing Systems, pp. 2845–2853 (2015)

Krizhevsky, A., Sutskever, I., Hinton, G.E.: Imagenet classification with deep convolutional neural networks. In: Advances in Neural Information Processing Systems, pp. 1097–1105 (2012)

LeCun, Y., Cortes, C., Burges, C.J.C.: The mnist database of handwritten digits (1998). http://yann.lecun.com/exdb/mnist 10:34 (1998)

Lee, K.-C., Ho, J., Kriegman, D.J.: Acquiring linear subspaces for face recognition under variable lighting. IEEE Trans. Pattern Anal. Mach. Intell. **27**(5), 684–698 (2005)

Li, G., Pong, T.K.: Global convergence of splitting methods for nonconvex composite optimization. SIAM J. Optim. **25**(4), 2434–2460 (2015)

Li, H., Schwab, J., Antholzer, S., Haltmeier, M.: Nett: Solving inverse problems with deep neural networks. Inverse Prob. **36**, 065005 (2020)

Lions, P.-L., Mercier, B.: Splitting algorithms for the sum of two nonlinear operators. SIAM J. Numer. Anal. **16**(6), 964–979 (1979)

Lorenz, D.A., Schöpfer, F., Wenger, S.: The linearized Bregman method via split feasibility problems: Analysis and generalizations. SIAM J. Imag. Sci. **7**(2), 1237–1262 (2014)

Lorenz, D.A., Wenger, S., Schöpfer, F., Magnor, M.: A sparse Kaczmarz solver and a linearized Bregman method for online compressed sensing. arXiv e-prints (2014)

Prasanta, P.C.: On the generalized distance in statistics. National Institute of Science of India (1936)

Matet, S., Rosasco, L., Villa, S., Vu, B.L.: Don't relax: Early stopping for convex regularization. arXiv preprint arXiv:1707.05422 (2017)

McLachlan, R.I., Quispel, G.R.W.: Six lectures on the geometric integration of ODEs, pp. 155–210. London Mathematical Society Lecture Note Series. Cambridge University Press, Cambridge (2001)

McLachlan, R.I., Quispel, G.R.W., Robidoux, N.: Geometric integration using discrete gradients. Philos. Trans. R. Soc. Lond. Ser. A Math. Phys. Eng. Sci. **357**(1754), 1021–1045 (1999)

Miyatake, Y., Sogabe, T., Zhang, S.-L.: On the equivalence between SOR-type methods for linear systems and the discrete gradient methods for gradient systems. J. Comput. Appl. Math. **342**, 58–69 (2018)

Moeller, M., Benning, M., Schönlieb, C., Cremers, D.: Variational depth from focus reconstruction. IEEE Trans. Image Process. **24**(12), 5369–5378 (2015)

Möllenhoff, T., Strekalovskiy, E., Moeller, M., Cremers, D.: The primal-dual hybrid gradient method for semiconvex splittings. SIAM J. Imag. Sci. **8**(2), 827–857 (2015)

Moreau, J.-J.: Proximité et dualité dans un espace hilbertien. Bulletin de la Société mathématique de France **93**, 273–299 (1965)

Morozov, V.A.: Regularization of incorrectly posed problems and the choice of regularization parameter. USSR Comput. Math. Math. Phys. **6**(1), 242–251 (1966)

Nair, V., Hinton, G.E.: Rectified linear units improve restricted Boltzmann machines. In: Proceedings of the 27th International Conference on Machine Learning (ICML-10), pp. 807–814 (2010)

Nemirovski, A., Juditsky, A., Lan, G., Shapiro, A.: Robust stochastic approximation approach to stochastic programming. SIAM J. Optim. **19**(4), 1574–1609 (2009)

Nemirovsky, A.S., Yudin, D.B.: Problem complexity and method efficiency in optimization (1983)

Nesterov, Y.: A method for unconstrained convex minimization problem with the rate of convergence $\mathcal{O}(1/k^2)$. In: Doklady AN USSR, Vol. 269, pp. 543–547 (1983)

Nesterov, Y.: Primal-dual subgradient methods for convex problems. Math. Program. **120**(1), 221–259 (2009)

Neubauer, A.: On Nesterov acceleration for Landweber iteration of linear ill-posed problems. J. Inverse Ill-posed Prob. **25**(3), 381–390 (2017)

Nielsen, F., Boltz, S.: The Burbea-Rao and Bhattacharyya centroids. IEEE Trans. Inf. Theory **57**(8), 5455–5466 (2011)

Ochs, P., Chen, Y., Brox, T., Pock, T.: iPiano: Inertial proximal algorithm for nonconvex optimization. SIAM J. Imag. Sci. **7**(2), 1388–1419 (2014)

Ochs, P., Ranftl, R., Brox, T., Pock, T.: Bilevel optimization with nonsmooth lower level problems. In: International Conference on Scale Space and Variational Methods in Computer Vision, pp. 654–665. Springer (2015)

Osher, S., Burger, M., Goldfarb, D., Xu, J., Yin, W.: An iterative regularization method for total variation-based image restoration. Multiscale Model. Simul. **4**(2), 460–489 (2005)

Oswald, P., Zhou, W.: Convergence analysis for Kaczmarz-type methods in a Hilbert space framework. Linear Algebra Appl. **478**, 131–161 (2015)

Ouyang, H., He, N., Tran, L., Gray, A.: Stochastic alternating direction method of multipliers. In: International Conference on Machine Learning, pp. 80–88 (2013)

Parikh, N., Boyd, S., et al.: Proximal algorithms. Found. Trends® Optim. **1**(3), 127–239 (2014)

Perona, P., Malik, J.: Scale-space and edge detection using anisotropic diffusion. IEEE Trans. Pattern Anal. Mach. Intell. **12**(7), 629–639 (1990)

Pock, T., Cremers, D., Bischof, H., Chambolle, A.: An algorithm for minimizing the Mumford-Shah functional. In: 2009 IEEE 12th International Conference on Computer Vision, pp. 1133–1140. IEEE (2009)

Resmerita, E., Scherzer, O.: Error estimates for non-quadratic regularization and the relation to enhancement. Inverse Prob. **22**(3), 801 (2006)

Riis, E.S., Ehrhardt, M.J., Quispel, G.R.W., Schönlieb, C.-B.: A geometric integration approach to nonsmooth, nonconvex optimisation. Foundations of Computational Mathematics (FOCM). ArXiv e-prints (2018)

Ringholm, T., Lazić, J., Schönlieb, C.-B.: Variational image regularization with Euler's elastica using a discrete gradient scheme. SIAM J. Imag. Sci. **11**(4), 2665–2691 (2018)

Robbins, H., Monro, S.: A stochastic approximation method. Ann. Math. Stat. **22**, 400–407 (1951)

Scherzer, O., Groetsch, C.: Inverse scale space theory for inverse problems. In: International Conference on Scale-Space Theories in Computer Vision, pp. 317–325. Springer (2001)

Marie Schmidt, F., Benning, M., Schönlieb, C.-B.: Inverse scale space decomposition. Inverse Prob. **34**(4), 179–212 (2018)

Schmidt, M., Le Roux, N., Bach, F.: Minimizing finite sums with the stochastic average gradient. Math. Program. **162**(1–2), 83–112 (2017)

Schöpfer, F., Lorenz, D.A.: Linear convergence of the randomized sparse Kaczmarz method. Math. Program. **173**(1), 509–536 (2019)

Schuster, T., Kaltenbacher, B., Hofmann, B., Kazimierski, K.S.: Regularization methods in Banach spaces, Vol. 10. Walter de Gruyter (2012)

Su, W., Boyd, S., Candes, E.J.: A differential equation for modeling Nesterov's accelerated gradient method: Theory and insights. J. Mach. Learn. Res. **17**(153), 1–43 (2016)

Teboulle, M.: Entropic proximal mappings with applications to nonlinear programming. Math. Oper. Res. **17**(3), 670–690 (1992)

Teboulle, M.: A simplified view of first order methods for optimization. Math. Program. **170**(1), 67–96 (2018)

Teboulle, M., Chen, G.: Convergence analysis of a proximal-like minimization algorithm using Bregman function. SIAM J. Optim. **3**(3), 538–543 (1993)

Valkonen, T.: A primal–dual hybrid gradient method for nonlinear operators with applications to MRI. Inverse Prob. **30**(5), 055012 (2014)

Wang, H., Banerjee, A.: Bregman alternating direction method of multipliers. In: Advances in Neural Information Processing Systems, pp. 2816–2824 (2014)

Widrow, B., Hoff, M.E.: Adaptive switching circuits. Technical report, Stanford Univ Ca Stanford Electronics Labs (1960)

Wright, S.J.: Coordinate descent algorithms. Math. Program. **1**(151), 3–34 (2015)

Xiao, L.: Dual averaging methods for regularized stochastic learning and online optimization. J. Mach. Learn. Res. **11**, 2543–2596 (2010)

Yin, W.: Analysis and generalizations of the linearized Bregman method. SIAM J. Imag. Sci. **3**(4), 856–877 (2010)

Yin, W., Osher, S., Goldfarb, D., Darbon, J.: Bregman iterative algorithms for \ell_1-minimization with applications to compressed sensing. SIAM J. Imag. Sci. **1**(1), 143–168 (2008)

Yosida, K.: Functional Analysis. Springer (1964)

Young, D.M.: Iterative Solution of Large Linear Systems. Computer Science and Applied Mathematics, 1st edn. Academic Press, Inc., Orlando (1971)

Zhang, H., Dai, Y.-H., Guo, L., Peng, W.: Proximal-like incremental aggregated gradient method with linear convergence under Bregman distance growth conditions. arXiv preprint arXiv:1711.01136 (2017)

Zhang, K.S., Peyré, G., Fadili, J., Pereyra, M.: Wasserstein control of mirror Langevin Monte Carlo. arXiv e-prints (2020)

Zhang, X., Burger, M., Osher, S.: A unified primal-dual algorithm framework based on Bregman iteration. J. Sci. Comput. **46**(1), 20–46 (2011)

Zhou, Z., Mertikopoulos, P., Bambos, N., Boyd, S., Glynn, P.W.: Stochastic mirror descent in variationally coherent optimization problems. In: Advances in Neural Information Processing Systems, pp. 7040–7049 (2017)

Zhu, M., Chan, T.: An efficient primal-dual hybrid gradient algorithm for total variation image restoration. UCLA CAM Report 34 (2008)

Fast Iterative Algorithms for Blind Phase Retrieval: A Survey

4

Huibin Chang, Li Yang, and Stefano Marchesini

Contents

Introduction... 140
Mathematical Formula and Nonlinear Optimization Model for BPR................... 141
    Mathematical Formula... 141
    Optimization Problems and Proximal Mapping 145
Fast Iterative Algorithms 147
    Alternating Projection (AP) Algorithms 147
    ePIE-Type Algorithms 149
    Proximal Algorithms.. 151
    ADMM... 153
    Convex Programming... 156
    Second-Order Algorithm Using Hessian...................................... 160
    Subspace Method... 162
Discussions.. 167
    Experimental Issues... 167
    Theoretical Analysis.. 168
    Further Discussions... 169
Conclusions... 170
References.. 171

Abstract

In nanoscale imaging technique and ultrafast laser, the reconstruction procedure is normally formulated as a blind phase retrieval (BPR) problem, where one has to recover both the sample and the probe (pupil) jointly from phaseless data. This

H. Chang (✉) · L. Yang
School of Mathematical Sciences, Tianjin Normal University, Tianjin, China

S. Marchesini
SLAC National Laboratory, Menlo Park, CA, USA

© Springer Nature Switzerland AG 2023
K. Chen et al. (eds.), *Handbook of Mathematical Models and Algorithms in Computer Vision and Imaging*, https://doi.org/10.1007/978-3-030-98661-2_116

survey first presents the mathematical formula of BPR and related nonlinear optimization problems and then gives a brief review of the recent iterative algorithms. It mainly consists of three types of algorithms, including the operator-splitting-based first-order optimization methods, second-order algorithm with Hessian, and subspace methods. The future research directions for experimental issues and theoretical analysis are further discussed.

Introduction

Phase retrieval (PR) plays a key role in nanoscale imaging technique (Pfeiffer 2018; Elser et al. 2018; Zheng et al. 2021; Gürsoy et al. 2022) and ultrafast laser (Trebino et al. 1997). Retrieving the images of the sample from phaseless data is a long-standing problem. Generally speaking, designing fast and reliable algorithms is challenging since directly solving the quadratic polynomials of PR is NP hard and the involved optimization problem is nonconvex and possibly nonsmooth. Thus, it has drawn the attentions of researchers for several decades (Luke 2005; Shechtman et al. 2015; Grohs et al. 2020; Fannjiang and Strohmer 2020). Among the general PR problems, besides the recovery of the sample, it is also of great importance to reconstruct the probes. The motivation of blind recovery is twofold: (1) characteristics of the probe (wave front sensing) and (2) improving the reconstruction quality of the sample. Essentially in practice, as the probe is almost never completely known, one has to solve such blind phase retrieval (BPR) problem, e.g., in coherent diffractive imaging (CDI) (Thibault and Guizar-Sicairos 2012), convention ptychography imaging (Thibault et al. 2009; Maiden and Rodenburg 2009), Fourier ptychography (Zheng et al. 2013; Ou et al. 2014), convolutional PR(Ahmed et al. 2018), frequency-resolved optical gating (Trebino et al. 1997), and others.

An early work by Chapman (1996) to solve the blind problem used the Wigner-distribution deconvolution method to retrieve the probe. In the optics community, alternating projection (AP) algorithms are very popular for nonblind PR problems (Marchesini 2007; Elser et al. 2018). Some AP algorithms have also been applied to BPR problems, e.g., Douglas-Rachford (DR)-based algorithm (Thibault et al. 2009), extended ptychographic engine (ePIE) and variants (Maiden and Rodenburg 2009; Maiden et al. 2017), and relaxed averaged alternating reflection (Luke 2005)-based projection algorithm (Marchesini et al. 2016). More advanced first-order optimization method includes proximal algorithms, Hesse et al. (2015), Yan (2020), and Huang et al. (2021), alternating direction of multiplier methods (ADMMs) (Chang et al. 2019a; Fannjiang and Zhang 2020), and convex programming method (Ahmed et al. 2018). To further accelerate the first-order optimization, several second-order algorithms utilizing the Hessian have also been developed (Qian et al. 2014; Yeh et al. 2015; Ma et al. 2018; Gao and Xu 2017; Kandel et al. 2021). Moreover, the subspace methods (Xin et al. 2021) were successfully applied to the BPR as Thibault and Guizar-Sicairos (2012), Chang et al. (2019a), and Fung and Wendy (2020).

ic algorithms for BPR problem, so as to provide instructions for practical use and
draw attentions of applied mathematician for further improvement. The remainder
of the survey is organized as follows: Section "Mathematical Formula and Nonlinear
Optimization Model for BPR" gives the mathematical formula for BPR and
related nonlinear optimization models, as well as the closed-form expression of the
proximal mapping. Fast iterative algorithms are reviewed in Section "Fast Iterative
Algorithms". Section "Discussions" further discusses the experimental issues and
theoretical analysis. Section "Conclusions" summarizes this survey.

Mathematical Formula and Nonlinear Optimization Model for BPR

First, introduce the general nonblind PR problem in the discrete setting. By
introducing a linear operator $A \in \mathbb{C}^{m,n}$, for the sample of interest $u \in \mathbb{C}^n$,
experimental instruments usually collect the quadratic phaseless data $f \in \mathbb{R}^m$ as
below:

$$f = |Au|^2, \qquad (1)$$

in the ideal situation. However, noise contamination is evitable in practice (Chang
et al. 2018b) as

$$f_{\text{noise}} = \text{Poi}(|Au|^2), \qquad (2)$$

where Poi denotes the random variable following i.i.d Poisson distribution. See
more advanced models for practical noise as outliers and structured and randomly
distributed uncorrelated noise sources in Godard et al. (2012), Reinhardt et al.
(2017), Wang et al. (2017), Odstrčil et al. (2018), Chang et al. (2019b), and
references therein.

Mathematical Formula

State the BPR problem starting from convention ptychography (Rodenburg 2008),
since the principle of other BPR problems can be explained in a similar manner, all
of which can be unified as the blind recovery problem.

As shown in Fig. 1, a detector in the far field measures a series of phaseless
intensities, by letting a localized coherent X-ray probe w scan through the sample u.
Let the 2D image and the localized 2D probe denote as $u \in \mathbb{C}^n$ with $\sqrt{n} \times \sqrt{n}$ pixels
and $w \in \mathbb{C}^{\bar{m}}$ with $\sqrt{\bar{m}} \times \sqrt{\bar{m}}$ pixels, respectively. Here both the sample and the probe
are rewritten as vectors by a lexicographical order. Let $f_j^P \in \mathbb{R}_+^{\bar{m}}$ $\forall 0 \leq j \leq J-1$
denote the phaseless measurements satisfying

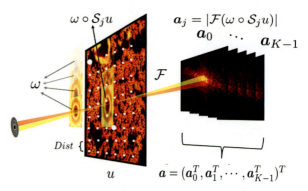

Fig. 1 Ptychographic PR (far field): A stack of phaseless data $f_j := a_j^2$ is collected, with w being the localized coherent probe and u being the image of interest (sample). The white dots represent the scanning lattice points, with $Dist$ denoting the sliding distance between centers of two adjacent frames

$$f_j^P = |\mathcal{F}(w \circ \mathcal{S}_j u)|^2, \qquad (3)$$

where the symbols $|\cdot|$, $(\cdot)^2$, and $\circ$ represent the element-wise absolute value and square of a vector and the element-wise multiplication of two vectors, respectively, the symbol $\mathcal{S}_j \in \mathbb{R}^{\bar{m} \times n}$ represents a matrix with binary elements extracting a patch (with the index j and size $\bar{m}$) from the entire sample, and the symbol $\mathcal{F}$ denotes the normalized discrete Fourier transformation (DFT). In practice, to get an accurate estimate of the probe, one has to solve a blind ptychographic PR problem. Note that the coherent CDI problem (Thibault and Guizar-Sicairos 2012) can be interpreted as a special blind ptychography problem with only one scanned frame ($J = 1$).

A recent super-resolution technique based on visible light called as the Fourier ptychography method (FP) has been developed by Zheng et al. (2013) and quickly spreads out for fruitful applications (Zheng et al. 2021). Letting w and u (here reuse the notations for simplicity) be the point spread function (PSF) of the imaging system and the sample of interest, the collected phaseless data f_j^{FP} of FP can be expressed as

$$f_j^{FP} = |\mathcal{F}^{-1}(\bar{w} \circ \mathcal{S}_j \bar{u})|^2 \text{ for } 0 \leq j \leq J-1$$

with $\bar{w} := \mathcal{F}w$ and $\bar{u} := \mathcal{F}u$.

Some similar problems dubbed as "convolutional PR" were recently studied (Qu et al. 2017, 2019; Ahmed et al. 2018). Given the sample u and the convolution kernel κ, the phaseless measurement f^{Cov} is given as

$$f^{Cov} = |\kappa \circledast u|^2 \qquad \text{Qu et al. (2017, 2019)}$$

or

$$f^{Cov} = |\mathcal{F}(\kappa \circledast u)|^2 \qquad \text{Ahmed et al. (2018)}, \qquad (4)$$

where the symbol $\circledast$ denotes the convolution.

Other interesting blind problem for full characterization of ultrashort optical pulses is to use frequency-resolved optical gating (FROG) (Trebino et al. 1997; Bendory et al. 2017; Kane and Vakhtin 2021). The phaseless measurement for a typical SHG-FROG can be obtained as

$$f_j^{\text{FROG}} = |\mathcal{F}(u \circ \mathcal{T}_j u)|^2,$$

where the symbol $\mathcal{T}_j$ denotes the translation. From the measurement $\{f_j^{\text{FROG}}\}_j$, one may also formulate it by BPR if assuming the element-wise multiplication for two independent variables.

All the mentioned problems can be unified as the BPR problem, i.e., to recover the probe (pupil, convolution kernel, or the signal itself) and the sample jointly. Essentially the relation between these two variables is bilinear. For conventional ptychography, the bilinear operators $\mathcal{A}: \mathbb{C}^{\tilde{m}} \times \mathbb{C}^n \to \mathbb{C}^m$ and $\mathcal{A}_j: \mathbb{C}^{\tilde{m}} \times \mathbb{C}^n \to \mathbb{C}^{\tilde{m}}\ \forall 0 \leq j \leq J-1$ are denoted as follows:

$$\mathcal{A}(w, u) := (\mathcal{A}_0^T(w, u), \mathcal{A}_1^T(w, u), \cdots, \mathcal{A}_{J-1}^T(w, u))^T, \quad (5)$$

with

$$\mathcal{A}_j(w, u) := \mathcal{F}(w \circ \mathcal{S}_j u)$$

and

$$f := (f_0^T, f_1^T, \cdots, f_{J-1}^T)^T \in \mathbb{R}_+^m.$$

Actually for all BPR problems, the bilinear operators can be unified as

$$\mathcal{A}_j(w, u) := \begin{cases} \mathcal{F}(w \circ \mathcal{S}_j u); & \text{Case I: CDI and ptychography} \\ \mathcal{F}^{-1}(\mathcal{F}w \circ \mathcal{S}_j(\mathcal{F}u)); & \text{Case II: Fourier ptychography} \\ \mathcal{F}(w \circ \mathcal{T}_j u); & \text{Case III: FROG} \\ w \circledast u, \text{ or } \mathcal{F}(w \circledast u); & \text{Case IV: Convolution PR} \end{cases} \quad (6)$$

where there are totally one frame as $J = 1$ for the last case for convolution PR. Hence, by introducing the general bilinear operator $\mathcal{A}(\cdot, \cdot)$, the BPR can be given below:

BPR: To find the "probe" w and the sample u, s.t. $|\mathcal{A}(w, u)|^2 = f$, (7)

where $\mathcal{A}$ is denoted as (5) and (6) and the per frame of phaseless measurements f_j represents the measurement from four cases. Note that the BPR problem is not limited to the cases with forward propagation as (6).

Denote two linear operators A_w, A_u as below:

$$A_w u = \mathcal{A}(w, u) \forall u;$$
$$A_u w = \mathcal{A}(w, u) \forall w; \qquad (8)$$

Then one can obtain the conjugate operators

$$A_w^* z = \begin{cases} \sum_j \mathcal{S}_j^T (\text{conj}(w) \circ \mathcal{F}^{-1} z_j); & \text{Case I} \\ \mathcal{F}^{-1} \sum_j \mathcal{S}_j^T (\text{conj}(\mathcal{F} w) \circ \mathcal{F} z_j); & \text{Case II} \\ \sum_j \mathcal{T}_j^T (\text{conj}(w) \circ \mathcal{F}^{-1} z_j); & \text{Case III} \\ \text{conj}(w) \circledast z, \text{ or } \text{conj}(w) \circledast \mathcal{F}^{-1} z; & \text{Case IV} \end{cases} \qquad (9)$$

and

$$A_u^* z = \begin{cases} \sum_j (\text{conj}(\mathcal{S}_j u) \circ \mathcal{F}^{-1} z_j); & \text{Case I} \\ \mathcal{F}^{-1} \sum_j (\text{conj}(\mathcal{S}_j \mathcal{F} u) \circ \mathcal{F} z_j); & \text{Case II} \\ \sum_j (\text{conj}(\mathcal{T}_j u) \circ \mathcal{F}^{-1} z_j); & \text{Case III} \\ \text{conj}(u) \circledast z, \text{ or } \text{conj}(u) \circledast \mathcal{F}^{-1} z; & \text{Case IV} \end{cases} \qquad (10)$$

$\forall z = (z_1^T, z_2^T, \cdots, z_{J-1}^T)^T \in \mathbb{C}^m$. Here $\sum_j$ is a simplified form of $\sum_{j=0}^{J-1}$. Consequently, one obtains

$$A_w^* A_w u = \begin{cases} \left(\sum_j \mathcal{S}_j^T |w|^2 \right) \circ u; & \text{Case I} \\ \mathcal{F}^{-1} \left(\left(\sum_j \mathcal{S}_j^T |\mathcal{F} w|^2 \right) \circ \mathcal{F} u \right); & \text{Case II} \\ \left(\sum_j \mathcal{T}_j^T |w|^2 \right) \circ u; & \text{Case III} \\ \text{conj}(w) \circledast w \circledast u; & \text{Case IV} \end{cases} \qquad (11)$$

and

$$\mathcal{A}_u^* \mathcal{A}_u w = \begin{cases} \left(\sum_j \mathcal{S}_j |u|^2\right) \circ w; & \text{Case I} \\ \mathcal{F}^{-1}\left(\left(\sum_j \mathcal{S}_j |\mathcal{F}u|^2\right) \circ \mathcal{F}w\right); & \text{Case II} \\ \left(\sum_j \mathcal{T}_j |u|^2\right) \circ w; & \text{Case III} \\ \text{conj}(u) \circledast u \circledast w. & \text{Case IV} \end{cases} \quad (12)$$

Optimization Problems and Proximal Mapping

Solving a nonblind problem may be NP hard if knowing w or u in advance. Other than directly solving equations as (7), one can solve the following nonlinear optimization problems in order to determine the underlying image u and probe w from noisy measurements f:

$$\min_{w,u} \mathcal{M}(|\mathcal{A}(w,u)|^2, f), \quad (13)$$

where the symbol $\mathcal{M}(\cdot,\cdot)$ represents the error between the unknown intensity $|\mathcal{A}(w,u)|^2$ and collected phaseless data f. Various metrics proposed under different noise settings include amplitude-based metric for Gaussian measurements (AGM) (Wen et al. 2012; Chang et al. 2016), intensity-based metric for Poisson measurements (IPM) (Thibault and Guizar-Sicairos 2012; Chen and Candes 2015; Chang et al. 2018b), and intensity-based metric for Gaussian measurements (IGM) (Qian et al. 2014; Candes et al. 2015; Sun et al. 2016), all of which can be expressed as

$$\mathcal{M}(g,f) := \begin{cases} \frac{1}{2}\|\sqrt{g} - \sqrt{f}\|^2; & \text{(AGM)} \\ \frac{1}{2}\langle g - f \circ \log(g), \mathbf{1}\rangle; & \text{(IPM)} \\ \frac{1}{2}\|g - f\|^2; & \text{(IGM)} \end{cases} \quad (14)$$

where the operations on vectors such as $\sqrt{\cdot}, \log(\cdot), |\cdot|, (\cdot)^2$ are all defined pointwisely in this survey, $\mathbf{1}$ denotes a vector whose entries all equal to ones, and $\|\cdot\|$ denotes the ℓ^2 norm in Euclidean space.

The proximal mapping for functions defined on complex Euclidean space is introduced below.

Definition 1. Given function $h : \mathbb{C}^N \to \mathbb{R} \bigcup \{+\infty\}$, the proximal mapping $\mathrm{Prox}_{h;\mu} : \mathbb{C}^N \to \mathbb{C}^N$ of h is defined by

$$\mathrm{Prox}_{h;\beta}(v) = \arg\min_{x}\left(h(x) + \frac{\beta}{2}\|x-v\|^2 \right), \tag{15}$$

with the symbol $\|\cdot\|$ denoted as the ℓ^2 norm of a complex vector on complex Euclidean space (use the same notation for real and complex spaces).

Namely, the proximal operator for the function $\mathcal{M}(|\cdot|^2, f)$ defined in (14) has a closed-form formula (Chang et al. 2018c) as below:

$$\mathrm{Prox}_{\mathcal{M}(|\cdot|^2, f);\beta}(z) = \begin{cases} \dfrac{\sqrt{f} + \beta|z|}{1+\beta} \circ \mathrm{sign}(z), & \text{for AGM}; \\[2mm] \dfrac{\beta|z| + \sqrt{(\beta|z|)^2 + 4(1+\beta)f}}{2(1+\beta)} \circ \mathrm{sign}(z), & \text{for IPM}; \\[2mm] \varpi_{\beta}(|z|) \circ \mathrm{sign}(z), & \text{for IGM}; \end{cases} \tag{16}$$

where $\forall z \in \mathbb{C}^m$, $(\mathrm{sign}(z))(t) := \mathrm{sign}(z(t))\ \forall\ 0 \leq t \leq m-1$, $\mathrm{sign}(x)$ for a scalar $x \in \mathbb{C}$ is denoted as $\mathrm{sign}(x) = \frac{x}{|x|}$ if $x \neq 0$, otherwise $\mathrm{sign}(0) := c$ with an arbitrary constant $c \in \mathbb{C}$ with unity length, and

$$\varpi_{\beta}(|z|)(t) = \begin{cases} \sqrt[3]{\frac{\beta|z(t)|}{4} + \sqrt{D(t)}} + \sqrt[3]{\frac{\beta|z(t)|}{4} - \sqrt{D(t)}}, & \text{if } D(t) \geq 0; \\[2mm] 2\sqrt{\frac{f(t) - \frac{\beta}{2}}{3}} \cos\left(\arccos \frac{\theta(t)}{3}\right), & \text{otherwise}, \end{cases} \tag{17}$$

for $0 \leq t \leq m-1$, with $D(t) = \dfrac{(\frac{\beta}{2}-f(t))^3}{27} + \dfrac{\beta^2|z(t)|^2}{16}$, and $\theta(t) = \dfrac{\beta|z(t)|}{4\sqrt{\frac{(f(t)-\frac{\beta}{2})^3}{27}}}$.

Note that the alternating direction method of multipliers (ADMM) was adopted in Wen et al. (2012) and Chang et al. (2016, 2018b) to solve the variational PR model in (13). However, due to the lack of the globally Lipschitz differentiable terms in the objective function, it seems difficult to guarantee its convergence. Some other variants of the metric have been recently proposed, such as the penalized metrics $\mathcal{M}(|\cdot|^2 + \epsilon \mathbf{1}, f + \epsilon \mathbf{1})$ by adding a small positive scalar ϵ as Guizar-Sicairos and Fienup (2008), Chang et al. (2019a), and Gao et al. (2020). Although it has simple form, the technique will make the related proximal mapping not have closed-form expression, such that additional computation cost as an inner loop may have to be introduced (Chang et al. 2019a). By cutting off the AGM near the origin, and then adding back a smooth function, one can keep the global minimizer unchanged. Hence, a novel smooth truncated AGM (ST-AGM) $\mathcal{G}_{\epsilon}(\cdot; f)$ with truncation parameter $\epsilon > 0$ (Chang et al. 2021) was designed below:

$$\mathcal{M}_\epsilon(z, f) := \sum_j M_\epsilon(z(j), f(j)), \tag{18}$$

where $\forall\ x \in \mathbb{C}, b \in \mathbb{R}^+,$

$$M_\epsilon(x, b) := \begin{cases} \dfrac{1-\epsilon}{2}\left(b - \frac{1}{\epsilon}|x|^2\right), & \text{if } |x| < \epsilon\sqrt{b}; \\ \dfrac{1}{2}\big||x| - \sqrt{b}\big|^2, & \text{otherwise}. \end{cases} \tag{19}$$

Readily its closed form of the corresponding proximal mapping can be found in Chang et al. (2021). More other elaborate metrics can be found in Luke (2005), Cai et al. (2021), and references therein.

Fast Iterative Algorithms

In this section, the main iterative algorithms for BPR will be introduced. Note that each algorithm may be designed originally for a specific case of (6). Hence, the basic idea based on the original case will be explained first, and the possible extensions to other cases will be discussed then.

Alternating Projection (AP) Algorithms

First consider BPR defined in (7) in the case of convention ptychography.
Given the exit wave in the far field $\Psi := (\Psi_0^T, \Psi_1^T, \cdots, \Psi_{J-1}^T)^T \in \mathbb{C}^m$, with

$$\Psi_j := \mathcal{F}(w \circ \mathcal{S}_j u)\ \forall 0 \leq j \leq J-1,$$

the optimal exit wave $\Psi^\star$ lies in the intersection of two following sets, i.e.,

$$\Psi^\star \in \widehat{\mathcal{X}_1} \bigcap \widehat{\mathcal{X}_2},$$

with

$$\widehat{\mathcal{X}_1} := \{\Psi := (\Psi_0^T, \Psi_1^T, \cdots, \Psi_{J-1}^T)^T \in \mathbb{C}^m : |\Psi_j| = \sqrt{f_j}\ \forall 0 \leq j \leq J-1\},$$
$$\widehat{\mathcal{X}_2} := \{\Psi \in \mathbb{C}^m : \exists w \in \mathbb{C}^{\bar{m}}, u \in \mathbb{C}^n, s.t.\ w \circ \mathcal{S}_j u = \mathcal{F}^{-1}\Psi_j\ \forall 0 \leq j \leq J-1\}. \tag{20}$$

The AP algorithm determining this intersection alternatively calculates the projections onto these two sets $\widehat{\mathcal{X}_1}$ and $\widehat{\mathcal{X}_2}$. Regarding the projection onto $\widehat{\mathcal{X}_1}$ as

$$\widehat{\mathcal{P}_1}(\Psi) := \arg\min_{\widehat{\Psi} \in \widehat{\mathcal{X}_1}} \|\widehat{\Psi} - \Psi\|^2,$$

one readily gets a closed-form solution

$$\widehat{\mathcal{P}}_1(\Psi) := \left((\widehat{\mathcal{P}}_1^0(\Psi))^T, \cdots, (\widehat{\mathcal{P}}_1^{J-1}(\Psi))^T\right)^T,$$

with

$$\widehat{\mathcal{P}}_1^j(\Psi) = \sqrt{f_j} \circ \text{sign}(\Psi_j) \; 0 \leq j \leq J-1.$$

For the projection onto $\widehat{\mathcal{X}_2}$, given Ψ^k as the solution in the kth iteration, one gets

$$\widehat{\mathcal{P}}_2(\Psi^k) := \left((\mathcal{F}(w^{k+1} \circ \mathcal{S}_0 u^{k+1}))^T, (\mathcal{F}(w^{k+1} \circ \mathcal{S}_1 u^{k+1}))^T, \cdots,\right.$$
$$\left.(\mathcal{F}(w^{k+1} \circ \mathcal{S}_{J-1} u^{k+1}))^T\right)^T,$$

where

$$(w^{k+1}, u^{k+1}) = \arg\min_{w,u} F(w, u, \Psi^k) := \tfrac{1}{2} \sum_j \|\mathcal{F}^{-1}\Psi_j^k - w \circ \mathcal{S}_j u\|^2. \qquad (21)$$

Unfortunately, it does not have a closed-form solution. One can solve (21) by alternating minimization (with T steps) as below:

$$w_{l+1} = \arg\min_w F(w, u_l, \Psi^k),$$
$$u_{l+1} = \arg\min_u F(w_{l+1}, u, \Psi^k) \; \forall l = 0, 1, \cdots, T-1. \qquad (22)$$

Readily one has

$$w_{l+1} \approx \frac{\sum_j \text{conj}(\mathcal{S}_j u_l) \circ \mathcal{F}^{-1}\Psi_j^k}{\sum_j |\mathcal{S}_j u_l|^2 + \bar{\alpha}_1};$$

$$u_{l+1} \approx \frac{\sum_j \mathcal{S}_j^T (\text{conj}(w_{l+1}) \circ \mathcal{F}^{-1}\Psi_j^k)}{\sum_j (\mathcal{S}_j^T |w_{l+1}|^2) + \bar{\alpha}_2} \; \forall l = 0, 1, \cdots, T-1, \qquad (23)$$

where the parameters $0 < \bar{\alpha}_1, \bar{\alpha}_2 \ll 1$ are introduced in order to avoid dividing by zeros.

Letting Ψ^k be iterative solution in the kth iteration, the standard AP for BPR can be directly given as below:

(1) Compute $\widehat{\Psi}^k$ by $\widehat{\Psi}_j^k = \mathcal{F}(w^{k+1} \circ \mathcal{S}_j u^{k+1})$, where the pair (w^{k+1}, u^{k+1}) is approximately solved by (23).
(2) Compute Ψ^{k+1} by $\Psi^{k+1} = \widehat{\mathcal{P}}_1(\widehat{\Psi}^k)$.

The DR algorithm for BPR can be formulated in two steps (Thibault et al. 2009), as follows:

4 Fast Iterative Algorithms for Blind Phase Retrieval: A Survey

(1) Compute $\widehat{\Psi}^k$ as the first step of AP.
(2) Compute Ψ^{k+1} by

$$\Psi^{k+1} = \Psi^k + \widehat{\mathcal{P}}_1(2\widehat{\Psi}^k - \Psi^k) - \widehat{\Psi}^k. \tag{24}$$

Note that the formula (24) utilizing DR operator is essentially Fienup's hybrid input–output map, which can also be derived with proper parameters from difference map (Elser 2003).

Since the fixed point of DR iteration may not exist, Marchesini et al. (2016) adopted the relaxed version of DR (dubbed as RAAR by Luke 2005) to further improve the stability of the reconstruction from noisy measurements, which simply takes weighted average of right term of (24) and $\widehat{\Psi}^k$ with a tunable parameter $\delta \in (0, 1)$ as

$$\Psi^{k+1} = \delta\big(\Psi^k + \widehat{\mathcal{P}}_1(2\widehat{\Psi}^k - \Psi^k) - \widehat{\Psi}^k\big) + (1-\delta)\widehat{\Psi}^k,$$

with $\widehat{\Psi}^k$ determined in a same manner as the first step of AP.

At the end of this part, extension of AP to general BPR problems will be discussed. Similarly as for the ptychography, introduce Ψ as

$$\Psi = \mathcal{A}(w, u), \text{ and } \Psi_j = \mathcal{A}_j(w, u).$$

In the same manner, one can define two constraint sets and establish the AP algorithms for the four cases of BPR. The only differences lie in the calculations of the projections onto the bilinear constraint set. As (21), consider

$$\min_{w,u} \|\Psi^k - \mathcal{A}(w, u)\| \tag{25}$$

by alternating minimization, where Ψ^k is the iterative solution. Then the scheme is given below:

$$w_{l+1} \approx (A_{u_l}^* A_{u_l} + \bar{\alpha}_1 \mathbf{I})^{-1} A_{u_l}^* \Psi^k,$$

$$u_{l+1} \approx (A_{w_{l+1}}^* A_{w_{l+1}} + \bar{\alpha}_2 \mathbf{I})^{-1} A_{w_{l+1}}^* \Psi^k \; \forall l = 0, 1, \cdots, T-1. \tag{26}$$

The detailed forms of these operators can be found in (9), (10), (11), and (12). Notably the inverse in (26) can be efficiently solved by pointwise division or DFT.

ePIE-Type Algorithms

This iterative algorithm can be expressed as an AP method for convention ptychography as follows: To find $\Psi_{n_k}^\star$ belonging to the intersection as

$$\Psi_{n_k}^\star \in \{|\Psi_{n_k}| = \sqrt{f_{n_k}}\} \cap \{\Psi_{n_k} : \exists w \in \mathbb{C}^{\bar{m}}, u \in \mathbb{C}^n, \quad s.t. \quad w \circ \mathcal{S}_{n_k} u = \mathcal{F}^{-1} \Psi_{n_k}\},$$

with a random frame index n_k. Let w^k, u^k be the iterative solutions in the kth iteration. By first computing the projection of $\psi_{n_k}^k := \mathcal{F}(w^k \circ \mathcal{S}_{n_k} u^k)$ by $\widehat{\mathcal{P}}_1^{n_k}(\psi_{n_k}^k)$, and then updating w^{k+1} and u^{k+1} by the gradient descent algorithm (inexact projection) for (21), the ePIE algorithm proposed by Maiden and Rodenburg (2009) can be expressed by updating w^{k+1} and u^{k+1} in parallel as

$$\begin{cases} w^{k+1} = w^k - \dfrac{d_2}{\|\mathcal{S}_{n_k} u^k\|_\infty^2} \mathcal{S}_{n_k} \mathrm{conj}(u^k) \circ \mathcal{F}^{-1}(\Psi_{n_k}^k - \widehat{\mathcal{P}}_1^{n_k}(\Psi_{n_k}^k)) \\ u^{k+1} = u^k - \dfrac{d_1}{\|\mathcal{S}_{n_k}^T w^k\|_\infty^2} \mathcal{S}_{n_k}^T \left(\mathrm{conj}(w^k) \circ \mathcal{F}^{-1}(\Psi_{n_k}^k - \widehat{\mathcal{P}}_1^{n_k}(\Psi_{n_k}^k)) \right), \end{cases} \quad (27)$$

with frame index $n_k \in \{0, 1, \cdots, J-1\}$ generated randomly and positive parameters d_1 and d_2 (default values are ones) and $\|w\|_\infty := \max_t |w(t)|$.

The regularized PIE (rPIE) was further proposed by Maiden et al. (2017) as

$$\begin{cases} w^{k+1} = w^k - \dfrac{1}{\delta\|\mathcal{S}_{n_k} u^k\|_\infty^2 + (1-\delta)\mathcal{S}_{n_k}|u^k|^2} \\ \qquad \circ \mathcal{S}_{n_k} \mathrm{conj}(u^k) \circ \mathcal{F}^{-1}(\Psi_{n_k}^k - \widehat{\mathcal{P}}_1^{n_k}(\Psi_{n_k}^k)), \\ u^{k+1} = u^k - \dfrac{1}{\delta\|\mathcal{S}_{n_k}^T w^k\|_\infty^2 + (1-\delta)\mathcal{S}_{n_k}^T|w^k|^2} \\ \qquad \circ \mathcal{S}_{n_k}^T \left(\mathrm{conj}(w^k) \circ \mathcal{F}^{-1}(\Psi_{n_k}^k - \widehat{\mathcal{P}}_1^{n_k}(\Psi_{n_k}^k)) \right), \end{cases} \quad (28)$$

with the scalar constant $\delta \in (0, 1)$. It can be interpreted as a hybrid scheme for the stepsize of gradient descent, which takes the weighted average of the denominator of the ePIE scheme (27) and first term in the denominator of AP scheme (23). The rPIE algorithm was further accelerated by momentum (Maiden et al. 2017).

One can directly get the ePIE and rPIE schemes for FP (Zheng et al. 2021) by replacing the variables w and u by $\mathcal{F}w$ and $\mathcal{F}u$. The ePIE-type algorithms are very popular in optics community, since it is enough to implement the algorithm if one knows how to calculate the gradient of the objective functions, and the memory footprint is much smaller than more advanced AP algorithm including DR and RAAR. However, it tends to unstable when the data redundancy is insufficient (e.g., noisy data, big-step scan) as reported in Chang et al. (2019a). Moreover, the theoretical convergence is unknown and seems challenging due to the relation with nonsmooth objective functions.

Note that if with totally $J = 1$ frame as CDI, the differences between the ePIE (with $d_1 = d_2 = 1$) and standard AP lie in the preconditioning matrices: AP utilizes the spatial weighted diagonal matrices $A_u^* A_u$ and $A_w^* A_w$, while ePIE utilizes the spatial-independent constant determined by the maximum of their diagonal matrices.

Proximal Algorithms

Proximal Heterogeneous Block Implicit-Explicit (PHeBIE) For convention ptychography, consider an optimization problem (to get rid of introducing redundant notations in this survey, slightly modify the constraint set of Ψ in Hesse et al. (2015) as $\widehat{\mathscr{X}_1}$ and adjust the notation of the first term of the following model accordingly in order to present an equivalent form) (Hesse et al. 2015) as follows:

$$\min_{w,u,\Psi} F(w, u, \Psi) + \mathbb{I}_{\widehat{\mathscr{X}_1}}(\Psi) + \mathbb{I}_{\mathscr{X}_1}(w) + \mathbb{I}_{\mathscr{X}_2}(u), \tag{29}$$

with $F(w, u, \Psi)$ and $\widehat{\mathscr{X}_1}$ denoted in (21) and (20), respectively, and the indicator function $\mathbb{I}_{\mathscr{X}}$ denoted as

$$\mathbb{I}_{\mathscr{X}}(\Psi) := \begin{cases} 0, & \text{if } \Psi \in \mathscr{X}, \\ +\infty, & \text{otherwise,} \end{cases}$$

where the amplitude constraints of the probe and image are incorporated (in Hesse et al. (2015), the authors further considered the compact support condition of the probes and image), where

$$\begin{aligned} \mathscr{X}_1 &:= \{w \in \mathbb{C}^{\bar{m}} : \|w\|_\infty \leq C_w\}; \\ \mathscr{X}_2 &:= \{u \in \mathbb{C}^n : \|u\|_\infty \leq C_u\} \end{aligned} \tag{30}$$

with two positive constants C_w, C_u. The projection operator onto $\mathscr{X}_1$ is readily obtained as

$$\text{Proj}(w; C_w) := \min\{C_w, |w|\} \circ \text{sign}(w) \; \forall w,$$

which is the closed-form expression for the minimizer to the problem

$$\min_{\|\tilde{w}\|_\infty \leq C_w} \tfrac{1}{2}\|\tilde{w} - w\|^2.$$

Similarly, one gets the projection onto $\mathscr{X}_2$ as $\text{Proj}(u; C_u)$.

Hesse et al. (2015) further adopted the proximal alternating linearized minimization (PALM) method (Bolte et al. 2014) for the BPR problem in the case of convention ptychography, such that the proximal heterogeneous block implicit-explicit (PHeBIE) (see Hesse et al. 2015, Algorithm 2.1) consists of two steps with three positive parameters d_1, d_2, and γ:

(1) Calculate w^{k+1}, u^{k+1} sequentially as

$$\begin{cases} w^{k+1} = \text{Proj}\left(w^k - \frac{1}{d_1^k} \sum_j \mathcal{S}_j \text{conj}(u^k) \circ \mathcal{F}^{-1}(w^k \circ \mathcal{S}_j u^k - \Psi_j^k); C_w\right); \\ u^{k+1} = \text{Proj}\left(u^k - \frac{1}{d_2^k} \sum_j \mathcal{S}_j^T (\text{conj}(w^{k+1}) \circ \mathcal{F}^{-1}(w^{k+1} \circ \mathcal{S}_j u^k - \Psi_j^k)), C_u\right), \end{cases} \quad (31)$$

with $d_1^k := d_1 \| \sum_j |\mathcal{S}_j u^k| \|_\infty^2$, $d_2^k := d_2 \| \sum_j \mathcal{S}_j^T |w^{k+1}| \|_\infty^2$.

(2) Calculate Ψ^{k+1} by

$$\Psi^{k+1} = \widehat{\mathcal{P}}_1\left(\frac{1}{1+\gamma}(\widehat{\Psi}^{k+1} + \gamma \Psi^k)\right),$$

with

$$\widehat{\Psi}_j^{k+1} := \mathcal{F}(w^{k+1} \circ \mathcal{S}_j u^{k+1}).$$

Under the assumption of boundedness of iterative sequences, the convergence of PALM to stationary points of (29) was proved (Hesse et al. 2015). To the knowledge, it is the first iterative algorithm with convergence guarantee for BPR problem. The PHeBIE has multiple similarities with ePIE. The main differences between them are that, for ePIE, only the gradient of F w.r.t. a randomly selected single frame is adopted to update w and u per outer loop as (27), while, for PHeBIE, each block of w and u can be updated in parallel by employing the gradient as (31) w.r.t. all adjacent frames. Therefore, PHeBIE is more stable than ePIE numerically.

One readily knows that the convergence rate relies on the Lipschitz constant of partial derivative of F. In order to get smaller constant, a direct way is to employ the derivative of a small block for unknowns. Hence, based on the partition of the sample and the probe, the parallel version of PHeBIE was also provided (Hesse et al. 2015) with convergence guarantee.

For more extensions to other cases of BPR, one can introduce a generalized nonlinear optimization model:

$$\min_{w,u,\Psi} \|\Psi - \mathcal{A}(w,u)\|^2 + \mathbb{I}_{\widehat{\mathscr{P}_1}}(\Psi) + \mathbb{I}_{\mathscr{P}_1}(w) + \mathbb{I}_{\mathscr{P}_2}(u),$$

where one adopts the same form as (25) for the first term. The detailed algorithms are omitted, since one only needs to update the gradient of first term following (9), (10), (11), and (12).

Variant of Proximal Algorithm Here introduce a general constraint set for the bilinear relation as

$$X := \{\Psi \in \mathbb{C}^m : \exists w \in \mathbb{C}^{\bar{m}}, u \in \mathbb{C}^n, s.t. \ \mathcal{A}(w,u) = \Psi\}. \quad (32)$$

Consider the optimization problem as

$$\min_z \mathbb{I}_{\widehat{\mathcal{X}_1}}(z) + \mathbb{I}_X(z). \tag{33}$$

By replacing the indicator function by the metrics, and further combining the alternating minimization with proximal algorithms, Han Yan (2020) derived a new proximal algorithm for the convention ptychography problem. Specifically, the proposed algorithm with a generalized form for BPR has the following steps:

Step 1: $z^{k+1} = \arg\min_z \mathcal{M}(|z|^2; f) + \frac{\beta}{2}\|z - \mathcal{A}(w^k, u^k)\|^2;$

Step 2: $w^{k+1} = \arg\min_w \|z^{k+1} - A_{u^k}w)\|^2.$

Step 3: $u^{k+1} = \arg\min_u \|z^{k+1} - A_{w^{k+1}}u)\|^2. \tag{34}$

Here the last two steps can be solved in a same manner as (26). The above algorithm has deep connections with the ADMM (Chang et al. 2019a). If removing the constraint of boundedness of two variables, and setting the penalization parameter to zero in (35), then by solving the constraint problem (37) by adding a penalization term $\|z - \mathcal{A}(w, u)\|^2$ without introducing the multiplier Λ, one can get exactly the same iterative scheme as (34). Besides, it was further improved by accelerated proximal gradient method in Yan (2020) and recently by stochastic gradient descent (Huang et al. 2021) for FP.

ADMM

As a typical operator-splitting algorithm, ADMM is very flexible and successfully applied to inverse and imaging problems (Wu and Tai 2010; Boyd et al. 2011), which is also adopted for classical and ptychographic PR problems (Wen et al. 2012; Chang et al. 2019a).

Consider the metrics using penalized-AGM (pAGM) and penalized-IPM (pIPM) as to measure the error of recovered intensity and the targets. A nonlinear optimization model (Chang et al. 2019a) was given as

$$\min_{w\in\mathbb{C}^{\bar{m}}, u\in\mathbb{C}^n} \mathcal{G}(\mathcal{A}(w, u)) + \mathbb{I}_{\mathcal{X}_1}(w) + \mathbb{I}_{\mathcal{X}_2}(u), \tag{35}$$

with $\mathcal{G}(z) := \mathcal{M}(|z|^2 + \epsilon\mathbf{1}, f + \epsilon\mathbf{1})$ and the constraint sets defined in (30). The authors further leveraged the additional data $c \in \mathbb{R}_+^{\bar{m}}$ to eliminate structural artifacts caused by grid scan and therefore obtained the following variant:

$$\min_{w\in\mathbb{C}^{\bar{m}}, u\in\mathbb{C}^n} \mathcal{G}(\mathcal{A}(w, u)) + \mathbb{I}_{\mathcal{X}_1}(w) + \mathbb{I}_{\mathcal{X}_2}(u) + \tau\widehat{\mathcal{G}}(\mathcal{F}w), \tag{36}$$

where τ is a positive parameter, the additional measurement c is the diffraction pattern (absolute value of Fourier transform of the probe) as $c := |\mathcal{F}u|$, and $\widehat{\mathcal{G}}(z) := \mathcal{B}(|z|^2 + \epsilon\mathbf{1}, c^2 + \epsilon\mathbf{1})$. For simplicity, assume that $\mathcal{G}$ and $\widehat{\mathcal{G}}$ adopt the same metric.

As the procedures for solving the two above models are quite similar using ADMM, only details for solving the first optimization model (35) are given below. By introducing an auxiliary variable $z = \mathcal{A}(w, u) \in \mathbb{C}^m$, an equivalent form of (35) is formulated below:

$$\min_{w,u,z} \mathcal{G}(z) + \mathbb{I}_{\mathscr{X}_1}(w) + \mathbb{I}_{\mathscr{X}_2}(u), \ s.t. \ z - \mathcal{A}(w, u) = 0. \tag{37}$$

The corresponding augmented Lagrangian reads

$$\Upsilon_\beta(w, u, z, \Lambda) := \mathcal{G}(z) + \mathbb{I}_{\mathscr{X}_1}(w) + \mathbb{I}_{\mathscr{X}_2}(u) + \Re(\langle z - \mathcal{A}(w, u), \Lambda \rangle) \\ + \frac{\beta}{2} \|z - \mathcal{A}(w, u)\|^2, \tag{38}$$

with the multiplier $\Lambda \in \mathbb{C}^m$ and a positive parameter β, where $\Re(\cdot)$ denotes the real part of a complex number. Then one considers the following problem:

$$\max_{\Lambda} \min_{w,u,z} \Upsilon_\beta(w, u, z, \Lambda). \tag{39}$$

Given the approximated solution $(w^k, u^k, z^k, \Lambda^k)$ in the kth iteration, the four-step iteration by the generalized ADMM (only the subproblems w.r.t. w or u have proximal terms) is given as follows:

$$\begin{cases} \text{Step 1:} & w^{k+1} = \arg\min_{w} \Upsilon_\beta(w, u^k, z^k, \Lambda^k) + \frac{\alpha_1}{2} \|w - w^k\|_{M_1^k}^2 \\ \text{Step 2:} & u^{k+1} = \arg\min_{u} \Upsilon_\beta(w^{k+1}, u, z^k, \Lambda^k) + \frac{\alpha_2}{2} \|u - u^k\|_{M_2^k}^2 \\ \text{Step 3:} & z^{k+1} = \arg\min_{z} \Upsilon_\beta(w^{k+1}, u^{k+1}, z, \Lambda^k) \\ \text{Step 4:} & \Lambda^{k+1} = \Lambda^k + \beta(z^{k+1} - \mathcal{A}(w^{k+1}, u^{k+1})), \end{cases} \tag{40}$$

with diagonal positive semidefinite matrices $M_1^k \in \mathbb{R}_+^{\bar{m} \times \bar{m}}$ and $M_2^k \in \mathbb{R}_+^{n \times n}$ and two penalization parameters $\alpha_1, \alpha_2 > 0$, where $\|w\|_{M_1^k}^2 := \langle M_1^k w, w \rangle$ and $\|u\|_{M_2^k}^2 := \langle M_2^k u, u \rangle$.

Detailed algorithms will be given focusing on convention ptychography as Chang et al. (2019a). Note that these two matrices M_1^k, M_2^k are assumed to be diagonal so that subproblems in Step 1 and Step 2 have closed-form solutions. Roughly speaking, based on splitting technique of proximal ADMM, subproblems of u, w, and z are element-wise optimization problems with closed-form solutions, such that each subproblem can be fast solved. In practice, these two matrices are chosen by hand, and an adaptive strategy was presented in Chang et al. (2019a) in order to guarantee the convergence. Letting

$$\hat{z}^k := z^k + \frac{\Lambda^k}{\beta}$$

and the diagonal matrices M_1^k and M_2^k satisfy

$$\begin{cases} \min_t \sum_j \left|\left(\mathcal{S}_j u^k\right)(t)\right|^2 + \frac{\alpha_1}{\beta}\mathrm{diag}(M_1^k)(t) > 0, \\ \min_t \sum_j \left|\left(\mathcal{S}_j^T w^{k+1}\right)(t)\right|^2 + \frac{\alpha_2}{\beta}\mathrm{diag}(M_2^k)(t) > 0, \end{cases} \quad (41)$$

the closed-form solutions of Step 1 and Step 2 are given as

$$\begin{cases} w^{k+1} = \mathrm{Proj}\left(\frac{\beta \sum_j \mathrm{conj}(\mathcal{S}_j u^k)\circ(\mathcal{F}^{-1}\hat{z}_j^k)+\alpha_1\mathrm{diag}(M_1^k)\circ w^k}{\beta \sum_j |\mathcal{S}_j u^k|^2+\alpha_1\mathrm{diag}(M_1^k)}; C_w\right); \\ u^{k+1} = \mathrm{Proj}\left(\frac{\beta \sum_j \mathcal{S}_j^T(\mathrm{conj}(w^{k+1})\circ\mathcal{F}^{-1}\hat{z}_j^k)+\alpha_2\mathrm{diag}(M_2^k)\circ u^k}{\beta \sum_j (\mathcal{S}_j^T|w^{k+1}|^2)+\alpha_2\mathrm{diag}(M_2^k)}; C_u\right). \end{cases} \quad (42)$$

For Step 3, denoting

$$z^+ = \mathcal{A}(w^{k+1}, u^{k+1}) - \frac{\Lambda^k}{\beta},$$

one has

$$z^{k+1} = \arg\min_z \frac{1}{2}\langle |z|^2 + \varepsilon\mathbf{1}_m - (f + \varepsilon\mathbf{1}_m)\circ\log(|z|^2 + \varepsilon\mathbf{1}_m), \mathbf{1}_m\rangle + \frac{\beta}{2}\|z - z^+\|^2.$$

The solution can be expressed as

$$z^{k+1} = \rho^\star \circ \mathrm{sign}(z^+), \quad (43)$$

where $\rho^\star(t)$ was solved by the gradient projection scheme expressed as

$$x_{l+1} = \max\left\{0, x_l - \delta\left((1 + \beta - \frac{f(t)+\varepsilon}{|x_l|^2+\varepsilon})x_l - \beta z^+(t)\right)\right\}, \forall l = 0, 1, \ldots, \quad (44)$$

if using the pIPM, or

$$x_{l+1} = \max\left\{0, x_l - \delta\left((1 + \beta - \frac{\sqrt{f(t)+\varepsilon}}{\sqrt{|x_l|^2+\varepsilon}})x_l - \beta z^+(t)\right)\right\}, \forall l = 0, 1, \ldots, \quad (45)$$

if using the pAGM with the stepsize $\delta > 0$, and $x_0 := |z^k(t)|$. Note that with the penalization parameter $\varepsilon = 0$, one can directly get the closed-form solution by (16) as Wen et al. (2012) and Chang et al. (2018b).

Under the condition of sufficient overlapping scan, and bounded preconditioning matrices M_1^k and M_2^k, the convergence of the ADMM can be derived on the sense that the iterative sequence generated by above algorithm converges to a stationary point of the augmented Lagrangian by letting the parameter β sufficiently large.

From the point of view of fixed point analysis, for nonblind problems (knowing the probe w), the authors Fannjiang and Zhang (2020) presented a variant ADMM to solve the following optimization problem:

$$\min_z \mathcal{M}(|z|^2; f) + \mathbb{I}_X(z), \tag{46}$$

with $X \subset \mathbb{C}^m$ defined in (32). By introducing the auxiliary variable $\bar{z} = z$ and decomposing the objective functions, the ADMM was proposed in Fannjiang and Zhang (2020) to solve

$$\min_{z,\bar{z}} \mathcal{M}(|z|^2; f) + \mathbb{I}_X(\bar{z}), \ s.t. \ z - \bar{z} = 0. \tag{47}$$

To further apply the idea to the BPR, alternating minimization was further adopted as

$$\begin{aligned} z^{k+1/2} &:= \arg\min_z \mathcal{M}(|z|^2; f) + \mathbb{I}_{X_1^k}(z); \\ z^{k+1} &:= \arg\min_z \mathcal{M}(|z|^2; f) + \mathbb{I}_{X_2^k}(z); \end{aligned} \tag{48}$$

where these two subproblems can be solved via ADMM as inner loop. Here one has to adjust the constraint sets with the update probe and sample, i.e.,

$$\begin{aligned} X_1^k &:= \{z : z = \mathcal{A}(w^k, u) \forall u \in \mathbb{C}^n\}, \\ X_2^k &:= \{z : z = \mathcal{A}(w, u^{k+1}) \forall w \in \mathbb{C}^{\bar{m}}\}. \end{aligned}$$

Note that the probe and sample can be readily determined by solving the least squares problem as

$$\begin{aligned} u^{k+1} &= \arg\min_u \|z^{k+1/2} - \mathcal{A}(w^k, u)\|^2, \\ w^{k+1} &= \arg\min_w \|z^{k+1} - \mathcal{A}(w, u^{k+1})\|^2, \end{aligned} \tag{49}$$

which can be solved by (23). Although the algorithms worked well with suitable initialization as reported in Fannjiang and Zhang (2020), the theoretical convergence for the blind recovery is still open.

Convex Programming

Ahmed et al. (2018) proposed a convex relaxation based on a lifted matrix recovery formulation that allows a nontrivial convex relaxation of the convolution PR.

Consider the convolution PR as

$$f^{\text{Cov}} = |\mathcal{F}\kappa \circ \mathcal{F}u|^2.$$

One basic assumption for unique recovery is that the variables κ and u belong to the subspace of $\mathbb{C}^n$, i.e.,

$$\kappa = \mathbf{B}h, \quad u = \mathbf{C}m,$$

where $h \in \mathbb{C}^{k_1}$ and $m \in \mathbb{C}^{k_2}$ with known matrices $\mathbf{B} \in \mathbb{C}^{n,k_1}$ and $\mathbf{C} \in \mathbb{C}^{n,k_2}$ ($k_1, k_2 \ll n$). Then one is concerned with the following problem with h, m as unknowns:

$$f^{\text{Cov}} = \frac{1}{\sqrt{n}} |\hat{\mathbf{B}}h \circ \hat{\mathbf{C}}m|^2 \tag{50}$$

with $\hat{\mathbf{B}} := \sqrt{n}\mathcal{F}\mathbf{B}, \hat{\mathbf{C}} := \sqrt{n}\mathcal{F}\mathbf{C}$. Further by the lifting technique in semidefinite programming (SDP), the above problem reduces to

$$f^{\text{Cov}}(l) = \frac{1}{n} \langle \mathbf{b}_l \mathbf{b}_l^*, \mathbf{H} \rangle \langle \mathbf{c}_l \mathbf{c}_l^*, \mathbf{M} \rangle, \tag{51}$$

where $\mathbf{H} := hh^*$, $\mathbf{M} := mm^*$ (rank 1 matrices), and $\langle \cdot, \cdot \rangle$ denotes the Frobenius inner product (trace of multiplication of two matrices). Here b_l^* and c_l^* are the rows of $\hat{\mathbf{B}}$ and $\hat{\mathbf{C}}$, respectively. By using a nuclear-norm minimization, to convexify the rank of matrix and further transform (51) to a convex constraint, then the following convex optimization model can be derived as

$$\min_{\mathbf{H} \succeq 0, \mathbf{M} \succeq 0} \quad \text{Tr}(\mathbf{H}) + \text{Tr}(\mathbf{M})$$

$$\text{s.t.} \quad \langle \mathbf{b}_l \mathbf{b}_l^*, \mathbf{H} \rangle \langle \mathbf{c}_l \mathbf{c}_l^*, \mathbf{M} \rangle \geq \bar{f}(l), \quad 0 \leq l \leq n-1, \tag{52}$$

with $\bar{f} := nf^{\text{Cov}}$.

An ADMM scheme was further developed (Ahmed et al. 2018) to solve (52). By introducing the convex constraint set

$$\mathscr{C} := \{(\mathbf{v}_1, \mathbf{v}_2) : \mathbf{v}_1(l)\mathbf{v}_2(l) \geq \bar{f}(l), \mathbf{v}_1(l) \geq 0, \mathbf{v}_2(l) \geq 0 \,\forall 0 \leq l \leq n-1\}$$

and $\mathbf{H}' = \mathbf{H}, \mathbf{M}' = \mathbf{M}$, an equivalent form can be given as

$$\min_{\mathbf{H},\mathbf{H}',\mathbf{M},\mathbf{M}',\mathbf{v}_1,\mathbf{v}_2} \mathbb{I}_{\mathscr{C}}(\mathbf{v}_1, \mathbf{v}_2) + \text{Tr}(\mathbf{H}) + \text{Tr}(\mathbf{M})$$

$$+ \mathbb{I}_{\{X \succeq 0\}}(\mathbf{H}') + \mathbb{I}_{\{X \succeq 0\}}(\mathbf{M}'),$$

$$\text{s.t.} \quad \mathbf{v}_1(l) - \langle \mathbf{b}_l \mathbf{b}_l^*, \mathbf{H} \rangle = 0, \quad \mathbf{v}_2(l) - \langle \mathbf{c}_l \mathbf{c}_l^*, \mathbf{M} \rangle = 0,$$

$$\mathbf{H}' - \mathbf{H} = 0, \quad \mathbf{M} - \mathbf{M}' = 0.$$

With the multipliers Λ_k for $k = 1, 2, 3, 4$ for the totally four constraints, the augmented Lagrangian with scalar form has the following form:

$$\mathcal{L}_c(\boldsymbol{H}, \boldsymbol{H}', \boldsymbol{M}, \boldsymbol{M}', \boldsymbol{v}_1, \boldsymbol{v}_2; \{\Lambda_k\}_{k=1}^4)$$
$$:= \mathbb{I}_\mathscr{C}(\boldsymbol{v}_1, \boldsymbol{v}_2) + Tr(\boldsymbol{H}) + Tr(\boldsymbol{M}) + \mathbb{I}_{\{X \succeq 0\}}(\boldsymbol{H}') + \mathbb{I}_{\{X \succeq 0\}}(\boldsymbol{M}')$$
$$+ \beta_1 \sum_l \left(\langle \Lambda_1(l), \boldsymbol{v}_1(l) - \langle \boldsymbol{b}_l \boldsymbol{b}_l^*, \boldsymbol{H} \rangle \rangle + \tfrac{1}{2} \| \boldsymbol{v}_1(l) - \langle \boldsymbol{b}_l \boldsymbol{b}_l^*, \boldsymbol{H} \rangle \|^2 \right)$$
$$+ \beta_1 \sum_l \left(\langle \Lambda_2(l), \boldsymbol{v}_2(l) - \langle \boldsymbol{c}_l \boldsymbol{c}_l^*, \boldsymbol{M} \rangle \rangle + \tfrac{1}{2} \| \boldsymbol{v}_2(l) - \langle \boldsymbol{c}_l \boldsymbol{c}_l^*, \boldsymbol{M} \rangle \|^2 \right)$$
$$+ \beta_2 \langle \Lambda_3, \boldsymbol{H}' - \boldsymbol{H} \rangle + \tfrac{\beta_2}{2} \| \boldsymbol{H}' - \boldsymbol{H} \|^2$$
$$+ \beta_2 \langle \Lambda_4, \boldsymbol{M}' - \boldsymbol{M} \rangle + \tfrac{\beta_2}{2} \| \boldsymbol{M}' - \boldsymbol{M} \|^2, \tag{53}$$

with two positive scalar parameters β_1, β_2. Then with alternating minimization and update of dual variables Λ_k, the iterative scheme is obtained. First, one can optimize the variables $\boldsymbol{H}$ and $\boldsymbol{M}$ in parallel and only consider

$$\boldsymbol{H}^\star := \arg\min_{\boldsymbol{H}} Tr(\boldsymbol{H}) + \beta_1 \sum_l \langle \Lambda_1(l), \boldsymbol{v}_1(l) - \langle \boldsymbol{b}_l \boldsymbol{b}_l^*, \boldsymbol{H} \rangle \rangle$$
$$+ \tfrac{\beta_1}{2} \| \boldsymbol{v}_1(l) - \langle \boldsymbol{b}_l \boldsymbol{b}_l^*, \boldsymbol{H} \rangle \|^2 + \beta_2 \langle \Lambda_3, \boldsymbol{H}' - \boldsymbol{H} \rangle + \tfrac{\beta_2}{2} \| \boldsymbol{H}' - \boldsymbol{H} \|^2.$$

By considering the first-order optimality condition (taking the derivative of the objective function w.r.t. $\boldsymbol{H}$), one obtains

$$\text{vec}(\boldsymbol{H}^\star) = \boldsymbol{T}_1^{-1} \text{vec}\big(\beta_1 \sum_l (\boldsymbol{v}_1(l) + \Lambda_1(l)) \boldsymbol{b}_l \boldsymbol{b}_l^* + \beta_2 (\boldsymbol{H}' - \Lambda_3) - \boldsymbol{I} \big),$$

with

$$\boldsymbol{T}_1 := \beta_1 \sum_l \text{vec}(\boldsymbol{b}_l \boldsymbol{b}_l^*) \text{vec}(\boldsymbol{b}_l \boldsymbol{b}_l^*)^* + \beta_2 \boldsymbol{I}.$$

Similarly, one can determine the optimal $\boldsymbol{M}^\star$ for the subproblem w.r.t. $\boldsymbol{M}$ by

$$\text{vec}(\boldsymbol{M}^\star) = \boldsymbol{T}_2^{-1} \text{vec}\big(\beta_1 \sum_l (\boldsymbol{v}_2(l) + \Lambda_2(l)) \boldsymbol{c}_l \boldsymbol{c}_l^* + \beta_2 (\boldsymbol{M}' - \Lambda_4) - \boldsymbol{I} \big),$$

with

$$\boldsymbol{T}_2 := \beta_1 \sum_l \text{vec}(\boldsymbol{c}_l \boldsymbol{c}_l^*) \text{vec}(\boldsymbol{c}_l \boldsymbol{c}_l^*)^* + \beta_2 \boldsymbol{I}.$$

For the H'−subproblem, denoting $\tilde{H} := H - \Lambda_3$, one considers the problem

$$H'^{\star} := \arg\min_{H'} \mathbb{I}_{\{X \succeq 0\}}(H') + \tfrac{1}{2}\|H' - \tilde{H}\|^2, \tag{54}$$

with the Hermitian matrix $\tilde{H}$(if initializing the multipliers Λ_3 and Λ_4 with Hermitian matrices, it can be readily guaranteed that all iterative sequences of these two multipliers are Hermitian). The closed-form solution of (54) can be directly given as

$$H'^{\star} = \text{Proj}_+(\tilde{H}),$$

with the operator Proj_+ defined as

$$\text{Proj}_+(\tilde{H}) := U \text{diag}(\max\{\text{diag}(\Sigma), 0\}) U^*$$

and $\tilde{H}$ has the eigen-decomposition as $\tilde{H} = U\Sigma U^*$ with unitary matrix U and diagonal matrix Σ.

Similarly,

$$M'^{\star} := \arg\min_{M'} \mathbb{I}_{\{X \succeq 0\}}(M') + \tfrac{1}{2}\|M' - (M - \Lambda_3)\|^2. \tag{55}$$

One can directly get the closed-form solution

$$M'^{\star} = \text{Proj}_+(M - \Lambda_3).$$

The subproblems w.r.t the variables v_1, v_2 can be solved in an element-wise manner, due to the independence of the optimization problem for each element of these two variables. Since they can be derived with standard discussion based on Karush-Kuhn-Tucker optimality conditions, the details here are omitted. Hence, all procedures to get the iterative scheme are summarized by further combining with the update of the multipliers as

$$\Lambda_1(l) \leftarrow \Lambda_1(l) + v_1(l) - \langle \mathbf{b}_l \mathbf{b}_l^*, H \rangle;$$
$$\Lambda_2(l) \leftarrow \Lambda_2(l) + v_2(l) - \langle \mathbf{c}_l \mathbf{c}_l^*, M \rangle;$$
$$\Lambda_3 \leftarrow \Lambda_3 + H' - H;$$
$$\Lambda_4 \leftarrow \Lambda_4 + M' - M.$$

Please see more details in the appendix of Ahmed et al. (2018).

As reported in Ahmed et al. (2018), this convex method showed excellent agreement with the theorem in the case of random subspaces. However, it was less effective on deterministic subspaces, including partial discrete cosine transforms or partial discrete wavelet transforms. One should also notice that although the model is convex, the lifting technique increased the dimension of original nonconvex optimization problem greatly, at the order of square of the original dimension, causing huge memory requirement as well as computational complexity. That may limit practical applications, especially for reconstructing 2D images or volumes.

It seems rather difficult to adopt the same convex method for other cases of BPR, since they cannot be rewritten as the same form as (50). Convexifying a general BPR problem should be an interesting research direction in the future.

Second-Order Algorithm Using Hessian

The second-order algorithms relying on the Hessian of the nonlinear optimization problems have also been developed for PR problem, such as using Newton method (NT) (Qian et al. 2014; Yeh et al. 2015), Levenberg-Marquardt method (LM) (Ma et al. 2018; Kandel et al. 2021), or Gauss-Newton algorithm (GN) (Gao and Xu 2017). Consider the following problem by rewriting (13)

$$\min_u \mathcal{Q}(u), \qquad (56)$$

with $\mathcal{Q}(u) := \mathcal{M}(|A_w u|^2, f)$. Given the initial guess u^0,

$$\mathcal{Q}(u) \approx \mathcal{Q}(u^0) + \Re\langle \nabla_u \mathcal{Q}(u^0), u - u^0 \rangle + \frac{1}{2}\Re(\langle \nabla_u^2 \mathcal{Q}(u^0)(u - u^0), u - u^0\rangle), \qquad (57)$$

where ∇_u^2 denotes the Hessian operator. Then a new estimate u^1 for the stationary point can be obtained by solving the following systems:

$$\nabla_u^2 \mathcal{Q}(u_0)(u^1 - u^0) = -\nabla_u \mathcal{Q}(u^0).$$

Assuming the Hessian matrix is nonsingular, the iterative scheme by NT is derived as

$$\text{Newton method:} \quad u^{k+1} = u^k - (\nabla_u^2 \mathcal{Q}(u^k))^{-1} \nabla_u \mathcal{Q}(u^k) \; \forall k. \qquad (58)$$

The gradient is given below:

$$\nabla_u \mathcal{Q}(u) = \begin{cases} A_w^*\left(A_w u - \frac{\sqrt{f}}{|A_w u|} \circ A_w u\right); & \text{(AGM)} \\ A_w^*\left(A_w u - \frac{f}{|A_w u|^2} \circ A_w u\right); & \text{(IPM)} \\ 2A_w^*\left(|A_w u|^2 \circ A_w u - f \circ A_w u\right); & \text{(IGM)} \end{cases} \qquad (59)$$

where the objective function $\mathcal{Q}(u)$ is rewritten as $\mathcal{Q}(u) = \mathcal{M}(|A_w u|^2, f)$ by denoting the matrix A_w as (8), and the detailed forms of the operators can be found in (9), (10), (11), and (12). The Hessian matrices for three metrics are complicated, and please see Appendix A of Yeh et al. (2015).

More efficient algorithms including LM and GN were developed, concerned with the nonlinear least squares problems (NLS) (56) with the AGM and IPM metrics (please see (14)). Namely, by denoting the residual function

$$r(u) = \begin{cases} |A_w u| - \sqrt{f}; & \text{(AGM)} \\ |A_w u|^2 - f; & \text{(IGM)} \end{cases}$$

consider the NLS problem below:

$$\min_u \mathcal{Q}(u) = \frac{1}{2}\|r(u)\|^2.$$

Then with Jacobian matrix as

$$J(u) := \nabla_u r(u) = \begin{cases} \text{diag}(\text{sign}(\text{conj}(A_w u))) A_w; & \text{(AGM)} \\ \text{diag}(\text{conj}(A_w u)) A_w & \text{(IGM)}; \end{cases}$$

the GN method considered

$$GN(u) := J^*(u) J(u),$$

as an estimate of the Hessian matrix, that leads to the following iterative scheme:

Gauss-Newton method:
$$\begin{aligned} u^{k+1} &= (J^*(u^k) J(u^k))^{-1} (u^k - \nabla_u \mathcal{Q}(u^k)) \\ &= (J^*(u^k) J(u^k))^{-1} (u^k - J^*(u^k) r(u^k)) \quad \forall k. \end{aligned}$$
(60)

Gao and Xu (2017) further proposed a global convergent GN algorithm with resampling for PR problem, which partial phaseless data was used to reformulate the GN matrix and the gradient per loop.

The Hessian matrix or the GN matrix cannot be guaranteed to be nonsingular practically. Hence, the LM method interpreted as a regularized variant of GN was proposed as

LM method: $\quad u^{k+1} = (J^*(u^k) J(u^k) + \mu^k \mathbf{I})^{-1} (u^k - J^*(u^k) r(u^k)) \quad \forall k$
(61)

with the adaptive parameter μ^k. Readily one knows μ^k cannot be too large; otherwise, the Hessian information is useless. To obtain fast convergence, Marquardt (1963) proposed the following strategy for μ^k depending on the diagonal matrix of GN matrix as

$$\mu^k = \mu_0 D_g(J^*(u^k) J(u^k))),$$

with $D_g(A)$ denoting the diagonal matrix with the elements from the main diagonal of the matrix A. Yamashita and Fukushima (2001) and Fan and Yuan (2005) proposed the scheme depending on the objective function value below:

$$\mu^k = (Q(u^k))^{\frac{\nu}{2}} \quad (62)$$

with $\nu \in [1, 2]$. Ma et al. (2018) further improved the scheme (62) as choosing a larger value when the iterative solution u^k is far away from the global minimizer, i.e.,

$$\mu^k = \text{Thresh}(u^k)(Q(u^k))^{\frac{\nu}{2}},$$

with

$$\text{Thresh}(u) = \begin{cases} \tau, & \text{if } Q(u^k) \geq c_0 \|u^k\|^2, \\ 1, & \text{otherwise,} \end{cases}$$

with $\tau \gg 1$ and parameter $c_0 > 0$.

The mentioned algorithms including Qian et al. (2014), Gao and Xu (2017), and Ma et al. (2018) focused on nonblind PR. With the generalized GN method and automatic differentiation, Kandel et al. (2021) proposed a variant LM algorithm for blind recovery, where, especially for IPM-based metric, it employed the generalized GN (GGN) as

$$GGN(u) := J^*(u^k) \nabla_g^2 \mathcal{M}(|A_{u^k} w|^2, f) J(u^k)$$

with $\mathcal{M}(g, f)$ defined in (14). Following the same manner with alternating minimization, one can easily derive the second-order algorithm for the blind problem as

$$\begin{aligned} u^{k+1} &= \arg\min_u \mathcal{M}(|A_{w^k} u|^2, f); \\ w^{k+1} &= \arg\min_w \mathcal{M}(|A_{u^{k+1}} u|^2, f); \end{aligned} \quad (63)$$

where both two subproblems are solved by NT, GN, or LM algorithms.

Subspace Method

The subspace method (Saad 2003; Xin et al. 2021) is a very powerful algorithm, iteratively refining the variable in the subspace of solution, which includes the Krylov subspace method as well-known conjugate gradient method, domain decomposition method, and multigrid method. It originally focused on solving the linear equations or least squares problems and now has been successfully extended to nonlinear equations or nonlinear optimization problems. In this part, the subspace methods for the PR and BPR problems will be reviewed.

Nonlinear Conjugate Gradient Algorithm Consider the following optimization problem:

$$\min f(x).$$

By the nonlinear conjugate gradient (NLCG) algorithm, the iterative scheme can be given below:

$$\begin{aligned} x^{k+1} &= x^k + \alpha^k d^k; \\ d^k &= -\nabla_x f(x^k) + \beta^{k-1} d^{k-1} \quad \forall k \geq 1, \end{aligned} \quad (64)$$

with the stepsize α^k and weight β^{k-1}, where d^k is the search direction. One may notice that the search direction d^k in NLCG is the combination of the gradient and the search direction d^{k-1} with the weight β^{k-1}. To get optimal parameters, the stepsize α^k is selected by the monotone line search procedures, while the weight β^k is determined based on the gradient $\nabla_x f(x^{k-1})$, $\nabla_x f(x^k)$ and the search direction d^{k-1} (typically five different formulas (Xin et al. 2021)).

The NLCG has been successfully applied to the BPR problem (Thibault and Guizar-Sicairos 2012; Qian et al. 2014). For example, Thibault and Guizar-Sicairos (2012) adopted the NLCG to solve the CDI problem. The iterative scheme can be given below:

$$\begin{aligned} (w^{k+1}, u^{k+1}) &= (w^k, u^k) + \alpha^k \Delta^k; \\ \Delta^k &= -g^k + \beta^{k-1} \Delta^{k-1} \quad \forall k \geq 1, \end{aligned} \quad (65)$$

with the gradient $g^k := (\nabla_w \mathcal{M}(|\mathcal{A}(w^k, u^k)|^2, f), \nabla_u \mathcal{M}(|\mathcal{A}(w^k, u^k)|^2, f))$ calculated by (59) and $\Delta := (\Delta_w, \Delta_u)$. The weight β^{k-1} is derived by the Polak-Ribére formula as

$$\beta^{k-1} = \frac{\langle g^k, g^k \rangle - \Re(\langle g^k, g^{k-1} \rangle)}{\langle g^{k-1}, g^{k-1} \rangle}.$$

To further get α^k, by estimating $\mathcal{M}(|\mathcal{A}(w^k + \alpha \Delta_w, u^k + \alpha \Delta_u)|^2, f)$ by the low-order polynomial as

$$\mathcal{M}(|\mathcal{A}(w^k + \alpha \Delta_w, u^k + \alpha \Delta_u)|^2, f) \approx \sum_{t=0}^{8} c_t \alpha^t,$$

the authors adopted the Newton–Raphson algorithm in order to minimize the following one-dimensional problem:

$$\alpha^k := \arg\min_{\alpha} \sum_{t=0}^{8} c_t \alpha^t.$$

Domain Decomposition Method The domain decomposition methods (DDMs) allow for highly parallel computing with good load balance, by decomposing the equations on whole domain to the problems on relatively small subdomains with information synchronization on the partition interfaces. They have played a great role in solving partial differential equations numerically and recently been successfully extended to large-scale image restoration, image reconstruction, and other inverse problems, e.g., Xu et al. (2010), Chang et al. (2015, 2021), Langer and Gaspoz (2019), Lee et al. (2019), and references therein. For ptychography imaging, several parallel algorithms (Nashed et al. 2014; Guizar-Sicairos et al. 2014; Marchesini et al. 2016; Enfedaque et al. 2019; Chang et al. 2021) have been developed. Specially, for convention ptychography, Chang et al. (2021) proposed an overlapping DDM with the ST-AGM as defined in (18), with fewer communication cost and theoretical convergence guarantee.

First, give the domain decomposition. Denote the whole region $\Omega := \{0, 1, 2, \cdots, n-1\}$ in the discrete setting. There exists the two-subdomain overlapping DD $\{\Omega_d\}_{d=1}^{2}$, such that

$$\Omega = \bigcup_{d=1}^{2} \Omega_d$$

with $\Omega_d := \{l_0^d, l_1^d, \cdots, l_{n_d-1}^d\}$, and the overlapping region is denoted as

$$\Omega_{1,2} := \Omega_1 \cap \Omega_2 = \{\hat{l}_0, \hat{l}_1, \cdots, \hat{l}_{\hat{n}-1}\}.$$

Here consider a special overlapping DD as shown in Fig. 2. Denote the restriction operators R_1, R_2 as

$$R_d u = u|_{\Omega_d}, \quad R_{1,2} u = u|_{\Omega_{1,2}},$$

i.e.,

$$(R_d u)(j) = u(l_j^d) \forall\, 0 \leq j \leq n_d - 1,$$
$$(R_{1,2} u)(j) = u(\hat{l}_j) \forall\, 0 \leq j \leq \hat{n} - 1.$$

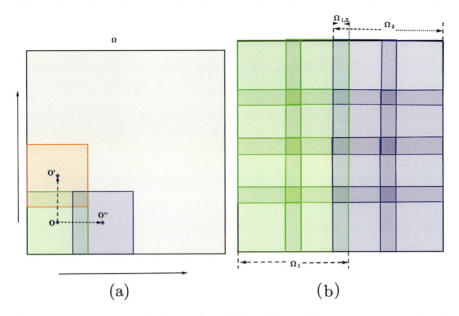

Fig. 2 (**a**) Ptychography scan in the domain Ω (grid scan): the starting scan centers at point **O** and then moves up (or to the right) with the center point **O'** (or **O''**); (**b**) two-subdomain DD (totally 4×4 frames): The subdomains Ω_1, Ω_2 are generated by two 4×2 scans, and the overlapping region $\Omega_{1,2} = \Omega_1 \cap \Omega_2$

Then two groups of localized shift operators can be introduced $\{\mathcal{S}^d_{j_d}\}_{j_d=0}^{J_d-1}$ for $d = 1, 2$ with $\sum_d J_d = J$.

For nonblind problem, denote the linear operators A_1, A_2 on the subdomains as

$$A_d u_d := \left((\mathcal{F}(w \circ \mathcal{S}^d_0 u_d))^T, (\mathcal{F}(w \circ \mathcal{S}^d_1 u_d)))^T, \cdots, (\mathcal{F}(w \circ \mathcal{S}^d_{J_d-1} u_d))^T \right)^T, \quad (66)$$

for $d = 1, 2$. Based on the continuity on the overlapping regions, one has

$$\pi_{1,2} u_1 = \pi_{2,1} u_2, \quad (67)$$

where the operators $\pi_{1,2}$ (restriction from Ω_1 into $\Omega_{1,2}$) and $\pi_{2,1}$ (restriction from Ω_2 into $\Omega_{1,2}$) are denoted as

$$\pi_{1,2} u_1 := R_{1,2} R_1^T u_1,$$

and

$$\pi_{2,1} u_2 := R_{1,2} R_2^T u_2.$$

Naturally, the measurement f is also decomposed to two nonoverlapping parts f_1, f_2, i.e.,

$$f_d := |A_d R_d u|^2.$$

Hereafter, consider the nonlinear optimization problem with ST-AGM. In order to enable the parallel computing of u_1 and u_2, introduce an auxiliary variable v which is only defined in the overlapping region $\Omega_{1,2}$, and then is concerned with the following model:

$$\min_{u_1, u_2, v} \sum_{d=1}^{2} \mathcal{G}_\epsilon(A_d u_d; f_d), \quad (68)$$
$$\text{s.t.} \quad \pi_{d, 3-d} u_d - v = 0, \ d = 1, 2.$$

In order to develop an iterative scheme without inner loop as well as with fast convergence for large-step scan, two auxiliary variables z_1, z_2 are introduced below:

$$\min_{u_1, u_2, v, z_1, z_2} \sum_{d=1}^{2} \mathcal{G}_\epsilon(z_d; f_d), \quad (69)$$
$$\text{s.t.} \quad \pi_{d, 3-d} u_d - v = 0, \quad A_d u_d - z_d = 0, \ d = 1, 2.$$

Then it is quite standard to solve the saddle point problem by ADMM. The details are omitted here, and please see more details in Chang et al. (2021).

Then for blind recovery, in order to reduce the grid pathology (Chang et al. 2019a) (ambiguity derived by the multiplication of any periodical function and the true solution) due to grid scan, introduce the support set constraint of the probe, i.e., $\mathcal{O} := \{w : (\mathcal{F}w)(j) = 0, \ j \in \mathscr{J}\}$, with the support set $\tilde{\mathscr{J}}$ denoted as the complement of the set $\mathscr{J}$ (index set for zero values for the Fourier transform of the probe). Then consider the blind ptychography problem for two-subdomain DD:

$$\min_{\{w, u_1, u_2, v\}} \sum_{d=1}^{2} \mathcal{G}_\epsilon(\mathcal{A}_d(w, u_d); f_d) + \mathbb{I}_\mathcal{O}(w),$$
$$\text{s.t.} \quad \pi_{d, 3-d} u_d - v = 0, d = 1, 2,$$

where the bilinear mapping $\mathcal{A}_d(w, u_d)$ is denoted as

$$(\mathcal{A}_d)_{j_d}(w, u_d) := \mathcal{F}(w \circ (S_{j_d}^d u_d)) \ \forall \ 0 \le j_d \le J_d - 1,$$

with $\sum_{d=1}^{2} J_d = J$, and the indicator function $\mathbb{I}_\mathcal{O}$. To enable parallel computing, consider the following constraint optimization problems:

$$\min_{\{w,w_1,w_2,u_1,u_2,v,z_1,z_2\}} \sum_{d=1}^{2} \mathcal{G}_\epsilon(z_d; f_d) + \mathbb{I}_\mathcal{O}(w)$$

$$\text{s.t.} \quad \pi_{1,2}u_1 - v = 0, \quad \pi_{2,1}u_2 - v = 0,$$

$$w_d = w, \ z_d = \mathcal{A}_d(w_d, u_d), d = 1, 2,$$

which was also efficiently solved by ADMM.

Multigrid Methods The multigrid method (MG) is a standard framework in order to accelerate solving partial differential equations (Hackbusch 1985), large-scale linear equations (Xu and Zikatanov 2017), and related optimization problems (Borzi and Schulz 2009) with the full approximation scheme (FAS) (Brandt and Livne 2011). A multigrid-based optimization framework based on Nash (2000) to reduce the computational for nonblind ptychographic PR was proposed by Fung and Wendy (2020), which utilized the hierarchical structures of the measured data.

Consider the following feasible problem (Fung and Wendy 2020) as

$$\min_u \sum_j \|\mathcal{F}^*(\sqrt{f_j} \circ \text{sign}(\mathcal{F}(w \circ \mathcal{S}_j u))) - w \circ \mathcal{S}_j u\|^2, \quad (70)$$

which is equivalent to the problem

$$\min_u \mathcal{M}(|\mathcal{A}(w, u)|^2, f)$$

using the AGM metric. Then the multigrid optimization framework based on FAS was further developed, where the coarse-grid subproblem was interpreted as a first-order approximation to the fine-grid problem. However, it is unclear how to extend the current algorithm to the blind problem.

Discussions

Experimental Issues

Probe Drift Probe drift happens in ptychography, when the data is very noisy. The mass center of the iterative probe will eventually touch the boundary such that the iterative algorithms fail eventually. Hence, the joint reconstruction will cause instability of the iterative algorithms from noisy experimental data. One simple strategy proposed by Marchesini et al. (2016) is to shift the probe to the mass center of the complex image periodically. Other possible way is to consider the compact support condition for the probe, or to get additional measurement for the probe by letting the light go through the vacuum as Marchesini et al. (2016) and Chang et al. (2019a). The related numerical stability shall be investigated, and one can refer to Huang and Xu (2020, 2021) for nonblind PR.

Flat Samples When the sample is nearly flat (such as weak absorption or scattering for biological specimens using hard X-ray sources), there will be no sufficient diversity of the measured phaseless data even by very dense scan. In such case, the iterative algorithms mentioned in this survey will become slow, and the recovered image quality gets worse. Acquiring of scattering map by linearization for large features of the sample (Dierolf et al. 2010b) or modeling with additional Kramers-Kronig relation (KKR) (Hirose et al. 2017) was exploited to improve the reconstruction quality. Besides, pairwise relations between adjacent frames were considered in Marchesini and Wu (2014) to accelerate projection algorithms for the flat sample.

Background Retrieval Parasitic scattering termed as background often happens experimentally, which may come from any element along the beam path other than the sample and the optical elements desired harmonic order (Chang et al. 2019b). Direct reconstruction without background removal will introduce structural artifacts to the reconstruction images. Several methods were designed, such as preconditioned gradient descent (Marchesini et al. 2013), preprocessing method (Wang et al. 2017), and ADMM for nonlinear optimization method with framewise-invariant background (Chang et al. 2019b). It is still a challenging problem since the practical background is sophisticated and cannot be assumed to be framewisely invariant.

High-Dimensional Problems The formula for all four cases for BPR holds for a thin (2D) object in paraxial approximation. For thick samples, the linear propagation as (7) will cause obvious errors, and one has to consider the nonlinear transform as Dierolf et al. (2010a). Other than the 3D imaging, high-dimensional problems may result from the spectromicroscopy (Maiden et al. 2013), multimode decomposition of partial coherence (Thibault and Menzel 2013; Chang et al. 2018a), and dichroic ptychography (Chang et al. 2020; Lo et al. 2021). Such strong nonlinearity coupling with the high-dimensional optimization causes difficulties for designing the stable and high-throughput algorithm.

Theoretical Analysis

Convergence of Iterative Algorithms Other than the projection onto nonconvex modulus constraint for nonblind PR, APs (Thibault et al. 2009; Marchesini et al. 2016) for BPR involve the bilinear constraint set. Some progress has been made for the general PR problem using projection algorithms (Hesse and Luke 2013; Marchesini et al. 2015; Chen and Fannjiang 2016). However, the corresponding convergence theories for BPR are still unclear. Moreover, only the PHeBIE- (Hesse et al. 2015) and ADMM-based algorithm (Chang et al. 2019a) for BPR provided rigorous convergence analysis. Hence, it is of great importance to either study the convergence of existing algorithms or develop new algorithms with clear convergence guarantee in the future.

Uniqueness Analysis Uniqueness can be guaranteed for 1D nonblind ptychographic PR for nonvanishing signals with the probe of proper size (Jaganathan et al. 2016). It can also be guaranteed for BPR (Bendory et al. 2019). By letting two signals lie in low-dimensional random subspaces, the uniqueness was obtained (Ahmed et al. 2018) with sufficient measurements. For 2D imaging problems, with a randomly phased probe, the uniqueness can be proved for the measurements which is strongly connected and possesses an anchor. See more discussions on more general cases together with sparse signals in Grohs et al. (2020). Readily for ptychography, nontrivial ambiguity including periodical function and linear phase exists for raster scan. Rigorous analysis about more general ambiguity was given (Fannjiang 2019). Experimentally more flexible spiral or random scan (Huang et al. 2014) has been exploited for stable recovery.

Further Discussions

Recently, some efficient algorithms have been developed for nonblind PR, such as the second-order algorithms including Ma et al. (2018), Gao and Xu (2017), and the multigrid method (Fung and Wendy 2020); however, it is not clear how they can be applied to the blind problem. Hence, we only list the algorithms for the BPR problem included in this survey, and please see the overview in Table 1.

Then we discuss the advantages and disadvantages of all listed algorithms. As the unique convex method, the convex programming (Ahmed et al. 2018) provided a convex relaxation such that it can gain the global minimizer. The dimension of the lifted matrix is much higher than that of the original form leading to the iterative algorithm with high complexity, and therefore it seems more impractical for real experimental analysis. Moreover, it is limited to the convolutional PR as the special case of BPR, since it relies on the structure as (50). All other listed algorithms designed based on the nonconvex optimization problem work well for perfect data (smaller scan stepsizes to guarantee enough redundancy and long exposure with high signal-to-noise ratio (SNR)). The AP, ePIE-type, proximal, and ADMM algorithms are of the first order and have closed-form expression for all iterative steps, all of which have already been efficiently implemented for practical ptychography and Fourier ptychography imaging instrument with low computational complexity. As reported in Chang et al. (2019a), the ePIE algorithm may get unstable for noisy measurements, and it seems more sensitive to the scan stepsizes for ptychography imaging, while the ADMM algorithm (Chang et al. 2019a) for ptychography imaging can offer promising performance even for noisy and insufficient data. The second-order algorithms utilizing the Hessian usually requires much more computation cost, and it can be accelerated by Gauss-Newton or Levenberg-Marquardt methods without direct calculation of the Hessian. Further requirement of parallel computing may consider the DDM (Chang et al. 2021).

Conclusions

In this survey, a short review of the iterative algorithms is provided for the nonlinear optimization problem arising from the BPR problem, mainly consisting of three types of algorithms as the first-order operator-splitting algorithms and second-order algorithms and subspace methods. There still exist sophisticated experimental issues and challenging theoretical analysis, which are further discussed in the last part. Learning-based methods have been a powerful tool for solving inverse problem and PR problems, which are not included in this survey.

This survey focuses on the BPR problems with forms expressed as (6). However, not all the BPR problem belongs to the categories of (6). Very recently, a resolution-enhanced parallel coded ptychography (CP) technique (Jiang et al. 2021, 2022) was reported which achieves the highest numerical aperture. With the sample u and the transmission profile of the engineered surface w, the phaseless data was generated as

$$f_j^{CP} = \left| \left(w \circ (\mathcal{S}_j u \circledast \kappa_1) \right) \circledast \kappa_2 \right|^2,$$

with κ_1, κ_2 as two known PSFs. Such advanced cases should be further investigated.

Table 1 Overview of all iterative algorithms for the blind phase retrieval (BPR) problem in this survey. "Y" and "N" are short for "yes" and "no", respectively

Name	Refs	Convex(Y/N)	Convergence(Y/N)
Alternating projection	Thibault et al. (2009); Marchesini et al. (2016)	N	N
ePIE-type algorithms	Maiden and Rodenburg (2009); Maiden et al. (2017)	N	N
Proximal algorithms	Hesse et al. (2015); Yan (2020)	N	Y (Hesse et al. 2015)
ADMM	Chang et al. (2019a); Fannjiang and Zhang (2020)	N	Y (Chang et al. 2019a)
Convex programming	Ahmed et al. (2018)	Y	Y
Second-order algorithms	Yeh et al. (2015); Kandel et al. (2021)	N	N
Subspace method	Thibault and Guizar-Sicairos (2012); Qian et al. (2014); Chang et al. (2021)	N	Y (Chang et al. 2021)

Acknowledgments The work of the first author was partially supported by the NSFC (Nos. 11871372, 11501413) and Natural Science Foundation of Tianjin (18JCYBJC16600). The authors would like to thank Prof. Guoan Zheng for the helpful discussions.

References

Ahmed, A., Aghasi, A., Hand, P.: Blind deconvolutional phase retrieval via convex programming (2018). NeurIPS (arXiv:1806.08091)

Bendory, T., Sidorenko, P., Eldar, Y.C.: On the uniqueness of frog methods. IEEE Sig. Process. Lett. **24**(5), 722–726 (2017)

Bendory, T., Edidin, D., Eldar, Y.C.: Blind phaseless short-time fourier transform recovery. IEEE Trans. Inf. Theory **66**(5), 3232–3241 (2019)

Bolte, J., Sabach, S., Teboulle, M.: Proximal alternating linearized minimization for nonconvex and nonsmooth problems. Math. Program. **146**(1–2), 459–494 (2014)

Borzì, A., Schulz, V.: Multigrid methods for pde optimization. SIAM Rev. **51**(2), 361–395 (2009)

Boyd, S., Parikh, N., Chu, E., Peleato, B., Eckstein, J.: Distributed optimization and statistical learning via the alternating direction method of multipliers. Found. Trends Mach. Learn. **3**(1), 1–122 (2011)

Brandt, A., Livne, O.E.: Multigrid Techniques: 1984 Guide with Applications to Fluid Dynamics, Revised Edition. SIAM, Philadelphia (2011)

Cai, J.-F., Huang, M., Li, D., Wang, Y.: The global landscape of phase retrieval II: quotient intensity models (2021). arXiv preprint arXiv:2112.07997

Candes, E.J., Li, X., Soltanolkotabi, M.: Phase retrieval via wirtinger flow: Theory and algorithms. IEEE Trans. Inf. Theory **61**(4), 1985–2007 (2015)

Chang, H., Tai, X.-C., Wang, L.-L., Yang, D.: Convergence rate of overlapping domain decomposition methods for the Rudin-Osher-Fatami model based on a dual formulation. SIAM J. Image Sci. **8**, 564–591 (2015)

Chang, H., Lou, Y., Ng, M.K., Zeng, T.: Phase retrieval from incomplete magnitude information via total variation regularization. SIAM J. Sci. Comput. **38**(6), A3672–A3695 (2016)

Chang, H., Enfedaque, P., Lou, Y., Marchesini, S.: Partially coherent ptychography by gradient decomposition of the probe. Acta Crystallogr. Sect. A: Found. Adv. **74**(3), 157–169 (2018a)

Chang, H., Lou, Y., Duan, Y., Marchesini, S.: Total variation–based phase retrieval for Poisson noise removal. SIAM J. Imaging Sci. **11**(1), 24–55 (2018b)

Chang, H., Marchesini, S., Lou, Y., Zeng, T.: Variational phase retrieval with globally convergent preconditioned proximal algorithm. SIAM J. Imaging Sci. **11**(1), 56–93 (2018c)

Chang, H., Enfedaque, P., Marchesini, S.: Blind ptychographic phase retrieval via convergent alternating direction method of multipliers. SIAM J. Imaging Sci. **12**(1), 153–185 (2019a)

Chang, H., Enfedaque, P., Zhang, J., Reinhardt, J., Enders, B., Yu, Y.-S., Shapiro, D., Schroer, C.G., Zeng, T., Marchesini, S.: Advanced denoising for x-ray ptychography. Opt. Express **27**(8), 10395–10418 (2019b)

Chang, H., Marcus, M.A., Marchesini, S.: Analyzer-free linear dichroic ptychography. J. Appl. Crystallogr. **53**(5), 1283–1292 (2020)

Chang, H., Glowinski, R., Marchesini, S., Tai, X.-C., Wang, Y., Zeng, T.: Overlapping domain decomposition methods for ptychographic imaging. SIAM J. Sci. Comput. **43**(3), B570–B597 (2021)

Chapman, H.N.: Phase-retrieval x-ray microscopy by wigner-distribution deconvolution. Ultramicroscopy **66**(3), 153–172 (1996)

Chen, Y., Candes, E.: Solving random quadratic systems of equations is nearly as easy as solving linear systems. In: Advances in Neural Information Processing Systems, pp. 739–747 (2015)

Chen, P., Fannjiang, A.: Fourier phase retrieval with a single mask by douglas–rachford algorithms. Appl. Comput. Harmon. Anal. **44**(3), 665–699 (2016)

Dierolf, M., Menzel, A., Thibault, P., Schneider, P., Kewish, C.M., Wepf, R., Bunk, O., Pfeiffer, F.: Ptychographic x-ray computed tomography at the nanoscale. Nature **467**(7314), 436–439 (2010a)

Dierolf, M., Thibault, P., Menzel, A., Kewish, C.M., Jefimovs, K., Schlichting, I., von König, K., Bunk, O., Pfeiffer, F.: Ptychographic coherent diffractive imaging of weakly scattering specimens. New J. Phys. **12**(3), 035017 (2010b)

Elser, V.: Phase retrieval by iterated projections. J. Opt. Soc. Am. A **20**(1), 40–55 (2003)

Elser, V., Lan, T.-Y., Bendory, T.: Benchmark problems for phase retrieval. SIAM J. Imaging Sci. **11**(4), 2429–2455 (2018)

Enfedaque, P., Chang, H., Enders, B., Shapiro, D., Marchesini, S.: High performance partial coherent x-ray ptychography. In: International Conference on Computational Science, pp. 46–59. Springer (2019)

Fan, J.-Y., Yuan, Y.-X.: On the quadratic convergence of the levenberg-marquardt method without nonsingularity assumption. Computing **74**(1), 23–39 (2005)

Fannjiang, A.: Raster grid pathology and the cure. Multiscale Model. Simul. **17**(3), 973–995 (2019)

Fannjiang, A., Strohmer, T.: The numerics of phase retrieval. Acta Numer. **29**, 125–228 (2020)

Fannjiang, A., Zhang, Z.: Fixed point analysis of douglas–rachford splitting for ptychography and phase retrieval. SIAM J. Imaging Sci. **13**(2), 609–650 (2020)

Fung, S.W., Wendy, Z.: Multigrid optimization for large-scale ptychographic phase retrieval. SIAM J. Imaging Sci. **13**(1), 214–233 (2020)

Gao, B., Xu, Z.: Phaseless recovery using the Gauss–Newton method. IEEE Trans. Sig. Process. **65**(22), 5885–5896 (2017)

Gao, B., Wang, Y., Xu, Z.: Solving a perturbed amplitude-based model for phase retrieval. IEEE Trans. Sig. Process. **68**, 5427–5440 (2020)

Godard, P., Allain, M., Chamard, V., Rodenburg, J.: Noise models for low counting rate coherent diffraction imaging. Opt. Express **20**(23), 25914–25934 (2012)

Grohs, P., Koppensteiner, S., Rathmair, M.: Phase retrieval: uniqueness and stability. SIAM Rev. **62**(2), 301–350 (2020)

Guizar-Sicairos, M., Fienup, J.R.: Phase retrieval with transverse translation diversity: a nonlinear optimization approach. Opt. Express **16**(10), 7264–7278 (2008)

Guizar-Sicairos, M., Johnson, I., Diaz, A., Holler, M., Karvinen, P., Stadler, H.-C., Dinapoli, R., Bunk, O., Menzel, A.: High-throughput ptychography using eiger: scanning x-ray nano-imaging of extended regions. Opt. Express **22**(12), 14859–14870 (2014)

Gürsoy, D., Chen-Wiegart, Y.-C.K., Jacobsen, C.: Lensless x-ray nanoimaging: revolutions and opportunities. IEEE Sig. Process. Mag. **39**(1), 44–54 (2022)

Hackbusch, W.: Multi-grid Methods and Applications, Springer, Berlin, Heidelberg (1985)

Hesse, R., Luke, D.R.: Nonconvex notions of regularity and convergence of fundamental algorithms for feasibility problems. SIAM J. Optim. **23**(4), 2397–2419 (2013)

Hesse, R., Luke, D.R., Sabach, S., Tam, M.K.: Proximal heterogeneous block implicit-explicit method and application to blind ptychographic diffraction imaging. SIAM J. Imaging Sci. **8**(1), 426–457 (2015)

Hirose, M., Shimomura, K., Burdet, N., Takahashi, Y.: Use of Kramers-Kronig relation in phase retrieval calculation in x-ray spectro-ptychography. Opt. Express **25**(8), 8593–8603 (2017)

Huang, M., Xu, Z.: The estimation performance of nonlinear least squares for phase retrieval. IEEE Trans. Inf. Theory **66**(12), 7967–7977 (2020)

Huang, M., Xu, Z.: Uniqueness and stability for the solution of a nonlinear least squares problem (2021). arXiv preprint arXiv:2104.10841

Huang, X., Yan, H., Harder, R., Hwu, Y., Robinson, I.K., Chu, Y.S.: Optimization of overlap uniformness for ptychography. Opt. Express **22**(10), 12634–12644 (2014)

Huang, Y., Jiang, S., Wang, R., Song, P., Zhang, J., Zheng, G., Ji, X., Zhang, Y.: Ptychography-based high-throughput lensless on-chip microscopy via incremental proximal algorithms. Opt. Express **29**(23), 37892–37906 (2021)

Jaganathan, K., Eldar, Y.C., Hassibi, B.: Stft phase retrieval: uniqueness guarantees and recovery algorithms. IEEE J. Sel. Top. Sig. Process. **10**(4), 770–781 (2016)

Jiang, S., Guo, C., Song, P., Zhou, N., Bian, Z., Zhu, J., Wang, R., Dong, P., Zhang, Z., Liao, J. et al.: Resolution-enhanced parallel coded ptychography for high-throughput optical imaging. ACS Photon. **8**(11), 3261–3271 (2021)

Jiang, S., Guo, C., Bian, Z., Wang, R., Zhu, J., Song, P., Hu, P., Hu, D., Zhang, Z., Hoshino, K. et al.: Ptychographic sensor for large-scale lensless microbial monitoring with high spatiotemporal resolution. Biosens. Bioelectron. **196**, 113699 (2022)

Kandel, S., Maddali, S., Nashed, Y.S., Hruszkewycz, S.O., Jacobsen, C., Allain, M.: Efficient ptychographic phase retrieval via a matrix-free levenberg-marquardt algorithm. Opt. Express **29**(15), 23019–23055 (2021)

Kane, D.J., Vakhtin, A.B.: A review of ptychographic techniques for ultrashort pulse measurement. Progress Quantum Electron.vol. 81, 100364 (2021)

Langer, A., Gaspoz, F.: Overlapping domain decomposition methods for total variation denoising. SIAM J. Numer. Anal. **57**(3), 1411–1444 (2019)

Lee, C.-O., Park, E.-H., Park, J.: A finite element approach for the dual Rudin–Osher–Fatemi model and its nonoverlapping domain decomposition methods. SIAM J. Sci. Comput. **41**(2), B205–B228 (2019)

Lo, Y.H., Zhou, J., Rana, A., Morrill, D., Gentry, C., Enders, B., Yu, Y.-S., Sun, C.-Y., Shapiro, D.A., Falcone, R.W., Kapteyn, H.C., Murnane, M.M., Gilbert, P.U.P.A., Miao, J.: X-ray linear dichroic ptychography. Proc. Natl. Acad. Sci. **118**(3), 2019068118 (2021)

Luke, D.R.: Relaxed averaged alternating reflections for diffraction imaging. Inverse Probl. **21**(1), 37–50 (2005)

Ma, C., Liu, X., Wen, Z.: Globally convergent levenberg-marquardt method for phase retrieval. IEEE Trans. Inf. Theory **65**(4), 2343–2359 (2018)

Maiden, A.M., Rodenburg, J.M.: An improved ptychographical phase retrieval algorithm for diffractive imaging. Ultramicroscopy **109**(10), 1256–1262 (2009)

Maiden, A., Morrison, G., Kaulich, B., Gianoncelli, A., Rodenburg, J.: Soft x-ray spectromicroscopy using ptychography with randomly phased illumination. Nat. Commun. **4**, 1669 (2013)

Maiden, A., Johnson, D., Li, P.: Further improvements to the ptychographical iterative engine. Optica **4**(7), 736–745 (2017)

Marchesini, S.: Invited article: a unified evaluation of iterative projection algorithms for phase retrieval. Rev. Sci. Instrum. **78**(1), 011301 (2007)

Marchesini, S., Wu, H.-T.: Rank-1 accelerated illumination recovery in scanning diffractive imaging by transparency estimation (2014). arXiv preprint arXiv:1408.1922

Marchesini, S., Schirotzek, A., Yang, C., Wu, H.-T., Maia, F.: Augmented projections for ptychographic imaging. Inverse Probl. **29**(11), 115009 (2013)

Marchesini, S., Tu, Y.-C., Wu, H.-T.: Alternating projection, ptychographic imaging and phase synchronization. Appl. Comput. Harmon. Anal. **41**(3), 815-851 (2015)

Marchesini, S., Krishnan, H., Shapiro, D.A., Perciano, T., Sethian, J.A., Daurer, B.J., Maia, F.R.: SHARP: a distributed, GPU-based ptychographic solver. J. Appl. Crystallogr. **49**(4), 1245–1252 (2016)

Marquardt, D.W.: An algorithm for least-squares estimation of nonlinear parameters. J. Soc. Ind. Appl. Math. **11**(2), 431–441 (1963)

Nash, S.G.: A multigrid approach to discretized optimization problems. Optim. Methods Softw. **14**(1–2), 99–116 (2000)

Nashed, Y.S., Vine, D.J., Peterka, T., Deng, J., Ross, R., Jacobsen, C.: Parallel ptychographic reconstruction. Opt. Express **22**(26), 32082–32097 (2014)

Odstrčil, M., Menzel, A., Guizar-Sicairos, M.: Iterative least-squares solver for generalized maximum-likelihood ptychography. Opt. Express **26**(3), 3108–3123 (2018)

Ou, X., Zheng, G., Yang, C.: Embedded pupil function recovery for fourier ptychographic microscopy. Opt. Express **22**(5), 4960–4972 (2014)

Pfeiffer, F.: X-ray ptychography. Nat. Photon **12**, 9–17 (2018)

Qian, J., Yang, C., Schirotzek, A., Maia, F., Marchesini, S.: Efficient algorithms for ptychographic phase retrieval. Inverse Probl. Appl. Contemp. Math. **615**, 261–280 (2014)

Qu, Q., Zhang, Y., Eldar, Y.C., Wright, J.: Convolutional phase retrieval. In: Proceedings of the 31st International Conference on Neural Information Processing Systems, pp. 6088–6098 (2017)

Qu, Q., Zhang, Y., Eldar, Y.C., Wright, J.: Convolutional phase retrieval via gradient descent. IEEE Trans. Inf. Theory **66**(3), 1785–1821 (2019)

Reinhardt, J., Hoppe, R., Hofmann, G., Damsgaard, C.D., Patommel, J., Baumbach, C., Baier, S., Rochet, A., Grunwaldt, J.-D., Falkenberg, G., Schroer, C.G.: Beamstop-based low-background ptychography to image weakly scattering objects. Ultramicroscopy **173**, 52–57 (2017)

Rodenburg, J.M.: Ptychography and related diffractive imaging methods. Adv. Imaging Electron Phys. **150**, 87–184 (2008)

Saad, Y.: Iterative Methods for Sparse Linear Systems, 2nd edn. Society for Industrial and Applied Mathematics (2003)

Shechtman, Y., Eldar, Y.C., Cohen, O., Chapman, H.N., Miao, J., Segev, M.: Phase retrieval with application to optical imaging: a contemporary overview. Sig. Process. Mag. IEEE **32**(3), 87–109 (2015)

Sun, J., Qu, Q., Wright, J.: A geometric analysis of phase retrieval. In: 2016 IEEE International Symposium on Information Theory (ISIT), pp. 2379–2383. IEEE (2016)

Thibault, P., Guizar-Sicairos, M.: Maximum-likelihood refinement for coherent diffractive imaging. New J. Phys. **14**(6), 063004 (2012)

Thibault, P., Menzel, A.: Reconstructing state mixtures from diffraction measurements. Nature **494**(7435), 68–71 (2013)

Thibault, P., Dierolf, M., Bunk, O., Menzel, A., Pfeiffer, F.: Probe retrieval in ptychographic coherent diffractive imaging. Ultramicroscopy **109**(4), 338–343 (2009)

Trebino, R., DeLong, K.W., Fittinghoff, D.N., Sweetser, J.N., Krumbügel, M.A., Richman, B.A., Kane, D.J.: Measuring ultrashort laser pulses in the time-frequency domain using frequency-resolved optical gating. Rev. Sci. Instrum. **68**(9), 3277–3295 (1997)

Wang, C., Xu, Z., Liu, H., Wang, Y., Wang, J., Tai, R.: Background noise removal in x-ray ptychography. Appl. Opt. **56**(8), 2099–2111 (2017)

Wen, Z., Yang, C., Liu, X., Marchesini, S.: Alternating direction methods for classical and ptychographic phase retrieval. Inverse Probl. **28**(11), 115010 (2012)

Wu, C., Tai, X.-C.: Augmented Lagrangian method, dual methods and split-Bregman iterations for ROF, vectorial TV and higher order models. SIAM J. Imaging Sci. **3**(3), 300–339 (2010)

Xin, L., Zaiwen, W., Ya-Xiang, Y.: Subspace methods for nonlinear optimization. CSIAM Trans. Appl. Math. **2**(4), 585–651 (2021)

Xu, J., Zikatanov, L.: Algebraic multigrid methods. Acta Numer. **26**, 591–721 (2017)

Xu, J., Tai, X.-C., Wang, L.-L.: A two-level domain decomposition method for image restoration. Inverse Probl. Imaging **4**(3), 523–545 (2010)

Yamashita, N., Fukushima, M.: On the rate of convergence of the levenberg-marquardt method. In: Alefeld, G., Chen, X. (eds.) Topics in Numerical Analysis, pp. 239–249. Springer, Vienna (2001)

Yan, H.: Ptychographic phase retrieval by proximal algorithms. New J. Phys. **22**(2), 023035.(2020)

Yeh, L.-H., Dong, J., Zhong, J., Tian, L., Chen, M., Tang, G., Soltanolkotabi, M., Waller, L.: Experimental robustness of fourier ptychography phase retrieval algorithms. Opt. Express **23**(26), 33214–33240 (2015)

Zheng, G., Horstmeyer, R., Yang, C.: Wide-field, high-resolution fourier ptychographic microscopy. Nat. Photon. **7**, 739–745 (2013)

Zheng, G., Shen, C., Jiang, S., Song, P., Yang, C.: Concept, implementations and applications of fourier ptychography. Nat. Rev. Phys. **3**(3), 207–223 (2021)

Modular ADMM-Based Strategies for Optimized Compression, Restoration, and Distributed Representations of Visual Data

5

Yehuda Dar and Alfred M. Bruckstein

Contents

Introduction	176
Modular ADMM-Based Optimization: General Construction and Guidelines	178
Unconstrained Lagrangian Optimizations via ADMM	178
Employing Black-Box Modules	180
Another Splitting Structure	181
Image Restoration Based on Denoising Modules	183
Modular Optimizations Based on Standard Compression Techniques	185
Preliminaries: Lossy Compression via Operational Rate-Distortion Optimization	185
Restoration by Compression	189
Modular Strategies for Intricate Compression Problems	191
Distributed Representations Using Black-Box Modules	198
The General Framework	198
Modular Optimizations for Holographic Compression of Images	199
Conclusion	202
References	205

Abstract

Iterative techniques are a well-established tool in modern imaging sciences, allowing to address complex optimization problems via sequences of simpler computational processes. This approach has been significantly expanded in recent years by iterative designs where explicit solutions of optimization subproblems were replaced by black-box applications of ready-to-use modules for

Y. Dar (✉)
Electrical and Computer Engineering Department, Rice University, Houston, TX, USA
e-mail: ydar@rice.edu

A. M. Bruckstein (✉)
Computer Science Department, Technion – Israel Institute of Technology, Haifa, Israel
e-mail: freddy@cs.technion.ac.il

denoising or compression. These modular designs are conceptually simple, yet often achieve impressive results. In this chapter, we overview the concept of modular optimization for imaging problems by focusing on structures induced by the alternating direction method of multipliers (ADMM) technique and their applications to intricate restoration and compression problems. We start by emphasizing general guidelines independent of the module type used and only then derive ADMM-based structures relying on denoising and compression methods. The wide perspective on the topic should motivate extensions of the types of problems addressed and the kinds of black boxes utilized by the modular optimization. As an example for a promising research avenue, we present our recent framework employing black-box modules for distributed representations of visual data.

Keywords

Modular optimization · Alternating direction method of multipliers (ADMM) · Inverse problems · Signal compression · Distributed representations

Introduction

During the last several decades, significant attention and efforts were invested in establishing solutions for a wide variety of imaging problems. The proposed methods often rely on models and techniques adapted to visual signals and the relevant problem settings. Naturally, along the contemporary challenges and open questions of the field, there are excellent solutions to various fundamental problems that were extensively studied throughout the years. This situation suggests addressing currently open problems by exploring their relations to existing methods developed for basic tasks.

A lot of work has been devoted to fundamental problems such as denoising of a single image contaminated by additive white Gaussian noise and lossy compression of still images with respect to squared errors as quality assessment measures. Persistent and thorough studies of such basic problems (in their classical settings) led to excellent solutions that are believed to be nearly perfect (see, e.g., Chatterjee and Milanfar 2009). However, the techniques for many other imaging tasks are in various degrees of maturity that leave room for possibly considerable improvements. Examples for types of currently active research lines include jointly addressing multiple imaging tasks (Burger et al. 2018; Corona et al. 2019a,b; Dar et al. 2018a,b,c,d), restoration with uncertainty about the degradation operator (Lai et al. 2016; Bahat et al. 2017), image compression with respect to modern perceptual quality measures (Ballé et al. 2017; Laparra et al. 2017), and tasks (also fundamental ones such as denoising and compression) involving visual data beyond a single natural image (this includes video, hyperspectral, medical, etc.).

In this chapter, we overview a recent and fascinating approach for elegant utilization of existing knowledge and available imaging tools for complex problems of interest. The general idea is to define an optimization problem such that when addressed using a specific iterative optimization technique, the resulting sequential algorithm calls for solving a subproblem corresponding to a fundamental task like denoising or compression. Then, the explicit solutions of the basic subproblem instances may be replaced by black-box applications of available methods, highly perfected over the years due to their prevalence and long-standing importance. Interestingly, the black-box modules utilized do not have to exactly match the formulation of the subproblems they replace, as long as they address the same fundamental task.

The main concept of *modular optimization strategies* described above was preceded by a line of optimization-based iterative algorithms including stages of *explicitly* solving regularized inverse problems, often associated with denoising or maximum a posteriori estimation tasks (see, e.g., Afonso et al. 2010; Zoran and Weiss 2011). Yet, actual employment of image denoisers as black boxes was explicitly proposed only later in the Plug-and-Play Priors framework (Venkatakrishnan et al. 2013; Sreehari et al. 2016), where the alternating direction method of multipliers (ADMM) (Boyd et al. 2011) was used to form iterative structures based on denoising modules to solve inverse imaging problems (specifically, demonstrated by Venkatakrishnan et al. (2013) and Sreehari et al. (2016) for tomographic reconstruction based on the BM3D denoiser (Dabov et al. 2007)). The Plug-and-Play Priors framework (based on ADMM and denoisers) proved very useful to a variety of practical inverse problems (Dar et al. 2016b; Rond et al. 2016; Brifman et al. 2016; Chan et al. 2017; Buzzard et al. 2018; Kwan et al. 2018; Yazaki et al. 2019; Brifman et al. 2019; Ahmad et al. 2019), and its convergence was analyzed for several particular cases (Chan et al. 2017; Chan 2019). Another prominent approach based on denoising modules is the Regularization-by-Denoising (RED) framework (Romano et al. 2017; Hong et al. 2019; Brifman et al. 2019), proposing to regularize the basic problem using the black-box denoising function. Then an efficient sequential procedure based on iterative optimization techniques of ADMM or a fixed-point strategy is called upon, thereby clarifying that modular optimizations can be constructed not only based on ADMM. Other non-ADMM methods using denoisers for restoration or reconstruction problems were proposed based on FISTA for addressing nonlinear problems (Kamilov et al. 2017; Ahmad et al. 2019), primal-dual splitting (Ono 2017), backward projections (Tirer and Giryes 2018a,b, 2019), and ISTA for online updates (Sun et al. 2019a,b). All of these firmly established the wide applicability of denoising-based modular approaches for inverse problems addressing restoration and reconstruction of visual data.

We here consider the modular optimization strategy as a general concept beyond the extensively studied aspect of using denoisers for solving inverse problems. The deviation from the denoising-based modular optimizations started by Dar et al. (2016a, 2018c), and also the related work of Beygi et al. (2017a,b), where inverse problems were addressed based on compression techniques, essentially functioning as complexity regularizers. Specifically, image deblurring and inpainting problems were addressed by Dar et al. (2018c) using JPEG2000 and the state-of-the-art image coding method of the High Efficiency Video Coding (HEVC) standard. Moreover, a shift-invariant regularizer was proposed by Dar et al. (2018c) to amend the limitations of the regular compression-based prior. All these complex problem structures were treated using the ADMM optimization tool in a Plug-and-Play manner.

Another important generalization is due to a recent research line (Dar et al. 2016a, 2018a,b,c,d), branching out from the original Plug-and-Plug Priors framework, suggesting to address intricate compression and restoration problems based on image and video compression modules applied as black boxes. Importantly, this framework shows that modularity is possible not only for priors and that the basic modules employed can be other than denoisers. Furthermore, using standard compression techniques in modular optimization frameworks extends the range of imaging problems addressed far outside the area of inverse problems. This extension is pursued in (Dar et al. 2018a,b,c,d) where systems involving acquisition, compression, and rendering processes are optimized based on ADMM and standard compression techniques. This established the ability to optimize complex systems while being compatible to prevalent compression standards and without using post-processing – thereby emphasizing on the usefulness of modular optimization strategies to much more than using denoisers as black-box priors or in ready-to-use modules. Specifically, the ADMM-based framework in (Dar et al. 2018a,b,c,d) also exhibits how to address intricate rate-distortion optimizations (a fundamental concept in modern compression techniques (Shoham and Gersho 1988; Ortega and Ramchandran 1998; Sullivan and Wiegand 1998)) by decoupling the challenging distortion metric from the actual compression task, consequently enabling the use of standard techniques as modules. Indeed, this idea inspired the nice work reported by Rott Shaham and Michaeli (2018) where an alternating minimization process is used to decouple a perceptual distortion metric from a standard compression technique – thus externally adding desired perceptual aspects into a standardized compression method.

We further point out here on a new direction of developing modular optimizations for distributed representations. In general, ADMM is a technique for distributed optimization, and, therefore, it is natural to utilize its valuable decoupling ability also for optimizations aimed at distributed representations. Specifically, we suggest to employ black-box modules for creating multiple descriptions of a given signal. Therefore, we overview our recent work (Dar and Bruckstein 2021) on holographic compression of images, where standard image compression techniques are adjusted to settings of duplication-based storage systems. The idea is to create a set of standard-compatible representations, all of them being equally important in refining the data reconstruction. We conclude by discussing the general implications of modular optimizations to distributed tasks.

Modular ADMM-Based Optimization: General Construction and Guidelines

Unconstrained Lagrangian Optimizations via ADMM

Consider an arbitrary optimization problem of an unconstrained Lagrangian form, namely,

$$\hat{\mathbf{v}} = \underset{\mathbf{v} \in \mathcal{M}}{\operatorname{argmin}} \ R(\mathbf{v}) + \lambda D(\mathbf{x}, \mathbf{v}) \tag{1}$$

to be solved for the optimization variable $\mathbf{v} \in \mathcal{M} \subset \mathbb{R}^N$, where $\mathcal{M}$ is a (continuous or discrete) subset of the N-dimensional real space. Moreover, the optimization (1) is defined for a given column vector $\mathbf{x} \in \mathbb{R}^M$. In this section, we refer to general scalar-valued functions satisfying $R : \mathcal{M} \to \mathbb{R}$ and $D : \mathbb{R}^M \times \mathbb{R}^N \to \mathbb{R}$. In the sequel discussing the applications for restoration and compression tasks, the general definitions given here take the following form. For restoration tasks, posed as inverse problems, $\mathcal{M}$ is set to be $\mathbb{R}^N$, and the functions R and D implement regularization and fidelity terms, respectively. In the case of compression, $\mathcal{M}$ is a discrete set of decompressed signals supported by the compression architecture, and the functions R and D measure bit-cost and distortion, respectively. While restoration and compression problems introduce various mathematical forms to the general optimization (1), we here address this general structure.

Particular instances of the problem (1) often take challenging forms that require significant engineering and/or computational resources. Addressing a new problem may require the design and implementation of a complete algorithm from scratch, ignoring existing knowledge and tools from potentially related problems. Then, computational difficulties may arise due to high dimensionality of specific instances of (1) such that direct solutions become very costly or even impractical. Such reasons motivate the translation of (1) into a tractable procedure addressing the original task, sometimes in an approximated manner, while avoiding the complications mentioned above. A prominent approach for such designs is described next.

The alternating direction method of multipliers (ADMM) technique (Boyd et al. 2011) is a popular tool for addressing the potentially challenging problem (1). For this we start by splitting the optimization variable such that (1) becomes

$$\hat{\mathbf{v}} = \underset{\mathbf{v} \in \mathcal{M}, \mathbf{z} \in \mathbb{R}^N}{\operatorname{argmin}} \underbrace{R(\mathbf{v}) + \lambda D(\mathbf{x}, \mathbf{z})}_{f} \tag{2}$$
$$\text{subject to} \quad \mathbf{v} = \mathbf{z}$$

where $\mathbf{z} \in \mathbb{R}^N$ is an auxiliary variable that is not directly constrained to the domain $\mathcal{M}$. Next, we apply the scaled form of the augmented Lagrangian and the method of multipliers (Boyd et al. 2011, Ch. 2) on (2) and obtain the iterative procedure

$$\left(\hat{\mathbf{v}}^{(t)}, \hat{\mathbf{z}}^{(t)}\right) = \underset{\mathbf{v} \in \mathcal{M}, \mathbf{z} \in \mathbb{R}^N}{\operatorname{argmin}} \ R(\mathbf{v}) + \lambda D(\mathbf{x}, \mathbf{z}) + \frac{\beta}{2} \left\| \mathbf{v} - \mathbf{z} + \mathbf{u}^{(t)} \right\|_2^2 \tag{3}$$

$$\mathbf{u}^{(t+1)} = \mathbf{u}^{(t)} + \left(\hat{\mathbf{v}}^{(t)} - \hat{\mathbf{z}}^{(t)}\right), \tag{4}$$

where t denotes the iteration index, $\mathbf{u}^{(t)} \in \mathbb{R}^N$ is the scaled dual variable, and β is an auxiliary parameter introduced by the augmented Lagrangian. Then, the ADMM form of the problem is derived by applying one iteration of alternating minimization on (3), yielding a series of simpler optimizations:

$$\hat{\mathbf{v}}^{(t)} = \underset{\mathbf{v} \in \mathcal{M}}{\operatorname{argmin}} \, R(\mathbf{v}) + \frac{\beta}{2} \left\| \mathbf{v} - \tilde{\mathbf{z}}^{(t)} \right\|_2^2 \qquad (5)$$

$$\hat{\mathbf{z}}^{(t)} = \underset{\mathbf{z} \in \mathbb{R}^N}{\operatorname{argmin}} \, \lambda D(\mathbf{x}, \mathbf{z}) + \frac{\beta}{2} \left\| \mathbf{z} - \tilde{\mathbf{v}}^{(t)} \right\|_2^2 \qquad (6)$$

$$\mathbf{u}^{(t+1)} = \mathbf{u}^{(t)} + \left(\hat{\mathbf{v}}^{(t)} - \hat{\mathbf{z}}^{(t)} \right) \qquad (7)$$

where $\tilde{\mathbf{z}}^{(t)} = \hat{\mathbf{z}}^{(t-1)} - \mathbf{u}^{(t)}$ and $\tilde{\mathbf{v}}^{(t)} = \hat{\mathbf{v}}^{(t)} + \mathbf{u}^{(t)}$. Importantly, in the last ADMM-based structure, the possibly nontrivial domain $\mathcal{M}$ and the related function R are decoupled from the second, perhaps intricate, function D. Accordingly, the new subtasks in the process are much simpler. Specifically, note that (6) is a continuous optimization problem over $\mathbb{R}^N$, regardless of the original domain of problem (1) that may be even discrete. Note that in the general case, where R, D, and $\mathcal{M}$ can induce non-convexity and discreteness to the problem, there are no convergence guarantees corresponding to the ADMM process formulated above, and its usefulness should be evaluated empirically. However, this common practice has already provided many useful methods for various applications, and selected examples of those are presented in sections "Image Restoration Based on Denoising Modules" and "Modular Optimizations Based on Standard Compression Techniques".

Employing Black-Box Modules

While the ADMM form in (5), (6), and (7) indeed seems easier to carry out than a complex instance of (1), the explicit definition and deployment of $\mathcal{M}$ and/or R in the optimization stage (5) may still require some engineering efforts (such as design, implementation, etc.). In the case of restoration tasks, this means detailed definitions and implementations of regularization functions. For compression architectures, one should establish binary compressed representations matching signal-domain reconstructions. As explained next, the fundamental idea of using black-box modules is to avoid explicit treatment of such details and still achieve excellent, or even state-of-the-art, results with respect to the actual goal.

The main guideline when addressing a problem based on modular optimization strategies is to *formulate the initial optimization problem* (in our case, an instance of (1)) and *choose an iterative optimization technique* (here, ADMM) *such that* the *resulting sequential process* includes:

- A stage corresponding to a basic problem, having well-established solutions readily available to use. In the developments presented here, we ask to replace the optimization stage (5) with a module applied as a black box and, by that, encapsulating the various aspects of the original problem domain $\mathcal{M}$ and function R. Now, if (5) can be identified as a prototype formulation corresponding to a fundamental problem (e.g., denoising, compression), then one can replace the direct treatments of (5) with application of a module addressing the same basic problem – possibly based on another formulation or even an algorithm that does not correspond to an explicit mathematical expression. Such module is applied as a black box and denoted here as

$$\hat{\mathbf{v}}^{(t)} = BlackBoxModule\left(\tilde{\mathbf{z}}^{(t)}; \theta\left(\beta\right)\right) \qquad (8)$$

where $\theta\left(\beta\right)$ (which will be denoted from now on as θ) is a parameter generalizing the role of the Lagrange multiplier β in determining the *implicit* trade-off between the components appeared in (5) before the replacement with the module. The generic method is summarized in Algorithm 1, where the number of parameters is reduced based on the relation $\tilde{\beta} \triangleq \frac{\beta}{2\lambda}$ such that only the parameters θ and $\tilde{\beta}$ are required as inputs for the method (for simplicity, we do not use the fact that both θ and $\tilde{\beta}$ originally depend on β).

- A subproblem considering the distance function D while having a form that can be practically solved. This refers here to subproblem (6). In many interesting applications, the distance function is a particular case of

$$D(\mathbf{x}, \mathbf{z}) = \sum_{j=1}^{K} \alpha_j \left\|\mathbf{A}_j \mathbf{x} - \mathbf{B}_j \mathbf{z}\right\|_2^2 \qquad (9)$$

for some positive integer K, positive real values $\{\alpha_j\}_{j=1}^{K}$, and matrices $\{\mathbf{A}_j\}_{j=1}^{K} \in \mathbb{R}^{\tilde{N} \times M}$, $\{\mathbf{B}_j\}_{j=1}^{K} \in \mathbb{R}^{\tilde{N} \times N}$. Then, for the form (9), the optimization step is a least squares problem that can be easily addressed for many structures of matrices $\{\mathbf{A}_j\}_{j=1}^{K} \in \mathbb{R}^{\tilde{N} \times M}$ and $\{\mathbf{B}_j\}_{j=1}^{K} \in \mathbb{R}^{\tilde{N} \times N}$.

One should note that the modular optimization process in Algorithm 1 provides a result that is an output of the black-box module applied in the last iteration. This eventual output can be the signal $\hat{\mathbf{v}}^{(t)} \in \mathcal{M}$ produced by the module at the last iteration and/or other relevant data possibly outputted by the module. This structure is useful, for example, in the case of compression where the important output is a binary compressed representation (i.e., a direct output of the module which is coupled with the signal $\hat{\mathbf{v}}^{(t)} \in \mathcal{M}$). Various applications may benefit from an alternative application that is described next.

Another Splitting Structure

We now turn to describe the construction of a process mirroring Algorithm 1 and utilized often for restoration and reconstruction problems. For the developments

Algorithm 1 General Modular Optimization – Type I: Overall Results Are Module Outputs
___
1: Inputs: $\mathbf{x}, \theta, \tilde{\beta}$.
2: Initialize $t = 0$, $\hat{\mathbf{z}}^{(0)} = \mathbf{x}$, $\mathbf{u}^{(1)} = \mathbf{0}$.
3: **repeat**
4: $\quad t \leftarrow t + 1$
5: $\quad \tilde{\mathbf{z}}^{(t)} = \hat{\mathbf{z}}^{(t-1)} - \mathbf{u}^{(t)}$
6: $\quad \hat{\mathbf{v}}^{(t)} = BlackBoxModule\left(\tilde{\mathbf{z}}^{(t)}; \theta\right)$
7: $\quad \tilde{\mathbf{v}}^{(t)} = \hat{\mathbf{v}}^{(t)} + \mathbf{u}^{(t)}$
8: $\quad \hat{\mathbf{z}}^{(t)} = \underset{\mathbf{z} \in \mathbb{R}^N}{\mathrm{argmin}}\, D(\mathbf{x}, \mathbf{z}) + \tilde{\beta} \left\| \mathbf{z} - \tilde{\mathbf{v}}^{(t)} \right\|_2^2$
9: $\quad \mathbf{u}^{(t+1)} = \mathbf{u}^{(t)} + \left(\hat{\mathbf{v}}^{(t)} - \hat{\mathbf{z}}^{(t)}\right)$
10: **until** stopping criterion is satisfied
11: Output: $\hat{\mathbf{v}}^{(t)}$ and/or other application-specific outputs of *BlackBoxModule*.
___

overviewed, here, we assume that the output domain of the basic optimization problem (1) satisfies $\mathcal{M} = \mathbb{R}^N$.

The alternative process stems from a delicate difference in the variable splitting applied on the basic problem, namely, instead of (2), we write

$$\hat{\mathbf{v}} = \underset{\mathbf{v} \in \mathbb{R}^N, \mathbf{z} \in \mathbb{R}^N}{\mathrm{argmin}}\, R(\mathbf{z}) + \lambda D(\mathbf{x}, \mathbf{v}) \tag{10}$$
$$\text{subject to} \quad \mathbf{v} = \mathbf{z}$$

where $\mathbf{z} \in \mathbb{R}^N$ is an auxiliary variable used here to replace the occurrence of $\mathbf{v}$ as the argument of R, whereas the function D still refers to $\mathbf{v}$ (note the difference from the variable splitting described in (2)). Then, similarly to section "Unconstrained Lagrangian Optimizations via ADMM", further developing (10) using the scaled form of the augmented Lagrangian, the method of multipliers, and alternating minimization gives

$$\hat{\mathbf{v}}^{(t)} = \underset{\mathbf{v} \in \mathbb{R}^N}{\mathrm{argmin}}\, \lambda D(\mathbf{x}, \mathbf{v}) + \frac{\beta}{2} \left\| \mathbf{v} - \tilde{\mathbf{z}}^{(t)} \right\|_2^2 \tag{11}$$

$$\hat{\mathbf{z}}^{(t)} = \underset{\mathbf{z} \in \mathbb{R}^N}{\mathrm{argmin}}\, R(\mathbf{z}) + \frac{\beta}{2} \left\| \mathbf{z} - \tilde{\mathbf{v}}^{(t)} \right\|_2^2 \tag{12}$$

$$\mathbf{u}^{(t+1)} = \mathbf{u}^{(t)} + \left(\hat{\mathbf{v}}^{(t)} - \hat{\mathbf{z}}^{(t)}\right) \tag{13}$$

where $\tilde{\mathbf{z}}^{(t)} = \hat{\mathbf{z}}^{(t-1)} - \mathbf{u}^{(t)}$ and $\tilde{\mathbf{v}}^{(t)} = \hat{\mathbf{v}}^{(t)} + \mathbf{u}^{(t)}$. Note that the current procedure in (11), (12), and (13) includes the same subproblems as in (5), (6), and (7) but in a different order (and also up to the setting $\mathcal{M} = \mathbb{R}^N$ used in this subsection).

Like in section "Employing Black-Box Modules", we identify the stage considering the function R, here in (12), as a solution to a fundamental problem that can be replaced by an available black-box implementation. This yields the process described in Algorithm 2. Note that the result of the procedure is not a direct output of the black-box module. This delicate change with respect to Algorithm 1 may lead to improved results in various applications such as image restoration (where the black-box module is utilized for regularization purposes and, in practice, it is often better not to use its output directly as the result of the entire procedure).

Algorithm 2 General Modular Optimization – Type II: Overall Results Are Not Module Outputs

1: Inputs: $\mathbf{x}, \theta, \tilde{\beta}$.
2: Initialize $t = 0$, $\hat{\mathbf{z}}^{(0)} = \mathbf{x}$, $\mathbf{u}^{(1)} = \mathbf{0}$.
3: **repeat**
4: $\quad t \leftarrow t + 1$
5: $\quad \tilde{\mathbf{z}}^{(t)} = \hat{\mathbf{z}}^{(t-1)} - \mathbf{u}^{(t)}$
6: $\quad \hat{\mathbf{v}}^{(t)} = \underset{\mathbf{v} \in \mathbb{R}^N}{\operatorname{argmin}}\, D(\mathbf{x}, \mathbf{v}) + \tilde{\beta} \left\| \mathbf{v} - \tilde{\mathbf{z}}^{(t)} \right\|_2^2$
7: $\quad \tilde{\mathbf{v}}^{(t)} = \hat{\mathbf{v}}^{(t)} + \mathbf{u}^{(t)}$
8: $\quad \hat{\mathbf{z}}^{(t)} = BlackBoxModule\left(\tilde{\mathbf{v}}^{(t)}; \theta\right)$
9: $\quad \mathbf{u}^{(t+1)} = \mathbf{u}^{(t)} + \left(\hat{\mathbf{v}}^{(t)} - \hat{\mathbf{z}}^{(t)}\right)$
10: **until** stopping criterion is satisfied
11: Output: $\hat{\mathbf{v}}^{(t)}$.

Image Restoration Based on Denoising Modules

In the previous section, we presented the modular optimization approach in its general form, independent of the type of tasks addressed and modules utilized. In this section, we focus on the prevalent application of denoising-based modular optimizations to image restoration problems.

The problem setting considered in this section is defined as follows. A signal $\mathbf{v}_0 \in \mathbb{R}^N$ is going through a degradation process, resulting in the observation $\mathbf{x} \in \mathbb{R}^M$ satisfying

$$\mathbf{x} = \mathbf{H}\mathbf{v}_0 + \mathbf{n}, \qquad (14)$$

where $\mathbf{H}$ is a $M \times N$ real matrix and $\mathbf{n}$ is a white Gaussian noise column-vector of length M (the noise components are zero mean and have variance σ_n^2). The restoration task is to estimate the unknown $\mathbf{v}_0$, given $\mathbf{x}$ and the knowledge of the degradation operator $\mathbf{H}$ and the noise variance σ_n^2. For the purpose of restoration, we define the function D from (1) as the fidelity term of the respective inverse problem, namely,

$$D(\mathbf{x}, \mathbf{v}) = \left\| \mathbf{x} - \mathbf{H}\mathbf{v} \right\|_2^2. \qquad (15)$$

The ADMM optimization structure based on black-box denoisers, first proposed in the Plug-and-Play Priors design (Venkatakrishnan et al. 2013; Sreehari et al. 2016), mainly stems from associating the function $R : \mathbb{R}^N \to \mathbb{R}$ with a regularizer implemented (explicitly or implicitly) in a ready-to-use denoising process. Then, the optimization for the $\hat{\mathbf{z}}^{(t)}$, appearing in (12), can be interpreted as an inverse problem for denoising $\tilde{\mathbf{v}}^{(t)}$ using the regularizer R. One can also perceive (12) as a maximum a posteriori (MAP) estimation of a signal from its noisy version $\tilde{\mathbf{v}}^{(t)}$, i.e.,

$$\hat{\mathbf{z}}^{(t)} = \operatorname*{argmax}_{\mathbf{z} \in \mathbb{R}^N} \log p_R(\mathbf{z}) + \log p_\eta \left(\tilde{\mathbf{v}}^{(t)} - \mathbf{z} \right) \tag{16}$$

where $p_R(\mathbf{z}) \triangleq exp(-R(\mathbf{z}))$ is the prior probability function assumed for the clean signal and p_η is the probability density function of an additive Gaussian noise vector η with i.i.d. components having zero mean and $1/\beta$ variance. Accordingly, the correspondence of (12) to denoising problems motivates the usage of black-box denoisers as the modules applied at stage 8 of Algorithm 2. These denoisers should be set to remove noise having variance of $1/\beta$ from the signal $\tilde{\mathbf{v}}^{(t)}$. Importantly, the substitution of (12) with applications of Gaussian denoisers was experimentally shown beneficial also for denoisers that do not follow the MAP estimation form or the regularized inverse problem approach. Specifically, one can even employ algorithmic denoisers that were designed based on completely different mindsets. The denoising-based restoration procedure for an arbitrary degradation operator $\mathbf{H}$ is summarized in Algorithm 3.

The decoupling induced by the ADMM structure leads to an additional conceptual simplification: stage 6 of Algorithm 3 can be interpreted as a ℓ_2-constrained deconvolution problem (or ℓ_2-regularized least squares computation) with respect to the degradation operator $\mathbf{H}$. Note that this is one of the simplest restoration formulations addressing the degradation process (14) from the regularized inverse-problem perspective. The corresponding analytic solution is

$$\hat{\mathbf{v}}^{(t)} = \left(\mathbf{H}^T \mathbf{H} + \tilde{\beta} \mathbf{I} \right)^{-1} \left(\mathbf{H}^T \mathbf{x} + \tilde{\beta} \tilde{\mathbf{z}}^{(t)} \right). \tag{17}$$

Alternatively, this computation can be numerically applied in various tractable ways (depending on the specific structure of $\mathbf{H}$). In summary, the overall modular restoration process relies on sequential application of conceptually simple tasks: Gaussian denoising and ℓ_2-constrained deconvolution.

Figures 1 and 2 show typical results obtained using the Plug-and-Play method, implemented in the code published with Chan et al. (2017), based on the BM3D denoiser (Dabov et al. 2007). The deblurring settings (Fig. 1) include a blur operator corresponding to a 9×9 pixels convolution kernel (Gaussian with standard deviation 1.75) and additive white Gaussian noise of standard deviation 10. The inpainting experiment (Fig. 2) considers 80% missing pixels and additive white Gaussian noise of standard deviation 10. Specifically, note the improvement in the PSNR of the intermediate estimates, $\hat{\mathbf{v}}^{(t)}$, evolving throughout the process iterations until

Algorithm 3 Restoration Based on Denoising Modules

1: Inputs: $\mathbf{x}, \theta, \tilde{\beta}$.
2: Initialize $t = 0$, $\hat{\mathbf{z}}^{(0)} = \mathbf{x}$, $\mathbf{u}^{(1)} = \mathbf{0}$.
3: **repeat**
4:     $t \leftarrow t + 1$
5:     $\tilde{\mathbf{z}}^{(t)} = \hat{\mathbf{z}}^{(t-1)} - \mathbf{u}^{(t)}$
6:     $\hat{\mathbf{v}}^{(t)} = \underset{\mathbf{v} \in \mathbb{R}^N}{\operatorname{argmin}} \left\| \mathbf{x} - \mathbf{H}\mathbf{v} \right\|_2^2 + \tilde{\beta} \left\| \mathbf{v} - \tilde{\mathbf{z}}^{(t)} \right\|_2^2$
7:     $\tilde{\mathbf{v}}^{(t)} = \hat{\mathbf{v}}^{(t)} + \mathbf{u}^{(t)}$
8:     $\hat{\mathbf{z}}^{(t)} = Denoiser\left(\tilde{\mathbf{v}}^{(t)}; \theta\right)$
9:     $\mathbf{u}^{(t+1)} = \mathbf{u}^{(t)} + \left(\hat{\mathbf{v}}^{(t)} - \hat{\mathbf{z}}^{(t)}\right)$
10: **until** stopping criterion is satisfied
11: Output: $\hat{\mathbf{v}}^{(t)}$.

practical convergence (Figs. 1d and 2d). See Chan et al. (2017) for details and analysis of the convergence appearing here.

Modular Optimizations Based on Standard Compression Techniques

In this section, we overview the utilization of compression modules for restoration and challenging compression purposes. The use of modules beyond denoisers further establishes the modularity property as a general idea, relevant to various tasks.

Preliminaries: Lossy Compression via Operational Rate-Distortion Optimization

Consider a signal, $\mathbf{x} \in \mathbb{R}^N$, to be compressed and represented as a sequence of bits. We describe a lossy compression procedure as the function

$$C : \mathbb{R}^N \to \mathcal{B}, \tag{18}$$

mapping the N-dimensional signal domain to a discrete set $\mathcal{B}$ of compressed representations in variable-length binary forms. The compression of $\mathbf{x}$ is

$$\mathbf{b} = C(\mathbf{x}), \tag{19}$$

where $\mathbf{b} \in \mathcal{B}$ is the binary compressed data useful for storage or transmission. Then, a matching decompression process gets the compressed data $\mathbf{b}$ as its input and reconstructs a signal-domain representation via

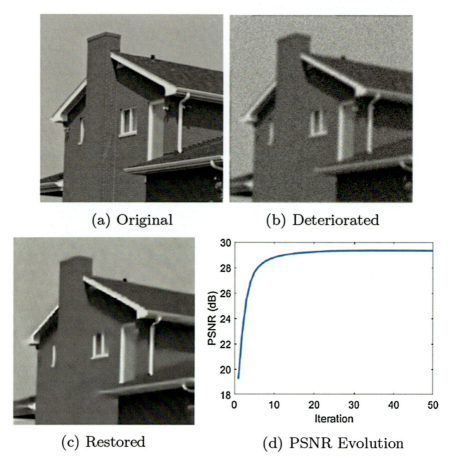

Fig. 1 Deblurring using denoising-based Plug-and-Play method (Chan et al. 2017). The utilized denoiser is BM3D (Dabov et al. 2007). The degradation includes a Gaussian blur (of 9×9 pixels kernel and 1.75 standard deviation), followed by additive noise with $\sigma_n = 10$, applied on the House image (256×256 pixels). (**a**) The original image. (**b**) Deteriorated image. (**c**) Restored image using the method by Chan et al. (2017) (29.33 dB). (**d**) The PSNR evolution of the intermediate estimate $\hat{\mathbf{v}}^{(t)}$ along the restoration-process iterations

$$\mathbf{v} = F(\mathbf{b}), \tag{20}$$

where

$$F : \mathcal{B} \to \mathcal{S} \tag{21}$$

maps the binary compressed representations in $\mathcal{B}$ to their corresponding decompressed signals from the discrete set $\mathcal{S} \subset \mathbb{R}^N$. The decompressed signal $\mathbf{v}$ can be

5 Modular ADMM-Based Strategies for Optimized Compression, Restoration,...

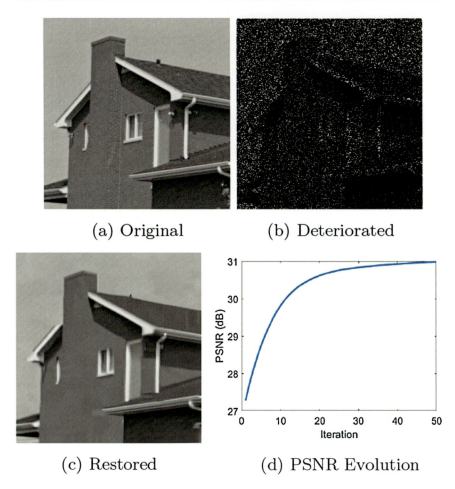

Fig. 2 Inpainting using denoising-based Plug-and-Play method (Chan et al. 2017). The employed denoiser is BM3D (Dabov et al. 2007). The degradation includes 80% missing pixels and additive noise with $\sigma_n = 10$, applied on the House image (256×256 pixels). (**a**) The original image. (**b**) Deteriorated image. (**c**) Restored image using the method by Chan et al. (2017) (30.98 dB). (**d**) The PSNR evolution of the intermediate estimate $\hat{\mathbf{v}}^{(t)}$ along the iterations

further processed or outputted to a user. For example, in the case of visual signals, $\mathbf{v}$ is usually displayed.

Modern compression architectures (see, e.g., Ortega and Ramchandran 1998; Sullivan and Wiegand 1998; Shukla et al. 2005; Sullivan et al. 2012) implement the compression function C using operational rate-distortion optimizations, a tool established by Shoham and Gersho (1988), Chou et al. (1989), and Ortega and Ramchandran (1998), and can be explained using our notions as follows. A given deterministic signal $\mathbf{x}$ is compressed based on an optimization process searching for its best compressed representation $\mathbf{b} \in \mathcal{B}$, coupled with the decompressed signal

$\mathbf{v} \in \mathcal{S}$. The optimization trades off two opposing aspects of the representation: bit-cost and reconstruction quality. The bit-cost of the binary representation $\mathbf{b} \in \mathcal{B}$ is its length. Since, by (20), each $\mathbf{b} \in \mathcal{B}$ corresponds to one decompressed signal $\mathbf{v} \in \mathcal{S}$, we define the bit-cost of a decompressed signal $\mathbf{v} \in \mathcal{S}$ as the length of its binary representation $\mathbf{b} = F^{-1}(\mathbf{v})$. We also define the function $R_\mathcal{S}(\mathbf{v})$ to evaluate the bit-cost of the compressed binary representation associated with $\mathbf{v}$. Specifically, for $\mathbf{v} \in \mathcal{S}$ that satisfies $\mathbf{v} = F(\mathbf{b})$, the bit-cost is

$$R_\mathcal{S}(\mathbf{v}) \triangleq \text{length}\{\mathbf{b}\}, \tag{22}$$

where length$\{\cdot\}$ counts the length of a binary description. The second part of the trade-off is the reconstruction distortion, $D(\mathbf{x}, \mathbf{v})$, evaluating the distance between the compression input $\mathbf{x}$ and its decompressed form $\mathbf{v}$. Note that the distortion value is real and nonnegative. Then, the optimization task including bit-cost constraints, corresponding to storage space or transmission bandwidth limitations, is

$$\hat{\mathbf{v}} = \underset{\mathbf{v} \in \mathcal{S}}{\text{argmin}}\ D(\mathbf{x}, \mathbf{v})$$
$$\text{subject to}\quad R_\mathcal{S}(\mathbf{v}) \leq r \tag{23}$$

where $r \geq 0$ is the maximal representation bit-cost allowed. Another relevant optimization problem, mirroring (23), is defined to minimize the compression bit-cost under a limited distortion amount, i.e.,

$$\hat{\mathbf{v}} = \underset{\mathbf{v} \in \mathcal{S}}{\text{argmin}}\ R_\mathcal{S}(\mathbf{v})$$
$$\text{subject to}\quad D(\mathbf{x}, \mathbf{v}) \leq d \tag{24}$$

where $d \geq 0$ is the tolerated distortion level. Without loss of generality, we consider here the optimization form in (24). The constrained optimization (24) is usually cast (see, e.g., Shoham and Gersho 1988, Chou et al. 1989, Ortega and Ramchandran 1998, Sullivan and Wiegand 1998, Shukla et al. 2005, and Sullivan et al. 2012) to its unconstrained Lagrangian form

$$\hat{\mathbf{v}} = \underset{\mathbf{v} \in \mathcal{S}}{\text{argmin}}\ R_\mathcal{S}(\mathbf{v}) + \lambda D(\mathbf{x}, \mathbf{v}) \tag{25}$$

where $\lambda \geq 0$ is a Lagrange multiplier corresponding to a distortion constraint $d_\lambda \geq 0$. Such compression without a prespecified distortion level is common, e.g., in video coding (Sullivan et al. 2012).

When working with high-dimensional signals (large N values), the discrete set $\mathcal{S}$ tends to be huge. Then, for arbitrarily structured distortion metrics $D(\mathbf{x}, \mathbf{v})$, one cannot directly solve the Lagrangian form in (25) via iterating over the elements in $\mathcal{S}$ and evaluating their corresponding costs (recall that (25) is a discrete optimization problem). Accordingly, compression methods are designed such that the combination of $D(\mathbf{x}, \mathbf{v})$, $\mathcal{S}$, and $\mathcal{B}$ leads to a computationally tractable task. This is often obtained using architectures where nonoverlapping signal segments are independently compressed with respect to the squared-error distortion measure (see details in the Appendix). However, while such computationally efficient architectures that rely on squared-error metrics are prevalent (we also refer them as *standard* compression techniques), they are often too simple and limit the compression performance one could wish for in various settings of interest. This will be further demonstrated in section "Modular Strategies for Intricate Compression Problems".

Restoration by Compression

Regularization of inverse problems based on complexity measures is a well-established approach for estimation tasks (see, e.g., Rissanen 2000). In a subclass of these methods, complexity is defined based on the number of bits required for the compressed representation of the candidate estimate. This motivated various studies of signal and image denoising using lossy compression techniques (see, for example, Natarajan 1995 and Liu and Moulin 2001). The extension of this idea to image restoration problems beyond Gaussian denoising was studied from a theoretical perspective by Moulin and Liu (2000), also including a limited experimental demonstration for Poisson denoising based on a particularly designed compression process. Implementing the compression-based approach for other image restoration problems (such as deblurring, inpainting, super resolution, etc.) was considered as impractical for a long while until the *Restoration by Compression* architecture (Dar et al. 2016a, 2018c) resolved the computational difficulties via ADMM-based modularity. Next, we overview the main construction and applicative aspects of the *Restoration by Compression* idea as a special case of the generic modular optimization designs presented above.

The core idea in the *Restoration by Compression* approach (Dar et al. 2016a, 2018c) is to exploit existing compression techniques such that their underlying signal models will be indirectly used for desired restoration purposes. For this, we define the function R in (1) as a complexity regularizer, measuring the likelihood of a signal based on its compression bit-cost (assuming that more probable signals receive shorter compressed representations). Specifically, the regularizer extends the bit-cost evaluation function (22) such that for any $\mathbf{z} \in \mathbb{R}^N$ it returns

$$R(\mathbf{z}) = \begin{cases} R_{\mathcal{S}}(\mathbf{z}) & \text{for } \mathbf{z} \in \mathcal{S} \\ \infty & \text{for } \mathbf{z} \notin \mathcal{S} \end{cases}, \tag{26}$$

where $\mathcal{S}$ and $R_{\mathcal{S}}$ are conceptually associated with an existing compression technique. Then, considering the complexity regularizer (26), the optimization for the $\hat{\mathbf{z}}^{(t)}$ in (12) becomes equivalent to an operational rate-distortion optimization

(25) with respect to the implicit architecture of the ready-to-use compression method. This motivates the replacement of (12) with an application of a black-box compression module, followed by its respective decompression process, i.e.,

$$\mathbf{b}^{(t)} = StandardCompression\left(\tilde{\mathbf{v}}^{(t)}; \theta\right) \tag{27}$$

$$\hat{\mathbf{z}}^{(t)} = StandardDecompression\left(\mathbf{b}^{(t)}\right). \tag{28}$$

Note that the application of both compression and decompression is in accordance with the optimization form in (25) that looks for the optimal decompressed signal corresponding to the given signal to compress. Interestingly, the utilized compression modules do not have to rely on rate-distortion optimizations (25), as in the case of transform coding architectures (such as the JPEG2000 method included in the following demonstrations). The Restoration by Compression procedure is summarized in Algorithm 4. See Dar et al. (2018c) for further detail on the parameters given to the compression modules in the iterative process. Moreover, the main concepts of the proposed algorithm are explained by Dar et al. (2018c) using rate-distortion theory for cyclo-stationary Gaussian signals.

Algorithm 4 Restoration by Compression: Basic Complexity Regularization

1: Inputs: $\mathbf{x}, \theta, \tilde{\beta}$.
2: Initialize $t = 0$, $\hat{\mathbf{z}}^{(0)} = \mathbf{x}$, $\mathbf{u}^{(1)} = \mathbf{0}$.
3: **repeat**
4: $\quad t \leftarrow t + 1$
5: $\quad \tilde{\mathbf{z}}^{(t)} = \hat{\mathbf{z}}^{(t-1)} - \mathbf{u}^{(t)}$
6: $\quad \hat{\mathbf{v}}^{(t)} = \underset{\mathbf{v} \in \mathbb{R}^N}{\operatorname{argmin}} \left\| \mathbf{x} - \mathbf{H}\mathbf{v} \right\|_2^2 + \tilde{\beta} \left\| \mathbf{v} - \tilde{\mathbf{z}}^{(t)} \right\|_2^2$
7: $\quad \tilde{\mathbf{v}}^{(t)} = \hat{\mathbf{v}}^{(t)} + \mathbf{u}^{(t)}$
8: $\quad \mathbf{b}^{(t)} = StandardCompression\left(\tilde{\mathbf{v}}^{(t)}; \theta\right)$
9: $\quad \hat{\mathbf{z}}^{(t)} = StandardDecompression\left(\mathbf{b}^{(t)}\right)$
10: $\quad \mathbf{u}^{(t+1)} = \mathbf{u}^{(t)} + \left(\hat{\mathbf{v}}^{(t)} - \hat{\mathbf{z}}^{(t)}\right)$
11: **until** stopping criterion is satisfied
12: Output: $\hat{\mathbf{v}}^{(t)}$.

Clearly, the artifacts introduced by the compression module participating in restoration process affect the produced estimate. Many compression artifacts originate in the common approach of independently coding nonoverlapping segments of the image. This block-based design also influences the complexity measure defining the regularizer in (26), essentially equivalent to summing the compression bit-costs of all the nonoverlapping blocks. This aspect was identified by Dar et al. (2018c) as introducing shift sensitivity into the regularizer (26). Accordingly, a shift-invariant complexity regularizer was proposed by Dar et al. (2018c), measuring the total bit-cost of *all the shifted versions* of the estimate evaluated. The shift operator $shift_j \{\cdot\}$

can be defined as a two-dimensional cyclical shift on an image or, alternatively, as returning the rectangular portion of the image that starts at a shifted coordinate from the upper-left corner pixel of the full image (see Dar et al. 2018c for details). For each $j \in \{1, \ldots, N_b\}$, the two-dimensional offset applied by $shift_j \{\cdot\}$ is different. This leads to an extended Restoration by Compression process, described in Algorithm 5, including ADMM-based decoupling of the compressions of various shifts of the processed signals. Further details on the shift-invariant regularizer and the algorithm development are provided in Dar et al. (2018c). The applications of Algorithm 5 for deblurring and inpainting of images are presented in Figs. 3 and 4, respectively. The compression modules employed are JPEG2000 and HEVC (in its BPG implementation for image coding (Bellard)). Since HEVC provides significantly better compression performance than JPEG2000, a corresponding gap in their restoration abilities is also evident.

Algorithm 5 Restoration by Compression: Shift-Invariant Complexity Regularization

---
1: Inputs: $\mathbf{y}, \theta, \tilde{\beta}$, and the number of shifts N_b.
2: Initialize $\left\{ \hat{\mathbf{z}}^{j,(0)} \right\}_{j=1}^{N_b}$ (depending on the deterioration type).
3: $t = 1$ and $\mathbf{u}^{j,(1)} = \mathbf{0}$ for $j = 1, \ldots, N_b$.
4: **repeat**
5: $\quad \tilde{\mathbf{z}}^{j,(t)} = \hat{\mathbf{z}}^{j,(t-1)} - \mathbf{u}^{j,(t)}$ for $j = 1, \ldots, N_b$
6: $\quad$ Solve the ℓ_2-constrained deconvolution:
$$\hat{\mathbf{v}}^{(t)} = \underset{\mathbf{v} \in \mathbb{R}^N}{\arg\min} \left\| \mathbf{x} - \mathbf{H}\mathbf{v} \right\|_2^2 + \tilde{\beta} \sum_{j=1}^{N_b} \left\| \mathbf{v} - \tilde{\mathbf{z}}^{j,(t)} \right\|_2^2$$
7: $\quad$ **for** $j = 1, \ldots, N_b$ **do**
8: $\quad\quad \tilde{\mathbf{v}}_{shifted}^{j,(t)} = shift_j \left\{ \hat{\mathbf{v}}^{(t)} + \mathbf{u}^{j,(t)} \right\}$
9: $\quad\quad \mathbf{b}^{j,(t)} = StandardCompression \left(\tilde{\mathbf{v}}_{shifted}^{j,(t)}; \theta \right)$
10: $\quad\quad \hat{\mathbf{z}}_{shifted}^{j,(t)} = StandardDecompression \left(\mathbf{b}^{j,(t)} \right)$
11: $\quad\quad \hat{\mathbf{z}}^{j,(t)} = shift_j^{-1} \left\{ \hat{\mathbf{z}}_{shifted}^{j,(t)} \right\}$
12: $\quad\quad \mathbf{u}^{j,(t+1)} = \mathbf{u}^{j,(t)} + \left(\hat{\mathbf{v}}^{(t)} - \hat{\mathbf{z}}^{j,(t)} \right)$
13: $\quad$ **end for**
14: $\quad t \leftarrow t + 1$
15: **until** stopping criterion is satisfied
16: Output: $\hat{\mathbf{v}}^{(t)}$.

Modular Strategies for Intricate Compression Problems

The utilization of available compression methods in modular restoration processes (Algorithms 4 and 5) naturally raises the question whether modular optimization strategies are relevant also to intricate compression problems. This is indeed the case, as established by the *System-Aware Compression* framework

Fig. 3 The deblurring experiment (settings #2 in Dar et al. 2018c) for the Cameraman image (256 × 256 pixels). (**a**) The underlying image. (**b**) Degraded image (20.76 dB). (**c**) Restored image using Algorithm 5 with JPEG2000 compression (28.10 dB). (**d**) Restored image using Algorithm 5 with HEVC compression (30.14 dB)

Fig. 4 The inpainting experiment (80% missing pixels) for the Barbara image (512 × 512 pixels). (**a**) The original image. (**b**) Deteriorated image. (**c**) Restored image using Algorithm 5 with JPEG2000 compression (24.83 dB). (**d**) Restored image using Algorithm 5 with HEVC compression (28.80 dB)

(Dar et al. 2018a,b,d), where ADMM-based modular strategies are employed for optimizing end-to-end performance of systems involving acquisition, compression, and rendering stages. The main idea is to decouple unusual distortion metrics from the actual compression tasks that, in turn, can be applied using black-box compression modules (which are operated with respect to the elementary squared-error metric). Hence, this methodology opens a new research path for addressing complex compression problems including, for example, optimizations for nonlocal processing/prediction architectures, enhancement filters or degradation processes, and perceptual metrics assessing subjective quality of audio/visual signals. Indeed, a successful implementation of this approach for perceptually oriented image compression (using an alternating minimization procedure) was proposed by Rott Shaham and Michaeli (2018).

In this section, we overview the *System-Aware Compression* concept (Dar et al. 2018a,b,d), demonstrating the main aspects of using modular optimizations for intricate compression problems. The motivation for the *System-Aware Compression* framework stems from a structure common to many imaging systems (see Fig. 5), where a source signal is first acquired, then compressed for its storage or transmission, and eventually decompressed and rendered back into a signal that can be displayed or further processed. Obviously, in such systems, the quality of the eventual output depends on the entire acquisition-rendering chain and not solely on the lossy compression component. Yet, the employed compression technique is often system independent, hence inducing suboptimal rate-distortion performance for the entire system. The *System-Aware Compression* architecture is a practical and modular way for optimizing the end-to-end performance (in its rate-distortion trade-off sense) of such acquisition-rendering systems.

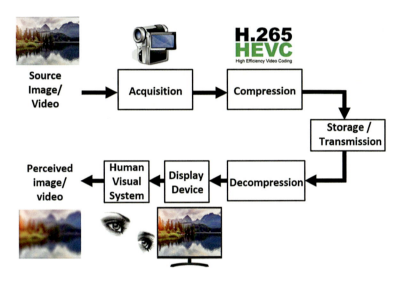

Fig. 5 The general imaging system structure motivating the System-Aware Compression approach

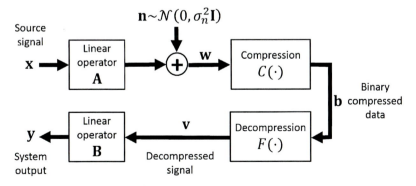

Fig. 6 The system model addressed by the System-Aware Compression framework

Let us describe the system structure considered for the mathematical development of the method (Fig. 6). A source signal, an N-length column vector $\mathbf{x} \in \mathbb{R}^N$, undergoes a linear processing represented by the $M \times N$ matrix $\mathbf{A}$ and, then, deteriorated by an additive white Gaussian noise vector $\mathbf{n} \sim \mathcal{N}\left(0, \sigma_n^2 \mathbf{I}\right)$, resulting in the signal

$$\mathbf{w} = \mathbf{A}\mathbf{x} + \mathbf{n} \qquad (29)$$

where $\mathbf{w}$ and $\mathbf{n}$ are M-length column vectors. We represent the lossy compression procedure via the mapping $C : \mathbb{R}^M \to \mathcal{B}$ from the M-dimensional signal domain to a discrete set $\mathcal{B}$ of binary compressed representations (which may have different lengths). The signal $\mathbf{w}$ is the input to the compression component of the system, producing the compressed binary data $\mathbf{b} = C(\mathbf{w})$ that can be stored or transmitted in an error-free manner. Then, on a device and settings depending on the specific application, the compressed data $\mathbf{b} \in \mathcal{B}$ is decompressed to provide the signal $\mathbf{v} = F(\mathbf{b})$ where $F : \mathcal{B} \to \mathcal{S}$ represents the decompression mapping between the binary compressed representations in $\mathcal{B}$ to the corresponding decompressed signals in the discrete set $\mathcal{S} \subset \mathbb{R}^M$. The decompressed signal $\mathbf{v}$ is further processed by the linear operator denoted as the $N \times M$ matrix $\mathbf{B}$, resulting in the system output signal

$$\mathbf{y} = \mathbf{B}\mathbf{v}, \qquad (30)$$

which is an N-length real-valued column vector.

As an example, consider an acquisition-compression-rendering system where the signal $\mathbf{w}$ is a sampled version of the source signal $\mathbf{x}$ and the system output $\mathbf{y}$ is the rendered version of the decompressed signal $\mathbf{v}$.

We assume here that the operators $\mathbf{A}$ and $\mathbf{B}$, as well as the noise variance σ_n^2, are known and fixed (i.e., cannot be optimized). Consequently, we formulate a new compression procedure in order to optimize the end-to-end rate-distortion performance of the entire system. Specifically, we want the system output $\mathbf{y}$ to be the best approximation of the source signal $\mathbf{x}$ under the bit-budget constraint. However, at the compression stage, we do not accurately know $\mathbf{x}$ but rather its degraded form $\mathbf{w}$ formulated in (29). This motivates us to suggest the following distortion metric with respect to the system output $\mathbf{y}$

$$D_s(\mathbf{w}, \mathbf{y}) = \frac{1}{M} \|\mathbf{w} - \mathbf{A}\mathbf{y}\|_2^2. \qquad (31)$$

This metric conforms with the fact that if $\mathbf{y}$ is close to $\mathbf{x}$, then, by (29), $\mathbf{w}$ will be close to $\mathbf{A}\mathbf{y}$ up to the noise $\mathbf{n}$. Indeed, for the ideal case of $\mathbf{y} = \mathbf{x}$, the metric (31) becomes

$$D_s(\mathbf{w}, \mathbf{x}) = \frac{1}{M} \|\mathbf{n}\|_2^2 \approx \sigma_n^2 \qquad (32)$$

where the last approximate equality is under the assumption of a sufficiently large M (the length of $\mathbf{n}$). Since $\mathbf{y} = \mathbf{B}\mathbf{v}$, we can rewrite the distortion $D_s(\mathbf{w}, \mathbf{y})$ in (31) as a function of the decompressed signal $\mathbf{v}$, namely,

$$D_c(\mathbf{w}, \mathbf{v}) = \frac{1}{M} \left\| \mathbf{w} - \mathbf{ABv} \right\|_2^2. \tag{33}$$

Since the operator $\mathbf{B}$ produces the output signal $\mathbf{y}$, an ideal result will be $\mathbf{y} = \mathbf{P}_B \mathbf{x}$, where $\mathbf{P}_B$ is the matrix projecting onto $\mathbf{B}$'s range. The corresponding ideal distortion is

$$d_0 \triangleq D_s(\mathbf{w}, \mathbf{P}_B \mathbf{x}) = \frac{1}{M} \left\| \mathbf{A}(\mathbf{I} - \mathbf{P}_B)\mathbf{x} + \mathbf{n} \right\|_2^2. \tag{34}$$

We use the distortion metric (33) to constrain the bit-cost minimization in the following rate-distortion optimization

$$\begin{aligned} \hat{\mathbf{v}} &= \underset{\mathbf{v} \in \mathcal{S}}{\operatorname{argmin}} \quad R(\mathbf{v}) \\ \text{subject to} &\quad d_0 \leq \frac{1}{M} \left\| \mathbf{w} - \mathbf{ABv} \right\|_2^2 \leq d_0 + d \end{aligned} \tag{35}$$

where $R(\mathbf{v})$ evaluates the length of the binary compressed description of the decompressed signal $\mathbf{v}$ and $d \geq 0$ determines the allowed distortion. By (34), the value d_0 depends on the operator $\mathbf{A}$, the null space of $\mathbf{B}$, the source signal $\mathbf{x}$, and the noise realization $\mathbf{n}$. Since $\mathbf{x}$ and $\mathbf{n}$ are unknown, d_0 cannot be accurately calculated in the operational case (in Dar et al. (2018d) we formulate the expected value of d_0 for the case of a cyclo-stationary Gaussian source signal). We address the optimization (35) using its unconstrained Lagrangian form

$$\hat{\mathbf{v}} = \underset{\mathbf{v} \in \mathcal{S}}{\operatorname{argmin}} \quad R(\mathbf{v}) + \lambda \frac{1}{M} \left\| \mathbf{w} - \mathbf{ABv} \right\|_2^2 \tag{36}$$

where $\lambda \geq 0$ is a Lagrange multiplier corresponding to some distortion constraint $d_\lambda \geq d_0$ (such optimization strategy with respect to some Lagrange multiplier is common, e.g., in video coding (Sullivan et al. 2012)). In the case of high-dimensional signals, the discrete set $\mathcal{S}$ is extremely large, and, therefore, it is impractical to directly solve the Lagrangian form in (36) for generally structured matrices $\mathbf{A}$ and $\mathbf{B}$. This difficulty vanishes, for example, when $\mathbf{A} = \mathbf{B} = \mathbf{I}$, reducing the Lagrangian optimization in (36) to the standard (system independent) compression form (see, e.g., Shoham and Gersho 1988 and Ortega and Ramchandran 1998).

The optimization (36) matches the generic template presented in section "Modular ADMM-Based Optimization: General Construction and Guidelines", and, therefore, we can formulate an ADMM-based modular procedure to address it. This modular optimization process is a special case of the generic procedure described in Algorithm 1, taking here the form of Algorithm 6. Note that we use the form of Algorithm 1 where the eventual output is the output of the module applied in the last iteration, which in our case corresponds to the output of the compression module in the last iteration (and this is the desired output because in this section we consider compression application, unlike the Restoration by Compression method in Algorithm 4 that its purpose is restoration by means of compression-based regularization). The interested reader is referred to Dar et al. (2018d) for a rate-distortion theoretic analysis for cyclo-stationary Gaussian signals and linear shift-invariant operators, explaining various aspects of the proposed procedure.

Algorithm 6 System-Aware Compression

1: Inputs: $\mathbf{w}, \theta, \tilde{\beta}$.
2: Initialize $t = 0$, $\hat{\mathbf{z}}^{(0)} = \mathbf{w}$, $\mathbf{u}^{(1)} = \mathbf{0}$.
3: **repeat**
4:     $t \leftarrow t + 1$
5:     $\tilde{\mathbf{z}}^{(t)} = \hat{\mathbf{z}}^{(t-1)} - \mathbf{u}^{(t)}$
6:     $\mathbf{b}^{(t)} = StandardCompression\left(\tilde{\mathbf{z}}^{(t)}, \theta\right)$
7:     $\hat{\mathbf{v}}^{(t)} = StandardDecompression\left(\mathbf{b}^{(t)}\right)$
8:     $\tilde{\mathbf{v}}^{(t)} = \hat{\mathbf{v}}^{(t)} + \mathbf{u}^{(t)}$
9:     $\hat{\mathbf{z}}^{(t)} = \underset{\mathbf{z} \in \mathbb{R}^N}{\operatorname{argmin}} \left\| \mathbf{w} - \mathbf{ABz} \right\|_2^2 + \tilde{\beta} \left\| \mathbf{z} - \tilde{\mathbf{v}}^{(t)} \right\|_2^2$
10:    $\mathbf{u}^{(t+1)} = \mathbf{u}^{(t)} + \left(\hat{\mathbf{v}}^{(t)} - \hat{\mathbf{z}}^{(t)}\right)$
11: **until** stopping criterion is satisfied
12: Output: $\mathbf{b}^{(t)}$, which is the binary compressed data obtained in the last iteration.

To demonstrate the essence of the System-Aware Compression approach, we provide here a representative example taken from Dar et al. (2018b), excluding the acquisition stage (i.e., $\mathbf{A} = \mathbf{I}$ and $\sigma_n^2 = 0$) while considering a post-decompression operator $\mathbf{B}$ implementing a shift-invariant Gaussian blur degradation. One can perceive this setting as optimizing image compression with respect to degradation occurring later (after decompression) at the display device, where no additional processing is done after decompression in order to counterbalance the degradation. To observe the gains achieved by the modular optimization approach, let us first examine the unoptimized (regular) compression process where the input image (Fig. 7a) is compressed using the state-of-the-art HEVC standard at a bit-rate of 3.75 bits per pixel (bpp), yielding the decompressed image in Fig. 7b (this is the image before blur degradation). Then, the decompressed image after degradation (Fig. 7c) is very blurry, as also reflected in the respective PSNR value (measured with respect to the image before compression). In the modular optimization approach, the input image is processed such that the compression in the last iteration is

5 Modular ADMM-Based Strategies for Optimized Compression, Restoration,... 197

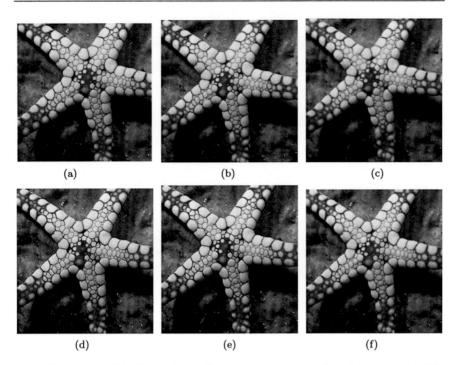

Fig. 7 Comparison of the System-Aware Compression approach and regular compression. The settings consider a Gaussian blur operator degrading the decompressed image. The intermediate and eventual images of the regular and the modular optimization process are presented. (**a**) Input. (**b**) Regular Decompression (3.75 bpp). (**c**) Regular Degraded Decompression (34.32 dB). (**d**) System-Aware Compression: Input to Last Iteration Compression. (**e**) System-Aware Compression: Decompression (2.21 bpp). (**f**) System-Aware Compression: Degraded Decompression (41.84 dB)

applied on a sharpened version (see Fig. 7d) adjusted to the known blur operator; then, the compressed image at bit-rate 2.21 bpp eventually results in a degraded decompression with moderate blur effects (Fig. 7f) and a PSNR gain of 7.52 dB with respect to the regular compression (which used even a higher bit-rate). See Dar et al. (2018b) for extensive experimental evaluation including PSNR-bitrate performance curves and comparison to additional alternatives. Furthermore, LCD display degradations associated with motion blur are also examined by Dar et al. (2018b).

Additional evaluations of the System-Aware Compression approach are provided by Dar et al. (2018d) for video compression settings including acquisition degradation of low-pass filtering and subsampling and post-decompression nearest-neighbor upsampling. In Dar et al. (2018a), the idea is demonstrated for a simplified model of multimedia distribution networks, where a set of possible degradation operators and their probabilities are considered by the optimized compression process.

Distributed Representations Using Black-Box Modules

All the above problems conduct optimizations for finding one signal (or compressed representation) that minimizes a Lagrangian cost of interest. As shown, these tasks are addressed very well by modular optimizations, relying on sequential black-box module applications. In this section, we demonstrate that the modular optimization approach is useful also to problems seeking for a set of signals (or representations) that collaboratively minimize a joint Lagrangian cost.

The General Framework

The following extends the settings and developments given in section "Unconstrained Lagrangian Optimizations via ADMM". The general optimization form for distributed representations broadens the single-representation problem in (1) to an unconstrained Lagrangian form optimizing several signals, i.e.,

$$\left(\hat{\mathbf{v}}_1, \ldots, \hat{\mathbf{v}}_K\right) = \underset{\mathbf{v}_1,\ldots,\mathbf{v}_K \in \mathcal{M}}{\operatorname{argmin}} \sum_{i=1}^{K} R\left(\mathbf{v}_i\right) + \lambda D\left(\mathbf{x}; \mathbf{v}_1, \ldots, \mathbf{v}_K\right) \qquad (37)$$

to be solved for the K representations $\mathbf{v}_1, \ldots, \mathbf{v}_K \in \mathcal{M} \subset \mathbb{R}^N$, where $\mathcal{M}$ is a (continuous or discrete) subset of the N-dimensional real space. While there are several optimization variables, they all intend to (possibly differently) represent the single given signal $\mathbf{x} \in \mathbb{R}^M$. The general scalar-valued function $R : \mathcal{M} \to \mathbb{R}$ is defined for individual representation as inputs, and D is a scalar-valued function receiving $\mathbf{x}$ and all the representations together as inputs.

The computational challenge of solving (37) is clear, as it is hard even in the case of optimizing one signal (as discussed in section "Modular Optimizations Based on Standard Compression Techniques"). Nevertheless, we can utilize variable splitting and ADMM techniques to develop a sequential optimization process addressing (37). Essentially, this is an extension of the ADMM constructions presented in section "Unconstrained Lagrangian Optimizations via ADMM". Here the developments originate in the variable splitting applied on (37) via

$$\left(\{\hat{\mathbf{v}}_i\}_{i=1}^{K}, \{\hat{\mathbf{z}}_i\}_{i=1}^{K}\right) = \underset{\substack{\{\mathbf{v}_i\}_{i=1}^{K} \in \mathcal{M} \\ \{\mathbf{z}_i\}_{i=1}^{K} \in \mathbb{R}^N}}{\operatorname{argmin}} \sum_{i=1}^{K} R\left(\mathbf{v}_i\right) + \lambda D\left(\mathbf{x}; \mathbf{z}_1, \ldots, \mathbf{z}_K\right) \qquad (38)$$

$$\text{subject to} \quad \mathbf{v}_i = \mathbf{z}_i \quad \text{for } i = 1, \ldots, K$$

where $\mathbf{z}_1, \ldots, \mathbf{z}_K \in \mathbb{R}^N$ are auxiliary variables that are not directly constrained to the eventual output domain $\mathcal{M}$ (similar to the developments in section "Unconstrained Lagrangian Optimizations via ADMM").

Then, the scaled form of the augmented Lagrangian and the method of multipliers (Boyd et al. 2011, Ch. 2) renders (38) into the sequential process

$$\left(\left\{\hat{\mathbf{v}}_i^{(t)}\right\}_{i=1}^K, \left\{\hat{\mathbf{z}}_i^{(t)}\right\}_{i=1}^K\right) = \qquad (39)$$

$$\operatorname*{argmin}_{\substack{\{\mathbf{v}_i\}_{i=1}^K \in \mathcal{M} \\ \{\mathbf{z}_i\}_{i=1}^K \in \mathbb{R}^N}} \sum_{i=1}^K R(\mathbf{v}_i) + \lambda D(\mathbf{x}; \mathbf{z}_1, \ldots, \mathbf{z}_K) + \frac{\beta}{2}\sum_{i=1}^K \left\|\mathbf{v}_i - \mathbf{z}_i + \mathbf{u}_i^{(t)}\right\|_2^2$$

$$\mathbf{u}_i^{(t+1)} = \mathbf{u}_i^{(t)} + \left(\hat{\mathbf{v}}_i^{(t)} - \hat{\mathbf{z}}_i^{(t)}\right) \quad \text{for } i=1,\ldots,K \qquad (40)$$

where t denotes the iteration index, $\left\{\mathbf{u}_i^{(t)}\right\}_{i=1}^K \in \mathbb{R}^N$ are the scaled dual variables, and β is an auxiliary parameter originating at the augmented Lagrangian (note that β is an intentionally joined parameter for the purpose of easing the parameter tuning process). The corresponding ADMM process is obtained by applying one iteration of alternating minimization on (39), leading to

$$\hat{\mathbf{v}}_i^{(t)} = \operatorname*{argmin}_{\mathbf{v}_i \in \mathcal{M}} R(\mathbf{v}_i) + \frac{\beta}{2}\left\|\mathbf{v}_i - \tilde{\mathbf{z}}_i^{(t)}\right\|_2^2 \quad \text{for } i=1,\ldots,K \qquad (41)$$

$$\hat{\mathbf{z}}_i^{(t)} = \operatorname*{argmin}_{\mathbf{z}_i \in \mathbb{R}^N} \lambda D\left(\mathbf{x}; \left\{\hat{\mathbf{z}}_j^{(t)}\right\}_{j=1}^{i-1}, \mathbf{z}_i, \left\{\hat{\mathbf{z}}_j^{(t-1)}\right\}_{j=i+1}^K\right) + \frac{\beta}{2}\left\|\mathbf{z}_i - \tilde{\mathbf{v}}_i^{(t)}\right\|_2^2$$

$$\text{for } i=1,\ldots,K \qquad (42)$$

$$\mathbf{u}_i^{(t+1)} = \mathbf{u}_i^{(t)} + \left(\hat{\mathbf{v}}_i^{(t)} - \hat{\mathbf{z}}_i^{(t)}\right) \quad \text{for } i=1,\ldots,K \qquad (43)$$

where $\tilde{\mathbf{z}}_i^{(t)} \triangleq \hat{\mathbf{z}}_i^{(t-1)} - \mathbf{u}_i^{(t)}$ and $\tilde{\mathbf{v}}_i^{(t)} \triangleq \hat{\mathbf{v}}_i^{(t)} + \mathbf{u}_i^{(t)}$. Nicely, the obtained process does not only decouple the treatment of $\{\mathcal{M}, R\}$ from D as before (see section "Unconstrained Lagrangian Optimizations via ADMM") but also separates the treatment of the various representations. Thus, (41), (42), and (43) simplify the challenging structure in (37). Moreover, the subproblems in (41) have the same form associated with black-box modules applied on individual signals (see section "Employing Black-Box Modules"). This casting leads us to the process summarized in Algorithm 7. Also note that in each iteration t the treatment of the K representations is sequential (this reordered procedure is equivalent to the form in (41), (42), and (43)).

Modular Optimizations for Holographic Compression of Images

We now turn to exemplify the generic approach in Algorithm 7 for the purpose of optimizing distributed representations in compressed, standard-compatible, forms.

Algorithm 7 General Modular Optimization of Multiple Representations
1: Inputs: $\mathbf{x}, \theta, \tilde{\beta}$.
2: Initialize $t = 0$, $\hat{\mathbf{z}}^{(0)} = \mathbf{x}$, $\mathbf{u}^{(1)} = \mathbf{0}$.
3: Initialize $t = 0$.
4: Initialize (for $i = 1, \ldots, K$) $\mathbf{u}_i^{(1)} = \mathbf{0}$ and $\hat{\mathbf{z}}_i^{(0)}$ according to the specific application.
5: **repeat**
6:     $t \leftarrow t + 1$
7:     **for** $i = 1, \ldots, K$ **do**
8:         $\tilde{\mathbf{z}}_i^{(t)} = \hat{\mathbf{z}}_i^{(t-1)} - \mathbf{u}_i^{(t)}$
9:         $\hat{\mathbf{v}}_i^{(t)} = BlackBoxModule\left(\tilde{\mathbf{z}}_i^{(t)}; \theta\right)$
10:        $\tilde{\mathbf{v}}_i^{(t)} = \hat{\mathbf{v}}_i^{(t)} + \mathbf{u}_i^{(t)}$
11:        $\hat{\mathbf{z}}_i^{(t)} = \underset{\mathbf{z}_i \in \mathbb{R}^N}{\mathrm{argmin}}\, \lambda D\left(\mathbf{x}; \left\{\hat{\mathbf{z}}_i^{(t)}\right\}_{j=1}^{i-1}, \mathbf{z}_i, \left\{\hat{\mathbf{z}}_i^{(t-1)}\right\}_{j=i+1}^{K}\right) + \tilde{\beta}\left\|\mathbf{z}_i - \tilde{\mathbf{v}}_i^{(t)}\right\|_2^2$
12:        $\mathbf{u}_i^{(t+1)} = \mathbf{u}_i^{(t)} + \left(\hat{\mathbf{v}}_i^{(t)} - \hat{\mathbf{z}}_i^{(t)}\right)$
13:     **end for**
14: **until** stopping criterion is satisfied
15: Output: $\hat{\mathbf{v}}_1^{(t)}, \ldots, \hat{\mathbf{v}}_K^{(t)}$ and/or other application-specific outputs of $BlackBoxModule$.

Our recent framework for holographic compression (Dar and Bruckstein 2021) represents a given signal using a set of distinct compressed descriptions, that any subset of them enables reconstruction of the signal at a quality determined solely by the number of compressed representations utilized. This property of holographic representations is useful for designing progressive refinement mechanisms independent of the order the representations are accessible (Bruckstein et al. 1998, 2000, 2018).

In Dar and Bruckstein (2021) we identified the shift sensitivity of standard compression techniques as a property useful for constructing holographic representations in binary compressed forms. Specifically, compressions of shifted versions of an image provide a set of distinct decompressed images of similar individual qualities, but combining subsets of them (by back-shifts and averaging) achieves remarkable quality gains (see details in Dar and Bruckstein 2021). While this architecture is new and interesting, it does not include optimization aspects. This led us to suggest an optimization procedure unleashing the potential benefits of the shift-based holographic compression settings. Here we can consider this optimization framework as a special case of the generic process described in Algorithm 7, described as follows. First, the general $\{\mathcal{M}, R\}$ notions are set to the respective components $\{\mathcal{S}, R_\mathcal{S}\}$ of a standard compression method (as defined in section "Preliminaries: Lossy Compression via Operational Rate-Distortion Optimization"). This makes the first component in (37) the accumulated bit-cost of all the compressed representations. In Dar and Bruckstein (2021) we set the function D to evaluate the average MSE of reconstructions formed using subsets of m out of the K representations, where $m \in \{2, \ldots, K\}$ and assuming $K > 1$. This improves the reconstruction quality for subsets of m representations, at the inevitable expense of reducing their individual qualities. Therefore, we also include in D a regularization term computing the average MSE of the single-representation reconstructions. We denote the sequence of integers from 1 to K as $[[K]] \triangleq \{1, \ldots, K\}$. For $m \in [[K]]$, an m-combination of the set $[[K]]$ is a subset of m distinct numbers from $[[K]]$. We denote the set of all m-combinations of $[[K]]$ as $\binom{[[K]]}{m}$, where the latter contains $\binom{K}{m}$ elements. The corresponding formulation of D is

$$D(\mathbf{x}; \mathbf{v}_1, \ldots, \mathbf{v}_K) = \frac{1}{\binom{K}{m}} \sum_{(i_1, \ldots, i_m) \in \binom{[[K]]}{m}} D^{(m)}\left(\mathbf{x}; \mathbf{v}_{i_1}, \ldots, \mathbf{v}_{i_m}\right) \quad (44)$$

$$+ \eta \frac{1}{K} \sum_{i=1}^{K} D^{(1)}(\mathbf{x}; \mathbf{v}_i)$$

The parameter η determines the regularization level of the individual representation quality. Moreover,

$$D^{(1)}(\mathbf{x}; \mathbf{v}_i) \triangleq \frac{1}{N} \left\| \mathbf{x} - \mathbf{S}_i^T \mathbf{v}_i \right\|_2^2 \quad (45)$$

is the reconstruction MSE corresponding to the single representation $\mathbf{v}_i$, and

$$D^{(m)}\left(\mathbf{x}; \mathbf{v}_{i_1}, \ldots, \mathbf{v}_{i_m}\right) \triangleq \frac{1}{N} \left\| \mathbf{x} - \frac{1}{m} \sum_{j=1}^{m} \mathbf{S}_{i_j}^T \mathbf{v}_{i_j} \right\|_2^2 \quad (46)$$

is the MSE of reconstruction using the m representations $\mathbf{v}_{i_1}, \ldots, \mathbf{v}_{i_m}$. The matrices $\mathbf{S}_i^T$ and $\mathbf{S}_{i_j}^T$ correspond to back shift operators matching the shift forms used to create the compressed representations (further details are available in Dar and Bruckstein 2021). Then, by the settings of $\mathcal{M}$, R, and D, Algorithm 7 is specified for optimizing shift-based holographic compressed representations – this process is described in Algorithm 8.

In Figs. 8 and 9, we provide representative results taken from Dar and Bruckstein (2021). First, Fig. 8 presents reconstructions obtained from JPEG2000-compatible holographic compressions optimized for using sets of four representations. Specifically note the similar quality obtained using the individual representations and how they collaboratively achieve progressive refinement. This behavior is also clearly demonstrated in Fig. 9 by the curves of PSNR versus number of representations (packets) used for reconstructions. In particular, Fig. 9 shows the curves obtained for all the subset combinations in each of the examined methods. This exhibits the ability of the proposed method for optimizing reconstructions that rely on a specified number of representations (independent of the actual participating signals). The interested reader is referred to Dar and Bruckstein (2021) for additional details and experimental demonstrations.

Algorithm 8 Modular Holographic Compression: Optimized for Reconstructions Using m Representations

1: Inputs: $\mathbf{x}, \tilde{\beta}, \lambda, \eta, \theta, m, K$.
2: Initialize $t = 0$.
3: Initialize (for $i = 1, \ldots, K$) $\hat{\mathbf{z}}_i^{(0)} = \mathbf{S}_i \mathbf{x}$ and $\mathbf{u}_i^{(1)} = \mathbf{0}$.
4: **repeat**
5: $\quad t \leftarrow t + 1$
6: $\quad$ **for** $i = 1, \ldots, K$ **do**
7: $\quad\quad \tilde{\mathbf{z}}_i^{(t)} = \hat{\mathbf{z}}_i^{(t-1)} - \mathbf{u}_i^{(t)}$
8: $\quad\quad \mathbf{b}_i^{(t)} = StandardCompression\left(\tilde{\mathbf{z}}_i^{(t)}, \theta\right)$
9: $\quad\quad \hat{\mathbf{v}}_i^{(t)} = StandardDecompression\left(\mathbf{b}_i^{(t)}\right)$
10: $\quad\quad \tilde{\mathbf{v}}_i^{(t)} = \hat{\mathbf{v}}_i^{(t)} + \mathbf{u}_i^{(t)}$
11: $\quad\quad \hat{\mathbf{z}}_i^{(t)} = \dfrac{N\tilde{\beta}\tilde{\mathbf{v}}_i^{(t)} + \frac{\eta}{\lambda K}\mathbf{S}_i\mathbf{x} + \frac{\lambda}{m^2 \cdot \binom{K}{m}}\mathbf{S}_i\mathbf{w}_i^{(m)}}{N\tilde{\beta} + \frac{\eta}{\lambda K} + \frac{\lambda}{m^2 \cdot \binom{K}{m}} \cdot |\mathcal{I}_i^{(m)}|}$
$\quad\quad$ where
$$\mathbf{w}_i^{(m)} \triangleq \sum_{(i_1,\ldots,i_m)\in \mathcal{I}_i^{(m)}} \left(m\mathbf{x} - \sum_{\substack{i_j < i \\ j \in \{1,\ldots,m\}}} \mathbf{S}_{i_j}^T \hat{\mathbf{z}}_{i_j}^{(t)} - \sum_{\substack{i_j > i \\ j \in \{1,\ldots,m\}}} \mathbf{S}_{i_j}^T \hat{\mathbf{z}}_{i_j}^{(t-1)} \right)$$
$\quad\quad$ and $\mathcal{I}_i^{(m)}$ contains all the m-combinations including the ith representation.
12: $\quad\quad \mathbf{u}_i^{[t+1]} = \mathbf{u}_i^{(t)} + \left(\hat{\mathbf{v}}_i^{(t)} - \hat{\mathbf{z}}_i^{(t)}\right)$
13: $\quad$ **end for**
14: **until** stopping criterion is satisfied
15: Output: The binary compressed packets $\mathbf{b}_1^{(t)}, \ldots, \mathbf{b}_K^{(t)}$.

Conclusion

In this chapter, we presented the recent methodology of modular optimizations, employing black-box modules in procedures addressing various imaging problems. The main idea is that fundamental tasks, such as denoising and compression, have excellent ready-to-use techniques that can be utilized for solving more intricate problems. We presented the developments of ADMM-based algorithms for modular optimizations, starting in general settings exhibiting the essence and prominent guidelines of the approach. Then, we overviewed settings where denoising and compression techniques are operated as stages in sequential procedures for restoration and intricate compression problems. We also outlined the extension of modular optimizations to formation of distributed representations and particularly exemplified it for the case of holographic compression of images. The perspectives emphasized in this chapter should motivate new ideas and settings extending the current class of module types used and problems addressed via modular optimization strategies.

5 Modular ADMM-Based Strategies for Optimized Compression, Restoration,...

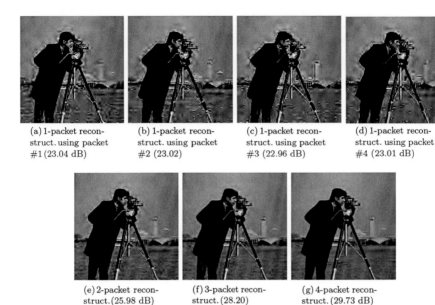

Fig. 8 Examples (taken from Dar and Bruckstein 2021) for reconstructions of the "Cameraman" image using several representations out of a set of four holographic compressed descriptions. The compression employed is JPEG2000 at a compression ratio of 1:50. (**a**)–(**d**) the 1-packet reconstructions using each of the individual packets. (**e**)–(**g**) examples for the m-packet reconstructions for $m = 2, 3, 4$

Appendix: Operational Rate-Distortion Optimizations in Block-Based Architectures

The computational challenge of operational rate-distortion optimizations (see section "Preliminaries: Lossy Compression via Operational Rate-Distortion Optimization") is often addressed via the squared-error metric

$$D(\mathbf{x}, \mathbf{v}) = \|\mathbf{x} - \mathbf{v}\|_2^2, \qquad (47)$$

leading to practical forms of the Lagrangian rate-distortion optimization (25). These useful structures also process the signal $\mathbf{x}$ based on its segmentation into a set of nonoverlapping blocks $\{\mathbf{x}_i\}_{i \in \mathcal{I}}$; here, each of them is a column vector of N_b samples, and $\mathcal{I}$ is the set of indices corresponding to the nonoverlapping blocks of the signal. Correspondingly, the decompressed signal $\mathbf{v}$ is decomposed into a set of nonoverlapping blocks $\{\mathbf{v}_i\}_{i \in \mathcal{I}}$. This lets us casting (47) into

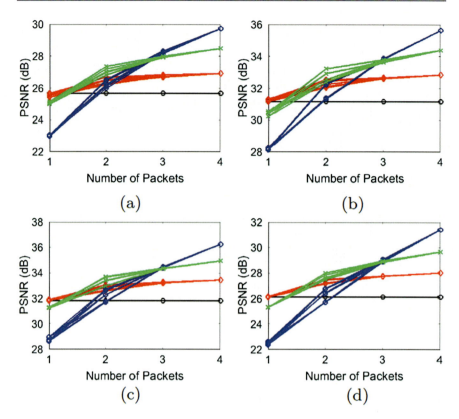

Fig. 9 PSNR versus the number of representations used for the reconstructions. The entire set contains four packets, each formed by JPEG2000 compression at 1:50 compression ratio. The black, red, green, and blue curves, respectively, represent the methods of exact duplications, baseline (unoptimized), optimized for reconstruction from pairs of packets, and optimized for reconstruction from four packets. (**a**) Cameraman. (**b**) House. (**c**) Lena. (**d**) Barbara

$$D(\mathbf{x}, \mathbf{v}) = \sum_{i \in \mathcal{I}} \|\mathbf{x}_i - \mathbf{v}_i\|_2^2, \tag{48}$$

exhibiting that, for squared-error measures, the total distortion can be computed as the sum of distortions associated with its nonoverlapping blocks. While this property is satisfied for any segmentation of the signal into nonoverlapping blocks, we will exemplify it here for blocks of equal sizes that allow using one block-level compression procedure for all the blocks.

Mirroring the definitions described in section "Preliminaries: Lossy Compression via Operational Rate-Distortion Optimization" for full-signal compression architectures, the block-level process corresponds to a function $C_b : \mathbb{R}^{N_b} \to \mathcal{B}_b$, mapping the N_b-dimensional signal-block domain to a discrete set $\mathcal{B}_b$ of binary compressed representations of blocks. The associated block decompression process is denoted by the function $F_b : \mathcal{B}_b \to \mathcal{S}_b$, mapping the binary compressed representations in $\mathcal{B}_b$ to their decompressed signal blocks from the discrete set $\mathcal{S}_b \subset \mathbb{R}^{N_b}$. The bit-cost evaluation function $R_b(\mathbf{v}_i)$ is defined to quantify the number of bits needed for the compressed representation matching the decompressed signal block $\mathbf{v}_i \in \mathbb{R}^{N_b}$. Then, the compression of the nonoverlapping signal blocks $\{\mathbf{x}_i\}_{i \in \mathcal{I}}$ producing the decompressed blocks $\{\mathbf{v}_i\}_{i \in \mathcal{I}}$ requires a total bit budget satisfying

$$R(\mathbf{v}) = \sum_{i \in \mathcal{I}} R_b(\mathbf{v}_i). \tag{49}$$

Plugging the block-based compression design into the Lagrangian form (25) gives

$$\{\hat{\mathbf{v}}_i\}_{i \in \mathcal{I}} = \underset{\{\mathbf{v}_i\}_{i \in \mathcal{I}} \in \mathcal{S}_b}{\operatorname{argmin}} \sum_{i \in \mathcal{I}} R_b(\mathbf{v}_i) + \lambda \sum_{i \in \mathcal{I}} \|\mathbf{x}_i - \mathbf{v}_i\|_2^2 \tag{50}$$

that reduces to a set of block-level rate-distortion Lagrangian optimizations, i.e.,

$$\text{For } i \in \mathcal{I}: \quad \hat{\mathbf{v}}_i = \underset{\mathbf{v}_i \in \mathcal{S}_b}{\operatorname{argmin}} \ R_b(\mathbf{v}_i) + \lambda \|\mathbf{x}_i - \mathbf{v}_i\|_2^2. \tag{51}$$

Note that the block-level optimizations in (51) are independent and refer to the same Lagrangian multiplier λ. Commonly, compression designs are based on processing of low-dimensional blocks, allowing to practically address the block-level optimizations in (51). For example, one can evaluate the Lagrangian cost for all the elements in $\mathcal{S}_b$ (since this set is sufficiently small).

References

Afonso, M.V., Bioucas-Dias, J.M., Figueiredo, M.A.: Fast image recovery using variable splitting and constrained optimization. IEEE Trans. Image Process. 19(9), 2345–2356 (2010)

Ahmad, R., Bouman, C.A., Buzzard, G.T., Chan, S., Reehorst, E.T., Schniter, P.: Plug and play methods for magnetic resonance imaging. arXiv preprint arXiv:1903.08616 (2019)

Bahat, Y., Efrat, N., Irani, M.: Non-uniform blind deblurring by reblurring. In: Proceedings of the IEEE International Conference on Computer Vision, pp. 3286–3294 (2017)

Ballé, J., Laparra, V., Simoncelli, E.P.: End-to-end optimized image compression. In: Proceedings of ICLR (2017)

F. Bellard, BPG 0.9.6. [Online]. Available: http://bellard.org/bpg/

Beygi, S., Jalali, S., Maleki, A., Mitra, U.: Compressed sensing of compressible signals. In: IEEE International Symposium on Information Theory (ISIT), pp. 2158–2162 (2017a)

Beygi, S., Jalali, S., Maleki, A., Mitra, U.: An efficient algorithm for compression-based compressed sensing. arXiv preprint arXiv:1704.01992 (2017b)

Boyd, S., Parikh, N., Chu, E., Peleato, B., Eckstein, J.: Distributed optimization and statistical learning via the alternating direction method of multipliers. Found. Trends Mach. Learn. 3(1), 1–122 (2011)

Brifman, A., Romano, Y., Elad, M.: Turning a denoiser into a super-resolver using plug and play priors. In: 2016 IEEE International Conference on Image Processing (ICIP), pp. 1404–1408. IEEE (2016)

Brifman, A., Romano, Y., Elad, M.: Unified single-image and video super-resolution via denoising algorithms. IEEE Trans. Image Process. **28**(12), 6063–6076 (2019)

Bruckstein, A.M., Holt, R.J., Netravali, A.N.: Holographic representations of images. IEEE Trans. Image Process. **7**(11), 1583–1597 (1998)

Bruckstein, A.M., Holt, R.J., Netravali, A.N.: On holographic transform compression of images. Int. J. Imag. Syst. Technol. **11**(5), 292–314 (2000)

Bruckstein, A.M., Ezerman, M.F., Fahreza, A.A., Ling, S.: Holographic sensing. arXiv preprint arXiv:1807.10899 (2018)

Burger, M., Dirks, H., Schonlieb, C.-B.: A variational model for joint motion estimation and image reconstruction. SIAM J. Imag. Sci. **11**(1), 94–128 (2018)

Buzzard, G.T., Chan, S.H., Sreehari, S., Bouman, C.A.: Plug-and-play unplugged: optimization-free reconstruction using consensus equilibrium. SIAM J. Imag. Sci. **11**(3), 2001–2020 (2018)

Chan, S.H.: Performance analysis of plug-and-play ADMM: a graph signal processing perspective. IEEE Trans. Comput. Imag. **5**(2), 274–286 (2019)

Chan, S.H., Wang, X., Elgendy, O.A.: Plug-and-play ADMM for image restoration: fixed-point convergence and applications. IEEE Trans. Comput. Imag. **3**(1), 84–98 (2017)

Chatterjee, P., Milanfar, P.: Is denoising dead? IEEE Trans. Image Process. **19**(4), 895–911 (2009)

Chou, P.A., Lookabaugh, T., Gray, R.M.: Optimal pruning with applications to tree-structured source coding and modeling. IEEE Trans. Inf. Theory **35**(2), 299–315 (1989)

Corona, V., Aviles-Rivero, A.I., Debroux, N., Graves, M., Le Guyader, C., Schönlieb, C.-B., Williams, G.: Multi-tasking to correct: motion-compensated mri via joint reconstruction and registration. In: International Conference on Scale Space and Variational Methods in Computer Vision, pp. 263–274. Springer (2019a)

Corona, V., Benning, M., Ehrhardt, M.J., Gladden, L.F., Mair, R., Reci, A., Sederman, A.J., Reichelt, S., Schönlieb, C.-B.: Enhancing joint reconstruction and segmentation with non-convex bregman iteration. Inverse Probl. **35**(5), 055001 (2019b)

Dabov, K., Foi, A., Katkovnik, V., Egiazarian, K.: Image denoising by sparse 3-D transform-domain collaborative filtering. IEEE Trans. Image Process. **16**(8), 2080–2095 (2007)

Dar, Y., Bruckstein, A.M.: Benefiting from duplicates of compressed data: shift-based holographic compression of images. J. Math. Imag. Vis. 1–14 **63**, 380–393 (2021)

Dar, Y., Bruckstein, A.M., Elad, M.: Image restoration via successive compression. In: Picture Coding Symposium (PCS), pp. 1–5 (2016a)

Dar, Y., Bruckstein, A.M., Elad, M., Giryes, R.: Postprocessing of compressed images via sequential denoising. IEEE Trans. Image Process. **25**(7), 3044–3058 (2016b)

Dar, Y., Elad, M., Bruckstein, A.M.: Compression for multiple reconstructions. In: IEEE International Conference on Image Processing (ICIP), pp. 440–444 (2018a)

Dar, Y., Elad, M., Bruckstein, A.M.: Optimized pre-compensating compression. IEEE Trans. Image Process. **27**(10), 4798–4809 (2018b)

Dar, Y., Elad, M., Bruckstein, A.M.: Restoration by compression. IEEE Trans. Sig. Process. **66**(22), 5833–5847 (2018c)

Dar, Y., Elad, M., Bruckstein, A.M.: System-aware compression. In: IEEE International Symposium on Information Theory (ISIT), pp. 2226–2230 (2018d)

Hong, T., Romano, Y., Elad, M.: Acceleration of red via vector extrapolation. J. Vis. Commun. Image Represent. **63**, 102575 (2019)

Kamilov, U.S., Mansour, H., Wohlberg, B.: A plug-and-play priors approach for solving nonlinear imaging inverse problems. IEEE Sig. Process. Lett. **24**(12), 1872–1876 (2017)

Kwan, C., Choi, J., Chan, S., Zhou, J., Budavari, B.: A super-resolution and fusion approach to enhancing hyperspectral images. Remote Sens. **10**(9), 1416 (2018)

Lai, W.-S., Huang, J.-B., Hu, Z., Ahuja, N., Yang, M.-H.: A comparative study for single image blind deblurring. In: Proceedings of the IEEE Conference on Computer Vision and Pattern Recognition, pp. 1701–1709 (2016)

Laparra, V., Berardino, A., Ballé, J., Simoncelli, E.P.: Perceptually optimized image rendering. J. Opt. Soc. Am. A **34**, 1511 (2017)

Liu, J., Moulin, P.: Complexity-regularized image denoising. IEEE Trans. Image Process. **10**(6), 841–851 (2001)

Moulin, P., Liu, J.: Statistical imaging and complexity regularization. IEEE Trans. Inf. Theory **46**(5), 1762–1777 (2000)

Natarajan, B.K.: Filtering random noise from deterministic signals via data compression. IEEE Trans. Sig. Process. **43**(11), 2595–2605 (1995)

Ono, S.: Primal-dual plug-and-play image restoration. IEEE Sig. Process. Lett. **24**(8), 1108–1112 (2017)

Ortega, A., Ramchandran, K.: Rate-distortion methods for image and video compression. IEEE Sig. Process. Mag. **15**(6), 23–50 (1998)

Rissanen, J.: MDL denoising. IEEE Trans. Inf. Theory **46**(7), 2537–2543 (2000)

Romano, Y., Elad, M., Milanfar, P.: The little engine that could: regularization by denoising (RED). SIAM J. Imag. Sci. **10**(4), 1804–1844 (2017)

Rond, A., Giryes, R., Elad, M.: Poisson inverse problems by the plug-and-play scheme. J. Vis. Commun. Image Represent. **41**, 96–108 (2016)

Rott Shaham, T., Michaeli, T.: Deformation aware image compression. In: Proceedings of the IEEE Conference on Computer Vision and Pattern Recognition, pp. 2453–2462 (2018)

Shoham, Y., Gersho, A.: Efficient bit allocation for an arbitrary set of quantizers. IEEE Trans. Acoust. Speech Sig. Process. **36**(9), 1445–1453 (1988)

Shukla, R., Dragotti, P.L., Do, M.N., Vetterli, M.: Rate-distortion optimized tree-structured compression algorithms for piecewise polynomial images. IEEE Trans. Image Process. **14**(3), 343–359 (2005)

Sreehari, S., Venkatakrishnan, S., Wohlberg, B., Buzzard, G.T., Drummy, L.F., Simmons, J.P., Bouman, C.A.: Plug-and-play priors for bright field electron tomography and sparse interpolation. IEEE Trans. Comput. Imag. **2**(4), 408–423 (2016)

Sullivan, G.J., Wiegand, T.: Rate-distortion optimization for video compression. IEEE Sig. Process. Mag. **15**(6), 74–90 (1998)

Sullivan, G.J., Ohm, J., Han, W.-J., Wiegand, T.: Overview of the high efficiency video coding (HEVC) standard. IEEE Trans. Circuits Syst. Video Technol. **22**(12), 1649–1668 (2012)

Sun, Y., Wohlberg, B., Kamilov, U.S.: An online plug-and-play algorithm for regularized image reconstruction. IEEE Trans. Comput. Imag.**5**, 395–408 (2019a)

Sun, Y., Xu, S., Li, Y., Tian, L., Wohlberg, B., Kamilov, U.S.: Regularized fourier ptychography using an online plug-and-play algorithm. In: ICASSP 2019-2019 IEEE International Conference on Acoustics, Speech and Signal Processing (ICASSP), pp. 7665–7669. IEEE (2019b)

Tirer, T., Giryes, R.: Image restoration by iterative denoising and backward projections. IEEE Trans. Image Process. **28**(3), 1220–1234 (2018a)

Tirer, T., Giryes, R.: An iterative denoising and backwards projections method and its advantages for blind deblurring. In: 2018 25th IEEE International Conference on Image Processing (ICIP), pp. 973–977. IEEE (2018b)

Tirer, T., Giryes, R.: Back-projection based fidelity term for ill-posed linear inverse problems. arXiv preprint arXiv:1906.06794 (2019)

Venkatakrishnan, S.V., Bouman, C.A., Wohlberg, B.: Plug-and-play priors for model based reconstruction. In: IEEE GlobalSIP (2013)

Yazaki, Y., Tanaka, Y., Chan, S.H.: Interpolation and denoising of graph signals using plug-and-play ADMM. In: ICASSP 2019-2019 IEEE International Conference on Acoustics, Speech and Signal Processing (ICASSP), pp. 5431–5435. IEEE (2019)

Zoran, D., Weiss, Y.: From learning models of natural image patches to whole image restoration. In: IEEE International Conference on Computer Vision (ICCV), pp. 479–486 (2011)

Connecting Hamilton-Jacobi Partial Differential Equations with Maximum a Posteriori and Posterior Mean Estimators for Some Non-convex Priors

Jérôme Darbon, Gabriel P. Langlois, and Tingwei Meng

Contents

Introduction	210
First-Order Hamilton-Jacobi PDEs and Optimization Problems	212
Single-Time HJ PDEs and Image Denoising Models	213
Multi-time HJ PDEs and Image Decomposition Models	214
Min-Plus Algebra for HJ PDEs and Certain Non-convex Regularizations	216
Application to Certain Decomposition Problems	220
Viscous Hamilton-Jacobi PDEs and Bayesian Estimation	224
Viscous HJ PDEs and Posterior Mean Estimators for Log-Concave Models	225
On Viscous HJ PDEs with Certain Non-log-Concave Priors	227
Conclusion	230
References	230

Abstract

Many imaging problems can be formulated as inverse problems expressed as finite-dimensional optimization problems. These optimization problems generally consist of minimizing the sum of a data fidelity and regularization terms. In Darbon (SIAM J. Imag. Sci. 8:2268–2293, 2015), Darbon and Meng, (On decomposition models in imaging sciences and multi-time Hamilton-Jacobi partial differential equations, arXiv preprint arXiv:1906.09502, 2019), connections between these optimization problems and (multi-time) Hamilton-Jacobi partial differential equations have been proposed under the convexity assumptions of both the data fidelity and regularization terms. In particular, under these convexity assumptions, some representation formulas for a minimizer can

---

J. Darbon (✉) · G. P. Langlois · T. Meng
Division of Applied Mathematics, Brown University, Providence, RI, USA
e-mail: jerome_darbon@brown.edu; gabriel_provencher_langlois@brown.edu; tingwei_meng@brown.edu

© Springer Nature Switzerland AG 2023
K. Chen et al. (eds.), *Handbook of Mathematical Models and Algorithms in Computer Vision and Imaging*, https://doi.org/10.1007/978-3-030-98661-2_56

be obtained. From a Bayesian perspective, such a minimizer can be seen as a maximum a posteriori estimator. In this chapter, we consider a certain class of non-convex regularizations and show that similar representation formulas for the minimizer can also be obtained. This is achieved by leveraging min-plus algebra techniques that have been originally developed for solving certain Hamilton-Jacobi partial differential equations arising in optimal control. Note that connections between viscous Hamilton-Jacobi partial differential equations and Bayesian posterior mean estimators with Gaussian data fidelity terms and log-concave priors have been highlighted in Darbon and Langlois, (On Bayesian posterior mean estimators in imaging sciences and Hamilton-Jacobi partial differential equations, arXiv preprint arXiv:2003.05572, 2020). We also present similar results for certain Bayesian posterior mean estimators with Gaussian data fidelity and certain non-log-concave priors using an analogue of min-plus algebra techniques.

Keywords

Hamilton–Jacobi partial differential equation · Maximum a posteriori estimation · Bayesian posterior mean estimation · Min-plus algebra · Imaging inverse problems

Introduction

Many low-level signal, image processing, and computer vision problems are formulated as inverse problems that can be solved using variational (Aubert and Kornprobst 2002; Scherzer et al. 2009; Vese et al. 2016) or Bayesian approaches (Winkler 2003). Both approaches have been very effective, for example, at solving image restoration (Bouman and Sauer 1993; Likas and Galatsanos 2004; Rudin et al. 1992), segmentation (Boykov et al. 2001; Chan et al. 2006; Chan and Vese 2001), and image decomposition problems (Aujol et al. 2005; Osher et al. 2003).

As an illustration, let us consider the following image denoising problem in finite dimension that formally reads as follows:

$$x = \bar{u} + \eta,$$

where $x \in \mathbb{R}^n$ is the observed image that is the sum of an unknown ideal image $\bar{u} \in \mathbb{R}^n$ and an additive perturbation or noise realization $\eta \in \mathbb{R}^n$. We aim to estimate $\bar{u}$.

A standard variational approach for solving this problem consists of estimating $\bar{u}$ as a minimizer of the following optimization problem:

$$\min_{u \in \mathbb{R}^n} \left\{ \lambda D(x - u) + J(u) \right\}, \tag{1}$$

where $D : \mathbb{R}^n \to \mathbb{R}$ is generally called the data fidelity term and contains the knowledge we have on the perturbation η while $J : \mathbb{R}^n \to \mathbb{R} \cup \{+\infty\}$ is called the regularization term and encodes the knowledge on the image we wish to reconstruct. The nonnegative parameter λ relatively weights the data fidelity and the regularization terms. Note that minimizers of (1) are called maximum a posteriori (MAP) estimators in a Bayesian setting. Also note that variational-based approaches for estimating $\bar{u}$ are particularly appealing when both the data fidelity and regularization terms are convex because (1) becomes a convex optimization problem that can be efficiently solved using convex optimization algorithms (see, e.g., Chambolle and Pock 2016). Many regularization terms have been proposed in the literature (Aubert and Kornprobst 2002; Winkler 2003). Popular choices for these regularization terms involve robust edge-preserving priors (Bouman and Sauer 1993; Charbonnier et al. 1997; Geman and Yang 1995; Geman and Reynolds 1992; Nikolova and Chan 2007; ?; Rudin et al. 1992) because they allow the reconstructed image to have sharp edges. For the sake of simplicity, we only describe in this introduction regularizations that are expressed using pairwise interactions which take the following form:

$$J(u) = \sum_{i,j=1}^{n} w_{ij} f(u_i - u_j), \quad (2)$$

where $f : \mathbb{R} \to \mathbb{R} \cup \{+\infty\}$ and $w_{i,j} \geq 0$. Note that our results that will be presented later do not rely on pairwise interaction-based models and work for more general regularization terms. A popular choice is the celebrated Total Variation (Bouman and Sauer 1993; Rudin et al. 1992), which corresponds to consider $f(z) = |z|$ in (2). The use of Total Variation as a regularization term has been very popular since the seminal works of Bouman and Sauer (1993); Rudin et al. (1992) because it is convex and allows the reconstructed image to preserve edges well. When the data fidelity D is quadratic, this model is known as the celebrated Rudin-Osher-Fatemi model (Rudin et al. 1992). Following the seminal works of Charbonnier et al. (1997), Geman and Yang (1995) and Geman and Reynolds (1992), another class of edge-preserving priors corresponds to half-quadratic-based regularizations that read as follows:

$$f(z) = \begin{cases} |z|^2 & \text{if } |z| \leq 1, \\ 1 & \text{otherwise.} \end{cases} \quad (3)$$

Note that the quadratic term above can be replaced by $|\cdot|$, i.e., we consider:

$$f(z) = \begin{cases} |z| & \text{if } |z| \leq 1, \\ 1 & \text{otherwise,} \end{cases} \quad (4)$$

which corresponds to the truncated Total Variation regularization (see Darbon et al. 2009; Dou et al. 2017 for instance).

There is a large body of literature on variational methods (e.g., Aubert and Kornprobst 2002; Chambolle et al. 2010; Chan and Shen 2005; Scherzer et al. 2009; Vese et al. 2016). In particular, in Darbon (2015) and Darbon and Meng (2020), connections between convex optimization problems of the form of (1) and Hamilton-Jacobi partial differential equations (HJ PDEs) were highlighted. Specifically, it is shown that the dependence of the minimal value of these problems with respect to the observed data x and the smoothing parameter λ is governed by HJ PDEs, where the initial data corresponds to the regularization term J and the Hamiltonian is related to the data fidelity (see section "First-Order Hamilton-Jacobi PDEs and Optimization Problems" for details). However, the connections between HJ PDEs and certain variational imaging problems described in Darbon (2015) and Darbon and Meng (2020) require the convexity of both the data fidelity and regularization terms. Note that these connections between HJ PDEs and imaging problems also hold for image decomposition models (see section "Multi-time HJ PDEs and Image Decomposition Models") using multi-time HJ PDEs (Darbon and Meng 2020).

Our goal is to extend the results of Darbon (2015) and Darbon and Meng (2020) to certain non-convex regularization terms using min-plus algebra techniques (Akian et al. 2006, 2008; Dower et al. 2015; Fleming and McEneaney 2000; Gaubert et al. 2011; Kolokoltsov and Maslov 1997; McEneaney 2006, 2007; McEneaney et al. 2008; McEneaney and Kluberg 2009) that were originally designed for solving certain HJ PDEs arising in optimal control problems. We also propose an analogue of this approach for certain Bayesian posterior mean estimators when the data fidelity is Gaussian.

The rest of this chapter is as follows. Section "First-Order Hamilton-Jacobi PDEs and Optimization Problems" reviews connections of image denoising and decomposition models with HJ PDEs under convexity assumptions. We then present a min-plus algebra approach for single-time and multi-time HJ PDEs that allows us to consider certain non-convex regularizations in these image denoising and decomposition models. In particular, this min-plus algebra approach yields practical numerical optimization algorithms for solving certain image denoising and decomposition models. Section "Viscous Hamilton-Jacobi PDEs and Bayesian Estimation" reviews connections between viscous HJ PDEs and posterior mean estimators with Gaussian data fidelity term and log-concave priors. We also present an analogue of the min-plus algebra technique for these viscous HJ PDEs with certain priors that are not log-concave. Finally, we draw some conclusions in section "Conclusion".

First-Order Hamilton-Jacobi PDEs and Optimization Problems

In this section, we discuss the connections between some variational optimization models in imaging sciences and HJ PDEs. In section "Single-Time HJ PDEs and Image Denoising Models", we consider the convex image denoising model (1) and the single-time HJ PDE. In section "Multi-time HJ PDEs and Image Decomposition Models", we review the connections between convex image decomposition models and the multi-time HJ PDE system. In section "Min-Plus Algebra for HJ PDEs and Certain Non-convex Regularizations", we use the min-plus algebra technique to solve certain optimization problems in which one regularization term is non-convex. In section "Application to Certain Decomposition Problems", we provide an application of the min-plus algebra technique to certain image decomposition problems, which yields practical numerical optimization algorithms.

Single-Time HJ PDEs and Image Denoising Models

As described in the introduction, an important class of optimization models in imaging sciences for denoising takes the form of (1), where $\lambda > 0$ is a positive parameter, $x \in \mathbb{R}^n$ is the observed image with n pixels, and $u \in \mathbb{R}^n$ is the reconstructed image. The objective function is the weighted sum of the convex regularization term J and the convex data fidelity term D.

The connection between the class of optimization models (1) and first-order HJ PDEs has been discussed in Darbon (2015). Specifically, if the data fidelity term λD can be written in the form of $tH^*\left(\frac{\cdot}{t}\right)$ (where H^* denotes the Legendre transform of a convex function H and $t > 0$ is a new parameter that depends on λ), then the minimization problem (1) defines a function $S \colon \mathbb{R}^n \times (0, +\infty) \to \mathbb{R}$ as follows:

$$S(x, t) = \min_{u \in \mathbb{R}^n} \left\{ J(u) + tH^*\left(\frac{x-u}{t}\right) \right\}. \tag{5}$$

For instance, if the noise is assumed to be Gaussian, independent, identically distributed, and additive, we impose the quadratic data fidelity $D(x) = \frac{1}{2}\|x\|_2^2$ for each $x \in \mathbb{R}^n$. Then D satisfies $\lambda D(x) = tH^*\left(\frac{x}{t}\right)$ where $H^*(x) = \frac{1}{2}\|x\|_2^2$ and $t = \frac{1}{\lambda}$.

Formula (5) is called the Lax-Oleinik formula (Bardi and Evans 1984; Evans 2010; Hopf 1965) in the PDE literature, and it solves the following first-order HJ PDE:

$$\begin{cases} \frac{\partial S}{\partial t}(x, t) + H(\nabla_x S(x, t)) = 0 & x \in \mathbb{R}^n, t > 0, \\ S(x, 0) = J(x) & x \in \mathbb{R}^n, \end{cases} \tag{6}$$

where the function $H \colon \mathbb{R}^n \to \mathbb{R}$ is called the Hamiltonian and $J \colon \mathbb{R}^n \to \mathbb{R} \cup \{+\infty\}$ is the initial data. In Darbon (2015), a representation formula for the minimizer of (5) is given, and we state it in the following proposition. Here and in the remainder of this chapter, we use $\Gamma_0(\mathbb{R}^n)$ to denote the set of convex, proper and lower semicontinuous functions from $\mathbb{R}^n$ to $\mathbb{R} \cup \{+\infty\}$.

Proposition 1. *Assume $J \in \Gamma_0(\mathbb{R}^n)$, and assume $H: \mathbb{R}^n \to \mathbb{R}$ is a differentiable, strictly convex, and 1-coercive function. Then the Lax-Oleinik formula (5) gives the differentiable and convex solution $S: \mathbb{R}^n \times (0, +\infty) \to \mathbb{R}$ to the HJ PDE (6). Moreover, for each $x \in \mathbb{R}^n$ and $t > 0$, the minimizer in (5) exists and is unique, which we denote by $u(x, t)$, and satisfies*

$$u(x, t) = x - t \nabla H(\nabla_x S(x, t)). \tag{7}$$

Equation (7) in this proposition gives the relation between the minimizer u in the Lax-Oleinik formula (5) and the spatial gradient of the solution to the HJ PDE (6). In other words, one can compute the minimizer in the corresponding denoising model (1) using the spatial gradient $\nabla_x S(x, t)$ of the solution, and vice versa.

There is another set of assumptions for the conclusion of the proposition above to hold. For the details, we refer the reader to Darbon (2015).

Multi-time HJ PDEs and Image Decomposition Models

In this subsection, we consider the following image decomposition models:

$$\min_{u_1, \ldots, u_N \in \mathbb{R}^n} \left\{ J\left(x - \sum_{i=1}^N u_i\right) + \sum_{i=1}^N \lambda_i f_i(u_i) \right\}, \tag{8}$$

where $\lambda_1, \ldots, \lambda_N$ are positive parameters, $x \in \mathbb{R}^n$ is the observed image with n pixels, and $u_1, \ldots, u_N \in \mathbb{R}^n$ correspond to the decomposition of the original image x. In Darbon and Meng (2020), the relation between the decomposition model (8) and the multi-time HJ PDE system has been proposed under the convexity assumptions of J and the functions $f_1, \ldots, f_N$.

In the decomposition model, an image is assumed to be the summation of $N + 1$ components, denoted as $u_1, \ldots, u_N$ and the residual $x - \sum_{i=1}^N u_i$. The feature of each part u_i is characterized by a convex function f_i, and the residual $x - \sum_{i=1}^N u_i$ is characterized by a convex regularization term J. If the function $\lambda_i f_i$ can be written in the form of $t_i H_i^*\left(\frac{\cdot}{t_i}\right)$ (where H_i^* denotes the Legendre transform of a convex function H_i and $t_i > 0$ is a new parameter which depends on λ_i) for each $i \in \{1, \ldots, N\}$, then the image decomposition model (8) defines a function $S: \mathbb{R}^n \times (0, +\infty)^N \to \mathbb{R}$ as follows:

$$S(x, t_1, \ldots, t_N) = \min_{u_1, \ldots, u_N \in \mathbb{R}^n} \left\{ J\left(x - \sum_{i=1}^N u_i\right) + \sum_{i=1}^N t_i H_i^*\left(\frac{u_i}{t_i}\right) \right\}. \tag{9}$$

This formula is called the generalized Lax-Oleinik formula (Lions and Rochet 1986; Tho 2005) which solves the following multi-time HJ PDE system:

$$\begin{cases} \frac{\partial S(x,t_1,\dots,t_N)}{\partial t_1} + H_1(\nabla_x S(x,t_1,\dots,t_N)) = 0 & x \in \mathbb{R}^n, t_1,\dots,t_N > 0, \\ \vdots \\ \frac{\partial S(x,t_1,\dots,t_N)}{\partial t_j} + H_j(\nabla_x S(x,t_1,\dots,t_N)) = 0 & x \in \mathbb{R}^n, t_1,\dots,t_N > 0, \\ \vdots \\ \frac{\partial S(x,t_1,\dots,t_N)}{\partial t_N} + H_N(\nabla_x S(x,t_1,\dots,t_N)) = 0 & x \in \mathbb{R}^n, t_1,\dots,t_N > 0, \\ S(x,0,\dots,0) = J(x) & x \in \mathbb{R}^n, \end{cases}$$
(10)

where $H_1,\dots,H_N \colon \mathbb{R}^n \to \mathbb{R}$ are called Hamiltonians and $J \colon \mathbb{R}^n \to \mathbb{R} \cup \{+\infty\}$ is the initial data. Under certain assumptions (see Prop. 2), the generalized Lax-Oleinik formula (9) gives the solution $S(x,t_1,\dots,t_N)$ to the multi-time HJ PDE system (10). In Darbon and Meng (2020), the relation between the minimizer in (9) and the spatial gradient $\nabla_x S(x,t_1,\dots,t_N)$ of the solution to the multi-time HJ PDE system (10) is studied. This relation is described in the following proposition.

Proposition 2. *Assume $J \in \Gamma_0(\mathbb{R}^n)$, and assume $H_j \colon \mathbb{R}^n \to \mathbb{R}$ is a convex and 1-coercive function for each $j \in \{1,\dots,N\}$. Suppose there exists $j \in \{1,\dots,N\}$ such that H_j is strictly convex. Then the generalized Lax-Oleinik formula (9) gives the differentiable and convex solution $S \colon \mathbb{R}^n \times (0,+\infty)^N \to \mathbb{R}$ to the multi-time HJ PDE system (10). Moreover, for each $x \in \mathbb{R}^n$ and $t_1,\dots,t_N > 0$, the minimizer in (9) exists. We denote by $(u_1(x,t_1,\dots,t_N),\dots,u_N(x,t_1,\dots,t_N))$ any minimizer of the minimization problem in (9) with parameters $x \in \mathbb{R}^n$ and $t_1,\dots,t_N \in (0,+\infty)$. Then, for each $j \in \{1,\dots,N\}$, there holds*

$$u_j(x,t_1,\dots,t_N) \in t_j \partial H_j(\nabla_x S(x,t_1,\dots,t_N)),$$

where ∂H_j denotes the subdifferential of H_j.

Furthermore, if all the Hamiltonians $H_1,\dots,H_N$ are differentiable, then the minimizer is unique and satisfies

$$u_j(x,t_1,\dots,t_N) = t_j \nabla H_j(\nabla_x S(x,t_1,\dots,t_N)), \tag{11}$$

for each $j \in \{1,\dots,N\}$.

As a result, when the assumptions in the proposition above are satisfied, one can compute the minimizer to the corresponding decomposition model (8) using equation (11) and the spatial gradient $\nabla_x S(x,t_1,\dots,t_N)$ of the solution to the multi-time HJ PDE (10).

Min-Plus Algebra for HJ PDEs and Certain Non-convex Regularizations

In the previous two subsections, we considered the optimization models (1) and (8) where each term was assumed to be convex. When J is non-convex, solutions to (6) may not be classical (in the sense that it is not differentiable). It is well-known that the concept of viscosity solutions (Bardi and Capuzzo-Dolcetta 1997; Barles 1994; Barron et al. 1984; Crandall et al. 1992; Evans 2010; Fleming and Soner 2006) is generally the appropriate notion of solutions for these HJ PDEs. Note that Lax-Oleinik formulas (1) and (8) yield viscosity solutions to their respective HJ PDEs (6) and (10). However, these Lax-Oleinik formulas result in non-convex optimization problems.

In this subsection, we use the min-plus algebra technique (Akian et al. 2006, 2008; Dower et al. 2015; Fleming and McEneaney 2000; Gaubert et al. 2011; Kolokoltsov and Maslov 1997; McEneaney 2006, 2007; McEneaney et al. 2008; McEneaney and Kluberg 2009) to handle the cases when the term J in (1) and (8) is assumed to be a non-convex function in the following form:

$$J(x) = \min_{i \in \{1,\dots,m\}} J_i(x) \text{ for every } x \in \mathbb{R}^n, \qquad (12)$$

where $J_i \in \Gamma_0(\mathbb{R}^n)$ for each $i \in \{1, \dots, m\}$.

First, we consider the single-time HJ PDE (6). By min-plus algebra theory, the semigroup of this HJ PDE is linear with respect to the min-plus algebra. In other words, under certain assumptions the solution S to the HJ PDE $\frac{\partial S}{\partial t}(x,t) + H(\nabla_x S(x,t)) = 0$ with initial data J is the minimum of the solution S_i to the HJ PDE $\frac{\partial S_i}{\partial t}(x,t) + H(\nabla_x S_i(x,t)) = 0$ with initial data J_i. Specifically, if the Lax-Oleinik formula (5) solves the HJ PDE (6) for each $i \in \{1, \dots, m\}$ and the minimizer u exists (for instance, when $J_i \in \Gamma_0(\mathbb{R}^n)$ for each $i \in \{1, \dots, m\}$, and $H \colon \mathbb{R}^n \to \mathbb{R}$ is a differentiable, strictly convex, and 1-coercive function), then we have:

$$\begin{aligned}
S(x,t) &= \min_{u \in \mathbb{R}^n} \left\{ J(u) + t H^*\left(\frac{x-u}{t}\right) \right\} \\
&= \min_{u \in \mathbb{R}^n} \left\{ \min_{i \in \{1,\dots,m\}} J_i(u) + t H^*\left(\frac{x-u}{t}\right) \right\} \\
&= \min_{u \in \mathbb{R}^n} \min_{i \in \{1,\dots,m\}} \left\{ J_i(u) + t H^*\left(\frac{x-u}{t}\right) \right\} \qquad (13) \\
&= \min_{i \in \{1,\dots,m\}} \left\{ \min_{u \in \mathbb{R}^n} \left\{ J_i(u) + t H^*\left(\frac{x-u}{t}\right) \right\} \right\} \\
&= \min_{i \in \{1,\dots,m\}} S_i(x,t).
\end{aligned}$$

Therefore, the solution $S(\boldsymbol{x}, t)$ is given by the pointwise minimum of $S_i(\boldsymbol{x}, t)$ for $i \in \{1, \ldots, m\}$. Note that the Lax-Oleinik formula (5) yields a convex problem for each $S_i(\boldsymbol{x}, t)$ with $i \in \{1, \ldots, m\}$. Therefore this approach seems particularly appealing to solve these non-convex optimization problems and associated HJ PDEs. Note that such an approach is embarrassingly parallel since we can solve the initial data J_i for each $i \in \{1, \ldots, m\}$ independently and compute in linear time the pointwise minimum. However, this approach is only feasible if m is not too big. We will see later in this subsection that robust edge-preserving priors (e.g., truncated Total Variation or truncated quadratic) can be written in the form of (12), but m is exponential in n.

We can also compute the set of minimizers $\boldsymbol{u}(\boldsymbol{x}, t)$ as follows. Here, we abuse notation and use $\boldsymbol{u}(\boldsymbol{x}, t)$ to denote the set of minimizers, which may be not a singleton set when the minimizer is not unique. We can write

$$\boldsymbol{u}(\boldsymbol{x}, t) = \arg\min_{\boldsymbol{u}\in\mathbb{R}^n} \left\{ \min_{i\in\{1,\ldots,m\}} J_i(\boldsymbol{u}) + tH^*\left(\frac{\boldsymbol{x}-\boldsymbol{u}}{t}\right) \right\}$$

$$= \arg\min_{\boldsymbol{u}\in\mathbb{R}^n} \min_{i\in\{1,\ldots,m\}} \left\{ J_i(\boldsymbol{u}) + tH^*\left(\frac{\boldsymbol{x}-\boldsymbol{u}}{t}\right) \right\} \quad (14)$$

$$= \bigcup_{i\in I(\boldsymbol{x},t)} \arg\min_{\boldsymbol{u}\in\mathbb{R}^n} \left\{ J_i(\boldsymbol{u}) + tH^*\left(\frac{\boldsymbol{x}-\boldsymbol{u}}{t}\right) \right\},$$

where the index set $I(\boldsymbol{x}, t)$ is defined by

$$I(\boldsymbol{x}, t) = \arg\min_{i\in\{1,\ldots,m\}} S_i(\boldsymbol{x}, t). \quad (15)$$

A specific example is when the regularization term J is the truncated regularization term with pairwise interactions in the following form:

$$J(\boldsymbol{x}) = \sum_{(i,j)\in E} w_{ij} f(x_i - x_j), \text{ for each } \boldsymbol{x} = (x_1, \ldots, x_n) \in \mathbb{R}^n, \quad (16)$$

where $w_{ij} \geqslant 0$, $f(x) = \min\{g(x), 1\}$ for some convex function $g\colon \mathbb{R} \to \mathbb{R}$ and $E = \{1, \ldots, n\} \times \{1, \ldots, n\}$. This function can be written as the minimum of a collection of convex functions $J_\Omega\colon \mathbb{R}^n \to \mathbb{R}$ as the following:

$$J(\boldsymbol{x}) = \min_{\Omega\subseteq E} J_\Omega,$$

with each J_Ω defined by

$$J_\Omega := \left\{ \sum_{(i,j)\in\Omega} w_{ij} + \sum_{(i,j)\notin\Omega} w_{ij} g(x_i - x_j) \right\},$$

where Ω is any subset of E. The truncated regularization term (16) can therefore be written in the form of (12), and hence the minimizer to the corresponding optimization problem (1) with the non-convex regularization term J in (16) can be computed using (14).

We give here two examples of truncated regularization term with pairwise interactions in the form of (16). First, let g be the ℓ^1 norm. Then J is the truncated discrete Total Variation regularization term defined by

$$J(x) = \sum_{(i,j)\in E} w_{ij} \min\{|x_i - x_j|, 1\}, \text{ for each } x = (x_1, \ldots, x_n) \in \mathbb{R}^n. \quad (17)$$

This function J can be written as the formula (16) with $f: \mathbb{R} \to \mathbb{R}$ given by Eq. (4). Second, let g be the quadratic function. Then J is the half-quadratic regularization term defined by

$$J(x) = \sum_{(i,j)\in E} w_{ij} \min\{(x_i - x_j)^2, 1\}, \text{ for each } x = (x_1, \ldots, x_n) \in \mathbb{R}^n. \quad (18)$$

This function J can be written as the formula (16) with $f: \mathbb{R} \to \mathbb{R}$ given by Eq. (3). This specific form of edge-preserving prior was investigated in the seminal works of Charbonnier et al. (1997), Geman and Yang (1995) and Geman and Reynolds (1992). Several algorithms have been proposed to solve the resultant non-convex optimization problem (13), i.e., the solution to the corresponding HJ PDE, for some specific choice of data fidelity terms (e.g., Allain et al. 2006; Idier 2001; Geman and Yang 1995; Geman and Reynolds 1992; Nikolova and Ng 2005; Champagnat and Idier 2004; Nikolova and Ng 2001).

Suppose now, for general regularization terms J in the form of (16), that we have Gaussian noise. Then the data fidelity term is quadratic and $H(p) = \frac{1}{2}\|p\|_2^2$ and $t = \frac{1}{\lambda}$. Hence, for this example, using (14), we obtain the set of minimizers:

$$u(x,t) = \bigcup_{\Omega \in I(x,t)} \arg\min_{u \in \mathbb{R}^n} \left\{ J_\Omega(u) + t H^*\left(\frac{x-u}{t}\right) \right\}$$

$$= \bigcup_{\Omega \in I(x,t)} \arg\min_{u \in \mathbb{R}^n} \left\{ \sum_{(i,j)\notin\Omega} w_{ij} g(u_i - u_j) + \frac{1}{2t}\|x - u\|_2^2 \right\}$$

$$= \bigcup_{\Omega \in I(x,t)} \{x - t\nabla_x S_\Omega(x,t)\}$$

where

$$S_\Omega(x,t) = \sum_{(i,j)\in\Omega} w_{ij} + \min_{u\in\mathbb{R}^n} \left\{ \sum_{(i,j)\notin\Omega} w_{ij} g(u_i - u_j) + \frac{1}{2t}\|x-u\|_2^2 \right\}$$

and

$$I(x,t) = \arg\min_{\Omega\subseteq E} S_\Omega(x,t).$$

The same result also holds for the multi-time HJ PDE system (10). Indeed, if J is a non-convex regularization term given by (12), and S, $S_j \colon \mathbb{R}^n \times (0,+\infty)^N \to \mathbb{R}$ are the solutions to the multi-time HJ PDE system (10) with initial data J and J_i, respectively, then similarly we have the min-plus linearity of the semigroup under certain assumptions. Specifically, if the Lax-Oleinik formula (9) solves the multi-time HJ PDE system (10) for each $i \in \{1,\ldots,m\}$ (for instance, when H and J_i satisfy the assumptions in Prop. 2 for each $i \in \{1,\ldots,m\}$), then there holds

$$\begin{aligned} S(x,t_1,\ldots,t_N) &= \min_{u_1,\ldots,u_N\in\mathbb{R}^n} \left\{ \min_{i\in\{1,\ldots,m\}} J_i\left(x - \sum_{j=1}^N u_j\right) + \sum_{j=1}^N t_j H_j^*\left(\frac{u_j}{t_j}\right) \right\} \\ &= \min_{i\in\{1,\ldots,m\}} \left\{ \min_{u_1,\ldots,u_N\in\mathbb{R}^n} \left\{ J_i\left(x - \sum_{j=1}^N u_j\right) + \sum_{j=1}^N t_j H_j^*\left(\frac{u_j}{t_j}\right) \right\} \right\} \\ &= \min_{i\in\{1,\ldots,m\}} S_i(x,t_1,\ldots,t_N). \end{aligned}$$
(19)

Let $M \subset \mathbb{R}^{n\times N}$ be the set of minimizers of (9) with J given by (12). Then M satisfies

$$\begin{aligned} M &= \arg\min_{u_1,\ldots,u_N\in\mathbb{R}^n} \left\{ \min_{i\in\{1,\ldots,m\}} J_i\left(x - \sum_{j=1}^N u_j\right) + \sum_{j=1}^N t_j H_j^*\left(\frac{u_j}{t_j}\right) \right\} \\ &= \bigcup_{i\in I(x,t_1,\ldots,t_N)} \arg\min_{u_1,\ldots,u_N\in\mathbb{R}^n} \left\{ J_i\left(x - \sum_{j=1}^N u_j\right) + \sum_{j=1}^N t_j H_j^*\left(\frac{u_j}{t_j}\right) \right\}, \end{aligned}$$
(20)

where the index set $I(x,t_1,\ldots,t_N)$ is defined by

$$I(x,t_1,\ldots,t_N) = \arg\min_{i\in\{1,\ldots,m\}} S_i(x,t_1,\ldots,t_N). \tag{21}$$

As a result, we can use (20) to obtain the minimizers of the decomposition model (8) with the non-convex regularization term J in the form of (12), such as the function in (16) and the truncated Total Variation function (17).

In summary, one can compute the minimizers of the optimization problems (1) and (8) with a non-convex function J in the form of (12) using the aforementioned min-plus algebra technique. Furthermore, this technique can be extended to handle other cases. For instance, in the denoising model (1), if the data fidelity term D is in the form of (12) and the prior term $\frac{J(u)}{\lambda}$ can be written as $tH^*\left(\frac{u}{t}\right)$, then one can still compute the minimizer of this problem using the min-plus algebra technique on the HJ PDE with initial data D. Similarly, because of the symmetry in the decomposition model (8), if there is only one non-convex term f_j and if it can be written in the form of (12), then one can apply the min-plus algebra technique to the multi-time HJ PDE with initial data f_j.

In general, however, there is a drawback to the min-plus algebra technique. To compute the minimizers using (14) and (20), we need to compute the index set $I(x,t)$ and $I(x,t_1,\ldots,t_N)$ defined in (15) and (21), which involves solving m HJ PDEs to obtain the solutions $S_1, \ldots, S_m$. When m is too large, this approach is impractical since it involves solving too many HJ PDEs. For instance, if J is the truncated Total Variation in (17), the number m equals the number of subsets of the set E, i.e., $m = 2^{|E|}$, which is computationally intractable. Hence, in general, it is impractical to use (14) and (20) to solve the problems (1) and (8) where the regularization term J is given by the truncated Total Variation. The same issue arises when the truncated Total Variation is replaced by half-quadratic regularization. Several authors attempted to address this intractability for half-quadratic regularizations by proposing heuristic optimization methods that aim to compute a global minimizer (Allain et al. 2006; Idier 2001; Geman and Yang 1995; Geman and Reynolds 1992; Nikolova and Ng 2005; Champagnat and Idier 2004; Nikolova and Ng 2001).

Application to Certain Decomposition Problems

In this section, we demonstrate how to use our formulation described in the previous sections to solve certain image decomposition problems. The variational formulation for image decomposition problems is in the form of (8), where the input image $x \in \mathbb{R}^n$ is decomposed into three components, which includes the geometrical part $x - u_1 - u_2$, the texture part u_1, and the noise u_2. The regularization function J for the geometrical part $x - u_1 - u_2$ is chosen to be the widely used Total Variation regularization function in order to preserve edges in the image. Here, we use the anisotropic Total Variation semi-norm (see, e.g., Darbon and Sigelle 2006; Darbon 2015) denoted by $|\cdot|_{TV}$. The noise is assumed to be Gaussian, and hence the data fidelity term f_2 is set to be the quadratic function. Many texture models have been proposed (see Aujol et al. 2003, 2005; Le Guen 2014; Winkler 2003 and the references in these papers). For instance, the indicator function of the unit ball with respect to Meyer's norm is used in Aujol et al. (2003, 2005), and the ℓ^1 norm is used in Le Guen (2014). Note that each texture model has some pros and cons and, to our knowledge, it remains an open problem whether one specific texture model is better than the others. In this example, we combine different texture regularizations proposed in the literature by taking the minimum of the indicator function of the unit ball with respect to Meyer's norm and the ℓ^1 norm. In other words, we consider the following variational problem:

$$\min_{\bm{u}_1, \bm{u}_2 \in \mathbb{R}^n} \left\{ J(\bm{x} - \bm{u}_1 - \bm{u}_2) + t_1 g\left(\frac{\bm{u}_1}{t_1}\right) + \frac{1}{2t_2}\|\bm{u}_2\|_2^2 \right\}, \qquad (22)$$

where $J \colon \mathbb{R}^n \to \mathbb{R}$ and $g \colon \mathbb{R}^n \to \mathbb{R} \cup \{+\infty\}$ are defined by

$$J(\bm{y}) := |\bm{y}|_{TV}, \qquad g(\bm{y}) := \min\{J^*(\bm{y}), \|\bm{y}\|_1\},$$

for each $\bm{y} \in \mathbb{R}^n$. Problem (22) is equivalent to the following mixed discrete-continuous optimization problem

$$\min_{\bm{u}_1, \bm{u}_2 \in \mathbb{R}^n} \min_{k \in \{1,2\}} \left\{ J(\bm{x} - \bm{u}_1 - \bm{u}_2) + t_1 g_k\left(\frac{\bm{u}_1}{t_1}\right) + \frac{1}{2t_2}\|\bm{u}_2\|_2^2 \right\}, \qquad (23)$$

where $g_1(\bm{y}) := J^*(\bm{y})$ and $g_2(\bm{y}) := \|\bm{y}\|_1$ for each $\bm{y} \in \mathbb{R}^n$. Note that solving mixed discrete-continuous optimization is hard in general (see Floudas and Pardalos 2009 for instance). However, we shall see that our proposed approach yields efficient optimization algorithms. Since the function g is the minimum of two convex functions, the problem (22) fits into our formulation, and can be solved using a similar idea as in (19) and (20). To be specific, define the two functions S_1 and S_2 by

$$S_1(\bm{x}, t_1, t_2) := \min_{\bm{u}_1, \bm{u}_2 \in \mathbb{R}^n} \left\{ J(\bm{x} - \bm{u}_1 - \bm{u}_2) + t_1 J^*\left(\frac{\bm{u}_1}{t_1}\right) + \frac{1}{2t_2}\|\bm{u}_2\|_2^2 \right\},$$

$$S_2(\bm{x}, t_1, t_2) := \min_{\bm{u}_1, \bm{u}_2 \in \mathbb{R}^n} \left\{ J(\bm{x} - \bm{u}_1 - \bm{u}_2) + \|\bm{u}_1\|_1 + \frac{1}{2t_2}\|\bm{u}_2\|_2^2 \right\}, \qquad (24)$$

where the sets of the minimizers in the two minimization problems above are denoted by $M_1(\bm{x}, t_1, t_2)$ and $M_2(\bm{x}, t_1, t_2)$, respectively. Using a similar argument as in (19) and (20), we conclude that the minimal value in (22) equals $\min\{S_1(\bm{x}, t_1, t_2), S_2(\bm{x}, t_1, t_2)\}$, and the set of minimizers in (22), denoted by $M(\bm{x}, t_1, t_2)$, satisfies

$$M(x, t_1, t_2) = \begin{cases} M_1(x, t_1, t_2) & S_1(x, t_1, t_2) < S_2(x, t_1, t_2), \\ M_2(x, t_1, t_2) & S_1(x, t_1, t_2) > S_2(x, t_1, t_2), \\ M_1(x, t_1, t_2) \cup M_2(x, t_1, t_2) & S_1(x, t_1, t_2) = S_2(x, t_1, t_2). \end{cases} \quad (25)$$

As a result, we solve the two minimization problems in (24) first, and then obtain the minimizers using (25) by comparing the minimal values $S_1(x, t_1, t_2)$ and $S_2(x, t_1, t_2)$.

Here, we present a numerical result. We solve the first optimization problem in (24) by a splitting method, where each subproblem can be solved using the proximal operator of the anisotropic Total Variation (for more details, see Darbon and Meng 2020). Similarly, a splitting method is used to split the second optimization problem in (24) to two subproblems, which are solved using the proximal operators of the anisotropic Total Variation and the ℓ^1-norm, respectively. To compute the proximal point of the anisotropic Total Variation, the algorithm in Chambolle and Darbon (2009), Darbon and Sigelle (2006), and Hochbaum (2001) is adopted, and it computes the proximal point without numerical errors. The input image x is the image "Barbara" shown in Fig. 1. The parameters are set to be $t_1 = 0.07$ and $t_2 = 0.01$. Let $(u_1, u_2) \in M_1(x, t_1, t_2)$ and $(v_1, v_2) \in M_2(x, t_1, t_2)$ be respectively the minimizers of the two minimization problems in (24) solved by the aforementioned splitting methods. We show these minimizers and the related images in Figs. 2 and 3. To be specific, the decomposition components $x - u_1 - u_2$, $u_1 + 0.5$, and $u_2 + 0.5$ given by the first optimization problem in (24) are shown in Fig. 2a, b, and c, respectively. The decomposition components $x - v_1 - v_2$, $v_1 + 0.5$, and $v_2 + 0.5$

Fig. 1 The input image x ("Barbara") in the example in section "Application to Certain Decomposition Problems"

Fig. 2 The minimizer of the first problem in (24). The output images $x - u_1 - u_2$, $u_1 + 0.5$, and $u_2 + 0.5$ are shown in (**a**), (**b**), and (**c**), respectively

given by the second optimization problem in (24) are shown in Fig. 3a, b, and c, respectively. We also compute the optimal values $S_1(x, t_1, t_2)$ and $S_2(x, t_1, t_2)$, and obtain

$$S_1(x, t_1, t_2) = 1832.81, \qquad S_2(x, t_1, t_2) = 4171.33.$$

Since $S_1(x, t_1, t_2) < S_2(x, t_1, t_2)$, we conclude that (u_1, u_2) is a minimizer in the decomposition problem (22), and the minimal value equals 1832.81. In other words, the optimal decomposition given by (22) is shown in Fig. 2.

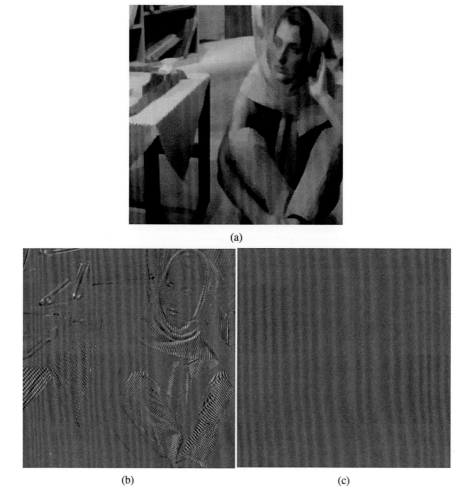

Fig. 3 The minimizer of the second problem in (24). The output images $x - v_1 - v_2$, $v_1 + 0.5$ and $v_2 + 0.5$ are shown in (**a**), (**b**) and (**c**), respectively

Viscous Hamilton-Jacobi PDEs and Bayesian Estimation

In contrast to variational approaches that frame imaging problems as optimization problems, Bayesian approaches frame them in a probabilistic framework. This framework combines observed data through a likelihood function (which models the noise corrupting the unknown image) and prior knowledge through a prior distribution (which models known properties of the image to reconstruct) to generate a posterior distribution from which an appropriate decision rule can select a meaningful image estimate. In this section, we present an analogue of the min-plus algebra technique discussed in section "Min-Plus Algebra for HJ PDEs and Certain Non-convex Regularizations" for certain Bayesian posterior mean estimators.

Viscous HJ PDEs and Posterior Mean Estimators for Log-Concave Models

Consider the following class of Bayesian posterior distributions:

$$q(u|(x,t,\epsilon)) := \frac{e^{-\left(J(u)+\frac{1}{2t}\|x-u\|_2^2\right)/\epsilon}}{\int_{\mathbb{R}^n} e^{-\left(J(u)+\frac{1}{2t}\|x-u\|_2^2\right)/\epsilon} du}, \tag{26}$$

where $x \in \mathbb{R}^n$ is the observed image with n pixels, and t and ϵ are positive parameters. The posterior distribution (26) is proportional to the product of a log-concave prior $u \mapsto e^{-J(u)/\epsilon}$ (possibly improper) and a Gaussian likelihood function $u \mapsto e^{-\frac{1}{2t\epsilon}\|x-u\|_2^2}$. This class of posterior distributions generates the family of Bayesian posterior mean estimators $u_{PM} : \mathbb{R}^n \times (0,+\infty) \times (0,+\infty) \to \mathbb{R}^n$ defined by

$$u_{PM}(x,t,\epsilon) := \int_{\mathbb{R}^n} u\, q(u|(x,t,\epsilon))\, du. \tag{27}$$

These are Bayesian estimators because they minimize the mean squared error (Kay 1993, pages 344–345):

$$u_{PM}(x,t,\epsilon) = \arg\min_{u\in\mathbb{R}^n} \int_{\mathbb{R}^n} \|\bar{u}-u\|_2^2\, q(\bar{u}|(x,t,\epsilon))\, d\bar{u}. \tag{28}$$

They are frequently called minimum mean squared error estimators for this reason.

The class of posterior distributions (26) also generates the family of maximum a posteriori estimators $u_{MAP} : \mathbb{R}^n \times (0,+\infty) \to \mathbb{R}^n$ defined by

$$u_{MAP}(x,t) = \arg\min_{u\in\mathbb{R}^n} \left\{ J(u) + \frac{1}{2t}\|x-u\|_2^2 \right\}, \tag{29}$$

where $u_{MAP}(x,t)$ is the mode of the posterior distribution (26). Note that the MAP estimator is also the minimizer of the solution (5) to the first-order HJ PDE (6) with Hamiltonian $H = \frac{1}{2}\|\cdot\|_2^2$ and initial data J.

There is a large body of literature on posterior mean estimators for image restoration problems (see e.g., Demoment 1989; Kay 1993; Winkler 2003). In particular, original connections between variational problems and Bayesian methods have been investigated in Louchet (2008), Louchet and Moisan (2013), Burger and Lucka (2014), Burger and Sciacchitano (2016), Gribonval (2011), Gribonval and Machart (2013), Gribonval and Nikolova (2018), and Darbon and Langlois

(2020). In particular, in Darbon and Langlois (2020), the authors described original connections between Bayesian posterior mean estimators and viscous HJ PDEs when $J \in \Gamma_0(\mathbb{R}^n)$ and the data fidelity term is Gaussian. We now briefly describe these connections here.

Consider the function $S_\epsilon : \mathbb{R}^n \times (0, +\infty) \to \mathbb{R}$ defined by

$$S_\epsilon(x, t) = -\epsilon \ln \left(\frac{1}{(2\pi t \epsilon)^{n/2}} \int_{\mathbb{R}^n} e^{-\left(J(u) + \frac{1}{2t}\|x-u\|_2^2\right)/\epsilon} \, du \right), \qquad (30)$$

which is proportional to the negative logarithm of the partition function of the posterior distribution (26). Under appropriate assumptions on the regularization term J (see Proposition 3), formula (30) corresponds to a Cole-Hopf transform (Evans 2010) and is the solution to the following viscous HJ PDE:

$$\begin{cases} \frac{\partial S_\epsilon}{\partial t}(x, t) + \frac{1}{2} \|\nabla_x S_\epsilon(x, t)\|_2^2 = \frac{\epsilon}{2} \Delta_x S_\epsilon(x, t) & x \in \mathbb{R}^n, t > 0, \\ S_\epsilon(x, 0) = J(x) & x \in \mathbb{R}^n, \end{cases} \qquad (31)$$

where J is the initial data. The solution to this PDE is also related to the first-order HJ PDE (6) when the Hamiltonian is $H = \frac{1}{2}\|\cdot\|_2^2$. The following proposition, which is given in Darbon and Langlois (2020), describes these connections.

Proposition 3. *Assume $J \in \Gamma_0(\mathbb{R}^n)$, int (dom J) $\neq \emptyset$, and $\inf_{u \in \mathbb{R}^n} J(u) = 0$. Then for every $\epsilon > 0$, the unique smooth solution $S_\epsilon : \mathbb{R}^n \times (0, +\infty) \to (0, +\infty)$ to the HJ PDE (31) is given by formula (30), where $(x, t) \mapsto S_\epsilon(x, t) - \frac{n\epsilon}{2} \ln t$ is jointly convex. Moreover, for each $x \in \mathbb{R}^n$, $t > 0$, and $\epsilon > 0$, the posterior mean estimator (27) and minimum mean squared error in (28) (with $u = u_{PM}(x, t, \epsilon)$) satisfy, respectively, the formulas:*

$$u_{PM}(x, t, \epsilon) = x - t \nabla_x S_\epsilon(x, t) \qquad (32)$$

and

$$\int_{\mathbb{R}^n} \|u_{PM}(x, t, \epsilon) - u\|_2^2 \, q(u|(x, t, \epsilon)) \, du = nt\epsilon - t^2 \epsilon \Delta_x S_\epsilon(x, t). \qquad (33)$$

In addition, for every $x \in \mathbb{R}^n$ and $t > 0$, the limits of $\lim_{\substack{\epsilon \to 0 \\ \epsilon > 0}} S_\epsilon(x, t)$ and $\lim_{\substack{\epsilon \to 0 \\ \epsilon > 0}} u_{PM}(x, t, \epsilon)$ exist and converge uniformly over every compact set of $\mathbb{R}^n \times (0, +\infty)$ in (x, t). Specifically, we have

$$\lim_{\substack{\epsilon \to 0 \\ \epsilon > 0}} S_\epsilon(x, t) = \min_{u \in \mathbb{R}^n} \left\{ J(u) + \frac{1}{2t} \|x - u\|_2^2 \right\}, \qquad (34)$$

where the right-hand side solves uniquely the first-order HJ PDE (6) with Hamiltonian $H = \frac{1}{2}\|\cdot\|_2^2$ and initial data J, and

$$\lim_{\substack{\epsilon \to 0 \\ \epsilon > 0}} \boldsymbol{u}_{PM}(\boldsymbol{x}, t, \epsilon) = \arg\min_{\boldsymbol{u} \in \mathbb{R}^n} \left\{ J(\boldsymbol{u}) + \frac{1}{2t} \|\boldsymbol{x} - \boldsymbol{u}\|_2^2 \right\}. \tag{35}$$

Under convexity assumptions on J, the representation formulas (32) and (33) relate the posterior mean estimate and the minimum mean squared error to the spatial gradient and Laplacian of the solution to the viscous HJ PDE (31), respectively. Hence one can compute the posterior mean estimator and minimum mean squared error using the spatial gradient $\nabla_{\boldsymbol{x}} S_\epsilon(\boldsymbol{x}, t)$ and the Laplacian $\Delta_{\boldsymbol{x}} S_\epsilon(\boldsymbol{x}, t)$ of the solution to the HJ PDE (31), respectively, or vice versa by computing the posterior mean and minimum mean squared error using, for instance, Markov chain Monte Carlo sampling strategies.

The limit (35) shows that the posterior mean $\boldsymbol{u}_{PM}(\boldsymbol{x}, t, \epsilon)$ converges to the maximum a posteriori $\boldsymbol{u}_{MAP}(\boldsymbol{x}, t)$ as the parameter $\epsilon \to 0$. A rough estimate of the squared Euclidean distance between the posterior mean estimator (27) and the maximum a posteriori (29) in terms of the parameters t and ϵ is given by

$$\|\boldsymbol{u}_{PM}(\boldsymbol{x}, t, \epsilon) - \boldsymbol{u}_{MAP}(\boldsymbol{x}, t)\|_2^2 \leqslant nt\epsilon. \tag{36}$$

On Viscous HJ PDEs with Certain Non-log-Concave Priors

So far, we have assumed that the regularization term J in the posterior distribution (26) and Proposition 3 is convex. Here, we consider an analogue of the min-plus algebra technique designed for certain first-order HJ PDEs tailed to viscous HJ PDEs, which will enable us to derive representation formulas for posterior mean estimators of the form of (27) whose priors are sums of log-concave priors, i.e., to certain mixture distributions.

Remember that the min-plus algebra technique for first-order HJ PDEs described in section "Min-Plus Algebra for HJ PDEs and Certain Non-convex Regularizations" involves initial data of the form $\min_{i \in \{1,...,m\}} J_i(\boldsymbol{x})$ where each $J_i \colon \mathbb{R}^n \to \mathbb{R} \cup \{+\infty\}$ is convex. Consider now initial data of the form

$$J(\boldsymbol{x}) = -\epsilon \ln \left(\sum_{i=1}^m e^{-J_i(\boldsymbol{x})/\epsilon} \right). \tag{37}$$

Note that formula (37) approximates the non-convex term (12) in that

$$\lim_{\substack{\epsilon \to 0 \\ \epsilon > 0}} -\epsilon \ln \left(\sum_{i=1}^m e^{-J_i(\boldsymbol{x})/\epsilon} \right) = \min_{i \in \{1,...,m\}} J_i(\boldsymbol{x}) \text{ for each } \boldsymbol{x} \in \mathbb{R}^n.$$

Now, assume int (dom J_i) $\neq \emptyset$ for each $i \in \{1, \dots, m\}$, and let

$$S_{i,\epsilon}(x, t) = -\epsilon \ln \left(\frac{1}{(2\pi t \epsilon)^{n/2}} \int_{\mathbb{R}^n} e^{-\left(J_i(u) + \frac{1}{2t}\|x-u\|_2^2\right)/\epsilon} du \right),$$

and

$$u_{i,PM}(x, t, \epsilon) = \frac{\int_{\mathbb{R}^n} u\, e^{-\left(J_i(u) + \frac{1}{2t}\|x-u\|_2^2\right)/\epsilon} du}{\int_{\mathbb{R}^n} e^{-\left(J_i(u) + \frac{1}{2t}\|x-u\|_2^2\right)/\epsilon} du}$$

denote, respectively, the solution to the viscous HJ PDE (31) with initial data J_i and its associated posterior mean. Then, a short calculation shows that for every $\epsilon > 0$, the function $S_\epsilon(x, t) : \mathbb{R}^n \times (0, +\infty) \to \mathbb{R}$ defined by

$$\begin{aligned} S_\epsilon(x, t) &= -\epsilon \ln \left(\sum_{i=1}^m \frac{1}{(2\pi t \epsilon)^{n/2}} \int_{\mathbb{R}^n} e^{-\left(J_i(u) + \frac{1}{2t}\|x-u\|_2^2\right)/\epsilon} du \right) \\ &= -\epsilon \ln \left(\sum_{i=1}^m e^{-S_{i,\epsilon}(x,t)/\epsilon} \right) \end{aligned} \quad (38)$$

is the unique smooth solution to the viscous HJ PDE (31) with initial data (37). As stated in section "Viscous HJ PDEs and Posterior Mean Estimators for Log-Concave Models", the posterior mean estimate $u_{PM}(x, t, \epsilon)$ is given by the representation formula:

$$u_{PM}(x, t, \epsilon) = x - t \nabla_x S_\epsilon(x, t), \quad (39)$$

which can be expressed in terms of the solutions $S_{i,\epsilon}(x, t)$, their spatial gradients $\nabla_x S_{i,\epsilon}(x, t)$, and posterior mean estimates $u_{i,PM}(x, t, \epsilon)$ as the weighted sums

$$\begin{aligned} u_{PM}(x, t, \epsilon) &= x - t \left(\frac{\sum_{i=1}^m \nabla_x S_{i,\epsilon}(x, t) e^{-S_{i,\epsilon}(x,t)/\epsilon}}{\sum_{i=1}^m e^{-S_{i,\epsilon}(x,t)/\epsilon}} \right) \\ &= \frac{\sum_{i=1}^m u_{i,PM}(x, t, \epsilon) e^{-S_{i,\epsilon}(x,t)/\epsilon}}{\sum_{i=1}^m e^{-S_{i,\epsilon}(x,t)/\epsilon}}. \end{aligned} \quad (40)$$

As an application of this result, we consider the problem of classifying a noisy image $x \in \mathbb{R}^n$ using a Gaussian mixture model (Duda et al. 2012): Suppose $J_i(u) = \frac{1}{2\sigma_i^2} \|u - \mu_i\|_2^2$, where $\mu_i \in \mathbb{R}^n$ and $\sigma_i > 0$. The regularized minimization problem (13) with quadratic data fidelity term $H = \frac{1}{2}\|\cdot\|_2^2$ is given by

$$S_0(x,t) = \min_{u \in \mathbb{R}^n} \left\{ \min_{i \in \{1,\ldots,m\}} \left\{ \frac{1}{2\sigma_i^2} \|u - \mu_i\|_2^2 + \frac{1}{2t} \|x - u\|_2^2 \right\} \right\}$$

$$= \min_{i \in \{1,\ldots,m\}} \left\{ \min_{u \in \mathbb{R}^n} \left\{ \frac{1}{2\sigma_i^2} \|u - \mu_i\|_2^2 + \frac{1}{2t} \|x - u\|_2^2 \right\} \right\} \quad (41)$$

$$= \min_{i \in \{1,\ldots,m\}} \left\{ \frac{1}{2(\sigma_i^2 + t)} \|x - \mu_i\|_2^2 \right\}.$$

Letting $I(x,t) = \arg\min_{i \in \{1,\ldots,m\}} \left\{ \frac{1}{2(\sigma_i^2+t)} \|x - \mu_i\|_2^2 \right\}$, the MAP estimator is then the collection:

$$u_{MAP}(x,t) = \bigcup_{i \in I(x,t)} \left\{ \frac{\sigma_i^2 x + t\mu_i}{\sigma_i^2 + t} \right\}.$$

Consider now the initial data (37):

$$J(u) = -\epsilon \ln \left(\sum_{i=1}^m e^{-\frac{1}{2\sigma_i^2 \epsilon} \|u - \mu_i\|_2^2} \right).$$

The solution $S_\epsilon(x,t)$ to the viscous HJ PDE (31) with initial data $J(x)$ is given by formula (38), which in this case can be computed analytically:

$$S_\epsilon(x,t) = -\epsilon \ln \left(\sum_{i=1}^m \left(\frac{\sigma_i^2}{\sigma_i^2 + t} \right)^{n/2} e^{-\frac{1}{2(\sigma_i^2+t)\epsilon} \|x - \mu_i\|_2^2} \right). \quad (42)$$

Since $e^{-S_{i,\epsilon}(x,t)/\epsilon} = \left(\frac{\sigma_i^2}{\sigma_i^2+t} \right)^{n/2} e^{-\frac{1}{2(\sigma_i^2+t)\epsilon} \|x - \mu_i\|_2^2}$, we can write the corresponding posterior mean estimator (40) using the representation formulas (39) and (40):

$$u_{PM}(x,t,\epsilon) = x - t \nabla_x S_\epsilon(x,t)$$

$$= \frac{\sum_{i=1}^m \left(\frac{\sigma_i^2 x + t\mu_i}{\sigma_i^2 + t} \right) \left(\frac{\sigma_i^2}{\sigma_i^2+t} \right)^{n/2} e^{-\frac{1}{2(\sigma_i^2+t)\epsilon} \|x - \mu_i\|_2^2}}{\sum_{i=1}^m \left(\frac{\sigma_i^2}{\sigma_i^2+t} \right)^{n/2} e^{-\frac{1}{2(\sigma_i^2+t)\epsilon} \|x - \mu_i\|_2^2}}. \quad (43)$$

Conclusion

In this chapter, we reviewed the connections of single-time HJ PDEs with image denoising models and the connections of multi-time HJ PDEs with image decomposition models under convexity assumptions. Specifically, under some assumptions, the minimizers of these optimization problems can be computed using the spatial gradient of the solution to the corresponding HJ PDEs. We also proposed a min-plus algebra technique to cope with certain non-convex regularization terms in imaging sciences problems. This suggests that certain non-convex optimization problem can be solved by computing several convex subproblems. For instance, if the denoising model (1) or the image decomposition model (8) involves a non-convex regularization term J that can be expressed as the minimum of m convex subproblems in the form of (12), then the minimizer of these non-convex problems can be solved using formulas (14) and (20). However, when m in (12) is too large, it is generally impractical to solve (14) and (20) using this min-plus technique because it involves solving too many HJ PDEs. However, our formulation yields practical numerical optimization algorithms for certain image denoising and decomposition problems.

We also reviewed connections between viscous HJ PDEs and a class of Bayesian methods and posterior mean estimators when the data fidelity term is Gaussian and the prior distribution is log-concave. Under some assumptions, the posterior mean estimator (27) and minimum mean squared error in (28) associated to the posterior distribution (26) can be computed using the spatial gradient and Laplacian of the solution to the viscous HJ PDE (31) via the representation formulas (32) and (33), respectively. We also proposed an analogue of the min-plus algebra technique designed for certain first-order HJ PDEs tailored to viscous HJ PDEs that enable us to compute posterior mean estimators with Gaussian fidelity term and prior that involves the sum of m log-concave priors, i.e., to certain mixture models. The corresponding posterior mean estimator with non-convex regularization J of the form of (37) can then be computed using the representation formulas (40) and posterior mean estimators (27) with convex regularization terms J_i.

Let us emphasize again that the proposed min-plus algebra technique for computations directly applies only for moderate m in (12). It would be of great interest to identify classes of non-convex regularizations for which novel numerical algorithms based on the min-plus algebra technique would not require to compute solutions to all m convex subproblems. To our knowledge, there is no available result in the literature on this matter.

Acknowledgments This work was funded by NSF 1820821.

References

Akian, M., Bapat, R., Gaubert, S.: Max-plus algebra. In: Handbook of Linear Algebra, 39 (2006)
Akian, M., Gaubert, S., Lakhoua, A.: The max-plus finite element method for solving deterministic optimal control problems: basic properties and convergence analysis. SIAM J. Control. Optim. **47**, 817–848 (2008)

Allain, M., Idier, J., Goussard, Y.: On global and local convergence of half-quadratic algorithms. IEEE Trans. Image Process. **15**, 1130–1142 (2006)

Aubert, G., Kornprobst, P.: Mathematical Problems in Image Processing. Springer (2002)

Aujol, J.-F., Aubert, G., Blanc-Féraud, L., Chambolle, A.: Image decomposition application to SAR images. In: L.D. Griffin, Lillholm, M. (eds.) Scale Space Methods in Computer Vision. Springer, Berlin/Heidelberg, pp. 297–312 (2003)

Aujol, J.-F., Aubert, G., Blanc-Féraud, L., Chambolle, A.: Image decomposition into a bounded variation component and an oscillating component. J. Math. Imaging Vision **22**, 71–88 (2005)

Bardi, M., Capuzzo-Dolcetta, I.: Optimal control and viscosity solutions of Hamilton-Jacobi-Bellman equations. Systems & Control: Foundations & Applications, Birkhäuser Boston, Inc., Boston (1997). With appendices by Maurizio Falcone and Pierpaolo Soravia

Bardi, M., Evans, L.: On Hopf's formulas for solutions of Hamilton-Jacobi equations. Nonlinear Anal. Theory Methods Appl. **8**, 1373–1381 (1984)

Barles, G.: Solutions de viscosité des équations de Hamilton-Jacobi. Mathématiques et Applications. Springer, Berlin/Heidelberg (1994)

Barron, E., Evans, L., Jensen, R.: Viscosity solutions of Isaacs' equations and differential games with Lipschitz controls. J. Differ. Equ. **53**, 213–233 (1984)

Bouman, C., Sauer, K.: A generalized gaussian image model for edge-preserving map estimation. IEEE Trans. Trans. Signal Process. **2**, 296–310 (1993)

Boykov, Y., Veksler, O., Zabih, R.: Fast approximate energy minimization via graph cuts. IEEE Trans. Pattern Anal. Mach. Intell. **23**, 1222–1239 (2001)

Burger, M., Lucka, F.: Maximum a posteriori estimates in linear inverse problems with log-concave priors are proper bayes estimators. Inverse Probl. **30**, 114004 (2014)

Burger, Y.D.M., Sciacchitano, F.: Bregman cost for non-gaussian noise. arXiv preprint arXiv:1608.07483 (2016)

Chambolle, A., Darbon, J.: On total variation minimization and surface evolution using parametric maximum flows. Int. J. Comput. Vis. **84**, 288–307 (2009)

Chambolle, A., Novaga, M., Cremers, D., Pock, T.: An introduction to total variation for image analysis. In: Theoretical Foundations and Numerical Methods for Sparse Recovery, De Gruyter (2010)

Chambolle, A., Pock, T.: An introduction to continuous optimization for imaging. Acta Numer. **25**, 161–319 (2016)

Champagnat, F., Idier, J.: A connection between half-quadratic criteria and em algorithms. IEEE Signal Processing Lett. **11**, 709–712 (2004)

Chan, T.F., Esedoglu, S., Nikolova, M.: Algorithms for finding global minimizers of image segmentation and denoising models. SIAM J. Appl. Math. **66**, 1632–1648 (2006)

Chan, T.F., Shen, J.: Image processing and analysis, Society for Industrial and Applied Mathematics (SIAM), Philadelphia (2005). Variational, PDE, wavelet, and stochastic methods

Chan, T.F., Vese, L.A.: Active contours without edges. IEEE Trans. Image Process. **10**, 266–277 (2001)

Charbonnier, P., Blanc-Feraud, L., Aubert, G., Barlaud, M.: Deterministic edge-preserving regularization in computed imaging. IEEE Trans. Image Process. **6**, 298–311 (1997)

Crandall, M.G., Ishii, H., Lions, P.-L.: User's guide to viscosity solutions of second order partial differential equations. Bull. Am. Math. Soc. **27**, 1–67 (1992)

Darbon, J.: On convex finite-dimensional variational methods in imaging sciences and Hamilton–Jacobi equations. SIAM J. Imag. Sci. **8**, 2268–2293 (2015)

Darbon, J., Ciril, I., Marquina, A., Chan, T.F., Osher, S.: A note on the bregmanized total variation and dual forms. In: 2009 16th IEEE International Conference on Image Processing (ICIP), Nov 2009, pp. 2965–2968

Darbon, J., Langlois, G.P.: On Bayesian posterior mean estimators in imaging sciences and Hamilton-Jacobi partial differential equations. arXiv preprint arXiv: 2003.05572 (2020)

Darbon, J., Meng, T.: On decomposition models in imaging sciences and multi-time Hamilton-Jacobi partial differential equations. SIAM Journal on Imaging Sciences. **13**(2), 971–1014 (2020). https://doi.org/10.1137/19M1266332

Darbon, J., Sigelle, M.: Image restoration with discrete constrained total variation part I: Fast and exact optimization. J. Math. Imaging Vision **26**, 261–276 (2006)

Demoment, G.: Image reconstruction and restoration: Overview of common estimation structures and problems. IEEE Trans. Acoust. Speech Signal Process. **37**, 2024–2036 (1989)

Dou, Z., Song, M., Gao, K., Jiang, Z.: Image smoothing via truncated total variation. IEEE Access **5**, 27337–27344 (2017)

Dower, P.M., McEneaney, W.M., Zhang, H.: Max-plus fundamental solution semigroups for optimal control problems. In: 2015 Proceedings of the Conference on Control and its Applications. SIAM, 2015, pp. 368–375

Duda, R.O., Hart, P.E., Stork, D.G.: Pattern Classification. Wiley (2012)

Evans, L.C.: Partial differential equations, vol. 19 of Graduate Studies in Mathematics, 2nd edn. American Mathematical Society, Providence (2010)

Fleming, W., McEneaney, W.: A max-plus-based algorithm for a Hamilton–Jacobi–Bellman equation of nonlinear filtering. SIAM J. Control. Optim. **38**, 683–710 (2000)

Fleming, W.H., Soner, H.M.: Controlled Markov Processes and Viscosity Solutions, vol. 25. Springer Science & Business Media (2006)

Floudas, C.A., Pardalos, P.M. (eds.): Encyclopedia of Optimization, 2nd edn. (2009)

Gaubert, S., McEneaney, W., Qu, Z.: Curse of dimensionality reduction in max-plus based approximation methods: Theoretical estimates and improved pruning algorithms. In: 2011 50th IEEE Conference on Decision and Control and European Control Conference. IEEE, 2011, pp. 1054–1061

Geman, D., Yang, C.: Nonlinear image recovery with half-quadratic regularization. IEEE Trans. Image Process. **4**, 932–946 (1995)

Geman, D., Reynolds, G.: Constrained restoration and the recovery of discontinuities. IEEE Trans. Pattern Anal. Mach. Intell. **14**, 367–383 (1992)

Gribonval, R.: Should penalized least squares regression be interpreted as maximum a posteriori estimation? IEEE Trans. Signal Process. **59**, 2405–2410 (2011)

Gribonval, R., Machart, P.: Reconciling" priors" &" priors" without prejudice? In: Advances in Neural Information Processing Systems, 2013, pp. 2193–2201

Gribonval, R., Nikolova, M.: On bayesian estimation and proximity operators, arXiv preprint arXiv:1807.04021 (2018)

Hochbaum, D.S.: An efficient algorithm for image segmentation, Markov random fields and related problems. J. ACM **48**, 686–701 (2001)

Hopf, E.: Generalized solutions of non-linear equations of first order. J. Math. Mech. **14**, 951–973 (1965)

Idier, J.: Convex half-quadratic criteria and interacting auxiliary variables for image restoration. IEEE Trans. Image Process. **10**, 1001–1009 (2001)

Kay, S.M.: Fundamentals of Statistical Signal Processing. Prentice Hall PTR (1993)

Kolokoltsov, V.N., Maslov, V.P.: Idempotent analysis and its applications, vol. 401 of Mathematics and its Applications. Kluwer Academic Publishers Group, Dordrecht (1997) Translation of ıt Idempotent analysis and its application in optimal control (Russian), "Nauka" Moscow, 1994 [MR1375021 (97d:49031)], Translated by V. E. Nazaikinskii, With an appendix by Pierre Del Moral

Le Guen, V.: Cartoon + Texture Image Decomposition by the TV-L1yModel. Image Process. Line **4**, 204–219 (2014)

Likas, A.C., Galatsanos, N.P.: A variational approach for bayesian blind image deconvolution. IEEE Trans. Signal Process. **52**, 2222–2233 (2004)

Lions, P.L., Rochet, J.-C.: Hopf formula and multitime Hamilton-Jacobi equations. Proc. Am. Math. Soc. **96**, 79–84 (1986)

Louchet, C.: Modèles variationnels et bayésiens pour le débruitage d'images: de la variation totale vers les moyennes non-locales. Ph.D. thesis, Université René Descartes-Paris V (2008)

Louchet, C., Moisan, L.: Posterior expectation of the total variation model: properties and experiments. SIAM J. Imaging Sci. **6**, 2640–2684 (2013)

McEneaney, W.: Max-plus methods for nonlinear control and estimation. Springer Science & Business Media (2006)

McEneaney, W.: A curse-of-dimensionality-free numerical method for solution of certain HJB PDEs. SIAM J. Control. Optim. **46**, 1239–1276 (2007)

McEneaney, W.M., Deshpande, A., Gaubert, S.: Curse-of-complexity attenuation in the curse-of-dimensionality-free method for HJB PDEs. In: 2008 American Control Conference. IEEE, 2008, pp. 4684–4690

McEneaney, W.M., Kluberg, L.J.: Convergence rate for a curse-of-dimensionality-free method for a class of HJB PDEs. SIAM J. Control. Optim. **48**, 3052–3079 (2009)

Nikolova, M., Chan, R.H.: The equivalence of half-quadratic minimization and the gradient linearization iteration. IEEE Trans. Image Process. **16**, 1623–1627 (2007)

Nikolova, M., Ng, M.: Fast image reconstruction algorithms combining half-quadratic regularization and preconditioning. In: Proceedings 2001 International Conference on Image Processing (Cat. No. 01CH37205), vol. 1. IEEE, 2001, pp. 277–280

Nikolova, M., Ng, M.K.: Analysis of half-quadratic minimization methods for signal and image recovery. SIAM J. Sci. Comput. **27**, 937–966 (2005)

Osher, S., A. Solé, and Vese, L.: , Image decomposition and restoration using total variation minimization and the H^{-1} norm, Multiscale Modeling & Simulation, 1 (2003), pp. 349–370.

Rudin, L., Osher, S., Fatemi, E.: Nonlinear total variation based noise removal algorithms. Physica D **60**, 259–268 (1992)

Scherzer, O., Grasmair, M., Grossauer, H., Haltmeier, M., Lenzen, F.: Variational methods in imaging, vol. 167 of Applied Mathematical Sciences. Springer, New York (2009)

Tho, N.: Hopf-Lax-Oleinik type formula for multi-time Hamilton-Jacobi equations. Acta Math. Vietnamica **30**, 275–287 (2005)

Vese, L.A., Le Guyader, C.: Variational methods in image processing, Chapman & Hall/CRC Mathematical and Computational Imaging Sciences. CRC Press, Boca Raton (2016)

Winkler, G.: Image Analysis, Random Fields and Dynamic Monte Carlo Methods. Applications of Mathematics. Springer, 2nd edn. (2003)

Multi-modality Imaging with Structure-Promoting Regularizers

7

Matthias J. Ehrhardt

Contents

Introduction	236
Application Examples	236
Variational Regularization	240
Contributions	241
Related Work	241
Mathematical Models for Structural Similarity	243
Measuring Structural Similarity	244
Structure-Promoting Regularizers	245
Isotropic Models	245
Anisotropic Models	247
Algorithmic Solution	250
Algorithm	250
Prewhitening	251
Numerical Comparison	252
Software, Data, and Parameters	252
Numerical Results	253
Discussion on Computational Cost	256
Conclusions	261
Open Problems	262
References	266

Abstract

Imaging with multiple modalities or multiple channels is becoming increasingly important for our modern society. A key tool for understanding and early diagnosis of cancer and dementia is PET-MR, a combined positron emission

---

M. J. Ehrhardt (✉)
Institute for Mathematical Innovation, University of Bath, Bath, UK
e-mail: m.ehrhardt@bath.ac.uk

© Springer Nature Switzerland AG 2023
K. Chen et al. (eds.), *Handbook of Mathematical Models and Algorithms in Computer Vision and Imaging*, https://doi.org/10.1007/978-3-030-98661-2_58

tomography and magnetic resonance imaging scanner which can simultaneously acquire functional and anatomical data. Similarly, in remote sensing, while hyperspectral sensors may allow to characterize and distinguish materials, digital cameras offer high spatial resolution to delineate objects. In both of these examples, the imaging modalities can be considered individually or jointly. In this chapter we discuss mathematical approaches which allow combining information from several imaging modalities so that multi-modality imaging can be more than just the sum of its components.

Introduction

Many tasks in almost all scientific fields can be posed as an inverse problem of the form

$$Ku = f \qquad (1)$$

where K is a mathematical model that connects an unknown quantity of interest u to measured data f. The task is to recover u from data f under the model K. In practice this task is difficult because of measurement errors in the data f and inaccuracies in the model K. Moreover, in many cases the model (1) lacks information we have at hand about the unknown quantity u such as its regularity. In this chapter we are interested in the situation when have a priori knowledge about the "structure" of u from a second measurement v which we want to exploit in the inversion. Throughout this chapter we will refer to v as the *side information*. Intuitively, this is the case when u and v describe different properties of the same geometry (in medicine: anatomy). We will be more precise in section "Mathematical Models for Structural Similarity" where we discuss mathematical models for structural similarity. The two notions we will discuss in detail are that the edges of the two images u and v have similar (1) locations (Arridge et al. 2008; Bresson and Chan 2008; Haber and Holtzman-Gazit 2013; Knoll et al. 2014; Ehrhardt et al. 2015) and (2) directions (Gallardo and Meju 2003, 2004; Haber and Holtzman-Gazit 2013; Ehrhardt and Arridge 2014; Ehrhardt et al. 2015; Rigie and La Riviere 2015; Ehrhardt and Betcke 2016; Ehrhardt et al. 2016; Knoll et al. 2016; Schramm et al. 2017; Bathke et al. 2017; Bungert et al. 2018; Kolehmainen et al. 2019). Real-world examples for these mathematical models are numerous as we will see in the next section.

Application Examples

Historically the first application where information from several modalities was combined was positron emission tomography (PET) and magnetic resonance imaging (MRI) in the early 1990s (Leahy and Yan 1991). Sharing information between two different imaging modalities is motivated by the fact that all images

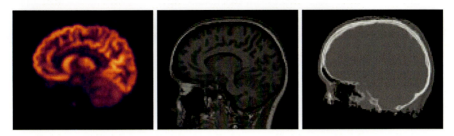

Fig. 1 PET-MR and PET-CT. A low resolution functional PET image (left) is to be reconstructed with the help of an anatomical MRI (middle) or CT image (right). As is evident from the images, all three images share many edges due to the same underlying anatomy. Note that the high soft tissue contrast in MRI makes it favorable over CT for this application. (Images curtesy of P. Markiewicz and J. Schott)

will be highly influenced by the same underlying anatomy; see Fig. 1. Since single-photon emission computed tomography (SPECT) imaging is both mathematically and physically similar to PET imaging, most of the proposed models can be directly translated and often models are proposed for both modalities simultaneously; see, e.g., Bowsher et al. (1996), Rangarajan et al. (2000), Chan et al. (2007) and Nuyts (4154). Over the years there always has been research in this direction (see, e.g., Bowsher et al. (1996), Rangarajan et al. (2000), Comtat et al. (2002), Bowsher et al. (2004), Baete et al. (2004), Chan et al. (2007), Chan et al. (2009), Tang and Rahmim (2009), Bousse et al. (2010), Pedemonte et al. (2011), Somayajula et al. (2011), Cheng-Liao and Qi (2011), Vunckx et al. (2012), Kazantsev et al. (2012), Bousse et al. (2012) and Bai et al. (2013)), which was intensified with the advent of the first simultaneous PET-MR scanner in 2011 (Delso et al. 2011); see, e.g., (Knoll et al. 2014; Ehrhardt et al. 2014, 2015; Tang and Rahmim 2015; Ehrhardt et al. 2016; Knoll et al. 2016; Schramm et al. 2017; Mehranian et al. 2018, 2017; Tsai et al. 2018; Zhang and Zhang 2018; Ehrhardt et al. 2019; Deidda et al. 2019).

The same motivation applies to other medical imaging techniques, for example, multi-contrast MRI; see, e.g., Bilgic et al. (2011), Ehrhardt and Betcke (2016), Huang et al. (2014), Sodickson et al. (2015), Song et al. (2018) and Xiang et al. (2019). In multi-contrast MRI multiple acquisition sequences are used to acquire data of the same patient; see Fig. 2 for a T_1- and a T_2-weighted image with shared anatomy. Other special cases are the combination of anatomical MRI (e.g., T_1-weighted) and magnetic particle imaging (Bathke et al. 2017), functional MRI (fMRI) and anatomical MRI (Rasch et al. 2018b), as well as anatomical (^{1}H) and fluorinated gas (^{19}F) MRI (Obert et al. 2020). A related imaging task is quantitative MRI (such as Magnetic Resonance Fingerprinting Ma et al. 7440) (Davies et al. 2013; Tang et al. 2018; Dong et al. 2019; Golbabaee et al. 2020) where one aims to reconstruct quantitative maps of tissue parameters (e.g., T_1, T_2, proton density, off-resonance frequency), but regularizers coupling these maps have not been used to date. The idea to couple channels has also been used for parallel MRI (Chen et al. 2013).

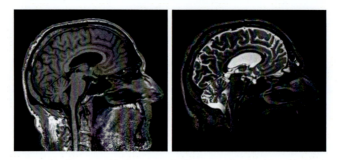

Fig. 2 Multi-contrast MRI. The same MRI scanner can produce different images depending on the acquisition sequence such as T_1-weighted (left) and T_2-weighted images (right). (Images courtesy of N. Burgos)

Fig. 3 Color imaging. The color image (left) is composed of three color channels (right) all of which show similar edges due to the same scenery. (Images courtesy of M. Ehrhardt)

Starting from the 1990s, mathematical models were developed that make use of the expected correlations between color channels of RGB images (Sapiro and Ringach 1996; Blomgren and Chan 1998; Sochen et al. 1998); see Fig. 3. Research in this field is still very active today; see, e.g., Tschumperlé and Deriche (2005), Bresson and Chan (2008), Goldluecke et al. (2012), Holt (2014), Ehrhardt and Arridge (2014), and Möller et al. (2014).

In remote sensing observations are often available from multiple sensors either mounted on a plane or on a satellite. For example, a hyperspectral camera with low spatial resolution and a digital camera with higher spatial resolution may be used simultaneously; see Fig. 4. This situation naturally invites for the fusion of information; see Ballester et al. (2006), Möller et al. (2012), Fang et al. (2013), Loncan et al. (2015), Yokoya et al. (2017), Duran et al. (2017), Bungert et al. (2018), Bungert et al. (2018) and references therein. In some situations the response of the cameras to certain wavelengths is (assumed to be) known such that the data can be fused making use of this knowledge. This is commonly referred to as *pansharpening* (Loncan et al. 2015; Yokoya et al. 2017; Duran et al. 2017). It is important to note that this assumption is sometimes not fulfilled, and many of the aforementioned algorithms are flexible enough to fuse data in this more general situation.

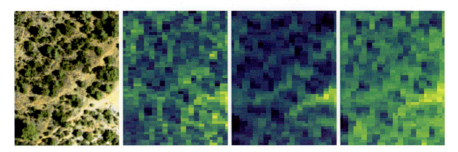

Fig. 4 Hyperspectral imaging + photography. A nowadays common scenario is that multiple cameras are mounted on a plane or satellite for remote sensing. While one camera carries spectral information (right), the other has high spatial resolution (left). (Images courtesy of D. Coomes)

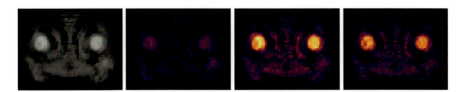

Fig. 5 Spectral CT. Standard (white-beam) CT on the left and three channels (28, 34, and 39 keV) of spectral CT on the right of an iodine-stained lizard head reconstructed by CIL (Ametova et al. 2019). The spectral channels clearly show a large increase in intensity from 28 to 34 keV, thereby revealing the presence, location, and concentration of iodine. (Images courtesy of J. Jorgensen and R. Warr)

Dual and spectral computed tomography (CT) is becoming increasingly popular in (bio-) medical imaging and material sciences due to its ability to distinguish different materials which would not be possible using a single energy; see Fig. 5. Since the energy channels have a very different signal-to-noise ratio, coupling them within the reconstruction allows to transfer information from high signal to low signal channels (Rigie and La Riviere 2015; Foygel Barber et al. 2016; Rigie et al. 2017; Kazantsev et al. 2018).

In geophysics, the coupling between modalities has been used to model similarity between electrical resistivity and seismic velocity (Gallardo and Meju 2003, 2004), estimating conductivity from multi-frequency data (Haber and Oldenburg 1997), inverting gravity and seismic tomography (Haber and Oldenburg 1997), and controlled-source electromagnetic resistivity inversion (Meju et al. 2019). For an overview and more details on examples in geophysics, see in Gallardo and Meju (2011) and Haber and Holtzman-Gazit (2013) and references therein.

Ideas from multi-modality imaging have recently also been used for art restoration. When a canvas is painted on both sides, an x-ray image shows the superposition of both paintings. The x-ray information can then be separated using photos of both sides of the canvas (Deligiannis et al. 2017).

Other examples that were considered in the literature are combining anatomical information and electrical impedance tomography (Kaipio et al. 1999; Kolehmainen et al. 2019), CT and MRI (Xi et al. 2015), photoacoustic and optical coherence tomography (Elbau et al. 2018), x-ray fluorescence and transmission tomography (Di et al. 2016), and various channels in multi-modal electron tomography (Huber et al. 2019). The combination of various imaging modalities into one system may eventually lead to what is sometimes referred to as *omni-tomography* (Wang et al. 2012).

Image reconstruction with side information is mathematically similar to multi-modal image registration, and thus it is not surprising that both fields share a lot of mathematical models; see, e.g., Wells III et al. (1996), Maes et al. (1997), Pluim et al. (2000), and Haber and Modersitzki (2006).

Variational Regularization

Inverse problems of the form (1) can be solved using variational regularization, i.e., framed as the optimization problem

$$u_\alpha \in \arg\min_u \mathcal{D}(Ku, f) + \alpha \mathcal{R}(u). \tag{2}$$

Here the *data fidelity* $\mathcal{D}: Y \times Y \to \mathbb{R}_\infty := \mathbb{R} \cup \{\infty\}$ measures how close the estimated data Ku fits the acquired data f. The *regularizer* (also referred to as the *prior*) $\mathcal{R}: X \to \mathbb{R}_\infty$ defines which properties of the image u we favor and which we do not. The trade-off between data fitting and regularization can be chosen using the *regularization parameter* $\alpha > 0$. Problems of this form have been extensively studied; see, for instance, (Engl et al. 1996; Scherzer et al. 2008; Ito and Jin 2014; Bredies and Lorenz 2018; Benning and Burger 2018) and references therein.

Three popular regularizers for imaging are the *squared H^1-semi norm* (H^1), the *total variation* (TV) (Rudin et al. 1992; Burger and Osher 2013), and the *total generalized variation* (TGV) (Bredies et al. 2010; Bredies and Holler 2014, 2015). It is common to model images as functions $u: \Omega \subset \mathbb{R}^d \to \mathbb{R}$. If u is smooth enough, then these regularizers are defined as

$$H^1(u) = \int_\Omega |\nabla u(x)|^2 \, dx \tag{3}$$

$$TV(u) = \int_\Omega |\nabla u(x)| \, dx \tag{4}$$

$$TGV(u) = \inf_\zeta \int_\Omega |\nabla u(x) - \zeta(x)| + \beta |E\zeta(x)| \, dx. \tag{5}$$

Here $\nabla u: \Omega \to \mathbb{R}^d, [\nabla u]_i = \partial_i u$ denotes the gradient of u, $E\zeta: \Omega \to \mathbb{R}^{d \times d}, [E\zeta]_{i,j} = (\partial_i \zeta_j + \partial_j \zeta_i)/2$ denotes the symmetrized gradient of a vector-field $\zeta: \Omega \to \mathbb{R}^d$ (see Bredies and Holler (2015) for more details), and $|\cdot|$ denotes the Euclidean/Frobenius norm. For TV and TGV it is of interest to develop other formulations which are well-defined even when u is not smooth. For simplicity, we do not go into more detail in this direction but refer the interested reader to the literature, e.g., Bredies et al. (2010) and Burger and Osher (2013).

All three regularizers promote solutions with different smoothness properties. H^1 promotes smooth solutions with small gradients everywhere, whereas TV promotes solutions which have sparse gradients, i.e., the images are piecewise constant and appear cartoon-like. The latter also leads to the staircase artifact which can be overcome by TGV which promotes piecewise linear solutions. None of these regularizers are able to encode additional information on the location or direction of edges.

Contributions

The contributions in this chapter are threefold.

Overview over existing methods We provide an overview on existing mathematical models for structural similarity which are related to the shared location or direction of edges. We then discuss various regularizers which promote similarity in this sense.

Higher order models Existing methods focus on incorporating additional information into regularizers modeling first-order smoothness. We extend existing methodology to second-order smoothness using the total generalized variation framework.

Extensive numerical comparison We highlight the properties of the discussed regularizers and the dependence on various parameters using two inverse problems: tomography and super-resolution.

Related Work

Joint Reconstruction

One can think of the setting (1) with extra information v as a special case when multiple measurements

$$K_i u_i = f_i \quad i = 1, \ldots, m \tag{6}$$

are taken. If $m = 2$ and one inverse problem is considerably less ill-posed, then this can be solved first to guide the inversion of the other. Some of the described models can be extended to the more general case (e.g., an arbitrary number of modalities) or the joint recovery of both/all unknowns (see, e.g., (Sapiro and Ringach 1996; Haber and Oldenburg 1997; Arridge and Simmons 1997; Gallardo and Meju 2003, 2004, 2011; Chen et al. 2013; Haber and Holtzman-Gazit 2013; Knoll et al. 2014;

Ehrhardt and Arridge 2014; Holt 2014; Ehrhardt et al. 2015; Rigie and La Riviere 2015; Knoll et al. 2016; Di et al. 2016; Mehranian et al. 2018; Zhang and Zhang 2018; Meju et al. 2019; Huber et al. 2019)), but it is out of the scope of this chapter to provide an overview on those. For an overview up to 2015, see Ehrhardt (2015). A few recent contributions are summarized in Arridge et al. (2020).

Model (6) may include several special cases: (i) multiple measurements of the same unknown, i.e., $u_i = u$, and (ii) measurements correspond to different states of the same unknown, e.g., in dynamic imaging $u_i = u(\cdot, t_i)$. The former case is covered by the standard literature when concatenating the measurements and the systems models, i.e., $(Ku)_i := K_i u$ and $f = (f_1, \ldots, f_m)$. The latter has been widely studied in the literature, too; see, e.g., (Schmitt and Louis 2002; Schmitt et al. 2002; Schuster et al. 2018) and references therein. Both of these are in general unrelated to multi-modality imaging.

Other Models for Similarity

The earliest contributions to structure-promoting regularizers for multi-modality imaging were made in the early 1990s by Leahy and Yan (1991) who used a segmentation of an anatomical MRI image to enhance PET reconstruction. This is achieved by carefully handcrafting a regularizer which can encode this information. In this chapter we will use the same strategy but in a continuous setting which is independent of the discretization and will not rely on a segmentation of the side information v. These ideas were subsequently refined in various directions (Bowsher et al. 1996; Rangarajan et al. 2000; Comtat et al. 2002; Bowsher et al. 2004; Baete et al. 2004; Chan et al. 2007, 2009; Bousse et al. 2010; Pedemonte et al. 2011; Bilgic et al. 2011; Bousse et al. 2012; Bai et al. 2013) of which Bowsher's prior (Bowsher et al. 2004) remains most popular today.

Other models that can combine information of multiple modalities are based on coupled diffusion (Arridge and Simmons 1997; Tschumperlé and Deriche 2005; Arridge et al. 2008), level sets (Cheng-Liao and Qi 2011), information theoretic priors (joint entropy, mutual information) (Nuyts 4154; Tang and Rahmim 2009; Somayajula et al. 2011; Tang and Rahmim 2015), Bregman distances (Ballester et al. 2006; Möller et al. 2012; Estellers et al. 2013; Kazantsev et al. 2014; Rasch et al. 2018b), Bregman iterations (Möller et al. 2014; Rasch et al. 2018a), the structure tensor (Estellers et al. 2015), joint dictionary learning (Deligiannis et al. 2017; Song et al. 2018, 2019), common edge weighting (Zhang and Zhang 2018), and deep learning (Xiang et al. 2019). Most of these methods are very different to what will be described in this chapter. There are some similarities between the methods of this chapter and methods which are based on the Bregman distance of the total variation (Ballester et al. 2006; Möller et al. 2012, 2014; Estellers et al. 2013; Kazantsev et al. 2014; Rasch et al. 2018a,b), but a detailed treatment is outside the scope of this section.

Mathematical Models for Structural Similarity

In this section we define mathematical models where we aim to capture the similarities as shown in Figs. 1, 2, 3, 4, and 5. We start by explicitly stating two definitions which capture structural similarity which have been used implicitly in the literature. The first is based on the location of edges or the edge set (Arridge et al. 2008; Bresson and Chan 2008; Haber and Holtzman-Gazit 2013; Chen et al. 2013; Knoll et al. 2014; Möller et al. 2014; Ehrhardt et al. 2015; Zhang and Zhang 2018), and the second is based on direction of edges or the shape of an object (Gallardo and Meju 2003, 2004; Haber and Holtzman-Gazit 2013; Ehrhardt and Arridge 2014; Ehrhardt et al. 2015; Rigie and La Riviere 2015; Knoll et al. 2016). The latter is essentially the same as Definition 5.1.6 in Ehrhardt (2015) except for the degenerate case when either $\nabla u(x) = 0$ or $\nabla v(x) = 0$.

Definition 1 (Structural similarity with edge sets). Two differentiable images $u, v : \Omega \to \mathbb{R}$ are said to be *structurally similar in the sense of edge sets* if

$$\mathcal{E}u = \mathcal{E}v \tag{7}$$

where $\mathcal{E}u = \{x \in \Omega \mid \nabla u(x) \neq 0\}$. We also write $u \stackrel{e}{\sim} v$ to denote that u and v are structurally similar in the sense of edge sets.

Definition 2 (Structural similarity with parallel level sets). Two differentiable images $u, v : \Omega \to \mathbb{R}$ are said to be *structurally similar in the sense of parallel level sets* if $u \stackrel{e}{\sim} v$ and for all $x \in \mathcal{E}u$ the gradients $\nabla u(x)$ and $\nabla v(x)$ are co-linear which we denote by $\nabla u(x) \parallel \nabla v(x)$, i.e., there exists $\alpha(x) \in \mathbb{R}$ such that

$$\nabla u(x) = \alpha(x) \nabla v(x). \tag{8}$$

We also write $u \stackrel{d}{\sim} v$ to denote that u and v are structurally similar in the sense of parallel level sets.

Remark 1. For smooth images u and v, their gradients are perpendicular to their level sets, i.e., $u^{-1}(s) = \{x \in \Omega \mid u(x) = s\}$. Thus parallel gradients are equivalent to parallel level sets which explains the naming. The notion that the structure of an image is contained in its level sets dates back to Caselles et al. (2002).

Remark 2. By definition, similarity with parallel level sets (Definition 2) is stronger than the definition that only involves edge sets (Definition 1). An example of two images u and v which have the same edge set but do not have parallel level sets is the following. $u, v : \Omega \subset \mathbb{R}^2 \to \mathbb{R}$, $u(x) = x_1$, $v(x) = x_2$. Clearly they have the same edge set since $\mathcal{E}u = \mathcal{E}v = \Omega$, but they do not have parallel level sets since $\nabla u(x) = [1, 0]$ but $\nabla v(x) = [0, 1]$.

Remark 3. Examples of images which have parallel level sets include:

1. *Function value transformations.* Let $f : \mathbb{R} \to \mathbb{R}$ be smooth and strictly monotonic, i.e., $f' > 0$ or $f' < 0$. Then $v := f \circ u \stackrel{d}{\sim} u$. This is readily to be seen from the fact that $\nabla v(x) = f'(u(x))\nabla u(x) \neq 0$ if and only if $\nabla u(x) \neq 0$.
2. *Local function value transformations.* Let $f_i : \mathbb{R} \to \mathbb{R}$ be smooth and strictly monotonic and $u = \sum_i u_i$ where u_i are smooth functions whose gradients have mutually disjoint support. Then $v := \sum_i f_i \circ u_i \stackrel{d}{\sim} u$.

Remark 4. It has been argued in the literature that many multi-modality images $z : \Omega \to \mathbb{R}^m$ essentially decompose as

$$z_i(x) = \tau_i(x)\rho(x) \tag{9}$$

where $\rho(x)$ describes its structure and τ is a material property; see, e.g., Kimmel et al. (2000) and Holt (2014). Since the material does not change arbitrarily, it is natural to assume that τ_i is slowly varying or even piecewise constant. In the latter case, if x is such that $\nabla \tau_i(x) = 0$, then we have

$$\nabla z_i(x) = \tau_i(x)\nabla \rho(x), \tag{10}$$

in particular if $\tau_i, \tau_j \neq 0$, then $z_i \stackrel{d}{\sim} z_j$. This property is also related to the material decomposition in spectral CT; see, e.g., Fessler et al. (2002), Heismann et al. (2012) and Long and Fessler (2014).

Measuring Structural Similarity

Measuring the degree of similarity with respect to the previous two definitions of structural similarity is not easy, and we will now discuss a couple of ideas from the literature. Here and for the rest of this chapter, we will make frequent use of the vector-valued representation of a set of images $z : \Omega \to \mathbb{R}^2$, $z(x) := [u(x), v(x)]$. We denote by J its Jacobian, i.e., $J : \Omega \to \mathbb{R}^{d \times 2}$, $J_{i,j} = \partial_i z_j$.

With the definition of the Jacobian, we see that $u \stackrel{e}{\sim} v$ if and only if

$$\int_\Omega |J(x)|_0 \, dx = \int_\Omega |\nabla u(x)|_0 \, dx = \int_\Omega |\nabla v(x)|_0 \, dx \tag{11}$$

where $|x|_0 := 1$ if $x \neq 0$ and 0 else.

Similarly, by definition $u \overset{d}{\sim} v$ if and only if $u \overset{e}{\sim} v$ and (a) rank $J(x) = 1$ for all $x \in \mathcal{E}u$. (a) is equivalent to (b) a vanishing determinant, i.e., $\det J^\top(x) J(x) = 0$. Simple calculations (see, e.g., Ehrhardt (2015)) show that

$$\det J^\top(x) J(x) = |\nabla u(x)|^2 |\nabla v(x)|^2 - \langle \nabla u(x), \nabla v(x) \rangle^2, \tag{12}$$

where we use the notation $\langle x, y \rangle = x^\top y$ for the inner product of two column vectors x and y. In order to get further equivalent statements, we turn to the singular values of the Jacobian which are given by

$$\sigma_1^2(x) = \frac{1}{2}\left[|J(x)|^2 + \sqrt{|J(x)|^4 - \det J^\top(x) J(x)}\right] \tag{13}$$

$$\sigma_2^2(x) = \frac{1}{2}\left[|J(x)|^2 - \sqrt{|J(x)|^4 - \det J^\top(x) J(x)}\right] \tag{14}$$

with $|J(x)|^2 = |\nabla u(x)|^2 + |\nabla v(x)|^2$; see, e.g., Ehrhardt (2015). Since $\sigma_1(x) \geq \sigma_2(x) \geq 0$ we have that (a) holds if and only if (c) the second singular value vanishes, i.e., $\sigma_2(x) = 0$ or (d) the vector of singular vectors $\sigma(x) = [\sigma_1(x), \sigma_2(x)]$ is 1-sparse.

Structure-Promoting Regularizers

Many of the abstract models from the previous section to measure the degree of similarity with respect to the previous two definitions of structural similarity are computationally challenging as they relate to non-convex constraints. In this section we will define convex structure-promoting regularizers which make them computationally tractable.

Isotropic Models

We first look at isotropic models which only depend on gradient magnitudes rather than directions, thus promoting structural similarity in the sense of edge sets, Definition 1.

First, based on (11) if we approximate $|J(x)|_0$ by $|J(x)|$, then

$$\mathrm{JTV}(u) = \int_\Omega |J(x)| \, \mathrm{d}x = \int_\Omega \sqrt{|\nabla u(x)|^2 + |\nabla v(x)|^2} \, \mathrm{d}x \tag{15}$$

$$\leq \int_\Omega |\nabla u(x)| + |\nabla v(x)| \, \mathrm{d}x = \mathrm{TV}(u) + \mathrm{TV}(v) \tag{16}$$

with equality if and only if $\mathcal{E}u \cap \mathcal{E}v = \emptyset$. This regularizer is called *joint total variation* in some communities (see, e.g., Chen et al. 2013; Haber and Holtzman-Gazit 2013; Ehrhardt et al. 2015, 2016) and *vectorial total variation* in others (see, e.g., Bresson and Chan 2008).

Remark 5. Note that JTV has the favorable property that if $\nabla v = 0$, then $\text{JTV}(u) = \text{TV}(u)$, so that it reduces to a well-defined regularization in u in this degenerate case. Note that this property also holds locally.

Remark 6. We would also like to note that there is a connection between JTV and the singular values of J. Let $\sigma_1, \sigma_2 : \Omega \to [0, \infty)$ be the two singular values of J, and then we have

$$\text{JTV}(u) = \int_\Omega \sqrt{\sigma_1^2(x) + \sigma_2^2(x)} \, dx \, . \tag{17}$$

Another strategy to favor edges at similar locations while reducing to a well-defined regularizer in the degenerate case is to introduce local weighting. Let $w : \Omega \to [0, 1]$ be an edge indicator function for v such that $w(x) = 1$ when $\nabla v(x) = 0$ and a small value whenever $|\nabla v(x)|$ is large. For example, choose

$$w(x) = \frac{\eta}{\sqrt{\eta^2 + |\nabla v(x)|^2}} \tag{18}$$

which is illustrated in Fig. 6. The figure shows that with a medium η the weight w in (18) shows the main structures of the images so that these can be promoted in the other image. If η is too small, then also unwanted structures are captured in w such as a smooth background variation. If η is too large, then the structures start to disappear.

For regularizers which are based on the image gradient ∇u, the weighting w can be used to favor edges at certain locations by replacing ∇ by $w\nabla$. For instance, for H^1 (3), TV (4), and TGV (5), this strategy results in

$$\text{wH}^1(u) = \int_\Omega |w(x)\nabla u(x)|^2 \, dx = \int_\Omega w^2(x)|\nabla u(x)|^2 \, dx \tag{19}$$

$$\text{wTV}(u) = \int_\Omega |w(x)\nabla u(x)| \, dx = \int_\Omega w(x)|\nabla u(x)| \, dx \tag{20}$$

$$\text{wTGV}(u) = \inf_\zeta \int_\Omega |w(x)\nabla u(x) - \zeta(x)| + \beta |E\zeta(x)| \, dx \tag{21}$$

which we will refer to as *weighted squared H^1-semi norm, weighted total variation*, and *weighted total generalized variation*. wTV was used in Arridge et al. (2008), Lenzen and Berger (2015) and Ehrhardt and Betcke (2016). A variant of wTV has been considered for single modality imaging in Hintermüller and Rincon-Camacho

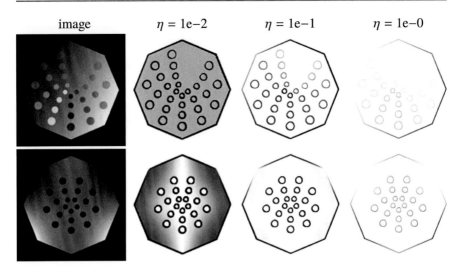

Fig. 6 Influence of the parameter η on estimation of **edge location**. The images on the right show the scalar field $w : \Omega \to [0, 1]$ which locally weights the influence of the regularizer; see (18). Here "black" denotes 0 and "white" denotes 1

(2010) and Dong et al. (2011) and extended to a variant of wTGV (Bredies et al. 2012).

Remark 7. The parameter η in w (see (18)) should be chosen in relation to $|\nabla v(x)|$. A common strategy is to normalize the side information first such that $\sup_{x \in \Omega} |\nabla v(x)| = 1$. Then desirable values of η are usually within the range $[0.01, 1]$.

Anisotropic Models

The same idea which resulted in isotropically "weighted" variants of common regularizers can be used anisotropically, i.e., by making the local weights vary with direction. Let us denote the anisotropic weighting by $D : \Omega \to \mathbb{R}^{d \times d}$. Similar to the isotropic variant, one would like the weight to become the identity matrix, i.e., $D(x) = I$, when $\nabla v(x) = 0$. In order to promote parallel level sets, it is desirable that $D(x)\nabla u(x)$ should be small if $\nabla u(x) \parallel \nabla v(x)$ and $D(x)\nabla u(x) = \nabla u(x)$ if $\nabla u(x) \perp \nabla v(x)$, i.e., $\langle \nabla u(x), \nabla v(x) \rangle = 0$. For example,

$$D(x) = I - \gamma \xi(x) \xi^\top(x), \quad \xi(x) = \frac{\nabla v(x)}{\sqrt{\eta^2 + |\nabla v(x)|^2}} \qquad (22)$$

for $\gamma \in (0, 1]$ (usually close to 1) and $\eta > 0$ satisfies all of these properties. Clearly if $\nabla v(x) = 0$, then $\xi = 0$ such that $D(x) = I$. Moreover, if $\nabla u(x) \parallel \nabla v(x)$, then there exists an α such that $\nabla u(x) = \alpha \nabla v(x)$ and

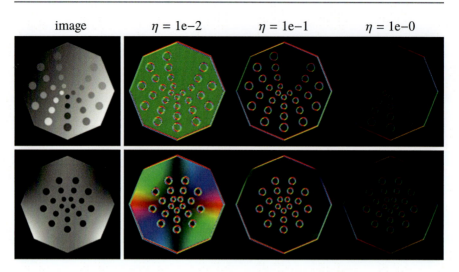

Fig. 7 Influence of the parameter η on estimation of **edge location and direction**. The images on the right show the vector field $\xi : \Omega \to \mathbb{R}^d$ which locally defines the influence of the regularizer; see, e.g., (22). Here "black" denotes that the magnitude of ξ, i.e., $|\xi(x)|$, is 0, and a bright color denotes that $|\xi(x)|$ is 1. The colors show the direction of the vector field ξ modulo its sign

$$D(x)\nabla u(x) = \left[I - \frac{\gamma}{\eta^2 + |\nabla v(x)|^2}\nabla v(x)\nabla v^\top(x)\right]\nabla u(x) \tag{23}$$

$$= \left[1 - \frac{\gamma|\nabla v(x)|^2}{\eta^2 + |\nabla v(x)|^2}\right]\nabla u(x). \tag{24}$$

The scalar weighting factor converges to $1 - \gamma$ for $|\nabla v(x)| \to \infty$. Finally, if $\nabla u(x) \perp \nabla v(x) = 0$, then clearly $D(x)\nabla u(x) = \nabla u(x)$.

The example of the matrix-field $D : \Omega \to \mathbb{R}^{d \times d}$ in (22) is determined by the vector-field $\xi : \Omega \to \mathbb{R}^d$ which we visualize in Fig. 7. The colors show the direction of the vector-field modulo its sign (since $\xi(x)\xi^\top(x)$ is invariant to a change of sign), and the brightness indicates its magnitude $|\xi(x)|$. Note that images appear as color versions of Fig. 6 which shows the isotropic weighting w.

Using a matrix-field in common regularizers leads to their "directional" variant

$$\mathrm{dH}^1(u) = \int_\Omega |D(x)\nabla u(x)|^2 \, dx \tag{25}$$

$$\mathrm{dTV}(u) = \int_\Omega |D(x)\nabla u(x)| \, dx \tag{26}$$

$$\mathrm{dTGV}(u) = \inf_\zeta \int_\Omega |D(x)\nabla u(x) - \zeta(x)| + \beta|E\zeta(x)| \, dx. \tag{27}$$

Remark 8. There is a connection between the particular choice of the matrix-field D in (22) and the Jacobian J.

$$|D(x)\nabla u(x)|^2 = |\nabla u(x) - \frac{\gamma}{\eta^2 + |\nabla v(x)|^2}\langle \nabla u(x), \nabla v(x)\rangle \nabla v(x)|^2 \qquad (28)$$

$$= |\nabla u(x)|^2 - \frac{2\gamma\eta^2 + \gamma(2-\gamma)|\nabla v(x)|^2}{(\eta^2 + |\nabla v(x)|^2)^2}\langle \nabla u(x), \nabla v(x)\rangle^2. \qquad (29)$$

For $\eta = 0$, $\gamma = 1$, and $|\nabla v(x)| = 1$, then with (12) we have

$$|D(x)\nabla u(x)|^2 = |\nabla u(x)|^2|\nabla v(x)|^2 - \langle \nabla u(x), \nabla v(x)\rangle^2 = \det J^\top(x) J(x). \qquad (30)$$

Thus, dH^1 corresponds to penalizing the determinant. This regularizer is widely used for joint reconstruction in geophysics under the name *cross-gradient* function since it is also the cross product of $\nabla u(x)$ and $\nabla v(x)$; see, e.g., (Gallardo and Meju 2003, 2004, 2011; Meju et al. 2019). Similarly the dTV used, for instance, in medical imaging (Ehrhardt and Betcke 2016; Ehrhardt et al. 2016; Bathke et al. 2017; Schramm et al. 2017; Kolehmainen et al. 2019; Obert et al. 2020) and remote sensing (Bungert et al. 2018) can be seen as penalizing the square root of the determinant.

Another strategy to promote parallel level sets is via nuclear norm of the Jacobian which is defined as $|J(x)|_* = \sum_{i=1}^{\min(d,2)} \sigma_i(x)$ where $\sigma_i(x)$ denotes the ith singular value of $J(x)$. Using the nuclear norm promotes sparse vectors of singular values $\sigma(x) = [\sigma_1(x), \sigma_2(x)]$ and thereby parallel level sets. As a regularizer

$$\text{TNV}(u) = \int_\Omega |J(x)|_* \, dx \qquad (31)$$

this strategy became known as *total nuclear variation*; see Holt (2014), Rigie and La Riviere (2015), Knoll et al. (2016), and Rigie et al. (2017).

All first-order regularizers of this section can be readily summarized in the following standard form

$$\mathcal{J}(u) = \int_\Omega \phi[B(x)\nabla u(x)] \, dx \qquad (32)$$

where $B(x) : \mathbb{R}^d \to \mathbb{R}^m$ is an affine transformation and $\phi : \mathbb{R}^m \to \mathbb{R}$. For details how B and ϕ can be chosen for specific regularizers to fit this framework, see Table 1. It is useful for Jacobian-based regularizers to use the reweighted Jacobian $[\nabla u(x), \xi(x)]$ with $\xi(x) = \eta \nabla v(x)$ instead.

Table 1 Examples of first-order structure-promoting regularizers; see (32)

Regularizer	Definition	$B(x)y$	m	$\phi(x)$
H^1	(3)	y	d	$|x|^2$
wH^1	(19)	$w(x)y$	d	$|x|^2$
dH^1	(25)	$D(x)y$	d	$|x|^2$
TV	(4)	y	d	$|x|$
wTV	(20)	$w(x)y$	d	$|x|$
dTV	(26)	$D(x)y$	d	$|x|$
JTV	(16)	$[y, \xi(x)]$	$d \times 2$	$|x|$
TNV	(31)	$[y, \xi(x)]$	$d \times 2$	$|x|_*$

Algorithmic Solution

Note that the solution to variational regularization (2) with either first- (32) or second-order structural regularization (5), (21), (27) can be cast into the general non-smooth composite optimization form

$$\min_x \mathcal{F}(Ax) + \mathcal{G}(x) \qquad (33)$$

with $\mathcal{F}(y) = \sum_{i=1}^n \mathcal{F}_i(y_i)$ and $Ax = [A_1 x, \ldots, A_n x]$; see Table 2. We denote by $\|\cdot\|_{2,1}, \|\cdot\|_2^2$ and $\|\cdot\|_{*,1}$ discretizations of

$$z \mapsto \int_\Omega |z(x)| \, dx, \quad z \mapsto \int_\Omega |z(x)|^2 \, dx \quad \text{and} \quad z \mapsto \int_\Omega |z(x)|_* \, dx. \qquad (34)$$

Note that all functionals $\mathcal{F}_i$ and $\mathcal{G}$ in Table 2 are proper, convex, and lower-semi continuous.

Algorithm

A popular algorithm to solve (33) and therefore (2) is the primal-dual hybrid gradient (PDHG) (Esser et al. 2010; Chambolle and Pock 2011); see Algorithm 1. It consists of two simple steps only involving basic linear algebra and the evaluation of the operator A and its adjoint A^*. Moreover, it involves the computation of the proximal operator of $\tau \mathcal{G}$ and the convex conjugate of $\sigma \mathcal{F}^*$ where τ and σ are scalar step sizes. The proximal operator of a functional $\mathcal{H}$ is defined as

$$\operatorname{prox}_\mathcal{H}(z) := \arg\min_x \left\{ \frac{1}{2} \|x - z\|_2^2 + \mathcal{H}(x) \right\}. \qquad (35)$$

The proximal operator can be computed in closed-form for $\|\cdot\|_{2,1}$ and $\|\cdot\|_2^2$. It also can be computed in closed-form for $\|\cdot\|_{*,1}$ if either the number channels

7 Multi-modality Imaging with Structure-Promoting Regularizers

Table 2 Mapping the variational regularization models into the composite optimization framework (33). In all cases we choose $A_1 x = Ku$, $\mathcal{F}_1(y_1) = \mathcal{D}(y_1, b)$, and $\mathcal{G}(x) = \iota_{\geq 0}(u)$.

Regularizer	Definition	x	$A_2 x$	$A_3 x$	$\mathcal{F}_2(y_2)$	$\mathcal{F}_3(y_3)$
H^1	(3)	u	∇u	–	$\alpha \|y_2\|_2^2$	–
wH^1	(19)	u	$w \nabla u$	–	$\alpha \|y_2\|_2^2$	–
dH^1	(19)	u	$D \nabla u$	–	$\alpha \|y_2\|_2^2$	–
TV	(4)	u	∇u	–	$\alpha \|y_2\|_{2,1}$	–
wTV	(20)	u	$w \nabla u$	–	$\alpha \|y_2\|_{2,1}$	–
dTV	(26)	u	$D \nabla u$	–	$\alpha \|y_2\|_{2,1}$	–
JTV	(16)	u	$[\nabla u, 0]$	–	$\alpha \|y_2 - [0, \xi]\|_{2,1}$	–
TNV	(31)	u	$[\nabla u, 0]$	–	$\alpha \|y_2 - [0, \xi]\|_{*,1}$	–
TGV	(5)	(u, ζ)	$\nabla u - \zeta$	$E\zeta$	$\alpha \|y_2\|_{2,1}$	$\alpha\beta \|y_3\|_{2,1}$
wTGV	(21)	(u, ζ)	$w\nabla u - \zeta$	$E\zeta$	$\alpha \|y_2\|_{2,1}$	$\alpha\beta \|y_3\|_{2,1}$
dTGV	(27)	(u, ζ)	$D\nabla u - \zeta$	$E\zeta$	$\alpha \|y_2\|_{2,1}$	$\alpha\beta \|y_3\|_{2,1}$

Algorithm 1 Primal-dual hybrid gradient (PDHG) to solve (33). Default values given in brackets

Input: iterates $x(=0)$, $y(=0)$, step size parameter $\rho(=1)$
Initialize: extrapolation $\bar{x} = x$, step sizes $\sigma = \rho/\|A\|$, $\tau = 0.999/(\rho\|A\|)$
1: **for** $k = 1, \ldots$ **do**
2: $\quad x^+ = \text{prox}_{\tau \mathcal{G}} (x - \tau A^* y)$
3: $\quad y^+ = \text{prox}_{\sigma \mathcal{F}^*} (y + \sigma A(2x^+ - x))$
4: **end for**

---

or the dimension of the domain are strictly less than 5, i.e., $m, d < 5$; see Holt (2014) for more details. Note also that the proximal operator of $\alpha \mathcal{F}(\cdot - \xi)$ can be readily computed based on the proximal operator of $\mathcal{F}$. More details on proximal operators, convex conjugates, and examples can be found, for example, in Bauschke and Combettes (2011), Combettes and Pesquet (2011), Parikh and Boyd (2014), and Chambolle and Pock (2016).

For some applications (e.g., x-ray tomography), a preconditioned (Pock and Chambolle 2011; Ehrhardt et al. 2019) or randomized (Chambolle et al. 2018; Ehrhardt et al. 2019) variant can be useful, but we will not consider these here for simplicity.

Prewhitening

Since the operator norms $\|A_i\|$, $i = 1, \ldots n$ can vary significantly, it is often advisable to "prewhiten" the problem by recasting it as

$$\min_x \tilde{\mathcal{F}}(\tilde{A}x) + \mathcal{G}(x). \tag{36}$$

with $\tilde{\mathcal{F}}(y) := \sum_{i=1}^{n} \mathcal{F}_i(\|A_i\| \cdot y_i)$ and $\tilde{A}_i x := A_i x / \|A_i\|$. Then trivially $\|\tilde{A}_i\| = 1$, $i = 1, \ldots, n$ so that all operator norms are equal. Note that the proximal operator of $\sigma\tilde{\mathcal{F}}$ is simple to compute if the proximal operators of $\sigma\mathcal{F}_i$, $i = 1, \ldots, n$ are simple to compute, since

$$[\operatorname{prox}_{\sigma\tilde{\mathcal{F}}}(y)]_i = \lambda_i^{-1}[\operatorname{prox}_{\sigma\lambda_i^2 \mathcal{F}_i}(\lambda_i y_i)], \tag{37}$$

for any $\lambda_i > 0$; see, for instance, Bredies and Lorenz (2018, Lemma 6.136).

Numerical Comparison

This section describes numerical experiments to compare first- and second-order structure-promoting regularizers.

Software, Data, and Parameters

Software The numerical computations are carried out in Python using ODL (version 1.0.0.dev0) (Adler et al. 2017) and ASTRA (van Aarle et al. 2015, 2016) for computing line integrals in the tomography example. The source code which reproduces all experiments in this chapter can be found at https://github.com/mehrhardt/Multi-Modality-Imaging-with-Structural-Priors.

Data We consider two test cases with different characteristics, both of which are visualized in Fig. 8. The first test case, later referred to as x-ray, is parallel beam x-ray reconstruction from only 15 views where additionally some detectors are broken. The latter is modeled by salt-and-pepper noise where 5% of all detectors are corrupted. We aim to recover an image with domain $[-1, 1]^2$ discretized with 200^2 pixels. The simulated x-ray camera has 100 detectors and a width of 3 in the same dimensions as the image domain. Therefore, the challenges are (1) sparse views, (2) small number of detectors, and (3) broken detectors.

The second test case, which we refer to as super-resolution, considers the task of super-resolution. Also here we aim to recover an image with domain $[-1, 1]^2$ discretized with 200^2 pixels. The forward operator is integrating over 5^2 pixels, thus mapping images of size 200^2 to images of size 40^2. In addition, Gaussian noise of mean zero and standard deviation of 0.01 is added.

Algorithmic parameters We chose the default value $\rho = 1$ for balancing the step sizes in PDHG and ran the algorithm for 3,000 iterations without choosing a specific stopping criterion.

7 Multi-modality Imaging with Structure-Promoting Regularizers

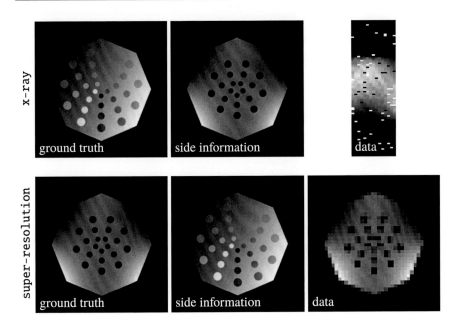

Fig. 8 Test cases for numerical experiments. Top: x-ray reconstruction from sparse views and failed detectors. Bottom: super-resolution by a factor of 5 and Gaussian noise

Numerical Results

The multiplicative scaling of an unconstrained optimization problem is arbitrary; nevertheless we report the absolute values here for completeness. For simplicity, all regularization parameters are shown as multiples of 1e−4. The figures at the bottom right of each image are PSNR and SSIM.

Test Case x-ray

Effect of edge weighting All structure-promoting regularizers described in section "Structure-Promoting Regularizers" have in common that they rely to some extent on the size of edges in the side information, i.e., $|\nabla v(x)|$. For JTV and TNV the actual values of $|\nabla v(x)|$ matter so that a parameter η is needed to correct for this. For all other regularizers a parameter η is needed to decide which edges to trust and which not. The effect of this edge weighting parameter η on all described regularizers is illustrated in Figs. 9, 10, and 11. The locally weighted regularizers (i.e., wH^1, wTV, and wTGV) and the directional regularizers (i.e., dH^1, dTV, and dTGV) have in common that if η is too small, then small artifacts around the edges appear. This effect is more pronounced in locally weighted regularizers. If η is too large, then the structure-promoting effect becomes too small. For joint total variation and total nuclear variation, similar effects exist with reverse relationship to η.

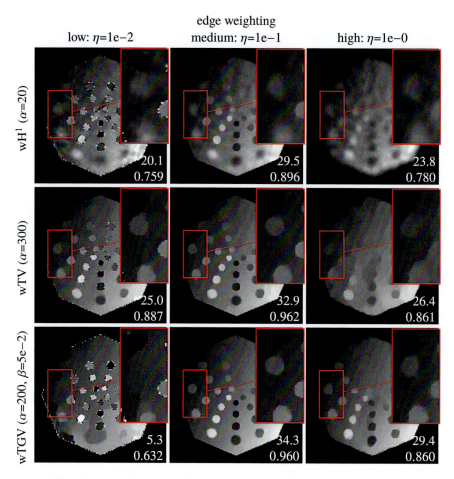

Fig. 9 Effect of edge weighting on locally weighted models for test case x-ray: increasing edge parameter η from left to right. All other parameters where tuned to maximize the PSNR and visual image quality

Effect of regularization The effect of the regularization parameter α on the solution is illustrated in Figs. 12, 13, and 14. All regularizers show the same behavior if α is too small or too large. If the regularization parameter is chosen too small, then artifacts from inverting an ill-posed operator are introduced, and if it is chosen too large, then all regularizers oversmooth the solution. Note that all structure-promoting regularizers have an increased robustness in areas of shared structures.

Comparison of regularizers All eleven regularizers are compared in Fig. 15. It can be seen that the structure-promoting regularizers perform much better in terms of PSNR and SSIM as their non-structure-promoting counterparts. Moreover, one can

7 Multi-modality Imaging with Structure-Promoting Regularizers

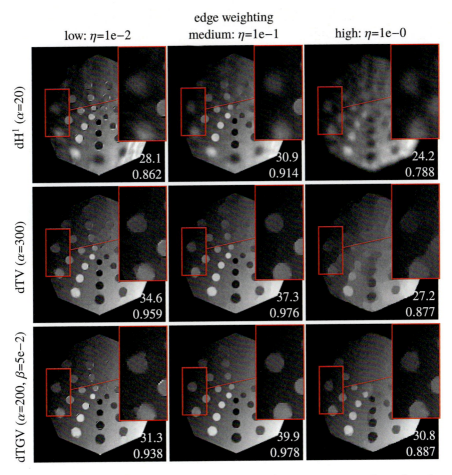

Fig. 10 Effect of edge weighting on directional models for test case x-ray: increasing edge parameter η from left to right. All other parameters where tuned to maximize the PSNR and visual image quality ($\gamma = 1$)

observe an interesting effect that the structure-promoting regularizers also perform visually better in regions where the structure is not shared, e.g., the outer ring of circles. This effect is most dominant for dTGV where the circle at the top left is clearly visible, while it is difficult to spot for many of the other regularizers.

Test Case Super-Resolution
Effect of edge weighting Figs. 16, 17, and 18 show the effect of the edge weighting parameter η. One can make similar observations as in Figs. 9, 10, and 11 for the test case x-ray. In addition, one can observe from the close-ups that if η is too small (or too large for JTV and TNV), then ghosting artifacts may appear. Note that these are present for TNV even for a moderate choice of η.

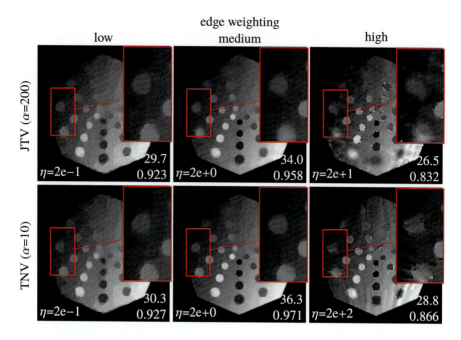

Fig. 11 Effect of edge weighting on joint total variation and total nuclear variation for test case x-ray: increasing edge parameter η from left to right. All other parameters where tuned to maximize the PSNR and visual image quality

Comparison of regularizers All regularizers are compared in Fig. 19 for the test case super-resolution. It can be noted from all images that introducing structural information allows to resolve some of the inner circles which have been merged for regularizers which are not structure-promoting. Moreover, all total generalized variation-based regularizers do not perform much better than the total variation-based regularizers. The directional regularizers as well as JTV and TNV perform best in terms of PSNR for this example.

Discussion on Computational Cost

The median computing times for the numerical experiments are reported in Table 3. The computing time of PDHG is mainly influenced by the dimensions of the models, the proximal operator, and the forward model. As can be seen from the table, H^1 and TV are roughly the same fast. TGV which uses a second primal variable in the space of the image gradient is significantly slower with about twice the computational cost. In all three cases, introducing isotropic weights (i.e., wH^1, wTV, and wTGV) increases the cost by about 6 seconds, and anisotropic weights (i.e., dH^1, dTV, and

7 Multi-modality Imaging with Structure-Promoting Regularizers

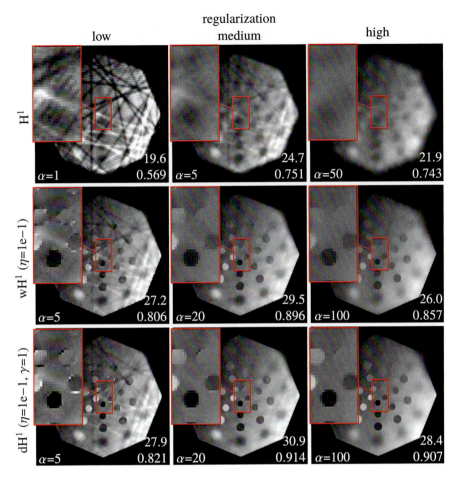

Fig. 12 H^1-semi norm-based structure-promoting regularizers for test case x-ray: increasing the regularization parameter α from left to right. All other parameters where tuned to maximize the PSNR and visual image quality. All regularizers in this figure reduce to the H^1-semi norm in areas where the side information is flat

dTGV) by about 12 s. JTV is more costly than dTV but not as costly as TGV. TNV is by far the most costly of all algorithms due to the need to compute singular value decompositions of 2×2-matrices for every pixel.

Since we run PDHG always for 3,000 iterations, we do not report computational time "till convergence" but computational cost for the full 3,000 iterations. It was observed at several occasions (see, e.g., Ehrhardt et al. 2019) that including side information into the regularizer not only improves the reconstruction but also

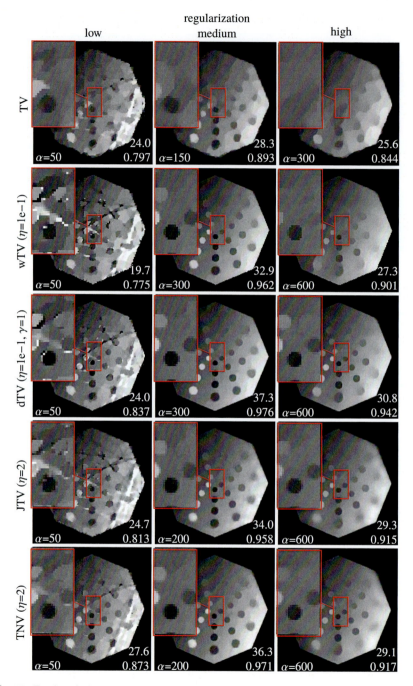

Fig. 13 Total variation based structure-promoting regularizers for test case x-ray: increasing the regularization parameter α from left to right. All other parameters where tuned to maximize the PSNR and visual image quality. All regularizers in this figure reduce to the total variation in areas where the side information is flat

7 Multi-modality Imaging with Structure-Promoting Regularizers

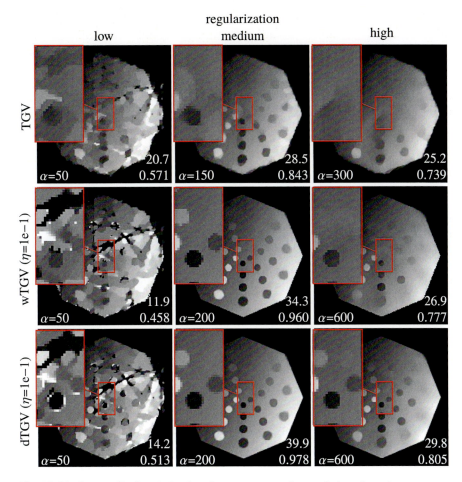

Fig. 14 Total generalized variation-based structure-promoting regularizers for test case x-ray: increasing the regularization parameter α from left to right. All other parameters where tuned to maximize the PSNR and visual image quality ($\beta = 5e-2$). All regularizers in this figure reduce to the total generalized variation in areas where the side information is flat

speeds up the algorithmic convergence. Intuitively this can be understood as more information is included into the optimization problem.

Comparing the regularizers regarding their computational time versus image quality trade-off, it can be noted that TNV should not be chosen since it is not better than dTV at 7-10x the computational cost. Whether H^1, TV, or TGV based regularizer is desirable depends on each individual application. For each of them, there is a clear trend that one achieves better image quality by introducing more information, i.e., first isotropic information and then anisotropic information, each of which increases their computational cost. However, the increase in computational cost is so small that for most applications the directional variant is likely to be favored.

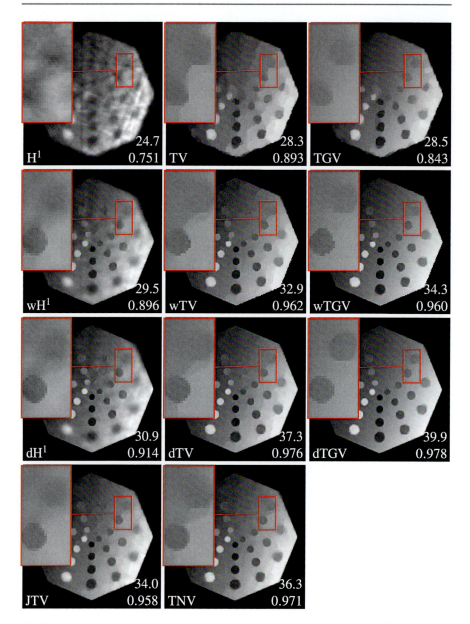

Fig. 15 Comparison of structure-promoting regularizers for test case x-ray. All parameters where tuned to maximize the PSNR and visual image quality

7 Multi-modality Imaging with Structure-Promoting Regularizers

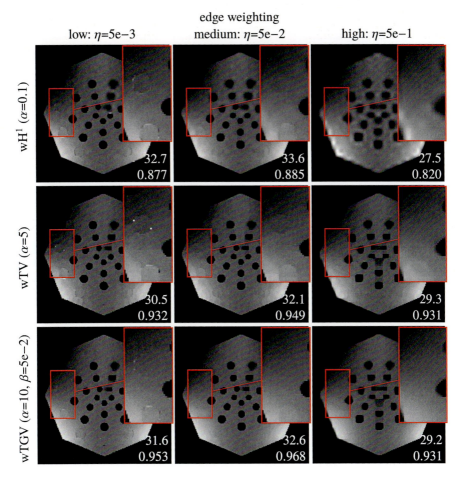

Fig. 16 Effect of edge weighting on locally weighted models for test case super-resolution: increasing edge parameter η from left to right. All other parameters where tuned to maximize the PSNR and visual image quality

Conclusions

This chapter introduced fundamental mathematical concepts on the structure of images and how structural similarity between images can be measured. The fundamental building blocks are the similarity based on edge sets and parallel level sets. These notions lead to several classes of structure-promoting regularizers all of which are convex and thereby lead to tractable optimization problems when used in variational regularization for linear inverse problems with convex data fits. While some of the regularizers are smooth and others are non-smooth, the resulting optimization problem for all of them can be efficiently computed by PDHG. The effectiveness

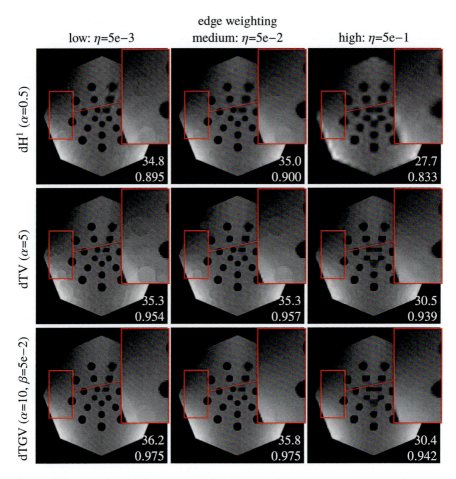

Fig. 17 Effect of edge weighting on directional models for test case `super-resolution`: increasing edge parameter η from left to right. All other parameters where tuned to maximize the PSNR and visual image quality ($\gamma = 0.9$)

of these regularizers for the promotion of structure has been observed in many applications and was also illustrated in this chapter on two simulation studies.

Open Problems

The mathematical framework for structure-promoting regularizers is by now well established and fairly mature. Open problems reside in practical problems in the translation of these techniques to applications which will also motivate further mathematical research.

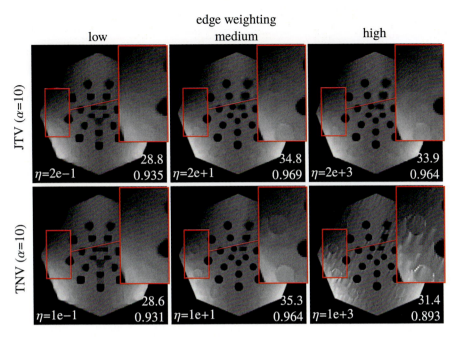

Fig. 18 Effect of edge weighting on joint total variation and total nuclear variation for test case super-resolution: increasing edge parameter η from left to right. All other parameters where tuned to maximize the PSNR and visual image quality

Misregistration The biggest open problem is misregistration. All of the described regularizers assume that both images are perfectly aligned. Even in scanners which have two imaging modalities in the same system such as PET-MR, this assumption is never perfectly fulfilled. This issue has not been addressed much in the literature. In Tsai et al. (2018), the authors proposed an alternating approach between image reconstruction and image registration with some success. In Bungert et al. (2018), the problem was formulated as a blind deconvolution problem so that translations can be compensated with a shifted kernel. A heuristic modification made this approach more robust to large translations (Bungert et al. 2018). A joint reconstruction and affine registration approach was proposed in Bungert and Ehrhardt (2020) which solves the misregistration problem in some cases, e.g., in neurology.

Extensions beyond two modalities It is natural to consider the case that more than one image is available as side information. For instance, in some remote sensing applications, a color photograph with high spatial resolution is available. Similarly, in PET-MR, images of more than one MR sequence might be available. This setting has also been considered in Mehranian et al. (2017) for a purely discrete model. Some of the regularizers to promote structural similarity in this chapter naturally

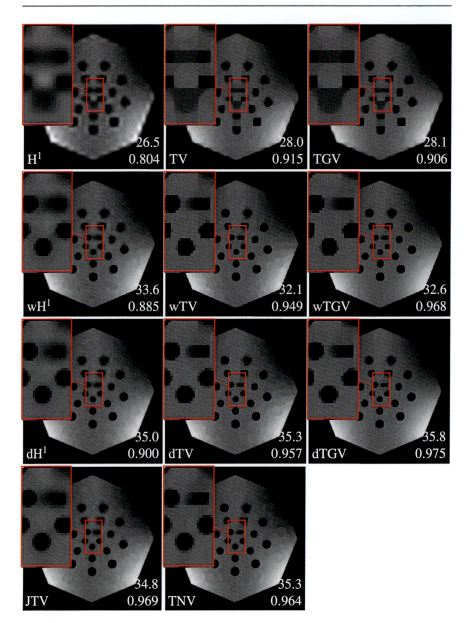

Fig. 19 Comparison of structure-promoting regularizers for test case super-resolution. All parameters where tuned to maximize the PSNR and visual image quality

Table 3 Computing times and PSNR for all tested regularizers

	Computing time		PSNR	
regularizer	x-ray	super-resolution	x-ray	super-resolution
H^1	30.43 s	18.34 s	24.7	26.5
wH^1	34.95 s	22.31 s	29.5	33.6
dH^1	40.87 s	27.80 s	30.9	35.0
TV	32.72 s	18.63 s	28.3	28.0
wTV	38.17 s	22.05 s	32.9	32.1
dTV	44.91 s	29.48 s	37.3	35.3
TGV	71.33 s	52.70 s	28.5	28.1
wTGV	77.67 s	58.44 s	34.3	32.6
dTGV	83.34 s	61.65 s	39.9	35.8
JTV	53.04 s	39.05 s	34.0	34.8
TNV	318.45 s	290.42 s	36.3	35.3

extend to multiple images as side information, but this has not yet been properly investigated.

Applications As we illustrated in section "Introduction," there are many applications where structure-promoting regularizers were already used or are on the horizon. The list of potential target applications grows steadily with more and more multi-modality scanners being introduced. Next to the misregistration mentioned before, the biggest hurdle in real applications is the interpretation of images that were created by fusing information from several modalities. A common question is "Which edges can I trust?" since often the reconstruction from multi-modality data would be performed on a finer resolution than for the single-modality case. For example, for PET-MR the reconstruction of PET data with an already reconstructed MR image as side information can be performed on the native MRI resolution. The answer might be that such an image should not be interpreted as a PET image, but in fact as a synergistic PET-MR image.

Joint reconstruction Throughout this chapter the focus was on improving the reconstruction of one image with the aid of another modality used as side information. Since the other image is rarely acquired directly, it is natural to aim to reconstruct both images simultaneously rather than sequentially. While conceptually appealing this strategy leads to many more complications than the approach discussed in this chapter which is sometimes referred to as one-sided reconstruction. While the mathematical framework for one-sided reconstruction is quite mature, the framework for joint reconstruction is despite a lot of research effort in the last 10 years still in its infancy. Fundamental problems like computationally tractable and efficient coupling of modalities are still unsolved. The appealing strategy of making use of the solid mathematical foundations of one-sided reconstruction for joint reconstruction in a mathematical sound and computationally tractable way is still not possible to date.

Acknowledgments The author acknowledges support from the EPSRC grant EP/S026045/1 and the Faraday Institution EP/T007745/1. Moreover, the author is grateful to all his collaborators which indirectly contributed to this chapter over the last couple of years.

References

van Aarle, W., Palenstijn, W.J., Cant, J., Janssens, E., Bleichrodt, F., Dabravolski, A., De Beenhouwer, J., Joost Batenburg, K., Sijbers, J.: Fast and flexible X-ray tomography using the ASTRA toolbox. Optics Express **24**(22), 25129 (2016). https://doi.org/10.1364/OE.24.025129

van Aarle, W., Palenstijn, W.J., De Beenhouwer, J., Altantzis, T., Bals, S., Batenburg, K.J., Sijbers, J.: The ASTRA Toolbox: A platform for advanced algorithm development in electron tomography. Ultramicroscopy **157**, 35–47 (2015). https://doi.org/10.1016/j.ultramic.2015.05.002

Adler, J., Kohr, H., Öktem, O.: Operator Discretization Library (ODL) (2017). https://doi.org/10.5281/zenodo.249479

Ametova, E., Fardell, G., Jørgensen, J.S., Lionheart, W.R.B., Papoutsellis, E., Pasca, E., Sykes, D., Turner, M., Warr, R., Withers, P.J.: Core Imaging Library (CIL) (2019). https://www.ccpi.ac.uk/cil

Arridge, S.R., Burger, M., Ehrhardt, M.J.: Preface to special issue on joint reconstruction and multi-modality/multi-spectral imaging. Inverse Prob. **36**, 020302 (2020)

Arridge, S.R., Kolehmainen, V., Schweiger, M.J.: Reconstruction and regularisation in optical tomography. In: Censor, A., Jiang, Y., Louis, M. (eds.) Mathematical Methods in Biomedical Imaging and Intensity-Modulated Radiation Therapy (IMRT). Scuola Normale Superiore (2008)

Arridge, S.R., Simmons, A.: Multi-spectral probabilistic diffusion using Bayesian classification. In: ter Haar Romeny, B.M., Florack, L., Koenderink, J.J., Viergever M.A. (eds.) Scale-Space Theories in Computer Vision, pp. 224–235. Springer, Berlin (1997). https://doi.org/10.1007/3-540-63167-4_53

Baete, K., Nuyts, J., Van Paesschen, W., Suetens, P., Dupont, P.: Anatomical-based FDG-PET reconstruction for the detection of hypo-metabolic regions in epilepsy. IEEE Trans. Med. Imaging **23**(4), 510–519 (2004). https://doi.org/10.1109/TMI.2004.825623

Bai, B., Li, Q., Leahy, R.M.: Magnetic resonance-guided positron emission tomography image reconstruction. Semin. Nucl. Med. **43**, 30–44 (2013). https://doi.org/10.1053/j.semnuclmed.2012.08.006

Ballester, C., Caselles, V., Igual, L., Verdera, J., Rougé, B.: A variational model for P+XS image fusion. Int. J. Comput. Vis. **69**(1), 43–58 (2006). https://doi.org/10.1007/s11263-006-6852-x

Bathke, C., Kluth, T., Maass, P.: Improved image reconstruction in magnetic particle imaging using structural a priori information. Int. J. Magn. Part. Imaging **3**(1) (2017)

Bauschke, H.H., Combettes, P.L.: Convex analysis and monotone operator theory in Hilbert spaces (2011). https://doi.org/10.1007/978-1-4419-9467-7

Benning, M., Burger, M.: Modern regularization methods for inverse problems. Acta Numerica **27**, 1–111 (2018). https://doi.org/10.1017/S0962492918000016

Bilgic, B., Goyal, V.K., Adalsteinsson, E.: Multi-contrast reconstruction with Bayesian compressed sensing. Magn. Reson. Med. **66**(6), 1601–1615 (2011). https://doi.org/10.1002/mrm.22956

Blomgren, P., Chan, T.F.: Color TV: Total variation methods for restoration of vector-valued images. IEEE Trans. Image Process. **7**(3), 304–309 (1998). https://doi.org/10.1109/83.661180

Bousse, A., Pedemonte, S., Kazantsev, D., Ourselin, S., Arridge, S.R., Hutton, B.F.: Weighted MRI-based Bowsher priors for SPECT brain image reconstruction. In: IEEE Nuclear Science Symposium and Medical Imaging Conference, pp. 3519–3522 (2010)

Bousse, A., Pedemonte, S., Thomas, B.A., Erlandsson, K., Ourselin, S., Arridge, S.R., Hutton, B.F.: Markov random field and Gaussian mixture for segmented MRI-based partial volume correction in PET. Phys. Med. Biol. **57**(20), 6681–6705 (2012). https://doi.org/10.1088/0031-9155/57/20/6681

Bowsher, J.E., Johnson, V.E., Turkington, T.G., Jaszczak, R.J., Floyd, C.E., Coleman, R.E.: Bayesian reconstruction and use of anatomical a priori information for emission tomography. IEEE Trans. Med. Imaging 15(5), 673–686 (1996). https://doi.org/10.1109/42.538945

Bowsher, J.E., Yuan, H., Hedlund, L.W., Turkington, T.G., Akabani, G., Badea, A., Kurylo, W.C., Wheeler, C.T., Cofer, G.P., Dewhirst, M.W., Johnson, G.A.: Utilizing MRI information to estimate F18-FDG distributions in rat flank tumors. In: IEEE Nuclear Science Symposium and Medical Imaging Conference, pp. 2488–2492 (2004). https://doi.org/10.1109/NSSMIC.2004.1462760

Bredies, K., Dong, Y., Hintermüller, M.: Spatially dependent regularization parameter selection in total generalized variation models for image restoration. Int. J. Comput. Math. 1–15 (2012). https://doi.org/10.1080/00207160.2012.700400

Bredies, K., Holler, M.: Regularization of linear inverse problems with total generalized variation. J. Inverse Ill-Posed Prob. 22(6), 871–913 (2014). https://doi.org/10.1515/jip-2013-0068

Bredies, K., Holler, M.: A TGV-based framework for variational image decompression, zooming, and reconstruction. Part II: Numerics. SIAM J. Imag. Sci. 8(4), 2851–2886 (2015). https://doi.org/10.1137/15M1023877

Bredies, K., Kunisch, K., Pock, T.: Total generalized variation. SIAM J. Imag. Sci. 3(3), 492–526 (2010). https://doi.org/10.1137/090769521

Bredies, K., Lorenz, D.A.: Mathematical Image Processing, 1 edn. Birkhäuser Basel (2018). https://doi.org/10.1007/978-3-030-01458-2

Bresson, X., Chan, T.F.: Fast dual minimization of the vectorial total variation norm and applications to color image processing. Inverse Prob. Imaging 2(4), 455–484 (2008). https://doi.org/10.3934/ipi.2008.2.455

Bungert, L., Coomes, D.A., Ehrhardt, M.J., Rasch, J., Reisenhofer, R., Schönlieb, C.B.: Blind image fusion for hyperspectral imaging with the directional total variation. Inverse Prob. 34(4), 044003 (2018). https://doi.org/10.1088/1361-6420/aaaf63

Bungert, L., Ehrhardt, M.J.: Robust image reconstruction with misaligned structural information (2020). http://arxiv.org/abs/2004.00589

Bungert, L., Ehrhardt, M.J., Reisenhofer, R.: Robust blind image fusion for misaligned hyperspectral imaging data. In: Proceedings in Applied Mathematics & Mechanics, vol. 18, p. e201800033 (2018). https://doi.org/10.1002/pamm.201800033

Burger, M., Osher, S.: A guide to the TV zoo. In: Level Set and PDE Based Reconstruction Methods in Imaging, Lecture Notes in Mathematics, vol. 2090, pp. 1–70. Springer (2013). https://doi.org/10.1007/978-3-319-01712-9

Caselles, V., Coll, B., Morel, J.M.: Geometry and color in natural images. J. Math. Imaging Vision 16(Section 2), 89–105 (2002). https://doi.org/10.1023/A:1013943314097

Chambolle, A., Ehrhardt, M.J., Richtárik, P., Schönlieb, C.B.: Stochastic primal-dual hybrid gradient algorithm with arbitrary sampling and imaging applications. SIAM J. Optim. 28(4), 2783–2808 (2018). https://doi.org/10.1007/s10851-010-0251-1

Chambolle, A., Pock, T.: A first-order primal-dual algorithm for convex problems with applications to imaging. J. Math. Imaging Vision 40(1), 120–145 (2011). https://doi.org/10.1007/s10851-010-0251-1

Chambolle, A., Pock, T.: An introduction to continuous optimization for imaging. Acta Numerica 25, 161–319 (2016). https://doi.org/10.1017/S096249291600009X

Chan, C., Fulton, R., Feng, D.D., Cai, W., Meikle, S.: An anatomically based regionally adaptive prior for MAP reconstruction in emission tomography. In: IEEE Nuclear Science Symposium and Medical Imaging Conference, pp. 4137–4141 (2007). https://doi.org/10.1109/NSSMIC.2007.4437032

Chan, C., Fulton, R., Feng, D.D., Meikle, S.: Regularized image reconstruction with an anatomically adaptive prior for positron emission tomography. Phys. Med. Biol. 54(24), 7379–400 (2009). https://doi.org/10.1088/0031-9155/54/24/009

Chen, C., Li, Y., Huang, J.: Calibrationless parallel MRI with joint total variation regularization. In: Medical Image Computing and Computer-Assisted Intervention, pp. 106–114 (2013). https://doi.org/10.1007/978-3-642-40760-4_14

Cheng-Liao, J., Qi, J.: PET image reconstruction with anatomical edge guided level set prior. Phys. Med. Biol. **56**, 6899–6918 (2011). https://doi.org/10.1088/0031-9155/56/21/009

Combettes, P.L., Pesquet, J.C.: Proximal splitting methods in signal processing. Springer Optim. Appl. **49**, 185–212 (2011). https://doi.org/10.1007/978-1-4419-9569-8_10

Comtat, C., Kinahan, P.E., Fessler, J.A., Beyer, T., Townsend, D.W., Defrise, M., Michel, C.J.: Clinically feasible reconstruction of 3D whole-body PET/CT data using blurred anatomical labels. Phys. Med. Biol. **47**(1), 1–20 (2002)

Davies, M., Puy, G., Vandergheynst, P., Wiaux, Y.: A compressed sensing framework for magnetic resonance fingerprinting. SIAM J. Imag. Sci. **7**(4), 2623–2656 (2013). https://doi.org/10.1137/130947246

Deidda, D., Karakatsanis, N.A., Robson, P.M., Tsai, Y.J., Efthimiou, N., Thielemans, K., Fayad, Z.A., Aykroyd, R.G., Tsoumpas, C.: Hybrid PET-MR list-mode kernelized expectation maximization reconstruction. Inverse Prob. **35**(4) (2019). https://doi.org/10.1088/1361-6420/ab013f

Deligiannis, N., Mota, J.F., Cornelis, B., Rodrigues, M.R., Daubechies, I.: Multi-modal dictionary learning for image separation with application in art investigation. IEEE Trans. Image Process. **26**(2), 751–764 (2017). https://doi.org/10.1109/TIP.2016.2623484

Delso, G., Furst, S., Jakoby, B., Ladebeck, R., Ganter, C., Nekolla, S.G., Schwaiger, M., Ziegler, S.I., Fürst, S., Jakoby, B., Ladebeck, R., Ganter, C., Nekolla, S.G., Schwaiger, M., Ziegler, S.I.: Performance measurements of the Siemens mMR integrated whole-body PET/MR scanner. J. Nucl. Med. **52**(12), 1914–22 (2011). https://doi.org/10.2967/jnumed.111.092726

Di, Z.W., Leyffer, S., Wild, S.M.: Optimization-based approach for joint X-Ray fluorescence and transmission tomographic inversion. SIAM J. Imag. Sci. **9**(1), 1–23 (2016)

Dong, G., Hintermüller, M., Papafitsoros, K.: Quantitative magnetic resonance imaging: From fingerprinting to integrated physics-based models. SIAM J. Imag. Sci. **12**(2), 927–971 (2019). https://doi.org/10.1137/18M1222211

Dong, Y., Hintermüller, M., Rincon-Camacho, M.M.: Automated regularization parameter selection in multi-scale total variation models for image restoration. J. Math. Imaging Vision **40**(1), 82–104 (2011). https://doi.org/10.1007/s10851-010-0248-9

Duran, J., Buades, A., Coll, B., Sbert, C., Blanchet, G.: A survey of pansharpening methods with a new band-decoupled variational model. ISPRS J. Photogramm. Remote Sens. **125**, 78–105 (2017). https://doi.org/10.1016/j.isprsjprs.2016.12.013

Ehrhardt, M.J.: Joint reconstruction for multi-modality imaging with common structure. Ph.d. thesis, University College London (2015)

Ehrhardt, M.J., Arridge, S.R.: Vector-valued image processing by parallel level sets. IEEE Trans. Image Process. **23**(1), 9–18 (2014). https://doi.org/10.1109/TIP.2013.2277775

Ehrhardt, M.J., Betcke, M.M.: Multi-contrast MRI reconstruction with structure-guided total variation. SIAM J. Imag. Sci. **9**(3), 1084–1106 (2016). https://doi.org/10.1137/15M1047325

Ehrhardt, M.J., Markiewicz, P.J., Liljeroth, M., Barnes, A., Kolehmainen, V., Duncan, J., Pizarro, L., Atkinson, D., Hutton, B.F., Ourselin, S., Thielemans, K., Arridge, S.R.: PET reconstruction with an anatomical MRI prior using parallel level sets. IEEE Trans. Med. Imaging **35**(9), 2189–2199 (2016). https://doi.org/10.1109/TMI.2016.2549601

Ehrhardt, M.J., Markiewicz, P.J., Schönlieb, C.B.: Faster PET reconstruction with non-smooth priors by randomization and preconditioning. Phys. Med. Biol. **64**(22), 225019 (2019). https://doi.org/10.1088/1361-6560/ab3d07

Ehrhardt, M.J., Thielemans, K., Pizarro, L., Atkinson, D., Ourselin, S., Hutton, B.F., Arridge, S.R.: Joint reconstruction of PET-MRI by exploiting structural similarity. Inverse Prob. **31**(1), 015001 (2015). https://doi.org/10.1088/0266-5611/31/1/015001

Ehrhardt, M.J., Thielemans, K., Pizarro, L., Markiewicz, P.J., Atkinson, D., Ourselin, S., Hutton, B.F., Arridge, S.R.: Joint reconstruction of PET-MRI by parallel level sets. In: IEEE Nuclear Science Symposium and Medical Imaging Conference (2014). https://doi.org/10.1109/NSSMIC.2014.7430895

Elbau, P., Mindrinos, L., Scherzer, O.: Quantitative reconstructions in multi-modal photoacoustic and optical coherence tomography imaging. Inverse Prob. **34**(1) (2018). https://doi.org/10.1088/1361-6420/aa9ae7

Engl, H.W., Hanke, M., Neubauer, A.: Regularization of Inverse Problems. Mathematics and Its Applications. Springer (1996)

Esser, E., Zhang, X., Chan, T.F.: A general framework for a class of first order primal-dual algorithms for convex optimization in imaging science. SIAM J. Imag. Sci. **3**(4), 1015–1046 (2010). https://doi.org/10.1137/09076934X

Estellers, V., Soatto, S., Bresson, X.: Adaptive regularization with the structure tensor. IEEE Trans. Image Process. **24**(6), 1777–1790 (2015). https://doi.org/10.1109/TIP.2015.2409562

Estellers, V., Thiran, J., Bresson, X.: Enhanced compressed sensing recovery with level set normals. IEEE Trans. Image Process. **22**(7), 2611–2626 (2013). https://doi.org/10.1109/TIP.2013.2253484

Fang, F., Li, F., Shen, C., Zhang, G.: A variational approach for pan-sharpening. IEEE Trans. Image Process. **22**(7), 2822–2834 (2013). https://doi.org/10.1109/TIP.2013.2258355

Fessler, J.A., Elbakri, I., Sukovic, P., Clinthorne, N.H.: Maximum-likelihood dual-energy tomographic image reconstruction. In: SPIE: Medical Imaging, vol. 4684, pp. 1–25 (2002). https://doi.org/doi:10.1117/12.467189

Foygel Barber, R., Sidky, E.Y., Gilat Schmidt, T., Pan, X.: An algorithm for constrained one-step inversion of spectral CT data. Phys. Med. Biol. **61**(10), 3784–3818 (2016). https://doi.org/10.1088/0031-9155/61/10/3784

Gallardo, L.A., Meju, M.A.: Characterization of heterogeneous near-surface materials by joint 2D inversion of DC resistivity and seismic data. Geophys. Res. Lett. **30**(13), 1658 (2003). https://doi.org/10.1029/2003GL017370

Gallardo, L.A., Meju, M.A.: Joint two-dimensional DC resistivity and seismic travel time inversion with cross-gradients constraints. J. Geophys. Res. **109**(B3), 1–11 (2004). https://doi.org/10.1029/2003JB002716

Gallardo, L.A., Meju, M.A.: Structure-coupled multiphysics imaging in geophysical sciences. Rev. Geophys. **49**, 1–19 (2011). https://doi.org/10.1029/2010RG000330.1.INTRODUCTION

Golbabaee, M., Chen, Z., Wiaux, Y., Davies, M.: CoverBLIP: accelerated and scalable iterative matched-filtering for magnetic resonance fingerprint reconstruction. Inverse Prob. **36**(1), 015003 (2020). https://doi.org/10.1088/1361-6420/ab4c9a

Goldluecke, B., Strekalovskiy, E., Cremers, D.: The natural vectorial total variation which arises from geometric measure theory. SIAM J. Imag. Sci. **5**(2), 537–563 (2012). https://doi.org/10.1137/110823766

Haber, E., Holtzman-Gazit, M.: Model fusion and joint inversion. Surv. Geophys. (34), 675–695 (2013). https://doi.org/10.1007/s10712-013-9232-4

Haber, E., Modersitzki, J.: Intensity gradient based registration and fusion of multi-modal images. In: Medical Image Computing and Computer-Assisted Intervention, vol. 46, pp. 726–733. Springer, Berlin/Heidelberg (2006). https://doi.org/10.1160/ME9046

Haber, E., Oldenburg, D.W.: Joint inversion: A structural approach. Inverse Prob. **13**, 63–77 (1997). https://doi.org/10.1088/0266-5611/13/1/006

Heismann, B., Schmidt, B., Flohr, T.: Spectral Computed Tomography. SPIE Press (2012)

Hintermüller, M., Rincon-Camacho, M.M.: Expected absolute value estimators for a spatially adapted regularization parameter choice rule in L1-TV-based image restoration. Inverse Prob. **26**(8), 085005 (2010). https://doi.org/10.1088/0266-5611/26/8/085005

Holt, K.M.: Total nuclear variation and jacobian extensions of total variation for vector fields. IEEE Trans. Image Process. **23**(9), 3975–3989 (2014). https://doi.org/10.1109/TIP.2014.2332397

Huang, J., Chen, C., Axel, L.: Fast Multi-contrast MRI reconstruction. Magn. Reson. Imaging **32**(10), 1344–52 (2014). https://doi.org/10.1016/j.mri.2014.08.025

Huber, R., Haberfehlner, G., Holler, M., Bredies, K.: Total generalized variation regularization for multi-modal electron tomography. Nanoscale 1–38 (2019). https://doi.org/10.1039/c8nr09058k

Ito, K., Jin, B.: Inverse Problems – Tikhonov Theory and Algorithms. World Scientific Publishing (2014). https://doi.org/10.1142/9120

Kaipio, J.P., Kolehmainen, V., Vauhkonen, M., Somersalo, E.: Inverse problems with structural prior information. Inverse Prob. **15**(3), 713–729 (1999). https://doi.org/10.1088/0266-5611/15/3/306

Kazantsev, D., Arridge, S.R., Pedemonte, S., Bousse, A., Erlandsson, K., Hutton, B.F., Ourselin, S.: An anatomically driven anisotropic diffusion filtering method for 3D SPECT reconstruction. Phys. Med. Biol. **57**(12), 3793–3810 (2012). https://doi.org/10.1088/0031-9155/57/12/3793

Kazantsev, D., Jørgensen, J.S., Andersen, M.S., Lionheart, W.R., Lee, P.D., Withers, P.J.: Joint image reconstruction method with correlative multi-channel prior for x-ray spectral computed tomography. Inverse Prob. **34**(6) (2018). https://doi.org/10.1088/1361-6420/aaba86

Kazantsev, D., Lionheart, W.R.B., Withers, P.J., Lee, P.D.: Multimodal image reconstruction using supplementary structural information in total variation regularization. Sens. Imaging **15**(1), 97 (2014). https://doi.org/10.1007/s11220-014-0097-5

Kimmel, R., Malladi, R., Sochen, N.: Images as embedded maps and minimal surfaces: movies, color, texture, and volumetric medical images. Int. J. Comput. Vis. **39**(2), 111–129 (2000). https://doi.org/10.1023/A:1008171026419

Knoll, F., Holler, M., Koesters, T., Otazo, R., Bredies, K., Sodickson, D.K.: Joint MR-PET reconstruction using a multi-channel image regularizer. IEEE Trans. Med. Imaging **36**(1) (2016). https://doi.org/10.1109/TMI.2016.2564989

Knoll, F., Koesters, T., Otazo, R., Boada, F., Sodickson, D.K.: Simultaneous MR-PET reconstruction using multi sensor compressed sensing and joint sparsity. In: International Society for Magnetic Resonance in Medicine, vol. 22 (2014)

Kolehmainen, V., Ehrhardt, M.J., Arridge, S.R.: Incorporating structural prior information and sparsity into EIT using parallel level sets. Inverse Prob. Imaging **13**(2), 285–307 (2019). https://doi.org/10.3934/ipi.2019015

Leahy, R.M., Yan, X.: Incorporation of anatomical MR data for improved functional imaging with PET. In: Information Processing in Medical Imaging, pp. 105–120. Springer (1991). https://doi.org/10.1007/BFb0033746

Lenzen, F., Berger, J.: Solution-driven adaptive total variation regularization. In: SSVM, pp. 203–215 (2015). https://doi.org/10.1007/978-3-642-24785-9

Loncan, L., De Almeida, L.B., Bioucas-Dias, J.M., Briottet, X., Chanussot, J., Dobigeon, N., Fabre, S., Liao, W., Licciardi, G.A., Simoes, M., Tourneret, J.Y., Veganzones, M.A., Vivone, G., Wei, Q., Yokoya, N.: Hyperspectral pansharpening: a review. IEEE Geosci. Remote Sens. Mag. **3**(3), 27–46 (2015). https://doi.org/10.1109/MGRS.2015.2440094

Long, Y., Fessler, J.A.: Multi-material decomposition using statistical image reconstruction for spectral CT. IEEE Trans. Med. Imaging **33**(8), 1614–1626 (2014). https://doi.org/10.1109/TMI.2014.2320284

Ma, D., Gulani, V., Seiberlich, N., Liu, K., Sunshine, J.L., Duerk, J.L., Griswold, M.A.: Magnetic resonance fingerprinting. Nature **495**(7440), 187–92 (2013). https://doi.org/10.1038/nature11971

Maes, F., Collignon, A., Vandermeulen, D., Marchal, G., Suetens, P.: Multimodality image registration by maximization of mutual information. IEEE Trans. Med. Imaging **16**(2), 187–98 (1997). https://doi.org/10.1109/42.563664

Mehranian, A., Belzunce, M., Prieto, C., Hammers, A., Reader, A.J.: Synergistic PET and SENSE MR image reconstruction using joint sparsity regularization. IEEE Trans. Med. Imaging **37**(1), 20–34 (2018). https://doi.org/10.1109/TMI.2017.2691044

Mehranian, A., Belzunce, M.A., Niccolini, F., Politis, M., Prieto, C., Turkheimer, F., Hammers, A., Reader, A.J.: PET image reconstruction using multi-parametric anato-functional priors. Phys. Med. Biol. (2017). https://doi.org/10.1042/BJ20101136>

Meju, M.A., Mackie, R.L., Miorelli, F., Saleh, A.S., Miller, R.V.: Structurally-tailored 3D anisotropic CSEM resistivity inversion with cross-gradients criterion and simultaneous model calibration. Geophysics **84**(6), 1–62 (2019). https://doi.org/10.1190/geo2018-0639.1

Möller, M., Brinkmann, E.M., Burger, M., Seybold, T.: Color Bregman TV. SIAM J. Imag. Sci. **7**(4), 2771–2806 (2014). https://doi.org/10.1137/130943388

Möller, M., Wittman, T., Bertozzi, A.L., Burger, M.: A variational approach for sharpening high dimensional images. SIAM J. Imag. Sci. **5**(1), 150–178 (2012). https://doi.org/10.1137/100810356

Nuyts, J.: The use of mutual information and joint entropy for anatomical priors in emission tomography. In: IEEE Nuclear Science Symposium and Medical Imaging Conference, pp. 4149–4154. IEEE (2007). https://doi.org/10.1109/NSSMIC.2007.4437034

Obert, A.J., Gutberlet, M., Kern, A.L., Kaireit, T.F., Grimm, R., Wacker, F., Vogel-Claussen, J.: 1H-guided reconstruction of 19F gas MRI in COPD patients. Magn. Reson. Med. 1–11 (2020). https://doi.org/10.1002/mrm.28209

Parikh, N., Boyd, S.P.: Proximal algorithms. Found Trends Optim **1**(3), 123–231 (2014). https://doi.org/10.1561/2400000003

Pedemonte, S., Bousse, A., Hutton, B.F., Arridge, S.R., Ourselin, S.: Probabilistic graphical model of SPECT/MRI. In: Machine Learning in Medical Imaging, pp. 167–174 (2011). https://doi.org/10.1007/978-3-642-24319-6_21

Pluim, J.P.W., Maintz, J.B.A., Viergever, M.A.: Image registration by maximization of combined mutual information and gradient information. IEEE Trans. Med. Imaging **19**(8), 809–14 (2000). https://doi.org/10.1109/42.876307

Pock, T., Chambolle, A.: Diagonal preconditioning for first order primal-dual algorithms in convex optimization. In: Proceedings of the IEEE International Conference on Computer Vision, pp. 1762–1769 (2011). https://doi.org/10.1109/ICCV.2011.6126441

Rangarajan, A., Hsiao, I.T., Gindi, G.: A Bayesian joint mixture framework for the integration of anatomical information in functional image reconstruction. J. Math. Imaging Vision **12**(3), 199–217 (2000). https://doi.org/10.1023/A:1008314015446

Rasch, J., Brinkmann, E.M., Burger, M.: Joint reconstruction via coupled bregman iterations with applications to PET-MR imaging. Inverse Prob. **34**(1), 014001 (2018a). https://doi.org/10.1088/1361-6420/aa9425

Rasch, J., Kolehmainen, V., Nivajarvi, R., Kettunen, M., Gröhn, O., Burger, M., Brinkmann, E.M.: Dynamic MRI reconstruction from undersampled data with an anatomical prescan. Inverse Prob. **34**(7) (2018b). https://doi.org/10.1088/1361-6420/aac3af

Rigie, D., La Riviere, P.: Joint reconstruction of multi-channel, spectral CT data via constrained total nuclear variation minimization. Phys. Med. Biol. **60**, 1741–1762 (2015). https://doi.org/10.1088/0031-9155/60/4/1741

Rigie, D.S., Sanchez, A.A., La Riviére, P.J.: Assessment of vectorial total variation penalties on realistic dual-energy CT data. Phys. Med. Biol. **62**(8), 3284–3298 (2017). https://doi.org/10.1088/1361-6560/aa6392

Rudin, L.I., Osher, S., Fatemi, E.: Nonlinear total variation based noise removal algorithms. Physica D: Nonlinear Phenom. **60**(1), 259–268 (1992). https://doi.org/10.1016/0167-2789(92)90242-F

Sapiro, G., Ringach, D.L.: Anisotropic diffusion of multivalued images with applications to color filtering. IEEE Trans. Image Process. **5**(11), 1582–1586 (1996). https://doi.org/10.1109/83.541429

Scherzer, O., Grasmair, M., Grossauer, H., Haltmeier, M., Lenzen, F.: Variational Methods in Imaging, vol. 167. Springer, New York/London (2008)

Schmitt, U., Louis, A.K.: Efficient algorithms for the regularization of dynamic inverse problems: I. Theory. Inverse Problems **18**(3), 645–658 (2002). https://doi.org/10.1088/0266-5611/18/3/308

Schmitt, U., Louis, A.K., Wolters, C., Vaukhonen, M.: Efficient algorithms for the regularization of dynamic inverse problems: II. Applications. Inverse Prob. **18**(3), 659–676 (2002). https://doi.org/10.1088/0266-5611/18/3/308

Schramm, G., Holler, M., Rezaei, A., Vunckx, K., Knoll, F., Bredies, K., Boada, F., Nuyts, J.: Evaluation of parallel level sets and Bowsher's method as segmentation-free anatomical priors for time-of-flight PET reconstruction. IEEE Trans. Med. Imaging **62**(2), 590–603 (2017). https://doi.org/10.1109/TMI.2017.2767940

Schuster, T., Hahn, B., Burger, M.: Dynamic inverse problems: Modelling – Regularization – numerics. Inverse Prob. **34**(4) (2018). https://doi.org/10.1088/1361-6420/aab0f5

Sochen, N., Kimmel, R., Malladi, R.: A general framework for low level vision. IEEE Trans. Image Process. **7**(3), 310–318 (1998). https://doi.org/10.1109/83.661181

Sodickson, D.K., Feng, L., Knoll, F., Cloos, M., Ben-Eliezer, N., Axel, L., Chandarana, H., Block, K.T., Otazo, R.: The rapid imaging renaissance: Sparser samples, denser dimensions, and glimmerings of a grand unified tomography. In: Proceedings of SPIE, vol. 9417, pp. 94170G1–9417014 (2015). https://doi.org/10.1117/12.2085033

Somayajula, S., Panagiotou, C., Rangarajan, A., Li, Q., Arridge, S.R., Leahy, R.M.: PET image reconstruction using information theoretic anatomical priors. IEEE Trans. Med. Imaging **30**(3), 537–549 (2011). https://doi.org/10.1109/TMI.2010.2076827

Song, P., Deng, X., Mota, J.F.C., Deligiannis, N., Dragotti, P.L., Rodrigues, M.: Multimodal image super-resolution via joint sparse representations induced by coupled dictionaries. IEEE Trans. Comput. Imaging 1–1 (2019). https://doi.org/10.1109/tci.2019.2916502

Song, P., Weizman, L., Mota, J.F., Eldar, Y.C., Rodrigues, M.R.: Coupled dictionary learning for multi-contrast MRI reconstruction. In: International Conference on Image Processing, 2, pp. 2880–2884 (2018). https://doi.org/10.1109/ICIP.2018.8451341

Tang, J., Rahmim, A.: Bayesian PET image reconstruction incorporating anato-functional joint entropy. Phys. Med. Biol. **54**(23), 7063–75 (2009). https://doi.org/10.1088/0031-9155/54/23/002

Tang, J., Rahmim, A.: Anatomy assisted PET image reconstruction incorporating multi-resolution joint entropy. Phys. Med. Biol. **60**(1), 31–48 (2015). https://doi.org/10.1088/0031-9155/60/1/31

Tang, S., Fernandez-Granda, C., Lannuzel, S., Bernstein, B., Lattanzi, R., Cloos, M., Knoll, F., Asslander, J.: Multicompartment magnetic resonance fingerprinting. Inverse Prob. **34**(9) (2018). https://doi.org/10.1088/1361-6420/aad1c3

Tsai, Y.J., Member, S., Bousse, A., Ahn, S., Charles, W., Arridge, S., Hutton, B.F., Member, S., Thielemans, K.: Algorithms for solving misalignment issues in penalized PET/CT reconstruction using anatomical priors. In: IEEE Nuclear Science Symposium and Medical Imaging Conference Proceedings (NSS/MIC). IEEE (2018)

Tschumperlé, D., Deriche, R.: Vector-valued image regularization with PDEs: A common framework for different applications. IEEE Trans. Pattern Anal. Mach. Intell. **27**(4), 506–517 (2005). https://doi.org/10.1109/TPAMI.2005.87

Vunckx, K., Atre, A., Baete, K., Reilhac, A., Deroose, C.M., Van Laere, K., Nuyts, J.: Evaluation of three MRI-based anatomical priors for quantitative PET brain imaging. IEEE Trans. Med. Imaging **31**(3), 599–612 (2012). https://doi.org/10.1109/TMI.2011.2173766

Wang, G., Zhang, J., Gao, H., Weir, V., Yu, H., Cong, W., Xu, X., Shen, H., Bennett, J., Furth, M., Wang, Y., Vannier, M.: Towards omni-tomography – grand fusion of multiple modalities for simultaneous interior tomography. PloS one **7**(6), e39700 (2012). https://doi.org/10.1371/journal.pone.0039700

Wells III, W.M., Viola, P., Atsumi, H., Nakajima, S., Kikinis, R.: Multi-modal volume registration by maximization of mutual information. Med. Image Anal. **1**(1), 35–51 (1996)

Xi, Y., Zhao, J., Bennett, J., Stacy, M., Sinusas, A., Wang, G.: Simultaneous CT-MRI reconstruction for constrained imaging geometries using structural coupling and compressive sensing. IEEE Trans. Biomed. Eng. (2015). https://doi.org/10.1109/TBME.2015.2487779

Xiang, L., Chen, Y., Chang, W., Zhan, Y., Lin, W., Wang, Q., Shen, D.: Deep-learning-based multi-modal fusion for fast MR reconstruction. IEEE Trans. Biomed. Eng. **66**(7), 2105–2114 (2019). https://doi.org/10.1109/TBME.2018.2883958

Yokoya, N., Grohnfeldt, C., Chanussot, J.: Hyperspectral and multispectral data fusion: A comparative review of the recent literature. IEEE Geosci. Remote Sens. Mag. **5**(2), 29–56 (2017). https://doi.org/10.1109/MGRS.2016.2637824

Zhang, Y., Zhang, X.: PET-MRI joint reconstruction with common edge weighted total variation regularization. Inverse Prob. **34**(6), 065006 (2018). https://doi.org/10.1088/1361-6420/aabce9

Diffraction Tomography, Fourier Reconstruction, and Full Waveform Inversion

8

Florian Faucher, Clemens Kirisits, Michael Quellmalz, Otmar Scherzer, and Eric Setterqvist

Contents

Introduction	274
Contribution and Outline	276
Experimental Setup	276
Forward Models	278
Incident Plane Wave	279
Modeling the Total Field Using Line and Point Sources	281
Numerical Comparison of Forward Models	282
Modeling the Scattered Field Assuming Incident Plane Waves	283

F. Faucher
Faculty of Mathematics, University of Vienna, Vienna, Austria

Project-Team Makutu, Inria Bordeaux Sud-Ouest, Talence, France
e-mail: florian.faucher@inria.fr; florian.faucher@univie.ac.at

C. Kirisits
Faculty of Mathematics, University of Vienna, Vienna, Austria
e-mail: clemens.kirisits@univie.ac.at

M. Quellmalz
Institute of Mathematics, Technical University Berlin, Berlin, Germany
e-mail: quellmalz@math.tu-berlin.de

O. Scherzer (✉)
Faculty of Mathematics, University of Vienna, Vienna, Austria

Johann Radon Institute for Computational and Applied Mathematics (RICAM), Linz, Austria

Christian Doppler Laboratory for Mathematical Modeling and Simulation of Next Generations of Ultrasound Devices (MaMSi), Vienna, Austria
e-mail: otmar.scherzer@univie.ac.at

E. Setterqvist
Johann Radon Institute for Computational and Applied Mathematics (RICAM), Linz, Austria
e-mail: eric.setterqvist@ricam.oeaw.ac.at

Modeling the Total Field Using Line and Point Sources............................. 283
Fourier Diffraction Theorem.. 284
  Rotating the Object... 286
  Varying Wave Number.. 287
  Rotating the Object with Multiple Wave Numbers................................. 288
Reconstruction Methods.. 289
  Reconstruction Using Full Waveform Inversion................................... 289
  Reconstruction Based on the Born and Rytov Approximations...................... 291
Numerical Experiments... 294
  Reconstruction of Circular Contrast with Various Amplitudes and Sizes.......... 294
  Reconstruction of Embedded Shapes: Phantom 1................................... 299
  Reconstruction of Embedded Shapes: Phantom 2................................... 304
  Computational Costs.. 305
Conclusion.. 308
References.. 309

Abstract

In this chapter, we study the mathematical imaging problem of diffraction tomography (DT), which is an inverse scattering technique used to find material properties of an object by illuminating it with probing waves and recording the scattered waves. Conventional DT relies on the Fourier diffraction theorem, which is applicable under the condition of weak scattering. However, if the object has high contrasts or is too large compared to the wavelength, it tends to produce multiple scattering, which complicates the reconstruction. In this chapter, we give a survey on diffraction tomography and compare the reconstruction of low- and high-contrast objects. We also implement and compare the reconstruction using the full waveform inversion method which, contrary to the Born and Rytov approximations, works with the total field and is more robust to multiple scattering.

Keywords

Diffraction tomography · Mathematical imaging · Fourier diffraction theorem · Full waveform inversion · Born approximation · Rytov approximation · Inverse problems

Introduction

Diffraction tomography (DT) is a technique for reconstructing the scattering potential of an object from measurements of waves scattered by that object. DT can be understood as an alternative to, or extension of, classical computerized tomography. In computerized tomography, a crucial assumption is that the radiation, X-rays, for instance, essentially propagates along straight lines through the object. The attenuated rays are recorded and can be related to material properties f of the object by means of the Radon, or X-ray, transform. A central result for the inversion of this relation is the Fourier slice theorem. Roughly speaking, it says that the Fourier transformed measurements are equal to the Fourier transform of f evaluated along slices through the origin (Natterer 1986).

The straight ray assumption of computerized tomography can be considered valid as long as the wavelength of the incident field is much smaller than the size of the relevant details in the object. As soon as the wavelength is similar to or greater than those details, for instance, in situations where X-rays are replaced by visible light, diffraction effects are no longer negligible. As an example of a medical application, an optical diffraction experiment in Sung et al. (2009) utilized a red laser of wavelength 633 nm to illuminate human cells of diameter around 10 μm, which include smaller subcellular organelles. One way to achieve better reconstruction quality in such cases is to drop the straight ray assumption and adopt a propagation model based on the wave equation instead.

The theoretical groundwork for DT was laid more than half a century ago (Wolf 1969). The central result derived there, sometimes called the *Fourier diffraction theorem*, says that the Fourier transformed measurements of the scattered wave are equal to the Fourier transform of the scattering potential evaluated along a hemisphere. This result relies on a series of assumptions: (i) the object is immersed in a homogeneous background, (ii) the incident field is a monochromatic plane wave, (iii) the scattered wave is measured on a plane in $\mathbb{R}^3$, and (iv) the first Born approximation of the scattered field is valid.

On the one hand, the Born approximation greatly simplifies the relationship between scattered wave and scattering potential. On the other hand, however, it generally requires the object to be weakly scattering, thus limiting the applicability of the Fourier diffraction theorem. An alternative is to assume validity of the first Rytov approximation instead (Iwata and Nagata 1975). While mathematically this amounts to essentially the same reconstruction problem, the underlying physical assumptions are not identical to those of the Born approximation, leading to a different range of applicability in general (Chen and Stamnes 1998; Slaney et al. 1984). Nevertheless, the restriction to weakly scattering objects remains.

Full waveform inversion (FWI) is a different approach that can overcome some of the limitations of the first-order methods, typically at the cost of being computationally more demanding. It relies on the iterative minimization of a cost functional which penalizes the misfit between measurements and forward simulations of the total field, cf. Bamberger et al. (1979), Lailly (1983), Pratt et al. (1998), Tarantola (1984), and Virieux and Operto (2009). Here, the forward model consists of the solution of the full wave equation, without simplification of first-order approximations. It results in a nonlinear minimization problem to be solved, typically with Newton-type methods (Virieux and Operto 2009; Nocedal and Wright 2006).

In practical experiments, there are sometimes only measurements of the intensity, i.e., the absolute value of the complex-valued wave, available. Different phase retrieval methods were investigated, e.g., in Maleki and Devaney (1993), Gbur and Wolf (2002), Horstmeyer et al. (2016), and Beinert and Quellmalz (2022). For this chapter, we assume that both the phase and amplitude information are present, which can be achieved by interferometry, cf. Wedberg and Stamnes (1995).

Contribution and Outline

In this chapter, we present a numerical comparison of three reconstruction approaches for diffraction tomography on simulated data, based on (i) the Born approximation, (ii) the Rytov approximation, and (iii) FWI. The setting we use for this comparison is 2D transmission imaging in a homogeneous background with (approximate) plane wave irradiation. The object is assumed to make a full turn during the experiment, providing measurements for a uniform set of incidence angles. In addition, we investigate how providing additional data by varying the wavelength affects the reconstruction. The scattering potentials considered here are test phantoms of varying sizes, shapes, and contrasts. Moreover, for data generation purposes, we compare several forward models.

For numerical reconstruction under the Born and Rytov approximations, a well-known method is the backpropagation algorithm (Devaney 1982), which is widely used in practice, cf. Müller et al. (2015) and also Fan et al. (2017). Our algorithms rely on the nonuniform discrete Fourier transform (NDFT), which was used in 3D Fourier diffraction tomography yielding better results than discrete backpropagation (Kirisits et al. 2021). Our FWI-based reconstruction uses an iterative Newton-type method on an L^2 distance between data and simulations. Here, the discretization of the partial differential equations associated with the wave propagation uses the hybridizable discontinuous Galerkin method (HDG), Cockburn et al. (2009) and Faucher and Scherzer (2020). It is implemented, together with the inverse procedure, in the open-source parallel software hawen,[1] Faucher (2021).

The outline of this chapter is as follows. The conceptual experiment is detailed in section "Experimental Setup". Forward models are presented in section "Forward Models", and their numerical performance is compared in section "Numerical Comparison of Forward Models". The Fourier diffraction theorem is formulated and discussed in section "Fourier Diffraction Theorem". Further, section "Reconstruction Methods" covers the reconstruction algorithms used for the numerical experiments, which are presented in section "Numerical Experiments". A concluding discussion of our findings is given in section "Conclusion".

Experimental Setup

We consider the tomographic reconstruction of a two-dimensional object taking into account diffraction of the incident field. The object is assumed to be embedded in a homogeneous background and illuminated or insonified by a monochromatic plane wave. In fact, for the computational experiments, we implement and compare several approaches to approximate the incident plane wave; see sections "Forward Models" and "Numerical Comparison of Forward Models". We restrict ourselves to transmission imaging, where the incident field propagates in direction $\mathbf{e}_2 = (0, 1)^\top$,

---

[1] https://ffaucher.gitlab.io/hawen-website/

8 Diffraction Tomography, Fourier Reconstruction, and Full Waveform Inversion

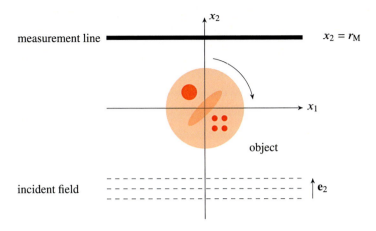

Fig. 1 Experimental setup

and the resulting field is measured on the line $x_2 = r_M$. The distance between the measurement line and the origin, $r_M > 0$, is sufficiently large so that it does not intersect the object. From the measurements, we aim to reconstruct the object's scattering properties. In order to improve the reconstruction quality, we generate additional data by rotating the object or changing the incident field's wavelength. See Fig. 1 for an illustration of the experimental setup.

We now introduce the physical quantities needed subsequently. Let $\lambda > 0$ denote the *wavelength* of the incident wave and $k_0 = 2\pi/\lambda$ the background *wave number*. Furthermore, let $n(\mathbf{x})$ denote the *refractive index* at position $\mathbf{x} \in \mathbb{R}^2$ and n_0 the constant refractive index of the background. From these quantities, we define the wave number

$$k(\mathbf{x}) := k_0 \frac{n(\mathbf{x})}{n_0}.$$

Furthermore, the wave number k can be equivalently expressed in terms of the angular frequency ω and the wave speed c such that

$$k(\mathbf{x}) = \frac{\omega}{c(\mathbf{x})} \quad \text{and} \quad k_0 = \frac{\omega}{c_0}, \qquad (1)$$

where c_0 is the constant wave speed in the background. The *scattering potential* f is obtained by subtracting the background wave number k_0:

$$f(\mathbf{x}) := k^2(\mathbf{x}) - k_0^2 = k_0^2 \left(\frac{n(\mathbf{x})^2}{n_0^2} - 1 \right). \qquad (2)$$

Note that, for all practical purposes, f can be assumed to be bounded and compactly supported in the disk $\mathcal{B}_{r_M} = \{\mathbf{x} \in \mathbb{R}^2 : \|\mathbf{x}\| < r_M\}$.

In our subsequent reconstructions with Born and Rytov approximations, f is the quantity to be reconstructed from the measured data and k_0 is known. On the other hand, with FWI, we reconstruct c; see Remark 2. These two quantities can be related to each other via

$$c(\mathbf{x}) = \sqrt{\frac{\omega^2}{k_0^2 + f(\mathbf{x})}}. \tag{3}$$

Forward Models

In this section, we propose several forward models for the experiment presented above. For all of them, the starting point is the system of equations

$$\left.\begin{aligned} \left(-\Delta - k(\mathbf{x})^2\right) u^{\text{tot}}(\mathbf{x}) &= g(\mathbf{x}), \\ \left(-\Delta - k_0^2\right) u^{\text{inc}}(\mathbf{x}) &= g(\mathbf{x}), \\ u^{\text{tot}}(\mathbf{x}) &= u^{\text{inc}}(\mathbf{x}) + u^{\text{sca}}(\mathbf{x}), \end{aligned}\right\} \quad \mathbf{x} \in \mathbb{R}^2. \tag{4}$$

Here, u^{inc} is the given *incident field*, the *total field* u^{tot} is what is recorded on the measurement line $\{\mathbf{x} \in \mathbb{R}^2 : x_2 = r_M\}$, and the difference between the two constitutes the *scattered field* u^{sca}. We describe different sources g in the following subsections. The scattered field u^{sca} is assumed to satisfy the Sommerfeld radiation condition which requires that

$$\lim_{\|\mathbf{x}\| \to \infty} \sqrt{\|\mathbf{x}\|} \left(\frac{\partial u^{\text{sca}}}{\partial \|\mathbf{x}\|} - ik_0 u^{\text{sca}} \right) = 0$$

uniformly for all directions $\mathbf{x}/\|\mathbf{x}\|$. It guarantees that u^{sca} is an outgoing wave. Further details concerning derivation and analytical properties of problems like Equation 4 can be found, for instance, in Colton and Kress (2013).

The models considered below are based on the following specifications of Equation 4. Their numbers agree with the corresponding subsection numbers, where the models will be explained in more detail.

1. Plane wave	No source ($g = 0$) and u^{inc} is an ideal plane wave; see Equation 5.
1.1 Born model	No source, u^{inc} is an ideal plane wave and u^{sca} satisfies the Born approximation.
1.2 Rytov model	No source, u^{inc} is an ideal plane wave, and u^{sca} satisfies the Rytov approximation.

8 Diffraction Tomography, Fourier Reconstruction, and Full Waveform Inversion

In addition, we propose the following two models with sources:

2.1 Point source g represents a point source located far from the object.
2.2 Line source g represents simultaneous point sources positioned along a straight line. We refer to this configuration as a "line source".

Section "Numerical Comparison of Forward Models" contains a numerical comparison of these forward models.

The proposed selection of equations is motivated in part by the availability of methods for their numerical inversion. While the Born and Rytov models can be inverted using nonuniform Fourier methods, the point and line source models are well-suited for FWI.

Incident Plane Wave

Monochromatic plane waves are basic solutions u of the homogeneous Helmholtz equation

$$\left(-\Delta - k_0^2\right) u = 0.$$

They take the form $u(\mathbf{x}) = e^{ik_0 \mathbf{x} \cdot \mathbf{s}}$, where the unit vector $\mathbf{s}$ specifies the direction of propagation of u. Plane waves are widely studied in imaging applications and theory, and we refer to Colton and Kress (2013), Devaney (2012), and Kak and Slaney (2001) for further information.

In the first model, we consider the incident field is a monochromatic plane wave propagating in direction $\mathbf{e}_2$

$$u^{\text{inc}}(\mathbf{x}) = e^{ik_0 x_2}. \tag{5}$$

In this case, we obtain from Equation 4 the following equation for the scattered field

$$\left(-\Delta - k(\mathbf{x})^2\right) u^{\text{sca}}(\mathbf{x}) = f(\mathbf{x}) e^{ik_0 x_2}. \tag{6}$$

The Born Approximation
Equation 6 can be written as

$$\left(-\Delta - k_0^2\right) u^{\text{sca}}(\mathbf{x}) = f(\mathbf{x}) \left(e^{ik_0 x_2} + u^{\text{sca}}(\mathbf{x})\right).$$

If the scattered field u^{sca} is negligible compared to the incident field $e^{ik_0 x_2}$, we can ignore u^{sca} on the right-hand side and obtain

$$(-\Delta - k_0^2) u^{\text{Born}}(\mathbf{x}) = f(\mathbf{x}) e^{ik_0 x_2}, \tag{7}$$

where u^{Born} is the *(first-order) Born approximation* to the scattered field. Supplementing this equation with the Sommerfeld radiation condition, we have a unique solution corresponding to an outgoing wave (Colton and Kress 2013). It can be written as a convolution

$$u^{\text{Born}}(\mathbf{x}) = \int_{\mathbb{R}^2} G(\mathbf{x} - \mathbf{y}) \, f(\mathbf{y}) \, e^{i k_0 y_2} \, d\mathbf{y}, \tag{8}$$

where G is the outgoing Green's function for the Helmholtz equation. In $\mathbb{R}^2$, it is given by

$$G(\mathbf{x}) = \frac{i}{4} H_0^{(1)}(k_0 \|\mathbf{x}\|), \qquad \mathbf{x} \in \mathbb{R}^2 \setminus \{\mathbf{0}\}, \tag{9}$$

where $H_0^{(1)}$ denotes the zeroth-order Hankel function of the first kind; see Colton and Kress (2013, Sect. 3.4). We note that, in spite of a singularity at the origin, G is locally integrable in $\mathbb{R}^2$.

The second-order Born approximation can be obtained by replacing the plane wave $e^{i k_0 y_2}$ in Equation 8 by the sum $e^{i k_0 y_2} + u^{\text{Born}}(\mathbf{y})$. Iterating this procedure yields Born approximations of arbitrary order. For more details, we refer to Kak and Slaney (2001, Sect. 6.2.1) and Devaney (2012).

The Rytov Approximation

In this subsection, we derive an alternative approximation for the scattered field. Introducing the complex phases φ^{inc}, φ^{tot}, and φ^{sca} according to

$$u^{\text{tot}} = e^{\varphi^{\text{tot}}}, \quad u^{\text{inc}} = e^{\varphi^{\text{inc}}}, \quad \varphi^{\text{tot}} = \varphi^{\text{inc}} + \varphi^{\text{sca}}, \tag{10}$$

one can derive from Equation 4, with $g = 0$, the following relation

$$(-\Delta - k_0^2)(u^{\text{inc}} \varphi^{\text{sca}}) = \left(f + (\nabla \varphi^{\text{sca}})^2 \right) u^{\text{inc}}, \tag{11}$$

where $(\nabla \varphi^{\text{sca}})^2 = (\partial \varphi^{\text{sca}} / \partial x_1)^2 + (\partial \varphi^{\text{sca}} / \partial x_2)^2$. The details of this derivation can be found, for instance, in Kak and Slaney (2001, Sect. 6.2.2). Neglecting $(\nabla \varphi^{\text{sca}})^2$ in Equation 11, we obtain

$$(-\Delta - k_0^2)(u^{\text{inc}} \varphi^{\text{Rytov}}) = f u^{\text{inc}}, \tag{12}$$

where φ^{Rytov} is the *Rytov approximation* to φ^{sca}. Note that we still assume u^{inc} to be a monochromatic plane wave, as given in Equation 5. Thus, the product $u^{\text{inc}} \varphi^{\text{Rytov}}$ solves the same equation as u^{Born}. If we define the Rytov approximation to the scattered field, u^{Rytov}, in analogy to Equation 10 via

$$u^{\text{Rytov}} = e^{\varphi^{\text{Rytov}}+\varphi^{\text{inc}}} - u^{\text{inc}},$$

and replace φ^{Rytov} by $\frac{u^{\text{Born}}}{u^{\text{inc}}}$, we obtain a relation between the two approximate scattered fields that can be expressed as

$$u^{\text{Born}} = u^{\text{inc}} \log\left(\frac{u^{\text{Rytov}}}{u^{\text{inc}}} + 1\right). \tag{13}$$

The relation between Born and Rytov in Equation 13 is not unique because of the multiple branches of the complex logarithm. In practical computations, this is addressed by a phase unwrapping as we will see in Equation 30.

Remark 1. There have been many investigations about the validity of the Born and Rytov approximations; see, e.g., Chen and Stamnes (1998), Slaney et al. (1984), or Kak and Slaney (2001, chap. 6). The Born approximation is reasonable only for a relatively (to the wavelength) small object. In particular, for a homogeneous cylinder of radius a, the Born approximation is valid if $a(n - n_0) < \lambda/4$, where λ is the wavelength of the incident wave and n is the constant refractive index inside the object. In contrast, the Rytov approximation only requires that $n - n_0 > (\nabla \varphi^{\text{sca}})^2/k_0^2$, i.e., the phase change of the scattered phase φ^{sca}, see Equation 10, is small over one wavelength, but it has no direct requirements on the object size and is therefore applicable for a larger class of objects. The latter is also observed in numerical simulations in Chen and Stamnes (1998). However, for objects that are small and have a low contrast $n - n_0$, the Born and Rytov approximation produce approximately the same results.

Modeling the Total Field Using Line and Point Sources

As an alternative to ideal incident plane waves, we consider models with one or several point sources. If arranged the right way, the resulting incident field can resemble a monochromatic plane wave. We refer to section "Numerical Comparison of Forward Models" for a numerical comparison of the different models presented here.

Point Source Far From Object
In this case, the right-hand side in Equation 4 is a Dirac delta function so that we obtain

$$\begin{cases} \left(-\Delta - k(\mathbf{x})^2\right) u_P^{\text{tot}}(\mathbf{x}) = \delta(\mathbf{x} - \mathbf{x}_0), \\ \left(-\Delta - k_0^2\right) u_P^{\text{inc}}(\mathbf{x}) = \delta(\mathbf{x} - \mathbf{x}_0). \end{cases} \tag{14}$$

If the position of the source is given by $\mathbf{x}_0 = -r_0 \mathbf{e}_2$ with $r_0 > 0$ sufficiently large, then, after appropriate rescaling, $u_{\mathrm{P}}^{\mathrm{inc}}$ approximates a plane wave with wave number k_0 and propagation direction $\mathbf{e}_2$ in a neighborhood of $\mathbf{0}$.

Line Source

Alternatively, we let g be a sum of Dirac functions and consider

$$\begin{cases} \left(-\Delta - k(\mathbf{x})^2\right) u_{\mathrm{L}}^{\mathrm{tot}}(\mathbf{x}) = \sum_{j=1}^{N_{\mathrm{sim}}} \delta(\mathbf{x} - \mathbf{x}_j), \\ \left(-\Delta - k_0^2\right) u_{\mathrm{L}}^{\mathrm{inc}}(\mathbf{x}) = \sum_{j=1}^{N_{\mathrm{sim}}} \delta(\mathbf{x} - \mathbf{x}_j), \end{cases} \quad (15)$$

where the number N_{sim} of simultaneous point sources should be sufficiently large. Moreover, the positions $\mathbf{x}_j$ should be arranged uniformly along a line perpendicular to the propagation direction $\mathbf{e}_2$ of the plane wave. This is illustrated in section "Modeling the Total Field Using Line and Point Sources".

Numerical Comparison of Forward Models

In this section, we numerically compare the forward models presented above. For the discretization of partial differential equations, several approaches exist, we mention, for instance, the finite differences that approximate the problem on a nodal grid (e.g., Virieux 1984), or methods that use the variational formulation, such as finite elements (Monk 2003) or discontinuous Galerkin methods (Hesthaven and Warburton 2007). In our work, we use the *hybridizable discontinuous Galerkin method* (HDG) for (HDG) the discretization and refer to Cockburn et al. (2009), Kirby et al. (2012), and Faucher and Scherzer (2020) for more details. The implementation precisely follows the steps prescribed in Faucher and Scherzer (2020), and it is carried out in the open-source parallel software hawen; see Faucher (2021) and Footnote 1. While the propagation is assumed on infinite space, the numerical simulations are performed on a finite discretization domain $\Omega \subset \mathbb{R}^2$, with absorbing boundary conditions (Engquist and Majda 1977) implemented to simulate free-space. It corresponds to the following Robin-type condition applied on the boundary Γ of the discretization domain Ω:

$$-\mathrm{i} k(\mathbf{x}) u(\mathbf{x}) + \partial_n u(\mathbf{x}) = 0, \qquad \text{for } x \text{ on } \Gamma. \quad (16)$$

where $\partial_n u$ denotes the normal derivative of u. The test sample used below is a homogeneous medium encompassing a circular object of radius 4.5 around the origin with contrast $f = 1$. This corresponds to the characteristic function

$$\mathbf{1}_a^{\text{disk}}(\mathbf{x}) := \begin{cases} 1, & \mathbf{x} \in \mathcal{B}_a, \\ 0, & \mathbf{x} \in \mathbb{R}^2 \setminus \mathcal{B}_a, \end{cases} \qquad (17)$$

of the disk $\mathcal{B}_a$ with radius $a > 0$. The incident plane wave has wave number $k_0 = 2\pi$.

Modeling the Scattered Field Assuming Incident Plane Waves

We consider the solutions u^{sca} of Equation 6 and u^{Born} of Equation 7, both satisfying boundary condition Equation 16, and simulated following the HDG discretization indicated above. As an alternative for computing the Born approximation u^{Born}, we discretize the convolution Equation 8 with the Green's function G given in Equation 9. In particular, applying an $N \times N$ quadrature on the uniform grid $\mathcal{R}_N = \{-r_M, -r_M + 2r_M/N, \ldots, r_M - 2r_M/N\}^2$ to Equation 8, we obtain

$$u^{\text{Born}}(\mathbf{x}) \approx u_{\text{conv},N}^{\text{Born}}(\mathbf{x}) := \left(\frac{2r_M}{N}\right)^2 \sum_{\mathbf{y} \in \mathcal{R}_N} G(\mathbf{x}-\mathbf{y}) f(\mathbf{y}) e^{ik_0 y_2}, \quad \mathbf{x} \in \mathbb{R}^2. \qquad (18)$$

In Fig. 2, we illustrate the solutions obtained with the different formulations. We observe that the transmission waves contain the most energy, that is, waves that "cross" the object. On the other hand, the solution has very low amplitude elsewhere, including the reflected waves. In Fig. 2d, we see that the Born approximation leads to an incorrect amplitude of the solution, in particular, the imaginary part. In addition, the imaginary part of $u_{\text{conv},200}^{\text{Born}}$ does not match the one of u^{Born}.

Modeling the Total Field Using Line and Point Sources

The objective is to evaluate how considering line and point sources differs from using incident plane waves and if the data obtained with both approaches are comparable. To compare the scattered fields obtained from Equations 14 and 15 with the solution u^{sca} of Equation 6, one needs to rescale according to

$$u_P^{\text{sca}} = \alpha_P \left(u_P^{\text{tot}} - u_P^{\text{inc}}\right),$$

$$u_L^{\text{sca}} = \alpha_L \left(u_L^{\text{tot}} - u_L^{\text{inc}}\right),$$

where α_P and α_L are constants depending only on the number and positions of the point sources $\mathbf{x}_j$. We illustrate in Fig. 3, where the line source is positioned at fixed height $x_2 = -15$ and composed of 441 points between $x_1 = -22$ and $x_1 = 22$. For the case of a point source, we have to consider a very wide domain, namely, $[-500, 500] \times [-500, 500]$, and the point source is positioned in

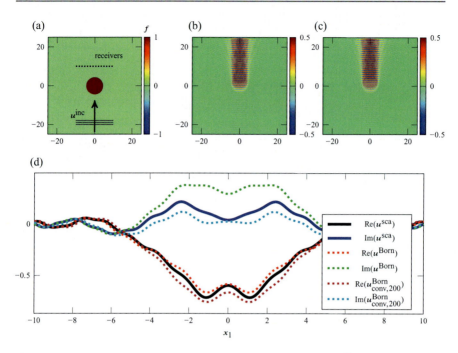

Fig. 2 Comparison of the scattered wave u^{sca} and the Born approximation u^{Born} The computations are performed on a domain $[-25, 25] \times [-25, 25]$ with boundary conditions given in Equation 16. Further, we display $u^{Born}_{conv,200}$ which is the Born approximation obtained by the convolution Equation 18. (**a**) Perturbation model $f = \mathbf{1}^{disk}_{4.5}$. (**b**) Real part of u^{sca}. (**c**) Real part of u^{Born}. (**d**) Comparison of the solutions at fixed height $z = 10$

$(x_1 = 0, x_2 = -480)$. In Fig. 3g, we plot the corresponding solutions on a line at height $x_2 = 10$, i.e., for measurements of transmission waves.

We see that the simulation using the line source is very close to the original solution u^{sca}; in fact, it is a more accurate representation than the Born approximation pictured in Fig. 2. The simulation using a point source positioned far away is also accurate, except for the middle area of the imaginary part. Furthermore, the major drawback of using a single point source is that it necessitates a very big domain, hence largely increasing the computational cost.

Fourier Diffraction Theorem

In this section, we discuss the inverse problem of recovering the scattering potential from measurements of the scattered wave under the Born or Rytov approximations. Before stating the fundamental result in this context, see Theorem 1 below, we have to introduce further notation.

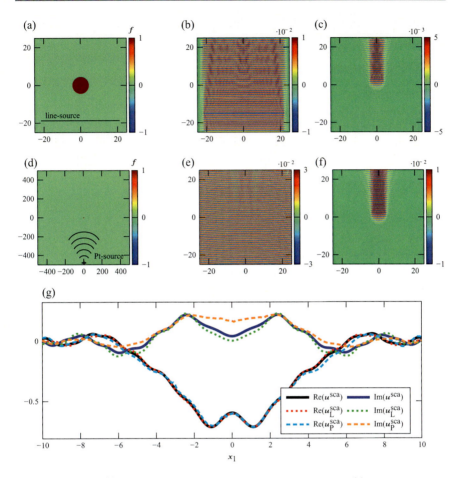

Fig. 3 Total field u_L^{tot} and scattered field u_L^{sca} (line source) and total field u_P^{tot} and scattered field u_P^{sca} (point source). The line source is composed of $N_{\text{sim}} = 441$ points at fixed height $x_2 = -15$. The computational domain for the point source is very large such that the perturbation is barely visible and the source is positioned in $\mathbf{x}_0 = (0, -480)^\top$. (**a**) Computational domain $[-25, 25]^2$ for line source with perturbation model $f = \mathbf{1}_{4.5}^{\text{disk}}$. (**b**) Real part of u_L^{tot}. (**c**) Real part of u_L^{sca}. (**d**) Computational domain $[-500, 500]^2$ for point source with perturbation model $f = \mathbf{1}_{4.5}^{\text{disk}}$. (**e**) Real part of u_P^{tot} near origin. (**f**) Real part of u_P^{sca} near origin. (**g**) Comparison of the solutions at fixed height $x_2 = 10$

We denote by $\mathcal{F}$ the two-dimensional Fourier transform and by $\mathcal{F}_1$ the partial Fourier transform with respect to the first coordinate,

$$\mathcal{F}\phi(\mathbf{k}) = (2\pi)^{-1} \int_{\mathbb{R}^2} \phi(\mathbf{x}) e^{-i\mathbf{x}\cdot\mathbf{k}} \, d\mathbf{x}, \qquad \mathbf{k} \in \mathbb{R}^2,$$

$$\mathcal{F}_1\phi(k_1, x_2) = (2\pi)^{-\frac{1}{2}} \int_{\mathbb{R}} \phi(\mathbf{x}) e^{-ik_1 x_1} \, dx_1, \qquad (k_1, x_2)^\top \in \mathbb{R}^2.$$

For $k_1 \in [-k_0, k_0]$, we define

$$\kappa(k_1) := \sqrt{k_0^2 - k_1^2}.$$

We can now formulate the Fourier diffraction theorem; see, for instance, Kak and Slaney (2001, Sect. 6.3), Natterer and Wübbeling (2001, Thm. 3.1), or Wolf (1969).

Theorem 1. *Let f be bounded with supp $f \subset \mathcal{B}_{r_M}$. For $k_1 \in (-k_0, k_0)$, we have*

$$\mathcal{F}_1 u^{\text{Born}}(k_1, r_M) = \sqrt{2\pi} \frac{i e^{i\kappa r_M}}{2\kappa} \mathcal{F}f(k_1, \kappa - k_0). \tag{19}$$

According to the Fourier diffraction theorem, Theorem 1, the measurements of the scattered wave u^{Born} can be related to the scattering potential f on a semicircle in k-space. Below we discuss how this so-called *k-space coverage* of the experiment is affected by (i) rotating the object and (ii) varying the wave number k_0 of the incident field u^{inc}.

Rotating the Object

Suppose the object rotates around the origin during the experiment. Then the resulting orientation-dependent scattering potential can be written as

$$f^\alpha(\mathbf{x}) = f(R_\alpha \mathbf{x}), \quad \mathbf{x} \in \mathbb{R}^2,$$

where α ranges over a (continuous or discrete) set of angles $A \subset [0, 2\pi]$ and

$$R_\alpha = \begin{pmatrix} \cos\alpha & -\sin\alpha \\ \sin\alpha & \cos\alpha \end{pmatrix}$$

is a rotation matrix. If we let u^α be the Born approximation to the wave scattered by f^α, the collected measurements are given by

$$u^\alpha(x_1, r_M), \quad x_1 \in \mathbb{R}, \quad \alpha \in A.$$

Exploiting the rotation property of the Fourier transform, $\mathcal{F}(f \circ R_\alpha) = (\mathcal{F}f) \circ R_\alpha$, we obtain from Equation 19

$$\mathcal{F}_1 u^\alpha(k_1, r_M) = \sqrt{2\pi} \frac{i e^{i\kappa r_M}}{2\kappa} \mathcal{F}f\left(R_\alpha(k_1, \kappa - k_0)^\top\right).$$

Thus, the k-space coverage, that is, the set of all spatial frequencies $\mathbf{y} \in \mathbb{R}^2$ at which $\mathcal{F}f$ is accessible via the Fourier diffraction theorem, is given by

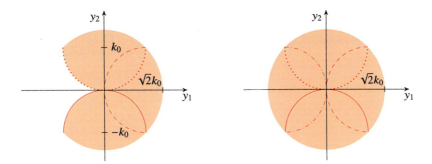

Fig. 4 k-space coverage for a rotating object. Left: half turn, $A = [0, \pi]$. Right: full turn, $A = [0, 2\pi]$. The k-space coverage (light red) is a union of infinitely many semicircles, each corresponding to a different orientation of the object. Some of the semicircles are depicted in red: solid arc ($\alpha = 0$), dashed ($\alpha = \pi/2$), dotted ($\alpha = \pi$), dash-dotted ($\alpha = 3\pi/2$)

$$\mathcal{Y} = \left\{ \mathbf{y} = R_\alpha(k_1, \kappa - k_0)^\top \in \mathbb{R}^2 : |k_1| < k_0,\ \alpha \in A \right\}.$$

It consists of rotated versions (around the origin) of the semicircle $(k_1, \kappa - k_0)^\top$, $|k_1| < k_0$, see Fig. 4.

Varying Wave Number

Now suppose the object is illuminated or insonified by plane waves with wave numbers ranging over a set $K \subset (0, +\infty)$. Recall the definition of the scattering potential $f_{k_0} = k_0^2(n^2/n_0^2 - 1)$ from Equation 2, but note that we have now added a subscript to indicate the dependence of f on k_0. If the variation of the object's refractive index n with $k_0 \in K$ is negligible, we can write

$$f_{k_0}(\mathbf{x}) = k_0^2 f_1(\mathbf{x}), \quad \mathbf{x} \in \mathbb{R}^2. \tag{20}$$

If no confusion arises, we may write $f = f_1$. Denoting by u_{k_0} the Born approximation to the wave scattered by f_{k_0}, the resulting collection of measurements is

$$u_{k_0}(x_1, r_M), \quad x_1 \in \mathbb{R}, \quad k_0 \in K.$$

Then, according to Equation 19, we have

$$\mathcal{F}_1 u_{k_0}(k_1, r_M) = \sqrt{2\pi}\, \frac{\mathrm{i} e^{\mathrm{i}\kappa r_M}}{2\kappa} k_0^2 \mathcal{F} f_1(k_1, \kappa - k_0).$$

Notice that now κ also varies with k_0. The resulting k-space coverage

Fig. 5 k-space coverage for k_0 covering the interval $[k_{\min}, k_{\max}]$

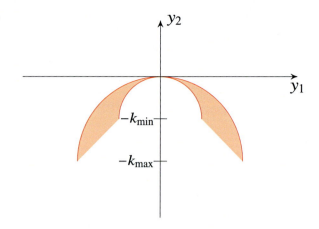

$$\mathcal{Y} = \left\{ \mathbf{y} = (k_1, \kappa - k_0)^\top \in \mathbb{R}^2 : k_0 \in K, \ |k_1| < k_0 \right\}$$

is a union of semicircles scaled and shifted (in direction of $-\mathbf{e}_2$) according to $k_0 \in K$.

Consider, for example, $K = [k_{\min}, k_{\max}]$. Then, the k-space coverage consists of all points $(y_1, y_2)^\top \in \mathbb{R}^2$ such that $|y_1| \leq k_{\max}$ and

$$\sqrt{k_{\max}^2 - y_1^2} - k_{\max} \geq y_2 \geq \begin{cases} -|y_1|, & |y_1| \geq k_{\min}, \\ \sqrt{k_{\min}^2 - y_1^2} - k_{\min}, & \text{otherwise,} \end{cases}$$

see Fig. 5. Note that, in contrast to the scenarios of section "Rotating the Object," there are large missing parts near the origin.

Rotating the Object with Multiple Wave Numbers

We combine the two previous observations by picking a finite set of wave numbers $K \subset (0, \infty)$ and performing a full rotation of the object for each $k_0 \in K$. Let $u_{k_0}^\alpha$ be the Born approximation to the wave scattered by $f_{k_0}^\alpha = k_0^2 f_1 \circ R_\alpha$. Then the full set of measurements is given by

$$u_{k_0}^\alpha(x_1, r_\mathrm{M}), \quad x_1 \in \mathbb{R}, \quad \alpha \in [0, 2\pi], \quad k_0 \in K, \tag{21}$$

and the Fourier diffraction theorem yields

$$\mathcal{F}_1 u_{k_0}^\alpha(k_1, r_\mathrm{M}) = \sqrt{2\pi} \frac{\mathrm{i} \mathrm{e}^{\mathrm{i}\kappa r_\mathrm{M}}}{2\kappa} k_0^2 \mathcal{F} f_1 \left(R_\alpha(k_1, \kappa - k_0)^\top \right). \tag{22}$$

We deduce from sections "Rotating the Object" and "Varying Wave Number" that the resulting k-space coverage $\mathcal{Y}$ is the union of disks with radii $\sqrt{2}k_0$, all centered at the origin. Hence, $\mathcal{Y}$ is just the largest disk, that is, the one corresponding to the largest wave number max K. However, smaller disks in k-space are covered more often, which might improve reliability of the reconstruction for noisy data.

Reconstruction Methods

In the following, we assume data generated by line sources according to the setup described in section "Modeling the Total Field Using Line and Point Sources". We simulate the total fields solutions to Equation 15, which are the synthetic data used for the reconstruction.

Reconstruction Using Full Waveform Inversion

For the identification of the physical properties of the medium, the *Full Waveform Inversion* (FWI) relies on an iterative minimization of a misfit functional which evaluates a distance between numerical simulation and measurements of the total field. The Full Waveform Inversion method arises in the context of seismic inversion for sub-surface Earth imaging, cf. Bamberger et al. (1979), Lailly (1983), Tarantola (1984), Pratt et al. (1998), and Virieux and Operto (2009), where the measured seismograms are compared to simulated waves.

With FWI, we invert with respect to the wave speed c, from which the wave number is defined according to Equation 1. It further connects with the model perturbation f according to Equation 3. In our experiment, c_0 is used as an initial guess (i.e., we start from constant background), and then c is inverted rather than f, as discussed in Remark 2. Given some measurements $\mathbf{d}$ of the total field, the quantitative reconstruction of the wave speed c is performed following the minimization of the misfit functional $\mathcal{J}$ such that

$$\min_{c} \mathcal{J}(c), \qquad \mathcal{J} = \mathrm{dist}\left(\mathcal{R}u^{\mathrm{tot}}, \mathbf{d}\right), \qquad \text{where } u \text{ solves Equation 15}. \qquad (23)$$

Here, dist($\cdot$) is a distance function to evaluate the difference between the measurements and the simulations, and $\mathcal{R}$ is a linear operator to restrict the solution to the positions of the receivers. For simplicity, we do not encode a regularization term in Equation 23 and refer the readers to, e.g., Faucher et al. (2020c), Kaltenbacher (2018), and the references therein.

Several formulations of the distance function have been studied for FWI (in particular, for seismic applications), such as a logarithmic criterion, Shin et al. (2007), the use of the signal's phase or amplitude, Bednar et al. (2007) and Pyun et al. (2007), the use of the envelope of the signal, Fichtner et al. (2008), criteria based upon cross-correlation, Luo and Schuster (1991), Van Leeuwen and Mulder (2010), Faucher et al. (2020a), and Faucher et al. (2021), or optimal transport distance, Métivier et al. (2016). Here, we rely on a least-squares approach where the misfit functional is defined as the L^2 distance between the data and simulations:

$$\mathcal{J}(c) := \frac{1}{2} \sum_{\omega \in c_0 K} \sum_{\alpha \in A} \| \mathcal{R} u^{\text{tot}}(c, \omega, \alpha) - \mathbf{d}(\omega, \alpha) \|^2_{L^2(-l_M, l_M)}, \qquad (24)$$

where $\mathbf{d}(\omega, \alpha)$ refers to the measurement data of the total field at the measurement plane with respect to the object rotated with angle α, and $u^{\text{tot}}(c, \omega, \alpha)$ is the solution of Equation 15 with $k(\mathbf{x}) = \omega/c(R_\alpha^\top \mathbf{x})$. The last term $R_\alpha^\top \mathbf{x}$ encodes the rotation of the object. We note that a rotation of the object is equivalent of the rotation of both the measurement line and the direction of the incident field. We have encoded a sum over the frequencies ω, which are chosen in accordance with the frequency content available in measurements. In the computational experiments, we further investigate uni- and multifrequency reconstructions.

The minimization of the misfit functional Equation 20 follows an iterative Newton-type method as depicted in Algorithm 1. Due to the computational cost, we use first-order information and avoid the Hessian computation (Virieux and Operto 2009): namely, we rely on the nonlinear conjugate gradient method for the model update, cf. Nocedal and Wright (2006) and Faucher (2017). Furthermore, to avoid the formation of the dense Jacobian, the gradient of the misfit functional

---

Algorithm 1 Iterative reconstruction of the wave speed model following the minimization of the misfit functional. At each iteration, the total field solution to Equation 15 is computed, and the gradient of the misfit functional is used to update the wave speed model. The algorithm stops when the prescribed number of iterations is performed for all of the frequencies of interest.

Input: Initial wave speed model c_0.
Initiate global iteration number $\ell := 1$;
for frequency $\omega \in c_0 K$ **do**
    for iteration $j = 1, \ldots, n_{iter}$ **do**
        Compute the solution to the wave equation using current wave speed model c_ℓ and frequency ω, that is, the solution to Equation 15 with $k(\mathbf{x}) = \omega/c_\ell(\mathbf{x})$;
        Evaluate the misfit functional $\mathcal{J}$ in Equation 24;
        Compute the gradient of the misfit functional using the adjoint-state method;
        Update the wave speed model using nonlinear conjugate gradient method to obtain $c_{\ell+1}$;
        Update global iteration number $\ell \leftarrow \ell + 1$;
    end
end
Output: Approximate wave speed c, from which the scattering potential f can be computed via Equations 2 and 1.

is computed using the adjoint-state method, cf. Pratt et al. (1998), Plessix (2006), Barucq et al. (2019), and Faucher and Scherzer (2020). In Algorithm 1, we further implement a progression in the frequency content, which is common to mitigate the ill-posedness of the nonlinear inverse problem, Bunks et al. (1995). We further invert each frequency independently, from low to high, as advocated by Barucq et al. (2019) and Faucher et al. (2020b). For the implementation details using the HDG discretization, we refer to Faucher and Scherzer (2020).

Remark 2. In the computational experiments, the reconstruction with FWI assumes the availability of the total fields which are solutions to Equation 15, and we invert with respect to the (frequency independent) wave speed c defined in Equation 3. We could instead use the representation with relation $k^2 = k_0^2 + f$ and invert with respect to the perturbation f, imposing the (known) smooth background c_0. Inverting with respect to c rather than f is mainly motivated by consistency with existing literature in FWI (Virieux and Operto 2009), in which the background model (c_0) is usually unknown. Nonetheless, reformulating the minimization with respect to f and imposing c_0 could improve the efficiency of FWI, as advocated by the data-space reflectivity inversion of Clément et al. (2001) and Faucher et al. (2020b).

Reconstruction Based on the Born and Rytov Approximations

In this section, we present numerical methods for the computation of the Born and Rytov approximations from Equations 7 and 12, respectively, as well as the reconstruction of the scattering potential. We concentrate on the case of full rotations of the object using incident waves with different wave numbers $k_0 \in K$; see section "Rotating the Object with Multiple Wave Numbers". The tomographic reconstruction is based on the Fourier diffraction theorem, Theorem 1, and the nonuniform discrete Fourier transform. Nonuniform Fourier methods have also been applied in computerized tomography (Potts and Steidl 2001), magnetic resonance imaging (Knopp et al. 2007), spherical tomography (Hielscher and Quellmalz 2015, 2016), or surface wave tomography (Hielscher et al. 2018).

In the following, we describe the discretization steps we apply. For $N \in 2\mathbb{N}$, let

$$\mathcal{I}_N := \left\{ -\frac{N}{2} + j : j = 0, \ldots, N-1 \right\}.$$

We sample the scattering potential f on the uniform grid $\mathcal{R}_N := \frac{2r_s}{N} \mathcal{I}_N^2$ in the square $[-r_s, r_s]^2$ for some $r_s > 0$. We assume that we are given measurements of the Born approximation

$$u_{k_0}^\alpha(x_1, r_\mathrm{M}), \qquad x_1 \in [-l_\mathrm{M}, l_\mathrm{M}],$$

for $\alpha \in A \subset [0, 2\pi]$ and $k_0 \in K$, cf. Equation 21. We want to reconstruct the scattering potential $f = f_1$; recall Equation 20, utilizing Equation 22. We adapt the reconstruction approach of Kirisits et al. (2021), which is written for the 3D case. First, we need to approximate the partial Fourier transform

$$\mathcal{F}_1 u_{k_0}^\alpha(k_1, r_\mathrm{M}) = \frac{1}{\sqrt{2\pi}} \int_\mathbb{R} u_{k_0}^\alpha(x_1, r_\mathrm{M}) \mathrm{e}^{-\mathrm{i} x_1 k_1} \, \mathrm{d}x_1, \qquad k_1 \in [-k_0, k_0]. \qquad (25)$$

The *discrete Fourier transform* (DFT) of $u(\cdot, r_\mathrm{M})$ on m equispaced points $x_1 \in (2l_\mathrm{M}/m)\mathcal{I}_m$ can be defined by

$$\mathbf{F}_{1,m} u(k_1, r_\mathrm{M}) := \frac{1}{\sqrt{2\pi}} \frac{2l_\mathrm{M}}{m} \sum_{x_1 \in \frac{2l_\mathrm{M}}{m} \mathcal{I}_m} u(x_1, r_\mathrm{M}) \mathrm{e}^{-\mathrm{i} x_1 k_1}, \qquad k_1 \in \frac{\pi}{l_\mathrm{M}} \mathcal{I}_m, \qquad (26)$$

which gives an approximation of Equation 25. Then, Equation 22 yields

$$k_0^2 \mathcal{F} f(R_\alpha(k_1, \kappa - k_0)^\top) = -\mathrm{i}\sqrt{\frac{2}{\pi}} \kappa \mathrm{e}^{-\mathrm{i}\kappa r_\mathrm{M}} \mathcal{F}_1 u_{k_0}^\alpha(k_1, r_\mathrm{M}) \qquad (27)$$

for $|k_1| \leq k_0$. Considering that we sample the angle α on the equispaced, discrete grid $A = (2\pi/n_A)\{0, 1, \ldots, n_A - 1\}$ and some finite set $K \subset (0, \infty)$, Equation 26 provides an approximation of $\mathcal{F} f$ on the non-uniform grid

$$\mathcal{Y}_{m,n_A} := \Big\{ R_\alpha(k_1, \kappa - k_0)^\top : \\ k_1 \in \frac{\pi}{l_\mathrm{M}} \mathcal{I}_m, \, |k_1| \leq k_0, \, \alpha \in \frac{2\pi}{n_A}\{0, 1, \ldots, n_A - 1\}, \, k_0 \in K \Big\}$$

in k-space, from which we want to reconstruct the scattering potential f.

Let M be the cardinality of $\mathcal{Y}_{m,n_A}$. The two-dimensional *nonuniform discrete Fourier transform* (NDFT) is the linear operator $\mathbf{F}_N : \mathbb{R}^{N^2} \to \mathbb{R}^M$ defined for the vector $\mathbf{f}_N := (f(\mathbf{x}))_{\mathbf{x} \in \mathcal{R}_N}$ elementwise by

$$\mathbf{F}_N \mathbf{f}_N(\mathbf{y}) := \frac{1}{2\pi} \frac{(2r_\mathrm{s})^2}{N^2} \sum_{\mathbf{x} \in \mathcal{R}_N} f(\mathbf{x}) \mathrm{e}^{-\mathrm{i}\mathbf{x} \cdot \mathbf{y}}, \qquad \mathbf{y} \in \mathcal{Y}_{m,n_A}, \qquad (28)$$

see Plonka et al. (2018, Section 7.1). It provides an approximation of the Fourier transform

$$\mathcal{F} f(\mathbf{y}) \approx \mathbf{F}_N \mathbf{f}_N(\mathbf{y}), \qquad \mathbf{y} \in \mathcal{Y}_{m,n_A}. \qquad (29)$$

solving an equation $\mathbf{F}_N \mathbf{f}_N(\mathbf{y}) = b$ for $\mathbf{f}_N$ amounts to applying an *inverse NDFT*, which usually utilizes an iterative method such as the conjugate gradient method on the normal equations (CGNE); see Kunis and Potts (2007) and Plonka et al. (2018, Section 7.6). One should be aware that the notation regarding conjugate gradient algorithms varies in the literature: the algorithm called CGNE in Hanke (1995) is known as CGNR in Kunis and Potts (2007). Conversely, the algorithm CGME in Hanke (1995) is known as CGNE in Kunis and Potts (2007).

In conclusion, our method for computing f given the Born approximation u^{Born} is summarized in Algorithm 2.

Algorithm 2 Iterative reconstruction of the scattering potential f based on the Born approximation using an inverse NDFT

Input: Measurement data

$$u_{k_0}^\alpha(x_1, r_{\text{M}}), \quad x_1 \in \frac{2l_{\text{M}}}{m} \mathcal{I}_m, \; \alpha \in A = \frac{2\pi}{n_A}\{0, \ldots, n_A - 1\}, \; k_0 \in K.$$

for $k_0 \in K$ **do**
    for $\alpha \in A$ **do**
        Compute $-i\sqrt{\frac{2}{\pi}}\kappa e^{-i\kappa r_{\text{M}}} \mathbf{F}_{1,m} u_{k_0}^\alpha(k_1, r_{\text{M}})$, $k_1 \in \frac{\pi}{l_{\text{M}}} \mathcal{I}_m$, with a DFT in Equation 26;
    end
end
Solve Equation 27 with Equation 29 for $\mathbf{f}_N$ using the conjugate gradient method;
Output: Approximate scattering potential $\mathbf{f}_N \approx (f(\mathbf{x}))_{\mathbf{x} \in \mathcal{R}_N}$.

The Rytov approximation u^{Rytov}, see Equation 12, is closely related to the Born approximation, but it has a different physical interpretation. Assuming that the measurements arise from the Rytov approximation, we apply Equation 13 to obtain u^{Born} from which we can proceed to recover f as shown above. We note that the actual implementation of Equation 13 requires a phase unwrapping because the complex logarithm is unique only up to adding $2\pi i$, cf. Müller et al. (2015). In particular, we use in the two-dimensional case

$$u^{\text{Born}} = u^{\text{inc}} \left(i \, \text{unwrap} \left(\arg \left(\frac{u^{\text{Rytov}}}{u^{\text{inc}}} + 1 \right) \right) + \ln \left| \frac{u^{\text{Rytov}}}{u^{\text{inc}}} + 1 \right| \right), \tag{30}$$

where arg denotes the principle argument of a complex number and unwrap denotes a standard unwrapping algorithm. For the reconstruction with the Rytov approximation, we can use Algorithm 2 as well, but we have to preprocess the data u by Equation 30.

Numerical Experiments

In this section, we carry out numerical experiments comparing the reconstruction obtained with FWI (section "Reconstruction Using Full Waveform Inversion") and Born and Rytov approximations (section "Reconstruction Based on the Born and Rytov Approximations"), using single and multi-frequency datasets. We consider different media with varying shapes and amplitude for the embedded objects. Our experiments use synthetic data with added noise: Firstly, synthetic simulations are carried out for the known wave speeds using the software hawen (Faucher 2021). The discretization relies on a fine mesh (usually a few hundred thousands cells in the discretized domain) and polynomials of order 5 to ensure accuracy. Then, white Gaussian noise is incorporated in the synthetic data, with a signal-to-noise ratio of 50 dB. The reconstruction with FWI also relies on software hawen, but uses different discretization setups to foster the computational time: the discretization mesh is coarser (usually a few tens of thousand cells) and the polynomial order varies with the cells, depending on the (local to the cell) wavelength, in order to remain as small as possible, as detailed in Faucher and Scherzer (2020). The computational cost of FWI is further discussed in section "Computational Costs".

We perform a full rotation of the object for a single or for multiple frequencies ω and thus wave numbers k_0, cf. section "Rotating the Object with Multiple Wave Numbers". For different frequencies, the scattering potential is scaled according to Equation 20. We always reconstruct the rescaled scattering potential f_1, which we will simply denote by f in the following. In all numerical experiments, we rely on forward data generated with the forward model of line sources in section "Line Source".

We compare the reconstruction quality based on the *peak signal-to-noise ratio* (PSNR) of the reconstruction $\mathbf{g}$ with respect to the ground truth $\mathbf{f}$ determined by

$$\text{PSNR}(\mathbf{f}, \mathbf{g}) := 10 \log_{10} \frac{\max_{\mathbf{x} \in \mathcal{R}_N} |\mathbf{f}(\mathbf{x})|^2}{N^{-2} \sum_{\mathbf{x} \in \mathcal{R}_N} |\mathbf{f}(\mathbf{x}) - \mathbf{g}(\mathbf{x})|^2},$$

where higher values indicate a better reconstruction quality.

Reconstruction of Circular Contrast with Various Amplitudes and Sizes

For the initial reconstruction experiments, we consider a circular object in a homogeneous background, namely, the scattering potential f of Equation 17. We investigate different sizes and contrasts for the object, as shown in Fig. 6. The data are generated for $n_A = 40$ angles of incidence equally partitioned between 0° to 351°, every 9°, and the measurement line is sampled on the 200 point uniform grid $10^{-1} \mathcal{I}_{200} \subset [-l_M, l_M]$ with $l_M = 10$. Let us note that in this context of a circularly

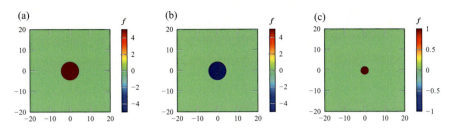

Fig. 6 Different perturbation models f used for the computational experiments, given for frequency $\omega/(2\pi) = 1$, with the relation to the wave speed given in Equation 3. Both the size and contrast vary: we consider two radii (4.5 and 2) and three contrasts (1, 5, and -5 with corresponding wave speeds $c = 0.9876$, $c = 0.9421$, and $c = 1.0701$, respectively), for a total of six configurations. The computations are carried out on the domain $[-50, 50] \times [-50, 50]$, i.e., a slightly larger setup than Fig. 3, and we only picture the area near the origin for clearer visualization. (**a**) Perturbation f for radius 4.5 and amplitude 5: model $f = 5 \cdot \mathbf{1}_{4.5}^{\text{disk}}$. (**b**) Perturbation f for radius 4.5 and amplitude -5: model $f = -5 \cdot \mathbf{1}_{4.5}^{\text{disk}}$. (**c**) Perturbation f for radius 2 and amplitude 1: model $f = \mathbf{1}_2^{\text{disk}}$

symmetric object, the data of each angle are similar and correspond to that of Fig. 3 for $f = \mathbf{1}_{4.5}^{\text{disk}}$.

Reconstruction Using FWI with Single-Frequency Datasets

We first only use data at frequency $\omega/(2\pi) = 1$, that is, wave number $k_0 = 2\pi$ for the reconstruction of the different perturbations illustrated in Fig. 6. With the background $k_0 = 2\pi$ (i.e., wave speed of 1), it means that we only rely on waves with wavelength 1 when propagating in the (homogeneous) background. Then all measurements in **x** are in multiples of the wavelength. In the case of a single frequency, only the inner loop remains in Algorithm 1, and we perform 50 iterations. In Fig. 7, we picture the reconstruction obtained for the six different perturbations f. We observe that the reconstructions of the smaller object of radius 2 (Fig. 7a, b and c) are more accurate, both in terms of the circular shape and in terms of amplitude. In the case of the larger object, the mild amplitude (Fig. 7d) is accurately recovered, while the stronger contrasts (Fig. 7e and f) are only partially retrieved. Here, the outer part of the disk appears, but the amplitude is incorrect with a ring effect and incorrect values in the inner area. Therefore, the reconstruction using single-frequency data is limited and its success depends on two factors: the size of the object and its contrast.

Reconstruction Using FWI with Multiple Frequency Datasets

The difficulty of recovering a large object with a strong contrast can be mitigated by the use of multi-frequency datasets, allowing a multiscale reconstruction (Bunks et al. 1995; Faucher et al. 2020b). We carry out the iterative reconstruction using increasing frequencies, starting with $\omega/(2\pi) = 0.2$ and up to $\omega/(2\pi) = 1$. Following Faucher et al. (2020b), we use a sequential progression, that is, every

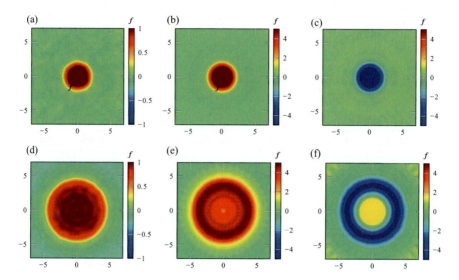

Fig. 7 Reconstruction using iterative minimization using data of frequency $\omega/(2\pi) = 1$ only. In each cases, 50 iterations are performed and the initial model consists in a constant background where $k_0 = 2\pi$. The data consist of $n_A = 40$ different angles of incidence from 0° to 351° (**a**) Reconstruction for model $f = \mathbf{1}_2^{\text{disk}}$ (PSNR 23.50). (**b**) Reconstruction for model $f = 5 \cdot \mathbf{1}_2^{\text{disk}}$ (PSNR 24.43). (**c**) Reconstruction for model $f = -5 \cdot \mathbf{1}_2^{\text{disk}}$ (PSNR 24.25). (**d**) Reconstruction for model $f = \mathbf{1}_{4.5}^{\text{disk}}$ (PSNR 14.79). (**e**) Reconstruction for model $f = 5 \cdot \mathbf{1}_{4.5}^{\text{disk}}$ (PSNR 15.71). (**f**) Reconstruction for model $f = -5 \cdot \mathbf{1}_{4.5}^{\text{disk}}$ (PSNR 10)

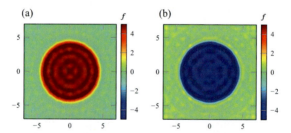

Fig. 8 Reconstruction using multi-frequency data from $\omega/(2\pi) = 0.2$ to $\omega/(2\pi) = 1$. The initial model consists in a constant wave speed $c_0 = 1$. The data consist of $n_A = 40$ different angles of incidence from 0° to 351° (**a**) Reconstruction for model $f = 5 \cdot \mathbf{1}_{4.5}^{\text{disk}}$ (PSNR 19.43). (**b**) Reconstruction for model $f = -5 \cdot \mathbf{1}_{4.5}^{\text{disk}}$ (PSNR 19.02)

frequency is inverted separately. The reconstructions for the object of radius 4.5 and contrast $f = \pm 5$ are pictured in Fig. 8. Contrary to the case of a single frequency (see Fig. 7d), the reconstruction is now accurate and stable: the amplitude is accurately retrieved and the circular shape is clear, avoiding the circular artifacts observed in Fig. 7e and f.

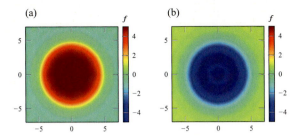

Fig. 9 Reconstruction using frequency $\omega/(2\pi) = 0.7$. The initial model consists in a constant wave speed $c_0 = 1$. The data consist of $n_A = 40$ different angles of incidence from $0°$ to $351°$. (**a**) Reconstruction for model $f = 5 \cdot \mathbf{1}_{4.5}^{\text{disk}}$ (PSNR 15.33). (**b**) Reconstruction for model $f = -5 \cdot \mathbf{1}_{4.5}^{\text{disk}}$ (PSNR 15.03)

Remark 3. It is possible to recover the model with a single frequency, which needs to be carefully chosen depending on the size of the object and the amplitude of the contrast. We have seen in Fig. 7 that the frequency $\omega/(2\pi) = 1$ is sufficient for the object of radius 2, but for the radius 4.5, we need a lower frequency (i.e., larger wavelength) to uncover the larger object. We illustrate in Fig. 9 the reconstruction using data at only $\omega/(2\pi) = 0.7$, where we see that the shape and contrast are retrieved accurately. Nonetheless, it is hard to predict this frequency a priori, and we believe it remains more natural to use multiple frequencies (when available in the data), to ensure the robustness of the algorithm.

Reconstruction Using Born and Rytov Approximations

For the reconstruction with Algorithm 2, which relies on the Born or Rytov approximation, we use the same data as in the above experiment. We use a grid with $N = 240$ and $r_s = N/(8\sqrt{2}) \approx 10$. The numerical results indicate that r_s should not be smaller than l_M. Since we have $k_1^2 \leq k_0^2$ and the distance between two grid points of k_1 is π/l_M, only around $2k_0 l_M/\pi \approx 40$ of them contribute to the data of the inverse NDFT.

In the following reconstructions, we use a fixed number of 20 iteration steps in the conjugate gradient method. Initially, we use the frequency $\omega/(2\pi) = 1$ of the incident wave; therefore, $k_0 = 2\pi$. Reconstructions of the circular model $f = \mathbf{1}_a^{\text{disk}}$ are shown in Fig. 10. We note that all reconstructions are reasonably good, where the Rytov reconstruction looks slightly better inside the object.

For a higher amplitude of the model function f, the limitations of the linear models become apparent. In Fig. 11, we see that the Born reconstruction of the larger object fails, and for the smaller object, only the Rytov approximation yields a good reconstruction, which is consistent with Remark 1. With the Born approximation, we recognize the object's shape but not its amplitude, which is consistent with the observations in Müller et al. (2016). However, as we see in Fig. 7, even the FWI reconstruction makes a considerable error in the object's interior, and we cannot expect the linear models to be better.

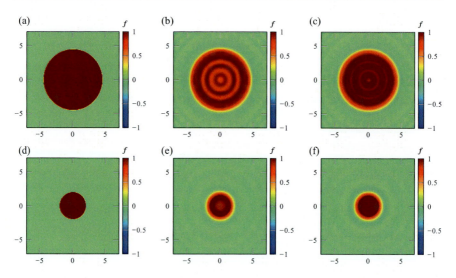

Fig. 10 Reconstructions with the Born and Rytov approximation, where the data $u(\cdot, r_M)$ is generated with the line source model. The incident field has the frequency $\omega/(2\pi) = 1$. Visible is only the cut out center, where we compute the PSNR. (**a**) Model $f = \mathbf{1}_{4.5}^{\text{disk}}$. (**b**) Born reconstruction (PSNR 19.35). (**c**) Rytov reconstruction (PSNR 19.28). (**d**) Model $f = \mathbf{1}_{2}^{\text{disk}}$. (**e**) Born reconstruction (PSNR 24.13). (**f**) Rytov reconstruction (PSNR 24.01)

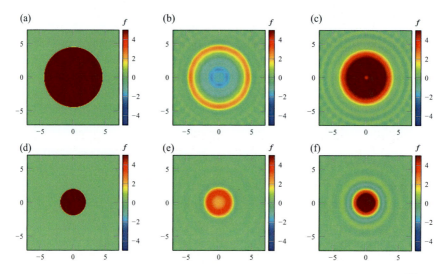

Fig. 11 Same setting as in Fig. 10, but with a higher amplitude of 5 (**a**) Model $f = 5 \cdot \mathbf{1}_{4.5}^{\text{disk}}$. (**b**) Born reconstruction (PSNR 4.68). (**c**) Rytov reconstruction (PSNR 11.91). (**d**) Model $f = 5 \cdot \mathbf{1}_{2}^{\text{disk}}$. (**e**) Born reconstruction (PSNR 19.39). (**f**) Rytov reconstruction (PSNR 21.31)

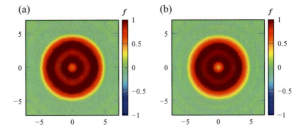

Fig. 12 Reconstructions of $1_{4.5}^{\text{disk}}$, where the incident field has the frequency $\omega/(2\pi) = 0.7$ instead of 1. The rest of the setting is from Fig. 10. (**a**) Born reconstruction (PSNR 18.05). (**b**) Rytov reconstruction (PSNR 16.52)

We see that the FWI and the Born/Rytov reconstructions contain different kinds of artifacts. Therefore, a comparison of the visual image quality perception does not necessarily yield the same conclusions as for the computed PSNR values. Furthermore, the size of the object has a considerable effect on the PSNR, e.g., the images in Fig. 11c and f show a comparable visual quality, but the latter's PSNR is considerably better because of the lower error in the background farther away from the object; see also Huynh-Thu and Ghanbari (2010) for a study on the PSNR.

In Fig. 12, we use the same setup as before, but with the frequency $\omega/(2\pi) = 0.7$ instead of $\omega/(2\pi) = 1$ and thus the wave number $k_0 = \omega$. Apparently, the reconstruction becomes worse with lower frequency, because it provides a smaller k-space coverage.

Reconstruction of Embedded Shapes: Phantom 1

We consider a more challenging scenario with shapes embedded in the background medium. In Fig. 13a, we picture the perturbation f consisting of a disk and heart included in an ellipse, with f varying from 0 to 0.5. The computational domain corresponds to $[-20, 20] \times [-20, 20]$, with line- sources positioned at a distance $R = 10$ and receivers in $r_M = 6$ to capture the data. The data are generated using $n_A = 100$ incidence angles α equispaced on $[0, 2\pi]$, following the steps described in section "Modeling the Total Field Using Line and Point Sources". This is illustrated in Fig. 13.

Reconstruction Using FWI

We carry out the reconstruction following Algorithm 1, and the results are pictured in Fig. 14, where we compare the use of single and multi-frequency data. In this example, we see that with relatively low-frequency data (i.e., relatively large wavelength), such as for frequency $\omega/(2\pi) = 0.7$ and $\omega/(2\pi) = 1$, the reconstruction is smooth; see Fig. 14a and b, and one needs to use smaller wavelengths to obtain a better reconstruction; see Fig. 14c and d. In Fig. 14e, we see that multi-frequency data gives the best reconstruction, it is also the most robust as one does not need to anticipate the appropriate wavelength before carrying out the reconstruction. Here both the shapes and contrast in amplitude are accurately obtained. We notice some

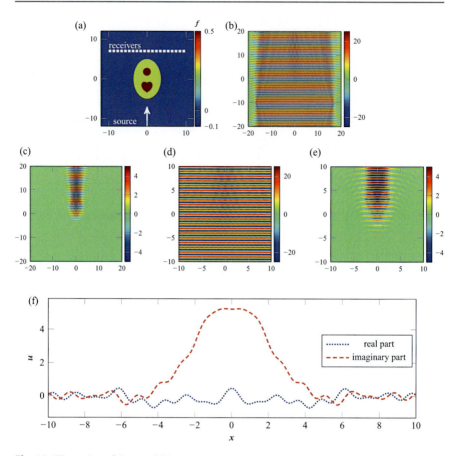

Fig. 13 Illustration of the acquisition setup and generated data. The computations are carried out on the domain $[-20, 20] \times [-20, 20]$. While FWI uses the total field, the reconstruction based upon Born and Rytov approximations use the scattered solutions, obtained after removing a reference solution corresponding to a propagation in an homogeneous medium, cf. section "Forward Models" (**a**) Perturbation model at frequency $\omega/(2\pi) = 1$, the wave speed is equal to 1 in the background. The positions of the source and the receivers recording transmission data are pictured in white. (**b**) Real part of the global solution to Equation 15 at frequency $\omega/(2\pi) = 1$, the source is discretized by $N_{\text{sim}} = 1361$ simultaneous excitations at fixed height $x_2 = -10$. (**c**) Real part of the scattering solution at frequency $\omega/(2\pi) = 1$. (**d**) Zoom near origin of figure panel (**b**). (**e**) Zoom near origin of figure panel (**c**). (**f**) Scattered solution measured at the 201 receivers positioned at fixed height $x_2 = 6$

oscillatory noise in the reconstructed models, which could certainly be reduced by incorporating a regularization criterion in the minimization (Faucher et al. 2020c).

In Fig. 15, we conduct a similar computational experiment, but increasing the contrast in the included heart shape where f has now a value of 2; see Fig. 15. We provide single and multi-frequency reconstructions and observe that large wavelengths still provide a smooth reconstruction. The high contrast in the heart is well recovered.

8 Diffraction Tomography, Fourier Reconstruction, and Full Waveform Inversion 301

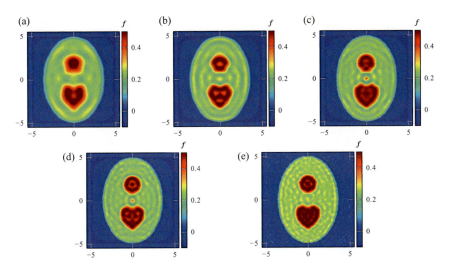

Fig. 14 Reconstruction of the model Fig. 13a with FWI and starting from a homogeneous background with $f = 0$. The models are given at frequency $\omega/(2\pi) = 1$ and the wave speed is equal to 1 in the background. (**a**) Using single-frequency, $\omega/(2\pi) = 0.7$ (PSNR 22.38). (**b**) Using single-frequency, $\omega/(2\pi) = 1$. (PSNR 22.91). (**c**) Using single-frequency, $\omega/(2\pi) = 1.2$. (PSNR 23.10). (**d**) Using single-frequency, $\omega/(2\pi) = 1.4$. (PSNR 23.31). (**e**) Using multi-frequency, $\omega/(2\pi) \in \{0.7, 1, 1.2, 1.4\}$. (PSNR 27.28)

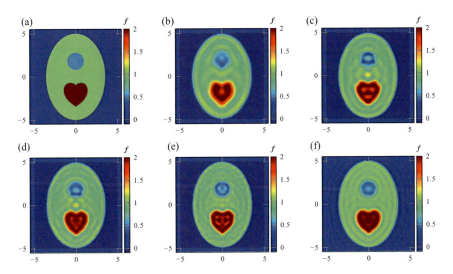

Fig. 15 Reconstruction with FWI starting from a homogeneous background with $f = 0$. The models are given at frequency $\omega/(2\pi) = 1$ and the wave speed is equal to 1 in the background (**a**) True model. (**b**) Using single-frequency, $\omega/(2\pi) = 0.7$ (PSNR 25.04). (**c**) Using single-frequency, $\omega/(2\pi) = 1$ (PSNR 25.91). (**d**) Using single-frequency, $\omega/(2\pi) = 1.2$ (PSNR 26.19). (**e**) Using single-frequency, $\omega/(2\pi) = 1.4$ (PSNR 26.75). (**f**) Using multi-frequency, $\omega/(2\pi) \in \{0.7, 1, 1.2, 1.4\}$ (PSNR 28.76)

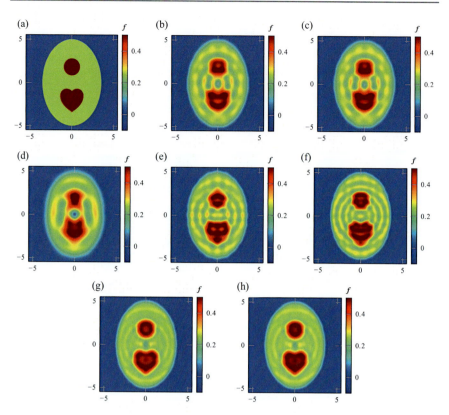

Fig. 16 Reconstructions for different frequencies of the incident wave. The PSNR is computed on the visible part of the grid for the real part of the reconstruction, since we know that f must be real (**a**) True model f. (**b**) Born reconstruction at frequency $\omega/(2\pi) = 1$. (PSNR 24.69). (**c**) Rytov reconstruction at frequency $\omega/(2\pi) = 1$. (PSNR 24.66). (**d**) Rytov reconstruction at frequency $\omega/(2\pi) = 0.7$. (PSNR 22.72). (**e**) Rytov reconstruction at frequency $\omega/(2\pi) = 1.2$. (PSNR 25.32). (**f**) Rytov reconstruction at frequency $\omega/(2\pi) = 1.4$. (PSNR 26.14). (**g**) Born reconstruction using multi-frequency, $\omega/(2\pi) \in \{0.7, 1, 1.2, 1.4\}$. (PSNR 26.37). (**h**) Rytov reconstruction using multi-frequency, $\omega/(2\pi) \in \{0.7, 1, 1.2, 1.4\}$. (PSNR 26.37)

Reconstruction Using Born and Rytov Approximations

We perform the reconstruction with Algorithm 2 of the test model f from Fig. 13. In the following tests, we discretize f on a finer grid of resolution $N = 720$, which covers the radius $r_s = 15/\sqrt{2}$. The PSNR is computed only on the central part of the grid that is visible in the image. Since we know that the f is real-valued, we take only the real part of the reconstruction. For simplicity, we use a constant number of 12 iterations in the conjugate gradient method of the inverse NDFT.

The Born and Rytov reconstructions are shown in Fig. 16, where the data u is the same as in section "Reconstruction Using FWI". The reconstruction with a higher frequency of the incident wave is more accurate, since it provides a larger k-space

8 Diffraction Tomography, Fourier Reconstruction, and Full Waveform Inversion

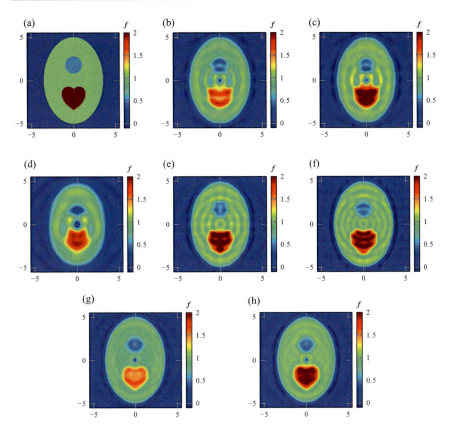

Fig. 17 Reconstructions with a higher contrast, where the rest of the setting is the same as in Fig. 16 (**a**) True model f. (**b**) Born reconstruction at frequency $\omega/(2\pi) = 1$. (PSNR 23.53). (**c**) Rytov reconstruction at frequency $\omega/(2\pi) = 1$. (PSNR 24.47). (**d**) Rytov reconstruction at frequency $\omega/(2\pi) = 0.7$. (PSNR 21.78). (**e**) Rytov reconstruction at frequency $\omega/(2\pi) = 1.2$. (PSNR 25.55). (**f**) Rytov reconstruction at frequency $\omega/(2\pi) = 1.4$. (PSNR 26.40). (**g**) Born reconstruction using multi-frequency, $\omega/(2\pi) \in \{0.7, 1, 1.2, 1.4\}$. (PSNR 23.77). (**h**) Rytov reconstruction using multi-frequency, $\omega/(2\pi) \in \{0.7, 1, 1.2, 1.4\}$. (PSNR 25.92)

coverage, which is the disk of radius $\sqrt{2}k_0 = \sqrt{2}\omega$, see section "Fourier Diffraction Theorem". Moreover, the multi-frequency reconstruction is shown in Fig. 16g and h. Even though the multi-frequency setup covers the same disk in k-space, it still seems superior because we have more data points of the Fourier transform $\mathcal{F}f$.

For the similar model from Fig. 15a with a higher contrast, the reconstructions with Born and Rytov approximation differ more from the FWI reconstruction because of the more severe scattering; see Fig. 17. In general, we can expect the FWI reconstruction to be better since it is a numerical approximation of the wave equation, of which the Born or Rytov approximations are just linearizations.

Reconstruction of Embedded Shapes: Phantom 2

We now consider the case with combinations of smaller convex and non-convex shapes included in the background medium.

Reconstruction Using FWI

In Figs. 18 and 19, we show the model perturbation, which consist in small objects buried in the background. FWI is carried out with single and multiple frequencies, while we investigate a mild contrast in Fig. 18 (where f is at most 0.5) and a stronger contrast in Fig. 18 (where f is at most 2). We see that the model can be recovered using a single frequency, which has to be selected depending on the contrast. As an alternative, multi-frequency data appears to be a robust candidate and always provides a good reconstruction, for both the object's shape and amplitude. The reconstruction quality with high contrast in Fig. 19 seems to be of a similar level as with low contrast.

Reconstruction Using Born and Rytov Approximations

In Fig. 20, we show the reconstruction using Born and Rytov approximations. Here, we can clearly see that we need a higher frequency in order to resolve small features of the object. However, the reconstructions are still inferior to the FWI.

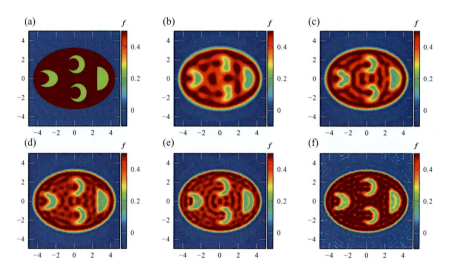

Fig. 18 Reconstruction with FWI starting from a homogeneous background with $f = 0$. The models are given at frequency $\omega/(2\pi) = 1$ and the wave speed is equal to 1 in the background (a) True contrast. (b) Using single-frequency, $\omega/(2\pi) = 0.7$ (PSNR 18.91). (c) Using single-frequency, $\omega/(2\pi) = 1$ (PSNR 19.50). (d) Using single-frequency, $\omega/(2\pi) = 1.2$ (PSNR 19.90). (e) Using single-frequency, $\omega/(2\pi) = 1.4$ (PSNR 20.10). (f) Using multi-frequency, $\omega/(2\pi) \in \{0.7, 1, 1.2, 1.4\}$ (PSNR 23.04)

8 Diffraction Tomography, Fourier Reconstruction, and Full Waveform Inversion

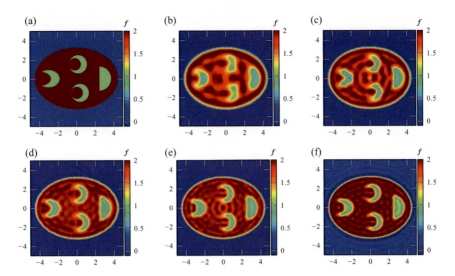

Fig. 19 Reconstruction with FWI starting from a homogeneous background with $f = 0$. The models are given at frequency $\omega/(2\pi) = 1$ and the wave speed is equal to 1 in the background (**a**) True contrast. (**b**) Using single-frequency, $\omega/(2\pi) = 0.7$ (PSNR 20.75). (**c**) Using single-frequency, $\omega/(2\pi) = 1$ (PSNR 21.50). (**d**) Using single-frequency, $\omega/(2\pi) = 1.2$ (PSNR 22.27). (**e**) Using single-frequency, $\omega/(2\pi) = 1.4$ (PSNR 22.72). (**f**) Using multi-frequency, $\omega/(2\pi) \in \{0.7, 1, 1.2, 1.4\}$ (PSNR 22.82)

With a higher contrast, the reconstruction with the Rytov approximation is considerably better than the one with the Born approximation; see Fig. 21. This is consistent with Remark 1. Interestingly, the shapes reconstruction in high contrast barely profits from taking frequencies higher than 1, even though the k-space coverage is larger. In this situation, the Rytov reconstruction is almost comparable to the one with lower contrast, but still worse than the FWI.

Computational Costs

Computational cost of FWI. The computational cost of FWI comes from the discretization and resolution of the wave problem Equation 15 for each of the sources in the acquisition, coupled with the iterative procedure of Algorithm 1. In our numerical experiments, we use the software hawen for the iterative inversion, Faucher (2021), Footnote 1, which relies on the Hybridizable discontinuous Galerkin discretization, Cockburn et al. (2009) and Faucher and Scherzer (2020). The number of degrees of freedom for the discretization depends on the number of cells in the mesh and the polynomial order. In the inversion experiments, we use a fixed mesh for all iterations, with about fifty thousand cells. On the other hand, the polynomial order is selected depending on the wavelength on each cell.

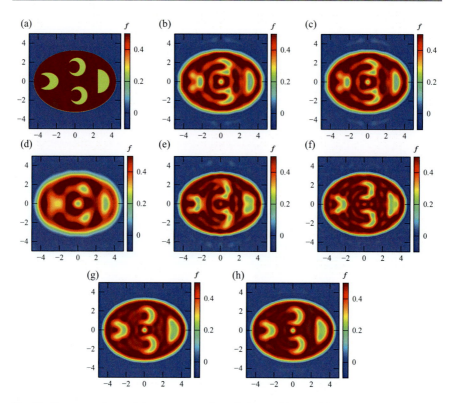

Fig. 20 Reconstructions of the more complicated shapes. The models are given at frequency $\omega/(2\pi) = 1$. (**a**) True model f. (**b**) Born reconstruction at frequency $\omega/(2\pi) = 1$. (PSNR 20.14). (**c**) Rytov reconstruction at frequency $\omega/(2\pi) = 1$. (PSNR 20.10). (**d**) Rytov reconstruction at frequency $\omega/(2\pi) = 0.7$. (PSNR 18.00). (**e**) Rytov reconstruction at frequency $\omega/(2\pi) = 1.2$. (PSNR 21.29). (**f**) Rytov reconstruction at frequency $\omega/(2\pi) = 1.4$. (PSNR 22.17). (**g**) Born reconstruction using multi-frequency, $\omega/(2\pi) \in \{0.7, 1, 1.2, 1.4\}$. (PSNR 21.91). (**h**) Rytov reconstruction using multi-frequency, $\omega/(2\pi) \in \{0.7, 1, 1.2, 1.4\}$. (PSNR 21.93)

That is, each of the cells in the mesh is allowed to have a different order (here between 3 to 7); see Faucher and Scherzer (2020). Then, when the frequency changes, while the mesh remains the same, the order of the polynomial evolves accordingly to the change of wavelength. Once the wave equation, Equation 15, is discretized, we obtain a linear system which size is the number of degrees of freedom that must be solved for the different sources (i.e., the different incident angles). We rely on the direct solver MUMPS, Amestoy et al. (2019), such that once the matrix factorization is computed, the numerical cost of having several sources (i.e., multiple right-hand sides in the linear system) is drastically mitigated, motivating the use of a direct solver instead of an iterative one.

8 Diffraction Tomography, Fourier Reconstruction, and Full Waveform Inversion

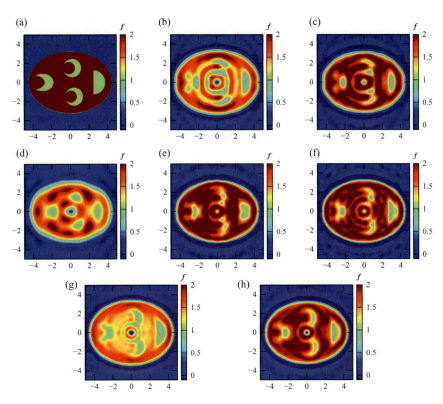

Fig. 21 Reconstructions of the more complicated shapes with a higher contrast than in Fig. 20. The models are given at frequency $\omega/(2\pi) = 1$. (**a**) True model f. (**b**) Born reconstruction at frequency $\omega/(2\pi) = 1$. (PSNR 17.16). (**c**) Rytov reconstruction at frequency $\omega/(2\pi) = 1$. (PSNR 18.73). (**d**) Rytov reconstruction at frequency $\omega/(2\pi) = 0.7$. (PSNR 16.10). (**e**) Rytov reconstruction at frequency $\omega/(2\pi) = 1.2$. (PSNR 20.14). (**f**) Rytov reconstruction at frequency $\omega/(2\pi) = 1.4$. (PSNR 21.15). (**g**) Born reconstruction using multi-frequency, $\omega/(2\pi) \in \{0.7, 1, 1.2, 1.4\}$. (PSNR 16.92). (**h**) Rytov reconstruction using multi-frequency, $\omega/(2\pi) \in \{0.7, 1, 1.2, 1.4\}$. (PSNR 20.33)

Our numerical experiments have been carried out on the Vienna Scientific Cluster VSC-4,[2] using 48 cores. For the reconstructions of Figs. 14, 15, 18, and 19, the size of the computational domain is 40×40, with about 350.000 degrees of freedom. Using single-frequency data, 50 iterations are performed in Algorithm 1, and the total computational time is of about 40 min. In the case of multiple frequencies, we have a total of 120 iterations, and the computational time is of about $1\,h\,45$ min.

Computational cost of Born and Rytov approximations. The conjugate gradient method used in the inverse NDFT requires in each iteration step the evaluation

---
[2] https://vsc.ac.at/

of an NDFT Equation 28 and its adjoint. We utilize the nonequispaced fast Fourier transform (NFFT) algorithm (Keiner et al. n.d.), implemented in the open-source software library nfft (Keiner et al. 2009), which can compute an NDFT in $O(N^2 \log N + M)$ arithmetic operations, which is considerably less than the $O(N^2 M)$ operations of a straightforward implementation of Equation 28.

The numerical simulations of section "Reconstruction Using Born and Rytov Approximations" have been carried out on a 4-core Intel Core i5-6500 processor. We used 12 iterations of the conjugate gradient method and noted that the reconstruction quality hardly benefits from a higher number of iterations. The reconstruction of an image took about 1 second, with a grid size 720×720 of f and 200×100 data points of u for each frequency. Therefore, the numerical computation of the Born and Rytov approximations is much faster than the FWI.

Conclusion

We study the imaging problem for diffraction tomography, where wave measurements are used to quantitatively reconstruct the physical properties, i.e., the refractive index. The forward operator that describes the wave propagation corresponds with the Helmholtz equation, which, under the assumption of small background perturbations, can be represented via the Born and Rytov approximations.

Firstly, we have compared different forward models in terms of the resulting measured data u. It highlights that, even in the case of a small circular object, the Born approximation is not entirely accurate to represent the total wave field given by the Helmholtz equation. In addition, the source that initiates the phenomenon (e.g., a point source located very far from the object, or simultaneous point source along a line) also plays an important role as it changes the resulting signals, hence leading to systematic differences depending on the choice of forward model. We found that the line source model approximates the plane wave pretty well.

Secondly, we have carried out the reconstruction using data from the total field u^{tot} and compared the efficiency of the Full Waveform Inversion method (FWI) with that of the Born and Rytov approximations. FWI works directly with the Helmholtz problem, Equation 15, hence giving a robust approach that can be implemented in all configurations, however at the cost of possibly intensive computations. On the other hand, the Born and Rytov are computationally cheap, but lack accuracy when the object is too large or when the contrast is too strong. We have also noted that the Rytov approximation gives better results than the Born one. Furthermore, for all reconstruction methods, we have shown that using data of multiple frequencies allows to improve the robustness of the reconstruction by providing information on multiple wavelengths.

Acknowledgments We thank the anonymous reviewer for carefully reading the manuscript and making various suggestions for its improvement. This work is supported by the Austrian Science Fund (FWF) within SFB F68 ("Tomography across the Scales"), Projects F68-06 and F68-07. FF is funded by the Austrian Science Fund (FWF) under the Lise Meitner fellowship M 2791-N.

References

Amestoy, P.R., Buttari, A., L'excellent, J.-Y., Mary, T.: Performance and scalability of the block low-rank multifrontal factorization on multicore architectures. ACM Trans. Math. Softw. (TOMS) **45**(1), 1–26 (2019). https://doi.org/10.1145/3242094

Bamberger, A., Chavent, G., Lailly, P.: About the stability of the inverse problem in the 1-d wave equation. J. Appl. Math. Optim. **5**, 1–47 (1979)

Barucq, H., Chavent, G., Faucher, F.: A priori estimates of attraction basins for nonlinear least squares, with application to Helmholtz seismic inverse problem. Inverse Probl. **35**(11), 115004 (2019). https://doi.org/10.1088/1361-6420

Bednar, J.B., Shin, C., Pyun, S.: Comparison of waveform inversion, part 2: phase approach. Geophys. Prospect. **55**(4), 465–475 (2007). ISSN: 1365-2478. https://doi.org/10.1111/j.1365-2478.2007.00618.x

Beinert, R., Quellmalz, M.: Total variation-based reconstruction and phase retrieval for diffraction tomography SIAM J. Imaging Sci. **15**(3), 1373–1399 (2022). ISSN: 1936-4954. https://doi.org/10.1137/22M1474382

Bunks, C., Saleck, F.M., Zaleski, S., Chavent, G.: Multiscale seismic waveform inversion. Geophysics **60**(5), 1457–1473 (1995). https://doi.org/10.1190/1.1443880

Chen, B., Stamnes, J.J.: Validity of diffraction tomography based on the first Born and the first Rytov approximations. Appl. Opt. **37**(14), 2996 (1998). https://doi.org/10.1364/ao.37.002996

Clément, F., Chavent, G., Gómez, S.: Migration-based traveltime wave-form inversion of 2-D simple structures: a synthetic example. Geophysics **66**(3), 845–860 (2001). https://doi.org/10.1190/1.1444974

Cockburn, B., Gopalakrishnan, J., Lazarov R.: Unified hybridization of discontinuous Galerkin, mixed, and continuous Galerkin methods for second order elliptic problems. SIAM J. Numer. Anal. **47**(2), 1319–1365 (2009). https://doi.org/10.1137/070706616

Colton, D., Kress, R.: Inverse Acoustic and Electromagnetic Scattering Theory. Applied Mathematical Sciences, vol. 93, 3rd edn. Springer, Berlin (2013). ISBN: 978-1-4614-4941-6. https://doi.org/10.1007/978-1-4614-4942-3

Devaney, A.: A filtered backpropagation algorithm for diffraction tomography. Ultrason. Imaging **4**(4), 336–350 (1982). https://doi.org/10.1016/0161-7346(82)90017-7

Devaney, A.: Mathematical Foundations of Imaging, Tomography and Wave-Field Inversion. Cambridge University Press (2012). https://doi.org/10.1017/CBO9781139047838

Engquist, B., Majda, A.: Absorbing boundary conditions for numerical simulation of waves. Proc. Natl. Acad. Sci. **74**(5), 1765–1766 (1977)

Fan, S., Smith-Dryden, S., Li, G., Saleh, B.E.A.: An iterative reconstruction algorithm for optical diffraction tomography. In: IEEE Photonics Conference (IPC), pp. 671–672 (2017). https://doi.org/10.1109/ipcon.2017.8116276

Faucher, F.: Contributions to seismic full waveform inversion for time harmonic wave equations: Stability estimates, convergence analysis, numerical experiments involving large scale optimization algorithms. PhD thesis. Université de Pau et Pays de l'Ardour, pp. 1–400 (2017)

Faucher, F.: Hawen: time-harmonic wave modeling and inversion using hybridizable discontinuous Galerkin discretization. J. Open Source Softw. **6**(57) (2021). https://doi.org/10.21105/joss.02699

Faucher, F., Scherzer, O.: Adjoint-state method for Hybridizable Discontinuous Galerkin discretization, application to the inverse acoustic wave problem. Comput. Methods Appl. Mech. Eng. **372**, 113406 (2020). ISSN: 0045-7825. https://doi.org/10.1016/j.cma.2020.113406

Faucher, F., Alessandrini, G., Barucq, H., de Hoop, M., Gaburro, R., Sincich, E.: Full Reciprocity-Gap Waveform Inversion, enabling sparse-source acquisition. Geophysics **85**(6), R461–R476 (2020a). https://doi.org/10.1190/geo2019-0527.1

Faucher, F., Chavent, G., Barucq, H., Calandra, H.: A priori estimates of attraction basins for velocity model reconstruction by time-harmonic Full Waveform Inversion and Data-Space Reflectivity formulation. Geophysics **85**(3), R223–R241 (2020b). https://doi.org/10.1190/geo2019-0251.1

Faucher, F., Scherzer O., Barucq, H.: Eigenvector models for solving the seismic inverse problem for the Helmholtz equation. Geophys. J. Int. (2020c). ISSN: 0956-540X. https://doi.org/10.1093/gji/ggaa009

Faucher, F., de Hoop, M.V., Scherzer, O.: Reciprocitygap misfit functional for Distributed Acoustic Sensing, combining data from passive and active sources. Geophysics **86**(2), R211–R220 (2021). ISSN: 0016-8033. https://doi.org/10.119/geo2020-0305.1

Fichtner, A., Kennett, B.L., Igel, H., Bunge, H.-P.: Theoretical back ground for continental- and global-scale full-waveform inversion in the time–frequency domain. Geophys. J. Int. **175**(2), 665–685 (2008). https://doi.org/10.1111/j.1365-246X.2008.03923.x

Gbur, G., Wolf, E.: Hybrid diffraction tomography without phase information. J. Opt. Soc. Am. A **19**(11), 2194–2202 (2002). https://doi.org/10.1364/OL27.001890

Hanke, M.: Conjugate Gradient Type Methods for Ill-Posed Problems. Pitman Research Notes in Mathematics Series, vol. 327. Longman Scientific & Technical, Harlow (1995)

Hesthaven, J.S., Warburton, T.: Nodal Discontinuous Galerkin Methods: Algorithms, Analysis, and Applications. Springer Science & Business Media (2007). https://doi.org/10.1007/978-0-387-72067-8

Hielscher, R., Potts, D., Quellmalz, M.: An SVD in spherical surface wave tomography. In: Hofmann, B., Leitao, A., Zubelli, J.P. (eds.) New Trends in Parameter Identification for Mathematical Models. Trends in Mathematics, pp. 121–144. Birkhäuser, Basel (2018). ISBN: 978-3-319-70823-2. https://doi.org/10.1007/978-3-319-70824-9_7

Hielscher, R., Quellmalz, M.: Optimal mollifiers for spherical de-convolution. Inverse Probl. **31**(8), 085001 (2015). https://doi.org/10.1088/02.665611/31/8/085001

Hielscher, R., Quellmalz, M.: Reconstructing a function on the sphere from its means along vertical slices. Inverse Probl. Imaging **10**(3), 711–739 (2016). ISSN: 1930-8337. https://doi.org/10.3934/ipi.2016018

Horstmeyer, R., Chung, J., Ou, X., Zheng, G., Yang, C.: Diffraction tomography with Fourier ptychography. Optica **3**(8), 827–835 (2016). https://doi.org/10.1364/OPTICA.3.000827

Huynh-Thu, Q., Ghanbari, M.: The accuracy of PSNR in predicting video quality for different video scenes and frame rates. Telecommun. Syst. **49**(1), 35–48 (2010). https://doi.org/10.1007/s112350109351x

Iwata, K., Nagata, R.: Calculation of refractive index distribution from interferograms using the Born and Rytov's approximation. Jpn. J. Appl. Phys. **14**(S1), 379–383 (1975). https://doi.org/10.7567/jjaps.14s1.379

Kak, A.C., Slaney M.: Principles of Computerized Tomographic Imaging. Classics in Applied Mathematics, vol. 33. Society for Industrial and Applied Mathematics (SIAM), Philadelphia (2001)

Kaltenbacher, B.: Minimization based formulations of inverse problems and their regularization. SIAM J. Optim. **28**(1), 620–645 (2018). https://doi.org/10.1137/17M1124036

Keiner, J., Kunis, S., Potts, D.: Using NFFT3 – a software library for various nonequispaced fast Fourier transforms. ACM Trans. Math. Softw. **36**, Article 19, 1–30 (2009). https://doi.org/10.1145/1555386.1555388

Keiner, J., Kunis, S., Potts, D.: NFFT 3.5, C subroutine library (n.d.). https://www.tu-chemnitz.de/~potts/nfft

Kirby, R.M., Sherwin, S.J., Cockburn, B.: To CG or to HDG: a comparative study. J. Sci. Comput. **51**(1), 183–212 (2012). https://doi.org/10.1007/s10915-011-9501-7

Kirisits, C., Quellmalz, M., Ritsch-Marte, M., Scherzer, O., Setterqvist, E., Steidl, G.: Fourier reconstruction for diffraction tomography of an object rotated into arbitrary orientations. Inverse Probl. (2021). ISSN: 0266-5611. https://doi.org/10.1088/1361-6420/ac2749

Knopp, T., Kunis, S., Potts, D.: A note on the iterative MRI reconstruction from nonuniform k-space data. Int. J. Biomed. Imag. (2007). https://doi.org/10.1155/2007/24727

Kunis, S., Potts, D.: Stability results for scattered data interpolation by trigonometric polynomials. SIAM J. Sci. Comput. **29**, 1403–1419 (2007). https://doi.org/10.1137/060665075

Lailly, P.: The seismic inverse problem as a sequence of before stack migrations. In: Bednar, J.B. (ed.) Conference on Inverse Scattering: Theory and Application, pp. 206–220. Society for Industrial and Applied Mathematics (1983)

Luo, Y., Schuster, G.T.: Wave-equation traveltime inversion. Geophysics **56**(5), 645–653 (1991). https://doi.org/10.1190/1.1443081

Maleki, M.H., Devaney, A.: Phase-retrieval and intensity-only recon-struction algorithms for optical diffraction tomography. J. Opt. Soc. Am. A **10**(5), 1086 (1993). https://doi.org/10.1364/josaa.10.001086

Métivier, L., Brossier, R., Mérigot, Q., Oudet, E., Virieux, J.: Measuring the misfit between seismograms using an optimal transport distance: application to full waveform inversion. Geophys. J. Int. **205**(1), 345–377 (2016). https://doi.org/10.1093/gji/ggw014

Monk, P.: Finite Element Methods for Maxwell's Equations. Oxford University Press, Oxford (2003)

Müller, P., Schürmann, M., Guck, J.: ODTbrain: a Python library for full-view, dense diffraction tomography. BMC Bioinform. **16**(367) (2015). https://doi.org/10.1186/s12859-015-0764-0

Müller, P., Schürmann, M., Guck, J.: The Theory of Diffraction Tomography (2016). arXiv: 1507.00466 [q-bio.QM]

Natterer, F.: The Mathematics of Computerized Tomography, x+222. B. G. Teubner, Stuttgart (1986). ISSN: 3-519-02103-X

Natterer, F., Wübbeling, F.: Mathematical Methods in Image Reconstruction. Monographs on Mathematical Modeling and Computation, vol. 5. SIAM, Philadelphia (2001)

Nocedal, J., Wright, S.J.: Numerical Optimization. Springer Series in Operations Research, 2nd edn. Springer, Berlin (2006)

Plessix, R.-E.: A review of the adjoint-state method for computing the gradient of a functional with geophysical applications. Geophys. J. Int. **167**(2), 495–503 (2006). https://doi.org/10.1111/j.1365-246X.2006.02978.x

Plonka, G., Potts, D., Steidl, G., Tasche, M.: Numerical Fourier Analysis. Applied and Numerical Harmonic Analysis. Birkhäuser (2018). ISSN: 978-3-030-04305-6. https://doi.org/10.1007/978-3-030-04306-3

Potts, D., Steidl, G.: A new linogram algorithm for computerized tomography. IMA J. Numer. Anal. **21**, 769–782 (2001). https://doi.org/10.1093/imanum/21.3.769

Pratt, R.G., Shin, C., Hick, G.J.: Gauss–Newton and full Newton methods in frequency–space seismic waveform inversion. Geophys. J. Int. **133**(2), 341–362 (1998). https://doi.org/10.1046/j.1365-246X.1998.00498.x

Pyun, S., Shin, C., Bednar, J.B.: Comparison of waveform inversion, part 3: amplitude approach. Geophys. Prospect. **55**(4), 477–485 (2007). ISSN: 1365-2478. https://doi.org/10.1111/j.1365-2478.2007.00619.x

Shin, C., Pyun, S., Bednar, J.B.: Comparison of waveform inversion, part 1: conventional wavefield vs logarithmic wavefield. Geophys. Prospect. **55**(4), 449–464 (2007). ISSN: 1365-2478. https://doi.org/10.1111/j.1365-2478.2007.00617.x

Slaney, M., Kak, A.C., Larsen, L.E.: Limitations of imaging with first-order diffraction tomography. IEEE Trans. Microw. Theory Techn. **32**(8), 860–874 (1984). https://doi.org/10.1109/TMTT.1984.1132783

Sung, Y., Choi, W., FangYen, C., Badizadegan, K., Dasari, R.R., Feld, M.S.: Optical diffraction tomography for high resolution live cell imaging. Opt. Express **17**(1), 266–277 (2009)

Tarantola, A.: Inversion of seismic reflection data in the acoustic approximation. Geophysics **49**, 1259–1266 (1984). https://doi.org/10.1190/1.1441754

Van Leeuwen, T., Mulder, W.: A correlation-based misfit criterion for wave-equation traveltime tomography. Geophys. J. Int. **182**(3), 1383–1394 (2010)

Virieux, J.: SH-wave propagation in heterogeneous media: velocity-stress finite-difference method. Geophysics **49**(11), 1933–1942 (1984)

Virieux, J., Operto, S.: An overview of full-waveform inversion in exploration geophysics. Geophysics **74**(6), WCC1–WCC26 (2009). https://doi.org/10.1190/1.3238367

Wedberg, T.C., Stamnes, J.J.: Comparison of phase retrieval methods for optical diffraction tomography. Pure Appl. Opt. **4**, 39–54 (1995). https://doi.org/10.1088/0963-9659/4/1/005

Wolf, E.: Three-dimensional structure determination of semi-transparent objects from holographic data. Opt. Commun. **1**, 153–156 (1969)

Models for Multiplicative Noise Removal

Xiangchu Feng and Xiaolong Zhu

Contents

Introduction	314
Variational Methods with Different Data Fidelity Terms	317
Statistical Property Based Models	318
MAP-Based Models	319
Root and Inverse Transformation-Based Models	320
Variational Methods with Different Regularizers	325
TV Regularization	325
Sparse Regularization	327
Nonconvex Regularization	330
Multitasks	334
Root Transformation	334
Fractional Transformation	335
Nonlocal Methods	335
Indirect Method	336
Direct Method	337
DNN Method	338
Indirect Method	339
Direct Method	342
Conclusion	343
References	343

Abstract

Image denoising is the most important step in image processing for further image analysis. It is an important topic in many applications, such as object recognition, digital entertainment, etc. The digital image can be corrupted with noise during

---

X. Feng (✉) · X. Zhu
School of Mathematics and Statistics, Xidian University, Xi'an, China
e-mail: xcfeng@mail.xidian.edu.cn; zxl001@aliyun.com

© Springer Nature Switzerland AG 2023
K. Chen et al. (eds.), *Handbook of Mathematical Models and Algorithms in Computer Vision and Imaging*, https://doi.org/10.1007/978-3-030-98661-2_60

acquisition, storage, and transmission. Noise can be classified as additive noise, multiplicative noise, and non-additive non-multiplicative noise (such as salt and pepper noise, Poisson noise). The main properties of a good image denoising model are that it will remove noise while preserving details of the image.

This chapter aims to present a review of multiplicative denoising models, especially for the multiplicative Gamma noise. Similar to denoising for additive Gaussian noise, these denoising approaches can be categorized as variational methods, non-local methods, and deep neural network-based methods. Due to space constraints, this chapter only discusses some of them. The rest of this chapter is organized as follows. Section "Introduction" is an introduction and section "Variational Methods with Different Data Fidelity Terms" describes variational methods with different data fidelity terms. Section "Variational Methods with Different Regularizers" introduces variational methods with different regularizers. Sections "Multitasks" to "DNN Method" describe multitasks, nonlocal, and deep neural network (DNN) methods. Finally, section "Conclusion" presents our conclusions.

Keywords

Multiplicative denoising · Variational methods · Nonlocal methods · Multitasks · DNN methods

Introduction

The most common noise encountered in real applications is thermal noise. It is additive and follows a Gaussian distribution with zero mean. Many image denoising approaches have been proposed for additive Gaussian noise (Shao et al. 2013; Lebrun et al. 2012). Generally, these approaches can be categorized as spatial domain, transform Domain, and learning-based method. Spatial domain methods include energy function methods, which exploit maximum a posteriori probability (MAP) estimation as the main tool, and nonlocal filters, which exploit the similarities between patches in an image. Transform domain methods consider transforming images into other domains, in which similarities of transformed coefficients are considered. Learning-based methods use sparse representations on a redundant dictionary or train a deep neural network through many training samples. In fact, after so many researches, the denoising results of these methods for additive Gaussian noise are close to the limitation (Chatterjee and Milanfar 2009).

This chapter focuses on reducing multiplicative speckle noise, especially for multiplicative Gamma noise. In many coherent imaging systems, digital images are usually accompanied by speckle noise (Singh and Jain 2016). It is caused by coherent processing of backscattered signals from multiple distributed targets. Speckle noise can be described as random multiplicative noise. It appears in many applications, e.g., in ultrasound imaging, where the noise follows a Rayleigh distribution; in electronic microscopy, where the multiplicative noise is

Poisson noise; and in synthetic aperture radar (SAR), where the noise follows a Gamma distribution. In fact, speckle in a SAR image is caused by constructive and destructive interference of coherent waves reflected by the many elementary scatterers contained within the imaged resolution cell. The magnitude of the complex observations of SAR can usually be modeled as corrupted by multiplicative Rayleigh noise. As a consequence, the noise present in the square of the magnitude, the so-called intensity, is exponentially distributed. To improve the quality of such data, a common approach in SAR imaging is to average independent intensity observations of the same scene to obtain so-called multi look data, which is then contaminated by multiplicative Gamma noise.

The Gamma noise model and Gamma distribution are given below. If we use f to denote the image intensity that the SAR measures for a given pixel whose backscattering coefficient is u, and assume that the SAR image represents an average of L looks (independent samples or pixels), then f is related to u by the multiplicative model

$$f = un \tag{1}$$

where n is the normalized fading random variable in the intensity image, following a Gamma distribution with unit mean and variance $1/L$. The probability density function (PDF) of n is given by

$$p_n(n) = \frac{L^L n^{L-1} e^{-Ln}}{\Gamma(L)}, \quad n \geq 0, \quad L \geq 1. \tag{2}$$

where $\Gamma(\cdot)$ denotes the gamma function.

The natural idea when dealing with this problem is to convert it into an additive problem by applying the logarithm while using a Gaussian distribution to approximate the distribution after the logarithm (Xie et al. 2002). The natural logarithmic transformation converts (1) into

$$\tilde{f} = \tilde{u} + \tilde{n} \tag{3}$$

where $\tilde{f} = \ln(f)$, $\tilde{u} = \ln(u)$ and $\tilde{n} = \ln(n)$. Owing to the monotonic of the logarithmic function, the probability density function of the random variable $\tilde{n}$ can be obtained from

$$p_{\tilde{n}}(\tilde{n}) = p_n(e^{\tilde{n}}) e^{\tilde{n}} \tag{4}$$

which leads to

$$p_{\tilde{n}}(\tilde{n}) = \frac{L^L e^{\tilde{n}L} e^{-Le^{\tilde{n}}}}{\Gamma(L)} \tag{5}$$

and the mean of $\tilde{n}$ is given by (Hoekman 1991)

Table 1 Cumulative distribution distance D

Looks L	1	2	3	4	5	6	9	12	16
Intensity	0.242	0.145	0.099	0.078	0.066	0.058	0.045	0.039	0.033
Log(Intensity)	0.070	0.050	0.040	0.034	0.030	0.028	0.023	0.019	0.016

$$E\left(\tilde{n}\right) = \psi\left(L\right) - \ln\left(L\right) \tag{6}$$

where $\psi\left(\cdot\right)$ is the Digamma function defined by

$$\psi\left(\cdot\right) = \frac{d}{dx}\ln\Gamma\left(x\right) \tag{7}$$

The new variance is given by

$$var\left(\tilde{n}\right) = \psi\left(1, L\right) \tag{8}$$

where $\psi\left(1, L\right)$ is known as the first-order Polygamma function of L.

The distance between cumulative distribution is defined as the maximum value of the absolute difference between the two cumulative distribution functions. For evaluating how close a general distribution $p(x)$ is to a Gaussian probability density function $g(x)$, their cumulative distributions are firstly calculated, denoted as $P(x)$ and $G(x)$. Then the distance D is given by (Table 1)

$$D = \max_{-\infty < x < \infty}\left|P\left(x\right) - G\left(x\right)\right| \tag{9}$$

As the number of looks increases, the probability density function of a speckle random variable approaches the Gaussian probability density function. When the number of looks is small, if Gaussian noise is used to approximate, the error is large. Therefore, it is necessary to directly process the multiplicative noise. Because speckle noise is non-Gaussian signal and spatially independent (Ullah et al. 2016; Abolhassani and Rostami 2012; Le and Vese 2003), noise removal is more complex and more challenging than additive noise removal.

As we know, the restoration process is to recover u from the degraded image $f = un$ and preserve image features including edges, point targets, textures, and so on. To achieve this objective, a variety of speckle noise removal methods has been proposed. Some methods have been well-known, such as variational methods, dictionary learning methods, nonlocal methods, deep neural network methods, etc.

A variational model for speckle noise removal normally consists of the regularization term (log-prior) and data fidelity term (log-likelihood). For Gamma noise, a variational model (AA-model) via the maximum a posteriori estimator was derived by (Aubert and Aujol 2008). Motivated by the effectiveness of the inverse scale space, Shi and Osher developed a strictly convex general model (SO-model) for speckle noise removal (Shi and Osher 2008). By applying I-divergence as a similarity term, an energy function (SST-model) was presented (Steidl and Teuber 2010). By applying an exponential transformation to the AA model, a globally convex model for speckle noise removal has been achieved in Jin and Yang (2010). The regularization term is commonly total variation (TV) and its variations (Xiao et al. 2010; Hu et al. 2013; Na et al. 2018).

Due to the sparse nature of the l_1 norm, TV requires the image to have some sparsity in the gradient domain. We know that the wavelet coefficients, ridgelet coefficients, or curvelet coefficients of a sharp image are sparse. Based on these, Durand et al. gave a hybrid method of curvelet field for removing multiplicative noise in Durand et al. (2010). A combination of total generalized variation filter (which has been proved to be able to reduce the blocky-effects by being aware of high-order smoothness) and shearlet transform (that effectively preserves anisotropic image features such as sharp edges, curves, and so on) was proposed in Ullah et al. (2017). In Huang et al. (2012), dictionary learning is used as a regularization term, and experimental results suggest that in terms of visual quality, peak signal-to-noise ratio, and mean absolute deviation error, the proposed algorithm outperforms many other methods. In addition to variational models, nonlocal methods (Teuber and Lang 2012; Huang et al. 2017) are also proposed. Recently, deep neural network methods (Wang et al. 2017, 2019) are presented, extensive experiments on synthetic and real images show that they achieve significant improvements over the state-of-the-art speckle reduction methods.

Variational Methods with Different Data Fidelity Terms

Usually, a variational method has two terms: a data fidelity term and a regularization term. More specifically, our interest is in recovering a true underlying image u from the noise corrupted observation $f = un$, where n is a random variable following Gamma distribution. To obtain an estimate $\hat{u}$, (10) is considered

$$\hat{u} = \arg\min_{u \in X} \left\{ E(u) := \phi(u, f) + \lambda \rho(u) \right\} \tag{10}$$

where $\lambda > 0$ is a tuning parameter, X is the space that the solution lies in. Depending on the model, X may be $L^2(\Omega)$, $BV(\Omega)$, etc. In the discrete case, usually $X = R^d$. In general, the data fidelity term ϕ reflects characteristics of the noise corrupting our observation, and the regularization term $\rho(\cdot)$ is a prior on the clean image u. A common choice for $\rho(\cdot)$ is total variation(TV)

$$\rho(u) = \int_\Omega |\nabla u| := J(u) \text{ or } |u|_{TV(\Omega)} \tag{11}$$

Statistical Property Based Models

(1) RLO-model
Under the assumption that the mean of the multiplicative noise is equal to 1 and the variance is known, Rudin, Lions, and Osher introduced the following denoising model (RLO) in Rudin et al. (2003):

$$\min_{u \in X} \left\{ J(u) + \lambda_1 \int_\Omega \frac{f}{u} dx + \lambda_2 \int_\Omega \left(\frac{f}{u} - 1\right)^2 dx \right\} \quad (12)$$

However, only basic statistical properties, the mean, and the variance of the noise are considered in the RLO model, which somehow limits its denoising ability. We know that the likelihoods of the multiplicative Poisson noise and the likelihood of the multiplicative Rayleigh noise (Setzer et al. 2010; Denis et al. 2009) are $\int (u - f \log u) dx$ and $\int \left(\frac{1}{2}\left(\frac{f}{u}\right)^2 + \log u\right) dx$, respectively.

Based on the MAP model of Poisson noise and Rayleigh noise, the above model (12) can be generalized into SO-model.

(2) General SO-model
In (2008), Shi and Osher proposed a new general model, which can be fitted in different areas by setting different parameters of a, b and c

$$\int_\Omega \left(a\frac{f}{u} + \frac{b}{2}\left(\frac{f}{u}\right)^2 + c \log u \right) dx \quad (13)$$

This is a nonconvex variational problem, coupled with the TV regularization term; it becomes the following problem:

$$\hat{u} = \arg\min_{u \in BV(\Omega)} \left\{ J(u) + \lambda \int \left(a\frac{f}{u} + \frac{b}{2}\left(\frac{f}{u}\right)^2 + c \log u \right) dx \right\} \quad (14)$$

where $c = a + b$ for the Gammer noise. By exponential transformation $u = e^w$, the problem is reduced to a convex one

$$\hat{w} = \arg\min_{w \in BV(\Omega)} \left\{ J(w) + \lambda \int \left(af \exp(-w) + \frac{b}{2}(f)^2 \exp(-2w) + (a+b)w \right) \right\},$$

$$\hat{u} = e^{\hat{w}}$$

$$(15)$$

The fidelity term $H(w, f) = \int \left(af \exp(-w) + \frac{b}{2}(f)^2 \exp(-2w) + (a+b)w \right)$ is globally strictly convex. Using gradient descent and the Euler-Lagrange equation for this total variation-based problem, (16) can be obtained:

$$w_t = \nabla \cdot \frac{\nabla w}{|\nabla w|} + \lambda \left(af \exp(-w) + b(f)^2 \exp(-2w) - (a+b) \right) \qquad (16)$$

Shi and Osher extended this convex model to obtain a nonlinear inverse scale space flow and its corresponding relaxed inverse scale space flow. The numerical results of SNR show significant improvement over the RLO model (Shi and Osher 2008).

MAP-Based Models

(3) AA-model
Based on the MAP estimator for multiplicative Gamma noise, Aubert and Aujol (2008) proposed to determine the denoised image as a minimizer in $S(\Omega) = \{u \in BV : u > 0\}$ of the following functional

$$\min_{u \in S(\Omega)} \lambda J(u) + \int_\Omega \left(\log u + \frac{f}{u} \right) dx \qquad (17)$$

The AA model (17) is nonconvex; finding its global solution is a challenging task. It is known that the convex optimization method has vast applications in image processing. Therefore, many works have been designed to relieve the nonconvex AA model.

(4) SO-model
In (2008), Shi and Osher suggested to keep the data fitting term in (17) but to replace the regularizer $|\nabla u|$ by $|\nabla \log u|$. Moreover, setting as in the log-model $w := \log u$, this results in the convex function

$$\hat{w} = \arg\min_{w \in BV(\Omega)} \int_\Omega f e^{-w} + w \, dx + \lambda J(w), \quad \hat{u} = e^{\hat{w}} \qquad (18)$$

In fact, it is the exponential form of the general SO-model (14) when $b = 0$. Furthermore, to better preserve textures and details, this model was extended by Chen and Cheng in 2011, incorporating it with a spatially dependent regularization

$$\min_{w \in BV(\Omega)} \int_\Omega \lambda(x) \left(w + f e^{-w} \right) dx + J(w) \qquad (19)$$

where $\lambda : \Omega \to R$ is a spatially varying parameter.

(5) I-divergence model

In connection with deblurring in the presence of multiplicative noise, the I-divergence, also called generalized Kullback-Leibler divergence

$$I(f, u) := \int_\Omega f \log \frac{f}{u} - f + u \, dx \tag{20}$$

is typically used as a data fitting term. The I-divergence is the Bregman distance of the function $F(u) := \int_\Omega u \log u - u \, dx$, i.e., $I(f, u) = F(f) - F(u) - \langle p, f - u \rangle$, where $p \in \partial F(u)$.

Therefore, it shares the useful properties of the Bregman distance, in particular, $I(f, u) \geq 0$. Ignoring the constant terms, the corresponding convex denoising model reads

$$\hat{u} = \underset{u \in BV(\Omega), u > 0}{\arg \min} \left\{ \int_\Omega u - f \log u \, dx + \lambda J(u) \right\} \tag{21}$$

The gradient of the data fitting terms in (18) and (21) coincide if we use again the relation $\log \hat{u} = \hat{w}$. Moreover, if we add TV-regularization, then both functions have the same minimizer. Since $\nabla e^w = e^w \nabla w$, for $u = e^w$, we have $\nabla u(x) = 0$ if and only if $\nabla w(x) = 0$. The minimizers $\hat{w}$ and $\hat{u}$ of functions (18) and (21) are unique and given by

$$0 = 1 - f e^{-\hat{w}} - \lambda \operatorname{div} \frac{\nabla \hat{w}}{|\nabla \hat{w}|} \quad for \quad |\nabla \hat{w}(x)| \neq 0 \tag{22}$$

$$0 = 1 - \frac{f}{\hat{u}} - \lambda \operatorname{div} \frac{\nabla \hat{u}}{|\nabla \hat{u}|} \quad for \quad |\nabla \hat{u}(x)| \neq 0 \tag{23}$$

Since $\frac{\nabla w}{|\nabla w|} = \frac{e^w \nabla w}{e^w |\nabla w|} = \frac{\nabla u}{|\nabla u|}$, we obtain the assertion.

Root and Inverse Transformation-Based Models

(6) m-V model

M-th root transformation was introduced to relax the nonconvexity of the AA model, which was referred to as the m-V model (Yun and Woo 2012). To relax the nonconvexity of the AA-model, they use the mth root transformation ($n_m = \sqrt[m]{n}$, $f_m = \sqrt[m]{f}$, $u_m = \sqrt[m]{u}$). Since the mth root function is monotonically increasing, the gradient operator is applied to mth root transformed images. Then the transformed new variational model is expressed as follows:

$$u^* = \arg\min_{u \in \sqrt[m]{U}} \langle m \log u + f u^{-m}, 1 \rangle + \lambda J(u)$$
$$\hat{u} = (u^*)^m \quad (24)$$

where $\langle \cdot, \cdot \rangle$ is a usual scalar product in Euclidean spaces, $m \geq 1$. The m-V model can be considered as a generalization of the AA-model (when $m = 1$) and the variational model based on Nakagami distribution (when $m = 2$).

The probability distribution of n_m, which is the mth root of the multiplicative noise n, becomes

$$p(n_m) = \frac{mL^L (n_m)^{mL-1}}{\Gamma(L)} e^{-L(n_m)^m} H(n_m) \quad (25)$$

The probability density function (25) is a special case of the generalized Gamma distribution. Hence, the mean value and the variance of n_m are

$$E(n_m) = \frac{\Gamma\left(L + \frac{1}{m}\right)}{\Gamma(L) \sqrt[m]{L}} \quad (26)$$

$$var(n_m) = \frac{\Gamma(L) \Gamma\left(L + \frac{2}{m}\right) - \Gamma\left(L + \frac{1}{m}\right)^2}{\Gamma(L)^2 \sqrt[m]{L^2}} \quad (27)$$

We know that if $u \in (0, C]$, then the objective function of the m-V model (24) is convex on the set $\left\{ u \mid 0 < u \leq \min\left\{ \sqrt[m]{(m+1)f}, \sqrt[m]{C1} \right\} \right\}$. We call this property as conditional convex, which is convex when $m \geq \frac{C}{\min f_j} - 1$.

(7) DZ-model

Since the performance of the m-V model critically depends on the choice of m, a relaxed method was proposed in Kang et al. (2013) to further relax the m-V model. Nevertheless, the method is convex only when m is large enough. In Dong and Zeng (2013), the authors suggested the following model:

$$\min_{u \in \bar{S}(\Omega)} E(u) := \int_\Omega \left(\log u + \frac{f}{u} \right) dx + \alpha \int_\Omega \left(\sqrt{\frac{u}{f}} - 1 \right)^2 dx + \lambda J(u) \quad (28)$$

with the penalty parameter $\alpha > 0$. $\bar{S}(\Omega) := \{ v \in BV(\Omega) : v \geq 0 \}$ is a closed and convex set, $\log 0 = -\infty$ and $\log \frac{1}{0} = +\infty$ in $\bar{S}(\Omega)$. They proved that if $\alpha \geq \frac{2\sqrt{6}}{9}$, the model (28) is strictly convex.

(8) Exp-model

It was pointed out that the model (28) is mainly suitable for a large value of L. Lu et al. (2016) replace $\sqrt{\frac{u}{f}} - 1$ in the DZ model with $\sqrt{\frac{u}{f}} - \beta 1$, yielding the following optimization problem:

$$\min_{u \in R_+^d} \left\langle \log u + \frac{f}{u}, 1 \right\rangle + \alpha \left\| \sqrt{\frac{u}{f}} - \beta 1 \right\|_2^2 + \lambda J(u) \tag{29}$$

The objective function of this model is strictly convex if $\alpha\beta \geq \frac{2\sqrt{6}}{9}$, where β is no less than 1 and varies with the level of the noise.

Furthermore, owing to the constraint $u > 0$ and the observation that exponent-like models usually provide better quality denoised images than their logarithm-like counterparts, the authors used the log transformation, $w = \log u$, and proposed the following model, called the exp model:

$$\min_{w \in BV(\Omega)} \lambda \int_\Omega \left[w + fe^{-w} + \alpha \left(\sqrt{\frac{e^w}{f}} - \beta \right)^2 \right] dx + J(w) \tag{30}$$

The objective function of this model is strictly convex if $\alpha\beta^4 \leq \frac{4096}{27}$.

With a spatially dependent regularization parameter λ, an energy function was presented in Na et al. (2018)

$$\int_\Omega w_r(x, y) \left[\frac{f}{\tilde{u}} - \log \frac{f}{\tilde{u}} + \alpha \left(\sqrt{\frac{\tilde{u}}{f}} - \beta \right)^2 \right] (y) \, dy \tag{31}$$

Next, the log transformation $u = \log(\tilde{u})$ is applied, resulting in

$$S_u^r(x) = \int_\Omega w_r(x, y) \bar{q}(u)(y) \, dy \tag{32}$$

which is the local expected value estimator of the function $\bar{q}(u)$

$$\bar{q}(u) = u + fe^{-u} - \log f + \alpha \left(\sqrt{\frac{e^u}{f}} - \beta 1 \right)^2 \tag{33}$$

Using (32), we can obtain the following TV minimization problem with local constraints:

$$\min_{u \in BV(\Omega)} J(u) = \int_\Omega |Du|, \ s.t. \ S_u^r(x) \leq C \ a.e. \ in \ \Omega \tag{34}$$

9 Models for Multiplicative Noise Removal

(9) Convex-model
Another work is the so-called discrete convex model (Zhao et al. 2014)

$$\min_{w, u \in R^d} \frac{1}{2} \|w - \mu e\|_2^2 + \alpha_1 \|Fw - u\|_1 + \alpha_2 J(u) \tag{35}$$

where μ is a constant, e is a vector of which all the components are valued one, $F = diag(f)$ is the diagonalization matrix of the noisy image f with main diagonal entries given by f_i, and w is expected as the inverse of the multiplicative noise: $w = \frac{1}{n}$. In fact, from $f = un$ we obtain $fw = u$, and using the matrix form $F = diag(f)$, we have $Fw = u$. The data fidelity term is $\|Fw - u\|_1$, i.e., $\|diag(f)w - u\|_1$, which is equivalent to $\|f - un\|_1$. It replaces the nonconvex data fidelity term in the AA model and leads to an unconditional convex problem. Except for the fidelity term and the TV regularization term, the third term $\|w - \mu e\|_2^2$ is introduced to avoid the trivial solution.

(10) mth root transformation model
Based on the statistical analysis of fractional transformation and root transformation, Zhao and Feng first take mth root transformation on the degradation problem $f = un$ (where n obeys the Gamma distribution, set $f_m = \sqrt[m]{f}$, $u_m = \sqrt[m]{u}$ and $\zeta_m = \sqrt[m]{\frac{1}{n}}$), and reformulate the degradation model (Zhao et al. 2018):

$$f_m \zeta_m = u_m \tag{36}$$

Then take L_1 norm $\int_\Omega |f_m \zeta_m - u_m| dx$ and TV semi-norm $\int_\Omega |\nabla u_m| dx$ as the data fidelity term and the regularization term, respectively, and introduce the quadratic penalty term $\int_\Omega (\zeta_m - u_m)^2 dx$ as the prior of noise. Consequently, the proposed model is formulated as (Zhao et al. 2018)

(a) $\{\zeta_m^*, u_m^*\} = \arg\min_{\zeta_m, u_m} \int_\Omega |f_m \zeta_m - u_m| dx + \frac{\alpha}{2} \int_\Omega |\zeta_m - u_m|^2 dx + \lambda \int_\Omega |\nabla u_m| dx$
(b) $\hat{u} = (u_m^*)^m$
$$\tag{37}$$

where $\{m : m \geq 1, m \in N\}$. α and λ are parameters to control the trade-off among three terms in the objective function. The model is based on the following theorems.

Theorem 1. *Suppose that n follows the Gamma distribution, set $\zeta_m = \frac{1}{\sqrt[m]{n}}$, $(m \geq 1, m \in N)$, then*

(i) *The probability density function (PDF) of ζ_m is*

$$p_{\zeta_m}(y) = \frac{L^L m}{\Gamma(L)} y^{-mL-1} e^{-\frac{L}{y^m}} \tag{38}$$

(ii) *The means of ζ_m is*

$$E(\zeta_m) = \frac{L^{\frac{1}{m}}\Gamma\left(L - \frac{1}{m}\right)}{\Gamma(L)} \tag{39}$$

and obtain the following trends with fixed $L(L \geq 3)$:

$$\lim_{m \to +\infty} E(\zeta_m) = 1 \tag{40}$$

$$\lim_{m \to +\infty} E\left((\zeta_m - 1)^2\right) = 0 \tag{41}$$

Theorem 2. *Suppose that the random variable n follows Gamma distribution, set $\zeta_m = \frac{1}{\sqrt[m]{n}}$, $(m \geq 1, m \in N)$; then the KL divergence of ζ_m and $N\left(\mu_{m,L}, \sigma_{m,L}^2\right)$ satisfies*

$$D_{KL}\left(\zeta_m \,\middle\|\, N\left(\mu_{m,L}, \sigma_{m,L}^2\right)\right) = o\left(\frac{1}{m}\right) + o\left(\frac{1}{L^2}\right) \tag{42}$$

where $\mu_{m,L} = E(\zeta_m)$, $\sigma_{m,L}^2 = E\left((\zeta_m - E(\zeta_m))^2\right)$.

The proposed model (37) is an unconditional convex problem with a parameter m. It is noted that it reduces to the work in Zhao et al. (2014) when $m = 1$. However, it is known from Fig. 1 that the probability density function of $\zeta = \frac{1}{n}(m = 1)$ is far away from the Gaussian distribution, especially for small L. That is to say, the model (Zhao et al. 2014) cannot describe the prior of the noise very well, which

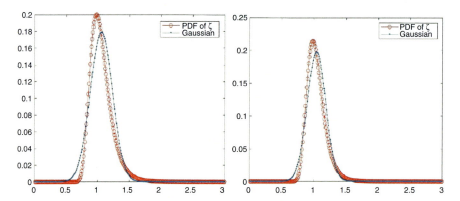

Fig. 1 Plots of the PDFs of ζ_m and $N\left(\mu_{m,L}, \sigma_{m,L}^2\right)$ with different m and L (**a**) $m = 4, L = 3$. (**b**) $m = 4, L = 4$

Fig. 2 Results of different methods when removing the multiplicative noise with $L = 4$. From the first to the last are original image, noisy image, the restored images of AA, convex, DZ, m-V, and the mth root transformation model, respectively

restricts its denoising performance. Comparatively, the new model is more flexible and extensible.

Moreover, it is worth noting that the data fidelity term in (37) is the L_1-norm $\int_\Omega |f_m \zeta_m - u_m| dx$. The main reason lies in that multiplicative noise mostly presents the corruption as the speckles or outliers onto the image, so L_1-norm outperforms L_2-norm or other convex representation as data fidelity term (Zhao et al. 2014) (Fig. 2).

Variational Methods with Different Regularizers

The regularizer $\rho(\cdot)$ has been extensively studied, and there are a few examples widely used in image recovery techniques. The choice of this regularizer depends on the assumptions made about the underlying image structure. Popular choices include the total variation (TV) semi-norm for image gradient sparsity, the l_1 norm for coefficient sparsity in a wavelet basis or other dictionary, and Huber-like functions which are akin to the l_1 norm but smooth. In general, image processors choose a regularizer according to two desiderata: one is that the objective function may be minimized efficiently and the other is that the regularizer accurately reflects image structure. The regularization term can be classified as TV, sparse, and nonconvex regularization.

TV Regularization

A frequently applied regularization term is the total variation (TV) semi-norm suggested in Rudin et al. (1992) by Rudin, Osher, and Fatemi (ROF), $|u|_{BV} := \sup_{p \in C_0^1, \|p\|_\infty \le 1} \int_\Omega u \operatorname{div} p \, dx$, which is formally (for sufficiently regular u)

$$J(u) = \int_\Omega |\nabla u| \, dx \qquad (43)$$

In the case of additive Gaussian noise, the minimizer $\hat{u}$ of the whole ROF function

$$\frac{1}{2}\int_{\Omega}(f-u)^2\,dx+\lambda J(u) \tag{44}$$

has many desirable properties. It preserves important structures such as edges, fulfills a maximum-minimum principle which reads in the discrete n-pixel setting as $f_{\min}\leq \hat{u}_i\leq f_{\max}$, where $f_{\min}$ and $f_{\max}$ denote the minimal and maximal coefficient of f, resp., and preserves the mean value, in the discrete case,

$$\sum_{i=1}^{n}\hat{u}_i=\sum_{i=1}^{n}f_i \tag{45}$$

The drawback of the model (44) consists of its staircasing effect so that meanwhile various alternative regularizers were considered.

(1) Non-local TV
The examples given above are standard TV regularization. Non-local TV is a promotion of NL-means. The idea of nonlocal means goes back to Buades et al. (2005) and was incorporated into the variational framework in Gilboa et al. (2006) and Gilboa and Osher (2009). We refer to these papers for further information on NL-means. Based on some pre-computed weights w, the regularization term is given by

$$\rho(u)=\int_{\Omega}|\nabla_w u|\,dx,\quad |\nabla_w u|:=\left(\int_{\Omega}\left(u(y)-u(x)\right)^2 w(x,y)\,dy\right)^{1/2}. \tag{46}$$

(2) Weberized TV
Inspiring from the Weberized TV regularization method, a nonconvex Weberized TV regularization-based multiplicative noise removal model was proposed in Xiao et al. (2010):

$$\hat{u}=\arg\min_{u\in X}\left\{E(u)=\int_{\Omega}\frac{|\nabla u|}{u}\,dx+\lambda\int_{\Omega}\left(\frac{f}{u}+\log u\right)dx\right\} \tag{47}$$

(3) Modified TV
Another variation of TV is proposed in Hu et al. (2013). When the gradient is small, a log is multiplied, and when the gradient is large, an affine transformation is made. Consider the following variational problem:

$$\min_{u\in X}E(u):=\min_{u\in X}\left\{\int_{\Omega}\rho(Du)+\lambda\int_{\Omega}\left(\log u+\frac{f}{u}\right)\right\} \tag{48}$$

where $\Omega \subset R^N$ is an open bounded open set with Lipschitz-regular boundary $\partial \Omega$, λ is a constant, $f : \Omega \to R^+$ is a given function, ρ is an even function from R^N to R having the linear growth

$$\rho(s) = \begin{cases} |s|\log(1+|s|), & |s| < M \\ b|s| - \frac{M^2}{1+M}, & |s| \geq M \end{cases} \tag{49}$$

where $b = M/(1+M) + \log(1+M)$, M is a positive constant, and its value is determined by the size of an image.

(4) TGV
To overcome these staircasing effects, higher-order regularization-based models were suggested in Chambolle and Lions (1997); Chan et al. (2000), and Li et al. (2007). As an early work, an inf-convolution TV (ICTV) model was proposed in Chambolle and Lions (1997), which takes the infimal convolution of TV and second-order TV. Moreover, Li et al. (2007) proposed a denoising model, involving a convex combination of TV and second-order TV as a regularizer. On the other hand, as a generalization of the ICTV, the TGV regularizer was proposed in Bredies et al. (2010). In particular, the second-order TGV is as follows:

$$TGV^2(u) = \min_{p \in P} \int_\Omega \alpha_1 |\nabla u - p| + \alpha_0 |\varepsilon(p)| \, dx \tag{50}$$

where $\varepsilon(p) = \frac{1}{2}\left(\nabla p + (\nabla p)^T\right)$ represents the distributional symmetrized derivative, and $\alpha_1, \alpha_0 > 0$ are the weighted parameters that control the balance between the first- and second-order terms. From the formulation (50) of TGV, it can be interpreted that $TGV^2(u)$ can automatically find an appropriate balance between the first- and the second-order derivative of u with respect to α_i.

Sparse Regularization

Due to the sparse nature of the l_1 norm, TV requires the image to have some sparsity in the gradient domain. We know that the wavelet coefficients, ridgelet coefficients, or curvelet coefficients of a sharp image are sparse. Based on these, Durand et al. gave a hybrid method of curvelet field for removing multiplicative noise in Durand et al. (2010).

(5) Curvelet Sparse
Durand et al. considered a hybrid model (DFN) by using the log-image data and a bias correction (Durand et al. 2010). Firstly, they studied the following sparse constraint problem:

$$\hat{\alpha} = \arg\min_{\alpha \in R^d} \left\| W\left(\log f\right) - \alpha \right\|_2^2 + \lambda \left\| \alpha \right\|_0 \qquad (51)$$

where W is the curvelet transform. Secondly, they proposed to minimize a specialized criterion composed of an L_1 data fidelity to $\hat{\alpha}$ and TV regularization in the log-image domain, i.e., the following problem was considered:

$$\hat{x} = \arg\min_{x \in R^d} \left\| \tilde{W} x \right\|_{TV} + \left\| \Lambda\left(x - \hat{\alpha}\right) \right\|_1 \qquad (52)$$

where $\tilde{W}$ is a left inverse of W, $\Lambda = diag\{\lambda_i\}$ is some weights. At last step, they restored the image by exponential transformation and bias correction according to the Gamma distribution, e.g.,

$$\hat{u} = \exp\left(\tilde{W}\hat{x}\right)\left(1 + \psi_1\left(L\right)/2\right) \qquad (53)$$

where $\psi_1(z) = \left(\frac{d}{dz}\right)^2 \log \Gamma(z)$, $\Gamma(z) = \int_0^{+\infty} \exp(-t) t^{z-1} dt$. In equation (51), they used curvelet and L_2 loss function to preserve the information of edges. Experimental results in Durand et al. (2010) show that the algorithm can obtain better results than SO (Shi and Osher 2008), AA (Aubert and Aujol 2008), and BS (Chesneau et al. 2010).

(6) Hybrid Model

Hao and Feng introduced dictionary learning instead of curvelet transform (Hao et al. 2012). The authors assumed that the log-image was sparse in the curvelet domain in Durand et al. (2010). However, in practice, it is very difficult to choose the correct dictionary on which the log-image is sparse. So instead of using the pre-selected basis, in the first stage, a dictionary is proposed to train by dictionary learning. Let $\omega : \omega = \log u$, and assume every patch in the log-image can be sparsely represented with the learned dictionary. Hao and Feng propose the following discrete model:

$$\min_{\alpha_{ij}, D, \omega} \left\{ \lambda \sum_{ij} \left(f \exp(-\omega) + \omega\right) + \frac{1}{2} \sum_{ij} \left\| R_{ij}\omega - D\alpha_{ij} \right\|_2^2 + \sum_{ij} \mu_{ij} \left\| \alpha_{ij} \right\|_0 \right\} \qquad (54)$$

where λ is a tuning parameter, R_{ij} is an $n \times N$ matrix that extracts the $(i, j)th$ block of size $n \times n$ from the $\sqrt{N} \times \sqrt{N}$ log-image ω, D is a dictionary, α_{ij} is the sparse representation coefficients of the $(i, j)th$ block with dictionary D, μ_{ij} are patch-specific weights, and $\left\| \alpha_{ij} \right\|_0$ stands for the count of nonzero entries in α_{ij}. Let ω^* be the minimizer of equation (54). It is amended by L_2 data fidelity and TV regularization in the log-image domain,

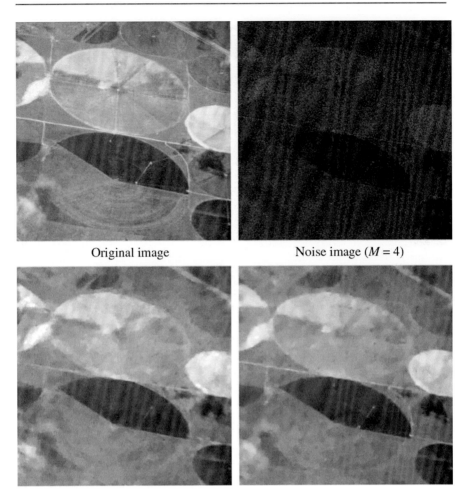

Fig. 3 The denoising experiment on Fields for $L = 4$

$$\min_{d} \frac{\delta}{2} \|d - \omega^*\|_2^2 + \|d\|_{TV} \qquad (55)$$

where δ is a turning parameter.

At the last stage, they transform the result obtained from the second step via an exponential function and bias correction. Let d^* be the solution to (55). d^* can be seen as the estimator of ω^*; it is prone to bias, which leads to the fact that the restored image is bias too. Using bias correction, we have (Figs. 3 and 4)

$$\hat{u} = \exp(d^*)\left(1 + \psi_1(L)/2\right) \qquad (56)$$

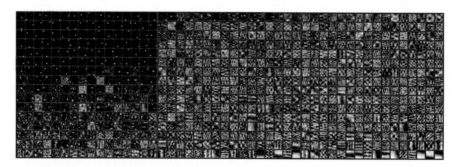

Fig. 4 The trained dictionary on fields for $L = 1, 4, 10$

Differ from this above approach, the following model adds a TV rule for the log domain (Huang et al. 2012).

(7) Dictionary Learning Plus Logarithmic Domain TV
Denoting by 1_Ω the constant 1 over the discrete image domain Ω (a $\sqrt{N} \times \sqrt{N}$ grid) and by $\langle \cdot, \cdot \rangle$ the usual scalar product in Euclidean spaces, the proposed model reads

$$\left\{\hat{D}, \hat{a}_{ij}, \hat{u}\right\} = \underset{\{D, a_{ij}, u > 0\}}{\arg\min} \lambda \left\langle \log u + \frac{f}{u}, 1_\Omega \right\rangle + \gamma \left\| \log u \right\|_{TV}$$
$$+ \frac{1}{2} \sum_{(i,j) \in P} \left\| D a_{ij} - R_{ij} \log u \right\|^2 + \sum_{(i,j) \in P} u_{ij} \left\| a_{ij} \right\|_0 \qquad (57)$$

where λ, γ are positive regularization parameters, $P = \left\{ 1, 2, \cdots, \sqrt{N} - \sqrt{n} + 1 \right\}^2$. $u \in R^N$ is the estimated image. The $\|\cdot\|_{TV}$ term is defined, in the discrete setting, by summing over the image domain Ω the norm of ∇u, the classical 2-neighbors discrete gradient estimate. $R_{i,j} \in R^{n \times N}$ is the matrix corresponding to the extraction of the patch located in (i, j), and $a_{i,j} \in R^K$ is the sparse vector of coefficients to represent the patch $R_{i,j} \log u$ with the dictionary $D \in R^{n \times K}$. The hidden parameters $(u_{i,j})_{(i,j) \in P}$ are determined by the optimization procedure described in Elad and Aharon (2006).

Nonconvex Regularization

(8) Fractional-Order TV
Toenhance the edge-preserving ability of TV, several nonconvex TV regularizers were proposed in Na et al. (2018); Krishnan and Fergus (2009), and Mei et al. (2018), which have the form $\rho(|\nabla u|) = \int_\Omega \varphi(|\nabla u|)\, dx$, where φ is the nonconvex function defined as

$$\varphi(s) = s^q \ (0 < q < 1), \quad \frac{\rho s^2}{1 + \rho s^2}, \quad \frac{1}{\rho}(1 + \rho s) \ (0 < q < 1) \tag{58}$$

Numerical results showed that the nonconvex TV regularizers were better at preserving edges and textures than TV (Nikolova et al. 2010). However, the nonconvex TV regularizers smooth homogeneous regions in the same way as TV. This indicates that they can yield some staircasing artifacts near smooth transition regions in the restored images.

SO model needs finally the nonlinear exponential transformation of the minimizing function, and the Weberized model is strongly dependent on the initialization and the numerical schemes. Tian et al. (2016) generalize the variation order from integer to fraction and obtain a fractional-order I-divergence model as follows:

$$\hat{u} = \arg\min_{u \in BV_\alpha} \left\{ \int_\Omega (u - f \log u) \, dx + \lambda \int_\Omega |\nabla^\alpha u| \, dx \right\} \tag{59}$$

where in the corresponding discrete form

$$\begin{cases} \int_\Omega |\nabla^\alpha u| \, dx = \sum\limits_{\substack{1 \le i \le M \\ 1 \le j \le N}} \left|(\nabla^\alpha u)_{i,j}\right| \\ \left|(\nabla^\alpha u)_{i,j}\right| = \sqrt{\left((\nabla_1^\alpha u)_{i,j}\right)^2 + \left((\nabla_2^\alpha u)_{i,j}\right)^2} \end{cases} \tag{60}$$

Note that the discrete form of the fractional-order gradient $\nabla^\alpha u$ can be evaluated by $(\nabla^\alpha u)_{i,j} = \left\langle (\nabla_1^\alpha u)_{i,j}, (\nabla_2^\alpha u)_{i,j} \right\rangle$ with $1 \le i \le M, 1 \le j \le N$, and

$$\begin{cases} (\nabla_1^\alpha u)_{i,j} = \sum\limits_{k=0}^{K-1} (-1)^k C_k^\alpha u_{i-k,j} \\ (\nabla_2^\alpha u)_{i,j} = \sum\limits_{k=0}^{K-1} (-1)^k C_k^\alpha u_{i,j-k} \end{cases}$$

where $K \ge 3$ is an integer constant, $C_k^\alpha = \Gamma(\alpha + 1) \big/ (\Gamma(k + 1) \Gamma(\alpha - k + 1))$, $\Gamma(\cdot)$ is the gamma function, and u is an image of size $M \times N$.

(9) Nonconvex Sparse Regularizer Model

Following the MAP estimation process, Han and Feng propose a new discrete minimization problem for removing speckle noise (Han et al. 2013):

$$\min_u \left\{ L \sum_{i=1}^n \left(u_i + e^{f_i - u_i}\right) + \lambda \sum_{i=1}^n \varphi(|\nabla_i u|) \right\} \tag{61}$$

where $\varphi(s) = \alpha s/(1+\alpha s)$, and λ is a constant parameter balancing the data term and the regularization term, which are based on the following two points:

Firstly, they point out the advantages of the proposed regularizer in the sparse framework. In fact, both the TV regularizer and the proposed regularizer can be seen as sparse measurements on the gradient modulus of u. The TV regularizer $\sum_{i=1}^{n} |\nabla_i u|$ is equals to the l_1-norm of the gradient modulus $\||\nabla_i u|\|_1$, while the proposed nonconvex regularizer $\sum_{i=1}^{n} \varphi(|\nabla_i u|)$ which can be converted into $\sum_{i=1}^{n} |\nabla_i u| / (|\nabla_i u| + \alpha^{-1})$ tends to be $\||\nabla_i u|\|_0$. Note that the parameter α used here should be set large enough. The approximating l_0-norm is a much sparser measurement than the l_1-norm. In sparse representation (Daubechies et al. 2010; Candes et al. 2008), the sparse property of the approximating l_0-norm has been widely used, which will lead to preserving edges of images.

Secondly, they present the underlying reason why the regularizer can protect edges from oversmoothing. This is equivalent to finding out what a good function $\varphi(\cdot)$ should be. On one hand, in order to protect edges from oversmoothing, $\varphi(s)$ should be imposed a "growth" condition of the type $\lim_{s \to +\infty} \varphi(s) = c$ (c is a constant) so that the contribution of the regularizer would not penalize the formation of strong gradients of u. In other words, the growth condition is used to protect large details of images. On the other hand, at near zero points ($s \to 0^+$), $\varphi(s)$ is preferable to have the same behavior as the TV regularizer so that u can be better smoothed in homogeneous regions of images. To make a balance between preserving edges and smoothing homogeneous regions, necessarily $\varphi(s)$ should have a nonconvex shape like the type $\varphi(s) = \alpha s/(1+\alpha s)$. Three different choices of $\varphi(s)$ are shown in Fig. 5. Therefore, the nonconvex sparse regularization is better than convex TV because the TV regularizer does not satisfy the growth condition (Fig. 6).

(10) Nonconvex TGV

Recently, Ochs et al. (2015) proposed a nonconvex extension of the TGV regularizer as follows:

$$NTGV(u) = \min_{p \in P} \int_{\Omega} \alpha_1 \varphi\left(|\nabla u - p|\right) + \alpha_0 \varphi\left(|\varepsilon(p)|\right) dx \qquad (62)$$

where $\varphi(x) = \frac{1}{\rho} \log(1 + \rho x)$ with the parameter $\rho > 0$ controlling the nonconvexity of the regularization term. This regularization takes advantage of both nonconvex regularization and TGV regularization.

The authors propose the following model Na et al. (2018) for the removal of heavy multiplicative noise, which utilizes an NTGV and $\lambda : \Omega \to R_+$:

9 Models for Multiplicative Noise Removal

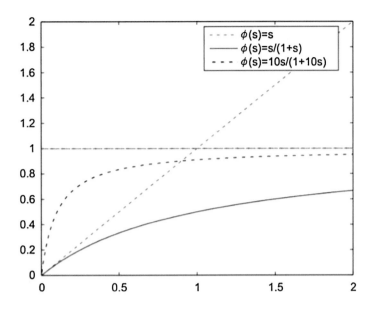

Fig. 5 Nonconvex and convex functions $\varphi(\cdot)$. The nonconvex function $\varphi(s) = s/(1+s)$ (resp. $\varphi(s) = 10s/(1+10s)$) corresponds to $\alpha = 1$ (resp. $\alpha = 10$). Both of their limits are 1 as $s \to +\infty$. The convex function $\varphi(s) = s$ corresponds to the case of the TV regularizer

Fig. 6 Local enlarged denoising results. From left to right, the clean image, the denoising results of the AA model, the BF model, and the nonconvex sparse regularizer model are listed

$$\min_{u \in X} \int_\Omega \lambda(x) \left[u + f e^{-u} + \alpha \left(\sqrt{\frac{e^u}{f}} - \beta 1 \right)^2 \right] dx + NTGV(u), \quad (63)$$

with the NTGV defined as

$$NTGV(u) = \min_{p \in P} \int_\Omega \alpha_1 \varphi_1 \left(|\nabla u - p| \right) + \alpha_0 \varphi_0 \left(|\varepsilon(p)| \right) dx, \quad (64)$$

where φ_i ($i = 0, 1$) are the nonconvex log functions, $\varphi_i(x) = \frac{1}{\rho_i} \log(1 + \rho_i x)$, where $\rho_i > 0$ control the nonconvexity of regularization terms. The parameters $\alpha > 0$ and $\beta \geq 0$ satisfy the condition $\alpha \beta^4 \leq \frac{4096}{27}$ to enforce the convexity of the data fidelity term. X and P are the corresponding solution spaces.

Multitasks

One of the advantages of using the variational method to build a model is that it can be easily extended to multitasking situations.

Root Transformation

The degraded image f is given by

$$f = (Au)n, \tag{65}$$

where A is a known linear and continuous blurring operator and $n \in L^2(\Omega)$ represents multiplicative noise with mean 1. Here, f is obtained from u, which is blurred by the blurring operator A and then is corrupted by the multiplicative noise n, assuming that $f > 0$. Until the past decade, a few variational methods have been proposed to handle the restoration problem with the multiplicative noise. Given the statistical properties of the multiplicative noise n, in Rudin et al. (2003) the recovery of the image $\hat{u}$ was based on solving the following constrained optimization problem:

$$\min_{u \in S(\Omega)} \int_\Omega |Du|$$

$$\text{s.t.} \int_\Omega \frac{f}{Au} dx = 1,$$

$$\int_\Omega \left(\frac{f}{Au} - 1 \right)^2 dx = \theta^2, \tag{66}$$

where θ^2 denotes the variance of n, $S(\Omega) = \{v \in BV(\Omega) : v \geq 0\}$, and $BV(\Omega)$ is the space of functions of bounded variation. In (66), only basic statistical properties, the mean and the variance, of the noise n are considered, which somehow limits the restored results. For this reason, based on the Bayes rule and Gamma distribution with mean 1, by using MAP estimator, Aubert and Aujol (2008) introduced a variational model as follows:

$$\min_{u \in S(\Omega)} \int_\Omega \left(\log(Au) + \frac{f}{Au} \right) dx + \lambda \int_\Omega |Du| \tag{67}$$

A quadratic penalty term is introduced in (67), which turns out to be

$$\min_{u \in \bar{S}(\Omega)} E_A(u) := \int_\Omega \left(\log(Au) + \frac{f}{Au}\right) dx + \alpha \int_\Omega \left(\sqrt{\frac{Au}{f}} - 1\right)^2 dx + \lambda \int_\Omega |Du| \tag{68}$$

where $\bar{S}(\Omega) := \{v \in BV(\Omega) : v \geq 0\}$.

Proposition 1. *If* $\alpha \geq \frac{2\sqrt{6}}{9}$, *then the model* (68) *is convex.*

Inspired by Dong and Zeng's model (Dong and Zeng 2013), the following TGV regularized model was presented in Shama et al. (2016)

$$\min_{u \in L^P(\Omega)} E(u) = \int_\Omega \left(\log Hu + \frac{f}{Hu}\right) dx + \beta \int_\Omega \left(\sqrt{\frac{Hu}{f}} - 1\right)^2 dx + TGV_\alpha^2(u) \tag{69}$$

where $\beta \geq \frac{2\sqrt{6}}{9}$, $p \in (1, \infty)$ and $p \leq d/(d-1)$, and $d = 2$ for the two-dimensional case.

Fractional Transformation

Zhao et al. (2014) introduced a new convex total variation-based model for restoring images contaminated with multiplicative noise and blur. The main notion is to reformulate a blur and multiplicative noise equation such that both the image variable and noise variable are decoupled. As a result, the concluding energy function involves the total variational filter, the term of the variance of the inverse of noise, the l_1-norm of the data fidelity term among the observed image, noise, and image variables. The convex optimization model is given by

$$\min_{w, u \in R^d} \frac{1}{2} \|w - \mu e\|_2^2 + \alpha_1 \|Fw - Hu\|_1 + \alpha_2 \|Du\|_2 \tag{70}$$

where α_1 and α_2 are two positive regularization parameters to control the balance between the three terms in the objective function, μ can be set to be the mean value of w, and e is a vector with all entries equal to 1.

Nonlocal Methods

Non-local means (NLM) is an algorithm in image processing for image denoising, which estimates each pixel based on the weighted average of all pixels inside a search window. The weight of a contributing pixel is evaluated on the basis of "similarity measure" of a neighborhood between the contributing and the target pixels. The NLM algorithm produces Gaussian denoised results with a higher peak signal-to-noise ratio (PSNR) value as well as good perceptual quality. It is natural to extend it to the non-Gaussian noise removal setting (Laus and Steidl 2019).

Indirect Method

In Huang et al. (2017), the Box-Cox transform is used to transform the random variable into an approximately normal distribution, and then the similar block BM3D method is used to denoise. We know Box-Cox transformation (Box and Cox 1964) can effectively transform a random variable and force it to follow normal distribution exactly or approximately if a suitable transformation parameter is selected. Furthermore, BM3D (Dabov et al. 2007) proposed by Dabov et al. is a rather novel method for additive Gaussian white noise removal. Therefore, inspired by the work proposed in Makitalo and Foi (2010, 2014), the authors proposed to transform the multiplicative noise removal to additive Gaussian noise removal by applying the Box-Cox transformation in Huang et al. (2017), and the images are finally recovered by an unbiased denoising algorithm. The Box-Cox transformation parameter is determined through a maximum likelihood method. After applying the Box-Cox transformation to the observed images, the BM3D method is utilized to restore the transformed image, and an unbiased improvement is performed so that the recovered image can finally be obtained.

Applying Box-Cox transformation with parameter λ to each pixel variable of $f = un$ to get

$$f^{(\lambda)} = \frac{(un)^\lambda - 1}{\lambda} \tag{71}$$

Suppose that u and n are independent, the expectation of $f^{(\lambda)}$ reads as

$$E\left(f^{(\lambda)}\right) = E\left(\frac{(un)^\lambda - 1}{\lambda}\right) = \frac{\Gamma(L + \lambda)}{\lambda L^\lambda \Gamma(L)} u^\lambda - \frac{1}{\lambda} \tag{72}$$

If λ is selected appropriately, $f^{(\lambda)}$ should follow or be close to Gaussian distribution, and it can be expressed as

$$f^{(\lambda)} = \frac{\Gamma(L + \lambda)}{\lambda L^\lambda \Gamma(L)} u^\lambda - \frac{1}{\lambda} + \varepsilon \tag{73}$$

where the random variable $\varepsilon \sim N\left(0, \sigma^2\right)$ is based on the assumptions in the Box-Cox transformation. In (73), if we consider $\frac{\Gamma(L+\lambda)}{\lambda L^\lambda \Gamma(L)} u^\lambda - \frac{1}{\lambda}$ as the original image and $f^{(\lambda)}$ as the observed image, the additive noise removal methods can be applied to (73) and a denoised approximation w of $\frac{\Gamma(L+\lambda)}{\lambda L^{\lambda}\Gamma(L)}u^{\lambda} - \frac{1}{\lambda}$ can be recovered. Finally, the reconstructed image can be obtained.

$$\hat{u} = L \left(\frac{\Gamma(L)(\lambda w + 1)}{\Gamma(L+\lambda)} \right)^{\frac{1}{\lambda}} \tag{74}$$

Direct Method

If we use nonlocal mean directly, the key is how to correctly estimate similar blocks under multiplicative noise. Now, to measure whether $u_1 = u_2$ by the noisy observations f_1, f_2, Deledalle et al. (2009) suggest using an approximate

$$s_{DDT}(f_1, f_2) := \int_S p_{f_1|u_1}(f_1|u) p_{f_2|u_2}(f_2|u) du \tag{75}$$

of the conditional density

$$p_{u_1-u_2|(f_1,f_2)}(0|f_1,f_2) = \frac{\int_S p_{u_1}(u) p_{u_2}(u) p_{f_1|u_1}(f_1|u) p_{f_2|u_2}(f_2|u) du}{p_{f_1}(f_1) p_{f_2}(f_2)} \tag{76}$$

as a measure of similarity. S_{DDT} is equal to the NL-mean filter under additive noise.

When the above method is generalized to multiplicative noise, the conditional density

$$p_{u_1-u_2|(f_1,f_2)}(0|f_1,f_2) = \int_S \frac{p_{u_1}(u) p_{u_2}(u)}{p_{f_1}(f_1) p_{f_2}(f_2)} p_{f_1|u_1}(f_1|u) p_{f_2|u_2}(f_2|u) du$$

$$\int_S \frac{p_{u_1}(u) p_{u_2}(u)}{p_{f_1}(f_1) p_{f_2}(f_2)} \frac{1}{u^2} p_{n_1}\left(\frac{f_1}{u}\right) p_{v_2}\left(\frac{f_2}{u}\right) du \tag{77}$$

is approximated by S_{DDT}.

For multiplicative Gamma noise,

$$s_{DDT}(f_1, f_2) = L \frac{\Gamma(2L-1)}{\Gamma(L)^2} \frac{(f_1 f_2)^{L-1}}{(f_1 + f_2)^{2L-1}}$$

$$= L \frac{\Gamma(2L-1)}{\Gamma(L)^2} \frac{1}{f_1 + f_2} \frac{1}{\left(2 + \frac{f_1}{f_2} + \frac{f_2}{f_1}\right)^{L-1}} \tag{78}$$

However, this measure does not seem to be optimal for multiplicative noise.

Next, a logarithmic transformation is done, and then the approximation of the conditional density is used. Considering the logarithmically transformed random variables $\tilde{f}_i = \ln(f_i)$, where

$$\underbrace{\ln(f_i)}_{\tilde{f}_i} = \ln(u_i n_i) = \underbrace{\ln(u_i)}_{\tilde{u}_i} + \underbrace{\ln(n_i)}_{\tilde{n}_i}, \quad i = 1, 2. \tag{79}$$

Lemma 1. For $f_1, f_2 > 0$ with $p_{f_i}(f_i)$ and $S = \text{supp}(p_{\tilde{u}_i})$, it holds that

$$p_{\tilde{u}_1 - \tilde{u}_2 | (\tilde{f}_1, \tilde{f}_2)} \left(0 | \ln(f_1), \ln(f_2) \right)$$

$$= \int_{\tilde{S}} \frac{p_{\tilde{u}_1}(t) \, p_{\tilde{u}_2}(t) \, p_{\tilde{f}_1 | \tilde{u}_1}\left(\ln(f_1) \big| t \right) p_{\tilde{f}_2 | \tilde{u}_2}\left(\ln(f_2) \big| t \right)}{p_{\tilde{f}_1}\left(\ln(f_1) \right) p_{\tilde{f}_2}\left(\ln(f_2) \right)} dt$$

$$= p_{\frac{u_1}{u_2} | (f_1, f_2)}\left(0 | f_1, f_2 \right) \tag{80}$$

Suppose n_i, $i = 1, 2$, be Gamma distributed random variables, for $f_1, f_2 > 0$, we can obtain

$$\begin{aligned} s(f_1, f_2) &= \frac{L^{2L}}{\Gamma(L)^2} (f_1 f_2)^L \int_0^{+\infty} \frac{1}{u^{2L+1}} \exp\left(-L \frac{f_1 + f_2}{u} \right) du \\ &= \frac{\Gamma(2L)}{\Gamma(L)^2} \frac{(f_1 f_2)^L}{(f_1 + f_2)^{2L}} = \frac{\Gamma(2L)}{\Gamma(L)^2} \frac{1}{\left(2 + \frac{f_1}{f_2} + \frac{f_2}{f_1} \right)^L} \end{aligned} \tag{81}$$

which has a maximum of $c = \frac{\Gamma(2L)}{\Gamma(L)^2} \frac{1}{4^L}$.

Then we can use the similarity (81) to calculate the weight function required by the NLM to determine Nonlocal filters for multiplicative Gamma noise.

DNN Method

At present, neural network-based methods have achieved great success in data processing and have also been applied to additive denoising and recovery, such as MLP, CSF, TNRD, and DnCNN (Chen and Pock 2016; Burger et al. 2012; Zhang et al. 2017). This induces us to generalize it to multiplicative noise removal.

Indirect Method

The splitting method is used to solve the variational problem, and one of the subproblems is replaced by DNN, which is essentially a plug-and-play model. Wang et al. (2019) propose a model for general multiplicative noise removal in (82).

$$\left(u^{k+1}, w^{k+1}\right) = \arg\min E(u, w)$$
$$= \left\{ \int_\Omega \left(afe^{-w} + \frac{b}{2}f^2 e^{-2w} + cw\right) dx + \frac{\theta_1}{2} \int_\Omega \left(u - e^w - d^k\right)^2 dx + \lambda \int_\Omega \Phi(u)\, dx \right\} \tag{82}$$

where θ_1 is the balance parameter. The second term $\int_\Omega \left(u - e^w - d^k\right)^2 dx$ makes $u = e^w$, and d^k is the Bregman distance. The last term is the deep CNN denoiser prior. Getting the solution of (82) directly is hard because of the term of $\Phi(u)$. Using the split method, we can import auxiliary variable $z = u$. Then (82) can be transformed into (83).

$$\left(u^{k+1}, w^{k+1}, z^{k+1}\right) = \arg\min E(u, w, z)$$
$$= \left\{ \int_\Omega \left(afe^{-w} + \frac{b}{2}f^2 e^{-2w} + cw\right) dx + \frac{\theta_1}{2} \int_\Omega \left(u - e^w - d^k\right)^2 dx \right.$$
$$\left. + \lambda \int_\Omega \Phi(z)\, dx + \frac{\theta_2}{2} \int_\Omega (z - u)^2 dx \right\} \tag{83}$$

For Gaussian noise, the parameters can be set as $c = 0, b = 1, a = -1$. Then (83) is changed into the form of (84).

$$\left(u^{k+1}, w^{k+1}, z^{k+1}\right) = \arg\min E(u, w, z)$$
$$= \left\{ \int_\Omega \left(\frac{1}{2}f^2 e^{-2w} - fe^{-w}\right) dx + \frac{\theta_1}{2} \int_\Omega \left(u - e^w - d^k\right)^2 dx \right.$$
$$\left. + \lambda \int_\Omega \Phi(z)\, dx + \frac{\theta_2}{2} \int_\Omega (z - u)^2 dx \right\} \tag{84}$$

Each variable will be solved separately after dissociation; by using the alternating optimization strategy, the optimization (84) can be divided into the following subproblems on (u, w, z):

$$w^{k+1} = \arg\min_{w} \left\{ E(w) = \int_{\Omega} \left(\frac{1}{2} f^2 e^{-2w} - f e^{-w} \right) dx + \frac{\theta_1}{2} \int_{\Omega} \left(u - e^w - d^k \right)^2 dx \right\}$$
(85)

$$z^{k+1} = \arg\min_{z} \left\{ E(z) = \lambda \int_{\Omega} \Phi(z) \, dx + \frac{\theta_2}{2} \int_{\Omega} (z - u)^2 \, dx \right\}$$
(86)

$$u^{k+1} = \arg\min_{u} \left\{ E(u) = \frac{\theta_1}{2} \int_{\Omega} \left(u - e^w - d^k \right)^2 dx + \frac{\theta_2}{2} \int_{\Omega} (z - u)^2 \, dx \right\}$$
(87)

For calculating w, we can deduce the corresponding Euler-Lagrange equation:

$$f e^{-w} - f^2 e^{-2w} - \theta_1 \left(u^{k+1} - e^w - d^k \right) = 0$$
(88)

We can solve w via gradient descent method as

$$w^{k+1} = w^k - \Delta t \, S^k$$
(89)

where Δt represents the time step, and

$$S^k = f e^{-w} - f^2 e^{-2w} - \theta_1 \left(u^{k+1} - e^w - d^k \right)$$
(90)

For calculating z, (86) can be changed into (91)

$$\min_{z} \left\{ E(z) = \int_{\Omega} \Phi(z) \, dx + \frac{1}{2 \left(\sqrt{\frac{\lambda}{\theta_2}} \right)^2} \int_{\Omega} (z - u)^2 \, dx \right\}$$
(91)

According to Bayesian theory, (91) is the Gaussian denoiser and the noise variance is λ/θ_2. In Wang et al. (2019), the authors use the CNN Gaussian denoiser for solving (91) by considering the performance and discriminative image prior modeling. The reason for using CNN is that it has achieved great success in Gaussian denoising and better performance (such as PSNR results outperforms BM3D's (Dabov et al. 2007)) than a model-based method. By incorporating CNN Gaussian denoiser into the model, we need not retrain the multiplicative noise removal model for different types of noise. We can deal with different types of noise only by changing the data fidelity.

9 Models for Multiplicative Noise Removal

For solving u, the corresponding Euler-Lagrange equation of (87) is shown in (92).

$$\theta_1 \left(u - e^w - d^{k+1} \right) + \theta_2 (u - z) = 0 \tag{92}$$

So, we can get u by using (93).

$$u^{k+1} = \frac{\theta_1 \left(e^{w^{k+1}} + d^{k+1} \right) + \theta_2 z^{k+1}}{\theta_1 + \theta_2} \tag{93}$$

The Bregman distance can be expressed as $d^{k+1} = d^k + e^{w^{k+1}} - u^{k+1}$.

Taking into account the above equations, we obtain the complete iteration used in the algorithm for multiplicative noise removal (Fig. 7).

Algorithm of multiplicative noise removal

1. Initialization: $u^0 = f$, $d^0 = 0$, $w^0 = \log u^0$ $k = 0$
2. Repeat
3. Compute w^k using Eq. (89);
4. Compute z^k using Eq. (91);
5. Compute u^k using Eq. (93);
6. Compute Bergman parameter d

using $d^{k+1} = d^k + e^{w^{k+1}} - u^{k+1}$;

1. $k = k + 1$;
2. Until k achieved the presetting value.

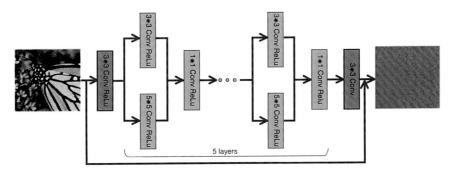

Fig. 7 The architecture of the proposed denoiser network

Direct Method

Let $F \in R^{W \times H}$ be the observed image intensity, $U \in R^{W \times H}$ be the noise free image, and $N \in R^{W \times H}$ be the speckle noise. Then assuming that the SAR image is an average of L looks, the observed image F is related to U by the following multiplicative model (Ulaby et al. 2019)

$$F = UN \tag{94}$$

One common assumption on N is that it follows a Gamma distribution with unit mean and variance $\frac{1}{L}$ and has the following probability density function (Ulaby et al. 2019)

$$p(N) = \frac{1}{\Gamma(L)} L^L N^{L-1} e^{-LN} \tag{95}$$

where $\Gamma(\cdot)$ denotes the Gamma function and $N \geq 0, L \geq 1$

The noise-estimating part of the ID-CNN network consists of eight convolutional layers (along with batch normalization and ReLU activation functions), with appropriate zero-padding to make sure that the output of each layer shares the same dimension with that of the input image (Wang et al. 2017). Each convolutional layer (except for the last convolutional layer) consists of 64 filters with the stride of one. Then the division residual layer with skip connection divides the input image by the estimated speckle noise. A hyperbolic tangent layer is stacked at the end of the network which serves as a nonlinear function. Here, L1 and L8 stand for the sequence of Conv-ReLU layers as depicted in Fig. 8. Similarly, L2 to L7 denote Conv-BNReLU layers. Some estimated results are shown in Table 2.

$$L_E(\varphi_E) = \frac{1}{WH} \sum_{w=1}^{W} \sum_{h=1}^{H} \left\| \varphi\left(F^{w,h}\right) - U^{w,h} \right\|_2^2 \tag{96}$$

$$L_{TV} = \sum_{w=1}^{W} \sum_{h=1}^{H} \sqrt{\left(\hat{U}^{w+1,h} - \hat{U}^{w,h}\right)^2 + \left(\hat{U}^{w,h+1} - \hat{U}^{w,h}\right)^2} \tag{97}$$

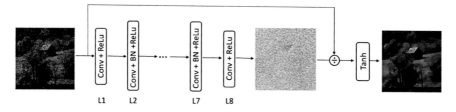

Fig. 8 Proposed ID-CNN network architecture for image despeckling

Table 2 The estimated results on real SAR images

# chip	PPB	SAR-BM3D	CNN	SAR-CNN	ID-CNN
1	42.49	69.26	32.32	50.76	**89.43**
2	8.63	10.95	7.50	8.93	**13.90**
3	103.25	127.38	31.65	99.13	**193.00**
4	34.84	63.83	7.65	43.13	**69.40**

Finally, the overall loss function is defined as follows:

$$L = L_E + \lambda_{TV} L_{TV} \tag{98}$$

Conclusion

This chapter presents a review of restoration models in the case of multiplicative noise. We introduce the main ideas for multiplicative denoising models and focus on the variational methods with different data fidelity terms, variant methods with different regularizers, multitasks methods, nonlocal methods, and DNN methods. We hope this chapter can provide some help for relevant researchers. We did not give the corresponding optimization method, although it is very important. The complete description probably needs twice as many pages as there are now. The interested reader can refer to the corresponding literature. We think Chambolle and Pock (2016) is a good review article for reference.

References

Abolhassani, M., Rostami, Y.: Speckle noise reduction by division and digital processing of a hologram. Optik **123**(10), 937–939 (2012)

Aubert, G., Aujol, J.-F.: A variational approach to removing multiplicative noise. SIAM J. Appl. Math. **68**(4), 925–946 (2008)

Box, G.E.P., Cox, D.R.: An analysis of transformations. J. R. Stat. Soc.: Ser. B (Methodol.) **26**(2), 211–243 (1964)

Bredies, K., Kunisch, K., Pock, T.: Total generalized variation. SIAM J. Imaging Sci. **3**(3), 492–526 (2010)

Buades, A., Coll, B., Morel, J.-M.: A non-local algorithm for image denoising. In: 2005 IEEE Computer Society Conference on Computer Vision and Pattern Recognition (CVPR'05), vol 2, pp. 60–65. IEEE (2005)

Burger, H.C., Schuler, C.J., Harmeling, S.: Image denoising: can plain neural networks compete with BM3D? In: 2012 IEEE Conference on Computer Vision and Pattern Recognition, pp. 2392–2399. IEEE (2012)

Candes, E.J., Wakin, M.B., Boyd, S.P.: Enhancing sparsity by reweighted l_1 minimization. J. Fourier Anal. Appl. **14**(5–6), 877–905 (2008)

Chambolle, A., Lions, P.-L.: Image recovery via total variation minimization and related problems. Numer. Math. **76**(2), 167–188 (1997)

Chambolle, A., Pock, T.: An introduction to continuous optimization for imaging. Acta Numer. **25**, 161–319 (2016)

Chan, T., Marquina, A., Mulet, P.: High-order total variation-based image restoration. SIAM J. Sci. Comput. **22**(2), 503–516 (2000)

Chatterjee, P., Milanfar, P.: Is denoising dead? IEEE Trans. Image Process. **19**(4), 895–911 (2009)

Chen, D.-Q., Cheng, L.-Z.: Spatially adapted total variation model to remove multiplicative noise. IEEE Trans. Image Process. **21**(4), 1650–1662 (2011)

Chen, Y., Pock, T.: Trainable nonlinear reaction diffusion: a flexible framework for fast and effective image restoration. IEEE Trans. Pattern Anal. Mach Intell. **39**(6), 1256–1272 (2016)

Chesneau, C., Fadili, J., Starck, J.-L.: Stein block thresholding for image denoising. Appl. Comput. Harmon. Anal. **28**(1), 67–88 (2010)

Dabov, K., Foi, A., Katkovnik, V., Egiazarian, K.: Image denoising by sparse 3-D transform-domain collaborative filtering. IEEE Trans. Image Process. **16**(8), 2080–2095 (2007)

Daubechies, I., DeVore, R., Fornasier, M., Güntürk, C.S.: Iteratively reweighted least squares minimization for sparse recovery. Commun. Pure Appl. Math.: J. Issued Courant Inst. Math. Sci. **63**(1), 1–38 (2010)

Deledalle, C.-A., Denis, L., Tupin, F.: Iterative weighted maximum likelihood denoising with probabilistic patch-based weights. IEEE Trans. Image Process. **18**(12), 2661–2672 (2009)

Denis, L., Tupin, F., Darbon, J., Sigelle, M.: SAR image regularization with fast approximate discrete minimization. IEEE Trans. Image Process. **18**(7), 1588–1600 (2009)

Dong, Y., Zeng, T.: A convex variational model for restoring blurred images with multiplicative noise. SIAM J. Imaging Sci. **6**(3), 1598–1625 (2013)

Durand, S., Fadili, J., Nikolova, M.: Multiplicative noise removal using L_1 fidelity on frame coefficients. J. Math. Imaging Vis. **36**(3), 201–226 (2010)

Elad, M., Aharon, M.: Image denoising via sparse and redundant representations over learned dictionaries. IEEE Trans. Image Process. **15**(12), 3736–3745 (2006)

Gilboa, G., Darbon, J., Osher, S., Chan, T.: Nonlocal convex functionals for image regularization. UCLA CAM-report, pp. 06–57 (2006)

Gilboa, G., Osher, S.: Nonlocal operators with applications to image processing. Multiscale Model. Simul. **7**(3), 1005–1028 (2009)

Han, Y., Feng, X.-C., Baciu, G., Wang, W.-W.: Nonconvex sparse regularizer based speckle noise removal. Pattern Recogn. **46**(3), 989–1001 (2013)

Hao, Y., Feng, X., Xu, J.: Multiplicative noise removal via sparse and redundant representations over learned dictionaries and total variation. Signal Process. **92**(6), 1536–1549 (2012)

Hoekman, D.H.: Speckle ensemble statistics of logarithmically scaled data (radar). IEEE Trans. Geosci. Remote Sens. **29**(1), 180–182 (1991)

Hu, X., Wu, Y.H., Li, L.: Analysis of a new variational model for image multiplicative denoising. J. Inequal. Appl. **2013**(1), 568 (2013)

Huang, Y.-M., Moisan, L., Ng, M.K., Zeng, T.: Multiplicative noise removal via a learned dictionary. IEEE Trans. Image Process. **21**(11), 4534–4543 (2012)

Huang, Y.-M., Yan, H.-Y., Zeng, T.: Multiplicative noise removal based on unbiased box-cox transformation. Commun. Comput. Phys. **22**(3), 803–828 (2017)

Jin, Z., Yang, X.: Analysis of a new variational model for multiplicative noise removal. J. Math. Anal. Appl. **362**(2), 415–426 (2010)

Kang, M., Yun, S., Woo, H.: Two-level convex relaxed variational model for multiplicative denoising. SIAM J. Imaging Sci. **6**(2), 875–903 (2013)

Krishnan, D., Fergus, R.: Fast image deconvolution using hyper-Laplacian priors. In: Advances in Neural Information Processing Systems, pp. 1033–1041 (2009)

Laus, F., Steidl, G.: Multivariate myriad filters based on parameter estimation of Student-t distributions. SIAM J. Imaging Sci. **12**(4), 1864–1904 (2019)

Le, T., Vese, L.: Additive and multiplicative piecewise-smooth segmentation models in a variational level set approach. UCLA CAM Report 03-52, University of California at Los Angeles, Los Angeles (2003)

Lebrun, M., Colom, M., Buades, A., Morel, J.-M.: Secrets of image denoising cuisine. Acta Numer. **21**, 475 (2012)

Li, F., Shen, C., Fan, J., Shen, C.: Image restoration combining a total variational filter and a fourth-order filter. J. Vis. Commun. Image Represent. **18**(4), 322–330 (2007)

Lu, J., Shen, L., Xu, C., Xu, Y.: Multiplicative noise removal in imaging: an exp-model and its fixed-point proximity algorithm. Appl. Comput. Harmon. Anal. **41**(2), 518–539 (2016)

Makitalo, M., Foi, A.: Optimal inversion of the Anscombe transformation in low-count Poisson image denoising. IEEE Trans. Image Process. **20**(1), 99–109 (2010)

Makitalo, M., Foi, A.: Noise parameter mismatch in variance stabilization, with an application to Poisson–Gaussian noise estimation. IEEE Trans. Image Process. **23**(12), 5348–5359 (2014)

Mei, J.-J., Dong, Y., Huang, T.-Z., Yin, W.: Cauchy noise removal by nonconvex ADMM with convergence guarantees. J. Sci. Comput. **74**(2), 743–766 (2018)

Na, H., Kang, M., Jung, M., Kang, M.: Nonconvex TGV regularization model for multiplicative noise removal with spatially varying parameters. Inverse Probl. Imaging **13**(1), 117 (2018)

Na, H., Kang, M., Jung, M., Kang, M.: An exp model with spatially adaptive regularization parameters for multiplicative noise removal. J. Sci. Comput. **75**(1), 478–509 (2018)

Nikolova, M., Ng, M.K., Tam, C.-P.: Fast nonconvex nonsmooth minimization methods for image restoration and reconstruction. IEEE Trans. Image Process. **19**(12), 3073–3088 (2010)

Ochs, P., Dosovitskiy, A., Brox, T., Pock, T.: On iteratively reweighted algorithms for nonsmooth nonconvex optimization in computer vision. SIAM J. Imaging Sci. **8**(1), 331–372 (2015)

Rudin, L., Lions, P.-L., Osher, S.: Multiplicative denoising and deblurring: theory and algorithms. In: Geometric Level Set Methods in Imaging, Vision, and Graphics, pp. 103–119. Springer, New York (2003)

Rudin, L.I., Osher, S., Fatemi, E.: Nonlinear total variation based noise removal algorithms. Phys. D: Nonlinear Phenom. **60**(1–4), 259–268 (1992)

Setzer, S., Steidl, G., Teuber, T.: Deblurring Poissonian images by split Bregman techniques. J. Vis. Commun. Image Represent. **21**(3), 193–199 (2010)

Shama, M.-G., Huang, T.-Z., Liu, J., Wang, S.: A convex total generalized variation regularized model for multiplicative noise and blur removal. Appl. Math. Comput. **276**, 109–121 (2016)

Shao, L., Yan, R., Li, X., Liu, Y.: From heuristic optimization to dictionary learning: a review and comprehensive comparison of image denoising algorithms. IEEE Trans. Cybern. **44**(7), 1001–1013 (2013)

Shi, J., Osher, S.: A nonlinear inverse scale space method for a convex multiplicative noise model. SIAM J. Imaging Sci. **1**(3), 294–321 (2008)

Singh, P., Jain, L.: A review on denoising of images under multiplicative noise. Int. Res. J. Eng. Technol. (IRJET) **03**(04), 574–579 (2016)

Steidl, G., Teuber, T.: Removing multiplicative noise by Douglas-Rachford splitting methods. J. Math. Imaging Vis. **36**(2), 168–184 (2010)

Teuber, T., Lang, A.: A new similarity measure for nonlocal filtering in the presence of multiplicative noise. Comput. Stat. Data Anal. **56**(12), 3821–3842 (2012)

Tian, D., Du, Y., Chen, D.: An adaptive fractional-order variation method for multiplicative noise removal. J. Inf. Sci. Eng. **32**(3), 747–762 (2016)

Ulaby, F., Dobson, M.C., Álvarez-Pérez, J.L.: Handbook of Radar Scattering Statistics for Terrain. Artech House, Norwood (2019)

Ullah, A., Chen, W., Khan, M.A.: A new variational approach for restoring images with multiplicative noise. Comput. Math. Appl. **71**(10), 2034–2050 (2016)

Ullah, A., Chen, W., Khan, M.A., Sun, H.: A new variational approach for multiplicative noise and blur removal. PloS One **12**(1), e0161787 (2017)

Wang, P., Zhang, H., Patel, V.M.: SAR image despeckling using a convolutional neural network. IEEE Signal Process. Lett. **24**(12), 1763–1767 (2017)

Wang, G., Pan, Z., Zhang, Z.: Deep CNN Denoiser prior for multiplicative noise removal. Multimed. Tools Appl. **78**(20), 29007–29019 (2019)

Xiao, L., Huang, L.-L., Wei, Z.-H.: A Weberized total variation regularization-based image multiplicative noise removal algorithm. EURASIP J. Adv. Signal Process. **2010**, 1–15 (2010)

Xie, H., Pierce, L.E., Ulaby, F.T.: Statistical properties of logarithmically transformed speckle. IEEE Trans. Geosci. Remote Sens. **40**(3), 721–727 (2002)

Yun, S., Woo, H.: A new multiplicative denoising variational model based on mth root transformation. IEEE Trans. Image Process. **21**(5), 2523–2533 (2012)

Zhao, X.-L., Wang, F., Ng, M.K.: A new convex optimization model for multiplicative noise and blur removal. SIAM J. Imaging Sci. **7**(1), 456–475 (2014)

Zhao, C.-P., Feng, X.-C., Jia, X.-X., He, R.-Q., Xu, C.: Root-transformation based multiplicative denoising model and its statistical analysis. Neurocomputing **275**, 2666–2680 (2018)

Zhang, K., Zuo, W., Chen, Y., Meng, D., Zhang, L.: Beyond a gaussian denoiser: residual learning of deep CNN for image denoising. IEEE Trans. Image Process. **26**(7), 3142–3155 (2017)

Recent Approaches to Metal Artifact Reduction in X-Ray CT Imaging

10

Soomin Jeon and Chang-Ock Lee

Contents

Introduction	348
Background: CT Image Formation and Metal Artifacts	350
Methods	353
Normalized Metal Artifact Reduction (NMAR)	353
Surgery-Based Metal Artifact Reduction (SMAR)	355
Convolutional Neural Network-Based MAR (CNN-MAR)	357
Industrial Application: 3D Cone Beam CT	360
Simulations and Results	365
Simulation Conditions	365
NMAR vs. SMAR: Patient Image Simulations	365
SMAR vs. CNN-MAR	369
NMAR vs. SMAR for 3D CBCT	371
Conclusion	374
References	375

Abstract

Metal artifacts severely degrade image quality by generating streak artifacts in X-ray computed tomography (CT) images. Metal artifact reduction (MAR) has long been an important issue because metal artifacts interfere with the acquisition of accurate contrast images, limiting the various applications of CT imaging. In this

---

S. Jeon
Department of Radiology, Massachusetts General Hospital and Harvard Medical School, Boston, MA, USA
e-mail: sjeon3@mgh.harvard.edu

C.-O. Lee (✉)
Department of Mathematical Sciences, KAIST, Daejeon, Republic of Korea
e-mail: colee@kaist.edu

© Springer Nature Switzerland AG 2023
K. Chen et al. (eds.), *Handbook of Mathematical Models and Algorithms in Computer Vision and Imaging*, https://doi.org/10.1007/978-3-030-98661-2_114

work, three recently developed CT MAR methods are introduced: normalized MAR, surgery-based MAR, and convolutional neural network-based MAR. Also, a MAR method for industrial cone beam CT is presented as an industrial application.

Keywords

Computed tomography (CT) · Convolutional neural network (CNN) · Normalized metal artifact reduction (NMAR) · Sinogram · Surgery based metal artifact reduction (SMAR)

Introduction

X-ray computed tomography (CT) is one of the most widely used tomographic imaging techniques for non-destructive visualization of structures inside objects. X-ray CT uses radiation from X-rays whose energy is absorbed according to the attenuation coefficients of the tissues in its path (Deans 2007). The cross-sectional image is reconstructed slice by slice from the measured X-ray data at different angles around the scanned object.

X-ray CT produces detailed, high-quality images, and its applicability is promising, but there are various artifacts that severely degrade the quality of CT images: beam hardening artifacts, scattering artifacts, and artifacts due to partial volume effects, photon starvation, undersampling, etc. (Barrett and Keat 2004). Artifacts in CT images are defined as system-induced discrepancies between the reconstructed CT image and the ground truth. These artifacts can be classified according to their causes: (i) physics-based artifacts arising from the physical processes involved during CT data acquisition; (ii) patient-based artifacts caused by factors such as patient movement or the presence of metallic objects in or on the patient; (iii) scanner-based artifacts due to defects in certain scanner functions; and (iv) others such as helical and multi-section artifacts. Among the various causes, implanted metals such as chest screws, dental fillings, and hip prostheses bring the most serious artifacts in CT images. They can also be classified according to their shape as streak artifacts, ring artifacts, cupping artifacts, etc.

The term metal artifact is a generic term for all artifacts caused by metallic objects such as dental implants and surgical clips which lead to various effects such as beam hardening, photon starvation, scattering, and noise increases (Boas and Fleischman 2012). Metal artifacts spread over the entire image in a bright and shadowy crown shape, damaging the quality of CT images and preventing accurate diagnoses. For this reason, as CT imaging becomes more popular, the importance of metal artifact reduction (MAR) technique increases.

Various studies have been attempted to understand metal artifacts, and several approaches have been proposed to reduce them. Existing MAR methods can be roughly classified into three categories: inpainting methods in the projection domain, iterative reconstruction methods, and other methods. For methods based on projection domain inpainting, sinogram data is calibrated with various types of inpainting techniques such as polynomial interpolation (Abdoli et al. 2010; Kalender et al. 1987; Klotz et al. 1990; Mahnken et al. 2003; Wei et al. 2004), wavelets (Zhao et al. 2002), Euler's elastica model (Gu et al. 2006), interpolation using adjacent pixel values (Kim et al. 2010), and forward projection (Bal and Spies 2006; Prell et al. 2009). However, these methods generate additional artifacts in the reconstructed CT image due to the inconsistency of the calibrated sinogram after inpainting. These extra artifacts deteriorate the quality of the X-ray CT image. In iterative reconstruction methods, the image is updated in a feedback manner through forward projection and back projection, for example, adding physics knowledge such as acquisition process and photon statistics (De Man et al. 2001; Kano and Koseki 2016; Lemmens et al. 2009; Wang et al. 1996). Iterative reconstruction methods achieve better image quality than inpainting-based methods. However, they are usually much slower due to the high computational cost. The category of other methods includes filtering methods (Bal et al. 2005; Kachelrieß et al. 2001), methods based on the wave front set (Park et al. 2016), and those with total variation minimization (Verburg and Seco 2012). There are also hybrid methods that combine different MAR methods (Watzke and Kalender 2004; Zhang et al. 2013). The first clinical application of the iterative MAR algorithm was achieved by Philips Health Care (Philips Healthcare 2012) though it was only applied to orthopedic implant cases (Zhang et al. 2020). The algorithm used by Phillips Health Care was based on the work by Timmer and Koehler (Koehler et al. 2012; Timmer 2008), whose methods were applied experimentally only for simple-shaped phantoms, with the result showing that residual artifacts still existed.

One of the state-of-the-art MAR algorithms is the normalized MAR (NMAR) algorithm (Meyer et al. 2010). It inpaints the corrupted part of the sinogram using normalization technique in conjunction with the prior image. In the first step, the metal is segmented in the image domain by a threshold. The forward projection then identifies the metal traces in the original projection. Before interpolation, the projection data is normalized based on the forward projection of the prior image. The prior image is acquired, for example, by a multi-threshold segmentation of the initial image. The original raw data is divided by the projection data of the prior image and then denormalized after interpolation.

Recently, a new metal artifact reduction algorithm based on sinogram surgery has been proposed to reduce metal artifacts without additional ones (Jeon and Lee 2018). The area around the metal region with similar CT numbers is extracted using the reconstructed CT numbers from the given sinogram. Then, the metal region and its surroundings are filled with the average CT number of the surrounding area to obtain a modified CT image. Using the forward projection of the modified CT image, a sinogram containing information about the anatomical structure is generated, and the sinogram surgery is performed using this and then back-projected to regenerate the CT image. The reconstructed CT image contains structural information around the metal region even if the original CT image includes severe artifacts near metallic objects. Unlike other interpolation-based MAR methods, the proposed algorithm uses this structural information to correct the corruption in the sinogram. The sinogram completion process is iteratively performed using the basic principles of CT image reconstruction to remove the metal effect from the sinogram.

Meanwhile, attempts are underway to exploit deep learning in almost all fields of science and technology. In particular, the concept of deep learning was also introduced to MAR in Ghani and Karl (2020), Gjesteby et al. (2017), Hwang et al. (2018), and Zhang and Yu (2018). Among these, the convolutional neural network (CNN)-based MAR (CNN-MAR) method (Zhang and Yu 2018) is best known as a general open framework. The CNN-MAR method consists of two phases: CNN phase and surgery phase with a prior image. In the CNN phase, CNN is used as an information fusion tool to produce a reduced artifact image by combining the uncorrected CT image and two pre-corrected ones from some model-based MAR methods as the input data of the neural network. The surgery phase further reduces the remaining artifacts by adding seamless surgery process with a prior image based on tissue classification.

This work introduces the NMAR algorithm (Meyer et al. 2010), the surgery-based MAR (SMAR) algorithm (Jeon and Lee 2018), and the CNN-MAR methods (Zhang and Yu 2018). It also reviews a methodology for reducing metal artifacts in three-dimensional industrial cone beam CT systems (Jeon et al. 2021).

Background: CT Image Formation and Metal Artifacts

In an X-ray CT system, the X-ray source and detector rotate simultaneously at regular angular intervals. A single projection data is obtained for each angle, and a stack of these projection data is called a sinogram. A cross-sectional image of an object can be obtained from the sinogram through a reconstruction process called filtered back projection (FBP). Depending on the geometry of photon spread, there are different types of X-ray CT systems, such as parallel beam CT, fan beam CT, cone beam CT, and helical CT. This work assumes using a parallel beam CT.

Let $f_E(\mathbf{x})$ denote the X-ray attenuation coefficient at a point $\mathbf{x}$ when the X-ray energy level is E. The Beer-Lambert law (Klotz et al. 1990) describes the attenuation of an X-ray along the path through which the X-ray passes a physical substance composed of a single species of uniform concentration by the first-order ordinary differential equation

$$\frac{dI}{dt}(\mathbf{x}) = -f_E(\mathbf{x})I(\mathbf{x}), \quad \mathbf{x} = s\Theta + t\Theta^\perp, \Theta = (\cos\theta, \sin\theta),$$

where I is the intensity of X-ray, s the distance along the detector, and t the distance along the path of X-ray; see Fig. 1. Solving the above equation gives the formula for I at the detector:

$$I_\theta(E, s) = I_0(E)e^{-\mathcal{R}_\theta f_E(s)},$$

10 Recent Approaches to Metal Artifact Reduction in X-Ray CT Imaging

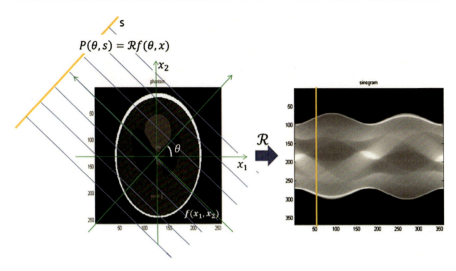

Fig. 1 Illustration of the Radon transform and the sinogram. (Reprinted from Jeon and Lee (2018) with permission from IOS Press)

where $I_0(E)$ is the initial intensity of the X-ray with energy level E. Here, the projection data $\mathcal{R}_\theta f_E(s)$ is the Radon transform of f_E defined by

$$\mathcal{R}_\theta f_E(s) := \int_{\mathbb{R}^2} f_E(\mathbf{x}) \delta(\Theta \cdot \mathbf{x} - s)\, d\mathbf{x},$$

which means the integral along the line

$$L_{\theta,s} := \left\{ \mathbf{x} \in \mathbb{R}^2 : \Theta \cdot \mathbf{x} = s,\ \Theta = (\cos\theta, \sin\theta) \right\},$$

where δ is the Dirac delta function.

Since the sinogram is a stack of projection data, it can be expressed as $\mathcal{R} f_E(\theta, s) = [\mathcal{R}_\theta f_E(s)]_\theta := [\mathcal{R}_{\theta_1} f_E(s), \mathcal{R}_{\theta_2} f_E(s), \ldots]$; see Fig. 1. Assuming that the X-ray is monochromatic, there is a relation

$$\mathcal{R} f_E(\theta, s) = \left[-\ln\left(\frac{I_\theta(E, s)}{I_0(E)} \right) \right]_\theta.$$

Then, a two-dimensional X-ray CT image is reconstructed by the inverse Radon transform of the sinogram $\mathcal{R} f_E$:

$$\begin{aligned} f_E(\mathbf{x}) &= \mathcal{R}^{-1}\{\mathcal{R} f_E\} \\ &= \frac{1}{4\pi^2} \int_0^\pi \int_{-\infty}^\infty |\omega| \mathcal{F}_s[\mathcal{R} f_E(\theta, s)](\omega) e^{i\omega \mathbf{x} \cdot \Theta}\, d\omega d\theta, \end{aligned} \quad (1)$$

where $\mathcal{F}_s$ is the 1D Fourier transform with respect to s.

In practice, since X-ray CT uses polychromatic X-rays, the measured X-ray intensity is given by

$$I_\theta(s) = \int_{E_{\min}}^{E_{\max}} I_\theta(E, s)\, dE = \int_{E_{\min}}^{E_{\max}} I_0(E) e^{-\mathcal{R}_\theta f_E(s)}\, dE, \qquad (2)$$

where $E_{\min}$ and $E_{\max}$ are the minimum and maximum energy levels of the X-ray, respectively. Then the sinogram $\mathcal{P} f_E$ is given by

$$\mathcal{P} f_E = [\mathcal{P}_\theta f_E]_\theta,$$

for

$$\mathcal{P}_\theta f_E(s) = -\ln\left(\frac{I_\theta(s)}{I_0}\right), \qquad (3)$$

where $I_0 = \int_{E_{\min}}^{E_{\max}} I_0(E)\, dE$. Then, the CT image is reconstructed from the sinogram using (1) with $\mathcal{P} f_E$ instead of $\mathcal{R} f_E$. The CT image reconstruction is shown in Fig. 2.

A smooth function is called a Schwartz function if all its derivatives including itself decay at infinity faster than the inverse of any polynomial. A function $g(\theta, s)$ defined on $[0, 2\pi) \times \mathbb{R}$ is said to satisfy the homogeneous polynomial condition if for $k = 1, 2, \ldots$, the integral

$$\int_{\mathbb{R}} g(\theta, s) s^k\, ds \qquad (4)$$

can be written as a k-th degree homogeneous polynomial in $\Theta = (\cos\theta, \sin\theta)$. By the Schwartz theorem for the Radon transform (Helgason 1965), $g = \mathcal{R} f$ for

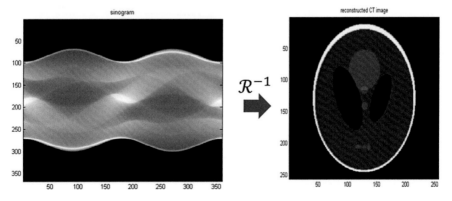

Fig. 2 Illustration of the reconstruction of CT image from sinogram with inverse Radon transform

some Schwartz function f if and only if g satisfies the homogeneous polynomial condition. In particular, when $k = 0$, (4) produces the same value regardless of θ for all sinograms generated by a single-energy X-ray CT machine. This is called the consistency condition of the sinogram.

Metal artifacts are mainly due to the beam hardening effect of polychromatic X-ray beam. When a polychromatic X-ray beam passes through an object, low-energy photons are preferentially absorbed, and thus the mean energy gradually increases (we say that the beam is hardened). The harder the beam, the less it attenuates. Therefore, the total attenuation is no longer proportional to the object thickness, unlike the monochromatic X-ray case. Hence, the generated sinogram becomes inconsistent. Because a monochromatic X-ray is assumed in the reconstruction (1), artifacts occur when a CT image is reconstructed from an inconsistent sinogram. If the target is a non-metal such as human tissue, these artifacts are not significant because the energy dependence of the attenuation coefficient is not high. However, the X-ray attenuation coefficient of materials such as metal with a high CT number is extremely energy dependent and produces erroneous projection data that is the source of metal artifacts.

Methods

Normalized Metal Artifact Reduction (NMAR)

Inpainting of Metal Traces in the Normalized Sinogram

The simplest inpainting-based MAR method is the LI method (Kalender et al. 1987) that fills the metal trace in the uncorrected sinogram by the linear interpolation of its neighboring unaffected projections in each projection view. In fact, the projection image called sinogram is made along the sine curve. Therefore, since the sinogram after interpolation is not consistent and even not smooth at the boundaries of the metal traces, new artifacts are necessarily introduced, and the structure near metals is distorted. However, such interpolation is less problematic in homogeneous regions, as interpolation on nearly flat sinogram provides a certain level of smoothness at the boundaries of the metal traces. The idea of normalization is to transform the sinogram so that it is comparatively flat.

Here, as a way to transform a sinogram into a more flat one, the method in Müller and Buzug (2009) is introduced. In the first step, the metal trace is determined by the forward projection of the metal extracted by the thresholding from the uncorrected CT image. A normalized sinogram is then created by dividing each pixel value of the given sinogram by the thickness of the object that the X-ray passes through. The metal trace determines where in the normalized sinogram is replaced by the inpainting (e.g., simple linear interpolation per projection view (Kalender et al. 1987)). Subsequently, the corrected sinogram is obtained by denormalizing the interpolated sinogram by multiplying it by the thickness of the object. Reconstructing a CT image with this corrected sinogram produces the corrected image.

This method gives excellent results in the absence of high contrast. If bones or metals are present, normalization with thickness cannot produce a very flat sinogram, resulting in new artifacts. To extend this idea to objects composed of bones, metals, and other high-contrast materials, NMAR uses a prior image that takes these materials into account. Through denormalization, NMAR restores traces of high-contrast objects buried in metal shadows. This is because the shape information of these objects is contained in the sinogram of the prior image. NMAR ensures a certain level of smoothness at the boundaries of the metal traces in the corrected sinogram and recovers traces of objects contained in the prior image.

NMAR Algorithm

Figure 3 provides a diagram of the different steps of NMAR algorithm. An uncorrected image is reconstructed from the original sinogram p. The metal image is then obtained by thresholding. The prior image f^{prior} is created by segmenting soft tissues and bones. Forward projection produces the corresponding sinograms. The original sinogram p is then normalized by division by p^{prior} projected from f^{prior}. The division is only performed on pixels where the divisor is greater than a

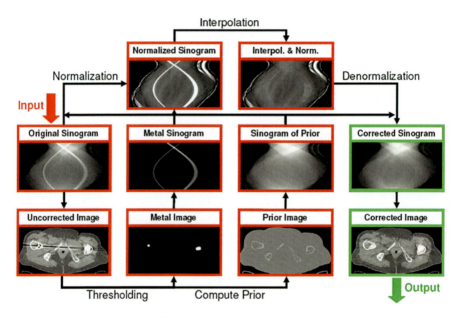

Fig. 3 Scheme of NMAR algorithm – from the original sinogram, an uncorrected image is reconstructed. By thresholding, the metal image and the prior image are obtained. Forward projection yields the corresponding sinograms. The normalized sinogram is then obtained by dividing the original sinogram by the sinogram of the prior image. The metal projections determine where data in the normalized sinogram are replaced by interpolation. The interpolated and normalized sinogram is denormalized by multiplying it with the sinogram of the prior image. Reconstruction yields the corrected image. (Reprinted from Meyer et al. (2010) with permission from John Wiley and Sons)

small positive value to avoid division by zero. A simple interpolation operation $\mathcal{M}_{int}$ is performed on the normalized sinogram p^{norm} to obtain a sinogram with metal traces removed. Subsequently, the corrected sinogram p^{corr} is obtained through denormalization, which multiplies $\mathcal{M}_{int} p^{norm}$ by p^{prior}:

$$p^{prior} = \mathcal{R} f^{prior},$$

$$p^{norm} = \frac{p}{p^{prior}},$$

$$p^{corr} = p^{prior} \mathcal{M}_{int} p^{norm}.$$

In this step, the structure information from the prior image is brought back into the metal trace because traces of high-contrast objects are included in the sinogram of the prior image. Normalization and multiplication procedures ensure that there is no difference between the original sinogram and corrected sinogram, except for metal trace. Hence, only sinogram values around metal traces are needed for normalization and denormalization. After reconstruction, the metal is inserted back into the corrected image.

An important step in NMAR algorithm is finding a good prior image. It should be modeled as close as possible to the uncorrected image, but should not contain artifacts. To achieve this, it is necessary to identify air regions, soft tissue regions, and bone regions. After smoothing the image with Gaussian, simple thresholding can be applied to segment air, soft tissue, and bone. It is also useful to smooth the steak structure as described in Müller and Buzug (2009) to reduce streak artifacts before segmentation. See Meyer et al. (2010) for more details.

Surgery-Based Metal Artifact Reduction (SMAR)

Even though NMAR algorithm removes metal artifacts very well, it still generates streaking artifacts because the corrected sinogram is not consistent. Recently, a new metal artifact reduction algorithm called SMAR, based on sinogram surgery, was proposed to reduce metal artifacts by calibrating the sinogram to be nearly consistent (Jeon and Lee 2018).

SMAR algorithm consists of two steps: a preprocessing step and an iterative reconstruction step. In the preprocessing step, the metal part from the given CT image is extracted, and then its metal trace is determined by the forward projection as in the NMAR algorithm. In the iterative reconstruction step, in order to moderate metal artifacts, several processes are performed such as average fill-in, sinogram surgery, and reconstruction from the updated sinogram. Detailed descriptions of each of these are given below.

Preprocessing Step
(1) Metal extraction: The metal region M can be extracted by simple thresholding.

(2) Surgery region designation: Once the metal region M has been extracted, its forward projection using the Radon transform $\mathcal{R}$ establishes the surgery region

$$M_{\text{proj}} = \text{supp}\{\mathcal{R}\chi_M\},$$

where χ is the characteristic function

$$\chi_M(x) = \begin{cases} 1 & \text{for } x \in M, \\ 0 & \text{otherwise.} \end{cases}$$

This region coincides with the corrupted part of the sinogram due to metal.

Iterative Reconstruction Step

The iterative reconstruction step has three steps: average fill-in, sinogram surgery, and reconstruction of the updated sinogram.

(1) Average fill-in: For the reconstructed CT image from the previous step, $f^{(n-1)}$, a connected region C is segmented which is surrounding M. Using $v^{(n-1)}$, the average of the attenuation coefficients $f^{(n-1)}$ of the region C, the average fill-in step is evaluated as

$$\widetilde{f}^{(n-1)} = v^{(n-1)}\chi_{C \cup M} + f^{(n-1)}(1 - \chi_{C \cup M}),$$

which moderates the streak structure of the CT image.

(2) Projection and sinogram surgery: By forward projection of $\widetilde{f}^{(n-1)}$, a new sinogram

$$\widetilde{p}^{(n-1)} = \mathcal{R}\widetilde{f}^{(n-1)}$$

is obtained. Using $\widetilde{p}^{(n-1)}$ a new sinogram

$$p^{(n)} = \widetilde{p}^{(n-1)}\chi_{M_{\text{proj}}} + p^{(0)}(1 - \chi_{M_{\text{proj}}}).$$

is produced. This is referred to as a sinogram surgery, where the corrupted sinogram part of a given sinogram $p^{(0)}$ is replaced with the newly generated sinogram $\widetilde{p}^{(n-1)}$ which is generated from the CT image with moderate metal artifact structure.

(3) CT image reconstruction: The new CT image is reconstructed by FBP

$$f^{(n)} = \mathcal{R}^{-1}p^{(n)}$$

The resulting image has less streak artifacts compared to $f^{(n-1)}$. Here, other sophisticated reconstruction methods can be also applied for the image quality improvement.

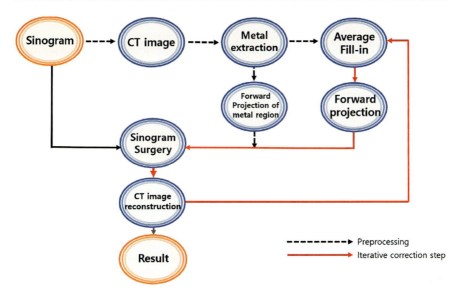

Fig. 4 Diagram for schematic description of SMAR algorithm

As the iterative reconstruction step is repeated, streak artifacts are reduced gradually because the missing data is complementarily replaced for both the sinogram and the reconstructed CT image. The iterative reconstruction step is terminated when the relative difference between the sinogram data becomes less than the tolerance level. The convergence of the SMAR algorithm is given empirically in the Appendix of Jeon and Lee (2018).

Figure 4 provides a schematic diagram of SMAR algorithm.

Convolutional Neural Network-Based MAR (CNN-MAR)

In this section, CNN-MAR method (Zhang and Yu 2018) is introduced as one of the most successful deep learning algorithms for metal artifact reduction. The CNN-MAR method uses two pre-corrected auxiliary images from the BHC method and the LI method, which were used in the hybrid MAR method in Zhang et al. (2013). BHC method is a model-based reconstruction method using total variation minimization (Verburg and Seco 2012). In this work, CNN-MAR methods with BHC and LI methods and with SMAR and LI methods are considered.

Training of the Convolutional Neural Network

The main goal of CNN training is to find optimal parameters that minimize the loss function. From a metal-free reference image, a set of images is generated: an image with metal inserted; an image with metal artifacts, which is used as a raw image;

and two pre-corrected auxiliary images. The loss function $Loss: \{\mathbf{U}, \mathbf{V}, W\} \to \mathbb{R}$ is defined by

$$Loss = \frac{1}{N} \sum_{n=1}^{N} \|C_L(\mathbf{u}_n, W) - \mathbf{v}_n\|_F^2,$$

where $\|\cdot\|_F$ denotes the Frobenius norm and N is the number of input data, $\mathbf{U} = \{\mathbf{u}_1, \cdots, \mathbf{u}_N\}$ the input data where each $\mathbf{u}_i$ consists of a raw image and two auxiliary images, $\mathbf{V} = \{\mathbf{v}_1, \cdots, \mathbf{v}_N\}$ a target data of reference images, and C_L a CNN containing a parameter set W. To optimize the loss function, stochastic gradient descent with momentum (SGDM) is used. SGDM is a variant of stochastic gradient descent (SGD) by adding momentum to accelerate the SGD algorithm which updates parameters randomly in order to avoid the situation "trap in local minima." SGD is based on traditional gradient descent (GD) algorithm. SGDM is formulated as

$$\Delta W^{(k)} = \mu \Delta W^{(k-1)} - \alpha \nabla Loss(W^{(k)}),$$
$$W^{(k+1)} = W^{(k)} + \Delta W^{(k)},$$

where μ is a momentum value (= 0.9), ΔW the direction vector, W the parameters, and α the learning rate and $\nabla Loss$ denotes the stochastic gradient. Therefore, ΔW is updated with remembering the past directions. Thanks to the momentum term, it is expected that the probability that W is trapped in local minima is reduced.

As in Zhang and Yu (2018), the CNN is constructed as follows: L convolutional layers are used with ReLU$(x) = \max(0, x)$ for nonlinear activation function. The first $L - 1$ layers are formulated as

$$C_0(\mathbf{u}) = \mathbf{u}_0,$$
$$C_m(\mathbf{u}) = \text{ReLU}(\mathbf{W}_m * C_{m-1}(\mathbf{u}) + \mathbf{b}_m), \quad m = 1, \ldots, L - 1,$$
$$C_L(\mathbf{u}) = \mathbf{W}_L * C_{L-1}(\mathbf{u}) + \mathbf{b}_L,$$

where $*$ stands for convolution, $\mathbf{W}_m$ is the m-th kernel, and $\mathbf{b}_m$ is the bias in the m-th layer. Each layer consists of 32 channels. The last layer generates an image that is close to the target. The convolutional kernel is 3×3 in each layer. Zero padding is used in each layer to maintain the size of output data as the same as input data. Whole architecture of the CNN is shown in Fig. 5.

CNN-MAR Method

The CNN-MAR algorithm consists of the following five steps:

(1) Two auxiliary MAR images are obtained.
(2) A CT image is obtained with reduced artifacts by the trained CNN.

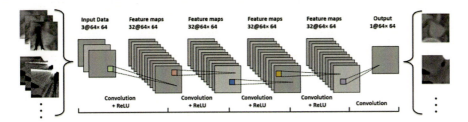

Fig. 5 Architecture of the CNN for metal artifact reduction. (Reprinted from Zhang and Yu (2018) with permission from IEEE)

(3) A CNN prior image is generated using tissue processing.
(4) A corrected sinogram is produced by replacing the metal traces in the sinogram of the CNN image using the sinogram of the CNN prior image.
(5) A corrected CT image is obtained using the inverse Radon transform.

Here, the details of Steps 2–4 are provided.

First, from the uncorrected image, two corrected auxiliary MAR images are obtained by the BHC and LI methods or by the SMAR and LI methods. Then these are combined as a three-channel image $\mathbf{u}^{\mathrm{input}}$, and the CNN-corrected image f^{CNN} is obtained through the CNN processing:

$$f^{\mathrm{CNN}} = C_L(\mathbf{u}^{\mathrm{input}}).$$

Here, the parameters in C_m have been found in advance from the CNN training. In implementation, $L = 5$ was used.

Even after the CNN processing, f^{CNN} still has considerable artifacts. Therefore, additional process is applied to reduce these artifacts; a prior image is generated from f^{CNN} by the tissue processing in Zhang and Yu (2018). First, because the water-equivalent tissues have similar attenuations and are accounted for a dominant proportion in a patient, the pixels corresponding to these tissues are assigned uniform values. For simple calculation, it is assumed that f^{CNN} consists of bone, water, and air. Using the k-means clustering on f^{CNN}, two thresholds are determined; one threshold is the bone-water threshold, and the other is the water-air threshold. Then, a binary image B is obtained with the water region set to 1 and the rest set to 0.

To replace the metal trace of the sinogram, a distance image D is introduced, which is made from the binary image B as follows. The pixel value of D is set to the distance between the pixel and its nearest 0 pixel if it is not greater than 5 and set to 5 if it is greater than 5. Hence, in the image $D = \{D_i\}$, most of the water pixels have the value 5, and there are 5-pixel transition regions, while the other pixels are zero. We compute the weighted average of the water pixel values:

$$\bar{f}^{\text{water}} = \frac{\sum_i D_i f_i^{\text{CNN}}}{\sum_i D_i}.$$

Then, the prior image is obtained:

$$f_i^{\text{prior}} = \frac{D_i}{5} \bar{f}^{\text{water}} + \left(1 - \frac{D_i}{5}\right) f_i^{\text{CNN}}.$$

This prior image $f^{\text{prior}} = \{f_i^{\text{prior}}\}$ is smoother than f^{CNN}. Using the prior image, sinogram correction and image reconstruction are performed as follows. First, let the metal trace occupy from the $(j_n + 1)$-th pixel to the $(j_n + \Delta_n)$-th pixel in the n-th projection view according to θ. Then the metal trace is replaced by the following:

$$p_{\theta,k_n}^{\text{corr}} = \frac{\left(\mathcal{R}_\theta f_{j_n+\Delta_n+1}^{\text{CNN}} - \mathcal{R}_\theta f_{j_n+\Delta_n+1}^{\text{prior}}\right) - \left(\mathcal{R}_\theta f_{j_n}^{\text{CNN}} - \mathcal{R}_\theta f_{j_n}^{\text{prior}}\right)}{\Delta_n + 1}(k_n - j_n)$$
$$+ \mathcal{R}_\theta f_{k_n}^{\text{prior}} + \left(\mathcal{R}_\theta f_{j_n}^{\text{CNN}} - \mathcal{R}_\theta f_{j_n}^{\text{prior}}\right), \quad j_n \leq k_n \leq j_n + \Delta_n + 1$$

and the other part of p_θ^{corr} is kept in $\mathcal{R}_\theta f^{\text{CNN}}$. This produces a new projection data $p^{\text{corr}} = [p_\theta^{\text{corr}}]_\theta$, which connects the correction of the metal trace to the surrounding unaffected projection data. It is kind of a seamless surgery of sinogram. Finally, a corrected CT image is reconstructed by the FBP algorithm for p^{corr}, and metals are inserted back into the corrected image. Note that this seamless surgery can also be used for the SMAR algorithm.

There are two key factors for the success of the CNN-MAR method: selection of the appropriate pre-corrected auxiliary CT images and preparation of training data. The first factor provides information to help CNN distinguish between tissue structures and artifacts. The second factor ensures the generality of the trained CNN by including as many kinds of metal artifact cases as possible.

Industrial Application: 3D Cone Beam CT

Industrial X-ray CT is used in various areas of industry as an internal inspection for manufactures such as flaw detection, failure analysis, metrology, and so on. Particularly, the reconstructed image obtained from three-dimensional cone beam CT (CBCT) provides ideal testing techniques to locate and measure volumetric details in three dimensions. However, a potential drawback with CT imaging is the possibility of artifacts due to physical phenomenon such as beam hardening effect.

In industry, computer-aided design (CAD) data is available for in-line inspection systems because most products are designed in the form of CAD data. In Jeon et al. (2021), a metal artifact reduction algorithm was proposed using the shape prior information given by CAD format. It is an extended version of SMAR algorithm in

Section "Surgery-Based Metal Artifact Reduction (SMAR)" to 3D, and CAD data is adopted as a shape prior information. In the SMAR algorithm, it is essential to accurately segment the average fill-in region for the success of the algorithm, and for this purpose, a registration algorithm is proposed to register the CAD data to the reconstructed CT volume.

Data Preparation
First, using the given CAD data, a binary volume data V_{CAD} such as

$$V_{CAD}(x) = \begin{cases} 1, & x \in \text{inside of the object} \\ 0, & x \in \text{outside of the object.} \end{cases}$$

is generated. Unlike the CAD data, the reconstructed CT volume is a three-dimensional image with a gray level with pixel values between 0 and 1. In order to distinguish between the inside and the outside of the scanned object, simple thresholding is applied after denoising. The level of the simple thresholding can be set considering the substance that makes up the scanned object. An anisotropic diffusion model called Perona-Malik (Perona et al. 1994) and the shock filter (Osher and Rudin 1990) are applied to the reconstructed CT volume for noise reduction. The resulting binarized CT volume is denoted as V_{CT}.

Registration via Shape Prior Chan-Vese Model
Since the reconstructed CT volume contains various artifacts, V_{CT} is not exactly the same with V_{CAD}.

With a given shape prior information, the following energy functional of the Chan-Vese model (Chan and Vese 2001) is considered for an image $I: \Omega \to \mathbb{R}$,

$$E(\phi, c_1, c_2) = \int_\Omega (I - c_1)^2 H(\phi) dx + \int_\Omega (I - c_2)^2 (1 - H(\phi)) \, dx, \qquad (5)$$

where H is the Heaviside function and ϕ is a level set function representation of the shape prior, whose zero level set is the boundary of the shape prior in the image I (Osher and Sethian 1988). Scalar values c_1 and c_2 become average intensities of I in the regions where ϕ is positive and negative, respectively. The level set-based approach (5) allows V_{CAD} to be registered to V_{CT}, the result being the closest in terms of volume.

Shape Prior SMAR Algorithm: Alignment and Registration
In the average fill-in step of SMAR, segmentation of the average fill-in region can be easily done by registering V_{CAD} into V_{CT}. Since the functional in (5) is non-convex, in order to avoid being trapped at local minima, V_{CAD} needs to be located as close as possible to V_{CT} while being resized to have the same size as V_{CT} before performing the minimization process.

For a three-dimensional binary object V, the moment tensor is defined by

$$T = \begin{bmatrix} T_{xx} & -T_{xy} & -T_{xz} \\ -T_{yx} & T_{yy} & -T_{yz} \\ -T_{zx} & -T_{zy} & T_{zz} \end{bmatrix},$$

where the moments T_{xx}, T_{yy}, T_{zz} and the products of moment T_{xy}, T_{xz}, T_{yz} are given by

$$T_{xx} = \int_V (x^2)\,dV,\; T_{yy} = \int_V (y^2)\,dV,\; T_{zz} = \int_V (z^2)\,dV,$$

and

$$T_{xy} = T_{yx} = -\int_V xy\,dV,\; T_{yz} = T_{zy} = -\int_V yz\,dV,\; T_{xz} = I_{zx} = -\int_V zx\,dV.$$

Since the moment tensor T is symmetric, by the principal axis theorem, the eigenvectors of T are the principal axes of V.

Let v_1, v_2, v_3 be unit eigenvectors of T. Then corresponding eigenvalues $\lambda_1, \lambda_2, \lambda_3$ satisfy the relation

$$Tv_i = \lambda_i v_i, \quad i = 1, 2, 3.$$

For the binary volumes V_1 and V_2, the ratio of object sizes can be obtained by using the relationship of the eigenvalues. If r denotes the ratio of sizes between V_1 and V_2, then

$$\begin{cases} T_{xx}^{V_1} = \int_{V_1} (x^2)\,dV \\ T_{xx}^{V_2} = \int_{V_2} (rx)^2 r^3\,dV \end{cases}$$

implies that $r^5 T_{xx}^{V_1} = T_{xx}^{V_2}$. Therefore, the scaling constant r becomes

$$r = \sqrt[5]{\frac{\lambda^{V_2}}{\lambda^{V_1}}}.$$

Using matrices of principal axes, Q_{CAD} and Q_{CT}, and scaling ratio r, the transformation matrix Q to align V_{CAD} to V_{CT} is expressed as

$$Q = r Q_{CT} Q_{CAD}^{-1} = r Q_{CT} Q_{CAD}^T.$$

Then, $V_{\text{align}} = Q V_{CAD}$ is closely aligned to V_{CT}.
Now, the functional in (5) for $I = V_{CT}$ is minimized with

$\phi(x, y, z)$
$$= \phi_0 \begin{bmatrix} (x-a)(n_1^2(1-\cos\theta)+\cos\theta)+(y-b)(n_1n_2(1-\cos\theta)-n_3\sin\theta)+(z-c)(n_1n_3(1-\cos\theta)+n_2\sin\theta) \\ (x-a)(n_1n_2(1-\cos\theta)+n_3\sin\theta)+(y-b)(n_2^2(1-\cos\theta)+\cos\theta)+(z-c)(n_2n_3(1-\cos\theta)-n_1\sin\theta) \\ (x-a)(n_1n_3(1-\cos\theta)-n_2\sin\theta)+(y-b)(n_2n_3(1-\cos\theta)+n_1\sin\theta)+(z-c)(n_3^2(1-\cos\theta)+\cos\theta) \end{bmatrix},$$
(6)

where ϕ_0 is a level set function representation of V_{align}. Here, (a, b, c) is the translation factor along x, y, z-axes, respectively, $\mathbf{n} = (n_1, n_2, n_3)$ a unit vector of rotation axis, and θ a rotation angle with respect to the rotation axis $\mathbf{n}$; see Fig. 6.

A particle swarm optimization (PSO) technique is used for the minimization process (Eberhart and Kennedy 1995; Jaberipour et al. 2011). To find a, b, c, n_1, n_2, n_3, and θ of (6) minimizing (5) by applying the modified PSO algorithm (Jaberipour et al. 2011), a strategy for efficient computation is adopted. As shown in Algorithm 1, the parameters are coupled into (a, b, c), (n_1, n_2, n_3), and θ depending on their meaning: translation, rotation axis, and rotation angle. First, fixing the center of mass of two volumes at the origin of the computation region, the initial seed $(a, b, c)^{(0)} = (0, 0, 0)$ is set. Then the principal axes of the moment matrices of the two volume data are aligned with the x, y, z-axes. At this time, the first principal

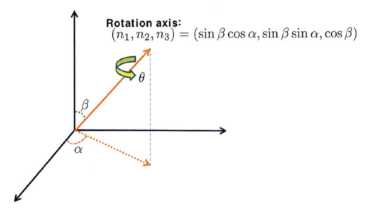

Fig. 6 Three-dimensional rotation. (Reprinted from Jeon et al. (2021) with permission from Taylor & Francis)

Algorithm 1 Finding a minimizer of E in (5)

An initial level set ϕ for aligned prior is given.
for $k = 1, 2, 3, \cdots$ **do**
  Update c_1, c_2 using $c_1 = \frac{\int_\Omega I(x) H(\phi) dx}{\int_\Omega H(\phi) dx}$ and $c_2 = \frac{\int_\Omega I(x)(1-H(\phi)) dx}{\int_\Omega 1-H(\phi) dx}$.
  For fixed $c_1, c_2, (a, b, c)$, and (n_1, n_2, n_3), update θ.
  For fixed $c_1, c_2, (n_1, n_2, n_3)$, and θ, update (a, b, c).
  For fixed c_1, c_2, θ, and (a, b, c), update normalized (n_1, n_2, n_3).
end
During the update process, reinitialization is applied first to avoid the numerical deterioration of the interface.

axis is aligned with z-axis and $(n_1, n_2, n_3)^{(0)} = (0, 0, 1)$ is set. Finally, the angle between the center slices of V_{CT} and V_{align} is computed and set as $\theta^{(0)}$. We generate particles in the proper intervals centered at $(a, b, c)^{(0)}$, $(n_1, n_2, n_3)^{(0)}$, and $\theta^{(0)}$. Which variable is updated first depends on how much the updated value affects other variables: updates in order of rotation angle, translation, and rotation axis.

The three-dimensional computation is highly time-consuming, and the most time-consuming part is the PSO process of finding parameters that minimize (5). As the number of particles increases, computation time is linearly increasing. Therefore, a two-resolution approach can be adopted to reduce the computation time of the registration process. For the down-sampled data, less particles can be used. The parameter obtained from the down-sampled data is used as an initial for the registration of the original sized data.

Shape Prior SMAR Algorithm: CT Volume Reconstruction

For sinogram surgery, the two-dimensional forward and backward projection operators, $\mathcal{R}$ and $\mathcal{R}^{-1}$, are straightforwardly extended to three-dimensional cone beam case. This approach is compatible with the filtered back projection (FBP), and the other sophisticated reconstruction methods can be also applied for the image quality improvement. The segmented region is obtained from the registration result of Algorithm 1 and is used as the average fill-in region. The flowchart of the whole shape prior SMAR algorithm is shown in Fig. 7.

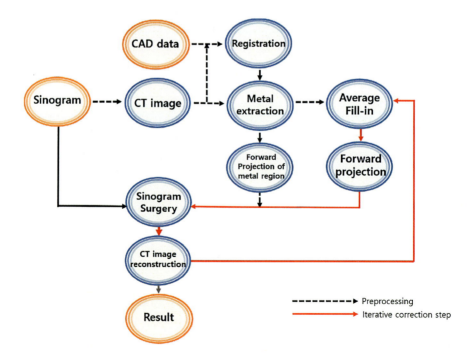

Fig. 7 The flowchart of the shape prior SMAR for 3D CBCT

Simulations and Results

Simulation Conditions

In the simulation study, to generate the polychromatic sinogram, the parallel beam were modeled with 512 channels per detector and 1800 views per half rotation. Seven discrete energy bins (10, 20, 30, 40, 60, 80, 100 keV) were defined (Table 1), and all X-ray coefficients were obtained from the National Institute of Standards and Technology (NIST) database (Hubbell and Seltzer 2004).

For a quantitative analysis, the metal effect-free CT images $f^\diamond$ were used as references, and the performance of MAR algorithms are measured with three error measurements: the relative l_2 error, the relative l_∞ error, and peak signal-to-noise ratio (PSNR). PSNR is defined as

$$\text{PSNR} = 20 \log \frac{\text{peak value}}{\text{RMSE}},$$

where peak value is the range of window and RMSE is the root mean squared error.

A simple notation is used for the relative error between a CT image f and the reference CT image $f^\diamond$,

$$\|f\|_{*,\diamond} := \frac{\|f - f^\diamond\|_*}{\|f^\diamond\|_*}, \quad * = 2, \infty,$$

where

$$\|f\|_2 := \sqrt{\sum_i |f_i|^2}, \quad \|f\|_\infty := \max_i |f_i|.$$

The iteration process of the SMAR algorithm was terminated when the relative difference of the sinogram data in the sinogram surgery region became less than Tol $= 10^{-4}$. In all reconstructed images, the window level with a width of 1000 centered at 0 (C/W = 0/1000 (HU)) is used.

NMAR vs. SMAR: Patient Image Simulations

In this section, the numerical results in Jeon and Lee (2018) are presented.

To compare NMAR and SMAR algorithms, patient images were tested (Fig. 8). Three cross-sectional images (pelvis, chest, and dental) were selected from a CT dataset acquired in a dosimetry study of ^{68}Ga- NOTA-RGD PET/CT (Kim et al. 2012). All study procedures were approved by the Institutional Review Board of Seoul National University Hospital, Seoul, Korea. Simulated metallic objects were inserted into the patient images while assuming that the metallic objects are titanium. The X-ray energy spectrum in Table 1 was used. Because it is difficult

Table 1 X-ray intensity and nominal Hounsfield units (HU) and linear attenuation coefficients for materials used in simulations

Energy [keV]	X-ray intensity	Air, dry [/mm] (−1000 HU)	Adipose [/mm] (−100 HU)	Water [/mm] (0 HU)	Tissue [/mm] (150 HU)	Bone [/mm] (1000 HU)	Iron [/mm] (1000 HU ≤)	Titanium [/mm] (1000 HU ≤)
10	0.0000	6.169E−04	0.3037	0.5329	0.6455	4.989	134.26	49.815
20	1.604	9.372E−05	0.0528	0.08096	0.09876	0.7002	20.21	7.1325
30	26.93	4.263E−05	0.02847	0.03756	0.04548	0.2329	6.435	2.2374
40	49.12	2.994E−05	0.02227	0.02683	0.03226	0.1165	2.856	0.9963
60	42.78	2.506E−05	0.01835	0.02059	0.02458	0.05509	0.9483	0.3447
80	46.31	2.002E−05	0.01673	0.01837	0.02188	0.03901	0.4684	0.1823
100	14.00	1.857E−05	0.01569	0.01707	0.02032	0.03246	0.2925	0.1224

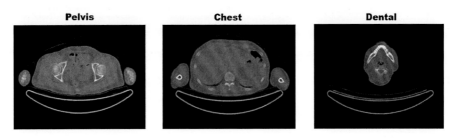

Fig. 8 Patient images (pelvis, chest, and dental). (Reprinted from Jeon and Lee (2018) with permission from IOS Press)

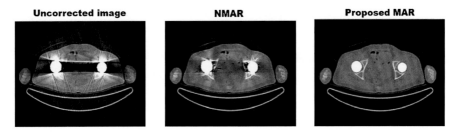

Fig. 9 Patient pelvis experiment. Uncorrected CT image (left) and results of NMAR (middle) and SMAR (right): C/W = 0/1000 (HU). (Reprinted from Jeon and Lee (2018) with permission from IOS Press)

to assign energy level-varying X-ray attenuation coefficients for each tissue type in patient images, it is assumed that only the X-ray attenuation coefficient of a metallic object depends on the X-ray energy level.

Figure 9 shows the simulation result for the pelvis of a patient with metallic hips. In the uncorrected CT image, there are streak artifacts between the metallic hips, and they corrupt the anatomical structure. NMAR reduces most of the streak artifacts; however, the resulting image contains bright and dark artifacts which blur the anatomical structure. In comparison, SMAR reduces streak artifacts effectively without generating such bright and dark artifacts. As a result, a clean CT image is obtained and the textures are also preserved well.

As shown in Table 2, the initial relative l_∞ and l_2 errors, 5.0085 and 0.7044, are decreased by nearly half to 3.5728 and 0.2341, respectively, for NMAR. The SMAR algorithm drops the errors more significantly, with the resulting relative l_∞ and l_2 errors becoming 0.2293 and 0.0269, respectively, the values which are decreased by a factor of 20 from the initial levels.

Figure 10 presents the experimental results for the chest of a patient. Two metallic screws inserted into the spine generate streak artifacts, which severely damage the anatomical structure near the spine. While NMAR does reduce the major part of the streak artifacts, it generates additional artifacts near the metallic objects,

Table 2 The performance comparison between NMAR and SMAR for the patient image simulations. The number of iterations is denoted by n. (Reprinted from Jeon and Lee (2018) with permission from IOS Press)

Phantom	Initial error		NMAR			SMAR	
	$\|f^{(0)}\|_{\infty,\diamond}$	$\|f^{(0)}\|_{2,\diamond}$	$\|\cdot\|_{\infty,\diamond}$	$\|\cdot\|_{2,\diamond}$	n	$\|f^{(n)}\|_{\infty,\diamond}$	$\|f^{(n)}\|_{2,\diamond}$
Pelvis	5.0085	0.7044	3.5728	0.2341	16	0.2293	0.0269
Chest	12.1957	0.9187	7.8182	0.3100	26	0.5878	0.0314
Dental	11.9255	1.7471	3.4632	0.4378	14	0.3734	0.0476

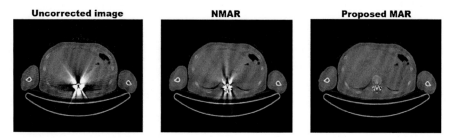

Fig. 10 Patient chest experiment. Uncorrected CT image (left) and results of NMAR (middle) and SMAR (right): C/W = 0/1000 (HU). (Reprinted from Jeon and Lee (2018) with permission from IOS Press)

resulting in bright and dark patterns. These newly generated artifacts appear near the metallic objects and thus corrupt the anatomical structure. Moreover, in the NMAR result, the metallic objects are thicker than the original metallic objects, whereas SMAR improves the image quality without generating additional artifacts, so that the anatomical structures near the spine can be successfully distinguished. As shown in Table 2, the initial relative l_∞ and l_2 errors, 12.1957 and 0.9187, are decreased in the NMAR result to 7.8182 and 0.3100, respectively. The resulting relative l_∞ and l_2 errors of SMAR are 0.5878 and 0.0314, respectively, values which are lower by a factor of 20 from the initial values.

In the dental image simulations, streak artifacts appear to connect three metallic objects, as shown in Fig. 11. As shown in the zoomed images of the solid boxes, both NMAR and SMAR reduce the streak artifacts. However, NMAR produces the shadow effects even in the region near the teeth and shows undulated artifacts across the entire image domain. Even in a region far from the metallic objects, undulated artifacts also appear, as shown in the zoomed images, and they degrade the image quality. As shown in Table 2, the initial relative l_∞ and l_2 errors, 11.9255 and 1.7471, are decreased for NMAR to 3.4632 and 0.4378, respectively. The resulting relative l_∞ and l_2 errors for SMAR are 0.3734 and 0.0476, respectively, showing decreases by a factor of 30 from the initial levels.

In patient image simulations, unlike NMAR, SMAR does not generate undulated artifacts. SMAR produces clear images and performs noticeably better than NMAR.

10 Recent Approaches to Metal Artifact Reduction in X-Ray CT Imaging 369

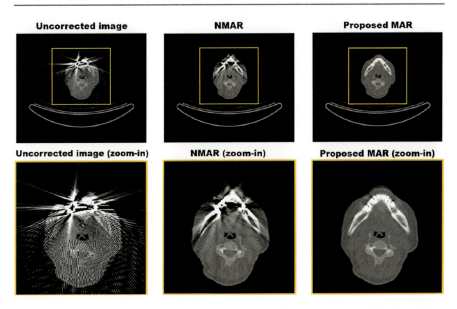

Fig. 11 Patient dental experiment. Uncorrected CT image (left) and results of NMAR (middle) and SMAR (right): C/W = 0/1000 (HU). (Reprinted from Jeon and Lee (2018) with permission from IOS Press)

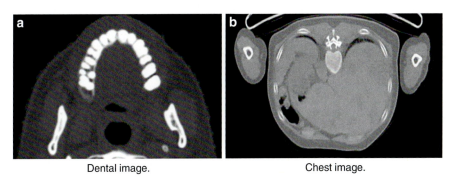

Dental image. Chest image.

Fig. 12 Reference images for the test of CNN-MAR. (**a**) Dental image. (**b**) Chest image

SMAR vs. CNN-MAR

Data Acquisition

For the training of CNN in the CNN-MAR method, data patches are obtained from dental, head, and pelvis images collected from "The 2016 Low-Dose CT Grand Challenge" training dataset (AAPM 2016). To show the performance of the CNN-MAR, a dental image and a chest image in Fig. 12 were chosen from a CT dataset acquired in a dosimetry study of ^{68}Ga-NOTA-RGD PET/CT (Kim et al. 2012). It is expected that the CNN works well when the dental image is used for test, but it will not work when the chest image is used since it is not in the training set.

Results

First, the CNN-MAR method in Zhang and Yu (2018) used BHC and LI results as auxiliary images for the input data, but in this work, SMAR and LI results are also provided as auxiliary images. Since the SMAR algorithm produces better results than the BHC method, it is expected that CNN-MAR will produce a better output if the BHC image input is replaced by an SMAR image. Furthermore, since the SMAR algorithm gives better results than LI, CNN-MAR with only one auxiliary image from the SMAR algorithm is considered.

Figure 13 is the results of the dental case. Indeed, streak artifacts between the teeth are reduced in all methods. However, there are big differences in the red-colored squares. From the numerical results in Table 3, CNN-MAR with SMAR and LI is the best.

Figure 14 is the results of the chest case. Note that the CNN did not learn the chest image patches. From the figure, it can be seen that CNN-MAR with SMAR and LI reduces the streak artifacts well compared with LI. However, comparing with SMAR, breastbone structure is not clearly reconstructed. In Table 4, SMAR shows

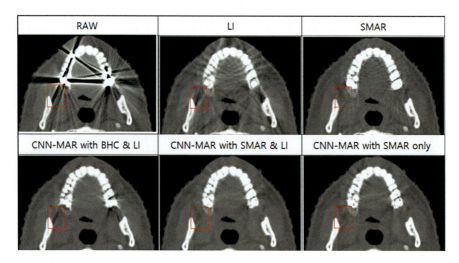

Fig. 13 Dental CT image results with various MAR methods

Table 3 Numerical results for the dental case

	Raw	CNN-MAR with BHC and LI	SMAR	CNN-MAR with SMAR and LI	CNN-MAR with SMAR only
PSNR	15.6867	30.6590	32.6508	32.9713	32.2920
Relative maximum error	6.1329	0.5401	0.7506	0.2812	0.3963
Relative L^2 error	0.4666	0.0832	0.0653	0.0638	0.0690

10 Recent Approaches to Metal Artifact Reduction in X-Ray CT Imaging

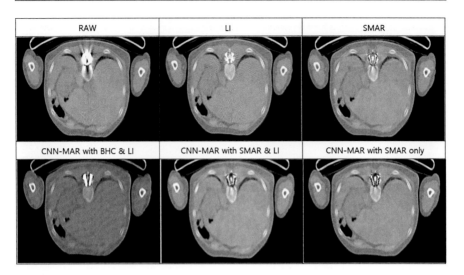

Fig. 14 Chest CT image results with various MAR methods

Table 4 Numerical results for the chest case

	Raw	CNN-MAR with BHC and LI	SMAR	CNN-MAR with SMAR and LI	CNN-MAR with SMAR only
PSNR	8.2912	23.8660	34.1109	29.1756	30.1433
Relative maximum error	12.7421	1.8401	0.4906	0.3192	0.2811
Relative l_2 error	1.0139	0.2812	0.0519	0.0916	0.0819

the best performance in terms of PSNR and the l_2 error. Furthermore, CNN-MAR with SMAR only shows the best performance in terms of the maximum error.

NMAR vs. SMAR for 3D CBCT

In this section, the numerical results in Jeon et al. (2021) are presented.

Phantoms and Hardware Specifications

Two real samples, Samples 1 and 2 in Figs. 15 and 16, respectively, are used for the simulations. Sample 1 consists of an acrylic body with 32 poles, and each pole is made of either Teflon or stainless steel. For the experiment, 4 stainless steel poles and 28 Teflon poles were used. Sample 2 consists of a cylindrical aluminum body with cylindrical holes of various sizes (six large, three middle, and three tiny cylindrical holes), and three lead poles are made to be inserted into the large cylindrical holes. Both samples are designed by CAD and each binary volume data is constructed from them.

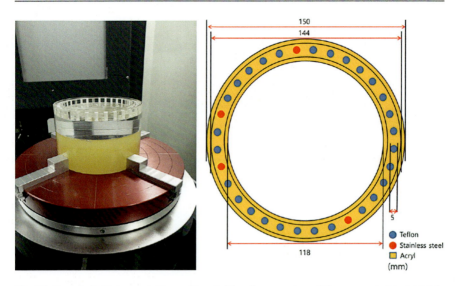

Fig. 15 Sample 1: Data acquisition setting (left) and upper view of the scanned object (right). (Reprinted from Jeon et al. (2021) with permission from Taylor & Francis)

Fig. 16 Sample 2: Data acquisition setting (left), upper view of sample body (middle), and lead (Pb) poles. (Reprinted from Jeon et al. (2021) with permission from Taylor & Francis)

The projection data of Sample 1 is acquired from an X-ray inspection system of *EB Tech Co., Ltd.*, and the scanning and reconstruction parameters are given in Table 5. The projection data of Sample 2 is acquired from an X-ray inspection system *Bright 240 450 Dual CTR* of *Dukin Co., Ltd.*; the specification of X-ray source is 450 kV and 700 W/1500 W, the focal spot is 0.4 mm/1.0 mm, the detector is a flat panel with size of 409.6 × 409.6 mm, and the resulting projection size per angle is 1024 × 1024.

Table 5 Parameters for Sample 1 data acquisition. (Reprinted from Jeon et al. (2021) with permission from Taylor & Francis)

	Values
Source-to-detector distance	2200 mm
Source-to-object distance	1500 mm
Tube voltage	80 kVp
Tube current	5 mA
Number of projection views	720
Scanning angle range	Full rotation (360°)
Detector pixel array size	1024×1024
Detector pitch	0.4 mm
Reconstructed volume size	$512 \times 512 \times 512$

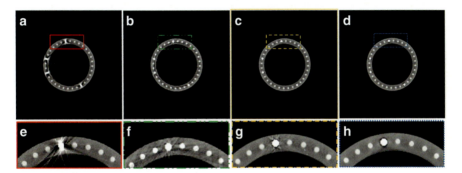

Fig. 17 MAR results for Sample 1: (**a**) uncorrected CT image, (**b**) NMAR, (**c**) SMAR, and (**d**) shape prior SMAR; (**e**), (**f**), (**g**), and (**h**) are zoomed-in images of (**a**), (**b**), (**c**), and (**d**), respectively. (Reprinted from Jeon et al. (2021) with permission from Taylor & Francis)

Test I: Performance Evaluation

The performance of the shape prior SMAR algorithm was evaluated and compared with NMAR. To demonstrate the benefits of the shape prior information, the SMAR algorithm was applied to two different situations: one is *with CAD* and the other is *without CAD*. In the *without CAD* case, the average fill-in regions are segmented with simple thresholding. Here, the real data for Sample 1 is used.

As shown in Fig. 17a and e, the inserted stainless poles generate severe metal artifacts. Although a considerable amount of artifacts has disappeared when NMAR is applied as shown in Fig. 17b and f, there are still streak shape of artifacts left. In the case of shape prior SMAR method (Fig. 17d), although slightly uneven parts are observed, streak shape of artifacts are almost eliminated, and the resulting image is very clean. Even though the shape prior information is not available, metal artifacts are reduced significantly as shown in Fig. 17c; however, the performance difference can be clearly seen in the zoomed-in images shown Fig. 17g and h.

Test II: Practical Application – Air Bubble Detection Simulation

Air bubble detection was simulated using Sample 2, where two lead poles are inserted into the aluminum body and about one-fourth of the pore size air bubble

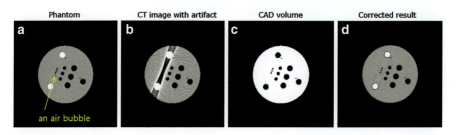

Fig. 18 Center slice views: (**a**) Sample 2 with an air bubble, (**b**) reconstructed CT image containing severe beam hardening artifacts, (**c**) V_{CAD}, and (**d**) corrected result. (Reprinted from Jeon et al. (2021) with permission from Taylor & Francis)

is included near the three tiny cylindrical pores as shown in Fig. 18a. However, as shown in Fig. 18c, the CAD data does not have the information about the air bubble. For implementation, three discrete bins (40, 60, 100 keV) are defined, and all X-ray attenuation coefficients are obtained from Table 1. For convenience, it is assumed that only the X-ray attenuation coefficient of the lead poles depends on the X-ray energy level. Using the registration results in the previous section, sinogram surgery is performed to reduce the metal artifacts due to two lead poles. As shown in Fig. 18c, an air bubble is not contained in the CAD data. Due to the severe artifacts, the bubble is hardly identified in Fig. 18b. The shape prior SMAR algorithm successfully reduces most of the streak artifacts, and it can also accurately detect the hidden air bubble as shown in Fig. 18d.

Conclusion

In this work, three recent approaches for metal artifact reduction in X-ray CT were investigated: NMAR, SMAR, and CNN-MAR.

NMAR has shown good performance for various types of metallic implants and thus been considered as one of the best currently available MAR algorithms. However, finding a good prior image is at the heart of this algorithm. Incorrect segmentation results can lead to residual artifacts. A more advanced segmentation algorithm will definitely improve the results compared to simple thresholding.

SMAR algorithm was applied to various patient images. As in other MAR approaches based on tissue classification, it is essential for the SMAR algorithm to find a good tissue classification. The average fill-in region is decided based on the tissue classification. The advantage of the SMAR algorithm stems from this point. By filling in the region surrounding metallic objects with the average values, a resulting image is obtained with less streak artifacts. Then this image is used as new input data for the next iteration. The SMAR algorithm tends to converge to a moderate value of the image intensity. Results can be improved when using a more sophisticated segmentation method rather than simple segmentation based on CT numbers.

From the aforementioned simulation results of CNN-MAR, it can be seen that CNN and tissue processing are two mutual beneficial steps. In the CNN step, useful information from pre-corrected auxiliary CT images is fused to avoid strong artifacts. However, mild artifacts typically remain. With the tissue processing, similar to other prior image-based MAR methods, it can remove such moderate artifacts and generates a prior image. Then the final result is produced by doing seamless surgery using the prior image.

In addition, for industrial cone beam CT, shape prior SMAR algorithm reduced metal artifacts using the shape prior information of the scanned object. For the segmentation task, a registration model was designed using level set approach. With CAD data, which is available in most cases in the manufacturing industry, the average fill-in region can be accurately segmented. Also, to overcome the non-convexity and nonlinearity of the energy functional for registration, an algorithm to find good initial parameters was proposed.

In the numerical section, the performance of the SMAR algorithm was compared with that of NMAR both qualitatively and quantitatively, and SMAR outperformed NMAR in the patient image simulation. Through numerical experiments, it was demonstrated that the SMAR algorithm reduces metal artifacts effectively without a loss of anatomical structures.

The CNN-MAR method used two pre-corrected auxiliary CT images to generate the output of CNN. Then, with the projection data of the prior image based on the tissue classification, the seamless surgery produced the final corrected projected data. The quality of the final corrected CT image is dependent upon the choice of the model-based reconstruction methods to generate the auxiliary images. CNN-MAR with SMAR and LI methods outperformed CNN-MAR with BHC and LI methods and with SMAR only. Note that CNN-MAR is not working well when it is applied for new types of images that are not in the training dataset; this result is expected in the general deep learning-based methods. The CNN-MAR method takes a long time for training; however, once trained, it produces outputs in a short time.

In the shape prior SMAR algorithm, through various experiments, the performance of the algorithm and the possibilities for the practical uses were investigated.

References

AAPM: Low dose CT grand challenge. Resource document. American Association of Physicists in Medicine (2016). http://www.aapm.org/GrandChallenge/LowDoseCT/

Abdoli, M., Ay, M.R., Ahmadian, A., Dierckx, R., Zaidi, H.: Reduction of dental filling metallic artifacts in CT-based attenuation correction of PET data using weighted virtual sinograms optimized by a genetic algorithm. Med. Phys. **37**(12), 6166–6177 (2010)

Bal, M., Spies, L.: Metal artifact reduction in CT using tissue-class modeling and adaptive prefiltering. Med. Phys. **33**(8), 2852–2859 (2006)

Bal, M., Celik, H., Subramanyan, K., Eck, K., Spies, L.: A radial adaptive filter for metal artifact reduction. Proc. SPIE **5747**, 2075–2082 (2005)

Barrett, J.F., Keat, N.: Artifacts in CT: recognition and avoidance. Radiographics **24**(6), 1679–1691 (2004)

Boas, F.E., Fleischmann, D.: CT artifacts: causes and reduction techniques. Imaging Med. **4**(2), 229–240 (2012)

Chan, T., Vese, L.: Active contours without edges. IEEE Trans. Image Process. **10**, 266–277 (2001)

De Man, B., Nuyts, J., Dupont, P., Marchal, G., Suetens, P.: An iterative maximum-likelihood polychromatic algorithm for CT. IEEE Trans. Med. Imaging **20**(10), 999–1008 (2001)

Deans, S.R.: The Radon Transform and Some of Its Applications. Dover, New York (2007)

Eberhart, R., Kennedy, J.: A new optimizer using particle swarm theory. In: MHS'95, Proceedings of the Sixth International Symposium on Micro Machine and Human Science, pp. 39–43 (1995). https://doi.org/10.1109/MHS.1995.494215

Ghani, M.U., Karl, W.C.: Fast enhanced CT metal artifact reduction using data domain deep learning. IEEE Trans. Comput. Imaging **6**, 181–193 (2020). https://doi.org/10.1109/TCI.2019.2937221

Gjesteby, L., Yang, Q., Xi, Y., Shan, H., Claus, B., Jin, Y., De Man, B., Wang, G.: Deep learning methods for CT image-domain metal artifact reduction. Proc. SPIE **10391**, 103910W (2017). https:doi.org/10.1117/12.2274427

Gu, J., Zhang, L., Yu, G., Xing, Y., Chen, Z.: X-ray CT metal artifacts reduction through curvature based sinogram inpainting. J. X-Ray Sci. Technol. **14**(2), 73–82 (2006)

Helgason, S.: The Radon transform on Euclidean spaces, compact two point homogeneous spaces and Grassmann manifolds. Acta Math. **113**, 153–180 (1965)

Huang, X., Wang, J., Tang, F., Zhong, T., Zhang, Y.: Metal artifact reduction on cervical CT images by deep residual learning. BioMed. Eng. OnLine **17**, 175 (2018). https://doi.org/10.1186/s12938-018-0609-y

Hubbell, J.H., Seltzer, S.M.: X-ray mass attenuation coefficients. Resource document. National Institute of Standards and Technology (2004). https://www.nist.gov/pml/x-ray-mass-attenuation-coefficients/

Jaberipour, M., Khorram, E., Karimi, B.: Particle swarm algorithm for solving systems of nonlinear equations. Comput. Math. Appl. **62**(2), 566–576 (2011). https://doi.org/10.1016/j.camwa.2011.05.031

Jeon, S., Lee, C.-O.: A CT metal artifact reduction algorithm based on sinogram surgery. J. X-Ray Sci. Technol. **26**, 413–434 (2018)

Jeon, S., Kim, S., Lee, C.-O.: Shape prior metal artefact reduction algorithm for industrial 3D cone beam CT. Nondestruct. Test. Eval. **36**(2), 176–194 (2021). https://doi.org/10.1080/10589759.2019.1709457

Kachelrieß, M., Watzke, O., Kalender, W.A.: Generalized multi-dimensional adaptive filtering (MAF) for conventional and spiral single-slice, multi-slice, and cone-beam CT. Med. Phys. **28**(4), 475–490 (2001)

Kalender, W.A., Hebel, R., Ebersberger, J.: Reduction of CT artifacts caused by metallic implants. Radiology **164**(2), 576–577 (1987)

Kano, T., Koseki, M.: A new metal artifact reduction algorithm based on a deteriorated CT image. J. X-Ray Sci. Technol. **24**(6), 901–912 (2016)

Kim, Y., Yoon, S., Yi, J.: Effective sinogram-inpainting for metal artifacts reduction in X-ray CT images. In: Proceedings of 2010 IEEE 17th International Conference on Image Processing, pp. 597–600 (2010)

Kim, J.H., Lee, J.S., Kang, K.W., Lee, H.-Y., Han, S.-W., Kim, T.-Y., Lee, Y.-S., Jeong, J.M., Lee, D.S.: Whole-body distribution and radiation dosimetry of ^{68}Ga-NOTA-RGD, a positron emission tomography agent for angiogenesis imaging. Cancer Biother. Radiopharm. **27**, 65–71 (2012)

Klotz, E., Kalender, W., Sokiranski, R., Felsenberg, D.: Algorithm for the reduction of CT artifacts caused by metallic implants. Proc. SPIE **1234**, 642–650 (1990)

Koehler, T., Brendel, B., Brown, K.: A new method for metal artifact reduction. In: The Second International Conference on Image Formation in X-Ray Computed Tomography, Salt Lake City (2012)

Lemmens, C., Faul, D., Nuyts, J.: Suppression of metal artifacts in CT using a reconstruction procedure that combines MAP and projection completion. IEEE Trans. Med. Imaging **28**(2), 250–260 (2009)

Mahnken, A.H., Raupach, R., Wildberger, J.E., Jung, B., Heussen, N., Flohr, T.G., Günther, R.W., Schaller, S.: A new algorithm for metal artifact reduction in computed tomography: in vitro and in vivo evaluation after total hip replacement. Investig. Radiol. **38**(12), 769–775 (2003)

Meyer, E., Raupach, R., Lell, M., Schmidt, B., Kachelrieß, M.: Normalized metal artifact reduction (NMAR) in computed tomography. Med. Phys. **37**(10), 5482–5493 (2010)

Müller, J., Buzug, T.M.: Spurious structures created by interpolation-based CT metal artifact reduction. Proc. SPIE **7258**, 72581Y (2009)

Osher, S., Rudin, L.I.: Feature-oriented image enhancement using shock filters. SIAM J. Numer. Anal. **27**, 919–940 (1990)

Osher, S., Sethian, J.A.: Fronts propagating with curvature-dependent speed: algorithms based on Hamilton-Jacobi formulations. J. Comput. Phys. **79**, 12–49 (1988)

Park, H.S., Hwang, D., Seo, J.K.: Metal artifact reduction for polychromatic X-ray CT based on a beam-hardening corrector. IEEE Trans. Med. Imaging **35**, 480–487 (2016)

Perona, P., Shiota, T., Malik, J.: Anisotropic diffusion. In: ter Haar Romeny, B.M. (ed.) Geometry-Driven Diffusion in Computer Vision, pp. 73–92. Springer, Dordrecht (1994)

Philips Healthcare: Metal artifact reduction for orthopedic implants (O-MAR), White Paper, Philips CT Clinical Science, Andover (2012)

Prell, D., Kyriakou, Y., Beister, M., Kalender, W.A.: A novel forward projection-based metal artifact reduction method for at-detector computed tomography. Phys. Med. Biol. **54**, 6575–6591 (2009)

Timmer, J.: Metal artifact correction in computed tomography. US Patent, 7,340,027 (2008)

Verburg, J.M., Seco, J.: CT metal artifact reduction method correcting for beam hardening and missing projections. Phys. Med. Biol. **57**(9), 2803–2818 (2012)

Wang, G., Snyder, D.L., O'Sullivan, J.A., Vannier, M.W.: Iterative deblurring for CT metal artifact reduction. IEEE Trans. Med. Imaging **15**(5), 657–664 (1996)

Watzke, O., Kalender, W.A.: A pragmatic approach to metal artifact reduction in CT: merging of metal artifact reduced images. Eur. J. Radiol. **14**(5), 849–856 (2004)

Wei, J., Chen, L., Sandison, G.A., Liang, Y., Xu, L.X.: X-ray CT high-density artifact suppression in the presence of bones. Phys. Med. Biol. **49**(24), 5407–5418 (2004)

Zhang, Y., Yu, H.: Convolutional neural network based metal artifact reduction in X-ray computed tomography. IEEE Trans. Med. Imaging **37**, 1370–1381 (2018)

Zhang, Y., Yan, H., Jia, X., Yang, J., Jiang, S.B., Mou, X.: A hybrid metal artifact reduction algorithm for X-ray CT. Med. Phys. **40**, 041910 (2013)

Zhang, K., Han, Q., Xu, X., Jiang, H., Ma, L., Zhang, Y., Yang, K., Chen, B., Wang, J.: Metal artifact reduction of orthopedics metal artifact reduction algorithm in total hip and knee arthroplasty. Medicine (Baltimore) **99**(11), e19268 (2020)

Zhao, S., Bae, K.T., Whiting, B., Wang, G.: A wavelet method for metal artifact reduction with multiple metallic objects in the field of view. J. X-Ray Sci. Technol. **10**, 67–76 (2002)

Domain Decomposition for Non-smooth (in Particular TV) Minimization

11

Andreas Langer

Contents

Introduction	380
Basic Idea of Domain Decomposition	380
Difficulty for Non-smooth and Non-separable Optimization Problems	385
Domain Decomposition for Smoothed Total Variation	390
Direct Splitting Approach	390
Decomposition Based on the Euler-Lagrange Equation	391
Decomposition for Predual Total Variation	391
Overlapping Domain Decomposition	392
Non-overlapping Domain Decomposition	397
Decomposition for Primal Total Variation	406
Basic Domain Decomposition Approach	406
Domain Decomposition Approach Based on the (Pre)Dual	412
Conclusion	421
References	422

Abstract

Domain decomposition is one of the most efficient techniques to derive efficient methods for large-scale problems. In this chapter such decomposition methods for the minimization of the total variation are discussed. We differ between approaches which directly tackle the (primal) total variation minimization and approaches which deal with their predual formulation. Thereby we mainly concentrate on the presentation of domain decomposition methods which guarantee to converge to a solution of the global problem.

A. Langer (✉)
Centre for Mathematical Sciences, Lund University, Lund, Sweden
e-mail: andreas.langer@math.lth.se

© Springer Nature Switzerland AG 2023
K. Chen et al. (eds.), *Handbook of Mathematical Models and Algorithms in Computer Vision and Imaging*, https://doi.org/10.1007/978-3-030-98661-2_104

Keywords

Domain decomposition · Schwarz method · Non-smooth optimisation · Total variation

Introduction

Due to the improvement of hardware, the dimensionality of measurements and in particular images is continuously increasing, resulting into large-scale data sets, which want to be processed further. For image processing, e.g., image restoration (denoising, deblurring, inpainting, etc.) and image analysis (segmentation, optical flow calculation, etc.), in the last decades non-smooth minimization problems such as total variation minimization became increasingly important. While being favorable due to the improved enhancement of images compared to smooth imaging approaches, non-smooth minimization problems typically scale badly with the dimension of the data. Hence, existing state-of-the-art standard methods for solving total variation minimization, as described in Burger et al. (2016) and Chambolle et al. (2010), perform well for small- and medium-scale problems. However, they are not able to perform in realistic CPU-time large imaging problems, such as 3D or even 4D imaging (spatial plus temporal dimensions) from functional magnetic resonance in nuclear medical imaging, astronomical imaging, or global terrestrial seismic imaging. Let us mention that with a clever implementation of these standard methods on a parallel architecture such as the graphics processing unit (GPU), one can accelerate them tremendously (Pock et al. 2008).

Here we are interested to address methods to large-scale total variation problems, which allow us to reduce the problem to a finite sequence of subproblems of a more manageable size by splitting the spatial domain into several smaller subdomains. Such methods are known under the name of *domain decomposition*.

Basic Idea of Domain Decomposition

Domain decomposition is a divide-and-conquer technique for solving partial differential equations by iteratively solving on each subdomain an appropriate defined subproblem. It has been shown multiple times (Quarteroni and Valli 1999; Toselli and Widlund 2006) that such methods are one of the most successful methods to construct efficient solvers for large-scale problems. The main reason for this is that they allow to reduce the dimension with the possibility for parallelization. In particular, domain decomposition is one of the most significant ways for devising parallel approaches that can benefit strongly from multiprocessor computers. Parallel approaches are mandatory when one has to solve large-scale numerical simulations, as they appear in a wide range of applications in physics and engineering. We summarize the main advantages of domain decomposition approaches, which include (i) dimension reduction; (ii) enhancement of parallelism;

11 Domain Decomposition for Non-smooth (in Particular TV) Minimization

Fig. 1 Overlapping decomposition into two domains

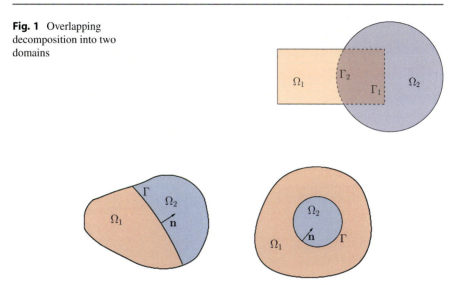

Fig. 2 Non-overlapping decomposition into two domains

(iii) localized treatment of complex and irregular geometries, singularities, and anomalous regions; (iv) and sometimes reduction of the computational complexity of the underlying solution method.

The first known domain decomposition has been proposed by Schwarz in (1869). In particular, he developed an overlapping domain decomposition method in order to show the existence of harmonic functions on irregular regions that are the union of overlapping subregions (Quarteroni and Valli 1999, p. 26). Since this pioneering work and due to the invention of computers and the need of fast computation, domain decomposition became one of the most successful numerical techniques.

When we talk about domain decomposition methods, we distinguish between an overlapping (see Fig. 1) and a non-overlapping (see Fig. 2) separation of the physical domain into two or more subdomains, as well as between successive and parallel computation of the subdomain problems.

Let us discuss the basic idea of domain decomposition methods for the *Poisson problem*, i.e., second-order self-adjoint elliptic problem,

$$\mathcal{L}u \equiv -\Delta u = f \quad \text{in } \Omega, \qquad u = 0 \quad \text{on } \partial\Omega, \tag{1}$$

for a decomposition of the spatial domain Ω into two subdomains. Here u is the unknown function; Δ denotes the Laplace operator; Ω is a two-dimensional domain, i.e., $\Omega \subset \mathbb{R}^2$, with Lipschitz boundary $\partial\Omega$; and f is a given function.

Non-overlapping Domain Decomposition

Let us start by splitting the spatial domain Ω into two non-overlapping subdomains Ω_1 and Ω_2 such that $\overline{\Omega} = \overline{\Omega_1 \cup \Omega_2}$ and $\Omega_1 \cap \Omega_2 = \emptyset$; cf. Fig. 2. We define the interface between these two regions by $\Gamma := \partial\Omega_1 \cap \partial\Omega_2$. In addition, we assume that the boundaries of the subdomains are Lipschitz continuous. Then problem (1) can be formulated as

$$\begin{cases} \mathcal{L}u_1 = f & \text{in } \Omega_1 \\ u_1 = 0 & \text{on } \partial\Omega_1 \cap \partial\Omega \\ u_1 = u_2 & \text{on } \Gamma \\ \frac{\partial u_1}{\partial \mathbf{n}} = \frac{\partial u_2}{\partial \mathbf{n}} & \text{on } \Gamma \\ \mathcal{L}u_2 = f & \text{in } \Omega_2 \\ u_2 = 0 & \text{on } \partial\Omega_2 \cap \partial\Omega \end{cases}, \qquad (2)$$

where each $\mathbf{n}$ is the outward pointed normal on Γ from Ω_1. Here we see that due to the partition of Ω, the original problem (1) is replaced by two subproblems on each subdomain by imposing both Neumann and Dirichlet conditions on Γ. These conditions transmit information from one domain patch to the other and therefore they are called *transmission conditions*. The equivalence between the Poisson problem (1) and the multi-domain problem (2) is in general not obvious, but can be shown under suitable regularity assumptions on f, typically $f \in L^2(\Omega)$, by considering the associated variational formulation; see, for example, Quarteroni and Valli (1999).

The successive Dirichlet-Neumann method We will now focus on solving the multi-domain problem (2) by an iterative method. Such methods typically introduce a sequence of subproblems on Ω_1 and Ω_2 for which boundary conditions at the internal boundary are provided, which play the role of the transmission conditions.

For a given λ^0, solve for each $k \geq 0$ with respect to u_1^{k+1} and u_2^{k+1}

$$\begin{cases} \mathcal{L}u_1^{k+1} = f & \text{in } \Omega_1 \\ u_1^{k+1} = 0 & \text{on } \partial\Omega_1 \setminus \Gamma \\ u_1^{k+1} = \lambda^k & \text{on } \Gamma \end{cases} \text{ and } \begin{cases} \mathcal{L}u_2^{k+1} = f & \text{in } \Omega_2 \\ u_2^{k+1} = 0 & \text{on } \partial\Omega_2 \setminus \Gamma \\ \frac{\partial u_2^{k+1}}{\partial \mathbf{n}} = \frac{\partial u_1^{k+1}}{\partial \mathbf{n}} & \text{on } \Gamma \end{cases} \qquad (3)$$

with

$$\lambda^{k+1} := \hat{\alpha} u_{2|\Gamma}^{k+1} + (1-\hat{\alpha})\lambda^k,$$

where $\hat{\alpha} > 0$ is an acceleration or relaxation parameter and $u_{|\Gamma}$ denotes the restriction of the function u to Γ. Note that the boundary conditions on the interface Γ are different for each subdomain problem.

We remark that this method does not necessarily converge, unless assumptions on the parameter $\hat{\alpha}$ or on Ω_1 and Ω_2 are made. However, if it is converging, then the rate of convergence is independent of the mesh size; see Marini and Quarteroni

(1989) for a convergence proof based on a functional analysis argument for partial differential equations.

Parallelism The Dirichlet-Neumann method (3) is generating at each step two boundary value problems, the first in Ω_1 and the second in Ω_2, to be solved successively. A simple modification frees these two subproblems from each other, which makes it more interesting in view of a parallel implementation. More precisely, when solving the boundary value problem in Ω_2 at the iteration step $k+1$, it is indeed enough to use u_1^k instead of u_1^{k+1}. That is, in (3) we simply replace the Neumann conditions on Γ by the new ones $\frac{\partial u_2^{k+1}}{\partial \mathbf{n}} = \frac{\partial u_1^k}{\partial \mathbf{n}}$.

Variational formulation For $i = 1, 2$ set $(w, v)_{\Omega_i} := \int_{\Omega_i} wv$, $a_i(w, v) := (\mathcal{L}w, v)_{\Omega_i}$, $W_i := \{w_i \in H^1(\Omega_i) : w_i|_{\partial\Omega\cap\partial\Omega_i} = 0\}$, and $W_\Gamma := \{\eta \in H^{1/2}(\Gamma) : \eta = w|_\Gamma$ for a suitable $w \in H_0^1(\Omega)\}$, where $H^{1/2}(\Gamma)$ is the trace space of $H^1(\Omega)$ on Γ. Then the variational formulation of (3) reads as follows:

$$\text{find } u_1^{k+1} \in W_1 : a_1(u_1^{k+1}, v_1) = (f, v_1)_{\Omega_1} \quad \forall v_1 \in H_0^1(\Omega_1)$$
$$u_1^{k+1} = \lambda^k \quad \text{on } \Gamma$$
$$\text{find } u_2^{k+1} \in W_2 : a_2(u_2^{k+1}, v_2) = (f, v_2)_{\Omega_2} \quad \forall v_2 \in H_0^1(\Omega_2)$$
$$a_2(u_2^{k+1}, R_2\mu) = (f, R_2\mu)_{\Omega_2} + (f, R_1\mu)_{\Omega_1} - a_2(u_1^{k+1}, R_1\mu) \ \forall \mu \in W_\Gamma$$

where $R_i : W_\Gamma \to W_i$, $i = 1, 2$, is some extension operator.

In a similar but different way, domain decomposition methods for an overlapping splitting of the spatial domain are derived.

Overlapping Domain Decomposition
The so-called *multiplicative* and *additive* Schwarz methods, whose terminology refers to successive and parallel overlapping domain decomposition methods, respectively, are shortly discussed next. Therefore, let us split the spatial domain Ω into two overlapping subdomains Ω_1 and Ω_2 such that $\Omega_1 \cap \Omega_2 \neq \emptyset$ and $\Omega = \Omega_1 \cup \Omega_2$; cf. Fig. 1. Further we denote $\Gamma_1 = \partial\Omega_1 \cap \Omega_2$ and $\Gamma_2 = \partial\Omega_2 \cap \Omega_1$ the interior boundaries of the subdomains.

Multiplicative Schwarz method The multiplicative Schwarz method starts with an initial value u^0 defined in Ω and vanishing on $\partial\Omega$ and computes a sequence of approximate solutions $u^1, u^2, \ldots$ by solving

$$\begin{cases} \mathcal{L}u_1^{k+1} = f & \text{in } \Omega_1 \\ u_1^{k+1} = u^k|_{\Gamma_1} & \text{on } \Gamma_1 \\ u_1^{k+1} = 0 & \text{on } \partial\Omega_1 \setminus \Gamma_1 \end{cases} \text{ and } \begin{cases} \mathcal{L}u_2^{k+1} = f & \text{in } \Omega_2 \\ u_2^{k+1} = u_1^{k+1}|_{\Gamma_2} & \text{on } \Gamma_2 \\ u_2^{k+1} = 0 & \text{on } \partial\Omega_2 \setminus \Gamma_2 \end{cases}, \quad (4)$$

with respect to u_1^{k+1} and u_2^{k+1}. That is, the subproblems are solved successively. The next approximate u^{k+1} is then defined by

$$u^{k+1}(x) = \begin{cases} u_2^{k+1}(x) & \text{if } x \in \Omega_2, \\ u_1^{k+1}(x) & \text{if } x \in \Omega \setminus \Omega_2. \end{cases}$$

It can be shown that the multiplicative Schwarz method (4) converges to a solution of problem (1); see Lions (1971, 1988) and for a variational based proof consult (Quarteroni and Valli 1999).

Variational formulation Set $(w, v) := \int_\Omega wv$, $a(w, v) := (\mathcal{L}w, v)$, and $W_i^0 := \{v \in H_0^1(\Omega): v = 0 \text{ in } \Omega \setminus \overline{\Omega_i}\}, i = 1, 2$, as closed subspaces of $H_0^1(\Omega)$ by extending their elements on Ω by 0. Moreover, we define the energy

$$\mathcal{J}(w, u) := \frac{1}{2}a(w, w) - (f, w) + a(u, w). \tag{5}$$

Let us rewrite (4) in the following form:

$$\begin{cases} \mathcal{L}(u^{k+1/2} - u^k) = f - \mathcal{L}u^k & \text{in } \Omega_1 \\ u^{k+1/2} - u^k \in W_1^0 \end{cases} \text{ and } \begin{cases} \mathcal{L}(u^{k+1} - u^{k+1/2}) = f - \mathcal{L}u^{k+1/2} & \text{in } \Omega_2 \\ u^{k+1} - u^{k+1/2} \in W_2^0. \end{cases}$$

The variational formulation of method (4) reads as follows: initialize $u^0 \in H_0^1(\Omega)$ and for $k \geq 0$ solve

$$\begin{cases} w_1^k \in W_1^0 : a(w_1^k, v_1) = (f, v_1) - a(u^k, v_1) & \text{for all } v_1 \in W_1^0 \\ u^{k+1/2} = u^k + w_1^k \\ w_2^k \in W_2^0 : a(w_2^k, v_2) = (f, v_2) - a(u^{k+1/2}, v_2) & \text{for all } v_2 \in W_2^0 \\ u^{k+1} = u^{k+1/2} + w_2^k \end{cases} \tag{6}$$

or equivalently

$$\begin{cases} w_1^k = \arg\min_{w_1 \in W_1^0} \mathcal{J}(w_1, u^k) \\ u^{k+1/2} = u^k + w_1^k \\ w_2^k = \arg\min_{w_2 \in W_2^0} \mathcal{J}(w_2, u^{k+1/2}) \\ u^{k+1} = u^{k+1/2} + w_2^k. \end{cases} \tag{7}$$

Additive Schwarz method If we make the two steps in (4) independent from each other, which allows for parallelization, then we obtain the additive alternating Schwarz method, which computes the sequence of approximations by solving

$$\begin{cases} \mathcal{L}u_1^{k+1} = f & \text{in } \Omega_1 \\ u_1^{k+1} = u_{|\Gamma_1}^k & \text{on } \Gamma_1 \\ u_1^{k+1} = 0 & \text{on } \partial\Omega_1 \setminus \Gamma_1 \end{cases} \quad \text{and} \quad \begin{cases} \mathcal{L}u_2^{k+1} = f & \text{in } \Omega_2 \\ u_2^{k+1} = u_{|\Gamma_2}^k & \text{on } \Gamma_2 \\ u_2^{k+1} = 0 & \text{on } \partial\Omega_2 \setminus \Gamma_2 \end{cases}. \qquad (8)$$

The next update u^{k+1} is then defined by

$$u^{k+1}(x) = \begin{cases} u_1^{k+1}(x) & x \in \Omega \setminus \Omega_2 \\ u_1^{k+1}(x) + u_2^{k+1}(x) - u^k(x) & x \in \Omega_1 \cap \Omega_2. \\ u_2^{k+1}(x) & x \in \Omega \setminus \Omega_1 \end{cases} \qquad (9)$$

Variational formulation The variational formulation of method (8) reads as

$$\begin{cases} w_1^k \in W_1^0 : a(w_1^k, v_1) = (f, v_1) - a(u^k, v_1) & \text{for all } v_1 \in W_1^0 \\ w_2^k \in W_2^0 : a(w_2^k, v_2) = (f, v_2) - a(u^k, v_2) & \text{for all } v_2 \in W_2^0 \\ u^{k+1} = u^k + w_1^k + w_2^k \end{cases}$$

or

$$\begin{cases} w_1^k = \arg\min_{w_1 \in W_1^0} \mathcal{J}(w_1, u^k) \\ w_2^k = \arg\min_{w_2 \in W_2^0} \mathcal{J}(w_2, u^k), \\ u^{k+1} = u^k + w_1^k + w_2^k \end{cases} \qquad (10)$$

where $\mathcal{J}$ is defined as in (5). By relation (9) we verify that the original formulation (8) is equivalent to the variational formulation.

Note that in the overlapping domain decomposition methods presented above, the subdomain problems are of the same type in each subdomain, while for the non-overlapping methods the subdomain problems differ due two interface conditions, which are distributed among the subdomain problems.

For a broader discussion on domain decomposition approaches for partial differential equations, we refer to Chan and Mathew (1994), Dolean et al. (2015), Mathew (2008), Quarteroni and Valli (1999), Toselli and Widlund (2006), and Smith et al. (2004).

Difficulty for Non-smooth and Non-separable Optimization Problems

Three main issues are of high interest when analyzing domain decomposition methods: (i) convergence, (ii) rate of convergence, and (iii) the independence of the rate of convergence on the mesh size, which can be interpreted as a preconditioning strategy. When talking about convergence, one usually means convergence to a solution of the global problem. However, we will also learn to know domain decomposition methods that do converge but not necessarily to a solution of the global problem. Hence, in the sequel when we talk about convergence, we distinguish between convergence to some point, which may not be a solution of the global problem, and convergence to a solution of the global problem. For smooth energies, the convergence to a solution of the global problem and the other two concerns are at large well established. We remark, that for non-smooth problems, decomposition algorithms may still work fine as long as the energy splits additively with respect to the domain decomposition. For such problems convergence to a solution of the original problem and sometimes even the rate of convergence are ensured; see, for example, Fornasier (2007), Tseng (2001), Tseng and Yun (2009), and Wright (2015). In (2009) Vonesch and Unser could provide preconditioning effects of a subspace correction algorithm for minimizing a non-smooth energy when applied to deblurring problems. Let us mention that there is a tremendous amount of literature devoted to splitting methods for non-smooth but separable problems in the context of coordinate descent methods (Wright 2015). We are not revising these methods, but concentrate on non-smooth and non-separable problems, where the situation to construct splitting methods that converge to the correct solution seems more complicated as the following counterexample by Warga (1963) indicates.

Example 1. Let $V := [0,1]^2$, $V_1 := \{(c,0) \colon c \in [0,1]\}$, $V_2 := \{(0,c) \colon c \in [0,1]\}$ and $\varphi \colon V \to \mathbb{R}$ given by $\varphi(x) = |x_1 - x_2| - \min\{x_1, x_2\}$, where $x = (x_1, x_2)$. We observe that φ is convex but non-smooth and non-additive with respect to the splitting, i.e., $\varphi(x) \neq \varphi((x_1, 0)) + \varphi((0, x_2))$. We have that $0 \in \arg\min_{x \in V_i} \varphi(x)$ for $i \in \{1,2\}$ and thus $x_2^k = x_1^k = 0$ for all $k \geq 0$. On the contrary $(1,1) \in \arg\min_{x \in V} \varphi(x)$.

While this example is more of an academic interest, non-smooth and non-separable problems often arise in image processing, where one is interested to obtain a non-smooth solution such that discontinuities (edges) are well represented. This may lead to the minimization of a functional that consists of a total variation term. Let $\Omega \subset \mathbb{R}^d$, $d = 1, 2$, be an open-bounded set with Lipschitz boundary $\partial \Omega$. For $u \in L^1(\Omega)$ we denote by

$$\int_\Omega |Du| := \sup \left\{ \int_\Omega u \operatorname{div} \mathbf{p} \, dx \colon \mathbf{p} \in C_0^1(\Omega, \mathbb{R}^d), |\mathbf{p}|_{\ell^2} \right.$$
$$\left. \leq 1 \text{ almost everywhere (a.e.) in } \Omega \right\} \qquad (11)$$

the total variation of u in Ω (Ambrosio et al. 2000; Giusti 1984), where $C_0^1(\Omega, \mathbb{R}^d)$ is the space of continuously differentiable vector-valued functions with compact support in Ω and $|\cdot|_{\ell^2}$ denotes the standard Euclidean norm. Here and in the rest of this chapter, bold letters indicate vector-valued functions. If $u \in W^{1,1}(\Omega)$, the

Sobolev space of L^1 functions with L^1 distributional derivative, then $\int_\Omega |Du| = \int_\Omega |\nabla u|_{\ell^2} \, dx$. Note that different vector norms may be used in the definition of the total variation. More precisely one may use $|\cdot|_{\ell^r}$ with $1 < r \leq \infty$. For example, the case $r = \infty$ is considered in Hintermüller and Kunisch (2004).

It is well established that the total variation preserves edges and discontinuities in images (Chambolle et al. 2010; Chan and Shen 2005), which is one of the reasons why it has been introduced to image processing as a regularization technique (Rudin et al. 1992). In this approach one typically minimizes an energy consisting of a data-fidelity term $\mathcal{D}$, which enforces the consistency between the observed image and the solution, a total variation term, as a regularizer, and a positive parameter λ weighting the importance of these two terms. That is, one solves

$$\min_u \mathcal{D}(u) + \lambda \int_\Omega |Du|.$$

The choice of the data term usually depends on the type of noise contamination. For example, in the case of Gaussian noise, a quadratic L^2 data fidelity term is used, while for impulsive noise an L^1 term is suggested (Alliney 1997) and seems more successful than an L^2 term (Nikolova 2002, 2004). Other and different fidelity terms have been considered in connection with other types of noise models as Poisson noise (Le et al. 2007), multiplicative noise (Aubert and Aujol 2008), and Rician noise (Getreuer et al. 2011). For images which are simultaneously contaminated by Gaussian and impulse noise (Cai et al. 2008), a combined L^1-L^2 data fidelity term has been suggested and demonstrated to work satisfactorily (Calatroni et al. 2017; Hintermüller and Langer 2013; Langer 2017b, 2019). We will restrict ourselves to Gaussian noise removal, i.e., L^2 data fidelity, as it will cover the fundamental domain decomposition approaches for total variation minimization proposed so far. That is, we consider the so-called L^2-TV model

$$\min_{u \in BV(\Omega)} \left\{ J(u) := \frac{1}{2} \|Tu - g\|_{L^2(\Omega)}^2 + \lambda \int_\Omega |Du| \right\}, \quad (12)$$

where $BV(\Omega) = \{u \in L^1(\Omega) : \int_\Omega |Du| < \infty\}$ is the space of bounded variation functions (Ambrosio et al. 2000), $g \in L^2(\Omega)$ is the observation, and $T \in \mathcal{L}(L^2(\Omega))$ is a bounded linear operator modeling the image formation device. Typical examples for T are (i) convolution operators, which describe blur in an image; (ii) the identity operator I, if an image is only corrupted by noise; (iii) the characteristic function of a subdomain marking missing parts, i.e., the inpainting domain; or (iv) the Fourier transform, if the observed data are given as corresponding frequencies. Since $\Omega \subset \mathbb{R}^d$, $d = 1, 2$, the embedding $BV(\Omega) \hookrightarrow L^2(\Omega)$ is continuous (Attouch et al. 2014, Theorem 10.1.3), and hence problem (12) is equivalent to $\min_{u \in L^2(\Omega)} J(u)$. In order to ensure the existence of a minimizer of J, we assume that J is coercive in $BV(\Omega)$, i.e., for every sequence $(u^n)_{n \in \mathbb{N}} \subset BV(\Omega)$ with $\|u^n\|_{BV(\Omega)} \to \infty$, we have $J(u^n) \to \infty$ or equivalently $\{u \in BV(\Omega) : J(u) \leq c\}$ is bounded in $BV(\Omega)$ for all constants $c > 0$. This condition holds if T does not annihilate constant functions, i.e., $1 \notin \ker(T)$ (Acar and Vogel 1994).

In the context of total variation minimization, the crucial difficulty in deriving suitable domain decomposition methods lies in the correct treatment of the interfaces of the domain decomposition patches, i.e., the preservation of crossing discontinuities and the correct matching where the solution is continuous. This difficulty is reflected by various effects of the total variation: (i) it is non-smooth, (ii) it preserves discontinuities and edges in images, and (iii) it is non-additive (non-separable) with respect to a non-overlapping domain decomposition, since the total variation of a function on the whole domain equals the sum of the total variation on the subdomains plus the size of the possible jumps at the interface. That is, let Ω_1 and Ω_2 be a disjoint (non-overlapping) decomposition of Ω, then the total variation has the following splitting property (cf. Ambrosio et al. (2000, Theorem 3.84, p. 177)):

$$\int_\Omega |D(u_{|\Omega_1} + u_{|\Omega_2})| = \int_{\Omega_1} |D(u_{|\Omega_1})| + \int_{\Omega_2} |D(u_{|\Omega_2})| \\ + \int_{\partial\Omega_1 \cap \partial\Omega_2} |u^+_{|\Omega_1} - u^-_{|\Omega_2}| \, d\mathcal{H}^{d-1}(x), \qquad (13)$$

where $\mathcal{H}^d$ denotes the Hausdorff measure of dimension d. The symbols u^+ and u^- denote the "interior" and "exterior" trace of u on $\partial\Omega_1 \cap \partial\Omega_2$, respectively.

For the L^2-TV model counterexamples of decomposition methods do exist, indicating failure of such splitting techniques. For example, in Fornasier et al. (2012) for a wavelet space decomposition method, a condition is derived which allows to establish global optimality of a limit point obtained by the decomposition method. Unfortunately, despite the good practical behavior of the method, this condition cannot be ensured to hold in general as shown by an example in Fornasier et al. (2012). Thus, the aforementioned condition may only be used in order to check a posteriori whether the algorithm indeed found a solution or failed to do so. A further counterexample for the L^2-TV model is presented in Lee and Nam (2017) for a decomposition of the spatial domain into two overlapping or non-overlapping domains; cf. Example 3 below.

We emphasize that for well-known approaches as those in Carstensen (1997), Chan and Mathew (1994), Tai and Tseng (2002), and Tai and Xu (2002), it is not clear yet whether they indeed converge to a global minimizer for non-smooth and non-additive problems, as any convergence theory in this direction is missing.

Instead of considering problem (12), one may tackle one of their dual or predual problems. In fact, if $T = I$, a predual of (12) is given by

$$\min \frac{1}{2} \|\operatorname{div} \mathbf{p} + g\|^2_{L^2(\Omega)} \quad \text{over } \mathbf{p} \in H_0(\operatorname{div}, \Omega) \qquad (14)$$

subject to (s.t.) $|\mathbf{p}(x)|_{\ell^2} \leq \lambda$ for almost all (f.a.a.) $x \in \Omega$, (see Hintermüller and Kunisch (2004) and Hintermüller and Rautenberg (2015)) where $H_0(\text{div}, \Omega) := \{\mathbf{v} \in L^2(\Omega)^d : \text{div } \mathbf{v} \in L^2(\Omega), \mathbf{v} \cdot \mathbf{n} = 0 \text{ on } \partial\Omega\}$ with $\mathbf{n}$ being the outward unit normal on $\partial\Omega$. Instead of (14), one may write equivalently

$$\min_{\mathbf{p} \in H_0(\text{div},\Omega)} \{F(\mathbf{p}) := \frac{1}{2}\|\text{div } \mathbf{p} + g\|_{L^2(\Omega)}^2 + \chi_K(\mathbf{p})\}, \tag{15}$$

or

$$\min_{\mathbf{p} \in H_0(\text{div},\Omega)} \{\mathfrak{F}(p) := \frac{1}{2}\|\text{div } \mathbf{p} + g\|_{L^2(\Omega)}^2 \, dx + \mathfrak{I}_\lambda(\mathbf{p})\}$$

where $K := \{\mathbf{p} \in H_0(\text{div}, \Omega) : |\mathbf{p}(x)|_{\ell^2} \leq \lambda \text{ f.a.a. } x \in \Omega\}$, χ_K being the characteristic function of the set K, i.e.,

$$\chi_K(\mathbf{p}) := \begin{cases} 0 & \text{if } \mathbf{p} \in K \\ \infty & \text{if } \mathbf{p} \notin K, \end{cases}$$

and

$$\mathfrak{I}_\lambda(\mathbf{p}) := \begin{cases} 0 & \text{if } |\mathbf{p}(x)|_{\ell^2} \leq \lambda \text{ f.a.a. } x \in \text{Dom}(\mathbf{p}) \\ \infty & \text{otherwise.} \end{cases}$$

with $\text{Dom}(\mathbf{p}) := \{x \in \Omega : \mathbf{p}(x) < \infty\}$ denoting the domain of $\mathbf{p}$. We note that if $T = I$, then (12) is strictly convex and hence possesses a unique minimizer, while its predual problem (14) may not have a unique solution, as it is "only" convex but not strictly convex. In the case of $T = I$, the solution u^* of (12) and a solution $\mathbf{p}^*$ of (14) are related by

$$u^* = \text{div } \mathbf{p}^* + g, \tag{16}$$

(see Hintermüller and Kunisch (2004)). Note that (14) is separable with respect to a disjoint decomposition of the spatial domain Ω. Let Ω be decomposed into M disjoint subdomains $(\Omega_j)_{j=1}^M$, then for $\mathbf{p} \in H_0(\text{div}, \Omega)$ we have

$$\int_\Omega |\text{div } \mathbf{p} + g|^2 \, dx + \mathfrak{I}_\lambda(\mathbf{p}) = \sum_{j=1}^M \int_{\Omega_j} |\text{div}(\mathbf{p}_{|\Omega_j}) + g|^2 \, dx + \mathfrak{I}_\lambda(\mathbf{p}_{|\Omega_j}). \tag{17}$$

Nevertheless, domain decomposition approaches as in Tai (2003) and Tai and Xu (2002), which may be utilized for obstacle problems, cannot be directly applied to (14), as the convergence theory used in Tai (2003) and Tai and Xu (2002) essentially relies on the strong convexity of the objective. For a class of non-smooth and non-separable minimization problems in Tseng and Yun (2009), a convergence theory for coordinate gradient descent methods is established. In that paper, convergence to a minimizer of the global problem could be proven only under the assumption of strict convexity of the objective, which does not hold for (14), and hence the convergence analysis in Tseng and Yun (2009) is not directly applicable to it.

Further basic terminology For a Banach space V we denote by V' its topological dual and $\langle \cdot, \cdot \rangle_{V' \times V}$ describes the bilinear canonical pairing over $V' \times V$. The norm of a Banach space V is written as $\|\cdot\|_V$. By $(\cdot, \cdot)$ we denote the standard inner product in $L^2(\Omega)$.

For a convex functional $\mathcal{F}: V \to \overline{\mathbb{R}}$, we define the *subdifferential* of $\mathcal{F}$ at $v \in V$ as the set valued function

$$\partial \mathcal{F}(v) := \begin{cases} \emptyset & \text{if } \mathcal{F}(v) = \infty, \\ \{v^* \in V': \langle v^*, u - v \rangle_{V' \times V} + \mathcal{F}(v) \leq \mathcal{F}(u) \quad \forall u \in V\} & \text{otherwise.} \end{cases}$$

It is clear from this definition, that $0 \in \partial \mathcal{F}(v)$ if and only if v is a minimizer of $\mathcal{F}$. Let V, W be two Banach spaces, then for any operator $\Lambda: V \to W$ we define by $\Lambda^*: W' \to V'$ its *adjoint*.

For ease of notation, in the sequel for any sequence $(v^n)_{n \in \mathbb{N}}$, we write $(v^n)_n$ instead.

Domain Decomposition for Smoothed Total Variation

If one seeks a minimizer of (12) in the Sobolev space $W^{1,1}(\Omega)$, then (12) becomes

$$\min_{u \in W^{1,1}(\Omega)} \{J(u) = \frac{1}{2}\|Tu - g\|^2_{L^2(\Omega)} + \lambda \int_\Omega |\nabla u| \, dx\}. \tag{18}$$

We note that the total variation of $u \in W^{1,1}(\Omega)$ is additive with respect to a disjoint decomposition of Ω, i.e., the interface term in (13) vanishes.

Direct Splitting Approach

By means of space decomposition, split $W^{1,1}(\Omega)$ into M subspaces $V_1, \ldots, V_M$ such that $W^{1,1}(\Omega) = \sum_{i=1}^M V_i$. Then following Chen and Tai (2007), Tai (2003), Tai and Tseng (2002), and Tai and Xu (2002) initialize $u^0 \in W^{1,1}(\Omega)$ and solve (18) successively by iterating for $n = 1, 2, \ldots$

$$v_i^n \in \arg\min_{v_i \in V_i} J(u^{n+(i-1)/M} + v_i), \quad u^{n+i/M} = u^{n+(i-1)/M} + v_i^n, \quad i = 1, \ldots, M.$$

Due to the optimality of v_i^n we get that $(J(u^n))_n$ is monotonically decreasing and hence $(u^n)_n \subset W^{1,1}(\Omega)$ is bounded, since J is coercive, i.e., $(u^n)_n \subset \{u \in$

11 Domain Decomposition for Non-smooth (in Particular TV) Minimization

$W^{1,1}(\Omega): J(u) \leq J(u^0)\}$. Note that $W^{1,1}(\Omega)$ is non-reflexive and (18) is convex but neither strongly nor strictly convex and still non-smooth, due to the presence of the L^1 term. Hence, the convergence theory of Tai (2003), Tai and Xu (2002), Tai and Tseng (2002), and Tseng and Yun (2009) does not cover this splitting algorithm. A similar decomposition method is considered in Chen and Tai (2007) but without any rigorous theoretical convergence analysis.

Decomposition Based on the Euler-Lagrange Equation

Assuming homogeneous Neumann boundary conditions, i.e., $\nabla u \cdot \mathbf{n} = 0$ on $\partial \Omega$ the Euler-Lagrange equation for (18) is

$$T^*(Tu - g) - \lambda \operatorname{div}\left(\frac{\nabla u}{|\nabla u|}\right) = 0.$$

Due to the presence of the term $\frac{1}{|\nabla u|}$, this equation is not well defined at points $\nabla u = 0$. To overcome this shortcoming, we introduce an additional small parameter $\epsilon > 0$ to slightly perturb the total variation semi-norm, such that (12) becomes

$$\min_{u \in W^{1,1}(\Omega)} \frac{1}{2}\|Tu - g\|_{L^2(\Omega)}^2 + \lambda \int_\Omega \sqrt{|\nabla u|^2 + \epsilon} \, dx. \tag{19}$$

The corresponding Euler-Lagrange equation is then

$$T^*(Tu - g) - \lambda \operatorname{div}\left(\frac{\nabla u}{\sqrt{|\nabla u|^2 + \epsilon}}\right) = 0. \tag{20}$$

Note that the functional in (19) is now strictly convex, Gâteaux differentiable, and separable. Hence, domain decomposition methods which converge to a solution of the global problem may be constructed following Tseng and Yun (2009). Domain decomposition methods for (19) and (20) have been considered, for example, in Chen and Tai (2007) and Xu et al. (2010, 2014).

While these smoothed problems possess the advantage that domain decomposition methods with desired convergence properties could be possibly designed, they do not generate solutions that preserve discontinuities and edges.

Decomposition for Predual Total Variation

In order to avoid the difficulties due to the minimization of a non-smooth and non-additive energy over a non-reflexive Banach space in (12), the predual problem (14) of (12) may be tackled instead. In particular the smooth objective and the box constraint in (14) seem more amenable to domain decomposition than the structure of (12). In fact, in Chang et al. (2015) overlapping and in Hintermüller and Langer (2015), Lee et al. (2019b), and Lee and Park (2019a,b) non-overlapping domain decomposition methods for (14) are proposed. Let us review the main ideas and results for these approaches.

Overlapping Domain Decomposition

Let Ω be partitioned into $M \in \mathbb{N}$ overlapping subdomains such that $\Omega = \bigcup_{j=1}^{M} \Omega_j$ and for any $j \in \{1, \ldots, M\}$ there is at least one $i \in \{1, \ldots, M\} \setminus \{j\}$ with $\Omega_i \cap \Omega_j \neq \emptyset$. Associated with this decomposition, we define subspaces $V_j := \{\mathbf{p} \in H_0(\text{div}, \Omega) \colon \text{supp}(\mathbf{p}) \subset \overline{\Omega}_j\}$, $j = 1, \ldots, M$. Based on this splitting, the fundamental idea of domain decomposition is to solve (14) or (15) by iteratively minimizing F on the subspaces V_j. Unfortunately $K \neq \sum_{j=1}^{M} K \cap V_j$ and hence there exist $\mathbf{p}_j \in K \cap V_j$, $j = 1, \ldots, M$, such that $\sum_{j=1}^{M} \mathbf{p}_j \notin K$, due to the overlapping region. This means that in general the subspaces V_j are too large. The introduction of a partition of unity, denoted by $(\theta_j)_{j=1}^{M}$, with the properties

$$\theta_j \in W^{1,\infty}(\Omega) \quad \text{for } j = 1, \ldots, M, \tag{21}$$

$$\sum_{j=1}^{M} \theta_j = 1, \tag{22}$$

$$\text{supp}(\theta_j) \subset \overline{\Omega}_j \quad \text{for } j = 1, \ldots, M, \tag{23}$$

allows us to define $K_j := \{\mathbf{p} \in H_0(\text{div}, \Omega) \colon |\mathbf{p}(x)|_{\ell^2} \leq |\theta_j(x)|\lambda \text{ f.a.a. } x \in \Omega\}$ for $j = 1, \ldots, M$. Properties (22) and (23) ensure that $(\theta_j \mathbf{p})_{j=1}^{M}$ is a partition of $\mathbf{p} \in H_0(\text{div}, \Omega)$ associated with the domain decomposition such that $\mathbf{p} = \sum_{j=1}^{M} \theta_j \mathbf{p}$ and $\text{supp}(\theta_j \mathbf{p}) \subset \overline{\Omega}_j$ for $j = 1, \ldots, M$. By (21) it is guaranteed that $\theta_j \mathbf{p} \in H_0(\text{div}, \Omega)$ provided that $\mathbf{p} \in H_0(\text{div}, \Omega)$ for $j = 1, \ldots, M$. This is easily seen by an application of the (generalized) Hölder inequality:

$$\|\theta_j \mathbf{p}\|_{L^2(\Omega)} \leq \|\theta_j\|_{L^\infty(\Omega)} \|\mathbf{p}\|_{L^2(\Omega)},$$

$$\|\text{div}(\theta_j \mathbf{p})\|_{L^2(\Omega)} \leq \|\nabla \theta_j\|_{L^\infty(\Omega)} \|\mathbf{p}\|_{L^2(\Omega)} + \|\theta_j\|_{L^\infty(\Omega)} \|\text{div } \mathbf{p}\|_{L^2(\Omega)}.$$

Moreover, one immediately sees that $K = \sum_{j=1}^{M} K_j$. In the case of a successive algorithm, this means (cf. Chang et al. (2015, Algorithm II)):

Note that, since $|\theta_j(\cdot)|\lambda : \overline{\Omega} \to \mathbb{R}_0^+$ is a bounded function, the existence of a minimizer of the subdomain problems in Algorithm 1 is ensured (Hintermüller and Rautenberg 2017, Proposition 3.2 (b)).

We are actually quite free in how to choose the partition of unity as long as the conditions (21), (22), and (23) hold. For example, one may additionally assume that $\theta_j \geq 0$ almost everywhere in Ω for $j = 1, \ldots, M$, as in Chang et al. (2015). One can

Algorithm 1 Basic successive overlapping algorithm for (14)

Pick an initial $\mathbf{p}^0 \in K$
for $n = 0, 1, \ldots$ do
 for $j = 1, \ldots, M$ do
  $\mathbf{p}_j^{n+1} \in \arg\min_{\mathbf{p}_j \in K_j} \mathfrak{F}(\mathbf{p}_j + \sum_{i<j} \mathbf{p}_i^{n+1} + \sum_{i>j} \theta_i \mathbf{p}^n)$
 end for
 $\mathbf{p}^{n+1} := \sum_{j=1}^M \mathbf{p}_j^{n+1}$
end for

also view it from the other way round, namely, that the partition of unity provides the overlapping splitting of the spatial domain. From a practical point of view, this has the advantage, that a partition of unity can always be easily constructed and hence an overlapping decomposition of the domain. In the case of a rectangle, which is a usual shape of an image, a simple example for a partition of unity for a splitting into three subdomains is shown in Fig. 3.

The first convergent overlapping domain decomposition method for the minimization of (14) is presented in Chang et al. (2015), here presented in Algorithm 2. There the partition of unity $(\theta_j)_j$ is chosen such that (21), (22), and (23), $\theta_j \geq 0$ and

$$\|\nabla \theta_j\|_{L^\infty(\Omega)} \leq \frac{C_\theta}{\delta}, \tag{24}$$

where $C_\theta > 0$ and $\delta > 0$ denotes the overlapping size, hold. The estimate (24) seems reasonable, as for small overlapping sizes we would expect a larger gradient and it may allow to get a feeling on how the convergence of the algorithm depends on the overlapping size; see Theorem 1.

Algorithm 2 Relaxed successive overlapping algorithm for (14)

Pick an initial $\mathbf{p}^0 \in K$.
Select a relaxation parameter $\hat{\alpha} \in (0, 1]$.
for $n = 0, 1, \ldots$ do
 for $j = 1, \ldots, M$ do
  $\hat{\mathbf{p}}_j^{n+1} \in \arg\min_{\mathbf{p}_j \in K_j} \mathfrak{F}(\mathbf{p}_j + \sum_{i<j} \mathbf{p}_i^{n+1} + \sum_{i>j} \theta_i \mathbf{p}^n)$
  $\mathbf{p}_j^{n+1} := (1 - \hat{\alpha}) \theta_j \mathbf{p}^n + \hat{\alpha} \hat{\mathbf{p}}_j^{n+1}$
 end for
 $\mathbf{p}^{n+1} := (1 - \hat{\alpha}) \mathbf{p}^n + \hat{\alpha} \sum_{j=1}^M \hat{\mathbf{p}}_j^{n+1}$
end for

In comparison to Algorithm 1, in Algorithm 2 a relaxation and associated parameter $\hat{\alpha} \in (0, 1]$ is introduced. This relaxation parameter weights the influence of the current (subspace) minimizer and a previous approximation. However, note that for $\hat{\alpha} = 1$ Algorithm 2 becomes Algorithm 1. Let $(\mathbf{p}_j^n)_n$ and $(\mathbf{p}^n)_n$ be

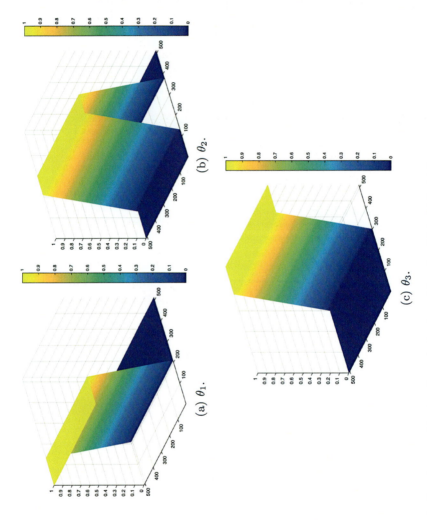

Fig. 3 Partition of unity for a decomposition into three overlapping subdomains. (**a**) θ_1 (**b**) θ_2. (**c**) θ_3

generated by Algorithm 2, then by a straightforward calculation one easily checks that $\mathbf{p}_j^{n+1} \in K_j$ for all $j \in \{1, \ldots, M\}$ and $\mathbf{p}^n \in K$.

By a simple modification, i.e., by replacing $\mathbf{p}_i^{n+1}$ by $\theta_i \mathbf{p}^n$ in the minimization problem in Algorithm 2, the subdomain problems are made independent from each other, which makes it more interesting in view of a parallel implementation. Its parallel version is presented in Algorithm 3.

Algorithm 3 Parallel overlapping algorithm for (14)

Pick an initial $\mathbf{p}^0 \in K$.
Select a relaxation parameter $\hat{\alpha} \in (0, \frac{1}{M}]$.
for $n = 0, 1, \ldots$ **do**
   for $j = 1, \ldots, M$ **do**
      $\hat{\mathbf{p}}_j^{n+1} \in \arg\min_{\mathbf{p}_j \in K_j} \mathfrak{F}(\mathbf{p}_j + \sum_{i \neq j} \theta_i \mathbf{p}^n)$
   end for
   $\mathbf{p}^{n+1} := (1 - \hat{\alpha})\mathbf{p}^n + \hat{\alpha} \sum_{j=1}^M \hat{\mathbf{p}}_j^{n+1}$
end for

Remark that the relaxation parameter $\hat{\alpha}$ is now only in the interval $(0, \frac{1}{M}]$, whose range is theoretically justified, in particular to guarantee the monotonic decay of $(F(\mathbf{p}^n))_n$; see Chang et al. (2015) for more details.

The convergence of Algorithms 2 and 3 to a solution of the global problem (15) with rate $\mathcal{O}(\frac{1}{n})$ is guaranteed. We recall this main result by referring to Chang et al. (2015) for its proof.

Theorem 1. *Let p^* be a minimizer of (14) and let $(p^n)_n$ be a sequence generated by Algorithm 2 or Algorithm 3. Due to (16) we set $u^n := g + \operatorname{div} p^n$ for all $n \in \mathbb{N}$ and $u^* := g + \operatorname{div} p^*$. Then for all $n \in \mathbb{N}$, we have*

$$\|u^n - u^*\|_{L^2(\Omega)}^2 \leq F(\mathbf{p}^n) - F(\mathbf{p}^*) \leq \frac{C^2}{n}$$

with

$$C := \sqrt{\zeta^0} \left(\frac{2}{\hat{\alpha}}(2M+1)^2 + 8\sqrt{2} C_\theta \lambda |\Omega|^{\frac{1}{2}} (\zeta^0)^{-\frac{1}{2}} \frac{M\sqrt{M}}{\delta \sqrt{\hat{\alpha}}} + \sqrt{2} - 1 \right)$$

where $\zeta^0 := |F(p^0) - F(p^*)|$.

We observe that the constant C in Theorem 1 depends on the tunable parameters $\hat{\alpha}$, δ, and M. Some comments according to these parameters are in order. Observe that if the number of subdomains M grows, C grows as well. In order to overcome this behavior, we may use a so-called coloring technique; see, e.g., Toselli and Widlund (2006). That is, Ω is partitioned into M_c classes of overlapping subdomains, where each class has a different color and each class is the union of disjoint subdomains with the same color. We note that in general the disjoint domains with the same color cannot be solved in parallel without introducing additional new constraints, as the following example, borrowed from Warga (1963), shows.

Example 2. Let $V := [0, 1] \times \{0\} \times [0, 1]$, $V_1 := \{(c, 0, 0) \; c \in [0, 1]\}$, $V_3 := \{(0, 0, c) \; c \in [0, 1]\}$, and $\varphi : V \to \mathbb{R}$ given by $\varphi(x) = |x_1 - x_3| - \min\{x_1, x_3\}$, where $x = (x_1, x_2, x_3)$. We have that $\mathbf{0} = \arg\min_{x_i \in V_i} \varphi(x)$ for $i \in \{1, 3\}$, while $(1, 0, 1) = \arg\min_{x \in V} \varphi(x)$.

However, if the problem is additively separable with respect to the considered disjoint decomposition, then it can be solved independently and in parallel with the disjoint domains. Since this property holds for the considered subdomain problems in Algorithms 2 and 3 with respect to the disjoint domains with the same color, we can replace M by M_c in Algorithms 2 and 3. Further let $N_0 \in \mathbb{N}$ be the maximum number of classes where a point $x \in \Omega$ can belong. A typical decomposition of a rectangular domain into a total of 16 subdomains colored by four colors is illustrated in Fig. 4a. In this example $M_c = 4 = N_0$. A splitting into overlapping stripes, as in Fig. 4b, would even reduce M_c and N_0 to 2. Then the constant C in Theorem 1 can be decreased to

$$C = \sqrt{\zeta^0} \left(\frac{2}{\hat{\alpha}} (2M_c + 1)^2 + 8\sqrt{2} C_\theta \lambda |\Omega|^{\frac{1}{2}} (\zeta^0)^{-\frac{1}{2}} \frac{M_c \sqrt{N_0}}{\delta \sqrt{\hat{\alpha}}} + \sqrt{2} - 1 \right),$$

where M_c and N_0 may be small, e.g., 2 or 4 (see above), even if the total number of subdomains grows. A complementary behavior is observed for the parameters $\hat{\alpha}$ and δ. That is, the smaller these parameters, the larger the constant C. Consequently one may choose $\hat{\alpha} = 1$ in Algorithm 2 and $\hat{\alpha} = \frac{1}{M}$ (or respectively $\hat{\alpha} = \frac{1}{M_c}$ when using a coloring technique) in Algorithm 3, which will lead to a faster convergence,

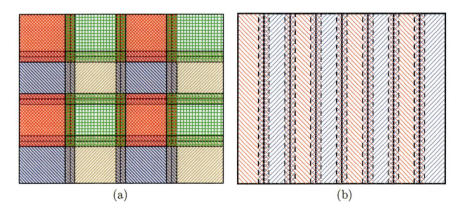

Fig. 4 Domain decomposition by coloring technique in M_c classes (colors) (**a**) $M_c = 4$. (**b**) $M_c = 2$

11 Domain Decomposition for Non-smooth (in Particular TV) Minimization

as also numerically observed in Chang et al. (2015). Note that $C \to \infty$ for $\delta \to 0$, which would be the non-overlapping case. Moreover, also the partition of unity cannot be carried over to the non-overlapping case, due to (21), as a non-overlapping decomposition would require a discontinuity at the interfaces of the patches. Hence, the here presented convergence results from Chang et al. (2015) do not imply the convergence of a non-overlapping domain decomposition method.

Subdomain problems Let $j \in \{1, \ldots, M\}$. Note that $\mathbf{p} \in K_j$ implies $\mathbf{p} \in H_0(\text{div}, \Omega)$. Hence, on a first sight each subproblem seems to be optimized on whole Ω, which would not be in the vein of domain decomposition. However, thanks to the partition of unity functions $(\theta_j)_{j=1}^M$ any $\mathbf{p} \in K_j$ has compact support in Ω_j solely. Consequently one only needs to compute a minimizer of the subproblems of Algorithms 2 and 3 in Ω_j. In order to compute a solution of the subdomain problems in practice, one may use, for example, the iterative scheme presented in Chambolle (2004) adapted to locally adaptive parameters $\theta_j \lambda$. Here for simplicity we assume that $\theta_j \geq 0$. In this situation the algorithm of Chambolle (2004) computes an approximation of $\hat{\mathbf{p}}_j^{n+1}$ in Algorithms 2 and 3 for $j = 1, \ldots, M$ by iterating

$$\hat{\mathbf{p}}_j^{n,0} \in K_j \text{ (e.g., } \hat{\mathbf{p}}_j^{n,0} = \hat{\mathbf{p}}_j^n\text{)}, \quad \hat{\mathbf{p}}_j^{n,\ell+1} = \frac{\theta_j \lambda \hat{\mathbf{p}}_j^{n,\ell} + \theta_j \lambda \tau \nabla(\text{div } \hat{\mathbf{p}}_j^{n,\ell} - g_j)}{\theta_j \lambda + \tau |\nabla(\text{div } \hat{\mathbf{p}}_j^{n,\ell} - g_j)|_{\ell^2}} \quad \text{for } \ell \geq 0,$$
(25)

where $g_j := g - \text{div}\left(\sum_{i<j} \mathbf{p}_i^{n+1} + \sum_{i>j} \theta_i \mathbf{p}^n\right)$ for Algorithm 2 and $g_j := g - \text{div}\left(\sum_{i \neq j} \theta_i \mathbf{p}^n\right)$ for Algorithm 3. For $0 < \tau \leq \frac{1}{8}$ one shows analogous to the proof of Chambolle (2004, Theorem 3.1) that the iterates $(\hat{\mathbf{p}}_j^{n,\ell})_\ell$ converge to a respective minimizer $\hat{\mathbf{p}}_j^{n+1}$ of the subdomain problems as $\ell \to \infty$. Due to the presence of θ_j in the nominator in (25), the update of $\hat{\mathbf{p}}_j^{n,\ell}$ is only performed in Ω_j and hence the subdomain problems are indeed restricted to the respective subdomains.

Non-overlapping Domain Decomposition

As already mentioned above, the convergence analysis carried out for the overlapping domain decomposition algorithms in Chang et al. (2015) leading to Theorem 1 cannot be directly applied to a non-overlapping splitting. In particular, till now it is still an open problem to construct a non-overlapping domain decomposition method for (14) in an infinite dimensional setting which is guaranteed to converge to a minimizer of the original global problem. However, for a finite difference and finite element discretization of (14), splitting methods which converge to the desired optimum are introduced in Hintermüller and Langer (2015), Lee et al. (2019b), and Lee and Park (2019a,b). The first method in this series has been proposed in Hintermüller and Langer (2015) for a finite difference discretization of (14), where instead of $|\mathbf{p}|_{\ell^2} \leq \lambda$ the constraint $|\mathbf{p}|_{\ell^\infty} \leq \lambda$ is originally used. Nevertheless, the algorithms in Hintermüller and Langer (2015) and its convergence results can be easily transformed into our setting, i.e., $|\mathbf{p}|_{\ell^2} \leq \lambda$, in which we will review them.

Finite Difference Setting

As our main application is image processing, in our discrete setting the spatial domain Ω^h is a mesh in $\mathbb{R}^2$ of size $N_1 \times N_2$, where $N_1, N_2 \in \mathbb{N}$ with mesh size $x_{i,j} - x_{i+1,j} = 1 = x_{i,j} - x_{i,j+1}$ for $x_{i,j} \in \Omega^h$, i.e., $\Omega^h = \{x_{i,j}\}_{\substack{1 \leq i \leq N_1 \\ 1 \leq j \leq N_2}}$. The respective function spaces are $X := \{u^h : \Omega^h \to \mathbb{R}\}$ and $Y = X^2$. For $u^h \in X$ and $\mathbf{p}^h = (p^{h,1}, p^{h,2}) \in Y$ we use the norms $\|u^h\|_X^2 := \|u^h\|_{\ell^2(\Omega^h)}^2 = \sum_{x \in \Omega^h} |u(x)|^2$ and $\|\mathbf{p}^h\|_Y^2 := \|p^{h,1}\|_X^2 + \|p^{h,2}\|_X^2$. On Ω^h the discrete gradient $\nabla_\Omega^h : X \to Y$ and the discrete $\mathrm{div}_\Omega^h : Y \to X$ are defined in a standard way by forward and backward differences such that $\mathrm{div}_\Omega^h = -(\nabla_\Omega^h)^*$; see, for example, Hintermüller and Langer (2015). Using this notation the discrete version of (15) is then written as

$$\min_{\mathbf{p}^h \in Y} F^h(\mathbf{p}^h) \qquad (26)$$

where $F^h(\mathbf{p}^h) := \|\mathrm{div}_\Omega^h \mathbf{p}^h + g^h\|_X^2 + \chi_{K^h}(\mathbf{p}^h)$ with $K^h := \{\mathbf{p}^h \in Y : |\mathbf{p}^h(x)|_{\ell^2} \leq \lambda \; \forall x \in \Omega^h\}$. Further let Ω^h be decomposed into $M \in \mathbb{N}$ overlapping or non-overlapping subdomains Ω_j^h such that $\Omega^h = \bigcup_{j=1}^M \Omega_j^h$. Associated with the subdomains we define $X_j := \{u^h : \Omega_j^h \to \mathbb{R}\}$ and $Y_j = X_j \times X_j$ together with the norms $\|u^h\|_{X_j}^2 := \sum_{x \in \Omega_j^h} |u^h(x)|^2$, $\|\mathbf{p}^h\|_{Y_j}^2 := \|p^{h,1}\|_{X_j}^2 + \|p^{h,2}\|_{X_j}^2$ for $u^h \in X_j$ and $\mathbf{p}^h \in Y_j$, $j = 1, \ldots, M$.

Approach via Finite Differences

Let Ω^h be decomposed into $M \in \mathbb{N}$ disjoint subdomains Ω_j^h such that $\Omega^h = \bigcup_{j=1}^M \Omega_j^h$ and $\Omega_j^h = \Omega^h \setminus (\bigcup_{i \neq j} \Omega_i^h)$ for $j = 1, \ldots, M$. Associated with this splitting, we set

$$\theta_j^h(x) := \begin{cases} 1 & \text{if } x \in \Omega_j^h \\ 0 & \text{if } x \in \Omega^h \setminus \Omega_j^h, \end{cases} \quad \text{for } j = 1, \ldots, M,$$

denoting a discrete partition of unity. In particular $\sum_{j=1}^M \theta_j^h(x) = 1$ for all $x \in \Omega^h$ and $\mathrm{supp}(\theta_j^h) = \Omega_j^h$ for $j = 1, \ldots, M$. For $\mathbf{p}^h \in Y$ we note that $\theta_j^h \mathbf{p}^h$ is an orthogonal projection of $\mathbf{p}^h$ onto Y_j, $j = 1, \ldots, M$, and $\mathbf{p}^h = \sum_{j=1}^M \theta_j^h \mathbf{p}^h$. Associated with the subdomains we define $K_j^h := \{\mathbf{p}^h \in Y : |\mathbf{p}^h(x)|_{\ell^2} \leq \lambda \theta_j(x) \; \forall x \in \Omega^h\}$ for $j = 1, \ldots, M$. With this splitting one may solve (26) by a successive domain decomposition algorithm (see Algorithm 4) or a parallel domain decomposition algorithm (see Algorithm 5; cf. Hintermüller and Langer (2015)).

Note that due to the disjoint decomposition of Ω^h we have that the sequences $(\mathbf{p}^{h,n})_n$ generated by Algorithms 4 and 5 are in K^h, as the constraint in K^h is

Algorithm 4 Successive non-overlapping algorithm for (14)

Pick an initial $\mathbf{p}^{h,0} \in K^h$.
for $n = 0, 1, \ldots$ do
  for $j = 1, \ldots, M$ do
   $\mathbf{p}_j^{h,n+1} \in \arg\min_{\mathbf{p}_j^h \in Y} \frac{1}{2} \| \operatorname{div}_\Omega^h (\mathbf{p}_j^h + \sum_{i<j} \mathbf{p}_i^{h,n+1} + \sum_{i>j} \mathbf{p}_i^{h,n}) + g^h \|_X^2 + \mathfrak{I}_{\lambda \theta_j^h}(\mathbf{p}_j^h)$
  end for
  $\mathbf{p}^{h,n+1} := \sum_{j=1}^M \mathbf{p}_j^{h,n+1}$
end for

Algorithm 5 Parallel non-overlapping algorithm for (14)

Pick an initial $\mathbf{p}^{h,0} \in K^h$.
for $n = 0, 1, \ldots$ do
  for $j = 1, \ldots, M$ do
   $\mathbf{p}_j^{h,n+1} \in \arg\min_{\mathbf{p}_j^h \in Y} \frac{1}{2} \| \operatorname{div}_\Omega^h (\mathbf{p}_j^h + \sum_{i \neq j} \theta_i^h \mathbf{p}^{h,n}) + g^h \|_X^2 + \mathfrak{I}_{\lambda \theta_j^h}(\mathbf{p}_j^h)$
  end for
  $\mathbf{p}^{h,n+1} := (1 - \frac{1}{M})\mathbf{p}^{h,n} + \frac{1}{M} \sum_{j=1}^M \mathbf{p}_j^{h,n+1}$
end for

pointwise and $|\mathbf{p}_j^{h,n} + \sum_{i<j} \mathbf{p}_i^{h,n} + \sum_{i>j} \mathbf{p}_i^{h,n-1}|_{\ell^2} \leq \lambda$ (Algorithm 4) as well as $|\mathbf{p}_j^{h,n} + \sum_{i \neq j} \theta_i^h \mathbf{p}^{h,n-1}|_{\ell^2} \leq \lambda$ (Algorithm 5). Similar as in Hintermüller and Langer (2015) one shows the following convergence results.

Theorem 2. *Let $(p^{h,n})_n$ be a sequence generated by Algorithm 4 or Algorithm 5. Then we have that*

(i) *$(F^h(p^{h,n}))_n$ is decreasing and converges.*
(ii) *The sequence $(p^{h,n})_n$ is bounded in Y and has an accumulation point which is a solution of (26).*

Additionally for the parallel non-overlapping domain decomposition method (Algorithm 5), a convergence order of $\mathcal{O}(\frac{1}{n})$ is ensured (Lee and Park 2019a).

Theorem 3. *Let $(p^{h,n})_n$ be a sequence generated by Algorithm 5 and $p^{h,*}$ a solution of (26), then for all $n \in \mathbb{N}$ we have*

$$F^h(p^{h,n}) - F^h(p^{h,*}) \leq \frac{C}{n},$$

where

$$C := M \left(\sum_{j=1}^M \frac{1}{2} \| \operatorname{div}_\Omega^h \theta_j^h (p^{h,*} - p^{h,0}) \|_X^2 \right) + (M-1)\left(F^h(p^{h,0}) - F^h(p^{h,*})\right). \tag{27}$$

As in the overlapping case, the constant in Theorem 3 can be reduced, if we use a coloring technique with $M_c \leq M$ classes. Then in (27) M can be replaced by M_c. However, C in Theorem 3 still depends on M as

$$\sum_{j=1}^{M_c} \frac{1}{2} \| \operatorname{div}_\Omega^h \theta_j^h (\mathbf{p}^{h,*} - \mathbf{p}^{h,0}) \|_X^2 \leq \| \operatorname{div}_\Omega^h (\mathbf{p}^{h,*} - \mathbf{p}^{h,0}) \|_X^2$$

$$+ c_1 \left(\max_{x \in \Omega^h} \left(|\mathbf{p}^{h,*}(x) - \mathbf{p}^{h,0}(x)|_{\ell^2} \right) \right)^2$$

where $c_1 \geq 0$ is a constant depending on M; see Lee and Park (2019a).

Algorithm 6 Accelerated parallel non-overlapping algorithm for (14)

Pick an initial $p^0 = q^0 \in K^h$ and $t^0 = 1$.
for $n = 0, 1, \ldots$ do
  for $j = 1, \ldots, M$ do
    $\mathbf{p}_j^{h,n+1} \in \arg\min_{\mathbf{p}_j^h \in Y} \frac{1}{2} \| \operatorname{div}_\Omega^h (M\mathbf{p}_j^h - (M-1)\theta_j^h \mathbf{q}^{h,n} + \sum_{i \neq j} \theta_i^h \mathbf{q}^{h,n}) + g^h \|_X^2$
    $+ \mathfrak{I}_{\lambda \theta_j^h}(M\mathbf{p}_j^h - (M-1)\theta_j^h \mathbf{q}^{h,n})$
  end for
  $\mathbf{p}^{h,n+1} := \sum_{j=1}^M \mathbf{p}_j^{h,n+1}$
  $t^{n+1} := \frac{1 + \sqrt{1 + 4(t^n)^2}}{2}$
  $\mathbf{q}^{h,n+1} := \mathbf{p}^{n+1} + \frac{t^n - 1}{t^{n+1}} (\mathbf{p}^{h,n+1} - \mathbf{p}^{h,n})$
end for

An accelerated version Instead of relaxing the global approximations of two consecutive iterations (see Algorithm 5), in Lee and Park (2019a) the relaxation step is put inside the local solution operator and is performed before the local solutions are computed. Then applying FISTA (Beck and Teboulle 2009) yields Algorithm 6, which converges with order $\mathcal{O}\left(\frac{1}{n^2}\right)$ (Lee and Park 2019a). Note that if in Algorithm 6 $t^n = 1$ for all iterations $n \in \mathbb{N}$, i.e., not using FISTA, then the algorithm still converges but only with order $\mathcal{O}(\frac{1}{n})$, as Algorithm 5.

Subdomain problems In order to restrict the subdomain problems in Algorithms 4, 5, and 6 to the respective subdomain plus a possible small stripe around the interface, a certain splitting property of the discrete divergence operator with respect to the disjoint decomposition of the spatial domain Ω^h is required, i.e.,

$$\sum_{x \in \Omega^h} \operatorname{div}_\Omega^h (\mathbf{p}_j^h + \mathbf{p}_{j^c}^h)(x) = \sum_{x \in \Omega_j^h \cup \widehat{\Omega}_j^h} \operatorname{div}_{\Omega_j \cup \widehat{\Omega}_j}^h (\mathbf{p}_j^h + \mathbf{p}_{j^c}^h)(x) + \zeta(\mathbf{p}_{j^c}^h),$$

Fig. 5 Non-overlapping domain decomposition of Ω^h into Ω_1^h and Ω_2^h with the small stripes $\widehat{\Omega}_i^h \subset (\Omega^h \setminus \Omega_i^h)$ for $i = 1, 2$

where $\mathbf{p}_j^h \in K_j^h$, $\mathbf{p}_{j^c}^h \in \sum_{i \neq j} K_i^h$, and ζ is a suitable function independent on $\mathbf{p}_j^h$, $j \in \{1, \ldots, M\}$. Here $\operatorname{div}_{\Omega_j \cup \widehat{\Omega}_j}^h$ is the usual discrete divergence on $\Omega_j^h \cup \widehat{\Omega}_j^h$, where $\widehat{\Omega}_j^h \subset \Omega^h \setminus \Omega_j^h$ is a small stripe around the interface between Ω_j^h and $\Omega^h \setminus \Omega_j^h$. A typical choice for $\widehat{\Omega}_j^h$ for which this splitting property holds is shown in Fig. 5 for a decomposition into two domains. Note that the stripe $\widehat{\Omega}_j^h$ may be arbitrarily small and hence in the limit case it may be viewed as the boundary of Ω_j^h inside Ω^h, i.e., $\widehat{\Omega}_j^h = \partial \Omega_j^h \setminus \partial \Omega^h$ and $\partial \Omega_j^h \cap \Omega_j^h = \emptyset$. Then using the above splitting property of the divergence operator, a solution $\mathbf{p}_j^{h,n+1}$ of the subspace minimization problem of Algorithms 5 and 6 in Ω_j is given as

$$\mathbf{p}_j^{h,n+1}\big|_{\Omega_j^h \cup \widehat{\Omega}_j^h} \in \arg\min_{\mathbf{p}_j^h \in \widehat{Y}_j} \frac{1}{2} \| \operatorname{div}_{\Omega_j \cup \widehat{\Omega}_j}^h \mathbf{p}_j^h + f_j \|_{\widehat{X}_j}^2 + \mathfrak{I}_{\lambda \theta_j^h}(\mathbf{p}_j^h)$$

$$\mathbf{p}_j^{h,n+1}\big|_{\Omega^h \setminus (\Omega_j^h \cup \widehat{\Omega}_j^h)} = 0,$$
(28)

where $f_j = \operatorname{div}_{\Omega_j \cup \widehat{\Omega}_j}^h ((1 - \theta_j^h) \mathbf{p}^{h,n})|_{\Omega_j^h \cup \widehat{\Omega}_j^h} + g^h|_{\Omega_j^h \cup \widehat{\Omega}_j^h} + \zeta((1 - \theta_j) \mathbf{p}^{h,n})|_{\Omega_j^h \cup \widehat{\Omega}_j^h}$, $\widehat{X}_j := \{u^h : \Omega_j^h \cup \widehat{\Omega}_j^h \to \mathbb{R}\}$, $\widehat{Y}_j := \widehat{X}_j \times \widehat{X}_j$, $\|u^h\|_{\widehat{X}_j}^2 := \sum_{x \in \Omega_j^h \cup \widehat{\Omega}_j^h} |u(x)|^2$ for $u^h \in \widehat{X}_j$, and $\widehat{K}_j^h := \{\mathbf{p}^h \in \widehat{Y}_i : |\mathbf{p}^h(x)|_{\ell^2} \leq \lambda \theta_j(x) \ \forall x \in \Omega_j^h \cup \widehat{\Omega}_j^h\}$. Hence, finding a solution of the subdomain problems reduces to solving an optimization problem on $\Omega_j^h \cup \widehat{\Omega}_j^h$ only. Note that due to the term $\mathfrak{I}_{\lambda \theta_j^h}(\mathbf{p}_j^h)$ the solution $\mathbf{p}_j^{h,n+1}(x) = 0$ for all $x \in \Omega^h \setminus \Omega_j^h$.

For $j = 1, \ldots, M$ we define $\widehat{V}_j := \{u^h : \widehat{\Omega}_j^h \to \mathbb{R}\}$ and rewrite the minimization problem in (28) as a constrained optimization problem in the following form:

$$\min_{\xi_j \in \widehat{Y}_j} \frac{1}{2} \| \operatorname{div}_{\Omega_j \cup \widehat{\Omega}_j}^h (\xi_j) + g^h|_{\Omega_j^h \cup \widehat{\Omega}_j^h} + \zeta((1 - \theta_j^h) \mathbf{p}^{h,n})|_{\Omega_j^h \cup \widehat{\Omega}_j^h} \|_{\widehat{X}_j}^2$$
(29)

s.t. $\operatorname{proj}_{\widehat{V}_j} \xi_j = \operatorname{proj}_{\widehat{V}_j} \mathbf{p}^{h,n}$ and $|\xi_j(x)|_{\ell^2} \leq \lambda$ for all $x \in \Omega_j^h \cup \widehat{\Omega}_j^h$, where $\text{proj}_{\widehat{V}_j}$ is the orthogonal projection onto $\widehat{V}_j$. In the case when $\widehat{\Omega}_j^h = \partial \Omega_j^h \setminus \partial \Omega^h$, the constraint $\text{proj}_{\widehat{V}_j} \boldsymbol{\xi}_j = \text{proj}_{\widehat{V}_j} \mathbf{p}^{h,n}$ is a respective inner boundary condition, which can be worked into the divergence operator. Let us define by $\widehat{\text{div}}^h$ the discrete divergence operator where this new boundary condition is considered. Then the optimization problem in the subdomains can be written as

$$\min_{\mathbf{p}_j^h \in Y_j} \frac{1}{2} \|\widehat{\text{div}}^h(\mathbf{p}_j^h) + g_{|\Omega_j^h}^h + \zeta((1 - \theta_j^h)\mathbf{p}^{h,n})_{|\Omega_j^h}\|_{X_j}^2 \tag{30}$$

$$\text{s.t. } |\mathbf{p}_j^h(x)|_{\ell^2} \le \lambda \text{ for all } x \in \Omega_j^h$$

or equivalently

$$\min_{\mathbf{p}_j^h \in Y_j} \frac{1}{2} \|\widehat{\text{div}}^h(\mathbf{p}_j^h) + g_{|\Omega_j^h}^h + \zeta((1 - \theta_j^h)\mathbf{p}^{h,n})_{|\Omega_j^h}\|_{X_j}^2 + \mathfrak{I}_\lambda(\mathbf{p}_j^h) \tag{31}$$

which is a minimization problem in Ω_j^h only. In a similar way the subdomain problems in Algorithm 6 can be restricted to the subdomains Ω_j^h, $j = 1, \ldots, M$.

In Hintermüller and Langer (2015) the augmented Lagrangian method (Bertsekas 2014; Ito and Kunisch 2008; Wu and Tai 2010) is used to solve (29). However, in view of (30) and (31), other known methods, as FISTA (Beck and Teboulle 2009) or a primal-dual algorithm (Chambolle and Pock 2011), may be utilized to solve the subspace minimization problems.

Finite Element Approach Based on FISTA

The main idea of this approach is based on the additivity property (17) of the objective of the predual problem (14). Remark that the discrete divergence operator designed in section "Finite Difference Setting" in a finite difference framework does not satisfy this splitting property. The difficulty in constructing a suitable domain decomposition method based on splitting (17) is that it has to be ensured that an approximation $\mathbf{p}$ obtained by a respective splitting method lies in $H_0(\text{div}, \Omega)$.

In this section we describe a domain decomposition method for (14) in a finite element setting, which is proposed in Lee and Park (2019b).

Let $\mathcal{T}$ be the set of all elements in Ω, e.g., the pixels, and $\mathcal{E}$ the set of edges between elements. Then we discretize (15) by using the lowest-order Raviart-Thomas element space (Raviart and Thomas 1977) defined as

$$Y := \{\mathbf{q} \in H_0(\text{div}, \Omega) \ : \ \mathbf{q}_{|T} \in \mathcal{RT}_0(T) \ \forall T \in \mathcal{T}, \ [\![\mathbf{q} \cdot \mathbf{n}]\!]_E = 0 \ \forall E \in \mathcal{E}\},$$

where $\mathcal{RT}_0(T) := \{\mathbf{q} : T \to \mathbb{R}^2 \ : \ \mathbf{q}(x) = a + bx, \ a, b \in \mathbb{R}^2\}$ being the smallest polynomial space with $(\mathbb{P}_0)^2 \subset \mathcal{RT}_0(T) \subset (\mathbb{P}_1)^2$ such that the divergence maps $\mathcal{RT}_0(T)$ onto $\mathbb{P}_0$ and $[\![\mathbf{q} \cdot \mathbf{n}]\!]_E$ denotes the jump across the edge E. Note that Y is a conforming approximation of $H_0(\text{div}, \Omega)$. Then $\mathbf{p} \in Y$ may be written as

11 Domain Decomposition for Non-smooth (in Particular TV) Minimization

$$\mathbf{p} = \sum_{i \in \mathcal{I}} (\mathbf{p})_i \psi_i,$$

where $\mathcal{I}$ is the set of indices of the basis functions $(\psi_i)_{i \in \mathcal{I}}$ of Y and $(\mathbf{p})_i$ denotes the respective degree of freedom. Based on these definitions, the finite element discretization of (15) is

$$\min_{\mathbf{p} \in Y} \frac{1}{2} \| \operatorname{div} \mathbf{p} + g \|_{L^2(\Omega)}^2 + \chi_C(\mathbf{p}), \tag{32}$$

where $C := \{ \mathbf{p} \in Y : |(\mathbf{p})_i|_{\ell^2} \leq \lambda \; \forall i \in \mathcal{I} \}$.

Associated with the non-overlapping decomposition $(\Omega_j)_{j=1}^M$ of Ω, we define the respective function spaces as

$$Y_j := \{ \mathbf{q} \in H_0(\operatorname{div}, \Omega_j) : \mathbf{q}_{|T} \in \mathcal{RT}_0(T) \; \forall T \in \mathcal{T}_j, \; [\![\mathbf{q} \cdot \mathbf{n}]\!]_E = 0 \; \forall E \in \mathcal{E}_j \},$$

where $\mathcal{T}_j$ and $\mathcal{E}_j$ are the collections of all elements and edges in $\overline{\Omega}_j$ for $j = 1, \ldots, M$. Let $\mathcal{I}_j$ be the set of indices of the basis functions for Y_j and $\mathcal{I}_\Gamma$ the set of indices of degree of freedom of Y on $\Gamma := \bigcup_{j<i} \partial\Omega_j \cap \partial\Omega_i$. By $Y_\Gamma = \operatorname{span}\{\psi_i\}_{i \in \mathcal{I}_\Gamma}$ we denote the interface function space. Further let $Y_I := \bigoplus_{j=1}^M Y_j$, $C_j := \{ \mathbf{p} \in Y_j : |(\mathbf{p})_i|_{\ell^2} \leq \lambda \; \forall i \in \mathcal{I}_j \}$, $C_I := \bigoplus_{j=1}^M C_j$ and $C_\Gamma := \{ \mathbf{p}_\Gamma \in Y_\Gamma : |(\mathbf{p}_\Gamma)_i|_{\ell^2} \leq \lambda \; \forall i \in \mathcal{I}_\Gamma \}$. Note that for $\mathbf{p} \in Y$ there exists a unique decomposition such that

$$\mathbf{p} = \mathbf{p}_I \oplus \mathbf{p}_\Gamma = \left(\bigoplus_{j=1}^M \mathbf{p}_j \right) \oplus \mathbf{p}_\Gamma$$

where $\mathbf{p}_j \in Y_j$ and $\mathbf{p}_\Gamma \in Y_\Gamma$. Define $\mathcal{H}_I : Y_\Gamma \to Y_I$ such that $\mathcal{H}_I \mathbf{p}_\Gamma$ solves

$$\min_{\mathbf{p}_I \in Y_I} \frac{1}{2} \| \operatorname{div}(\mathbf{p}_I \oplus \mathbf{p}_\Gamma) + g \|_{L^2(\Omega)}^2 + \chi_{C_I}(\mathbf{p}_I) \tag{33}$$

for a fixed $\mathbf{p}_\Gamma \in C_\Gamma$. We remark that thanks to the splitting property (17) a solution of (33) can be obtained by independently solving on each subspace

$$\min_{\mathbf{p}_j \in Y_j} \frac{1}{2} \| \operatorname{div}(\mathbf{p}_j \oplus \mathbf{p}_\Gamma|_{\Omega_j}) + g \|_{L^2(\Omega)}^2 + \chi_{C_j}(\mathbf{p}_j).$$

With these definitions instead of minimizing (32), one solves

$$\mathbf{p}_\Gamma \in \arg\min_{\mathbf{p}_\Gamma \in Y_\Gamma} \frac{1}{2} \| \operatorname{div}(\mathcal{H}_I \mathbf{p}_\Gamma \oplus \mathbf{p}_\Gamma) + g \|_{L^2(\Omega)}^2 + \chi_{C_\Gamma}(\mathbf{p}_\Gamma). \tag{34}$$

It can be shown that (i) if $\mathbf{p}^* \in Y$ is a solution of (32), then $\mathbf{p}^*_\Gamma = \mathbf{p}^*|_\Gamma$ is a solution of (34) and (ii) if $\mathbf{p}^*_\Gamma \in Y_\Gamma$ is a solution of (34), then $\mathbf{p}^* = \mathcal{H}_I \mathbf{p}^*_\Gamma \oplus \mathbf{p}^*_\Gamma$ is a solution of (32) (Lee and Park 2019a). Using FISTA to solve (34) the domain decomposition algorithm presented in Algorithm 7 is obtained (Lee and Park 2019a), where proj_{C_Γ} is the orthogonal projection onto C_Γ.

Algorithm 7 Parallelizable FISTA for (14)

Choose $L \geq 4$. Pick an initial $\mathbf{p}^0_\Gamma = \mathbf{q}^0_\Gamma = 0_\Gamma$ and $t^0 = 1$.
for $n = 0, 1, \ldots$ do
$\quad \mathcal{H}_I \mathbf{q}^n_\Gamma \in \arg\min_{\mathbf{p}_I \in Y_I} \frac{1}{2} \| \text{div}(\mathbf{p}_I \oplus \mathbf{p}^n_\Gamma) + g \|^2_{L^2(\Omega)} + \chi_{C_I}(\mathbf{p}_I)$
$\quad \mathbf{p}^{n+1}_\Gamma := \text{proj}_{C_\Gamma} \left(\mathbf{q}^n_\Gamma - \frac{1}{L} \text{div}^* \left(\text{div} \left(\mathcal{H}_I \mathbf{q}^n_\Gamma \oplus \mathbf{q}^n_\Gamma \right) + g \right)_{|Y_\Gamma} \right)$
$\quad t^{n+1} := \frac{1+\sqrt{1+4(t^n)^2}}{2}$
$\quad \mathbf{q}^{n+1}_\Gamma := \mathbf{p}^{n+1}_\Gamma + \frac{t^n - 1}{t^{n+1}} (\mathbf{p}^{n+1}_\Gamma - \mathbf{p}^n_\Gamma)$
end for

We remark once more that the minimizer $\mathcal{H}_I \mathbf{q}^n_\Gamma$ in Algorithm 7 may be obtained by solving independently on each subdomain

$$\mathbf{p}^n_j \in \arg\min_{\mathbf{p}_j \in Y_j} \frac{1}{2} \| \text{div}(\mathbf{p}_j \oplus \mathbf{p}^n_{\Gamma|\Omega_j}) + g \|^2_{L^2(\Omega)} + \chi_{C_j}(p_j)$$

and setting $\mathcal{H}_I \mathbf{q}^n_\Gamma = \bigoplus_{j=1}^M \mathbf{p}^n_j$. Due to the utilization of FISTA, Algorithm 7 converges with order $\mathcal{O}(1/n^2)$ to a solution $\mathbf{p}^*_\Gamma \in Y_\Gamma$ of (34).

This approach relies on a splitting into a problem defined on Y_I and a subdomain problem defined on the interface Y_Γ, which are alternately solved. A similar decoupling approach is presented in Lee et al. (2019a), where the functional to be minimized is additively separated with respect to a finite difference discretization into a problem on disjoint subdomains and one interface problem. By utilizing the primal-dual algorithm of Chambolle and Pock (2011) these two problems are successively solved. Note that due to a disjoint splitting, a parallelization of the problem on these disjoint subdomains is possible. This method is used to minimize a functional consisting of a total variation term and an L^1 date fidelity term with applications to image denoising, inpainting, and deblurring. For block coordinate descent methods, a similar splitting approach is presented in Chambolle and Pock (2015).

A FETI Approach

In contrary to the above finite element approach, in Lee et al. (2019b) a further and different domain decomposition method is proposed, where the local function spaces $\tilde{Y}_j$ are defined in the tearing-and-interconnecting fashion by

$$\tilde{Y}_j := \{ \mathbf{q} \in H_0(\text{div}, \Omega_j) \ : \ \mathbf{q}_{|T} \in \mathcal{RT}_0(T) \ \forall T \in \mathcal{T}_j, \ [\![\mathbf{q} \cdot \mathbf{n}]\!]_E = 0 \ \forall E \in \mathcal{E} \setminus \Gamma \}.$$

For $\mathbf{p} \in \tilde{Y}_j$ the jump $[\![\mathbf{p} \cdot n]\!]_\Gamma$ might be non-zero, which is related to tearing the subdomain solutions apart. Further let $\tilde{\mathcal{I}}_j$ be the set of indices of the basis functions $\tilde{Y}_j$ and $\tilde{Y} = \bigoplus_{j=1}^M \tilde{Y}_j$. Then based on the splitting (17) on each subdomain Ω_j, $j = 1, \ldots, M$, the following optimization problem might be solved:

$$\tilde{\mathbf{p}}_j \in \arg\min_{\mathbf{p}_j \in \tilde{Y}_j} \frac{1}{2} \| \operatorname{div} \mathbf{p}_j + g \|_{L^2(\Omega_j)}^2 + \chi_{\tilde{C}_j}(\mathbf{p}_j), \tag{35}$$

where $\tilde{C}_j := \{\mathbf{p} \in \tilde{Y}_j \,:\, |(\mathbf{p})_i|_{\ell^2} \leq \lambda \ \forall i \in \tilde{\mathcal{I}}_j\}$. Then in order to ensure that $\tilde{\mathbf{p}} := \bigoplus_{j=1}^M \tilde{\mathbf{p}}_j \in Y$, where $\tilde{\mathbf{p}}_j$ is a solution of (35), we need to enforce that $[\![\tilde{\mathbf{p}} \cdot n]\!]_E = 0$ for all $E \in \mathcal{E}$, i.e., we interconnect the subdomain solutions. For this purpose we define the operator $B : \tilde{Y} \to \mathbb{R}^{|\mathcal{I}_\Gamma|}$ such that $B\tilde{\mathbf{p}}_{|E} = [\![\tilde{\mathbf{p}} \cdot n]\!]_E$ for E being an edge between two domain patches. Then (32) is equivalent to

$$\min_{\tilde{\mathbf{p}} \in \tilde{Y}} \sum_{j=1}^M \frac{1}{2} \| \operatorname{div} \tilde{\mathbf{p}}_j + g \|_{L^2(\Omega_j)}^2 + \chi_{\tilde{C}_j}(\tilde{\mathbf{p}}_j) \quad \text{s.t.} \quad B\tilde{\mathbf{p}} = 0.$$

Utilizing the method of Lagrange multiplier, this optimization problem can be formulated as a saddle point problem:

$$\min_{\tilde{\mathbf{p}} \in \tilde{Y}} \max_{\mu \in \mathbb{R}^{|\mathcal{I}_\Gamma|}} \sum_{j=1}^M \frac{1}{2} \| \operatorname{div} \tilde{\mathbf{p}}_j + g \|_{L^2(\Omega_j)}^2 + \chi_{\tilde{C}_j}(\tilde{\mathbf{p}}_j) + \langle B\tilde{\mathbf{p}}, \mu \rangle_{\mathbb{R}^{|\mathcal{I}_\Gamma|}}. \tag{36}$$

Since B is bounded, the saddle point problem (36) can be solved by the primal-dual algorithm proposed in Chambolle and Pock (2011) which yields Algorithm 8.

Algorithm 8 Primal-dual FETI algorithm for (14)

Choose $L \geq 2$, $\tau, \sigma > 0$ with $\tau\sigma = \frac{1}{L}$. Let $\tilde{\mathbf{p}}^0 = 0$ and $\lambda^0 = 0$.
for $n = 0, 1, \ldots$ **do**
$\quad \lambda^{n+1} = \lambda^n + \sigma B(2\tilde{\mathbf{p}}^n - \tilde{\mathbf{p}}^{n-1})$
$\quad \tilde{\mathbf{p}}^{n+1} := \min_{\tilde{\mathbf{p}} \in \tilde{Y}} \sum_{j=1}^M \frac{1}{2} \| \operatorname{div} \tilde{\mathbf{p}}_j + g \|_{L^2(\Omega_j)}^2 + \chi_{\tilde{C}_j}(\tilde{\mathbf{p}}_j) + \frac{1}{2\tau}\|\tilde{\mathbf{p}}_j - \tilde{\mathbf{p}}_j^n + (\tau B^* \lambda^{n+1})_{|\Omega_j}\|_{L^2(\Omega_j)}^2$
end for

Note that the minimization problem in Algorithm 8 can be solved in parallel independently on each subdomain Ω_j, $j = 1, \ldots, M$. Moreover, in Lee et al. (2019b) it is stated that this algorithm converges with $\mathcal{O}(1/n)$ to a primal solution $\tilde{\mathbf{p}}^* \in \tilde{Y}$ of (36), which follows from Chambolle and Pock (2016, Theorem 5.1). Moreover, there is an isomorphism $\phi : Y \to \ker(B) \subset \tilde{Y}$ such that $\phi^{-1}\tilde{\mathbf{p}}^*$ solves (32) (Lee et al. 2019b). This primal-dual domain decomposition approach has been extended to other but related functionals in Lee and Park (2019a), where also image inpainting and segmentation problems with either L^2 or L^1 date fidelity terms are considered. Also for these applications, the convergence of the splitting method to a minimizer of the global problem is ensured. A similar tearing-and-interconnecting strategy together with the primal-dual algorithm (Chambolle and Pock 2011) has been used in Duan et al. (2016) for image segmentation, more precisely for the convex Chan-Vese model (Chan et al. 2006). However, in this setting the convergence of the algorithm to a minimizer of the global problem seems unclear, as the existence of an isomorphism, similar to the one above, is not shown. Note that in Duan et al. (2016) the minimization of the total variation is directly considered and not its predual counterpart.

Decomposition for Primal Total Variation

In this section we review domain decomposition methods which directly tackle the L^2-TV model (12). Historically such methods for the L^2-TV model were considered before its predual problem was suggested to be solved by splitting methods. In particular several domain decomposition methods for tackling directly total variation minimization are presented in Duan et al. (2016), Duan and Tai (2012), Fornasier et al. (2009, 2010), Fornasier and Schönlieb (2009), Hintermüller and Langer (2013, 2014), Lee et al. (2016), and Schönlieb (2009), which are not proven to converge to a solution of the global problem (12). Although in some of these works the proposed methods are theoretically investigated with respect to their convergence properties, in the best case only the criterion is derived under which the convergence to a minimizer of the global problem is achieved. We will review these decomposition methods and their convergence properties in section "Basic Domain Decomposition Approach". In particular we even present an example for the minimization of the L^2-TV model which shows that in general these methods cannot be guaranteed to converge to a minimizer of the global problem. After this quite negative result, we turn to domain decomposition methods for the L^2-TV model which indeed converge to a solution of the original problem. These methods are based on the splitting methods for the predual problem (14) presented in section "Decomposition for Predual Total Variation". We recall that the decomposition methods in section "Decomposition for Predual Total Variation" converge to a minimizer of the original global problem. Hence, they serve us as a role model for deriving domain decomposition methods for the L^2-TV model with this desired convergence property. In particular by transforming the decomposition methods of the predual problem via dualization into the function space of the L^2-TV model, such methods could be constructed, as we will discuss in section "Domain Decomposition Approach Based on the (Pre)Dual".

Basic Domain Decomposition Approach

Following the general philosophy of subspace correction and inspired by the variational formulations (7) and (10), we seek to minimize J by decomposing

$L^2(\Omega)$ into $M \in \mathbb{N}$ appropriate subspaces U_j such that $L^2(\Omega) = \sum_{j=1}^{M} U_j$. In terms of domain decomposition, let Ω be separated into M subdomains Ω_j, $j = 1, \ldots, M$. Here the decomposition of the domain may be overlapping or non-overlapping. Then $U_j := \{u \in L^2(\Omega) : \mathrm{supp}(u) \subset \Omega_j\}$ for $j = 1, \ldots, M$. With this splitting we aim to solve (12) by Algorithm 9.

Algorithm 9 Basic parallel domain decomposition algorithm for (12)

Initialise: $u^0 \in L^2(\Omega)$
for $n = 0, 1, \ldots$ do
  for $j = 1, \ldots, M$ do
    $u_j^{n+1} \in \arg\min_{u_j \in U_j} J(u_j + (1 - \theta_j)u^n)$
  end for
  $u^{n+1} := \frac{(M-1)u^n + \sum_{j=1}^{M} u_j^{n+1}}{M}$
end for

Here $(\theta_j)_{j=1}^{M} \subset L^\infty(\Omega)$ is a partition of unity with the properties (i) $\sum_{i=j}^{M} \theta_j = 1$ and (ii) $\theta_j \in U_j$ for $j = 1, \ldots, M$. From the assumptions on θ_j we obtain $u^n = \sum_{j=1}^{M} (\theta_j u^n)$. Further, if the U_js are orthogonal, i.e., $U = \bigoplus_{j=1}^{M} U_j$, then $\theta_j u^n = u_j^n$ for all $n \in \mathbb{N}$ and hence there is no need to introduce a partition of unity. The successive version of Algorithm 9 is stated in Algorithm 10.

Algorithm 10 Basic successive domain decomposition algorithm for (12)

Initialise: $u^0 \in U$
for $n = 0, 1, \ldots$ do
  for $j = 1, \ldots, M$ do
    $u_j^{n+1} \in \arg\min_{u_j \in U_j} J(u_j + \sum_{i<j} u_i^{n+1} + \sum_{i>j} (\theta_i)u^n)$
  end for
  $u^{n+1} := \sum_{j=1}^{M} u_j^{n+1}$
end for

We define the orthogonal complement of U_j in $L^2(\Omega)$ by U_j^c, i.e., $L^2(\Omega) = U_j \oplus U_j^c$, and we denote by proj_{U_j} the corresponding orthogonal projection onto U_j for $j = 1, \ldots, M$. Moreover, we define the domain of a functional $\mathcal{J}: L^2(\Omega) \to \bar{\mathbb{R}}$ as the set $\mathrm{Dom}(\mathcal{J}) = \{v \in L^2(\Omega) : \mathcal{J}(v) \neq \infty\}$.

Note that the subspace minimization problems in Algorithm 9 and in Algorithm 10 can be written as constrained optimization problems of the form

$$\min_{v \in L^2(\Omega)} J(v) \quad \text{s.t.} \ Av = b,$$

where $A : L^2(\Omega) \to L^2(\Omega)$ is a linear and continuous operator on $L^2(\Omega)$ and $b \in L^2(\Omega)$. In particular, we have

$$\min_{v \in L^2(\Omega)} J(v+b) \quad \text{s.t. } \text{proj}_{U_j^c} v = 0,$$

or equivalently

$$\min_{v \in L^2(\Omega)} J(v) \quad \text{s.t. } \text{proj}_{U_j^c}(v) = \text{proj}_{U_j^c}(b), \tag{37}$$

where $b = \sum_{i<j} u_i^{n+1} + \sum_{i>j} \theta_i u^n$ in Algorithm 10 and $b = (1 - \theta_j) u^n$ for the minimization problem in Algorithm 9 for $j = 1, \ldots, M$. For any *attainable* $b \in U_i$, i.e., there exists an $u \in \text{Dom}(J)$ such that $\text{proj}_{U_j^c}(u) = \text{proj}_{U_j^c}(b)$, we observe that $\{u \in L^2(\Omega) : \text{proj}_{U_j^c}(u) = \text{proj}_{U_j^c}(b), J(u) \leq c\} \subset \{J \leq c\}$ for all $c > 0$, $j = 1, \ldots, M$, and $i \in \{1, \ldots, M\} \setminus \{j\}$. Hence, by the coercivity of J, the former set is bounded and thus (37) has a solution, as every u_j^n in Algorithms 9 and 10 is attainable.

Let us mention that such domain decomposition algorithms for (12) have been first considered in Fornasier and Schönlieb (2009) for a non-overlapping and in Fornasier et al. (2009, 2010) for an overlapping decomposition of the spatial domain in the context of image reconstruction.

Convergence Properties

It can be shown that Algorithms 9 and 10 generate sequences $(u^n)_n$ in $L^2(\Omega)$, which have subsequences that weakly converge in $L^2(\Omega)$ and $BV(\Omega)$, such that $(J(u^n))_n$ is non-increasing for all $n \in \mathbb{N}$ (Hintermüller and Langer 2013, Proposition 3.1). As a consequence $(J(u^n))_n$ is also convergent, since it is bounded from below. Unfortunately the limit point of such subsequences is not guaranteed to be a solution of the global problem (12), as the following one-dimensional ($d = 1$) counterexample demonstrates:

Example 3. Let $\Omega \subset \mathbb{R}^1$ be the interval (a_1, a_2), $a_1 < a_2$, decomposed into two subintervals Ω_1 and Ω_2 such that $\Omega = \Omega_1 \cup \Omega_2$ and $|\Omega_j| = l$, $0 < l < a_2 - a_1$, for $j = 1, 2$. Further let $g = 1$ and $T = I$ in (12). We initialize Algorithms 9 and 10 with $u^0 = 0$. In the first iteration for the subspace minimization in U_1, we solve

$$\min_{u_1 \in U_1} \frac{1}{2} \int_\Omega |u_1 + b - 1|^2 \, dx + \lambda \int_\Omega |D(u_1 + b)|,$$

where $b = 0$ since $u^0 = 0$. As $\text{proj}_{U_1^c} u_1 = 0$, one can reason that every minimizer has to be of the form $c 1_{\Omega_1}$ for $c \in [0, 1]$, where $1_{\Omega_1}(x) = 1$ if $x \in \Omega_1$ and $1_{\Omega_1}(x) = 0$ otherwise. Therefore, we just need to solve

$$\min_{c \in [0,1]} \frac{1}{2} \int_{\Omega_1} |c - 1|^2 \, dx + \lambda c.$$

The associated optimality condition for c is

$$l(c - 1) + \lambda = 0$$

which is equivalent to

$$c = 1 - \frac{\lambda}{l}.$$

Hence, for $\lambda = l$ (in particular for $\lambda \geq l$) the minimizer is $c = 0$ and hence $u_1^1 = 0$. In this situation $b = 0$ for all $j \in \{1, \ldots, M\}$ and both algorithms. Consequently $u_j^1 = 0$ for all $j \in \{1, \ldots, M\}$ and hence $u^1 = 0 = u^0$. If $\lambda = l$, a repetition of these steps shows that $u^n = 0$ for all $n \in \mathbb{N}$.

On the contrary the minimizer of the global optimization problem (12) is $u^* = 1$ for any $\lambda \geq 0$.

Note that this example works for an overlapping as well as for a non-overlapping decomposition of the spatial domain Ω. Moreover, this counterexample can be easily extended to a multi-domain decomposition and to $\mathbb{R}^2$ by letting $\Omega \subset \mathbb{R}^2$ be a rectangle decomposed into stripes, for example, as in Fig. 4b. A similar counterexample has been presented in Lee and Nam (2017) for a finite difference discretization by using the relation to the predual problem.

Despite this quite negative result, in a finite difference setting in Hintermüller and Langer (2013) an estimate of the distance of a limit point $u^{h,\infty}$ obtained by discrete version of Algorithm 9 or Algorithm 10 to the true global minimizer $u^{h,*}$ is obtained. Let us use the finite difference setting of section "Finite Difference Setting", define $X_j^c := \sum_{i \neq j} X_i$ for $j \in \{1, \ldots, M\}$, and consider the discrete version of J defined as

$$J^h(u^h) := \|T^h u^h - g^h\|_X^2 + \sum_{x \in \Omega} |\nabla_\Omega^h u^h(x)|,$$

where $T^h : X \to X$ is a bounded linear operator. Then, if $T^{h*}T^h$ is positive definite in the direction $u^{h,\infty} - u^{h,*}$ with smallest eigenvalue $\sigma > 0$ and $\hat{\eta}^h \in \arg\min_{\eta^h \in \bigcup_{j=1}^M \left(\partial J^h(u^{h,\infty}) \cap X_j^c\right)} \|\eta^h\|_X$, then

$$\|u^{h,\infty} - u^{h,*}\|_X \leq \frac{\|\hat{\eta}^h\|_X}{\alpha_2 \sigma}. \tag{38}$$

Note that the Lagrange multiplier $\hat{\eta}^h$ indicates the influence of the constraint on the solution. If $\hat{\eta}^h = 0$, then the minimizer of the discrete version of (37) is equivalent to the minimizer of J^h in X and hence is indeed a solution of the global problem. On the contrary, if $\hat{\eta}^h \neq 0$, then the discrete version of the constraint in

(37) has influence on the solution, which consequently does not coincide with the global solution. Hence, this estimate does not contradict with the counterexample, but instead provides an a posteriori upper bound to check whether the algorithm is indeed converged for a considered example. In particular if $\|\hat{\eta}_j^{h,n_k}\|_X \to 0$ for $k \to \infty$ along a suitable subsequence $(n_k)_k$ for at least one $j \in \{1, \ldots, M\}$, then any accumulation point of the sequence $(u^{h,n})_n$ generated by the discrete version of Algorithm 9 or Algorithm 10 minimizes J^h. By this observation, with the help of this estimate in Hintermüller and Langer (2013), it is demonstrated by numerical experiments that Algorithms 9 and 10 generate sequences which seem to converge to the global minimizer, because $\|\hat{\eta}^h\|_X$ tends to zero.

It is worth mentioning that Algorithms 9 and 10 have not only been proposed for the L^2-TV model but also for total variation minimization with a combined L^1/L^2 data fidelity term, which seems in particular suitable for removing simultaneously Gaussian and impulsive noise in images (Hintermüller and Langer 2013). For a non-overlapping decomposition of the domain Ω, these algorithms have been also utilized for total variation minimization with an H^{-1} constraint, i.e., for solving

$$\min_{u \in BV(\Omega)} \frac{1}{2}\|Tu - g\|_{-1}^2 + \int_\Omega |Du|,$$

where $\|\cdot\|_{-1}$ denotes the $H^{-1}(\Omega)$ norm (Schönlieb 2009). In Chang et al. (2014) a similar splitting method for minimizing the nonlocal total variation (see Gilboa and Osher (2009), Peyré et al. (2008), Zhang et al. (2010), and the references therein for more information on nonlocal total variation) is described without any rigorous theoretical analysis. For total variation image segmentation in Duan et al. (2016) and Duan and Tai (2012), the domain decomposition methods based on an additive decomposition of the objective have been proposed. Nevertheless, a proof of convergence of these methods to a solution of the global problem is missing.

Subspace Minimization

Algorithms 9 and 10 require that the subspace minimization problems are solved exactly, which is in general not easily possible. Moreover, due to the presence of the operator T, which acts on the variable to be minimized, a restriction of the subspace minimization problems to the respective subdomains and subspaces seems in general difficult, in particular if T is a global operator. Therefore, in Fornasier et al. (2010), Fornasier and Schönlieb (2009), and Hintermüller and Langer (2014) the subproblems are approximated by the so-called *surrogate* functionals (Daubechies et al. 2004): assume $a, u_j \in U_j$, $b \in \sum_{i \neq j} U_i$ and define

$$J_j^s(u_j + b, a + b) := J(u_j + b) + \frac{1}{2}\left(\delta\|u_j + b - (a+b)\|_{L^2(\Omega)}^2\right.$$
$$\left. - \|T(u_j + b - (a+b))\|_{L^2(\Omega)}^2\right)$$
$$= \frac{\delta}{2}\|u_j - \left(a + \frac{1}{\delta}T^*\left(g - T(a+b)\right)\right)\|_{L^2(\Omega)}^2$$
$$+ \lambda \int_\Omega |D(u_j + b)| + \Phi(a, b, g)$$

for $j = 1, \ldots, M$, where $\delta > \|T\|^2$ and Φ is a function of a, b, g and independent of u_j. Now note that u_j is not anymore effected by T and J_j^s is strictly convex. Then a solution u_j^{n+1} of the subspace minimization problems in Algorithms 9 and 10 is realized by the following algorithm: for $u_j^{n,0} \in U_j$

$$u_j^{n,k+1} = \arg\min_{u_j \in U_j} J_j^s(u_j + b, u_j^{n,k} + b), \qquad k \geq 0, \tag{39}$$

where $b = \sum_{i<j} u_i^{n+1} + \sum_{i>j}(\theta_i)u^n$ for the alternating algorithm (cf. Algorithm 10) and $b = (1 - \theta_j)u^n$ for the parallel version (cf. Algorithm 9) for $j = 1, \ldots, M$. Note that the sequence $(u_j^{n,k})_k$ generated by (39) converges to a minimizer u_j^{n+1} of the corresponding subproblems of Algorithms 9 and 10 (Daubechies et al. 2007).

By introducing small stripes around the interfaces of the subdomains as in Fig. 5, i.e., $\widehat{\Omega}_j \subset \Omega \setminus \Omega_j$ is a small stripe around the interface between Ω_j and $\Omega \setminus \Omega_j$ and by the splitting property of the total variation

$$\int_\Omega |D(u_j + b)| = \int_{\Omega_j \cup \widehat{\Omega}_j} |D(u_j + b)|_{\Omega_j \cup \widehat{\Omega}_j}| + \int_{\Omega \setminus (\Omega_j \cup \widehat{\Omega}_j)} |D(b)|_{\Omega \setminus (\Omega_j \cup \widehat{\Omega}_j)}|$$
$$+ \int_{\partial(\Omega_j \cup \widehat{\Omega}_j) \cap \partial(\Omega \setminus (\Omega_j \cup \widehat{\Omega}_j))} |b^+ - b^-| \, d\mathcal{H}^{d-1}(x),$$

where b is understood as above, we can restrict the minimization problem in (39) to the domain $\Omega_j \cup \widehat{\Omega}_j$ for $j = 1, \ldots, M$, respectively. Then the respective subdomain problems can be written as constrained minimization problems of the form

$$\min_{u_j \in U_j \oplus \widehat{U}_j} \frac{\delta}{2}\|u_j - z_j\|_{L^2(\Omega_j \cup \widehat{\Omega}_j)}^2 + \lambda \int_{\Omega_j \cup \widehat{\Omega}_j} \left|D(u_j + b)|_{\Omega_j \cup \widehat{\Omega}_j}\right| \tag{40}$$

s.t. $\operatorname{proj}_{\widehat{U}_j} u_j = 0$

where $\widehat{U}_j := \{u \in L^2(\Omega): \mathrm{supp}(u) \subset \widehat{\Omega}_j\}$, $z_j = \left(u_j^{n,k} + \frac{1}{\delta}T^*\left(g - T(u_j^{n,k} + b)\right)\right)\Big|_{\Omega_j \cup \widehat{\Omega}_j}$

and $b \in \sum_{i \neq j} U_i$ as above. Note that such a splitting holds for overlapping and non-overlapping domain decompositions. Moreover, in case of an overlapping domain decomposition in Fornasier et al. (2010) for a discrete setting, the subproblems are completely restricted to Ω_j, $j = 1, \ldots, M$, respectively, due to an induced trace condition, i.e., $\widehat{\Omega}_j$ is replaced by $\Gamma_j := \partial\Omega_j \setminus \partial\Omega$ and the constraint in (40) is then a trace condition on Γ_j. In Fornasier et al. (2010) and Fornasier and Schönlieb (2009) the resulting subspace minimization problems are solved by *oblique thresholding*, which is based on an iterative proximity map algorithm and the computation of a Lagrange multiplier by a fixed point iteration. In order to speed up the computation, in Langer et al. (2013) each subproblem is suggested to be solved by a *Bregmanized operator splitting – split Bregman* algorithm.

In practice in order to obtain an approximation of the subspace minimization problems of Algorithms 9 and 10, only a finite number of (inner) iterations of (39) can be performed. Nevertheless, the respective generated sequence $(u^n)_n$ of Algorithms 9 and 10 still satisfies the following convergence properties:

(i) $J(u^n) \geq J(u^{n+1})$ for all $n \in \mathbb{N}$.
(ii) $\lim_{n \to \infty} \|u^{n+1} - u^n\|_{L^2(\Omega)} = 0$.
(iii) The sequence $(u^n)_n$ has subsequences that converge weakly in $L^2(\Omega)$ and $BV(\Omega)$.

Of course this does not imply the convergence of the sequence $(u^n)_n$ to a minimizer of J; cf. Example 3. Nevertheless, it means that independently how accurately the subdomain problems are solved, the overall convergence is untouched. In a finite difference setting a similar estimate as the one in (38) can be shown (see Hintermüller and Langer (2014)), which again provides an upper bound of the distance between the obtained limit and a minimizer of the global problem.

Domain Decomposition Approach Based on the (Pre)Dual

We have seen that for the predual problem (14), the domain decomposition methods, which are guaranteed to converge to a minimizer of the original global problem, can be constructed. Based on these methods one can pursue the following strategy in order to design a domain decomposition method for problem (12): The domain decomposition methods in Algorithms 1, 2, 3, 4, and 5 are constituted by its subdomain problems. Then the dual problems of these subdomain problems are computed, yielding a sequence of subdomain problems of the primal problem. Due to predualization and dualization, the final constituted domain decomposition methods of the primal problem (12) look different than the splitting strategies presented in section "Basic Domain Decomposition Approach". Using this idea in Langer and Gaspoz (2019) and Lee and Nam (2017) overlapping and non-overlapping domain decomposition methods that converge to the minimizer of J with $T = I$ are designed. In particular, in Langer and Gaspoz (2019) overlapping domain decomposition methods in an infinite dimensional setting are proposed, while non-overlapping domain decomposition methods in a finite difference setting are constructed in Lee and Nam (2017). It turns out that in a discrete setting in the limit case when the overlapping size tends to 0, i.e., in the case of a non-overlapping decomposition, the approach in Langer and Gaspoz (2019) becomes the one in Lee and Nam (2017). In this vein in the following we concentrate on describing the derivation of the overlapping methods, as the construction of the non-overlapping methods runs analogously and is a special case of the overlapping method.

Derivation of the Methods

Let us focus now on the derivation of overlapping domain decomposition methods for solving (12) with $T = I$, i.e.,

$$\min_{u \in L^2(\Omega)} \frac{1}{2} \|u - g\|_{L^2(\Omega)}^2 + \lambda \int_\Omega |Du|. \tag{41}$$

Therefore, partition Ω into $M \in \mathbb{N}$ overlapping subdomains as in section "Overlapping Domain Decomposition". For deriving the decomposition methods, we need to compute the dual problems of the subdomain problems of Algorithms 2 and 3. The subdomain problem in Ω_j, $j = \{1, \ldots, M\}$, of these algorithms may be rewritten as

$$\arg\min \left\{ \frac{1}{2} \|\operatorname{div} \mathbf{v} + f\|_{L^2(\Omega)}^2 : \mathbf{v} \in H_0(\operatorname{div}, \Omega), |\mathbf{v}(x)|_{\ell^2} \leq \beta(x) \text{ f.a.a. } x \in \Omega \right\} \tag{42}$$

where $\beta := \lambda \theta_j$ with $\theta_j \geq 0$ defined as in (21), (22), and (23), $f = \operatorname{div}\left(\sum_{i<j} \mathbf{p}_i^{n+1} + \sum_{i>j} \theta_i \mathbf{p}^n\right) + g$ for Algorithm 2 and $f = \operatorname{div}\left(\sum_{i \neq j} \theta_i \mathbf{p}^n\right) + g$ for Algorithm 3 for any $n \geq 0$. If $\beta : \overline{\Omega} \to \mathbb{R}_0^+$, $\beta \in H^1(\Omega) \cap C(\overline{\Omega})$, $\|\nabla \beta\|_{L^\infty(\Omega)} < \infty$, and $\operatorname{supp}(\beta) \subseteq \overline{\Omega}$, then a Fenchel dual of (42) is given by

$$\arg\min_{u \in L^2(\Omega)} \left\{ \frac{1}{2} \|u - f\|_{L^2(\Omega)}^2 + \int_\Omega \beta |Du| \right\}, \tag{43}$$

whose minimizer is unique (Langer and Gaspoz 2019). Here and in the sequel, the expression $\int_\Omega \beta |Du|$ describes the integral of β on Ω with respect to the measure $|Du|$, where Du is the distributional gradient of u. Hence, the subdomain problems of our domain decomposition method are of the form (43). In order that (43) is well defined, a partition of unity function needs to have the following properties:

$$\sum_{i=1}^M \theta_j \equiv 1 \text{ and } \theta_j \geq 0 \text{ a.e. on } \overline{\Omega} \text{ for } j = 1, 2, \ldots, M, \tag{44}$$

$$\operatorname{supp} \theta_j \subset \overline{\Omega}_j \text{ for } j = 1, 2, \ldots, M, \tag{45}$$

$$\theta_j \in H^1(\Omega) \cap C(\overline{\Omega}) \text{ and } \|\nabla \theta_j\|_{\mathbb{L}^\infty(\Omega)} < \infty \text{ for } j = 1, 2, \ldots, M, \tag{46}$$

as $\beta = \lambda \theta_j$ for the subproblem in Ω_j, $j \in \{1, \ldots, M\}$. In comparison to (21), (22), and (23) the additional requirements $\theta_j \in C(\overline{\Omega})$ and $\theta_j \geq 0$ a.e. on $\overline{\Omega}$ are needed such that $\int_\Omega \beta |Du|$ is well defined for $u \in L^2(\Omega)$. In the sequel of this section, we will only use a partition of unity function with the properties (44), (45), and (46) and denote it by $(\theta_j)_{j=1}^M$.

Now let us turn to the choice of f in (43). For this purpose we consider the basic successive algorithm (Algorithm 1) in domain Ω_M, where we have from the (predual) subdomain problem (42) that $f_M^{n+1} := f = \operatorname{div}\left(\sum_{i=1}^{M-1} \mathbf{p}_i^{n+1}\right) + g$, where we introduced the subscript M and the superscript $n + 1$ to make the dependency of f on the domain and iteration visible. As we are designing a decomposition method for the L^2-TV model, we do not want to compute in each iteration the dual variables $\mathbf{p}_i^{n+1}$, since then we could stick directly to Algorithm 1 of the predual problem. Note that for the solution u^* of (43) and a solution $\mathbf{p}^*$ of (42), the relation (16) still holds, i.e.,

$$u^* = \operatorname{div} \mathbf{p}^* + f.$$

Consequently, let $\mathbf{p}_j^{n+1}$ be a solution of the predual subproblem in iteration $n + 1$, then $u_j^{n+1} = \operatorname{div} \mathbf{p}_j^{n+1} + f_j^{n+1}$, $j = 1, \ldots, M$. Plugging this into the definition of f_M^{n+1}, we obtain

$$f_M^{n+1} = \sum_{j=1}^{M-1} \left(u_j^{n+1} - f_j^{n+1}\right) + g.$$

This motivates the choice of f_j^{n+1} for all $j \in \{1, \ldots, M\}$ as

$$f_j^{n+1} = \sum_{i>j}(u_i^n - f_i^n) + \sum_{i<j}(u_i^{n+1} - f_i^{n+1}) + g$$

for a successive algorithm; see Algorithm 11.

Let the partition of unity $(\theta_j)_{j=1}^M$ be as above, then we have the following convergence result (Langer and Gaspoz 2019).

11 Domain Decomposition for Non-smooth (in Particular TV) Minimization

---

Algorithm 11 Successive overlapping algorithm for (41)

Initialize: $u_j^0 (= 0) \in L^2(\Omega)$, $f_j^0 = 0 \in L^2(\Omega)$, $j = 1, \ldots, M$
for $n = 0, 1, 2, \ldots$ **do**
  for $j = 1, \ldots, M$ **do**
    $f_j^{n+1} = \sum_{i > j}(u_i^n - f_i^n) + \sum_{i < j}(u_i^{n+1} - f_i^{n+1}) + g$
    $u_j^{n+1} = \arg\min_{u_j \in L^2(\Omega)} \frac{1}{2}\|u_j - f_j^{n+1}\|_{L^2(\Omega)}^2 + \lambda \int_\Omega \theta_j |Du_j|$
  end for
  $u^{n+1} = g + \sum_{j=1}^M u_j^{n+1} - f_j^{n+1} (= u_M^{n+1})$
end for

Theorem 4. *Assume that $(f_j^n)_n$ is bounded in $L^2(\Omega)$ for $j = 1, \ldots, M$, then Algorithm 11 generates a sequence $(u^n)_n$ which converges strongly in $L^2(\Omega)$ to a unique minimizer of (12) with $T = I$.*

The strong convergence is due to the fact that $(\|u^n\|_{L^2(\Omega)})_n$ is monotonically decreasing. We remark that the boundedness assumption on $(f_j^n)_n$, $j = 1, \ldots, M$, is essential for the convergence proof, but this assumption automatically holds in a finite dimensional setting, which is, for example, the situation when the considered problem is discretized.

The parallel version of Algorithm 11 is presented in Algorithm 12.

Algorithm 12 Parallel overlapping algorithm for (41)

Initialize: $v_j^0 = 0$ for $j = 1, \ldots, M$
for $n = 0, 1, 2, \ldots$ **do**
  $f_j^{n+1} = \sum_{i \neq j} v_i^n + g$, $j = 1, \ldots, M$
  $u_j^{n+1} = \arg\min_{u_j \in L^2(\Omega)} \frac{1}{2}\|u_j - f_j^{n+1}\|_{L^2(\Omega)}^2 + \lambda \int_\Omega \theta_j |Du_j|$, $j = 1, \ldots, M$
  $v_j^{n+1} = \frac{(M-1)v_j^n + u_j^{n+1} - f_j^{n+1}}{M}$, $j = 1, \ldots, M$
  $u^{n+1} = g + \sum_{j=1}^M v_j^{n+1} (= \frac{\sum_{j=1}^M u_j^{n+1}}{M})$
end for

Note that here for the update of f_j^{n+1} an averaging (relaxation) is introduced, which is necessary for theoretical reasons in order to guarantee a similar convergence result as for the successive algorithm.

Theorem 5. *Assume that $(f_j^n)_n$ is bounded in $L^2(\Omega)$ for $j = 1, \ldots, M$, then Algorithm 12 generates a sequence $(u^n)_n$ which converges strongly in $L^2(\Omega)$ to a unique minimizer of (12) with $T = I$*

Proof. Since $f_j^{n+1} = \sum_{i \neq j} v_i^n + g$ we get that

$$f_j^{n+1} - f_1^{n+1} = v_1^n - v_j^n \tag{47}$$

for all $j = 1, \ldots, M$ and $n \geq 0$. Hence, this yields

$$f_M^{n+1} + \sum_{j=2}^{M-1} \left(f_j^{n+1} - f_1^{n+1} \right) = \sum_{j=1}^{M-1} v_j^n + g + \sum_{j=2}^{M-1} \left(v_1^n - v_j^n \right) = (M-1)v_1^n + g$$

for $n \geq 0$. Due to the boundedness of $(f_j^{n+1})_n$ for $j = 1, \ldots, M$ in $L^2(\Omega)$, also $(v_1^n)_n$ is bounded in $L^2(\Omega)$. Consequently by (47) the sequence $(v_j^n)_n$ for $j = 1, \ldots, M$ is bounded in $L^2(\Omega)$.

The rest of the proof follows the lines of the proof of Langer and Gaspoz (2019, Theorem 2.12) by straightforwardly adjusting the arguments to a splitting into $M \in \mathbb{N}$ domains.

Subspace Minimization

Let us turn now to the question how to realize the subspace minimization problems of Algorithms 11 and 12 and restrict them to the respective subdomains. We consider, for example, the subspace minimization with respect to u_1, i.e.,

$$u_1^{n+1} = \arg\min_{u_1 \in L^2(\Omega)} \frac{1}{2} \|u_1 - f_1^{n+1}\|_{L^2(\Omega)}^2 + \lambda \int_\Omega \theta_1 |Du_1|, \tag{48}$$

by anticipating that the arguments are analogue for the other subdomain problems. There are two different approaches on how to compute the solution of (48) by solving a minimization on Ω_1 only. These two approaches relate to "First optimize then discretize" and "First discretize then optimize," where the optimization part allows to restrict the problem to the subdomain. Hence, the first approach restricts the minimization problem in an infinite dimensional setting before discretization, while the second approach first discretizes (48) and then restricts the optimization process to the subdomain Ω_1.

First optimize then discretize The restriction of the subproblem is based on the following statement, cf. Langer and Gaspoz (2019, Lemma 2.2).

Lemma 1. *Let* $u \in L^2(\Omega)$, $\theta : \overline{\Omega} \to \mathbb{R}_0^+$, $\theta \in H^1(\Omega) \cap C(\overline{\Omega})$, $\|\nabla \theta\|_{\mathbb{L}^\infty(\Omega)} < \infty$, $\mathrm{supp}(\theta) \subseteq \overline{\Omega}$, *and* $K := \{p \in H_0(\mathrm{div}, \Omega) : |p(x)|_{\ell^2} \leq \theta(x) \text{ f.a.a. } x \in \Omega\}$ *then*

$$\int_\Omega \theta |Du| = \int_{\mathrm{supp}(\lambda)} \theta |Du|.$$

11 Domain Decomposition for Non-smooth (in Particular TV) Minimization

Proof. Let $\Omega_0 := \Omega \setminus \mathrm{supp}(\theta)$, then we get

$$\int_{\mathrm{supp}(\theta)} \theta|Du| = \int_{\Omega \setminus \Omega_0} \theta|Du| = \sup_{\mathbf{p} \in \mathcal{K}(1,C_0(\Omega \setminus \Omega_0, \mathbb{R}^2))} \langle \theta Du, \mathbf{p} \rangle_{C_0(\Omega \setminus \Omega_0, \mathbb{R}^2)' \times C_0(\Omega \setminus \Omega_0, \mathbb{R}^2)}$$

$$= \sup_{\mathbf{p} \in \mathcal{K}(1,C_0(\Omega, \mathbb{R}^2))} \langle \theta Du, \mathbf{p} \rangle_{C_0(\Omega, \mathbb{R}^2)' \times C_0(\Omega, \mathbb{R}^2)}$$

$$= \int_\Omega \theta|Du|,$$

since $\theta \in C(\overline{\Omega})$ and $\theta(x) = 0$ f.a.a. $x \in \Omega_0$, where $\mathcal{K}(\theta, C_0(\Omega, \mathbb{R}^2)) := \{\mathbf{p} \in C_0(\Omega, \mathbb{R}^2) : |\mathbf{p}(x)|_{\ell^2} \leq \theta(x)$ f.a.a. $x \in \Omega\}$ with $C_0(\Omega, \mathbb{R}^2)$ denoting the space of $\mathbb{R}^2$-valued continuous functions with compact support in Ω.

Utilizing Lemma 1 one can show that the minimizer of (48) can be computed by solving a minimization problem in Ω_1 only.

Proposition 1. *The solution $u_1^{n+1} \in L^2(\Omega)$ of the minimization problem in (48) is given by*

$$u_1^{n+1} = \begin{cases} f_1^{n+1} & \text{in } \Omega \setminus \Omega_1 \\ \arg\min_{u_1 \in L^2(\Omega_1)} \frac{1}{2}\|u_1 - f_1^{n+1}\|_{L^2(\Omega_1)}^2 + \lambda \int_{\Omega_1} \theta_1|Du_1| & \text{in } \Omega_1. \end{cases} \quad (49)$$

Proof. Since the partition of unity is such that $\mathrm{supp}(\theta_1) \subseteq \Omega_1$, we have due to Lemma 1 that $\int_\Omega \theta_1|Du_1| = \int_{\Omega_1} \theta_1|Du_1|$. Hence, by the optimality of u_1^{n+1} we get $f_1^{n+1} - u_1^{n+1} \in \partial\lambda \int_{\Omega_1} \theta_1|Du_1^{n+1}|$. That is,

$$(f_1^{n+1} - u_1^{n+1}, v - u_1^{n+1}) + \lambda \int_{\Omega_1} \theta_1|Du_1^{n+1}| \leq \lambda \int_{\Omega_1} \theta_1|Dv| \quad \forall v \in L^2(\Omega).$$

This inequality holds if

$$\int_{\Omega \setminus \Omega_1} (f_1^{n+1} - u_1^{n+1})(v - u_1^{n+1})dx \leq 0 \quad \text{and}$$

$$\int_{\Omega_1} (f_1^{n+1} - u_1^{n+1})(v - u_1^{n+1})dx + \lambda \int_{\Omega_1} \theta_1|Du_1^{n+1}| \leq \lambda \int_{\Omega_1} \theta_1|Dv|$$

for all $v \in L^2(\Omega)$. Hence, u_1^{n+1} fulfilling these two latter inequalities is a minimizer of the subspace minimization problem (48). By the uniqueness of the minimizer, we therefore obtain (49).

Due to the presence of the function θ_1, the usual total variation minimization techniques cannot be used directly to compute a minimizer of the optimization problem in (49), but may be used after being adapted to locally weighted total variation minimization. We note that the minimization of locally weighted total variation has been already considered in the literature (see, for example, Langer (2017a)), where an algorithm for solving a minimization problem of the type (48) is already presented. An alternative method modifying the split Bregman algorithm (Goldstein and Osher 2009) to locally weighted total variation minimization is proposed in Langer and Gaspoz (2019). Utilizing one of these methods for a practical implementation would then require a suitable discretization.

First discretize then optimize Since Algorithms 11 and 12 are designed for an overlapping splitting, let Ω^h be a discrete rectangular image domain containing $N_1 \times N_2$ pixels, $N_1, N_2 \in \mathbb{N}$, and decomposed into overlapping subdomains Ω_i^h, $i = 1, \ldots, M$ such that $\Omega^h = \bigcup_{i=1}^{M} \Omega_i^h$ and for any $i \in \{1, \ldots, M\}$ there exists at least one $j \in \{1, \ldots, M\} \setminus \{i\}$ such that $\Omega_i^h \cap \Omega_j^h \neq \emptyset$. Moreover, we use the finite difference discretization introduced in section "Finite Difference Setting". Then the discretized version of (48) is written as

$$u_1^{h,n+1} = \arg\min_{u_1^h \in X} \frac{1}{2} \|u_1^h - f_1^{h,n+1}\|_X^2 + \lambda \sum_{x \in \Omega^h} \theta_1^h(x) |\nabla_\Omega^h u_1^h(x)|_{\ell^2}, \qquad (50)$$

where $\theta_1^h \in X$ is the discrete version of the above introduced θ_1 satisfying (44), (45), and (46). Since $\theta_1^h(x) = 0$ for all $x \in \Omega^h \setminus \Omega_1^h$ we can write the above minimization problem as

$$u_1^{h,n+1} = \begin{cases} f_1^{h,n+1} & \text{in } \Omega^h \setminus \Omega_1^h \\ \arg\min_{u_1^h|_{\Omega_1^h} \in X_1} \frac{1}{2} \|u_1^h - f_1^{h,n+1}\|_{X_1}^2 + \lambda \sum_{x \in \Omega_1^h} \theta_1^h(x) |\nabla_\Omega^h u_1^h(x)|_{\ell^2} & \text{in } \Omega_1^h, \end{cases}$$

(51)

where $u_1^h \in X$ is such that $u_1^h(x) = f_1^{h,n+1}(x)$ for $x \in \Omega^h \setminus \Omega_1^h$. Hence, in order to obtain $u_1^{h,n+1}$, only a minimization problem in Ω_1^h has to be solved, i.e.,

$$\arg\min_{u_1^h|_{\Omega_1^h} \in X_1} \frac{1}{2} \|u_1^h - f_1^{h,n+1}\|_{X_1}^2 + \lambda \sum_{x \in \Omega_1^h} \theta_1^h(x) |(\nabla_\Omega^h u_1^h)|_{\Omega_1^h}(x)|_{\ell^2}.$$

Note that ∇_Ω^h is not a local operator, but nonetheless quite local, i.e., it affects only the neighboring pixels. Hence, by carefully considering the restriction to Ω_1^h (i.e., we use Dirichlet boundary conditions on the interface between Ω_1^h and $\Omega^h \setminus \Omega_1^h$), $u_{1,\Omega_1^h}^{h,n+1} \in X_1$ is obtained by solving an optimization in Ω_1^h only. Consequently locally weighted total variation minimization techniques may be used by carefully adjusting the gradient operator of the total variation term. An implementation based on the split Bregman algorithm is presented in Langer and Gaspoz (2019), which allows to obtain $u_{1,\Omega_1^h}^{h,n+1}$ by solving a linear system only of size $|\Omega_1^h|$.

Let us mention that all the results presented in this section hold symmetrically for the minimization with respect to u_i, $i = 2, \ldots, M$ and that the notations should be just adjusted accordingly.

Limit Case: Non-overlapping Decomposition

We remark that in a discrete setting the continuity assumption on θ_j^h, for $j = 1, \ldots, M$, is obsolete. Hence, we may let the overlapping size go to 0, yielding a non-overlapping decomposition. That is,

$$\theta_j^h(x) = \begin{cases} 1 & \text{if } x \in \Omega_i^h \\ 0 & \text{else} \end{cases}$$

for $j = 1, \ldots, M$. Then the subspace minimization problems read as

$$\arg\min_{u_j \in X_j} \frac{1}{2}\|u_j^h - f_j^{h,n+1}\|_{X_j}^2 + \lambda \sum_{x \in \Omega_j^h} |\nabla_\Omega^h u_j^h(x)|_{\ell^2},$$

$j = 1, \ldots, M$. Thus, in a discrete setting, using this discretization and restriction approach, in the limit case of a non-overlapping decomposition, Algorithms 11 and 12 become the successive domain decomposition (block Gauss-Seidel) and parallel domain decomposition (relaxed block Jacobi) method of Lee and Nam (2017), respectively. Moreover, in Lee and Nam (2017) these methods have been extended to (12) with $T \neq I$ by using the surrogate functional idea (cf. section "Subspace Minimization"), on J^h, i.e., for $u^h, a^h \in X$ we define

$$J^{h,s}(u^h, a^h) := J^h(u^h) + \frac{1}{2}\left(\delta\|u^h - a^h\|_X^2 - \|T^h(u^h - a^h)\|_X^2\right),$$

where $\delta > \|T^h\|^2$. Then we have

$$\arg\min_{u^h \in X} J^{h,s}(u^h, a^h) = \arg\min_{u^h \in X} \frac{1}{2}\|u^h - \frac{1}{\delta}(T^{h*}g^h + (\delta - T^{h*}T^h)a^h)\|_X^2$$

$$+ \frac{\lambda}{\delta} \sum_{x \in \Omega^h} |\nabla u^h(x)|_{\ell^2}$$

and an approximation of the minimizer of J is obtained by iteratively minimizing

$$u^{h,0} = 0, \quad u^{h,n+1} = \arg\min_{u^h \in X} J^{h,s}(u^h, u^{h,n}) \quad n \geq 0. \tag{52}$$

Since in each iteration we have to solve a problem which is of the same type as (12) with $T = I$, we may use Algorithm 11 or Algorithm 12 now in a non-overlapping and finite difference setting to speed up the solution process, leading to Algorithms 13 and 14.

Algorithm 13 Successive non-overlapping algorithm for (12)

Initialize: $u_j^{h,0} := 0$, $v_j^{h,0} := 0$ for $j = 1, \ldots, M$
for $n = 0, 1, 2, \ldots$ **do**
$\quad f^{h,n+1} = \frac{1}{\delta}(T^{h*}g^h + (\delta - T^{h*}T^h)u^{h,n}$, $q_i^{h,0} = v_j^{h,n}$ for $j = 1, \ldots, M$ and $k = 1$
$\quad$ **while** $J^{h,s}(f^{h,n+1} - \sum_{j=1}^{M} q_j^{h,k}, u^{h,n}) \leq J^h(u^{h,n})$ **do**
$\quad\quad$ **for** $j = 1, \ldots, M$ **do**
$\quad\quad\quad f_j^{h,k} = f^{h,n+1} - \sum_{i>j} q_i^{h,k-1} - \sum_{i<j} q_i^{h,k}$ in Ω_j^h
$\quad\quad\quad u_j^{h,k} = \arg\min_{u_j^h \in X} \frac{1}{2}\|u_j^h - f_j^{h,k}\|_X^2 + \frac{\lambda}{\delta} \sum_{x \in \Omega_j^h} |\nabla u_j^h(x)|$
$\quad\quad\quad q_j^{h,k} = f_j^{h,k} - u_j^{h,k}$
$\quad\quad\quad k = k + 1$
$\quad\quad$ **end for**
$\quad$ **end while**
$\quad u^{h,n+1} = f^{h,n+1} - \sum_{j=1}^{M} q_j^{h,k}$ and $v_j^{h,n+1} = q_j^{h,k}$ for $j = 1, \ldots, M$
end for

Algorithm 14 Parallel non-overlapping algorithm for (12)

Initialize: $u_j^{h,0} := 0$, $v_j^{h,0} := 0$ for $j = 1, \ldots, M$
for $n = 0, 1, 2, \ldots$ **do**
$\quad f^{n+1} = \frac{1}{\delta}(T^{h*}g^h + (\delta - T^{h*}T^h)u^{h,n}$, $q_i^{h,0} = v_j^{h,n}$ for $j = 1, \ldots, M$ and $k = 1$
$\quad$ **while** $J^{h,s}(f^{h,n+1} - \sum_{j=1}^{M} q_j^{h,k}, u^{h,n}) \leq J^h(u^{h,n})$ **do**
$\quad\quad$ **for** $j = 1, \ldots, M$ **do**
$\quad\quad\quad f_j^{h,k} = g^h - \sum_{i \neq j} q_i^{h,k-1}$ in Ω_j^h
$\quad\quad\quad u_j^{h,k} = \arg\min_{u_j^h \in X} \frac{1}{2}\|u_j^h - f_j^{h,k}\|_X^2 + \frac{\lambda}{\delta} \sum_{x \in \Omega_j^h} |\nabla u_j^h(x)|$
$\quad\quad\quad q_j^{h,k} = \frac{(M-1)q_j^{h,k-1} + f_j^{h,k} - u_j^{h,k}}{M}$
$\quad\quad\quad k = k + 1$
$\quad\quad$ **end for**
$\quad$ **end while**
$\quad u^{h,n+1} = f^{h,n+1} - \sum_{j=1}^{M} q_j^{h,k}$ and $v_j^{h,n+1} = q_j^{h,k}$ for $j = 1, \ldots, M$
end for

In Lee and Nam (2017) it is shown for $M = 2$ that these algorithms produce sequences $(u^n)_n$ whose accumulation points are minimizers of J^h.

Conclusion

Domain decomposition methods are known to be one of the most successful methods to construct efficient solvers for large-scale problems. Nevertheless, only quite recently such methods are developed for total variation minimization. Therefore, the research in this direction is far from being complete, as only very little is known yet. We summarize that the domain decomposition algorithms for total variation minimization with a theoretical guarantee to convergence to the minimizer of the global problem are till now given for (i) the discrete predual problem with a non-overlapping decomposition using finite differences (Hintermüller and Langer 2015) or finite elements (Lee et al. 2019b; Lee and Park 2019b), (ii) the continuous predual problem with an overlapping decomposition (Chang et al. 2015), (iii) the discrete primal problem with a non-overlapping decomposition (Lee and Nam 2017), (iv) and the continuous primal problem with an overlapping decomposition (Langer and Gaspoz 2019). This list of achievements indicates that constructing overlapping domain decomposition methods in an infinite dimensional setting seems easier than non-overlapping domain decomposition methods. A reason for this may be guessed when one looks at the Poisson problem (see section "Basic Idea of Domain Decomposition"). There one sees that in order to construct convergent non-overlapping methods, the subdomain problems differ in each subdomain due to the interface conditions, while in the overlapping situation all subdomain problems are of the same type. This ostensible flexibility in creating subdomain problems for a non-overlapping splitting may lead to additional difficulties for problems where the solution is discontinuous, as the interface conditions are not clear. In particular, neither of the interface conditions in (2) are suitable.

For the domain decomposition methods tackling the predual problem (14), not only the convergence but also the convergence order is known. We note that the decomposition methods for the continuous problems only cover the image denoising case, i.e., the L^2-TV model with $T = I$, while the methods for the discretized objectives can also handle image inpainting and image segmentation problems. The primal-dual approach in Lee et al. (2019a) is even successfully applied to image deblurring. Of course, by using the surrogate idea (also called operator splitting (Combettes and Wajs 2005)), the L^2-TV model can be cast to an image denoising type of problem for any operator T. But it is in general unclear how accurately the solution of the domain decomposition iteration has to be computed in order to guarantee the convergence of the outer surrogate iteration. Interesting tasks arising, for example, in medical imaging where T might be a sampled Fourier transform or Radon transform, which are very global operators, have not yet been thoroughly considered.

References

Acar, R., Vogel, C.R.: Analysis of bounded variation penalty methods for ill-posed problems. Inverse Probl. **10**(6), 1217–1229 (1994)

Alliney, S.: A property of the minimum vectors of a regularizing functional defined by means of the absolute norm. IEEE Trans. Signal Process. **45**(4), 913–917 (1997)

Ambrosio, L., Fusco, N., Pallara, D.: Functions of Bounded Variation and Free Discontinuity Problems. Oxford Mathematical Monographs. The Clarendon Press/Oxford University Press, New York (2000)

Attouch, H., Buttazzo, G., Michaille, G.: Variational Analysis in Sobolev and BV Spaces. MOS-SIAM Series on Optimization, 2nd edn. Society for Industrial and Applied Mathematics (SIAM)/Mathematical Optimization Society, Philadelphia (2014). Applications to PDEs and optimization

Aubert, G., Aujol, J.-F.: A variational approach to removing multiplicative noise. SIAM J. Appl. Math. **68**(4), 925–946 (2008)

Beck, A., Teboulle, M.: A fast iterative shrinkage-thresholding algorithm for linear inverse problems. SIAM J. Imaging Sci. **2**(1), 183–202 (2009)

Bertsekas, D.P.: Constrained Optimization and Lagrange Multiplier Methods. Academic Press, New York (2014)

Burger, M., Sawatzky, A., Steidl, G.: First order algorithms in variational image processing. In: Splitting Methods in Communication, Imaging, Science, and Engineering. Scientific Computation, pp. 345–407. Springer, Cham (2016)

Cai, J.-F., Chan, R.H., Nikolova, M.: Two-phase approach for deblurring images corrupted by impulse plus Gaussian noise. Inverse Probl. Imaging **2**(2), 187–204 (2008)

Calatroni, L., De Los Reyes, J.C., Schönlieb, C.-B.: Infimal convolution of data discrepancies for mixed noise removal. SIAM J. Imaging Sci. **10**(3), 1196–1233 (2017)

Carstensen, C.: Domain decomposition for a non-smooth convex minimization problem and its application to plasticity. Numer. Linear Algebra Appl. **4**(3), 177–190 (1997)

Chambolle, A.: An algorithm for total variation minimization and applications. J. Math. Imaging Vis. **20**(1–2), 89–97 (2004). Special issue on mathematics and image analysis

Chambolle, A., Pock, T.: A First-order Primal-dual Algorithm for Convex Problems with Applications to Imaging. J. Math. Imaging Vis. **40**(1), 120–145 (2011)

Chambolle, A., Pock, T.: A remark on accelerated block coordinate descent for computing the proximity operators of a sum of convex functions. SMAI J. Comput. Math. **1**, 29–54 (2015)

Chambolle, A., Pock, T.: An introduction to continuous optimization for imaging. Acta Numer. **25**, 161–319 (2016)

Chambolle, A., Caselles, V., Cremers, D., Novaga, M., Pock, T.: An introduction to total variation for image analysis. Theor. Found. Numer. Methods Sparse Recovery **9**, 263–340 (2010)

Chan, T.F., Mathew, T.P.: Domain decomposition algorithms. In: Acta Numerica, pp. 61–143. Cambridge University Press, Cambridge (1994)

Chan, T.F., Shen, J.J.: Image Processing and Analysis: Variational, PDE, Wavelet, and Stochastic Methods. SIAM, Philadelphia (2005)

Chan, T.F., Esedoglu, S., Nikolova, M.: Algorithms for finding global minimizers of image segmentation and denoising models. SIAM J. Appl. Math. **66**(5), 1632–1648 (2006)

Chang, H., Zhang, X., Tai, X.-C., Yang, D.: Domain decomposition methods for nonlocal total variation image restoration. J. Sci. Comput. **60**(1), 79–100 (2014)

Chang, H., Tai, X.-C., Wang, L.-L., Yang, D.: Convergence rate of overlapping domain decomposition methods for the Rudin–Osher–Fatemi model based on a dual formulation. SIAM J. Imaging Sci. **8**(1), 564–591 (2015)

Chen, K., Tai, X.-C.: A nonlinear multigrid method for total variation minimization from image restoration. J. Sci. Comput. **33**(2), 115–138 (2007)

Combettes, P.L., Wajs, V.R.: Signal recovery by proximal forward-backward splitting. Multiscale Model. Simul. **4**(4), 1168–1200 (electronic) (2005)

Daubechies, I., Defrise, M., De Mol, C.: An iterative thresholding algorithm for linear inverse problems with a sparsity constraint. Commun. Pure Appl. Math. **57**(11), 1413–1457 (2004)

Daubechies, I., Teschke, G., Vese, L.: Iteratively solving linear inverse problems under general convex constraints. Inverse Probl. Imaging **1**(1), 29–46 (2007)

Dolean, V., Jolivet, P., Nataf, F.: An Introduction to Domain Decomposition Methods: Algorithms, Theory, and Parallel Implementation, vol. 144. SIAM, Philadelphia (2015)

Duan, Y., Tai, X.-C.: Domain decomposition methods with graph cuts algorithms for total variation minimization. Adv. Comput. Math. **36**(2), 175–199 (2012)

Duan, Y., Chang, H., Tai, X.-C.: Convergent non-overlapping domain decomposition methods for variational image segmentation. J. Sci. Comput. **69**(2), 532–555 (2016)

Fornasier, M.: Domain decomposition methods for linear inverse problems with sparsity constraints. Inverse Probl. Int. J. Theory Pract. Inverse Probl. Inverse Methods Comput. Inversion Data **23**(6), 2505–2526 (2007)

Fornasier, M., Schönlieb, C.-B.: Subspace correction methods for total variation and l_1-minimization. SIAM J. Numer. Anal. **47**(5), 3397–3428 (2009)

Fornasier, M., Langer, A., Schönlieb, C.-B.: Domain decomposition methods for compressed sensing. In: Proceedings of the International Conference of SampTA09, Marseilles, arXiv preprint arXiv:0902.0124 (2009)

Fornasier, M., Langer, A., Schönlieb, C.-B.: A convergent overlapping domain decomposition method for total variation minimization. Numerische Mathematik **116**(4), 645–685 (2010)

Fornasier, M., Kim, Y., Langer, A., Schönlieb, C.: Wavelet decomposition method for L_2/TV-image deblurring. SIAM J. Imaging Sci. **5**(3), 857–885 (2012)

Getreuer, P., Tong, M., Vese, L.A.: A variational model for the restoration of mr images corrupted by blur and rician noise. In: International Symposium on Visual Computing, pp. 686–698. Springer (2011)

Gilboa, G., Osher, S.: Nonlocal operators with applications to image processing. Multiscale Model. Simul. **7**(3), 1005–1028 (2009)

Giusti, E.: Minimal Surfaces and Functions of Bounded Variation. Monographs in Mathematics, vol. 80. Birkhäuser Verlag, Basel (1984)

Goldstein, T., Osher, S.: The split Bregman method for $L1$-regularized problems. SIAM J. Imaging Sci. **2**(2), 323–343 (2009)

Hintermüller, M., Kunisch, K.: Total bounded variation regularization as a bilaterally constrained optimization problem. SIAM J. Appl. Math. **64**(4), 1311–1333 (2004)

Hintermüller, M., Langer, A.: Subspace correction methods for a class of nonsmooth and nonadditive convex variational problems with mixed L^1/L^2 data-fidelity in image processing. SIAM J. Imaging Sci. **6**(4), 2134–2173 (2013)

Hintermüller, M., Langer, A.: Surrogate functional based subspace correction methods for image processing. In: Domain Decomposition Methods in Science and Engineering XXI, pp. 829–837. Springer, Cham (2014)

Hintermüller, M., Langer, A.: Non-overlapping domain decomposition methods for dual total variation based image denoising. J. Sci. Comput. **62**(2), 456–481 (2015)

Hintermüller, M., Rautenberg, C.: On the density of classes of closed convex sets with pointwise constraints in sobolev spaces. J. Math. Anal. Appl. **426**(1), 585–593 (2015)

Hintermüller, M., Rautenberg, C.N.: Optimal selection of the regularization function in a weighted total variation model. Part I: Modelling and theory. J. Math. Imaging Vis. **59**(3), 498–514 (2017)

Ito, K., Kunisch, K.: Lagrange Multiplier Approach to Variational Problems and Applications, vol. 15. SIAM, Philadelphia (2008)

Langer, A.: Automated parameter selection for total variation minimization in image restoration. J. Math. Imaging Vis. **57**(2), 239–268 (2017a)

Langer, A.: Automated parameter selection in the L^1-L^2-TV model for removing Gaussian plus impulse noise. Inverse Probl. **33**(7), 74002 (2017b)

Langer, A.: Locally adaptive total variation for removing mixed Gaussian–impulse noise. Int. J. Comput. Math. **96**(2), 298–316 (2019)

Langer, A., Gaspoz, F.: Overlapping domain decomposition methods for total variation denoising. SIAM J. Numer. Anal. **57**(3), 1411–1444 (2019)
Langer, A., Osher, S., Schönlieb, C.-B.: Bregmanized domain decomposition for image restoration. J. Sci. Comput. **54**(2–3), 549–576 (2013)
Le, T., Chartrand, R., Asaki, T.J.: A variational approach to reconstructing images corrupted by poisson noise. J. Math. Imaging Vis. **27**(3), 257–263 (2007)
Lee, C.-O., Nam, C.: Primal domain decomposition methods for the total variation minimization, based on dual decomposition. SIAM J. Sci. Comput. **39**(2), B403–B423 (2017)
Lee, C.-O., Park, J.: Fast nonoverlapping block Jacobi method for the dual Rudin–Osher–Fatemi model. SIAM J. Imaging Sci. **12**(4), 2009–2034 (2019a)
Lee, C.-O., Park, J.: A finite element nonoverlapping domain decomposition method with lagrange multipliers for the dual total variation minimizations. J. Sci. Comput. **81**(3), 2331–2355 (2019b)
Lee, C.-O., Lee, J.H., Woo, H., Yun, S.: Block decomposition methods for total variation by primal–dual stitching. J. Sci. Comput. **68**(1), 273–302 (2016)
Lee, C.-O., Nam, C., Park, J.: Domain decomposition methods using dual conversion for the total variation minimization with L^1 fidelity term. J. Sci. Comput. **78**(2), 951–970 (2019a)
Lee, C.-O., Park, E.-H., Park, J.: A finite element approach for the dual Rudin–Osher–Fatemi model and its nonoverlapping domain decomposition methods. SIAM J. Sci. Comput. **41**(2), B205–B228 (2019b)
Lions, J.L.: Optimal Control of Systems Governed by Partial Differential Equations. Die Grundlehren der mathematischen Wissenschaften, vol. 170. Springer (1971)
Lions, P.-L.: On the Schwarz alternating method. I. In: First International Symposium on Domain Decomposition Methods for Partial Differential Equations, Paris, pp. 1–42 (1988)
Marini, L.D., Quarteroni, A.: A relaxation procedure for domain decomposition methods using finite elements. Numerische Mathematik **55**(5), 575–598 (1989)
Mathew, T.: Domain Decomposition Methods for the Numerical Solution of Partial Differential Equations, vol. 61. Springer Science & Business Media, Berlin (2008)
Nikolova, M.: Minimizers of cost-functions involving nonsmooth data-fidelity terms. Application to the processing of outliers. SIAM J. Numer. Anal. **40**(3), 965–994 (electronic) (2002)
Nikolova, M.: A variational approach to remove outliers and impulse noise. J. Math. Imaging Vis. **20**(1–2), 99–120 (2004)
Peyré, G., Bougleux, S., Cohen, L.: Non-local regularization of inverse problems. In: European Conference on Computer Vision, pp. 57–68. Springer (2008)
Pock, T., Unger, M., Cremers, D., Bischof, H.: Fast and exact solution of total variation models on the gpu. In: 2008 IEEE Computer Society Conference on Computer Vision and Pattern Recognition Workshops, pp. 1–8. IEEE (2008)
Quarteroni, A., Valli, A.: Domain Decomposition Methods for Partial Differential Equations. Oxford University Press, New York (1999)
Raviart, P.-A., Thomas, J.-M.: A mixed finite element method for 2-nd order elliptic problems. In: Mathematical Aspects of Finite Element Methods, pp. 292–315. Springer, Berlin (1977)
Rudin, L.I., Osher, S., Fatemi, E.: Nonlinear total variation based noise removal algorithms. Phys. D: Nonlinear Phenom. **60**(1), 259–268 (1992)
Schönlieb, C.-B.: Total variation minimization with an H^{-1} constraint. CRM Ser. **9**, 201–232 (2009)
Schwarz, H.A.: Über einige Abbildungsaufgaben. Journal für die reine und angewandte Mathematik **1869**(70), 105–120 (1869)
Smith, B., Bjorstad, P., Gropp, W.: Domain Decomposition: Parallel Multilevel Methods for Elliptic Partial Differential Equations. Cambridge University Press, Dordrecht (2004)
Tai, X.-C.: Rate of convergence for some constraint decomposition methods for nonlinear variational inequalities. Numerische Mathematik **93**(4), 755–786 (2003)
Tai, X.-C., Tseng, P.: Convergence rate analysis of an asynchronous space decomposition method for convex minimization. Math. Comput. **71**(239), 1105–1135 (2002)
Tai, X.-C., Xu, J.: Global and uniform convergence of subspace correction methods for some convex optimization problems. Math. Comput. **71**(237), 105–124 (2002)

Toselli, A., Widlund, O.: Domain Decomposition Methods: Algorithms and Theory, vol. 34. Springer Science & Business Media, Dordrecht (2006)

Tseng, P.: Convergence of a block coordinate descent method for nondifferentiable minimization. J. Optim. Theory Appl. **109**(3), 475–494 (2001)

Tseng, P., Yun, S.: A coordinate gradient descent method for nonsmooth separable minimization. Math. Prog. **117**(1–2), 387–423 (2009)

Vonesch, C., Unser, M.: A fast multilevel algorithm for wavelet-regularized image restoration. IEEE Trans. Image Process. **18**(3), 509–523 (2009)

Warga, J.: Minimizing Certain Concex Functions. J. Soc. Indust. Appl. Math. **11**, 588–593 (1963)

Wright, S.J.: Coordinate descent algorithms. Math. Prog. **151**(1), 3–34 (2015)

Wu, C., Tai, X.-C.: Augmented lagrangian method, dual methods, and split bregman iteration for rof, vectorial tv, and high order models. SIAM J. Imaging Sci. **3**(3), 300–339 (2010)

Xu, J., Tai, X.-C., Wang, L.-L.: A two-level domain decomposition method for image restoration. Inverse Probl. Imaging **4**(3), 523–545 (2010)

Xu, J., Chang, H.B., Qin, J.: Domain decomposition method for image deblurring. J. Comput. Appl. Math. **271**, 401–414 (2014)

Zhang, X., Burger, M., Bresson, X., Osher, S.: Bregmanized nonlocal regularization for deconvolution and sparse reconstruction. SIAM J. Imaging Sci. **3**(3), 253–276 (2010)

Fast Numerical Methods for Image Segmentation Models

12

Noor Badshah

Contents

Introduction	428
Mathematical Models for Image Segmentation	429
Two-Phase Segmentation Models	429
Snakes: Active Contour Model	429
Geodesic Active Contour Model (GAC)	430
Chan-Vese Model	431
Fast Numerical Methods:	433
Multigrid Solver for Solving a Class of Variational Problems with Application to Image Segmentation	446
Sobolev Gradient Minimization of Curve Length in Chan-Vese Model	449
Multiphase Image Segmentation	452
Multigrid Method for Multiphase Segmentation Model	452
Multigrid Method with Typical and Modified Smoother	454
Local Fourier Analysis and a Modified Smoother	455
Convex Multiphase Image Segmentation Model	460
A Three-Stage Approach for Multiphase Segmentation Degraded Color Images	466
Stage 2: Dimension Lifting with Secondary Color Space	468
Selective Segmentation Models	469
Image Segmentation Under Geometrical Conditions	469
Active Contour-Based Image Selective Model	471
Dual-Level Set Selective Segmentation Model	475
One-Level Selective Segmentation Model	477
Reproducible Kernel Hilbert Space-Based Image Segmentation	479
An Optimization-Based Multilevel Algorithm for Selective Image Segmentation Models	485

N. Badshah (✉)
Department of Basic Sciences, University of Engineering and Technology, Peshawar, Pakistan

© Springer Nature Switzerland AG 2023
K. Chen et al. (eds.), *Handbook of Mathematical Models and Algorithms in Computer Vision and Imaging*, https://doi.org/10.1007/978-3-030-98661-2_121

Machine/Deep Learning Techniques for Image Segmentation........................	492
Machine Learning with Region-Based Active Contour Models in Medical Image Segmentation..	492
ResBCU-Net: Deep Learning Approach for Segmentation of Skin Images...........	494
Conclusion...	498
References...	499

Abstract

In this chapter, three different types of segmentation problems are studied, namely, two-phase segmentation problems, multiphase segmentation problems, and selective segmentation problems. Three types of numerical methods are discussed here as well. Some of them are time marching schemes, multigrid methods, and multilevel methods. Two types of minimization techniques are discussed, like L^2 gradient minimization and Sobolev gradient-based minimization techniques. At the end two deep/machine learning approaches for segmentation of images are also presented.

Keywords

Image segmentation · Euler-Lagrange's equations · Sobolev gradient · Finite differences · Machine learning · Deep learning

Introduction

Image segmentation is one of the fundamental tasks in image analysis and computer vision. The purpose of image segmentation is to partition a given image into different meaningful regions based on the intensity homogeneity, pattern similarities, colors similarities, etc. The goal of image segmentation is to represent an image in such a way that could be easily analyzed. There are two main concerns related to image segmentation: (i) modeling image segmentation problems and (ii) fast and advanced numerical methods for the solution of partial differential equations arising from the minimization of these models. There are many algorithms/models present in the literature for the solution of image segmentation problems. Among these, some of them use edge or region information of the image for segmentation purpose. The most basic edge-based model is the geodesic snake model (Kass et al. 1988; Caselles et al. 1997), which is based on edge information in the image, and a gradient flow is used as a stopping term to get correct boundaries with sudden changes in the gradient for attracting the contour to the object boundary. The Chan-Vese model (Chan and Vese 2001) is based on the variation in regions. For that purpose it uses region statistics as a stopping criterion.

The Allen-Cahn (AC) equation was originally introduced as a phenomenological model for antiphase domain coarsening in a binary alloy (Allen and Cahn 1979). This equation can be used to model flow problems based on mean curvature. This type of flow is one of the important element for active contour-based image segmentation models. For these types of methods, there exists a very fast computational method such as multigrid method (Badshah and Chen 2008). This chapter is dedicated to the minimization techniques of various models developed for the segmentation of images, which leads to a highly nonlinear partial differential equation. Some well-known numerical methods for the solution of these partial differential equations have been discussed.

Mathematical Models for Image Segmentation

Image segmentation links low-level vision with high-level vision. It is the process of partitioning an image into a collection of objects which can, later on, be used for performing high-level tasks like object detection, tracking, recognition, etc. The current section is about the existing mathematical models developed for image segmentation. Active contour models have attained much attention in image segmentation nowadays. These models for segmentation of images are divided into two groups, namely, (i) edge-based segmentation models and (ii) region-based segmentation models. In the next section, edge-based active models are discussed in detail.

Two-Phase Segmentation Models

Snakes: Active Contour Model

A snake is an energy-based active contour model which minimizes the deformable curve combined with some constraints and/or drag or pull forces that will pull the contour toward object boundaries, whereas the internal energies will resist the deformation in the contour. The first active contour model was developed by Kass et al. (1988). This type of model locates abrupt changes in the intensity through the deformation of a curve Υ in the image z. Those type of abrupt changes in the intensities usually occur at the edges of objects in an image z. The energy functional of the snake model has external and internal forces. The image energy/force is responsible to push the contour/snake toward image features like lines, edges, etc. Whereas the internal energy works for the smoothness of the contour, the external energy pulls/drags the contour/snake toward the desired boundary of the object (local minima of the functional). For a parametric planar curve $\Upsilon(p) = (x(p), y(p)) \in \Omega, 0 \leq p \leq 1$, the following energy functional is proposed:

$$F^K(\Upsilon(p)) = \varpi \int_0^1 \left| \frac{\partial \Upsilon(p)}{\partial p} \right|^2 dp + \beta \int_0^1 \left| \frac{\partial^2 \Upsilon(p)}{\partial^2 p} \right|^2 dp \\ + \lambda \int_0^1 e^2(\nabla(z * K_\sigma)(\Upsilon(p))) dp, \quad (1)$$

where $\varpi > 0, \beta > 0$ and $\lambda > $ are the trade-off parameters. Also e is the edge detector function and is given by the following:

$$e(\nabla(z * K_\sigma)) = \frac{1}{1 + \gamma |\nabla(z * K_\sigma)|^2}, \tag{2}$$

where $K_\sigma(x, y) = \frac{1}{2\pi\sigma^2} \exp\left((x - \mu_x)^2 + (y - \mu_y)^2 / 2\sigma^2\right)$ is the well-known Gaussian kernel and γ is a positive parameter. F^K is nonconvex functional (Kass et al. 1988) and can be easily stuck at local minima. The local minima of F^K can be the solution of the following Euler-Lagrange's equation:

$$-\varpi \frac{\partial^2 \Upsilon}{\partial p^2} + \beta \frac{\partial^4 \Upsilon}{\partial p^4} + \lambda \nabla e^2 = 0. \tag{3}$$

The numerical solution of this fourth-order partial differential equation can be found by using finite difference method (Kass et al. 1988).

Geodesic Active Contour Model (GAC)

In 1997, Casselles et al. proposed another edge-based model by using a new type of curve parametrization. This is an improvement in snake energy functional (Kass et al. 1988). The energy functional of the GAC model is given by the following:

$$F^C(\Upsilon(p)) = \int_0^1 e(|\nabla z(\Upsilon(p))|)|\Upsilon'(p)|dp. \tag{4}$$

Given that $L(\Upsilon)$ represents the Euclidean length of the moving contour Υ and since $L(\Upsilon) = \int_0^1 |\Upsilon'(p)|dp = \int_0^{L(\Upsilon)} ds$, where ds is the Euclidean length element, Eq. (4) may be written as follows:

$$F^C(\Upsilon(p)) = \int_0^{L(\Upsilon)} e(|\nabla z(\Upsilon(p))|) ds \tag{5}$$

This energy functional introduces a new length through weighted Euclidean differential length ds by the edge detector e which uses edge information (Aubert and Kornprobst 2002). The function e is the same as given in (2). The equivalence between minimizing F^C and minimizing F^K at $\beta = 0$ was studied in Caselles et al. (1997). Hence the direction for which F^C decreases most rapidly provides us the following minimization flow: more details of its derivation can be found in Caselles et al. (1997):

$$\frac{\partial \Upsilon}{\partial t} = e\kappa \vec{N} - (\nabla e \cdot \vec{N})\vec{N}, \tag{6}$$

where κ represents curvature and $\mathcal{N}$ is the unit normal vector. This equation leads toward the optimal length of the contour. The steady-state solution of (6) will be the solution of Euler-Lagrange's equation for the energy functional given in Eq. (5). By introducing level set idea, the evolution equation takes the following form:

$$\frac{\partial \phi}{\partial t} = |\nabla \phi|(\nabla \cdot (e \frac{\nabla \phi}{|\nabla \phi|}) + v_1 e), \qquad (7)$$

ϕ is a level set function and the contour Υ is the zero level set $\phi(x, y) = 0$. A balloon term $v_1 e$, $v_1 > 0$ is included to speed up the convergence.

These models are based on the edge detector e which uses the gradient of the image so these models can only detect objects whose boundaries are defined by gradient. Also, in practice, the discrete gradients are bounded, and hence the stopping function e may not vanish on the boundaries, and the contour may leak through the image edges (Chan and Vese 2001). These models may not work very well in noisy images.

Chan-Vese Model

In 2001, Chan and Vese proposed a region-based energy functional which uses data fitting statistics as a stopping process and is a special case of piecewise constant Mumford-Shah model (Mumford and Shah 1989). Let z be the known bounded function (image data) and assume that z has two regions (say foreground and background) of approximately constant intensities z_i and z_o. Assume that the object to be detected is represented by the region with intensity z_i and its boundary is Γ_0. Let the average intensity approximating z_i and z_o be c_1 and c_2, respectively. Let Γ by the interface separating the regions where the average intensities are c_1 and c_2. Based on constant average intensities in two different regions, the following energy is introduced:

$$F^{CV}(\Gamma, c_1, c_2) = \mu \cdot (\text{len}(\Gamma)) + \nu \cdot \text{area}(\text{inside}(\Gamma))$$
$$+ \eta \int_{\text{inside}(\Gamma)} |z - c_1|^2 d\Omega + \gamma \int_{\text{outside}(\Gamma)} |z - c_2|^2 d\Omega, \qquad (8)$$

where c_1 and c_2 are unknown constants and $\mu \geq 0$, $\nu \geq 0$, $\eta, \gamma > 0$ are fixed parameters. In Chan and Vese (2001) $\eta = \gamma = 1$, Γ is generally a hypersurface in $\mathbb{R}^n$, and "len(Γ)" is the length in $\mathcal{H}^{n-1}(\Gamma)$. In most of the cases, $\nu = 0$ is taken and only length constraint is imposed. Thus Chan and Vese in (2001) proposed the following energy functional for minimization:

$$\inf_{\Gamma, c_1, c_2} F^{CV}(\Gamma, c_1, c_2). \qquad (9)$$

where F^{CV} is given in Eq. (8). This functional is a special case of the piecewise constant Mumford and Shah energy functional (Mumford and Shah 1989).

Level Set Representation of the Model

Consider a Lipschitz function $\phi : \mathbb{R}^2 \to \mathbb{R}$, whose zero level set represents the region interface Γ and has opposite signs in different regions (Osher and Sethian 1988). In the level set representation of an unknown curve Γ, transform it from lower low dimension into higher dimension.

So by using level set representation, the Eq. (8) becomes the following:

$$F^{CV}(\phi, c_1, c_2) = \mu \int_\Omega |\nabla H(\phi)| d\Omega + \nu \int_\Omega H(\phi) d\Omega$$
$$+ \eta \int_\Omega |z - c_1|^2 H(\phi) d\Omega$$
$$+ \gamma \int_\Omega |z - c_2|^2 (1 - H(\phi)) d\Omega. \quad (10)$$

Once the optimal value ϕ is obtained, the final solution (segmented image) can be found by using the following:

$$u = c_1 H(\phi) + c_2 (1 - H(\phi)).$$

For the existence of minimizers and its relation with the Mumford and Shah model, please see Chan and Vese (2001). It must be noted that c_1, c_2 are the optimal average constant intensities inside and outside curve $\phi = 0$. $H(\phi)$ is the Heaviside function and is used as region descriptor. Due to discontinuity of Heaviside function at origin, a regularized Heaviside function $H_\epsilon(\phi)$ is introduced, and the above functional (10) is minimized with respect to ϕ to the get the following differential equation:

$$\begin{cases} \delta_\epsilon(\phi) \left[\mu \nabla \cdot \left(\frac{\nabla \phi}{|\nabla \phi|} \right) - \nu - \eta(z - c_1)^2 + \gamma(z - c_2)^2 \right] = 0 & \text{in } \Omega, \\ \frac{\delta_\epsilon(\phi)}{|\nabla \phi|} \frac{\partial \phi}{\partial n} = 0 & \text{on } \partial \Omega. \end{cases} \quad (11)$$

The corresponding unsteady parabolic partial differential equation is considered (Chan and Vese 2001) by introducing an artificial time t.

$$\begin{cases} \frac{\partial \phi}{\partial t} = \delta_\epsilon(\phi) \left[\mu \nabla \cdot \left(\frac{\nabla \phi}{|\nabla \phi|} \right) - \nu - \eta(z - c_1)^2 + \gamma(z - c_2)^2 \right] & \text{in } \Omega, \\ \phi(t, x, y) = \phi_0(x, y) & \text{in } \Omega \\ \frac{\partial \phi}{\partial n} = 0 & \text{on } \partial \Omega. \end{cases} \quad (12)$$

12 Fast Numerical Methods for Image Segmentation Models

Note that the steady-state solution of this parabolic partial differential equation will give solution of the corresponding elliptic partial differential equation given in Eq. (11). This is a nonlinear partial differential equation whose solution is done through fast numerical methods which are discussed in the next section.

Fast Numerical Methods:

Solution of nonlinear partial differential equations is a challenging task and is an open problem. In this section, a brief survey on some well-known fast numerical methods for the solution of partial differential equations arising from the minimization of mathematical models for segmentation problems are given. One of the simplest and easy to implement method for this task is explicit method, but this method is stable for small time step, which leads toward a large number of iterations for convergence. Here some well-known stable methods are discussed.

Semi-implicit Method

Consider the following evolution problem which is obtained from minimization of Chan-Vese model:

$$\begin{cases} c_1(\phi) = \dfrac{\int_\Omega z H_\epsilon(\phi) d\Omega}{\int_\Omega H_\epsilon(\phi) dx dy d\Omega} \qquad c_2(\phi) = \dfrac{\int_\Omega z(1 - H_\epsilon(\phi)) d\Omega}{\int_\Omega (1 - H_\epsilon(\phi)) d\Omega} \\ \dfrac{\partial \phi}{\partial t} = \delta_\epsilon(\phi) \left[\mu \nabla \cdot \left(\dfrac{\nabla \phi}{|\nabla \phi|} \right) - \nu - \eta (z - c_1)^2 + \gamma (z - c_2)^2 \right] \quad \text{in } \Omega, \\ \phi(0, x, y) = \phi_0(x, y) \qquad \text{in } \Omega, \\ \dfrac{\partial \phi}{\partial n} = 0 \qquad \text{on } \partial\Omega. \end{cases}$$

(13)

For given initial ϕ, the constant average intensities $c_1(\phi)$ and $c_2(\phi)$ are computed first. And then ϕ is computed by solving the nonlinear PDE given in Eq. (13). Steps of the semi-implicit method for solution of this equation are given here. Suppose that the size of given input image z is $m_1 \times m_2$. Finite difference scheme is used for discretization. Let $x, y \in \Omega$ be the spatial variables, h_1, h_2 be the horizontal and vertical space step size, and Δt be the time step. Divide the image domain into $m_1 \times m_2$ grid points, and let $(x_i, y_j) = (ih_1, jh_2)$, for $i = 1, 2, \ldots, m_1$ and $j = 1, 2, \ldots, m_2$. Also let $\phi_{i,j}^k = \phi(k\Delta t, x_i, y_j)$ be an approximation of $\phi(t, x, y)$ in the kth iteration, where $k \geq 0$ and $\phi^0 = \phi_0$ be the initial value.

Discretize the parabolic PDE given in Eq. (13) by using finite differences to get the following nonlinear difference equation to be used for updating $\phi^{(k)}$:

$$\frac{\phi_{ij}^{k+1} - \phi_{ij}^k}{\Delta t} = \delta_\epsilon(\phi_{ij}^k) \left[\frac{\mu}{h_1^2} \Delta_-^x \left(\frac{\Delta_+^x \phi_{ij}^{k+1}}{\sqrt{(\Delta_+^x \phi_{ij}^k / h_1)^2 + ((\phi_{i,j+1}^k - \phi_{i,j-1}^k)/2h_2)^2 + \beta_1}} \right) \right.$$

$$+ \frac{\mu}{h_2^2} \Delta_-^y \left(\frac{\Delta_+^y \phi_{ij}^{k+1}}{\sqrt{((\phi_{i+1,j}^k - \phi_{i-1,j}^k)/2h_1)^2 + (\Delta_+^y \phi_{ij}^k / h_2)^2 + \beta_1}} \right)$$

$$\left. - \nu - \eta (z_{ij} - c_1(\phi^k))^2 + \gamma (z_{ij} - c_2(\phi^k))^2 \right].$$

Here $\beta_1 > 0$ is a parameter which avoid singularity. Let $h_1 = h_2 = h = 1$ for simplicity but this is not fixed; different values may be used. Linearizing the above difference equation and denoting the coefficients of $\phi_{i+1,j}^{k+1}, \phi_{i-1,j}^{k+1}, \phi_{i,j+1}^{k+1}, \phi_{i,j-1}^{k+1}$ by A_1, A_2, A_3, A_4, respectively, lead to the following system of linear equations:

$$\phi_{ij}^{k+1} \left[1 + \mu \delta_\epsilon \left(\phi_{ij}^k \right) (A_1 + A_2 + A_3 + A_4) \right]$$
$$= \phi_{ij}^k + \Delta t \delta_\epsilon \left(\phi_{ij}^k \right) \left[\mu \left(A_1 \phi_{i+1,j}^{k+1} + A_2 \phi_{i-1,j}^{k+1} + A_3 \phi_{i,j+1}^{k+1} \right. \right. \quad (14)$$
$$\left. \left. + A_4 \phi_{i,j-1}^{k+1} \right) - \nu - \eta \left(z_{ij} - c_1 \left(\phi^k \right) \right)^2 + \gamma \left(z_{ij} - c_2 \left(\phi^k \right) \right)^2 \right].$$

If the coefficients A_1, A_2, A_3, A_4 are frozen on the previous iteration, then the above system of nonlinear equations will become a linear system of equation:

$$A\phi^{(k+1)} = f^{(k)},$$

where A is a block tri-diagonal matrix, which can be solved by using any iterative method.

Re-initialization of the level set function is done to prevent the level set function from becoming too flat. This may be done by solving the following initial value problem; see for reference Sussman et al. (1994):

$$\begin{cases} \dfrac{\partial \xi}{\partial t} = \text{sgn}(\phi(t))(1 - |\nabla \xi|) \\ \xi(0, t) = \phi(t), \end{cases} \quad (15)$$

where ϕ is obtained from solution of Eq. (14) (Chan and Vese 2001).

12 Fast Numerical Methods for Image Segmentation Models

Algorithm 1 Chan-Vese (CV) algorithm for two-phase image segmentation

$$\phi^{k+1} \to CV(\phi^k, \mu, tol)$$

1. For given ϕ_0, calculate average intensities c_1 and c_2 using first two formulas in Eq. *(13)*.
2. Keep c_1 and c_2 fixed, and find numerical solution of the PDE in Eq. *(13)*, to have ϕ^{k+1}.
3. Compute c_1 and c_2 using ϕ^{k+1}.
4. If $|\phi^{k+1} - \phi^k| < tol$ stop else.
5. Re-initialize ϕ, by solving Equation *(15)*, and do step 2.

Note that the semi-implicit method for the solution of parabolic partial differential equations is unconditionally stable (Weickert and Kühne 2002) so will be convergent for large time steps in lower-dimensional problems. As the dimension of the problem increases, the bandwidth of the system matrix becomes much larger and results in a big condition number if the time step is taken larger, whereas the small time step, in that case, would require a large number of iterations, which lead toward slow convergence. This drawback of semi-implicit method was also observed in experimental results; see for details Badshah and Chen (2008) and Weickert et al. (1997).

Additive Operator Splitting (AOS) Method

Operator splitting methods for the solution of PDE have attained much attention from researchers in recent times. Some of these operator splitting methods are additive operator splitting (AOS) (Weickert et al. 1997; Lu et al. 1992), multiplicative operator splitting (MOS) (Barash et al. 2003), and additive+multiplicative operator splitting (AMOS) (Geiser and Bartecki 2017). Only AOS method is discussed here in detail. Weickert et al. (1997) solved the nonlinear diffusion problem by using an additive operator splitting (AOS) method. In this method, a m-dimensional differential operator is converted into m one-dimensional differential operators, and each one-dimensional problem is solved by using the semi-implicit method. The solution in m dimension is the simple algebraic mean of m one-dimensional solutions. Jeon et al. (2005) used AOS method solution of parabolic PDE obtained in minimization of Chan-Vese model for image segmentation. Let us consider the PDE (13):

$$\frac{\partial \phi}{\partial t} = \delta_\epsilon(\phi) \left[\mu \nabla \cdot \left(\frac{\nabla \phi}{|\nabla \phi|} \right) - \nu - \eta(z - c_1)^2 + \gamma(z - c_2)^2 \right]. \tag{16}$$

Corresponding one-dimensional PDE is considered to be discretized. Let k and i represent time and spatial indices, respectively, and $h = 1$ is the spatial step size. Let $\phi_i^k = \phi(i, k)$, and then at the ith grid, the one-dimensional semi-implicit discretization of Eq. (16) is given by the following:

$$\frac{\phi_i^{k+1} - \phi_i^k}{\Delta t} = \delta_\epsilon(\phi_i^k) \left(\frac{\phi_{i+1}^{k+1} - \phi_i^{k+1}}{|\Delta_+^x \phi_i^k|} - \frac{\phi_i^{k+1} - \phi_{i-1}^{k+1}}{|\Delta_+^x \phi_{i-1}^k|} + F_i \right), \quad (17)$$

where $F_i = [-\nu - \eta(z_i - c_1)^2 + \gamma(z_i - c_2)^2]$. Let:

$$A_1 = \frac{1}{|\Delta_+^x \phi_i^k|} \quad \text{and} \quad A_2 = \frac{1}{|\Delta_+^x \phi_{i-1}^k|},$$

so Equation (17) becomes the following:

$$\phi_i^{k+1} = \phi_i^k + \Delta t \delta_\epsilon(\phi_i^k)(A_1 \phi_{i+1}^{k+1} - (A_1 + A_2)\phi_i^{k+1} + A_2 \phi_{i-1}^{k+1} + F_i). \quad (18)$$

Thus with AOS method, solve problems in x- and y-directions with double time step to get two separate solutions say ϕ_1 and ϕ_2, and then find the average as follows:

$$\phi = \frac{1}{2}(\phi_1 + \phi_2).$$

Although no stability constraint on the time step is present when the AOS scheme is utilized, the size of the time step cannot be very large because splitting-related artifacts associated with loss of rotational invariance will emerge. The practical implication of this is that the number of iterations needed for the contour to converge remains quite large. For images of large sizes, the methods discussed in this chapter are very slow in convergence. To avoid this problem, multigrid method is the best option.

Multigrid Method

A multigrid method for the Chan-Vese model proposed by Badshah and Chen (2008) is presented here. The proposed method is based on the global formulation of the Chan-Vese model proposed by Chan et al. (2006). Consider Euler-Lagrange's equation deduced from the minimization of Chan-Vese energy functional given in (11):

$$\delta_\epsilon(\phi) \left[\mu \operatorname{div}\left(\frac{\nabla \phi}{|\nabla \phi|} \right) - \eta(z(x,y) - c_1)^2 + \gamma(z(x,y) - c_2)^2 \right] = 0,$$

$\delta_\epsilon(\phi)$ has non-compact support, so the above equation may be written as follows:

$$\mu \operatorname{div}\left(\frac{\nabla \phi}{|\nabla \phi|} \right) - \eta(z(x,y) - c_1)^2 + \gamma(z(x,y) - c_2)^2 = 0. \quad (19)$$

Equation (19) is Euler-Lagrange's equation of the following functional:

$$\mu \int_\Omega |\nabla \phi| d\Omega + \int_\Omega (\eta(z(x,y) - c_1)^2 - \gamma(z(x,y) - c_2)^2)\phi(x,y) d\Omega. \quad (20)$$

This is the convex formulation of the Chan-Vese model (Chan and Vese 2001) proposed by Chan et al. in (2006). But the functional given in Equation (20) is

12 Fast Numerical Methods for Image Segmentation Models

homogenous in ϕ of degree 1 Chan et al. (2006). This means that this functional has no stationary point, so it needs to impose some extra constraints on ϕ that is $0 \le |\phi| \le 1$.

Use finite difference scheme to discretize Equation (19) for ϕ. The corresponding discrete equation at a grid point (i, j) is given by the following:

$$\left[\mu \left\{ \frac{\Delta_-^x}{h_1} \left(\frac{\Delta_+^x \phi_{i,j}/h_1}{\sqrt{(\Delta_+^x \phi_{i,j}/h_1)^2 + (\Delta_+^y \phi_{i,j}/h_2)^2 + \beta_1}} \right) \right. \right.$$
$$\left. + \frac{\Delta_-^y}{h_2} \left(\frac{\Delta_+^y \phi_{i,j}/h_2}{\sqrt{(\Delta_+^x \phi_{i,j}/h_1)^2 + (\Delta_+^y \phi_{i,j}/h_2)^2 + \beta_1}} \right) \right\} \tag{21}$$
$$\left. - \eta(z_{i,j} - c_1)^2 + \gamma(z_{i,j} - c_2)^2 \right] = 0,$$

where $\beta_1 > 0$ is a small parameter to avoid zero denominator. Equation (21) may be written in the following way:

$$\left[\mu \left\{ \Delta_-^x \left(\frac{\Delta_+^x \phi_{i,j}}{\sqrt{(\Delta_+^x \phi_{i,j})^2 + (\lambda \Delta_+^y \phi_{i,j})^2 + \bar{\beta}}} \right) \right. \right. \tag{22}$$
$$\left. + \lambda^2 \Delta_-^y \left(\frac{\Delta_+^y \phi_{i,j}}{\sqrt{(\Delta_+^x \phi_{i,j})^2 + (\lambda \Delta_+^y \phi_{i,j})^2 + \bar{\beta}}} \right) \right\}$$
$$\left. - \eta(z_{i,j} - c_1)^2 + \gamma(z_{i,j} - c_2)^2 \right] = 0,$$

$$\Longrightarrow \mu \left\{ \Delta_-^x \left(\frac{\Delta_+^x \phi_{i,j}}{\sqrt{(\Delta_+^x \phi_{i,j})^2 + (\lambda \Delta_+^y \phi_{i,j})^2 + \bar{\beta}}} \right) \right.$$
$$\left. + \lambda^2 \Delta_-^y \left(\frac{\Delta_+^y \phi_{i,j}}{\sqrt{(\Delta_+^x \phi_{i,j})^2 + (\lambda \Delta_+^y \phi_{i,j})^2 + \bar{\beta}}} \right) \right\}$$
$$= \eta(z_{i,j} - c_1)^2 - \gamma(z_{i,j} - c_2)^2, \tag{23}$$

where $\underline{\mu} = \mu/h_1$, $\bar{\beta} = h_1^2 \beta_1$, and $\lambda = h_1/h_2$, with Neumann's boundary conditions:

$$\phi_{i,0} = \phi_{i,1}, \quad \phi_{i,m_2+1} = \phi_{i,m_2}, \quad \phi_{0,j} = \phi_{1,j}, \quad \phi_{m_1+1,j} = \phi_{m_1,j}, \quad (24)$$
$$\text{for } i = 1, \ldots, m_1, \quad j = 1, \ldots, m_2 \text{ and } 0 \leq |\phi_{i,j}| \leq 1.$$

Note that the left side of Eq. (23) resembles with the denoising model by Rudin et al. (1992) using the total variation (TV) regularization. The parameter $\beta > 0$ should be a small quantity to avoid the singularities.

The Full Approximation Scheme

Multigrid scheme usually known as full approximation scheme (FAS) constitutes three main steps, namely, smoothers, interpolation, and coarse grid solvers; for details see Brandt (1977) and Briggs (1999). Denote the system of nonlinear equations given in Equation (23) and (24) by the following:

$$N^h(\phi^h) = f^h \qquad (25)$$

where $h_1 = h_2 = h$, ϕ^h and f^h are grid functions on a $m_1 \times m_2$ cell-centered rectangular grid Ω^h with spacing (h_1, h_2). Let Ω^{2h} denote the $m_1/2 \times m_2/2$ cell-centered grid which results from standard coarsening of Ω^h. Let $e^h = \phi^h - \Phi^h$ be the solution's error, where Φ^h is a good approximation to solution of (25) in the sense that e^h is smooth. Such smoothness can only be achieved by a careful choice of suitable smoothers – a major task in developing a working multigrid method.

Let $r^h = f^h - N^h(\Phi^h)$ be the residual. Then the nonlinear residual equation will be as follows:

$$N^h(\Phi^h + e^h) - N^h(\Phi^h) = r^h. \qquad (26)$$

Smooth components of error e^h may not be visible on fine gird Ω^h and hence cannot be well approximated. But that can be well approximated on coarse grid Ω^{2h}. Therefore any iterative method which smooths the error on the fine grid can be further well approximated by the use of the coarse grid correction. Note that on coarse grid the residual equation is solved which is less expansive as there will be half the number of grid points. Once a coarse grid approximation of the error is obtained, then it will be transferred back to the fine grid to correct the approximation Φ^h. This is known as a two-grid cycle, and the recursive use of two-grid cycle is termed as a multigrid method. Restriction and interpolation operators for transferring grid functions between Ω^h and Ω^{2h} for cell-centered discretization are defined here:

Restriction

$$I_h^{2h}\Psi^h = \Psi^{2h}$$

where:

$$\Psi^{2h}_{\ell,m} = \frac{1}{4}\left(\Psi^h_{2\ell-1,2m-1} + \Psi^h_{2\ell-1,2m} + \Psi^h_{2\ell,2m-1} + \Psi^h_{2\ell,2m}\right),$$

$$1 \leq \ell \leq m_1/2, \quad 1 \leq m \leq m_2/2.$$

is a full weighting operator (Chen 2005; Trottenberg and Schuller 2001).

Interpolation

$$I_{2h}^h \Psi^{2h} = \Psi^h$$

where:

$$\Psi^h_{2\ell,2m} = \frac{1}{16}\left(9\Psi^{2h}_{\ell,m} + 3\Psi^{2h}_{\ell+1,m} + 3\Psi^{2h}_{\ell,m+1} + \Psi^{2h}_{\ell+1,m+1}\right),$$

$$\Psi^h_{2\ell-1,2m} = \frac{1}{16}\left(9\Psi^{2h}_{\ell,m} + 3\Psi^{2h}_{\ell-1,m} + 3\Psi^{2h}_{\ell,m+1} + \Psi^{2h}_{\ell-1,m+1}\right),$$

$$\Psi^h_{2\ell,2m-1} = \frac{1}{16}\left(9\Psi^{2h}_{\ell,m} + 3\Psi^{2h}_{\ell+1,m} + 3\Psi^{2h}_{\ell,m-1} + \Psi^{2h}_{\ell+1,m-1}\right),$$

$$\Psi^h_{2\ell-1,2m-1} = \frac{1}{16}\left(9\Psi^{2h}_{\ell,m} + 3\Psi^{2h}_{\ell-1,m} + 3\Psi^{2h}_{\ell,m-1} + \Psi^{2h}_{\ell-1,m-1}\right),$$

$$\text{for} \quad 1 \leq \ell \leq m_1/2, \quad 1 \leq m \leq m_2/2.$$

is known as a bilinear interpolation operator.

It remains to discuss the most important ingredient of a MG: smoothing. Two types of smoothers, namely, local and global smoothers, are discussed here in detail.

Smoother I: Local Smoother

This smoother is proposed in Badshah and Chen (2008). In this method the system of nonlinear equations is linearized locally, by computing differential coefficients $D(\phi)$ on each grid (i, j) locally to get a system of linear equations. Note that the Gauss-Seidel has the best smoothing property, so apply the Gauss-Seidel method to derive system of linear equations to smooth the error. Using a few steps of this smoother to smooth the error will ensure a convergent nonlinear multigrid. Equation (23) can be written as follows:

$$\mu\left\{\left[\frac{\Delta^x_+\phi_{i,j}}{\sqrt{(\Delta^x_+\phi_{i,j})^2+(\lambda\Delta^y_+\phi_{i,j})^2+\bar{\beta}}}-\frac{\Delta^x_+\phi_{i-1,j}}{\sqrt{(\Delta^x_+\phi_{i-1,j})^2+(\lambda\Delta^y_+\phi_{i-1,j})^2+\bar{\beta}}}\right]\right.$$

$$\left.+\lambda^2\left[\frac{\Delta^y_+\phi_{i,j}}{\sqrt{(\Delta^x_+\phi_{i,j})^2+(\lambda\Delta^y_+\phi_{i,j})^2+\bar{\beta}}}-\frac{\Delta^y_+\phi_{i,j-1}}{\sqrt{(\Delta^x_+\phi_{i,j-1})^2+(\lambda\Delta^y_+\phi_{i,j-1})^2+\bar{\beta}}}\right]\right\}$$

$$=\eta(z_{i,j}-c_1)^2-\gamma(z_{i,j}-c_2)^2.$$

Denoting the differential coefficients in the above equation (intended below to be frozen in local linearization) by $D(\phi)_{i,j}$, $D(\phi)_{i-1,j}$, $D(\phi)_{i,j-1}$ gives the following linear equation:

$$\mu\left\{\left[D(\phi)_{i,j}(\phi_{i+1,j}-\phi_{i,j})-D(\phi)_{i-1,j}(\phi_{i,j}-\phi_{i-1,j})\right]\right.$$

$$\left.+\lambda^2\left[D(\phi)_{i,j}(\phi_{i,j+1}-\phi_{i,j})-D(\phi)_{i,j-1}(\phi_{i,j}-\phi_{i,j-1})\right]\right\} \quad (27)$$

$$=\eta(z_{i,j}-c_1)^2-\gamma(z_{i,j}-c_2)^2=f_{i,j}.$$

Note that all differential coefficients $D(\phi)_{i,j}$, $D(\phi)_{i-1,j}$, and $D(\phi)_{i,j-1}$ contain $\phi_{i,j}$, which will be evaluated at previous time step, and the same values will be used in the rest of the process. Let $\widetilde{\varphi}$ be an approximation to ϕ. By putting the values of $\widetilde{\varphi}$ at each grid point in Eq. (27) other than the grid point (i, j) and also computing D at each grid point (i, j), a linear equation in $\phi_{i,j}$ will be obtained:

$$\left\{\left[D(\widetilde{\varphi})_{i,j}(\widetilde{\varphi}_{i+1,j}-\phi_{i,j})-D(\widetilde{\varphi})_{i-1,j}(\phi_{i,j}-\widetilde{\varphi}_{i-1,j})\right]\right.$$

$$\left.+\lambda^2\left[D(\widetilde{\varphi})_{i,j}(\widetilde{\varphi}_{i,j+1}-\phi_{i,j})-D(\widetilde{\varphi})_{i,j-1}(\phi_{i,j}-\widetilde{\varphi}_{i,j-1})\right]\right\}\equiv f_{i,j}/\underline{\mu}\equiv \bar{f}_{i,j}.$$

Algorithm for solving this equation for $\phi_{i,j}$ to update the approximation at each pixel (i, j):

Smoother II: Global Smoother
This smoother is proposed in Savage and Chen (2005) for image denoising model and extended to segmentation model in Badshah and Chen (2008). In this method the system of nonlinear equations is linearized globally at each step by computing differential coefficients $D(\phi)$ on each grid point (i, j). To the resulting system of linear equations, Gauss-Seidel relaxation is applied. Note that the global smoother is different from the local smoother defined above. The algorithm is given as follows:

Algorithm 2 Algorithm for smoother I

$$\phi^h \longleftarrow Smoother1(\phi^h, \bar{f}^h, ITER, tol)$$

where $ITER$ is the maximum number of inner iterations.
for $i = 1 : m_1$
    for $j = 1 : m_2$
        for iter=1:ITER
        $\widetilde{\varphi}^h \leftarrow \Phi^h$

$$\phi_{i,j} = \frac{\left[\left\{D(\widetilde{\varphi}^h)_{i,j}\widetilde{\varphi}^h_{i+1,j} + D(\widetilde{\varphi}^h)_{i-1,j}\widetilde{\varphi}^h_{i-1,j} + \lambda^2 D(\widetilde{\varphi}^h)_{i,j}\widetilde{\varphi}^h_{i,j+1} + \lambda^2 D(\widetilde{\varphi}^h)_{i,j-1}\widetilde{\varphi}^h_{i,j-1}\right\} - \bar{f}_{i,j}\right]}{D(\widetilde{\varphi}^h)_{i,j} + D(\widetilde{\varphi}^h)_{i-1,j} + \lambda^2(D(\widetilde{\varphi}^h)_{i,j} + D(\widetilde{\varphi}^h)_{i,j-1})}$$

        if $|\phi_{i,j} - \widetilde{\varphi}^h_{i,j}| < tol$ then stop
        end
    end
end

Algorithm 3 Algorithm for smoother II

$$\phi^h \longleftarrow Smoother2(\phi^h, \bar{f}^h, ITER, tol)$$

for $i = 1 : m_1$
    for $j = 1 : m_2$

$$D(\phi^h)_{i,j} = \sqrt{[(\Delta^x_+\phi_{i,j})^2 + (\lambda \Delta^y_+\phi_{i,j})^2 + \bar{\beta}]}$$

    end
end
$\varphi^h = \phi^h$
for $iter = 1 : maxit$
    for $i = 1 : n$
        for $j = 1 : m$
        $\widetilde{\varphi}^h \leftarrow \varphi^h$

$$\varphi_{i,j} = \frac{\left[\left\{D(\phi^h)_{i,j}\widetilde{\varphi}^h_{i+1,j} + D(\phi^h)_{i-1,j}\widetilde{\varphi}^h_{i-1,j} + \lambda^2 D(\phi^h)_{i,j}\widetilde{\varphi}^h_{i,j+1} + \lambda^2 D(\phi^h)_{i,j-1}\widetilde{\varphi}^h_{i,j-1}\right\} - \bar{f}_{i,j}\right]}{D(\phi^h)_{i,j} + D(\phi^h)_{i-1,j} + \lambda^2(D(\phi^h)_{i,j} + D(\phi^h)_{i,j-1})}$$

        end
    end
end
$\phi^h \leftarrow \varphi$

Here updating of the coefficients needs to be done at the beginning of each smoothing step globally and to be stored for relaxation use. This was found to be necessary for the total variation denoising model of Rudin et al. (1992).

The Multigrid Algorithm

The algorithm for solving equation given in Eq. (25) by using the multigrid method is given here. For further details see Chen (2005), Trottenberg and Schuller (2001) and references therein:

Algorithm 4 Multigrid algorithm

Set up the following multigrid parameters:
it_1 Number of steps required for pre-smoothing on each level
it_2 Number of steps required for post-smoothing on each level
$\gamma = 1$ or   2   Selection of V-cycle or W-cycle
rr: Relative residual
For given Φ^h compute $\bar{f}^h$ and keep it fixed. One-step V-cycle of nonlinear multigrid method for CV model is presented here.
FAS
Start

$$\phi^h \longleftarrow FASCYC(\phi^h, \bar{f}^h, ITER, it_1, it_2, \gamma, tol)$$

$\Phi_0 = \Phi^h$

1. On the coarsest grid, solve Eq. (25) by using SI or AOS methods (Weickert et al. 1997) and then stop.
   On finer grids do smoothing, i.e.:

   $$\phi^h \longleftarrow Smoother^{it_1}(\phi^h, \bar{f}^h, ITER, it_1, it_2, \gamma). \qquad \text{(Pre-Smoothing)}$$

2. **Restriction:**

   $$\phi^{2h} = I_h^{2h}\phi^h, \quad \bar{\phi}^{2h} = \phi^{2h}.$$

   $$\bar{f}^{2h} = I_h^{2h}(\bar{f}^h - H^h\phi^h) + N^{2h}(\phi^{2h})$$

   $$\phi^{2h} \longleftarrow FASCYC_\gamma^{2h}(\phi^h, \bar{f}^h, ITER, it_1, it_2, \gamma)$$

3. **Interpolation:**

   $$\phi^h \longleftarrow \phi^h + I_{2h}^h(\phi^{2h} - \bar{\phi}^{2h})$$

4.
   $$\phi^h \longleftarrow Smoother^{it_2}(\phi^h, \bar{f}^h, ITER, it_1, it_2, \gamma). \qquad \text{(Post-Smoothing)}$$

Update $\bar{f}^h$.
If $rr = \dfrac{\|\phi^h - \phi_0\|_2}{\|\phi_0\|_2} < tol$, stop.
Else go to **Start**.

12 Fast Numerical Methods for Image Segmentation Models

Local Fourier Analysis of Smoothers

The standard FAS multilevel algorithm (such as Algorithm 4) does not automatically converge for many problems if simple smoothers are used. The key for convergence lies in effective smoothers or reduction of residuals to a smoothed form (Chen 2005; Trottenberg and Schuller 2001). Here local Fourier analysis (LFA) is done to check the effectiveness of the smoothers (say smoother I and smoother II).

Note that LFA cannot be applied to nonlinear smoothers in general. However, for linearized smoothers, the analysis can only be done for each individual smoothing iteration, and the obtained smoothing rates change from iteration to iteration. However, the general trends, e.g., if the three consecutive smoothing rates are 0.58, 0.60, and 0.45 (instead of a constant rate say 0.4), the underlying smoother is effective. Likewise, if the consecutive rates are such that 1.4, 0.99, and 1.09, then the smoother may not be that much effective.

Let us assume that the image domain is a square say $m = m_1 = m_2$. Denote $h = h_1 = h_2$. The typical grid equation on Ω^h is as follows:

$$D(\phi_{i,j})(\phi_{i+1,j} - \phi_{i,j}) - D(\phi_{i-1,j})(\phi_{i,j} - \phi_{i-1,j})$$
$$+ \lambda^2 [D(\phi_{i,j})(\phi_{i,j+1} - \phi_{i,j}) - D(\phi_{i,j-1})(\phi_{i,j} - \phi_{i,j-1})] = \bar{f}_{i,j}.$$

For the local smoother, introduce the following notations $g_1 = D(\widetilde{\phi})_{i-1,j} = D(\phi^{(k)})_{i-1,j}$, $g_2 = D(\widetilde{\phi})_{i,j} = D(\phi^{(k)})_{i,j}$, and $g_3 = D(\widetilde{\phi})_{i,j-1} = D(\phi^{(k)})_{i,j-1}$, and similarly for the global smoother, g_1, g_2, g_3 will be computed globally as follows: $g_1 = D(\widetilde{\Phi})_{i-1,j}$, $g_2 = D(\widetilde{\Phi})_{i,j}$, and $g_3 = D(\widetilde{\Phi})_{i,j-1}$ where $\widetilde{\Phi}$ is the iterate at the previous sweep (global fixed point). Also as $h_1 = h_2$, so $\lambda^2 = 1$. So:

$$-(g_1 + 2g_2 + g_3)\phi_{i,j}^{k+1} + g_1 \phi_{i-1,j}^{k+1} + g_3 \phi_{i,j-1}^{k+1} + g_2(\phi_{i,j+1}^{k} + \phi_{i+1,j}^{k}) = \bar{f}_{i,j}.$$

The corresponding error equation will be as follows:

$$-(g_1 + 2g_2 + g_3)e_{i,j}^{k+1} + g_1 e_{i-1,j}^{k+1} + g_3 e_{i,j-1}^{k+1} + g_2(e_{i,j+1}^{k} + e_{i+1,j}^{k}) = 0, \quad (28)$$

where $e_{i,j}^{k+1} = \phi_{i,j} - \phi_{i,j}^{k+1}$ and $e_{i,j}^{k} = \phi_{i,j} - \phi_{i,j}^{k}$ are the local error functions after and before the pre(post) smoothing step, respectively.

Recall that the local Fourier analysis (LFA) measures the largest amplification factor in a relaxation scheme (Brandt 1977; Chen 2005; Trottenberg and Schuller 2001). Let the general Fourier component be as follows:

$$B_{\theta_1, \theta_2}(x_i, y_j) = \exp\left(i\alpha_1 \frac{x_i}{h} + i\alpha_2 \frac{y_j}{h}\right) = \exp\left(\frac{2i\theta_1 i \pi}{m} + \frac{2i\theta_2 j \pi}{m}\right), \quad \mathbf{i} = \sqrt{-1}.$$

Here $\alpha_1 = \dfrac{2\theta_1 \pi}{m}, \alpha_2 = \dfrac{2\theta_2 \pi}{m} \in [-\pi, \pi]$. The LFA involves expanding the following:

$$e^{k+1} = \sum_{\theta_1,\theta_2=-m/2}^{m/2} \psi_{\theta_1,\theta_2}^{k+1} B_{\theta_1,\theta_2}(x_i, y_j), \quad e^k = \sum_{\theta_1,\theta_2=-m/2}^{m/2} \psi_{\theta_1,\theta_2}^k B_{\theta_1,\theta_2}(x_i, y_j)$$

in Fourier components. Now estimate the maximum ratio:

$$\bar{\mu} = \max_{\theta_1,\theta_2} \mu(\theta_1, \theta_2) = |\psi_{\theta_1,\theta_2}^{k+1} / \psi_{\theta_1,\theta_2}^k|$$

in the high-frequency range $(\alpha_1, \alpha_2) \in [-\pi, \pi] \setminus \left[-\frac{\pi}{2}, \frac{\pi}{2}\right]$ which defines the smoothing rate (Trottenberg and Schuller 2001). Now replace all grid functions by their Fourier series and essentially consider the so-called amplification factor, i.e., the ratio between ψ_θ^{k+1} and ψ_θ^k for each θ where $\theta = (\theta_1, \theta_2)$. Then for the Fourier component of the error functions $e_{i,j}^k$ and $e_{i,j}^{k+1}$ before and after relaxation sweep, consider the following:

$$e_{i,j}^k = \psi_\theta^k e^{\mathbf{i}(2\pi\theta_1 i + 2\pi\theta_2 j)/m} \quad \text{and} \quad e_{i,j}^{k+1} = \psi_\theta^{k+1} e^{\mathbf{i}(2\pi\theta_1 i + 2\pi\theta_2 j)/m}, \tag{29}$$

putting these values in Equation (28) and defining the following:

$$\mu(\theta) = \left|\frac{\psi_\theta^{k+1}}{\psi_\theta^k}\right|$$

and introducing $|\theta| = \max(|\theta_1|, |\theta_2|)$; the smoothing factor $\bar{\mu}$ is then obtained as follows:

$$\bar{\mu} = \max_{\hat{\rho}\pi \leq |\theta| \leq \pi} \mu(\theta),$$

where $\hat{\rho}$ is the mesh size ratio and the range $\hat{\rho}\pi \leq |\theta| \leq \pi$ is the suitable range of high-frequency components, i.e., the range of components that cannot be approximated on the coarser grid. For standard coarsening $\hat{\rho} = \frac{1}{2}$, Brandt (1977). The smoothing factor $\bar{\mu}$ is computed for both smoothers. To proceed with an analysis, compute g_1, g_2 and g_3 or the following function:

$$D(\phi) = \sqrt{(\Delta_+^x \phi)^2 + (\Delta_+^y \phi)^2 + \bar{\beta}},$$

Numerically, and work out the smoothing factor $\bar{\mu}$ for each set of coefficients g_1, g_2, and g_3 within a smoother. Use the complete set of coefficients g_1, g_2 and g_3 for computing the smoothing factors $\bar{\mu}$, and display the maximum of such factors:

$$\hat{\mu} = \max_{g_1,g_2,g_3} \bar{\mu} = \max_{g_1,g_2,g_3} \max_\theta \mu(\theta).$$

12 Fast Numerical Methods for Image Segmentation Models

Table 1 $\hat{\mu}$ in the first 4 cycles of out MG algorithm

MG cycle	Smoothing steps	Rate I:$\hat{\mu}^I$	Rate II:$\hat{\mu}^{II}$
1	Pre-smoothing-1	0.4942	0.6776
	Pre-smoothing-2	0.4941	0.9317
	Post-smoothing-1	0.4942	0.9135
	Post-smoothing-2	0.4942	0.9427
2	Pre-smoothing-1	0.6003	0.9561
	Pre-smoothing-2	0.6003	0.9174
	Post-smoothing-1	0.6003	0.9581
	Post-smoothing-2	0.6003	0.9577
3	Pre-smoothing-1	0.7760	0.9533
	Pre-smoothing-2	0.7760	0.9193
	Post-smoothing-1	0.7757	0.9092
	Post-smoothing-2	0.7749	0.9040
4	Pre-smoothing-1	0.6025	0.9594
	Pre-smoothing-2	0.6026	0.9456
	Post-smoothing-1	0.6026	0.9286
	Post-smoothing-2	0.6026	0.9678

As such a linear analysis is based on freezing the nonlinear coefficients, the results should be viewed only as a guide to smoother's effectiveness and a way to distinguish smoothers.

Take a test example of size to 32×32, and display $\hat{\mu}$ in the first four cycles of the MG algorithm as in Table 1 where Pre-1 refers to the case of "pre-smoothing" and Post-1 to "post-smoothing," etc. By considering the average rate from all pixels, the averages are, respectively, 0.49 and 0.71 for smoothers I and II. Clearly in this example smoother I appears to be more effective than smoother II in terms of rates. For experimental results and comparison, the readers are referred to Badshah and Chen (2008).

In Table 2, the comparison of multigrid, SI, and AOS methods is given. The terms used in the heading of Table 2 have the following meanings:

Size: The size of given image $m \times n$.
Itr: Number of iterations used to get the required result.
CPU(s): Time in seconds required for CPU to perform these iterations.
SI: Semi-implicit method.
AOS: Additive operator splitting.
MG: Multigrid method.
ART: Artificial image and **REAL:** Real image like MRI.
**: Results with high CPU or out of memory.
S-I: Smoother I.
S-II: Smoother II.

Table 2 Comparison of MG with SI and AOS methods

Prob.	Size	AOS method Itr	CPU(s)	AOS multi-resolution Itr	CPU(s)	SI method Itr	CPU(s)	MG method (S-I) Itr	CPU(s)	MG method (S-II) Itr	CPU(s)
ART	128^2	60	4.8	60	4.8	80	16.5	2	8.5	2	8
	256^2	140	50	80	34	100	90.3	2	9.4	3	13.4
	512^2	280	421	170	277	439	1.3×10^4	2	13	3	17
	1024^2	1200	7661	240	1630	**	**	2	27	3	32
	2048^2	**	**	**	**	**	**	2	90	3	100
REAL	128^2	100	10.5	100	10.5	130	32.2	3	12.8	4	15
	256^2	280	110.5	156	68	180	450	3	14	4	22.2
	512^2	800	1230	312	503	**	1×10^4	3	19.2	4	22.2
	1024^2	**	**	**	**	**	**	3	40.7	4	42
	2048^2	**	**	**	**	**	**	3	133	4	136.9

AOS multi-resolution: AOS method is implemented in coarse to fine-level manner, i.e., AOS method is used to solve the problem on the coarsest level and interpolate the solution to the fine level and use it as initial guess, to solve the problem on fine level using AOS method and so on until the finest level is reached.

From Table 2, it can be observed that the MG method is as fast as the SI method and AOS method for images of small sizes, but it is more efficient for images having large sizes, where the abovementioned methods are very slow or not working.

Multigrid Solver for Solving a Class of Variational Problems with Application to Image Segmentation

In section "Multigrid Method", a multigrid method is discussed in detail for a specific type of image segmentation model, namely, Chan-Vese two-phase model (Chan and Vese 2001). In Roberts et al. (2019), the author proposed a new multigrid method for the following unconstraint model:

$$\min_u \left\{ \mu \int_\Omega g(|\nabla z(\boldsymbol{x})|)|\nabla u| d\Omega + \lambda \int_\Omega \mathcal{F} u \, d\Omega + \theta \int_\Omega \mathcal{D} u \, d\Omega + \alpha \int_\Omega v_{\varepsilon_2}(u) d\Omega \right\} \tag{30}$$

where $\mathcal{F}$ is the data fitting term, $\mathcal{D}$ is the distance metric, and v_{ε_2} is the convex-relaxation penalty term which enforces the constraint that $0 \leq u \leq 1$; see Chan et al. (2006) for choice of v_{ε_2}. The corresponding Euler-Lagrange equation is obtained by minimizing the above functional and is given by the following:

$$\mu \nabla \cdot \left(g(|\nabla z(x)|) \frac{\nabla u}{|\nabla u|_{\varepsilon_1}} \right) - \lambda \mathcal{F} - \theta \alpha \mathcal{D} - v'_{\varepsilon_2}(u) = 0 \qquad (31)$$

with Neumann boundary conditions and where $\varepsilon_1, \varepsilon_2$ are small positive parameters. Multigrid methods discussed in sections "Multigrid Method" and "Multigrid Method for Multiphase Segmentation Model" may not be applied for solution of type of PDE given in (31), due to the following reasons:

1. In the PDE given in (31), the Euler-Lagrange equation arose from minimization of convex formulation of CV model, which has an extra constraint of $0 \leq u \leq 1$, which means that the solution of the PDE will be a binary function everywhere. And hence there will be significant jumps in the values of $v'_{\varepsilon_2}(u)$; this leads to instability of pixel-wise fixed point smoother, and hence the basic multigrid method fails.
2. Small value of ε_2 can lead to the divergence of the algorithm due to discontinuity of the function $v'_{\varepsilon_2}(u)$, whereas large value of ε_2 may guarantee the convergence of the algorithm but change the nature of the problem.
3. ε_1 is the parameter which avoids singularity in the PDE. Most of the multigrid method's convergence depends on the value of ε_1; small value can lead to the nonconvergence of the algorithm, and large value changes the nature of the problem.
4. In the discretization step, all functions will be approximated at the half pixels and due to nonsmoothness of the edge function, its approximation at the half pixel may be very inaccurate.
5. Divergence term in the PDE (31) is highly nonlinear. Approximation of this term around the interfaces in g and u may be inaccurate due to the use of singularity parameter ε_1 as discussed above.

To address these bullets and to apply multigrid methods, the authors in Roberts et al. (2019) introduced a new formulation of the models given in (30).

First Algorithm
Model in (30) is reformulated by removing the penalty term $v_{\varepsilon_2}(u)$ which is done by introducing a new variable v. The new reformulated model becomes the following:

$$\min_{u,v} \left\{ \mu \int_\Omega g(|\nabla z(x)|) |\nabla u| d\Omega + \lambda \int_\Omega \mathcal{F} v \, d\Omega \right.$$
$$\left. + \theta \int_\Omega \mathcal{D} v \, d\Omega + \alpha \int_\Omega v_{\varepsilon_2}(v) d\Omega + \frac{\theta_B}{2} \|u - v\|_{L^2}^2 \right\}, \qquad (32)$$

where θ_B is a tuning parameter. This model will be minimized with respect to u and v. To minimize with respect to u, the above model reduces to the following:

$$\min_{u}\left\{\mu\int_{\Omega}g(|\nabla z(\boldsymbol{x})|)|\nabla u|d\Omega+\frac{\theta_B}{2}\|u-v\|_{L^2}^2\right\}. \tag{33}$$

Minimization with respect to u leads to the following PDE:

$$\mu\nabla\cdot\left(g(|\nabla z|)\frac{\nabla u}{|\nabla u|_{\varepsilon_1}}\right)=0 \tag{34}$$

With Neumann boundary condition. In the minimization problem for v, the following minimization problem is considered:

$$\min_{v}\left\{\lambda\int_{\Omega}\mathcal{F}v\,d\Omega+\theta\int_{\Omega}\mathcal{D}v\,d\Omega+\alpha\int_{\Omega}v_{\varepsilon_2}(v)d\Omega+\frac{\theta_B}{2}\|u-v\|_{L^2}^2\right\}, \tag{35}$$

whose solution is as follows:

$$v^{(k+1)}=v=\min\left\{\max\left\{u-\frac{\lambda\mathcal{F}+\theta\mathcal{D}}{\theta_B},0\right\},1\right\}. \tag{36}$$

It can be noted that both PDEs do not contain v'_{ε_2}, which is the one of the achievement of the proposed algorithm. For detailed steps of the algorithm, see Roberts et al. (2019).

Furthermore, the authors introduced Split-Bregman iterations for removing nonlinearity in the weighted TV term. This is done by introducing a new variable **d** for the weighted TV, and hence the minimization problem given in (30) becomes the following:

$$\min_{u,\mathbf{d}}\left\{\mu\int_{\Omega}|\mathbf{d}|_g d\Omega+\lambda\int_{\Omega}\mathcal{F}u\,d\Omega+\theta\int_{\Omega}\mathcal{D}u\,d\Omega\right.$$
$$\left.+\alpha\int_{\Omega}v_{\varepsilon_2}(u)d\Omega+\frac{\lambda_B}{2}\|\mathbf{d}-\nabla u-b\|_{L^2}^2\right\}, \tag{37}$$

where $|\mathbf{d}|_g=g(|\nabla z|)|\nabla u|$ and $\lambda_B\geq 0$. Note that b is the Bregman update which has a simple update formula. To find optimal value of u, the following minimization problem will be solved:

$$\min_{u}\left\{\lambda\int_{\Omega}\mathcal{F}u\,d\Omega+\theta\int_{\Omega}\mathcal{D}u\,d\Omega+\alpha\int_{\Omega}v_{\varepsilon_2}(u)d\Omega+\frac{\lambda_B}{2}\|\mathbf{d}-\nabla u-b\|_{L^2}^2\right\}. \tag{38}$$

Minimization problem for **d** takes the following form:

12 Fast Numerical Methods for Image Segmentation Models

$$\min_{\mathbf{d}} \left\{ \mu \int_\Omega |\mathbf{d}|_g d\Omega + \frac{\lambda_B}{2} \|\mathbf{d} - \nabla u - b\|_{L^2}^2 \right\}, \tag{39}$$

whose closed form solution is given as follows:

$$\mathbf{d} = \text{shrink}\left(\nabla u + b \frac{\mu g(|\nabla z|)}{\lambda_B}\right), \tag{40}$$

where $\text{shrink}(a, b) = sgn(a) \max\{|a| - b, 0\}$. b given in Equation (37) can be updated as follows:

$$b^{(k+1)} = b^{(k)} + \nabla u^{(k+1)} - \mathbf{d}^{(k+1)}. \tag{41}$$

In the Bregman iterations, Equation (38) is remaining to be solved which is still nonlinear and is not amenable to fast multigrid method. To solve this problem, the authors reformulate the minimization problem (33) to the following:

$$\min_{u,d} \left\{ \mu \int_\Omega |\mathbf{d}|_g \, d\Omega + \frac{\theta_B}{2} \|u - v\|_{L^2}^2 + \frac{\lambda_B}{2} \|\mathbf{d} - \nabla u - b\|_{L^2}^2 \right\} \tag{42}$$

using Bregman splitting where θ_B and λ_B are fixed nonnegative parameters. The following subproblem is considered for u:

$$u^{(k+1)} = \arg\min_u \left\{ \frac{\theta_B}{2} \left\|u - v^{(k)}\right\|_{L^2}^2 + \frac{\lambda_B}{2} \left\|\mathbf{d}^{(k)} - \nabla u - b^{(k)}\right\|_{L^2}^2 \right\} \tag{43}$$

and the minimizer is the solution of the following:

$$-\lambda_B \Delta u + \theta_B u = \theta_B v^{(k)} - \lambda_B \nabla \cdot \left(\mathbf{d}^{(k)} - b^{(k)}\right) \tag{44}$$

with Neumann boundary conditions. This is a linear PDE which can be solved by a multigrid method. $\mathbf{d}$, b, and v will be updated as given in (40), (41), and (36), respectively. PDEs obtained from minimization of various subproblems are solved by using additive operator splitting method and multigrid methods; for detail see Sect. 5 in Roberts et al. (2019).

Sobolev Gradient Minimization of Curve Length in Chan-Vese Model

In Yuan and He (2012), the Sobolev gradient is used to minimize the length term in the Chan-Vese segmentation model. Denote the length term in Chan-Vese model by $Ł(\phi) = \int_\Omega \delta_\epsilon(\phi)|\nabla \phi| d\Omega$. The Sobolev gradient of the curve length functional $L(\phi)$ may be represented through L^2 gradient. As done earlier, the Gáteaux derivative of $Ł(\phi)$ in the direction of a test function $h \in C_0^\infty$ is given by the following:

$$Ł'(\phi)h = \lim_{\epsilon \to 0} \frac{Ł(\phi + \epsilon h) - Ł(\phi)}{\epsilon} = \left\langle \delta(\phi) \frac{\nabla \phi}{|\nabla \phi|}, \nabla h \right\rangle_{L^2(\Omega)^2} + \int_\Omega \delta'(\phi)|\nabla \phi| h d\Omega. \tag{45}$$

The inner product can be simplified by using integration by parts, which will happen if ϕ belongs to Sobolev space $H^{2,2}(\Omega)$. The Gâteaux derivative of length term $Ł'(\phi)h$ is defined to be the unique element that represents the bounded linear functional $Ł'(\phi)$ in $L^2(\Omega)$ as follows:

$$Ł'(\phi)h = \langle \nabla Ł(\phi), h \rangle_{L^2(\Omega)} \tag{46}$$

where $\nabla Ł(\phi)$ is the gradient of $Ł(\phi)$ in L^2 space. Integration by parts is applied on Eq. (45) to get the following:

$$\nabla Ł(\phi) = -\delta_\epsilon(\phi) \left[\mu \operatorname{div} \left(\frac{\nabla \phi}{|\nabla \phi|} \right) \right] \tag{47}$$

with Neumann boundary conditions.

To find the Sobolev gradient of $L(\phi)$, integration by parts will not be used to integrate the inner product term in Eq. (45). Define the following:

$$D\phi = \begin{pmatrix} \phi \\ \nabla \phi \end{pmatrix}$$

where $\phi \in H^{1,2}(\Omega)$. In $\phi \in H^{1,2}(\Omega)$, the inner product may be defined as follows:

$$\langle \phi, h \rangle_{H^{1,2}(\Omega)} = \int_\Omega \phi h + \langle \nabla \phi, \nabla h \rangle_{H^{1,2}(\Omega)^2} = \langle D\phi, Dh \rangle_{L^2(\Omega)^3}, \ h \in H^{1,2}(\Omega).$$

For any function $\phi, h \in H^{1,2}(\Omega)$, it is well known that the Gâteaux derivative $Ł'(\phi)$ which is given in Eq. (45) exists and is a bounded linear functional on $H^{1,2}(\Omega)$. By the Riesz theorem, the Gâteaux derivative $Ł'(\phi)h$ is defined to be the unique element $R(\phi)$ that represents the bounded linear functional $Ł'(\phi)$ on $H^{1,2}(\Omega)$ as follows:

$$Ł'(\phi)h = \langle R(\phi), h \rangle_{H^{1,2}(\Omega)}. \tag{48}$$

Here, $R(phi)$ is the Sobolev gradient which is denoted by $\nabla_s Ł(\phi)$. For $\phi \in H^{2,2}(\Omega)$, using integration by parts on Eq. (45), $Ł'(\phi)$ can be represented by L^2 gradient as follows:

$$Ł'(\phi)h = \langle R(\phi), h \rangle_{H^{1,2}(\Omega)} = \langle D(\nabla_s Ł(\phi)), Dh \rangle_{L^2(\Omega)^3} = \langle D^* D(\nabla_s Ł(\phi)), h \rangle_{L^2(\Omega)} \tag{49}$$

12 Fast Numerical Methods for Image Segmentation Models

where $D^* = (I, -\nabla)$ is the adjoint of D. The two gradients may be related in the following way:

$$D^*D(\nabla_s Ł(\phi)) = \nabla Ł(\phi) \text{ or } \nabla_s Ł(\phi) = (D^*D)^{-1} \nabla Ł(\phi) \quad (50)$$

it can be noted that

$$D^*D = (I, -\nabla)\begin{pmatrix} I \\ \nabla \end{pmatrix} = I - \Delta.$$

Combine all these results to get the Sobolev gradient of the length term, which is given as follows:

$$\nabla Ł(\phi) = -(I - \Delta)^{-1}\left(\delta_\epsilon(\phi)\left[\mu \text{div}\left(\frac{\nabla \phi}{|\nabla \phi|}\right)\right]\right). \quad (51)$$

The data fitting term of the Chan-Vese model:

$$E(\phi) = \eta \int_\Omega |z - c_1|^2 H(\phi) d\Omega + \gamma \int_\Omega |z - c_2|^2 (1 - H(\phi)) d\Omega$$

is minimized by using L^2 gradient $\nabla E(\phi)$. Thus the combined evolution equation is given by the following:

$$\begin{cases} \frac{\partial \phi}{\partial t} = (I - \Delta)^{-1} \delta_\epsilon(\phi)\left[\mu \nabla \cdot \left(\frac{\nabla \phi}{|\nabla \phi|}\right) - \nu - \eta(z - c_1)^2 + \gamma(z - c_2)^2\right] & \text{in } \Omega, \\ \phi(t, x, y) = \phi_0(x, y) & \text{in } \Omega \\ \frac{\partial \phi}{\partial n} = 0 & \text{on } \partial \Omega. \end{cases} \quad (52)$$

Numerical Method

The evolution equation given in Eq. (52) is solved in the following way: the Sobolev gradient term is computed by introducing an intermediate variable say Φ, i.e.:

$$\Phi = (I - \Delta)^{-1} \delta_\epsilon(\phi)\left[\mu \nabla \cdot \left(\frac{\nabla \phi}{|\nabla \phi|}\right)\right] \quad (53)$$

or:

$$(I - \Delta)\Phi = \delta_\epsilon(\phi)\left[\mu \nabla \cdot \left(\frac{\nabla \phi}{|\nabla \phi|}\right)\right]. \quad (54)$$

Table 3 Speed comparison of L^2 and Sobolev gradients

Prob.	Prob 1		Prob 2		Prob 3		Prob 4		Prob 5	
L^2 gradient	Itr	CPU(s)	Itr	CPU(s)	Itr	CPU(s)	Itr	CPU(s)	Itr	CPU(s)
	400	313	60	13	50	9	700	632	100	70
L^2+Sobolev grads	40	8	28	7	17	4	88	64	60	12

For given value of $\phi_{i,j}^{(k)}$, the above equation will be solved by using fast Poisson solver to get $\Phi(\phi_{i,j}^{(k)}, \phi_{i,j}^{(k+1)})$. To find numerical solution of evolution equation given in Eq. (52), the following procedure will be followed. Starting with the initial value of ϕ, compute c_1 and c_2. Then the numerical approximation of the Eq. (52) can be found by solving the following discrete equation:

$$\frac{\phi_{i,j}^{(k+1)} - \phi_{i,j}^{(k)}}{\Delta t} = \mu \Phi(\phi_{i,j}^{(k)}, \phi_{i,j}^{(k+1)}) + \delta(\phi_{i,j}^{(k)}) \left[-\lambda_1 (z_{i,j} - c_1)^2 + \lambda_2 (z_{i,j} - c_2)^2 \right]. \tag{55}$$

For more details and algorithm, please see Yuan and He (2012).

Speed comparison of both type gradients, i.e., L^2 and L^2 combined with Sobolev gradients in Table 3. Both methods are tested on five different type of problems, and their number of iterations and CPU time in seconds is recorded. It is seen from the table that L^2 combined with Sobolev gradients showed good results compared to L^2 gradient only.

Multiphase Image Segmentation

Multigrid Method for Multiphase Segmentation Model

In the previous section, Chan-Vese model was discussed which divides a gray image into two phases say foreground and background. Another model proposed by Vese and Chan (2002), which divides an image into four phases, will be discussed here. By using one level set function, an image will be divided into two phases, whereas increasing the number of level set functions will increase the number of phases. To segment an image into n phases, $\log_2 n$ number of level set functions will be required.

Consider $p = \log_2 n$ level set function $\phi_\ell : \Omega \to \mathbb{R}$ for $\ell = 1, 2, \ldots, p$. The union of the zero level sets of all ϕ_ℓ will determine the set of edges in the segmented image. For $1 \leq s \leq n = 2^p$, denote by $c_s = \text{mean}(z)$ the average value of image gray scales in phase s and by χ_s the characteristic function for phase s. The following energy functional is proposed; see for detail Vese and Chan (2002):

$$F_n(c, \Phi) = \sum_{1 \leq s \leq n} \int_\Omega (z(x, y) - c_s)^2 \chi_s \, dx \, dy$$

$$+ \mu \sum_{1 \leq \ell \leq p} \int_\Omega |\nabla H(\phi_\ell)| \, dx \, dy \qquad (56)$$

where $c = (c_1, c_2, \ldots, c_n)$ and $\Phi = (\phi_1, \phi_2, \ldots, \phi_p)$; note $n = 2^p$. In this section, main focus will be on the four-phase segmentation, i.e., $n = 4$ or $p = 2$.

Consider the following minimization problem for four-phase segmentation:

$$\min_{c, \Phi} F_4(c, \Phi), \qquad (57)$$

where:

$$F_4(c, \Phi) = \int_\Omega (z(x, y) - c_{11})^2 H(\phi_1) H(\phi_2) \, dx \, dy$$

$$+ \int_\Omega (z(x, y) - c_{10})^2 H(\phi_1)(1 - H(\phi_2)) \, dx \, dy$$

$$+ \int_\Omega (z(x, y) - c_{01})^2 (1 - H(\phi_1)) H(\phi_2) \, dx \, dy$$

$$+ \mu \int_\Omega |\nabla H(\phi_1)| \, dx \, dy$$

$$+ \int_\Omega (z(x, y) - c_{00})^2 (1 - H(\phi_1))(1 - H(\phi_2)) \, dx \, dy$$

$$+ \mu \int_\Omega |\nabla H(\phi_2)| \, dx \, dy \qquad (58)$$

where $c = (c_{11}, c_{10}, c_{01}, c_{00})$ is the vector of average intensities in different phases of the given image and $\Phi = (\phi_1, \phi_2)$ is the vector of level sets used for segmentation of an image into various phases. Minimization of (57) with respect to Φ leads to the following system of equations:

$$\begin{cases} \delta_\epsilon(\phi_1) \left[\mu \nabla \cdot \frac{\nabla \phi_1}{|\nabla \phi_1|} - [T_1 H_\epsilon(\phi_2) + T_2(1 - H_\epsilon(\phi_2))] \right] = 0, \\ \delta_\epsilon(\phi_2) \left[\mu \nabla \cdot \frac{\nabla \phi_2}{|\nabla \phi_2|} - [T_1 H_\epsilon(\phi_1) + T_2(1 - H_\epsilon(\phi_1))] \right] = 0, \end{cases} \qquad (59)$$

with Neumann boundary conditions, where $T_1 = (z - c_{11})^2 - (z - c_{01})^2$ and $T_2 = (z - c_{10})^2 - (z - c_{00})^2$. This system of coupled partial differential equations is usually solved by introducing artificial time variable and using well-known time marching schemes like semi-implicit and additive operator splitting methods which

Multigrid Method with Typical and Modified Smoother

Using finite difference schemes to discretize (59) for ϕ_ℓ, the equations at a pixel point (i, j) are given by the following:

$$\begin{cases} \delta_\epsilon(\phi_1)_{i,j} \left\{ \dfrac{\Delta_-^x}{h_1} \dfrac{\mu \Delta_+^x(\phi_1)_{i,j}/h_1}{\sqrt{(\Delta_+^x(\phi_1)_{i,j}/h_1)^2 + (\Delta_+^y(\phi_1)_{i,j}/h_2)^2 + \beta}} - (T_1)_{i,j} H_\epsilon(\phi_2)_{i,j} + \right. \\ \left. \dfrac{\Delta_-^y}{h_2} \dfrac{\mu \Delta_+^y(\phi_1)_{i,j}/h_2}{\sqrt{(\Delta_+^x(\phi_1)_{i,j}/h_1)^2 + (\Delta_+^y(\phi_1)_{i,j}/h_2)^2 + \beta}} - (T_2)_{i,j}(1 - H_\epsilon(\phi_2)_{i,j}) \right\} = 0, \\ \delta_\epsilon(\phi_2)_{i,j} \left\{ \dfrac{\Delta_-^x}{h_1} \dfrac{\mu \Delta_+^x(\phi_2)_{i,j}/h_1}{\sqrt{(\Delta_+^x(\phi_2)_{i,j}/h_1)^2 + (\Delta_+^y(\phi_2)_{i,j}/h_2)^2 + \beta}} - (T_1)_{i,j} H_\epsilon(\phi_1)_{i,j} + \right. \\ \left. \dfrac{\Delta_-^y}{h_2} \dfrac{\mu \Delta_+^y(\phi_2)_{i,j}/h_2}{\sqrt{(\Delta_+^x(\phi_2)_{i,j}/h_1)^2 + (\Delta_+^y(\phi_2)_{i,j}/h_2)^2 + \beta}} - (T_2)_{i,j}(1 - H_\epsilon(\phi_1)_{i,j}) \right\} = 0, \end{cases} \tag{60}$$

Let $\bar{\mu} = \mu/h_1$, $\bar{\beta} = h_1^2 \beta$ and $\lambda = h_1/h_2$. Also denote $(f_1)_{i,j} = (T_1)_{i,j} H_\epsilon(\phi_2)_{i,j} + T_2)_{i,j}(1 - H_\epsilon(\phi_2)_{i,j})$ and $(f_2)_{i,j} = (T_1)_{i,j} H_\epsilon(\phi_1)_{i,j} + T_2)_{i,j}(1 - H_\epsilon(\phi_1)_{i,j})$.

For $\ell = 1, 2$, introducing the following notation for the differential coefficients as follows:

$$D_\ell(\phi_\ell)_{i,j} = \dfrac{1}{\sqrt{(\Delta_+^x(\phi_\ell)_{i,j})^2 + (\lambda \Delta_+^y(\phi_\ell)_{i,j})^2 + \bar{\beta}}},$$

$$D_\ell(\phi_\ell)_{i-1,j} = \dfrac{1}{\sqrt{(\Delta_+^x(\phi_\ell)_{i-1,j})^2 + (\lambda \Delta_+^y(\phi_\ell)_{i-1,j})^2 + \bar{\beta}}},$$

$$D_\ell(\phi_\ell)_{i,j-1} = \dfrac{1}{\sqrt{(\Delta_+^x(\phi_\ell)_{i,j-1})^2 + (\lambda \Delta_+^y(\phi_\ell)_{i,j-1})^2 + \bar{\beta}}}.$$

Thus locally linearized form of Equation (60) is given by the following:

$$\left[D_\ell(\phi_\ell)_{i,j}((\phi_\ell)_{i+1,j} - (\phi_\ell)_{i,j}) - D_\ell(\phi_\ell)_{i-1,j}((\phi_\ell)_{i,j} - (\phi_\ell)_{i-1,j}) \right]$$
$$+ \lambda^2 \left[D_\ell(\phi_\ell)_{i,j}((\phi_\ell)_{i,j+1} - (\phi_\ell)_{i,j}) - D_\ell(\phi_\ell)_{i,j-1}((\phi_\ell)_{i,j} - (\phi_\ell)_{i,j-1}) \right]$$
$$= (\bar{f}_\ell)_{i,j}, \tag{61}$$

where $\bar{f}_\ell = f_\ell / \bar{\mu}$.

12 Fast Numerical Methods for Image Segmentation Models

Let $\widetilde{\phi}_\ell$ be the approximation to ϕ_ℓ at the current iteration. Then from Equation (61), pursuing only local unknowns ϕ_ℓ at (i, j) in the following linear equations:

$$\left[D_\ell(\widetilde{\phi}_\ell)_{i,j}((\widetilde{\phi}_\ell)_{i+1,j} - (\phi_\ell)_{i,j}) - D_\ell(\widetilde{\phi}_\ell)_{i-1,j}((\phi_\ell)_{i,j} - (\widetilde{\phi}_\ell)_{i-1,j}) \right]$$
$$+ \lambda^2 \left[D_\ell(\widetilde{\phi}_\ell)_{i,j}((\widetilde{\phi}_\ell)_{i,j+1} - (\phi_\ell)_{i,j}) - D_\ell(\widetilde{\phi}_\ell)_{i,j-1}((\phi_\ell)_{i,j} - (\widetilde{\phi}_\ell)_{i,j-1}) \right]$$
$$= (\bar{f}_\ell)_{i,j}. \tag{62}$$

These equations will be solved for $(\phi_\ell)_{i,j}$, and store their values in $(\widetilde{\phi}_\ell)_{i,j}$, to use it in the next iteration. This equation is used as a smoother in the multigrid Algorithm 4. For further details, see Badshah and Chen (2009). Local Fourier analysis is usually used to check the convergence of the smoother, and this is discussed in the next section.

Local Fourier Analysis and a Modified Smoother

Local Fourier analysis (LFA) is a suitable tool to analyze the convergence rate of any iterative method for linear equations. However, the underlying equations are nonlinear, so LFA will consider a linearized equation, and as linearization occurs locally at each pixel, the maximum rate from all pixel locations will be considered.

Consider a square image with $m = m_1 = m_2$ and $h_1 = h_2 = h$ for simplicity, then $\lambda = 1$. Given the previous iterate at step k, $\widetilde{\phi}_\ell = \phi_\ell^{(k)}$, denote $a_1 = D_1(\widetilde{\phi}_1)_{i-1,j}$, $a_2 = D_1(\widetilde{\phi}_1)_{i,j}$, $a_3 = D_1(\widetilde{\phi}_1)_{i,j-1}$, $b_1 = D_2(\widetilde{\phi}_2)_{i-1,j}$, $b_2 = D_2(\widetilde{\phi}_2)_{i,j}$, $b_3 = D_2(\widetilde{\phi}_2)_{i,j-1}$ which are to be considered as local constants. From (61), the grid equation at (i, j) is the following local smoother:

$$\begin{cases} -(a_1 + 2a_2 + a_3)(\phi_1)_{i,j}^{(k+1)} + a_1(\phi_1)_{i-1,j}^{(k+1)} + a_3(\phi_1)_{i,j-1}^{(k+1)} \\ + a_2[(\phi_1)_{i+1,j}^{(k)} + (\phi_1)_{i,j+1}^{(k)}] = (\bar{f}_1)_{i,j}, -(b_1 + 2b_2 + b_3)(\phi_2)_{i,j}^{(k+1)} \\ + b_1(\phi_2)_{i-1,j}^{(k+1)} + b_3(\phi_2)_{i,j-1}^{(k+1)} + b_2[(\phi_2)_{i+1,j}^{(k)} + (\phi_2)_{i,j+1}^{(k)}] = (\bar{f}_2)_{i,j}. \end{cases} \tag{63}$$

Define the error functions by $e_1^{(k)} = \phi_1 - \phi_1^{(k)}$ and $e_2^{(k)} = \phi_2 - \phi_2^{(k)}$. Then using (127) and (63) with frozen $(\bar{f}_1)_{i,j}$ and $(\bar{f}_2)_{i,j}$, the error equations are as follows:

$$\begin{cases} a_1(e_1)_{i-1,j}^{(k+1)} + a_3(e_1)_{i,j-1}^{(k+1)} + a_2[(e_1)_{i+1,j}^{(k)} + (e_1)_{i,j+1}^{(k)}] \\ -(a_1 + 2a_2 + a_3)(e_1)_{i,j}^{(k+1)} = 0 b_1(e_2)_{i-1,j}^{(k+1)} \\ +b_3(e_2)_{i,j-1}^{(k+1)} + b_2[(e_2)_{i+1,j}^{(k)} + (e_2)_{i,j+1}^{(k)}] - (b_1 + 2b_2 + b_3)(e_2)_{i,j}^{(k+1)} = 0. \end{cases} \quad (64)$$

Recall that the LFA measures the largest amplification factor in a relaxation scheme (Brandt 1977; Chen 2005; Trottenberg and Schuller 2001). Let a general Fourier component be the following:

$$\Theta_{\alpha,\beta}(x_i, y_j) = \exp\left(i\theta_\alpha \frac{x_i}{h} + i\theta_\beta \frac{y_j}{h}\right) = \exp\left(\frac{2i\alpha\pi}{m} + \frac{2i\beta j\pi}{m}\right).$$

Note that $\theta_\alpha, \theta_\beta \in [-\pi, \pi]$. The LFA expands:

$$e_1^{(k)} = \sum_{\alpha,\beta=-m/2}^{m/2} \left(\psi_1^{(k)}\right)_{\alpha,\beta} \Theta_{\alpha,\beta}(x_i, y_j), \quad e_2^{(k)} = \sum_{\alpha,\beta=-m/2}^{m/2} \left(\psi_2^{(k)}\right)_{\alpha,\beta} \Theta_{\alpha,\beta}(x_i, y_j)$$

in Fourier components. Taking the largest spectral radius (maximum eigenvalue) of the amplification matrix $\mathcal{A}_{\alpha,\beta}$ (Trottenberg and Schuller 2001):

$$\begin{bmatrix} (\psi_1^{(k+1)})_{\alpha,\beta} \\ (\psi_2^{(k+1)})_{\alpha,\beta} \end{bmatrix} = \mathcal{A}_{\alpha,\beta} \begin{bmatrix} (\psi_1^{(k)})_{\alpha,\beta} \\ (\psi_2^{(k)})_{\alpha,\beta} \end{bmatrix}.$$

After substituting these components into (64) for $e_1^{(k+1)}$, $e_1^{(k)}$ and $e_2^{(k+1)}$, $e_2^{(k)}$:

$$\mathcal{A}_{\alpha,\beta} = \begin{bmatrix} \dfrac{a_2\left(e^{\frac{2i\alpha\pi}{m}} + e^{\frac{2i\beta\pi}{m}}\right)}{\left(a_1+2a_2+a_3-a_1 e^{\frac{-2i\alpha\pi}{m}} - a_3 e^{\frac{-2i\beta\pi}{m}}\right)} & 0 \\ 0 & \dfrac{b_2\left(e^{\frac{2i\alpha\pi}{m}} + e^{\frac{2i\beta\pi}{m}}\right)}{\left(b_1+2b_2+b_3-b_1 e^{\frac{-2i\alpha\pi}{m}} - b_3 e^{\frac{-2i\beta\pi}{m}}\right)} \end{bmatrix}.$$

At the kth iteration, each rate $\bar{\mu}^{(k)}(i, j) = \max_{\alpha,\beta} \rho(\mathcal{A}_{\alpha,\beta})$ in the high-frequency range $(\theta_\alpha, \theta_\beta) \in [-\pi, \pi] \setminus [-\frac{\pi}{2}, \frac{\pi}{2}]$, measuring the effectiveness of a smoother (Brandt 1977), is dependent on $a_\ell, b_\ell, \ell = 1, 2, 3$, which in turn depends on the pixel location (I, j). Therefore looking for the largest smoothing rate for all i, j (i.e., among all such pixels):

$$\hat{\mu} = \max_{a_1,a_2,a_3,b_1,b_2,b_3} \bar{\mu}^{(k)}(i, j).$$

12 Fast Numerical Methods for Image Segmentation Models

Table 4 The smoothing rate for a local smoother with 3 inner iterations

Outer iterations s	The smoothing rate $\hat{\mu}_s$	The smoothing rate taking out "odd pixels" $\hat{\mu}_s^*$
1	0.6862	0.5720
2	0.6861	0.3170
3	0.6861	0.2747

However, due to the high nonlinearity, it is useful to define the smoothing rate as the maximum of the above-accumulated rates out of all s relaxation steps by the following:

$$\hat{\mu}_s = \max_{i,j} \bar{\mu}^{(1)}(i,j)\bar{\mu}^{(2)}(i,j)\cdots\bar{\mu}^{(s)}(i,j).$$

Clearly for linear equations, where a_ℓ, b_ℓ are constants, $\bar{\mu} = \bar{\mu}^{(k)}$ is a constant so $\hat{\mu}_s = \bar{\mu}^{(s)}$. Here, as a_ℓ, b_ℓ are not constants, with this particular definition, and allowing the possibility of $\bar{\mu}^{(k)}(i,j) \approx 1$ for some i,j,k. As long as $\hat{\mu}_s \ll 1$, then a smoother will be effective. In Table 4, smoothing rates for an artificial image of size $m = 32$ are given; note that similar results are obtained with $m = 64$. Here, in Table 4, the "odd pixels" refer to positions where the relative ratios between a_2 and $\max(a_1, a_3)$, or the ratios between b_2 and $\max(b_1, b_3)$, are quite large. Clearly our smoother is ineffective overall due to these odd pixels. This prompted to consider how to improve the overall smoothing rate (column 2 in Table 4).

A modified smoother. To motivate the idea, consider the particular case of an odd pixel assigned with the following:

$$a_1 = 0.3536, a_2 = 10{,}000, a_3 = 0.3536, b_1 = 0.3536, b_2 = 10{,}000, b_3 = 0.3536 \tag{65}$$

for which LFA as described above gives a local (large) rate of $\mu = 0.99996$. An alternative to (63), the following under-relaxation smoothing scheme at these odd pixels:

$$\begin{cases} a_1(\phi_1)_{i-1,j}^{(k+1)} + a_3(\phi_1)_{i,j-1}^{(k+1)} + a_2\left[(\phi_1)_{i+1,j}^{(k)} + (\phi_1)_{i,j+1}^{(k)}\right] \\ -(a_1 + 2a_2 + a_3)(1+\omega)(\phi_1)_{i,j}^{(k+1)} + \omega(a_1 + 2a_2 + a_3)(\phi_1)_{i,j}^{(k)} = (\bar{f}_1)_{i,j}, \\ b_1(\phi_2)_{i-1,j}^{(k+1)} + b_3(\phi_2)_{i,j-1}^{(k+1)} + b_2\left[(\phi_2)_{i+1,j}^{(k)} + (\phi_2)_{i,j+1}^{(k)}\right] \\ -(b_1 + 2b_2 + b_3)(1+\omega)(\phi_2)_{i,j}^{(k+1)} + \omega(b_1 + 2b_2 + b_3)(\phi_2)_{i,j}^{(k)} = (\bar{f}_2)_{i,j}, \end{cases} \tag{66}$$

for some $0 \leq \omega \leq 1$ (note $\omega = 0$ reduces to the previous local smoother). The new error equation is as follows:

$$\begin{cases} a_1(e_1)_{i-1,j}^{(k+1)} + a_3(e_1)_{i,j-1}^{(k+1)} + a_2\left[(e_1)_{i+1,j}^{(k)} + (e_1)_{i,j+1}^{(k)}\right] \\ \qquad - (a_1 + 2a_2 + a_3)(1+\omega)(e_1)_{i,j}^{(k+1)} + \omega(a_1 + 2a_2 + a_3)(e_1)_{i,j}^{(k)} = 0, \\ b_1(e_2)_{i-1,j}^{(k+1)} + b_3(e_2)_{i,j-1}^{(k+1)} + b_2\left[(e_2)_{i+1,j}^{(k)} + (e_2)_{i,j+1}^{(k)}\right] \\ \qquad - (1+\omega)(b_1 + 2b_2 + b_3)(e_2)_{i,j}^{(k+1)} + \omega(b_1 + 2b_2 + b_3)(e_2)_{i,j}^{(k)} = 0. \end{cases} \quad (67)$$

Then the corresponding new Fourier amplification matrix is as follows:

$$\mathcal{A}_{\alpha,\beta} = \begin{bmatrix} \dfrac{a_2\left(e^{\frac{2i\alpha\pi}{m}} + e^{\frac{2i\beta\pi}{m}}\right) + \omega(a_1 + 2a_2 + a_3)}{(1+\omega)(a_1 + 2a_2 + a_3) - a_1 e^{\frac{-2i\alpha\pi}{m}} - a_3 e^{\frac{-2i\beta\pi}{m}}} & 0 \\ 0 & \dfrac{b_2\left(e^{\frac{2i\alpha\pi}{m}} + e^{\frac{2i\beta\pi}{m}}\right) + \omega(b_1 + 2b_2 + b_3)}{(1+\omega)(b_1 + 2b_2 + b_3) - b_1 e^{\frac{-2i\alpha\pi}{m}} - b_3 e^{\frac{-2i\beta\pi}{m}}} \end{bmatrix}.$$

Equation (66) with $\omega = 0.7$, this new scheme yields a much better rate of $\mu = 0.75026$. The choice of $\omega = 0$ is based on numerical experience.

Therefore, the modified smoother will be (66) using a variable ω written in a form similar to (62) as follows:

$$D_\ell(\widetilde{\phi}_\ell)_{i,j}\left[(\widetilde{\phi}_\ell)_{i+1,j} - (1+\omega)(\phi_\ell)_{i,j} + \omega(\widetilde{\phi}_\ell)_{i,j}\right]$$
$$- D_\ell(\widetilde{\phi}_\ell)_{i-1,j}\left[(1+\omega)(\phi_\ell)_{i,j} - \omega(\widetilde{\phi}_\ell)_{i,j} - (\widetilde{\phi}_\ell)_{i-1,j}\right]$$
$$+ \lambda^2 D_\ell(\widetilde{\phi}_\ell)_{i,j}\left[(\widetilde{\phi}_\ell)_{i,j+1} - (1+\omega)(\phi_\ell)_{i,j} + \omega(\widetilde{\phi}_\ell)_{i,j}\right]$$
$$- \lambda^2 D_\ell(\widetilde{\phi}_\ell)_{i,j-1}\left[(1+\omega)(\phi_\ell)_{i,j} - \omega(\widetilde{\phi}_\ell)_{i,j} - (\widetilde{\phi}_\ell)_{i,j-1}\right] = (\bar{f}_\ell)_{i,j}. \quad (68)$$

An algorithm for the modified smoother is given by the following:

Table 5 The smoothing rate for a modified local smoother

Outer iterations s	The smoothing rate $\hat{\mu}_s$
1	0.5720
2	0.3170
3	0.2747

Algorithm 5 Modified smoother for multiphase model

Implementation steps of the modified smoother given in Eq. (68) are demonstrated here as an algorithm:

$$\phi_\ell^h \longleftarrow Smoother(\phi_\ell^h, \bar{f}_\ell^h, \text{maxit}, \omega, K, tol)$$

where $\ell = 1, 2$ and h is the step size on level Ω^h. Set $K = 100$.
for $i = 1 : m_1$
    for $j = 1 : m_2$
        for $iter = 1 : \text{maxit}$
            if $|D_\ell(\widetilde{\phi}_\ell)_{i,j}^h| \geq K \max(|D_\ell(\widetilde{\phi}_\ell)_{i-1,j}^h|, |D_\ell(\widetilde{\phi}_\ell)_{i,j-1}^h|)$ for any ℓ, set $\omega = 0.7$;
            otherwise set $\omega = 0$.
$$\widetilde{\phi}_\ell^h \leftarrow \phi_\ell^h,$$
$$A_\ell = D_\ell(\widetilde{\phi}_\ell)_{i,j}^h((\widetilde{\phi}_\ell)_{i+1,j}^h + \omega(\widetilde{\phi}_\ell)_{i,j}^h) + D_\ell(\widetilde{\phi}_\ell)_{i-1,j}^h((\widetilde{\phi}_\ell)_{i-1,j}^h + \omega(\widetilde{\phi}_\ell)_{i,j}^h),$$
$$B_\ell = D_\ell(\widetilde{\phi}_\ell)_{i,j}^h((\widetilde{\phi}_\ell)_{i,j+1}^h + \omega(\widetilde{\phi}_\ell)_{i,j}^h) + D_\ell(\widetilde{\phi}_\ell)_{i,j-1}^h((\widetilde{\phi}_\ell)_{i,j-1}^h + \omega(\widetilde{\phi}_\ell)_{i,j}^h),$$

$$(\phi_\ell)_{i,j}^h = \frac{A_\ell + \lambda^2 B_\ell - \bar{f}_{\ell i,j}}{(1+\omega)(D_\ell(\widetilde{\phi}_\ell)_{i,j}^h + D_\ell(\widetilde{\phi}_\ell)_{i-1,j}^h + \lambda^2(D_\ell(\widetilde{\phi}_\ell)_{i,j}^h + D_\ell(\widetilde{\phi}_\ell)_{i,j-1}^h))}$$

            if $|(\phi_\ell)_{i,j}^h - (\widetilde{\phi}_\ell)_{i,j}^h| < tol$ Stop
            end
        end
    end
end

The smoothing analysis of the improved smoother is done again in the same steps and is given in Table 4. Clearly, the smoothing rates of the modified smoother are much more acceptable (note the accumulated number of smoothing steps is $3s$ since 3 inner iterations for each outer step are used) (Table 5).

In Table 6, speed comparison of the multigrid with typical local smoother (MG1), multigrid with modified smoother (MG1m), and additive operator splitting method (AOS) in terms of the number of iterations and CPU time is given. Fast convergence of the MG method can clearly be observed from the table. MG algorithms yield a computation time of $O(N \log N)$ where $N = m_1 \times m_2$.

Table 6 Speed comparison of MG1 (multigrid with typical local smoother), MG1m (multigrid with modified smoother), and AOS methods in terms of number of iterations ("Itr") and CPU time ("CPU"). Here "–" implies no convergence (to the tolerance) or very slow convergence

Image size	AOS Itr	AOS CPU	MG1 Itr	MG1 CPU	MG1m Itr	MG1m CPU
128 × 128	80	22	3	5	2	2
256 × 256	150	193	4	13	2	7
512 × 512	1500	42,600	4	74	2	33
1024 × 1024	–	–	4	525	2	148

Convex Multiphase Image Segmentation Model

The Vese-Chan model (Vese and Chan 2002) discussed in previous section has the advantage that the segmented phases cannot produce vacuum or overlap by construction. Moreover, it considerably reduces the number of level set functions needed and can represent complex boundaries. One of the drawbacks of the Vese-Chan model is that the energy functional of the model is a nonconvex and hence may stuck at local minima. This local minima may lead toward wrong segmentation. In Yang et al. (2014), a convex formulation of the Vese-Chan model (Vese and Chan 2002) is proposed. The energy functional of the Vese-Chan model is given in Equation 57. The convex model is then solved by using the Bregman iterations (Bregman 1967), which are discussed here.

The Bregman Iterations

Some basic definitions and theorems related to Bregman distance and Bregman iterations (Bregman 1967) are given here.

Definition 1. For an energy functional $E(\cdot)$, the Bregman distance between two functions say u and v is given by the following:

$$D_E^q(u, v) = E(u) - E(v) - \langle q, u - v \rangle,$$

where q is in the sub-gradient of $E(\cdot)$, i.e., $\partial E(v)$ at v.

To solve a minimization problem of the following form:

$$\min_u E(u) + \beta W(u), \qquad \beta > 0 \qquad (69)$$

where $W(\cdot)$ is convex with $\min_u W(u) = 0$, Bregman iterations are defined in the following way:

Definition 2. For given parameter $\beta > 0$, the Bregman iterations are defined as follows:

$$u^{(k+1)} = \arg\min_u D_E^{q^{(k)}}(u, u^{(k)}) + \beta W(u), \qquad q^{(k)} \in \partial E(u^{(k)}).$$

The next theorem plays an important role in minimization of the problem types given in (69).

Theorem 1. *The minimization problem given in (69) can be solved by the following Bregman iterations:*

$$u^{(k+1)} = \arg\min_u D_E^{q^{(k)}}(u, u^{(k)}) + \beta W(u) \tag{70}$$

$$= \arg\min_u E(u) - \langle q^{(k)}, u - u^{(k)} \rangle + \beta W(u) \tag{71}$$

where $q^{(k)} \in \partial E(u^{(k)})$. *Suppose that* $W(\cdot)$ *is differentiable; then:*

$$q^{(k+1)} = q^{(k)} - \beta \nabla W(u^{(k+1)}). \tag{72}$$

Convergence of the Bregman iterations is proven by Osher et al. in 2005 by stating the following convergence theorem:

Theorem 2. *Consider a minimization problem of type given in (69) and satisfying the condition given therein. Then $u^{(k)}$ obtained by Bregman iterations will satisfy the following conditions:*

1. $u^{(k)}$ *decreases monotonically on* W: $W(u^{(k+1)}) \leq W(u^{(k)})$.
2. *If u^* is a minimizer of W, then* $W(u^{(k)}) \leq W(u^*) + \frac{D_E^{q^{(0)}}(u^*, u^{(0)})}{\beta^{(k)}}$.

Convex Multiphase Model

In 2014, Yang et al. proposed a convex formulation of the Vese-Chan four-phase model. For this reconsider Equation (59):

$$\begin{cases} \delta_\epsilon(\phi_1)\left[\mu\nabla \cdot \frac{\nabla\phi_1}{|\nabla\phi_1|} - [T_1 H_\epsilon(\phi_2) + T_2(1 - H_\epsilon(\phi_2))]\right] = 0, \\ \\ \delta_\epsilon(\phi_2)\left[\mu\nabla \cdot \frac{\nabla\phi_2}{|\nabla\phi_2|} - [T_1 H_\epsilon(\phi_1) + T_2(1 - H_\epsilon(\phi_1))]\right] = 0, \end{cases} \tag{73}$$

with Neumann boundary conditions, where $T_1 = (z - c_{11})^2 - (z - c_{01})^2$ and $T_2 = (z - c_{10})^2 - (z - c_{00})^2$. Note that $H_\epsilon(w)$ has a non-compact support, so its derivative $\delta'_\epsilon(w) \neq 0$ for all $w \in R$. Thus the above system of equations may be written as follows:

$$\begin{cases} \left[\mu \nabla \cdot \dfrac{\nabla \phi_1}{|\nabla \phi_1|} - [T_1 H_\epsilon(\phi_2) + T_2(1 - H_\epsilon(\phi_2))]\right] = 0, \\ \left[\mu \nabla \cdot \dfrac{\nabla \phi_2}{|\nabla \phi_2|} - [T_1 H_\epsilon(\phi_1) + T_2(1 - H_\epsilon(\phi_1))]\right] = 0. \end{cases} \quad (74)$$

For simplification, suppose the following:

$$\begin{cases} r_1 = [T_1 H_\epsilon(\phi_2) + T_2(1 - H_\epsilon(\phi_2))], \\ r_2 = [T_1 H_\epsilon(\phi_1) + T_2(1 - H_\epsilon(\phi_1))]. \end{cases} \quad (75)$$

Thus the simplified gradient flow equation for (74) becomes the following:

$$\begin{cases} \dfrac{\partial \phi_1}{\partial t} = \mu \nabla \cdot \dfrac{\nabla \phi_1}{|\nabla \phi_1|} - r_1, \\ \dfrac{\partial \phi_2}{\partial t} = \mu \nabla \cdot \dfrac{\nabla \phi_2}{|\nabla \phi_2|} - r_2. \end{cases} \quad (76)$$

These partial differential equations are Euler-Lagrange's equation of the following energy functional:

$$\widetilde{F}(\phi_1, \phi_2) = \mu |\nabla \phi_1|_1 + \mu |\nabla \phi_2|_1 + \langle \phi_1, r_1 \rangle + \langle \phi_2, r_2 \rangle, \quad (77)$$

where $|\nabla(\cdot)|_1$ is the total variation (TV) norm and $\langle \cdot, \cdot \rangle$ is the inner product, respectively, and may be written as follows:

$$\begin{cases} |\nabla \phi_i|_1 = \displaystyle\int_\Omega |\nabla \phi_i(x)| dx = TV(\phi_i) \\ \langle \phi_i, r_i \rangle = \displaystyle\int_\Omega \phi_i(x) r_i(x) dx. \end{cases} \quad (78)$$

The energy functional given in Equation (77) is homogeneous in ϕ_i and does not have a minimizer in general. In order to make the minimizer well defined, introduction of some extra constraints on ϕ_i is necessary. As a result the following functional will be considered for minimization:

$$\min_{0 \leq \phi_1, \phi_2 \leq 1} \widetilde{F}(\phi_1, \phi_2) = \min_{0 \leq \phi_1, \phi_2 \leq 1} (\mu |\nabla \phi_1|_1 + \mu |\nabla \phi_2|_1 + \langle \phi_1, r_1 \rangle + \langle \phi_2, r_2 \rangle). \quad (79)$$

As $0 \leq \phi_1, \phi_2 \leq 1$, there is no need of using Heaviside function H_i.

12 Fast Numerical Methods for Image Segmentation Models

$$\begin{cases} \bar{r}_1 = [T_1\phi_2 + T_2(1-\phi_2)], \\ \bar{r}_2 = [T_1\phi_1 + T_2(1-\phi_1)]. \end{cases} \tag{80}$$

To use edge information of the image, they used weighted TV norm, which is given by the following:

$$TV_g(\phi_i) = \int_\Omega g(|\nabla z|)|\nabla \phi_i|dx = |\nabla \phi_i|_g, \tag{81}$$

where $g(w) = \frac{1}{1+c|z|^2}$ is the edge detector function. Now by using these terms, the energy functional given in Equation (79) becomes the following:

$$\min_{0 \leq \phi_1, \phi_2 \leq 1} \widetilde{F}(\phi_1, \phi_2) = \min_{0 \leq \phi_1, \phi_2 \leq 1} (\mu|\nabla\phi_1|_g + \mu|\nabla\phi_2|_g + \langle\phi_1, \bar{r}_1\rangle + \langle\phi_2, \bar{r}_2\rangle). \tag{82}$$

After solving this minimization problem, the following four-phase segmentation domains can be defined by thresholding the level set functions ϕ_1 and ϕ_2:

$$\begin{cases} \Omega_1 = \{x : \phi_1(x) > 0.5, \phi_2(x) > 0.5\} \\ \Omega_2 = \{x : \phi_1(x) > 0.5, \phi_2(x) < 0.5\} \\ \Omega_3 = \{x : \phi_1(x) < 0.5, \phi_2(x) > 0.5\} \\ \Omega_4 = \{x : \phi_1(x) < 0.5, \phi_2(x) < 0.5\}. \end{cases} \tag{83}$$

Also note that $\mathbf{c} = [c_{11}, c_{10}, c_{01}, c_{00}]$ is the vector of average intensities of image inside $\Omega_1, \Omega_2, \Omega_3, \Omega_4$, respectively. The model given in Equation (82) will give four-phase segmentation and can be extended to n phases, for which $m = \log_2 n$ level set functions will be required. The functional can be written as follows:

$$\min_{0 \leq \phi_i \leq 1} \widetilde{F}_n(\phi_1, \phi_2, \ldots, \phi_m) = \min_{0 \leq \phi_i \leq 1} \left(\mu \sum_{i=1}^m |\nabla\phi_i|_g + \sum_{i=1}^m \langle\phi_i, \bar{r}_i\rangle \right). \tag{84}$$

Split Bregman Method for the Model

The minimization problem given in (79) can be solved by using the split Bregman method. For this one must introduce two new auxiliary variables $\mathbf{p_i} = \nabla\phi_i, i = 1, 2$. Thus the minimization problem given in Equation (79) can be converted into the following equivalent constrained minimization problem:

$$\min_{\substack{0 \leq \phi_1, \phi_2 \leq 1 \\ \mathbf{p_1}, \mathbf{p_2}}} (\mu |\mathbf{p_1}|_g + \mu |\mathbf{p_2}|_g + \langle \phi_1, \bar{r}_1 \rangle + \langle \phi_2, \bar{r}_2 \rangle)),$$

$$\text{such that } \mathbf{p_i} = \nabla \phi_i, \quad i = 1, 2. \tag{85}$$

Corresponding unconstrained minimization problem can be obtained by introducing two quadratic penalty terms $\|\mathbf{p_i} - \nabla \phi_i\|^2$, $i = 1, 2$, which is given by the following:

$$(\phi_1^*, \phi_2^*, \mathbf{p_{1*}}, \mathbf{p_{2*}}) = \arg \min_{\substack{0 \leq \phi_1, \phi_2 \leq 1 \\ \mathbf{p_1}, \mathbf{p_2}}} \Big(\mu |\mathbf{p_1}|_g + \mu |\mathbf{p_2}|_g + \langle \phi_1, \bar{r}_1 \rangle$$
$$+ \langle \phi_2, \bar{r}_2 \rangle + \tfrac{\alpha}{2} \|\mathbf{p_1} - \nabla \phi_1\|^2 + \tfrac{\alpha}{2} \|\mathbf{p_2} - \nabla \phi_2\|^2 \Big), \tag{86}$$

where $\alpha > 0$ is a constant. Bregman iterations for the solution; this unconstrained minimization problem is given in the following theorem:

Theorem 3. *The minimization problem (79) of the proposed model can be converted to a series of optimization problems:*

$$(\phi_1^{(k+1)}, \phi_2^{(k+1)}, \mathbf{p_1}^{(k+1)}, \mathbf{p_2}^{(k+1)}) = \arg \min_{\substack{0 \leq \phi_1, \phi_2 \leq 1 \\ \mathbf{p_1}, \mathbf{p_2}}} \Big(\mu |\mathbf{p_1}|_g + \mu |\mathbf{p_2}|_g + \langle \phi_1, \bar{r}_1 \rangle$$
$$+ \langle \phi_2, \bar{r}_2 \rangle + \frac{\alpha}{2} \|\mathbf{p_1} - \nabla \phi_1 - \mathbf{b_1}^{(k)}\|^2$$
$$+ \frac{\alpha}{2} \|\mathbf{p_2} - \nabla \phi_2 - \mathbf{b_2}^{(k)}\|^2 \Big), \tag{87}$$

where $\mathbf{b_i} = (b_{ix}, b_{iy})$, $i = 1, 2$ are the Bregman variables, which can be updated by the following Bregman iterations with initial values $\mathbf{b}_i^0 = (0, 0)$, $i = 1, 2$:

$$\mathbf{b}_i^{(k+1)} = \mathbf{b}_i^{(k)} + \nabla \phi_i^{(k+1)} - \mathbf{p}_i^{(k+1)}, \text{ for } i = 1, 2. \tag{88}$$

To solve the minimization problem given in Equation (79), it is enough to solve the minimization problem given in Equation (87). The iterative minimization scheme can be achieved through the following two steps for solution of Equation (87).

- Keeping $\mathbf{p}_1^{(k)}$ and $\mathbf{p}_2^{(k)}$ and minimizing Equation (87) with respect to ϕ_1 and ϕ_2 give the following:

$$(\phi_1^{(k+1)}, \phi_2^{(k+1)}) = \arg \min_{0 \leq \phi_1, \phi_2 \leq 1} \Big(\langle \phi_1, \bar{r}_1^{(k)} \rangle + \langle \phi_2, \bar{r}_2^{(k)} \rangle + \frac{\alpha}{2} \|\mathbf{p_1} - \nabla \phi_1 - \mathbf{b}_1^{(k)}\|^2$$
$$+ \frac{\alpha}{2} \|p_2 - \nabla \phi_2 - b_2^{(k)}\|^2 \Big). \tag{89}$$

- Secondly, keeping $\phi_1^{(k+1)}$ and $\phi_2^{(k+1)}$ fixed and minimizing Equation (87) with respect to $\mathbf{p}_1$ and $\mathbf{p}_2$ give the following:

$$\left(\mathbf{p}_1^{(k+1)}, \mathbf{p}_2^{(k+1)}\right) = \arg\min_{\mathbf{p}_1, \mathbf{p}_2} \left(\mu |\mathbf{p}_1|_g + \mu |\mathbf{p}_2|_g + \frac{\alpha}{2} \|\mathbf{p}_1 - \nabla \phi_1^{(k+1)} - \mathbf{b}_1^{(k)}\|^2 \right.$$
$$\left. + \frac{\alpha}{2} \|\mathbf{p}_2 - \nabla \phi_2^{(k+1)} - \mathbf{b}_2^{(k)}\|^2 \right). \tag{90}$$

Theorem 4. *For fixed $\mathbf{b}_1^{(k)}$ and $\mathbf{b}_2^{(k)}$, the minimizer $(\phi_1^{(k+1)}, \phi_2^{(k+1)})$ of the minimization problem (89) will satisfy the following equations:*

$$\Delta \phi_1^{(k+1)} = \frac{1}{\alpha} \bar{r}_1^{(k)} + \nabla \cdot \left(\mathbf{p}_1^{(k)} - \mathbf{b}_1^{(k)}\right) \qquad 0 \leq \phi_1^{(k+1)} \leq 1. \tag{91}$$

$$\Delta \phi_2^{(k+1)} = \frac{1}{\alpha} \bar{r}_2^{(k)} + \nabla \cdot \left(\mathbf{p}_2^{(k)} - \mathbf{b}_2^{(k)}\right) \qquad 0 \leq \phi_2^{(k+1)} \leq 1. \tag{92}$$

These Laplace equations are solved by using Gauss-Seidel method and obtained the following relation for $\phi_\ell^{(k+1)}$:

$$\begin{cases} \gamma_{\ell,i,j}^{(k)} = p_{\ell,x,i-1,j}^{(k)} - p_{\ell,x,i,j}^{(k)} + p_{\ell,y,i,j-1}^{(k)} - p_{\ell,y,i,j}^{(k)} - \left(b_{\ell,x,i-1,j}^{(k)} - b_{\ell,x,i,j}^{(k)} \right. \\ \qquad\qquad \left. + b_{\ell,y,i,j-1}^{(k)} - b_{\ell,y,i,j}^{(k)}\right) \\ \tau_{\ell,i,j}^{(k)} = \frac{1}{4}\left(\phi_{\ell,i-1,j}^{(k)} + \phi_{\ell,i+1,j}^{(k)} + \phi_{\ell,i,j-1}^{(k)} + \phi_{\ell,i,j+1}^{(k)} - \frac{1}{\alpha}r_{\ell,i,j}^{(k)} + \gamma_{\ell,i,j}^{(k)}\right) \\ \phi_{\ell,i,j}^{(k+1)} = \max\left\{\min\left\{\tau_{\ell,i,j}^{(k)}, 1\right\}, 0\right\} \end{cases} \tag{93}$$

where $\ell = 1, 2$.

Now to find $\mathbf{b}_1$ and $\mathbf{b}_1$, the following theorem is very useful to note:

Theorem 5. *For fixed $\phi_1^{(k+1)}$ and $\phi_1^{(k+1)}$, the minimizer $(\mathbf{p}_1^{(k+1)}, \mathbf{p}_2^{(k+1)})$ of the minimization problem given in Equation (90) will satisfy the following vector shrinkage operator:*

$$\mathbf{p}_1^{(k+1)} = \text{shrinkage}_g\left(\mathbf{b}_1^{(k)} + \nabla \phi_1^{(k+1)}, \frac{1}{\alpha}\right) = \text{shrinkage}\left(\mathbf{b}_1^{(k)} + \nabla \phi_1^{(k+1)}, \frac{\rho}{\alpha}\right) \tag{94}$$

$$p_2^{(k+1)} = \text{shrinkage}_g\left(b_2^{(k)} + \nabla\phi_2^{(k+1)}, \frac{1}{\alpha}\right) = \text{shrinkage}\left(b_2^{(k)} + \nabla\phi_2^{(k+1)}, \frac{\rho}{\alpha}\right) \tag{95}$$

where the vector shrinkage operator is given by the following:

$$\text{shrinkage}(x, \xi) = \begin{cases} \frac{x}{|x|} \max(|x| - \xi, 0), & x \neq 0 \\ 0, & x = 0 \end{cases} \tag{96}$$

For further details and experimental results of the proposed model and method, see Yang et al. (2014). In Fig. 1, the proposed method is tested on an artificial image. In Fig. 2, results of the proposed model on a real MRI image are given.

A Three-Stage Approach for Multiphase Segmentation Degraded Color Images

In 2017, Cai et al. proposed a smoothing, lifting, and thresholding method with three stages for multiphase segmentation of color images corrupted by different degradations: noise, information loss, and blur. The proposed method works in the following steps: in step one, a smooth restored image is obtained by applying the convex models Cai et al. (2013) and Chan et al. (2014) on each channel of original color image space. In the second stage, the smooth color image is transformed to

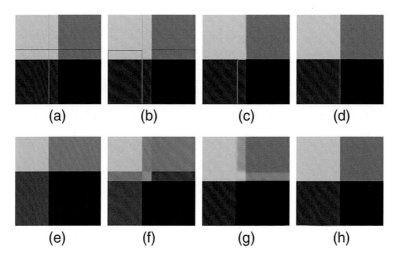

Fig. 1 Application of the proposed model to a simple synthetic image. (**a**)–(**d**): The active contour evolving process from the initial contour to the final contour. (**e**)–(**h**): The corresponding fitting images z at different iterations

12 Fast Numerical Methods for Image Segmentation Models

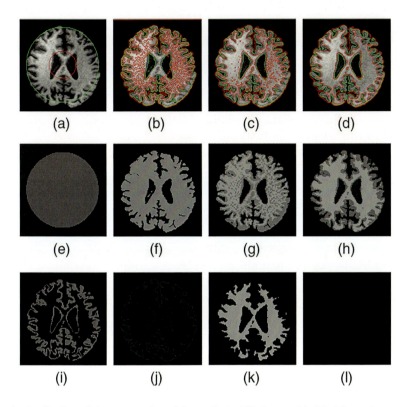

Fig. 2 Application of the proposed model to a brain MR image. (**a**)–(**d**): The active contour evolving process. (**e**)–(**h**): The evolution process of the fitting image z. (**i**)–(**l**): The final four segments with four averages $c_{11} = 113.1278$, $c_{10} = 48.3514$, $c_{01} = 167.2793$, and $c_{00} = 4.0692$

a secondary color space, which provides complementary information, and then a new vector-valued image is formed by using all channels from both color spaces (original and transformed). In stage 3, a multichannel thresholding is applied on the obtained vector-valued image to get segmented image.

Stage 1: Restoration and Smoothing of Given Image

Let z be a color image with d channels say z_i for $i = 1, 2, \ldots, d$;, the following energy functional is considered:

$$E(z_i) = \frac{\lambda}{2} \int_\Omega \Psi_i (z_i - K u_i)^2 dx + \frac{\mu}{2} \int_\Omega |\nabla u_i|^2 dx + \int_\Omega |\nabla u_i| dx, \quad i = 1, 2, \ldots, d, \tag{97}$$

where $\Psi_i(\cdot)$ is the characteristic function and is a region descriptor. For existence and uniqueness of the minimizer of the above functional, see Cai et al. (2017).

The above model (97) is considered in discrete setting and is solved for the unique minimizer $\bar{u}_i$ for each channel i by using different methods such as primal-dual method (Chambolle and Pock 2011; Chen et al. 2014), alternating direction method (Boyd et al. 2010), and split Bregman method (Goldstein and Osher 2009; Bregman 1967). Once $\bar{u}_i$ is found, it is rescaled onto [0, 1] and hence $\{\bar{u}_i\}_{i=1}^{d} \in [0, 1]^d$.

Stage 2: Dimension Lifting with Secondary Color Space

In first stage, a restored smooth image $\bar{u}_i$ is obtained, whereas in this stage, the dimension lifting is performed on $\bar{u}_i$ to extract more additional information from a different color space that help the segmentation in the later stage. Popular choices for other less correlated color spaces are the HSV (hue, saturation, and value), the CB (chromaticity-brightness), HSI (hue, saturation, and intensity), and the Lab (perceived lightness, red-green, and yellow-blue). Note that the Lab is a better color space than RGB, HSV, and HSI for segmentation. In this stage the authors created the Lab color space with the aim to be perceptually uniform in the sense that the numerical difference between two colors is proportional to perceived color difference. Here, the Lab is used as the additional color space, where the L channel correlates to perceived lightness, while the a and b channels correlate approximately with red-green and yellow-blue, respectively.

Let $\bar{u}'$ denote Lab transform of $\bar{u}$, rescaling all the channels of $\bar{u}'$ on the interval [0, 1] to yield an image denoted by $\bar{u}^t \in [0, 1]^3$. Introduce a new image $\bar{u}*$ by stacking $\bar{u}$ and $\bar{u}^t$ having six channels as follows:

$$\bar{u}* = (\bar{u}_1, \bar{u}_2, \bar{u}_3, \bar{u}_1^t, \bar{u}_2^t, \bar{u}_3^t).$$

This image will be used for segmentation in the next stage.

Stage 3: Segmentation

Segmentation of the vector-valued image $\bar{u}*$, obtained from the second stage in K segments, is done by using thresholding. This is based on the K-means algorithm (Kanungo et al. 2002) because of its simplicity and good asymptotic properties. According to the value of K, the algorithm clusters all points of $\{\bar{u} * (x) : x \in \Omega\}$ into K Voronoi-shaped cells, say $\Omega_1 \cup \Omega_2 \cup \ldots \cup \Omega_K = \Omega$. The mean vector $c_k \in \Omega^6$ on each cell Ω_k by the following:

$$c_k = \frac{\int_{\Omega_k} \bar{u} * (x) dx}{\int_{\Omega_k} dx}, \quad k = 1, 2, \ldots, K. \tag{98}$$

Recall that each entry $c_k[i]$ for $i = 1, 2, \ldots, 6$ is a value belonging to $\{R, G, B, L, a, b\}$, respectively. Using $c_k[i]$, $\bar{u}*$ can be divided into K phases by the following:

$$S_k := \{x \in \Omega : \|\bar{u} * -c_k\|_2 = \min_{1 \leq j \leq K} \|\bar{u} * -c_j\|_2\}, \qquad k = 1, 2, \ldots, K. \tag{99}$$

Clearly $\cup_{k=1}^{K} S_k = \Omega$ and $\cap_{k=1}^{K} S_k = $. For further details see Cai et al. (2013, 2017).

Selective Segmentation Models

Usually, two types of image segmentation problems are discussed in image processing: one is global segmentation, in which the complete image is segmented into all possible segments/regions, and the other one is the selective segmentation, in which a region of interest is segmented in an image. In previous sections, all discussions were about global segmentation. Another possible name used for selective segmentation in literature is interactive segmentation. This section is mainly devoted to selective segmentation.

Image Segmentation Under Geometrical Conditions

A model which is used for selective segmentation based on some geometrical constraints (like a set of points near the region of interest ROI) is proposed by Guyader and Gout (2008). The proposed model is based on the geodesic active contour model (Caselles et al. 1997) and geometrical constraints. Let $B = \{(x_i^*, y_i^*) \in \Omega, \ 1 \leq i \leq n_p\} \subset \Omega$ be the set of n_p distinct points near the boundary of the region of interest in the given image $z(x, y)$. The aim is to find an optimal contour $\Gamma \subset \Omega$ that best approaches the points from the set B while detecting the desired object in an image. The model works in the following way: let g be an edge detector function defined as follows:

$$g(w) = \frac{1}{1 + w^2}.$$

It must be noted that $g(|\nabla z(x, y)|)$ approaches to zero near edges in an image as discussed earlier. The purpose of the edge detector function g is to stop the evolving curve on edges/boundaries of the objects (ROI). A function $d(x, y)$ (distance metric) is introduced to stop the evolving curve near the geometrical points given in set B. This function $d(x, y)$ can be defined in the following way (Guyader and Gout 2008):

$$\forall (x, y) \in \Omega, \qquad d(x, y) = \prod_{i=1}^{n_p} \left(1 - e^{-\frac{(x - x_i^*)^2}{2\sigma^2}} e^{-\frac{(y - y_i^*)^2}{2\sigma^2}} \right). \tag{100}$$

There exist other distance metrics d as well like the following:

$$d(x, y) = distance((x, y), B) = \min_{(x_i^*, y_i^*) \in B} \left| (x, y) - (x_i^*, y_i^*) \right|$$

for all $(x, y) \in \Omega$ and $i = 1, 2, \ldots n_p$; see for others (Gout et al. 2005). Clearly $d(x, y)$ acts locally and will be approximately 0 in the neighborhood of points of set B. The aim is to find a contour Γ along which either $d \simeq 0$ or $g \simeq 0$. The following energy functional is proposed:

$$F(\Gamma) = \int_\Gamma d(x, y) g(|\nabla z(x, y)|) ds. \tag{101}$$

The contour Γ will stop at local minima where $d \simeq 0$ (in the neighborhood of points for B) or $g \simeq 0$ (near object boundaries). By introducing level set function ϕ, functional given in Equation (101) becomes the following:

$$F_\epsilon(\phi(x, y)) = \int_\Omega d(x, y) g(|\nabla z(x, y)|) \delta_\epsilon(\phi) |\nabla \phi(x, y)| dx dy, \tag{102}$$

where $\delta_\epsilon(\phi)$ is the regularized delta function. The functional $F_\epsilon(\phi(x, y))$ will be minimized with respect to $\phi(x, y)$, by considering the following minimization problem:

$$\min_{\phi(x,y)} F_\epsilon(\phi(x, y)), \tag{103}$$

where $F_\epsilon(\phi(x, y))$ is given in Equation (102). First variation of the functional given in Equation (103) leads to the following Euler-Lagrange's equation:

$$-\delta_\epsilon(\phi(x, y)) \nabla \cdot \left(d(x, y) g(|\nabla z(x, y)|) \frac{\nabla \phi(x, y)}{|\nabla \phi(x, y)|} \right) = 0.$$

Guyader and Gout (2008) solved the following evolution equation by introducing artificial time step t:

$$\frac{\partial \phi(x, y)}{\partial t} = \delta_\epsilon(\phi(x, y)) \nabla \cdot \left(d(x, y) g(|\nabla z(x, y)|) \frac{\nabla \phi(x, y)}{|\nabla \phi(x, y)|} \right) \tag{104}$$

with the boundary condition:

$$\frac{\partial \phi(x, y)}{\partial \vec{n}} = 0,$$

where $\vec{n}$ is the outward unit normal to the boundary $\partial \Omega$. Clearly the quantity $\frac{\partial \phi(x, y)}{\partial t}$ tends to 0 when a local minimum is achieved. In other words, if the model converges, the curve will not evolve any more since a steady state has been reached. A rescaling can be made so that the motion is applied to all level sets by replacing $\delta_\epsilon(\phi(x, y))$ by $|\nabla \phi(x, y)|$. Furthermore, it makes the flow independent of the scaling of ϕ (Alvarez et al. 1992; Zhao et al. 2000). Thus they considered the following evolution problem:

$$\phi(x, y, 0) = \phi_0(x, y),$$

$$\frac{\partial \phi(x, y)}{\partial t} = |\nabla \phi(x, y)| \nabla \cdot \left(d(x, y) g(|\nabla z(x, y)|) \frac{\nabla \phi(x, y)}{|\nabla \phi(x, y)|} \right), \quad (105)$$

$$\frac{\partial \phi(x, y)}{\partial \vec{n}} = 0 \quad \text{on } \partial \Omega,$$

where $\phi_0(x, y)$ is the initial value of $\phi(x, y)$. To avoid the evolving curve to stuck at local minima, an extra term known as "balloon term" is given by $\alpha d(x, y) g(|\nabla z(x, y)|)$, where $\alpha > 0$. Thus the following evolution problem is considered for solution:

$$\phi(x, y, 0) = \phi_0(x, y)$$

$$\frac{\partial \phi(x, y)}{\partial t} = |\nabla \phi(x, y)| \nabla \cdot \left(d(x, y) g(|\nabla z(x, y)|) \frac{\nabla \phi(x, y)}{|\nabla \phi(x, y)|} \right)$$

$$+ \alpha d(x, y) g(|\nabla z(x, y)|) |\nabla \phi(x, y)| \quad (106)$$

$$\frac{\partial \phi(x, y)}{\partial n} = 0 \quad \text{on } \partial \Omega.$$

After some manipulations:

$$\frac{\partial \phi(x, y)}{\partial t} = |\nabla \phi(x, y)| d(x, y) g(|\nabla z(x, y)|) \nabla \cdot \left(\frac{\nabla \phi(x, y)}{|\nabla \phi(x, y)|} \right) \quad (107)$$

$$+ \nabla (d(x, y) g(|\nabla z(x, y)|)) \cdot \nabla \phi + \alpha d(x, y) g(|\nabla z(x, y)|) |\nabla \phi(x, y)|.$$

This elliptic-type partial differential equation can be solved by using any time marching scheme. One of the best among those is the additive operator splitting (AOS) method (Weickert et al. 1997), which is discussed earlier.

Active Contour-Based Image Selective Model

Badshah and Chen in (2010) proposed a model for selective segmentation of gray images, which is the extension of Gout model (Guyader and Gout 2008) by using region information of the image combined with geodesic contour model. The following minimization problem was proposed:

$$\min_{\phi(x,y), c_1, c_2} F(\phi(x, y), c_1, c_2), \quad (108)$$

where:

$$F(\Gamma, c_1, c_2) = \mu \int_\Gamma d(x,y) g(|\nabla z(x,y)|) ds + \lambda_1 \int_{inside(\Gamma)} |z(x,y) - c_1|^2 dx dy$$
$$+ \lambda_2 \int_{outside(\Gamma)} |z(x,y) - c_2|^2 dx dy, \tag{109}$$

where μ is a positive parameter. Clearly if $\lambda_1 = \lambda_2 = 0$ and $\mu = 1$, then minimization problem (109) reduces to minimization problem (101).

Using level set function and introducing regularized Heaviside function, the energy functional given in Equation (109) becomes the following:

$$\min_{\phi(x,y), c_1, c_2} F_\epsilon(\phi(x,y), c_1, c_2), \tag{110}$$

where:

$$F_\epsilon(\phi(x,y), c_1, c_2)$$
$$= \mu \int_\Omega d(x,y) g(|\nabla z(x,y)|) \delta_\epsilon(\phi(x,y)) |\nabla \phi(x,y)| dx dy$$
$$+ \lambda_1 \int_\Omega |z(x,y) - c_1|^2 H_\epsilon(\phi(x,y)) dx dy + \lambda_2 \int_\Omega |z(x,y) - c_2|^2 (1 - H_\epsilon(\phi(x,y))) dx dy. \tag{111}$$

Note that c_1 and c_2 are the average intensities as discussed earlier. Introducing $G(x,y) = d(x,y) g(|\nabla z(x,y)|)$ and then taking first variation of the proposed functional with respect to ϕ through Gâteaux derivatives lead to the following Euler-Lagrange's equation:

$$\delta_\epsilon(\phi) \mu \nabla \cdot \left(G(x,y) \frac{\nabla \phi}{|\nabla \phi|} \right)$$
$$- \delta_\epsilon(\phi)(\lambda_1 (z(x,y) - c_1)^2 - \lambda_2 (z(x,y) - c_2)^2) = 0, \quad \text{on } \Omega$$
$$G(x,y) \frac{\delta_\epsilon(\phi)}{|\nabla \phi|} \frac{\partial \phi}{\partial \vec{n}} = 0, \quad \text{on } \partial \Omega. \tag{112}$$

Solution of this elliptic PDE is the steady-state solution of the following evolution equation (parabolic PDE):

$$\frac{\partial \phi}{\partial t} = \delta_\epsilon(\phi) \mu \nabla \cdot \left(G(x,y) \frac{\nabla \phi}{|\nabla \phi|} \right)$$
$$- \delta_\epsilon(\phi)(\lambda_1 (z(x,y) - c_1)^2 - \lambda_2 (z(x,y) - c_2)^2) \tag{113}$$

12 Fast Numerical Methods for Image Segmentation Models

with the boundary condition:

$$G(x,y)\frac{\delta_\epsilon(\phi)}{|\nabla\phi|}\frac{\partial\phi}{\partial\vec{n}}\bigg|_{\partial\Omega} = 0,$$

where $\vec{n}$ is the unit normal vector to the boundary of Ω. At steady state $\frac{\partial\phi}{\partial t} = 0$, which means the local minimum has been reached. After some manipulation, the above equation becomes the following:

$$\begin{cases} \phi(x,y,0) = \phi_0(x,y) \\ \frac{\partial\phi}{\partial t} = \mu\delta_\epsilon(\phi(x,y))\nabla\cdot\left(G(x,y)\frac{\nabla\phi}{|\nabla\phi|}\right) \\ \quad -\delta_\epsilon(\phi)(\lambda_1(z(x,y)-c_1)^2 - \lambda_2(z(x,y)-c_2)^2), \\ G(x,y)\frac{\delta_\epsilon(\phi)}{|\nabla\phi|}\frac{\partial\phi}{\partial n}\bigg|_{\partial\Omega} = 0. \end{cases} \qquad (114)$$

A term $\alpha G(x,y)|\nabla\phi|$ (known as a balloon term) could be added to speed up the convergence of the evolution equation as discussed in the previous section, where α is a positive constant. This term prevents the curve from stopping on a nonsignificant local minimum and is also of importance when initializing the process with a curve inside the object to be detected (Guyader and Gout 2008). Thus Equation (114) with balloon term can be written as follows:

$$\begin{cases} \phi(x,y,0) = \phi_0(x,y) \\ \frac{\partial\phi}{\partial t} = \mu\delta_\epsilon(\phi(x,y))\nabla\cdot\left(G(x,y)\frac{\nabla\phi}{|\nabla\phi|}\right) \\ \quad -\delta_\epsilon(\phi)(\lambda_1(z(x,y)-c_1)^2 - \lambda_2(z(x,y)-c_2)^2) + \alpha G(x,y)|\nabla\phi|, \\ G(x,y)\frac{\delta_\epsilon(\phi)}{|\nabla\phi|}\frac{\partial\phi}{\partial n}\bigg|_{\partial\Omega} = 0, \end{cases} \qquad (115)$$

after some manipulation leads to the following:

$$\begin{cases} \phi(x,y,0) = \phi_0(x,y) \\ \frac{\partial\phi}{\partial t} = \mu\delta_\epsilon(\phi(x,y))G(x,y)\nabla\cdot\left(\frac{\nabla\phi}{|\nabla\phi|}\right) + \mu\delta_\epsilon(\phi(x,y))\nabla G(x,y)\cdot\left(\frac{\nabla\phi}{|\nabla\phi|}\right) \\ \quad -\delta_\epsilon(\phi)(\lambda_1(z-c_1)^2 - \lambda_2(z-c_2)^2) + \alpha G(x,y)|\nabla\phi|, \\ G(x,y)\frac{\delta_\epsilon(\phi)}{|\nabla\phi|}\frac{\partial\phi}{\partial n}\bigg|_{\partial\Omega} = 0. \end{cases} \qquad (116)$$

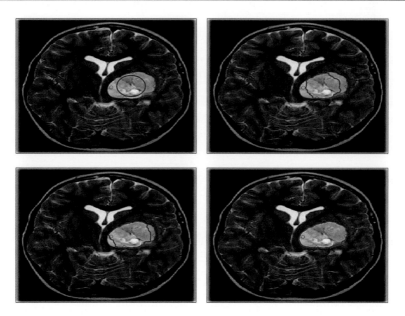

Fig. 3 To detect a tumor in a real brain MRI image with 4 markers with initial guess, $\phi_0 = \sqrt{(x - x_0)^2 + (y - y_0)^2} - r_0$, where x_0 and y_0 are the average of x, y-components of the markers, respectively. $\mu = $ (size of $z)^2/10$, $\lambda_1 = 0.0001$, $\lambda_2 = 0.0001$, $\alpha = -1.51 \times 10^{-2}$, and $\sigma = 4$

Existence and uniqueness of the solution can be proven along similar lines to Guyader and Gout (2008). This Equation (116) is solved by using time marching scheme like semi-implicit and additive operator splitting methods, which are discussed in the previous sections.

In Fig. 3, the proposed model is tested on a real brain MRI image to detect a tumor by taking four marker points near tumor in brain MR image. The initial condition is $\phi_0 = \sqrt{(x - x_0)^2 + (y - y_0)^2} - r_0$, where x_0 and y_0 are the average of x, y-components of the markers, respectively. The other parameters used are $\mu = $ (size of $z)^2/10$, $\lambda_1 = 0.0001$, $\lambda_2 = 0.0001$, $\alpha = -1.51 \times 10^{-2}$, and $\sigma = 4$. Top left figure is the original image with initial data, and top right figure is the result after 10 iterations. Bottom left figure is the result after 40 iterations, and bottom right figure is the final result after 200 iterations.

Parameter's selection. Initialization of the level set $\phi_0 = \sqrt{(x - x_0)^2 + (y - y_0)^2} - r_0$ is done automatically by taking x_0 and y_0 as the average of x, y-components of the marker's points, and r_0 is the minimum distance of the center from all marker's points. In most of the cases, $\lambda_1 = \lambda_2$ and may be taken small values of them. α controls the expanding of contour near edges of the object region whose values are near to zero and can be positive or negative. And μ is usually taken as multiple of the size of the given image, and this parameter must be chosen very carefully as the model is very sensitive with the selection of this parameter.

Dual-Level Set Selective Segmentation Model

In 2012, Rada and Chen proposed a selective segmentation model, in which two level sets (global and local) are constructed. Global-level set ϕ_G caries out global segmentation and local-level set ϕ_L carries out local selective segmentation. Introduce the following:

$$\begin{cases} \Gamma_L = \partial \Omega_L = \{(x, y) \in \Omega_L \mid \phi_L(x, y) = 0\} \\ \text{inside } (\Gamma_L) = \Omega_L = \{(x, y) \in \Omega_L \mid \phi_L(x, y) > 0\} \\ \text{outside } (\Gamma_L) = \Omega \backslash \overline{\Omega_L} = \{(x, y) \in \Omega_L \mid \phi_L(x, y) < 0\} \end{cases} \quad (117)$$

$$\begin{cases} \Gamma_G = \partial \Omega_G = \{(x, y) \in \Omega \mid \phi_G(x, y) = 0\} \\ \text{inside } (\Gamma_G) = \Omega_G = \{(x, y) \in \Omega \mid \phi_G(x, y) > 0\} \\ \text{outside } (\Gamma_G) = \Omega \backslash \overline{\Omega_G} = \{(x, y) \in \Omega \mid \phi_G(x, y) < 0\} \end{cases} \quad (118)$$

Note that $\Omega_L \subset \Omega_G \subset \Omega$. To look for all features Ω_G in the whole image domain Ω and the selective features Ω_L in the local domain Ω_G, they proposed the following energy functional by using regularized Heaviside function:

$$\min_{\phi_L(x,y), \phi_G(x,y), c_1, c_2} F_\epsilon \left(\phi_L(x, y), \phi_G(x, y) c_1, c_2 \right)$$

$$= \mu_1 \int_\Omega d(x, y) g(|\nabla z(x, y)|) \delta_\epsilon \left(\phi_L(x, y) \right) |\nabla \phi_L(x, y)| H_\epsilon \left(\phi_G(x, y) + \gamma \right) dx dy$$

$$+ \frac{\mu_L}{2} \int_\Omega \left(|\nabla \phi_L(x, y)| - 1 \right)^2 dx dy$$

$$+ \mu_2 \int_\Omega g(|\nabla z(x, y)|) \delta_\epsilon \left(\phi_G(x, y) \right) |\nabla \phi_G(x, y)| dx dy$$

$$+ \frac{\mu_G}{2} \int_\Omega \left(|\nabla \phi_G(x, y)| - 1 \right)^2 dx dy + \lambda_{1G} \int_\Omega |z(x, y) - c_1|^2 H_\epsilon \left(\phi_G(x, y) dx dy \right)$$

$$+ \lambda_{2G} \int_\Omega |z(x, y) - c_2|^2 \left(1 - H_\epsilon \right) \left(\phi_G(x, y) \right) dx dy$$

$$+ \lambda_1 \int_\Omega |z(x, y) - c_1|^2 H_\epsilon \left(\phi_L(x, y) dx dy \right)$$

$$+ \lambda_2 \int_\Omega |z(x, y) - c_1|^2 \left(1 - H_\epsilon \left(\phi_L(x, y) \right) \right) H \left(\phi_G(x, y) dx dy \right)$$

$$+ \lambda_3 \int_\Omega |z(x, y) - c_2|^2 \left(1 - H_\epsilon \left(\phi_L(x, y) \right) \right) \left(1 - H_\epsilon \left(\phi_G(x, y) \right) \right) dx dy \quad (119)$$

Here μ_L, μ_G are positive. Keeping ϕ fixed and minimizing with respect to c_1 and c_2 lead the following:

$$c_1 = \frac{\lambda_{1G} \int_\Omega z H_\epsilon(\phi_G) \, dxdy + \lambda_1 \int_\Omega z H_\epsilon(\phi_L) \, dxdy + \lambda_2 \int_\Omega z \left(1 - H_\epsilon(\phi_L)\right) H_\epsilon(\phi_G) \, dxdy}{\lambda_{1G} \int_\Omega H_\epsilon(\phi_G) \, dxdy + \lambda_1 \int_\Omega H_\epsilon(\phi_L) \, dxdy + \lambda_2 \int_\Omega \left(1 - H_\epsilon(\phi_L)\right) H_\epsilon(\phi_G) \, dxdy}$$

$$c_2 = \frac{\lambda_{2G} \int_\Omega z \left(1 - H_\epsilon(\phi_G)\right) dxdy + \lambda_3 \int_\Omega z \left(1 - H_\epsilon(\phi_L)\right)\left(1 - H_\epsilon(\phi_G)\right) dxdy}{\lambda_{2G} \int_\Omega \left(1 - H_\epsilon(\phi_G)\right) dxdy + \lambda_3 \int_\Omega \left(1 - H_\epsilon(\phi_L)\right)\left(1 - H_\epsilon(\phi_G)\right) dxdy}$$

First variation of the functional given in Equation (119) with respect to ϕ_L and letting $G(x, y) = d(x, y) g(|\nabla z(x, y)|)$ lead the following:

$$\begin{cases} \mu_1 \delta_\epsilon(\phi_L) \nabla \cdot \left(G(x, y) H_\epsilon(\phi_G + \gamma) \frac{\nabla \phi_L}{|\nabla \phi_L|}\right) + \mu_L \nabla \cdot \left(\left(1 - \frac{1}{|\nabla \phi_L|}\right) \nabla \phi_L\right) \\ + \delta_\epsilon(\phi_L) \left(-\lambda_1 (z(x,y) - c_1)^2 + \lambda_2 (z(x,y) - c_1)^2 H_\varepsilon(\phi_G) \right. \\ \left. + \lambda_3 (z(x,y) - c_2)^2 (1 - H_\varepsilon(\phi_G))\right) = 0, \quad \text{in } \Omega \\ \frac{\partial \phi_L}{\partial n} = 0, \quad \text{on } \partial \Omega \end{cases}$$

(120)

with Neumann boundary conditions. In similar way, Euler-Lagrange's equation can be derived for ϕ_G. Introducing balloon terms as discussed earlier leads to the following equations:

$$\begin{cases} \mu_1 \delta_\epsilon(\phi_L) \nabla \cdot \left(G(x, y) H_\epsilon(\phi_G + \gamma) \frac{\nabla \phi_L}{|\nabla \phi_L|}\right) + \mu_L \nabla \cdot \left(\left(1 - \frac{1}{|\nabla \phi_L|}\right) \nabla \phi_L\right) \\ + \delta_\epsilon(\phi_L) \left(-\lambda_1 (z(x,y) - c_1)^2 + \lambda_2 (z(x,y) - c_1)^2 H_\varepsilon(\phi_G) \right. \\ \left. + \lambda_3 (z(x,y) - c_2)^2 (1 - H_\varepsilon(\phi_G))\right) + \alpha G(x, y) |\nabla \phi_L| = 0, \quad \text{in } \Omega \\ \frac{\partial \phi_L}{\partial n} = 0, \quad \text{on } \partial \Omega \end{cases}$$

(121)

and:

$$\begin{cases} \mu_2 \delta_\epsilon(\phi_G) \nabla \cdot \left(g(x, y) \frac{\nabla \phi_G}{|\nabla \phi_G|}\right) + \mu_G \nabla \cdot \left(\left(1 - \frac{1}{|\nabla \phi_G|}\right) \nabla \phi_G\right) \\ + \delta_\epsilon(\phi_G + \gamma) \left(-\mu_1 G(x, y) |\nabla H_\varepsilon(\phi_L)|\right) + \delta_\epsilon(\phi_G) \left(-\lambda_{1G} (z(x,y) - c_1)^2 \right. \\ \left. + \lambda_{2G} (z(x,y) - c_2)^2 - \lambda_2 (z(x,y) - c_1)^2 (1 - H(\phi_L)) \right. \\ \left. + \lambda_3 (z(x,y) - c_2)^2 (1 - H(\phi_L))\right) + \alpha g(x, y) |\nabla \phi_G| = 0, \quad \text{in } \Omega \\ \frac{\partial \phi_G}{\partial n} = 0, \quad \text{on } \partial \Omega \end{cases}$$

(122)

12 Fast Numerical Methods for Image Segmentation Models

An additive operator splitting method (time marching scheme) is used to the respective parabolic partial differential equation:

$$\begin{cases} \frac{\partial \phi_G}{\partial t} = \mu_1 \delta_\epsilon (\phi_L) \nabla \cdot \left(G(x,y) H_\epsilon (\phi_G + \gamma) \frac{\nabla \phi_L}{|\nabla \phi_L|} \right) \\ \quad + \mu_L \nabla \cdot \left(\left(1 - \frac{1}{|\nabla \phi_L|} \right) \nabla \phi_L \right) \\ \quad + \delta_\epsilon (\phi_L) \left(-\lambda_1 (z(x,y) - c_1)^2 + \lambda_2 (z(x,y) - c_1)^2 H_\epsilon (\phi_G) \right. \\ \quad \left. + \lambda_3 (z(x,y) - c_2)^2 (1 - H_\epsilon (\phi_G)) \right) + \alpha G(x,y) |\nabla \phi_L|, \quad \text{in } \Omega \\ \frac{\partial \phi_L}{\partial n} = 0, \quad \text{on } \partial \Omega \end{cases} \tag{123}$$

and:

$$\begin{cases} \frac{\partial \phi_G}{\partial t} = \mu_2 \delta_\epsilon (\phi_G) \nabla \cdot \left(g(x,y) \frac{\nabla \phi_G}{|\nabla \phi_G|} \right) + \mu_G \nabla \cdot \left(\left(1 - \frac{1}{|\nabla \phi_G|} \right) \nabla \phi_G \right) \\ \quad + \delta_\epsilon (\phi_G + \gamma) \left(-\mu_1 G(x,y) |\nabla H_\epsilon (\phi_L)| \right) + \delta_\epsilon (\phi_G) \left(-\lambda_{1G} (z(x,y) - c_1)^2 \right. \\ \quad \left. + \lambda_{2G} (z(x,y) - c_2)^2 - \lambda_2 (z(x,y) - c_1)^2 (1 - H(\phi_L)) \right. \\ \quad \left. + \lambda_3 (z(x,y) - c_2)^2 (1 - H(\phi_L)) \right) + \alpha g(x,y) |\nabla \phi_G|, \quad \text{in } \Omega \\ \frac{\partial \phi_G}{\partial n} = 0, \quad \text{on } \partial \Omega \end{cases} \tag{124}$$

For further solution steps and experimental results, see Rada and Chen (2012). The model produces good and accurate results in hard images and images having overlapped regions but has high computational cost due to solution of system of PDEs for updating two level sets.

One-Level Selective Segmentation Model

In Rada and Chen (2013), proposed a one-level selective segmentation model. Consider the set of some geometrical points in the image domain as discussed earlier. They proposed the following energy functional:

$$\min_{\Gamma, c_2} F(\Gamma, c_2) = \min_{\Gamma, c_2} \left\{ \mu \int_\Gamma g\left(|\nabla z(x,y)| \right) dxdy \right.$$

$$\left. + \lambda_1 \int_{\text{inside}(\Gamma)} |z(x,y) - c_1|^2 \, dxdy \right.$$

$$+ \lambda_2 \int_{\text{outside}(\Gamma)} |z(x, y) - c_2|^2 \, dxdy$$

$$+ \nu \left\{ \left(\int_{\text{inside}(\Gamma)} dxdy - A_1 \right)^2 + \left(\int_{\text{outsite}(\Gamma)} dxdy - A_2 \right)^2 \right\}, \quad (125)$$

where $\lambda_1, \lambda_2, \mu, \nu$ are positive constants and g is the edge detector function which was defined earlier. Note that c_1 is known, which is the average intensity of the polygon constructed in the image by using the marker points. c_2 and Γ are unknown and need to found by minimizing the functional in (125). A_1 and A_2 are the areas of the region inside and outside polygon constructed from the marker points. Using level set function and regularized Heaviside function, the functional given in (125) takes the following form:

$$\min_{\phi(x,y),c_2} F_\epsilon \left(\phi(x, y), c_2 \right) = \mu \int_\Omega g \left(| \nabla z(x, y) | \right) \delta_\epsilon(\phi(x, y)) |\nabla(\phi(x, y))|$$

$$+ \, dxdy + \lambda_1 \int_\Omega |z(x, y) - c_1|^2 H_\epsilon(\phi(x, y)) dxdy$$

$$+ \lambda_2 \int_\Omega |z(x, y) - c_2|^2 \left(1 - H_\epsilon(\phi(x, y)) \right) dxdy$$

$$+ \nu \left\{ \left(\int_\Omega H_\epsilon(\phi(x, y)) dxdy - A_1 \right)^2 \right.$$

$$\left. + \left(\int_\Omega (1 - H_\epsilon(\phi(x, y)) dxdy) - A_2 \right)^2 \right\} dxdy.$$

$$(126)$$

Keeping ϕ fixed and minimizing this functional with respect to c_2 give the following:

$$c_2(\phi(x, y)) = \frac{\int_\Omega z(x, y) \left(1 - H_\epsilon(\phi(x, y)) \right) dxdy}{\int_\Omega \left(1 - H_\epsilon(\phi(x, y)) \right) dxdy}$$

and keeping c_2 fixed and if the marker points are not near to the boundary of the region of interest. Thus first variation with respect to ϕ gives the following Euler-Lagrange's equation:

$$\delta_\epsilon(\phi) \left\{ \mu \nabla \cdot | g(|\nabla z(x, y)|) \frac{\nabla \phi}{|\nabla \phi|} \right) - \left[\lambda_1 \left(z(x, y) - c_1 \right)^2 - \lambda_2 \left(z(x, y) - c_2 \right)^2 \right]$$

$$- \nu \left[\left(\int_\Omega H \, dxdy - A_1 \right) - \left(\int_\Omega (1 - H) dxdy - A_2 \right) \right] \right\} = 0 \text{ in } \Omega, \quad (127)$$

with Neumann boundary condition. If the marker points are near the boundary of the ROI, then Equation (127) becomes after introducing balloon term the following:

$$\delta_\epsilon(\phi)\left\{\mu\nabla\cdot|d(x,y)g(|\nabla z(x,y)|))\frac{\nabla\phi}{|\nabla\phi|}\right)$$
$$-\left[\lambda_1\left(z(x,y)-c_1\right)^2-\lambda_2\left(z(x,y)-c_2\right)^2\right]$$
$$-\nu\left[\left(\int_\Omega H dxdy-A_1\right)-\left(\int_\Omega(1-H)dxdy-A_2\right)\right]\right\}$$
$$-\alpha d(x,y)g(x,y)|\nabla\phi|=0. \qquad(128)$$

Corresponding unsteady partial differential equation is solved by using additive operator splitting method which is discussed earlier; for reference see Badshah and Chen (2010) and Rada and Chen (2012, 2013). For experimental results of the model, see Rada and Chen (2013).

Reproducible Kernel Hilbert Space-Based Image Segmentation

One of the basic problems in image segmentation is to handle low contrast and missing edge information. This problem is addressed in many papers. One of that is given in Burrows et al. (2021), in which Burrows et al. proposed methods for segmentation of images having objects with low contrast by making weak edges more prominent. To make the unclear/weak edges more prominent, the authors used reproducible kernel Hilbert space (RKHS) and approximated Heaviside functions.

Deng et al. in (2016) used RKHS and approximated Heaviside functions for another type of imaging problem, namely, image super resolution. RKHS models the smooth parts of an image, while edges may be represented by a set of approximated Heaviside functions. For details about RKHS and approximated Heaviside functions, see Deng et al. (2016) and Burrows et al. (2021).

Global Segmentation Model
This is a two-stage model for segmentation of images with low contrast and noise. In the first stage, RKHS-based model is used to get clean approximation of the original noisy image, and then edge components are separated from the smooth components. In the second stage, a suitable segmentation model is used on the clean image. The following model is proposed for separating edge features and removing noise:

$$\min_{d,\beta}\frac{1}{2}\|z-(Kd+\Psi\beta)\|^2+p_1 d^T K d+p_2\|\beta\|_1+p_3 g^T|\nabla(Kd+\Psi\beta)|, \qquad(129)$$

where Ψ collects values of the variation $\psi(v\cdot x+c)$ with v as the orientation at position c. ψ is the one-dimensional approximated Heaviside function:

$$\psi(t) = \frac{1}{2} + \frac{1}{\pi}\arctan(\frac{t}{\delta}).$$

β is a vector of all weights used for computing the edge part of an image, which is modeled from the set of $\psi(t)$. K is a $\ell \times N$ matrix with $K_{j,k} = K(x_j, x_k)$; g is the edge detector function based on $\Psi\beta$, performing better than a gradient-based one. The final term encourages the contrast to be low in homogeneous regions and high near edges.

The model given in Eq. 129 is solved by introducing auxiliary variables say $\theta = \beta$, $W = Kd + \Psi\beta$, and $v = \nabla W$, to have the following scheme:

$$\min_{d,\beta,\theta,W,v} \frac{1}{2}\|z - (Kd + \Psi\beta)\|^2 + p_1 d^T Kd + p_2\|\theta\|_1 + p_3 g^T |v| + \frac{\rho_1}{2}\|\theta$$
$$-\beta + b_1\|^2 + \frac{\rho_2}{2}\|W - (Kd + \Psi\beta) + b_2\|^2 + \frac{\rho_3}{2}\|v - \nabla W + b_3\|^2. \quad (130)$$

To implement a block coordinate descent scheme, take the following initial approximations: $d^{(0)}$, $\beta^{(0)}$, $\theta^{(0)}$, $W^{(0)}$, $v^{(0)}$, and update them alternatively and iteratively as follows:

The d problem in proximal form:

$$d^{(k)} = \arg\min d \frac{1}{2}\|z - (Kd + \Psi\beta^{(k-1)})\|^2 + p_1 d^T Kd + \frac{\zeta_1}{2}\|d - d^{(k-1)}\|^2$$
$$+ \frac{\rho_2}{2}\|W^{(k-1)} - (Kd + \Psi\beta^{(k-1)}) + b_2^{(k-1)}\|^2, \quad (131)$$

solution of this problem is obtained after some manipulation as follows:

$$d^{(k-1)} = A^{-1}\left(K^T z - (1+\rho_2)K^T \Psi\beta^{(k-1)} + \zeta_1 d^{(k-1)} + \rho_2 K^T (W^{(k-1)} + b_2^{(k-1)})\right), \quad (132)$$

Where:

$$A = (1 + \rho_2)K^T K + 2p_1 K + \zeta_1 I, \qquad (I \text{ is the identity matrix}).$$

Linearizing β problem and solving give the following proximal linear form:

$$\beta^{(k)} = \arg\min_{\beta} \langle \hat{p}^{(k)}, \beta - \hat{\beta}^{(k-1)}\rangle + \frac{\rho}{2}\|\theta^{(k-1)} - \beta + b_1^{(k-1)}\|^2 + \frac{\zeta_2}{2}\|\beta - \hat{\beta}^{(k-1)}\|^2, \quad (133)$$

where $\hat{\beta}^{(k-1)} = \beta^{(k-1)} + \omega^{(k-1)}(\beta^{(k-1)} - \beta^{(k-2)})$ and $\hat{p}^{(k)} = \nabla f(\hat{\beta}^{(k-1)})$, with:

$$f(\hat{\beta}^{(k-1)}) = \frac{1}{2}\|z - (Kd^{(k)} + \Psi\hat{\beta}^{(k-1)})\|^2 + \mu g^T |v^{(k-1)}|$$
$$+ \frac{\rho_2}{2}\|W^{(k-1)} - (Kd^{(k)} + \Psi\hat{\beta}^{(k-1)}) + b_2^{(k-1)}\|^2, \quad (134)$$

12 Fast Numerical Methods for Image Segmentation Models

$$\beta^{(k)} = \frac{1}{(\rho_1 + \zeta_2)} (\rho_1(\theta^{(k-1)} + b_1^{(k-1)}) + \zeta_2 \hat{\beta}^{(k-1)} - \hat{p}^{(k)}). \tag{135}$$

Subproblems for θ, W and v, are given as follows:

$$\theta^{(k)} = \arg\min_{\theta} \alpha \|\theta\|_1 + \frac{\rho_1}{2} \left\| \theta - \beta^{(k)} + b_1^{(k-1)} \right\|_2^2,$$

$$W^{(k)} = \arg\min_{W} \frac{\rho_2}{2} \left\| W - \left(Kd^{(k)} + \Psi\beta^{(k)} \right) + b_2^{(k-1)} \right\|^2 \tag{136}$$
$$+ \frac{\rho_3}{2} \left\| v^{(k-1)} - \nabla W + b_3^{(k-1)} \right\|^2,$$

$$v^{(k)} = \arg\min_{v} g^\top |v| + \frac{\rho_3}{2} \left\| v - \nabla W^{(k)} + b_3^{(k-1)} \right\|^2$$

Corresponding solutions are given by the following:

$$\theta^{(k)} = \text{shrink}\left(\beta^{(k)} - b_1^{(k-1)}, \frac{\rho_2}{\rho_1} \right), \tag{137}$$

$$W^{(k)} = \Re \left[\mathcal{F}^* \left(\frac{\rho_3 \mathcal{F}\left(\nabla^* \left(v^{(k-1)} + b_3^{(k-1)} \right) \right) + \rho_2 \mathcal{F}\left(Kd^{(k)} + \Psi\beta^{(k)} - b_2^{(k)} \right)}{\rho_2 + \rho_3 \mathcal{F}(\nabla^2)} \right) \right]$$
$$\tag{138}$$

$$v^{(k)} = \text{shrink}\left(\nabla W^{(k)} - b_3^{(k-1)}, \frac{v}{\rho_3} \cdot g \right) \tag{139}$$

Bregman parameters are updated as follows:

$$b_1^{(k)} = b_1^{(k-1)} + \theta^{(k)} - \beta^{(k)} \tag{140}$$

$$b_2^{(k)} = b_2^{(k-1)} + W^{(k)} - \left(Kd^{(k)} + \Psi\beta^{(k)} \right) \tag{141}$$

$$b_3^{(k)} = b_3^{(k-1)} + v^{(k)} - \nabla W^{(k)}. \tag{142}$$

The first-stage model given in Eq. 129 gives us separation edges from the rest and gives us a clean image say $M = Kd + \Psi\beta$. This clean image M is used in the next stage as an input in the segmentation model (Chan et al. 2006) and is given by the following:

$$F(u) = \int_\Omega g(|\Psi\beta|)|\nabla u|d\mathbf{x} + \lambda_1 \int_\Omega (M - c_1)^2 u d\mathbf{x}$$
$$+ \lambda_2 \int_\Omega (M - c_2)^2 (1 - u)d\mathbf{x} + \xi \int_\Omega v(u)d\mathbf{x} \qquad (143)$$

Using similar framework, the authors proposed a combined model which combines RKHS model with convex CV model. As a result the following model is proposed:

$$\min_{d,\beta,0\leq u\leq 1,c_1,c_2} F(d,\beta,u,c_1,c_2) \qquad (144)$$

where:

$$F(d,\beta,u,c_1,c_2) = \frac{1}{2}\|z - (Kd + \Psi\beta)\|^2 + \gamma d^T Kd + \alpha\|\beta\|_1 + \mu g^T |\nabla u|$$
$$+ \lambda[(Kd - \Psi\beta - c_1)^2 u + (Kd - \Psi\beta - c_2)^2 (1-u)]. \qquad (145)$$

To avoid non-differentiability of ℓ_1 norm, the following auxiliary variables are done before, $\theta = \beta$ and $w = (w_1, w_2) = \nabla u$. Thus the minimization problem becomes the following:

$$\min_{d,\beta,\theta,w,0\leq u\leq 1,c_1,c_2} F(d,\beta,\theta,w,u,c_1,c_2) \qquad (146)$$

where

$$F(d,\beta,\theta,w,u,c_1,c_2) = \frac{1}{2}\|z - (Kd + \Psi\beta)\|^2 + \gamma d^T Kd + \alpha\|\beta\|_1 + \mu g^T |w|$$
$$+ \lambda[(Kd - \Psi\beta - c_1)^2 u + (Kd - \Psi\beta - c_2)^2 (1-u)]$$
$$+ \frac{\rho_1}{2}\|\theta - \beta + b_1\|_2^2 + \frac{\rho_2}{2}\|w - \nabla u + b_2\|_2^2. \qquad (147)$$

This equation leads to subproblems for $d, \beta, \theta, c_1, c_2, u, w$, for the solution BCD scheme is used as follows:

Subproblem 1.

$$d^{(k)} = \arg\min_d \frac{1}{2}\|z - \left(Kd + \Psi\beta^{(k-1)}\right)\|^2 + \gamma(d)^T Kd + \frac{\zeta_1}{2}\|d - d^{(k-1)}\|^2$$
$$+ \lambda\left[\left(u^{(k-1)}\right)^T \left(Kd + \Psi\beta^{(k-1)} - c_1^{(k-1)}\right)^2\right.$$
$$+ \left.\left(1 - u^{(k-1)}\right)^T \left(Kd + \Psi\beta^{(k-1)} - c_2^{(k-1)}\right)^2\right]. \qquad (148)$$

The solution is given by the following:

$$d^{(k)} = A^{-1}\left(K^T z - (1+2\lambda)K^T \Psi \beta^{(k-1)} + \zeta_1 d^{(k-1)} \right.$$
$$\left. +2\lambda K^T \left[c_1^{(k-1)} u^{(k-1)} + c_2^{(k-1)}\left(1 - u^{(k-1)}\right)\right]\right) \quad (149)$$

where $A = (1+2\lambda)K^T K + 2\gamma K + \zeta_1 I$.

Subproblem 2. To get optimal value of β, the following subproblem will be solved:

$$\beta^{(k)} = \arg\min_{\beta} \langle \hat{p}^{(k)}, \beta - \hat{\beta}^{(k-1)} \rangle + \rho_1 over2 \|\theta^{(k-1)} - \beta + b_1^{(k-1)}\| + \frac{\zeta}{2}\|\beta - \hat{\beta}(k-1)\|^2, \quad (150)$$

where $\hat{\beta}^{(k-1)} = \beta^{(k-1)} + \omega^{(k-1)}\left(\beta^{(k-1)} - \beta^{(k-2)}\right)$ and $\hat{p}^{(k)} = \nabla f\left(\hat{\beta}^{(k-1)}\right)$, where f is given by the following:

$$f\left(\hat{\beta}^{(k-1)}\right) = \tfrac{1}{2}\|z - \left(Kd^{(k)} + \Psi\hat{\beta}^{(k-1)}\right)\|^2 + \mu g^T \left|w^{(k-1)}\right|$$
$$+\lambda\left[\left(u^{(k-1)}\right)^T \left(Kd^{(k)} + \Psi\hat{\beta}^{(k-1)} - c_1^{(k-1)}\right)^2 \right.$$
$$\left. + \left(1 - u^{(k-1)}\right)^T \left(Kd^{(k)} + \Psi\hat{\beta}^{(k-1)} - c_2^{(k-1)}\right)^2\right] \quad (151)$$

$$\nabla f\left(\hat{\beta}^{(k-1)}\right) = -\Psi^T\left(z - \left(Kd^{(k)} + \Psi\hat{\beta}^{(k-1)}\right)\right)$$
$$-2\mu\iota\Psi^T\left(\left|w^{(k-1)}\right| \odot \left(g\left(\Psi\hat{\beta}^{(k-1)}\right)\right)^2 \odot \left(\Psi\hat{\beta}^{(k-1)}\right)\right) \quad (152)$$
$$+2\lambda\Psi^T\left[Kd^{(k)} + \Psi\hat{\beta}^{(k-1)} - c_1^{(k-1)} u^{(k-1)} - c_2^{(k-1)}\left(1 - u^{(k-1)}\right)\right],$$

where $\odot$ denotes the Hadamard product between vectors (component-wise multiplication). Thus the β update is given as follows:

$$\beta^{(k)} = \frac{1}{(\rho_1 + L_2)}\left(\rho_1\left(\theta^{(k-1)} + b_1^{(k-1)}\right) + \zeta_2 \hat{\beta}^{(k-1)} - \hat{p}^{(k)}\right).$$

Subproblem 3. For the optimal solution of θ, the following minimization subproblem is solved:

$$\theta^{(k)} = \arg\min_{\theta} \alpha\|\theta\|_1 + \frac{\rho_1}{2}\|\theta - \beta^{(k)} + b_1^{(k-1)}\|_2^2, \quad (153)$$

whose solution is given by the following:

$$\theta^{(k)} = \mathrm{shrink}(\beta^{(k)} - b_1^{(k-1)}, \frac{\alpha}{\rho_1}), \tag{154}$$

and the Bregman parameter is updated in the following way:

$$b_1^{(k)} = b_1^{(k-1)} + \theta^{(k)} - \beta^{(k)} \tag{155}$$

Subproblem 4. This subproblem is solved for finding c_1 and c_2, for which the following minimization problem is solved:

$$c_1^{(k)} = \arg\min_{c_1} \lambda \left(u^{(k-1)}\right)^\top \left(Kd^{(k)} + \Psi\beta^{(k)} - c_1\right)^2 + \frac{\zeta_3}{2}\left\|c_1 - c_1^{(k-1)}\right\|^2, \tag{156}$$

$$c_2^{(k)} = \arg\min_{c_2} \lambda \left(1 - u^{(k-1)}\right)^\top \left(Kd^{(k)} + \Psi\beta^{(k)} - c_2\right)^2 + \frac{\zeta_4}{2}\left\|c_2 - c_2^{(k-1)}\right\|^2, \tag{157}$$

and the solutions are given by the following:

$$c_1^{(k)} = \frac{\zeta_3 c_1^{(k-1)} + 2\lambda \left(u^{(k-1)}\right)^\top \left(Kd^{(k)} + \Psi\beta^{(k)}\right)}{\zeta_3 + 2\lambda \left(u^{(k-1)}\right)^\top I}, \tag{158}$$

$$c_2^{(k)} = \frac{\zeta_4 c_2^{(k-1)} + 2\lambda \left(1 - u^{(k)}\right)^\top \left(Kd^{(k)} + \Psi\beta^{(k)}\right)}{\zeta_4 + 2\lambda \left(1 - u^{(k)}\right)^\top I}. \tag{159}$$

Subproblem 5. In this subproblem, the following minimization problem is solved for optimal value of u:

$$u^{(k)} = \arg\min_{u \in [0,1]} \frac{\rho_2}{2} \left\|\mathbf{w}^{(k-1)} - \nabla u + \mathbf{b}_2^{(k-1)}\right\|^2 + \frac{\zeta_5}{2}\left\|u - u^{(k-1)}\right\|^2$$

$$\lambda \left[(u)^\top \left(Kd^{(k)} + \Psi\beta^{(k)} - c_1^{(k)}\right)^2 \right. \tag{160}$$

$$\left. + (1-u)^\top \left(Kd^{(k)} + \Psi\beta^{(k)} - c_2^{(k)}\right)^2 \right].$$

The solution to this is given by the following:

$$u^{(k)} = \Re\left[\mathcal{F}^*\left(\frac{\rho_2 \mathcal{F}\left(\nabla^*\left(\mathbf{w}^{(k-1)} + \mathbf{b}_2^{(k-1)}\right)\right) - \lambda \mathcal{F}\left(r^{(k)}\right) + \zeta_5 \mathcal{F}\left(u^{(k-1)}\right)}{\zeta_5 + \rho_2 \mathcal{F}\left(\nabla^2\right)}\right)\right], \tag{161}$$

where $r^{(k)} = \left(Kd^{(k)} + \Psi\beta^{(k)} - c_1^{(k)}\right)^2 - \left(Kd^{(k)} + \Psi\beta^{(k)} - c_2^{(k)}\right)^2$ and $\mathcal{F}$ is the fast Fourier transform operator and $\mathcal{F}^*$ is its inverse.

Subproblem 6. In this subproblem, the following minimization problem is solved to update w:

$$w^{(k)} = \arg\min_{w} \mu g^\top |w| + \frac{\rho_2}{2} \|w - \nabla u^{(k)} + b_2^{(k-1)}\|_2^2. \tag{162}$$

Solving this minimization problem leads to the following solution:

$$w^{(k)} = \text{shrink}(\nabla u^{(k)} - b_2^{(k-1)}, \frac{\mu}{\rho_2} \cdot g), \tag{163}$$

the Bregman parameter is updated as follows:

$$b_2^{(k)} = b_2^{(k-1)} + w^{(k)} - \nabla u^{(k)}. \tag{164}$$

For experimental results, comparison, and extension of the model to selective segmentation, the readers are advised to see Burrows et al. (2021).

An Optimization-Based Multilevel Algorithm for Selective Image Segmentation Models

In 2017, Jumaat and Chen proposed a multilevel method for solution of Badshah-Chen selective segmentation model discussed in section "Active Contour-Based Image Selective Model" and Rada-Chen selective segmentation discussed in section "One-Level Selective Segmentation Model".

Multilevel Algorithm for Badshah-Chen (BC) Model

Consider energy functional of the Badshah-Chen model given in Equation (110):

$$\min_{\phi(x,y), c_1, c_2} F_\epsilon(\phi(x,y), c_1, c_2) = \mu \int_\Omega G(x,y)|\nabla H_\epsilon(\phi(x,y))|dxdy$$

$$+ \lambda_1 \int_\Omega |z(x,y) - c_1|^2 H_\epsilon(\phi(x,y))dxdy$$

$$+ \lambda_2 \int_\Omega |z(x,y) - c_2|^2 (1 - H_\epsilon(\phi(x,y)))dxdy.$$

where $G(x,y) = d(x,y)g(|\nabla z(x,y)|)$ and $|\nabla H_\epsilon(\phi(x,y))| = \delta_\epsilon(\phi)|\nabla\phi(x,y)|$. Suppose that the average intensities c_1 and c_2 are found at the start by using (13), and to update ϕ, the following minimization problem will be considered:

$$\min_{\phi(x,y)} F_\epsilon(\phi(x,y)) = \mu \int_\Omega G(x,y)\delta_\epsilon(\phi)|\nabla\phi(x,y)|dxdy$$
$$+ \lambda_1 \int_\Omega |z(x,y) - c_1|^2 H_\epsilon(\phi(x,y))dxdy \qquad (165)$$
$$+ \lambda_2 \int_\Omega |z(x,y) - c_2|^2 (1 - H_\epsilon(\phi(x,y)))dxdy.$$

Here assume that given image $z(x,y)$ has size $n \times n$ where $n = 2^L$. The standard coarsening defines $L+1$ levels where $k = 1$(finest level), $2, \ldots, L, L+1$(coarsest level); furthermore, k-th level has $\tau_k \times \tau_k$ "superpixels," and each "superpixel" has $b_k \times b_k$ pixels where $\tau_k = \frac{n}{b_k}$ and $b_k = 2^{k-1}$. By using discrete form of TV $|\nabla\phi|$, Equation (165) can be written as follows:

$$\min_{\{\phi_{i,j}\}'s} F(\phi_{1,1}, \phi_{2,1}, \ldots, \phi_{m_1-1,m_2}, \phi_{m_1,m_2})$$
$$= \mu \sum_{i=1}^{m_1-1} \sum_{j=1}^{m_2-1} G_{i,j} \sqrt{\left(\frac{\phi_{i+1,j} - \phi_{i,j}}{h}\right)^2 + \left(\frac{\phi_{i,j+1} - \phi_{i,j}}{h}\right)^2} . \delta_\epsilon(\phi_{i,j})h^2$$
$$+ \sum_{i=1}^{m_1-1} \sum_{j=1}^{m_2-1} [\lambda_1(z_{i,j} - c_1)^2 H_\epsilon(\phi_{i,j}) + \lambda_2(z_{i,j} - c_2)^2(1 - H_\epsilon(\phi_{i,j}))].h^2.$$
$$= \underline{\mu} \sum_{i=1}^{m_1-1} \sum_{j=1}^{m_2-1} G_{i,j} \sqrt{\left(\phi_{i+1,j} - \phi_{i,j}\right)^2 + \left(\phi_{i,j+1} - \phi_{i,j}\right)^2} . \delta_\epsilon(\phi_{i,j}) \qquad (166)$$
$$+ \sum_{i=1}^{m_1-1} \sum_{j=1}^{m_2-1} \underbrace{[\lambda_2(z_{i,j} - c_1)^2 - \lambda_2(z_{i,j} - c_2)^2]}_{r(x,y)} H_\epsilon(\phi_{i,j}) + \text{terms independent of } \phi,$$

where $\underline{\mu} = \mu/h$ and the minimization is done with respect to ϕ, so the last term will not be considered from here onward. Consider fine-level local minimization first, which is done by using coordinate descent method.

The Finest-Level Local Minimization ($k = 1$)
Let $\widetilde{\phi}$ be the current iterate. Then our idea is to solve a series of subproblems of the following form:

$$\min_C F_\epsilon(\widetilde{\phi} + C)$$

where C is a local and piecewise constant function. Consider a particular pixel (i, j). Clearly if only $\phi_{i,j}$ is allowed to vary, we simply consider the local subproblem:

$$\min_{\phi_{i,j}} F^{\text{local}}(\phi_{i,j}) = \mu\bigg[G_{ij}\sqrt{(\phi_{ij}-\widetilde{\phi}_{i+1,j})^2 + (\phi_{ij}-\widetilde{\phi}_{i,j+1})^2}\delta_\epsilon(\phi_{i,j})$$
$$+ G_{i-1,j}\sqrt{(\phi_{ij}-\widetilde{\phi}_{i-1,j})^2 + (\widetilde{\phi}_{i-1,j}-\widetilde{\phi}_{i-1,j+1})^2}\delta_\epsilon(\widetilde{\phi}_{i-1,j})$$
$$+ G_{i,j-1}\sqrt{(\phi_{ij}-\widetilde{\phi}_{i,j-1})^2 + (\widetilde{\phi}_{i,j-1}-\widetilde{\phi}_{i+1,j-1})^2}\delta_\epsilon(\widetilde{\phi}_{i,j-1})\bigg]$$
$$+ r_{ij} H(\widetilde{\phi}_{ij}),$$

where $r_{i,j} = \lambda_1(z_{i,j}-c_1)^2 - \lambda_2(z_{i,j}-c_2)^2$. Starting from $\phi_{i,j}^{\text{old}} = \widetilde{\phi}_{i,j}$, we can iterate the following (Richardson type) scheme to obtain an approximation for $\phi_{i,j}$:

$$\phi_{i,j}^{\text{new}} = RHS/LHS, \qquad (167)$$

where:

$$RHS = \mu\bigg[G_{ij}\frac{(\widetilde{\phi}_{i+1,j}+\widetilde{\phi}_{i,j+1})}{L_1}\delta_\epsilon(\phi_{i,j}^{\text{old}}) + G_{i-1,j}\frac{\widetilde{\phi}_{i-1,j}\cdot\delta_\epsilon(\widetilde{\phi}_{i-1,j})}{L_2}$$
$$+ G_{i,j-1}\frac{\widetilde{\phi}_{i,j-1}\cdot\delta_\epsilon(\widetilde{\phi}_{i,j-1})}{L_3}\bigg] + r_{i,j}\delta_\epsilon(\widetilde{\phi}_{i,j}),$$

$$LHS = \mu\bigg[\frac{2\delta_\epsilon(\phi_{i,j}^{\text{old}})}{L_1} + \frac{2\epsilon L_1}{\pi(\epsilon^2+\phi_{i,j}^{\text{old}\,2})^2} + \frac{\delta_\epsilon(\widetilde{\phi}_{i-1,j})}{L_2} + \frac{\delta_\epsilon(\widetilde{\phi}_{i,j-1})}{L_3}\bigg]$$

and

$$L_1 = \sqrt{(\phi_{ij}^{\text{old}}-\widetilde{\phi}_{i+1,j})^2 + (\phi_{ij}^{\text{old}}-\widetilde{\phi}_{i,j+1})^2 + \beta}$$
$$L_2 = \sqrt{(\phi_{ij}^{\text{old}}-\widetilde{\phi}_{i-1,j})^2 + (\widetilde{\phi}_{i-1,j}-\widetilde{\phi}_{i-1,j+1})^2 + \beta}$$
$$L_3 = \sqrt{(\phi_{ij}^{\text{old}}-\widetilde{\phi}_{i,j-1})^2 + (\widetilde{\phi}_{i,j-1}-\widetilde{\phi}_{i+1,j-1})^2 + \beta},$$

and $\gamma > 0$ is a regularizing parameter. Equation (167) is usually done for few steps only to update $\widetilde{\phi}_{i,j}$.

The General-Level k Local Minimization ($1 < k \leq L$)

On a general-level k, consider the following minimization subproblem:

$$\min_C F(\widetilde{\phi}+C), \qquad (168)$$

where C is a local and piecewise constant function of support $\tau_k \times \tau_k = 2^{k-1} \times 2^{k-1}$ at each block (i,j) of pixels. On kth level, the subproblem may be taken as follows:

$$\hat{c} = \arg\min_{c \in \mathbb{R}^{\tau_k \times \tau_k}} F(\widetilde{\phi} + I_k B_k c), \qquad C_k = I_k B_k \hat{c}, \qquad (169)$$

where $B_k : \mathbb{R} \to \mathbb{R}^{\tau_k \times \tau_k}$ duplicates a constant to a block of constants and $I_k : \mathbb{R}^{\tau_k \times \tau_k} \to \mathbb{R}^{n \times n}$ is the interpolation operator so $C_k \in \mathbb{R}^{n \times n}$. Here we may illustrate $C_k = I_k B_k \hat{c}$ as follows (Chan and Chen 2006):

$$C_k = \begin{bmatrix} 0 & 0 & \cdots\cdots\cdots & 0 & 0 \\ \cdots\cdots & \cdots\cdots & \cdots\cdots\cdots & \cdots\cdots & \cdots\cdots \\ 0 & \cdots & c & \cdots & c & \cdots & 0 \\ \cdots\cdots & \cdots\cdots & \cdots\cdots & \cdots\cdots & \cdots\cdots \\ 0 & \cdots & c & \cdots & c & \cdots & 0 \\ \cdots\cdots & \cdots\cdots & \cdots\cdots & \cdots\cdots & \cdots\cdots \\ 0 & 0 & \cdots\cdots\cdots & 0 & 0 \end{bmatrix} \text{ to approximate } \begin{bmatrix} c_{11} & \cdots & \cdots\cdots\cdots & \cdots & c_{1n} \\ \cdots\cdots & \cdots\cdots & \cdots\cdots & \cdots\cdots & \cdots\cdots \\ c_{i1} & \cdots & c_{ii} & \cdots & c_{ij} & \cdots & c_{in} \\ \cdots\cdots & \cdots\cdots & \cdots\cdots & \cdots\cdots & \cdots\cdots \\ c_{j1} & \cdots & c_{ji} & \cdots & c_{jj} & \cdots & c_{jn} \\ \cdots\cdots & \cdots\cdots & \cdots\cdots & \cdots\cdots & \cdots\cdots \\ c_{n1} & \cdots & \cdots\cdots\cdots & \cdots & c_{nn} \end{bmatrix}.$$

Details of solving the local minimization subproblem (169) are here. Set on level k, $b = \tau_k = 2^{k-1}$, $k_1 = (i-1)b + 1$, $k_2 = ib$, $\ell_1 = (j-1)b + 1$, $\ell_2 = jb$. Firstly, note that on level k, there are only $m_1/\tau_k \times m_2/\tau_k$ subproblems each of which is essentially one dimensional (mimicking a coarse grid of a geometric multigrid method). Secondly, introduce the Richardson-type iterative method adopted for each subproblem.

At each block (i, j) of pixels, solve (169) for $c_{i,j}$. Observe that each TV term $|\nabla \phi|$ does not change within the interior pixels of each block on level k because of the following:

$$\sqrt{[(c_{i,j} + \widetilde{\phi}_{k,\ell}) - (c_{i,j} + \widetilde{\phi}_{k+1,\ell})]^2 + [(c_{i,j} + \widetilde{\phi}_{k,\ell}) - (c_{i,j} + \widetilde{\phi}_{k,\ell+1})]^2}$$
$$= \sqrt{[\widetilde{\phi}_{k,\ell} - \widetilde{\phi}_{k+1,\ell}]^2 + [\widetilde{\phi}_{k,\ell} - \widetilde{\phi}_{k,\ell+1}]^2} \equiv T_{k,\ell}.$$

So it remains to consider the contribution to the TV term stemming from the boundary pixels (of the block) and the contribution of all interior pixels to the δ_ϵ term. Thus solving (169) is equivalent to solving the following (i, j) block local minimization problem:

$$\min_{c_{i,j}} F(\widetilde{\phi}_{i,j} + I_k B_k c_{i,j})$$

$$= \mu \sum_{\ell=\ell_1}^{\ell_2} G_{k_1-1,\ell} \sqrt{[c_{i,j} - (\widetilde{\phi}_{k_1-1,\ell} - \widetilde{\phi}_{k_1,\ell})]^2 + [\widetilde{\phi}_{k_1-1,\ell} - \widetilde{\phi}_{k_1-1,\ell+1}]^2} \cdot \delta_\epsilon(c_{i,j} + \widetilde{\phi}_{k_1-1,\ell})$$

$$+ \mu \sum_{k=k_1}^{k_2-1} G_{k,\ell_2} \sqrt{[c_{i,j} - (\widetilde{\phi}_{k,\ell_2+1} - \widetilde{\phi}_{k,\ell_2})]^2 + [\widetilde{\phi}_{k,\ell_2} - \widetilde{\phi}_{k+1,\ell_2}]^2} \cdot \delta_\epsilon(c_{i,j} + \widetilde{\phi}_{k,\ell_2})$$

$$+ \mu G_{k_2,\ell_2} \sqrt{[c_{i,j} - (\widetilde{\phi}_{k_2,\ell_2+1} - \widetilde{\phi}_{k_2,\ell_2})]^2 + [c_{i,j} - (\widetilde{\phi}_{k_2+1,\ell_2} - \widetilde{\phi}_{k_2,\ell_2})]^2} \cdot \delta_\epsilon(c_{i,j} + \widetilde{\phi}_{k_2,\ell_2})$$

$$+\underline{\mu}G_{k_2,\ell}\sum_{\ell=\ell_1}^{\ell_2-1}\sqrt{[c_{i,j}-(\widetilde{\phi}_{k_2+1,\ell}-\widetilde{\phi}_{k_2,\ell})]^2+[\widetilde{\phi}_{k_2,\ell}-\widetilde{\phi}_{k_2,\ell+1}]^2}.\delta_\epsilon(c_{i,j}+\widetilde{\phi}_{k_2,\ell})$$

$$+\underline{\mu}G_{k,\ell_1}\sum_{k=k_1}^{k_2}\sqrt{[c_{i,j}-(\widetilde{\phi}_{k,\ell_1-1}-\widetilde{\phi}_{k,\ell_1})]^2+[\widetilde{\phi}_{k,\ell_1-1}-\widetilde{\phi}_{k+1,\ell_1-1}]^2}.\delta_\epsilon(c_{i,j}+\widetilde{\phi}_{k,\ell_1})$$

$$+\sum_{k=k_1+1}^{k_2-1}\sum_{\ell=\ell_1+1}^{\ell_2-1}T_{k,\ell}.\delta_\epsilon(c_{i,j}+\widetilde{\phi}_{k,\ell})+\sum_{\ell=\ell_1}^{\ell_2}\sum_{k=k_1}^{k_2}r(k,\ell)H_\epsilon(c_{i,j}+\widetilde{\phi}_{k,\ell}). \qquad (170)$$

To simplify the formulae, let:

$$\Phi_{k,\ell}=\widetilde{\phi}_{k,\ell+1}-\widetilde{\phi}_{k,\ell}, \qquad \Theta_{k,\ell}=\widetilde{\phi}_{k+1,\ell}-\widetilde{\phi}_{k,\ell},$$

and:

$$P_{k,\ell}=\frac{\Phi_{k,\ell}+\Theta_{k,\ell}}{2}, \qquad Q_{k,\ell}=\frac{\Phi_{k,\ell}-\Theta_{k,\ell}}{2}.$$

Using the identity:

$$\sqrt{(c-a)^2+(c-b)^2}=\sqrt{2}\sqrt{\left(c-\frac{a+b}{2}\right)^2+\left(\frac{a-b}{2}\right)^2},$$

we may rewrite (170) as the following problem:

$$\mathcal{F}(c_{i,j})=\underline{\mu}G_{k_1-1,\ell}\sum_{\ell=\ell_1}^{\ell_2}\sqrt{(c_{i,j}-\Theta_{k_1-1,\ell})^2+\Phi_{k_1-1,\ell}^2}.\delta_\epsilon(c_{i,j}+\widetilde{\phi}_{k_1-1,\ell})$$

$$+\underline{\mu}G_{1,\ell_2}\sum_{k=k_1}^{k_2-1}\sqrt{(c_{i,j}-\Phi_{k,\ell_2})^2+\Theta_{k,\ell_2}^2}\delta_\epsilon(c_{i,j}+\widetilde{\phi}_{k,\ell_2})$$

$$+\underline{\mu}\sum_{\ell=\ell_1}^{\ell_2-1}G_{k_2,\ell}\sqrt{(c_{i,j}-\Theta_{k_2,\ell})^2+\Phi_{k_2,\ell}^2}\delta_\epsilon(c_{i,j}+\widetilde{\phi}_{k_2,\ell})$$

$$+\underline{\mu}\sum_{k=k_1}^{k_2}G_{k,\ell_1}\sqrt{(c_{i,j}-\Phi_{k,\ell_1})^2+\Theta_{k,\ell_1}^2}\delta_\epsilon(c_{i,j}+\widetilde{\phi}_{k,\ell_1})$$

$$+\underline{\mu}G_{k_2,\ell_2}\sqrt{2}\sqrt{(c_{i,j}-P_{k_2,\ell_2})^2+(Q_{k_2,\ell_2})^2}\delta_\epsilon(c_{i,j}+\widetilde{\phi}_{k_2,\ell_2})$$

$$+\underline{\mu}\sum_{k=k_1+1}^{k_2-1}\sum_{\ell=\ell_1+1}^{\ell_2-1}T_{k,\ell}.\delta_\epsilon(c_{i,j}+\widetilde{\phi}_{k,\ell})+\sum_{k=k_1}^{k_2}\sum_{\ell=\ell_1}^{\ell_2}r_{k,\ell}H_\epsilon(c_{i,j}+\widetilde{\phi}_{k,\ell}).$$

The first-order condition for $\mathcal{F}'(c_{i,j}) = 0$ and doing some manipulations, the following iterative scheme for c_{ij} will be achieved:

$$c_{i,j}^{\text{new}} = RHS^{\text{old}}/LHS^{\text{old}}, \qquad (171)$$

starting from $c_{i,j}^{\text{old}} = 0$:

$$RHS^{\text{old}} = 2\mu \sum_{\ell=\ell_1}^{\ell_2} G_{k_1-1,\ell} \frac{\widetilde{\phi}_{k_1-1,\ell}\sqrt{(c_{i,j}^{\text{old}} - \Theta_{k_1-1,\ell})^2 + \Phi_{k_1-1,\ell}^2}}{(\epsilon^2 + (c_{i,j}^{\text{old}} + \widetilde{\phi}_{k_1-1,\ell})^2)^2}$$

$$+ \mu \sum_{\ell=\ell_1}^{\ell_2} \frac{\Theta_{k_1-1,\ell}}{(\epsilon^2 + (c_{i,j}^{\text{old}} + \widetilde{\phi}_{k_1-1,\ell})^2)^2 \sqrt{(c_{i,j}^{\text{old}} - \Theta_{k_1-1,\ell})^2 + \Phi_{k_1-1,\ell}^2}}$$

$$+ \ldots + 2\mu \sum_{k=k_1+1}^{k_2-1} \sum_{\ell=\ell_1+1}^{\ell_2-1} T_{k,\ell} \cdot \frac{\widetilde{\phi}_{k,\ell}}{(\epsilon^2 + (c_{i,j}^{\text{old}} + \widetilde{\phi}_{k,\ell})^2)^2}$$

$$- \sum_{k=k_1}^{k_2} \sum_{\ell=\ell_1}^{\ell_2} r_{k,\ell} \cdot \left[\frac{2c_{i,j}^{\text{old}} \widetilde{\phi}_{k,\ell}}{(\epsilon^2 + \phi_{k,\ell}^2)^2} + \frac{1}{(\epsilon^2 + (c_{i,j}^{\text{old}} + \widetilde{\phi}_{k,\ell})^2)} \right],$$

and:

$$LHS^{\text{old}} = -2\mu \sum_{\ell=\ell_1}^{\ell_2} G_{k_1-1,\ell} \frac{\sqrt{(c_{i,j}^{\text{old}} - \Theta_{k_1-1,\ell})^2 + \Phi_{k_1-1,\ell}^2}}{(\epsilon^2 + (c_{i,j}^{\text{old}} + \widetilde{\phi}_{k_1-1,\ell})^2)^2}$$

$$+ \mu \sum_{\ell=\ell_1}^{\ell_2} \frac{1}{(\epsilon^2 + (c_{i,j}^{\text{old}} + \widetilde{\phi}_{k_1-1,\ell})^2)^2 \sqrt{(c_{i,j}^{\text{old}} - \Theta_{k_1-1,\ell})^2 + \Phi_{k_1-1,\ell}^2}}$$

$$+ \cdots - 2\mu \sum_{k=k_1+1}^{k_2-1} \sum_{\ell=\ell_1+1}^{\ell_2-1} \frac{T_{k,\ell}}{(\epsilon^2 + (c_{i,j}^{\text{old}} + \widetilde{\phi}_{k,\ell})^2)^2}$$

$$- 2 \sum_{k=k_1}^{k_2} \sum_{\ell=\ell_1}^{\ell_2} r_{k,\ell} \frac{\widetilde{\phi}_{k,\ell}}{(\epsilon^2 + \widetilde{\phi}_{k,\ell}^2)^2}.$$

Once $c_{i,j}$ is obtained, $\widetilde{\phi}_{k,\ell}$ is updated as follows:

$$\phi_{k,l} = \widetilde{\phi}_{k,\ell} + c_{i,j}$$

to the full (i, j) block.

12 Fast Numerical Methods for Image Segmentation Models

The Coarsest-Level Minimization ($k = L + 1$)

On the coarsest level, the whole image is considered to be a single block, so contribution for updating the constant will only come from the delta function term $\delta_\epsilon(\phi)$, i.e., no contribution from the TV term. Thus consider the following local minimization problem on the coarsest level:

$$\min_c F(\widetilde{\phi} + I_k B_k c) = \min_c \mu \sum_{i=1}^{m_1} \sum_{j=1}^{m_2} T_{i,j} \delta_\epsilon(\widetilde{\phi}_{i,j} + c) + \sum_{i=1}^{m_1} \sum_{j=1}^{m_2} r_{i,j} H_\epsilon(\widetilde{\phi}_{i,j} + c).$$

Taking variation with respect to c and equating to 0 leads to the following:

$$-\mu \sum_{i=1}^{m_1} \sum_{j=1}^{m_2} G_{ij} G_{i,j} T_{i,j} \frac{\widetilde{\phi}_{i,j} + c^{\text{new}}}{(\epsilon^2 + (\widetilde{\phi}_{i,j} + c)^2)^2}$$

$$+ \sum_{i=1}^{m_1} \sum_{j=1}^{m_2} r_{i,j} \left[\frac{2c^{\text{old}} \widetilde{\phi}_{i,j}}{(\epsilon^2 + \widetilde{\phi}_{i,j}^2)^2} + \frac{1}{(\epsilon^2 + (c^{\text{old}} + \widetilde{\phi}_{i,j})^2)} \right.$$

$$\left. - \frac{2c^{\text{new}} \widetilde{\phi}_{i,j}}{(\epsilon^2 + \widetilde{\phi}_{i,j}^2)^2} \right] = 0. \tag{172}$$

Linearizing and solving this equation for c^{new} and then updating $\widetilde{\phi}$ will be similarly done as above.

Here is the algorithm for multilevel method for BC model:

Algorithm 6 2D multilevel algorithm (ML1)

$[\phi, c_1, c_2] \leftarrow Opt\ Multilevel1(\phi, z)$ Given the image z and an initial guess $\phi = \widetilde{\phi}$ with $L + 1$ levels, our multilevel algorithm proceeds as follows:
Start
set $\phi_0 = \widetilde{\phi}$ and compute c_1, c_2.
for level $k = 1, 2, \ldots, L + 1$.
    If $k = 1$, for finest level   solve *(167)*.
    Elseif $k = L + 1$ i.e. on coarsest level.   solve *(172)* to find c
    Else on all other levels   solve *(171)*.
    Update $\phi = \widetilde{\phi} + I_k B_k c$.
end
Go to Start with $\widetilde{\phi} = \phi$ unless $\|\phi - \phi_0\| < tol$.

Exactly in same lines, multilevel method for RC model can be derived, and it is left as an exercise for the reader. For comparison and experimental results, see Jumaat and Chen (2017).

Machine/Deep Learning Techniques for Image Segmentation

In this section, a survey of deep/machine learning techniques for image segmentation is given.

Machine Learning with Region-Based Active Contour Models in Medical Image Segmentation

In 2017, Pratando et al. proposed an architecture which integrates machine learning with a region-based active contour model.

Proposed Framework
The proposed framework can be constructed from any algorithm used for classification, which is combined with a region-based model with a level set method. The matrix of classifier probability scores is generated by using KNN and support vector machine (SVM). The matrix is then regularized and combined with Chan-Vese (CV) active contour model (Chan and Vese 2001) which is discussed in section "Chan-Vese Model" in detail.

Classifier Probability Scores
For a given image z, a matrix of classifier probability scores is generated from the classification algorithms. Here two classification algorithms, namely, KNN and SVM, are investigated.

KNN. KNN provides scores in the range [0, 1] which can be implemented easily using the fuzzy KNN rule. This rule is derived from the fuzzy set and the KNN classifier in machine learning. For a reference set $X_R = \{x_i : 1 \leq i \leq m_R\}$ and a set of l-dimensional vectors $W = \{w_i : 1 \leq i \leq m_R\}$, $w_i = (w_{i,1}, w_{i,2}, \ldots, w_{i,l},)$, l and m_R are the number of classes and the number of elements in the reference set X_R, respectively. Due to fuzziness of the vectors, the following condition must be satisfied:

$$\sum_{j=1}^{l} w_{i,j} = 1, \qquad 0 \leq w_{i,j} \leq 1. \tag{173}$$

The value of $w_{i,j}$ for $1 \leq i \leq m_R$ and $1 \leq j \leq l$ is the membership value of the i-th object to class j. For a particular x to be classified, the set K of indices corresponding to the classes of k-nearest neighbors of x in X_R is obtained. The fuzzy decision vector v in the fuzzy KNN is computed in the following way:

$$v = \frac{1}{k} \sum_{s \in K} w_s. \tag{174}$$

The maximum v_j, $1 \leq j \leq l$ where $v = (v_1, v_2, \ldots, v_l)$ is used to define the object class in the original KNN.

Support vector machine SVM. In support vector machine, the given data is divided into two classes by finding a hyperplane between the classes with largest margin. This is done by using a sign function $class(x) = sgn(h(x))$, where $h(x)$ is the separating hyperplane for the two classes and is given by the following:

$$h(x) = \mathbf{w}_0^T \mathbf{x} + b_0 \tag{175}$$

where $\mathbf{w}_0$ is a d-dimensional optimal weight vector, $\mathbf{x}$ is the given data, and b_0 is the optimal bias. Since it may be difficult to separate the data in the original input space, a transformation of the data into a higher dimensional space through function φ is introduced. The $h(x)$ takes the following form:

$$h(x) = \mathbf{w}_0^T \varphi(\mathbf{x}) + b_0. \tag{176}$$

It is still hard to find φ explicitly, so a kernel $K(\mathbf{x}, \mathbf{x}_i)$ is introduced and thus (176) may be written as follows:

$$h(x) = \sum_{i=1}^{N} \alpha_i y_i K(\mathbf{x}, \mathbf{x}_i) + b_0, \tag{177}$$

where α_i is the estimated SVM parameter and $y_i \in \{1, -1\}$ is the desired class for the corresponding $\mathbf{x}_i$. The value of $h(x)$ is the SVM evaluation score and the sign is the predicted class. Note that the scores of KNN falls in the range $[0, 1]$ and that of SVM in the range $(-\infty, \infty)$, which can be converted to a prior probability score.

Regularization for Classifier Probability Score

Classifiers generate binary results by applying a hard limiter function to the probability scores. Let $s \in [0, 1]$ be a probability score and ρ be a regularization function that maps s to a real value in $[0, 1]$. The traditional classifiers generate binary results by the following:

$$\rho_1(s) = \begin{cases} 1 & \text{if } s \geq \frac{1}{2} \\ 0 & \text{if } s < \frac{1}{2} \end{cases} \tag{178}$$

Instead of refining these binary scores using machine learning algorithms. To retain the probability scores which are processed further by applying any region-based active contour model. This aims to find an optimal solution where the function $\rho(s)$ can be simply expressed by the following:

$$\rho_2(s) = s. \tag{179}$$

A nonlinear function ρ approximately lying under ρ_2 for $s > 0.5$ and above ρ_2 for $s < 0.5$ leads to better results. The regularization function in general should satisfy the following conditions:

1. The domain and the range, ρ, should be [0, 1].
2. It should be increasing.
3. The following equations must hold:

$$\lim_{s \to 0} \rho(s) = 0 \tag{180}$$

$$\lim_{s \to 0.5} \rho(s) = 0.5 \tag{181}$$

$$\lim_{s \to 1} \rho(s) = 1. \tag{182}$$

4. It should be close to 0.5 when s is in the vicinity of 0.5.

There are some more options for taking regularization functions $\rho(s)$; for details see Pratondo et al. (2017).

The map of ρ is then fed to a region-based active contour model. Through energy minimization using the level set method, the optimum solution for the desired region can be obtained. For experimental results, data set utilization, and comparisons, see Pratondo et al. (2017).

ResBCU-Net: Deep Learning Approach for Segmentation of Skin Images

In 2022, Badshah and Ahmad proposed a new architecture based on CNNs, namely, ResBCU-Net for segmentation of skin images/medical images. The network, ResBCU-Net, is an extension of the U-Net which utilizes residual blocks, batch normalization, and bidirectional ConvLSTM. In addition, we present an extended form of ResBCU-Net, ResBCU-Net(d = 3), which takes advantage of densely connected layers in its bottleneck section.

Proposed Work

Based on U-Net (Olaf et al. 2015) and inspired by residual blocks (He et al. 2016), batch normalization (Ioffe and Szegedy 2015), and bidirectional convolutional long-short-term memory (BConvLSTM) network (Song et al. 2018), a neural network, named as ResBCU-Net, shown in the Fig. 4 was proposed for segmentation of skin/medical images. The authors have made changes in the encoding path and decoding path of the classical U-Net, which is explained here in detail by considering encoding and decoding separately.

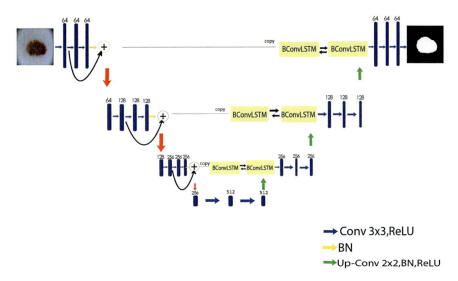

Fig. 4 ResBCU-Net architecture with residual blocks in the encoding path and BConvLSTM in the decoding path. The numbers on top of the rectangles show number of channels

Encoding

Unlike to the U-Net (Olaf et al. 2015), encoding/contracting path of ResBCU-Net consists of residual blocks (He et al. 2016) and batch normalization layers (Ioffe and Szegedy 2015) with nine convolution layers. The path consists of three blocks; each block contains three convolution layers followed by a batch normalization layer. The output of first convolution layer in each block is added with the output of the batch normalization layer, which is then followed by a max pooling layer. At the same time, before the max pooling layer, the output of each block is passed for concatenation with the corresponding output of the decoding/expanding path.

Residual Blocks

Successive sequences of convolution layers lead to learning of different features; in some cases it may also lead to learning of redundant features; and adding more layers lead to higher training error. To solve this problem in such deeper models, residual blocks are introduced in He et al. (2016). An input to some convolution layers is added to the output of the layers; the resultant is again fed to the successive convolution layers; an example of residual block is shown in the Fig. 5.

The authors utilized this approach for ResBCU-Net encoding path. Instead of blocks of two convolution layers in the encoding path, three convolutions blocks each followed by a batch normalization layer are introduced. Each block is then converted to residual blocks by adding the output of the first convolution layer to the output of the batch normalization layer in the block, as shown in the Fig. 6.

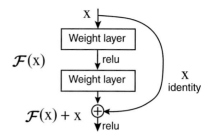

Fig. 5 ResNet residual block

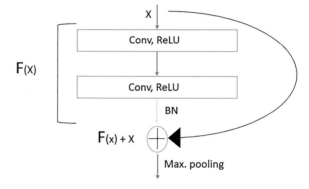

Fig. 6 ResBCU-Net residual block

Batch Normalization

To avoid over-fitting and for acceleration of the training process, batch normalization layers (Ioffe and Szegedy 2015) in the encoding and decoding path of ResBCU-Net are included. The batch normalization layer controls variation in distribution by calculating mean and standard deviation values of the data set as a whole by adjusting the mean to 0 and variance to 1; the equation for batch normalization (BN) is given below:

$$BN = \gamma_c \left[\frac{I_{n,c,h,w} - \mu_c}{\sqrt{\sigma_c^2 - \epsilon}} \right] + \beta_c$$

where $I_{n,c,h,w}$ represents n-number of images provided to a neural network at a time with c channels, h heights, and w widths. μ_c and σ_c^2 are channel-wise global mean and variance of the images, respectively. β_c and γ_c are learnable mean and standard deviation, respectively, while ϵ is kept constant as 0.00001.

Batch normalization layers in each block of the encoding and decoding path are introduced. In the encoding path, BN layers are used at the end of each block just before max pooling layer. After max pooling layer of the third block of convolution layers, bottleneck section of the network starts, where only two convolution layers each followed by an activation function, ReLU, are used. While in the decoding path, the batch normalization layers after each up-sampling layer are used, which are then followed by activation functions, ReLU, before proceeding to the next block.

Decoding

The decoding/expanding path of ResBCU-Net, inspired by BCDU-Net (Guo et al. 2019), contains convolution layers, up-sampling layers, batch normalization layers (Ioffe and Szegedy 2015), and bidirectional LSTM convolutions (BConvLSTM) (Song et al. 2018). Right after the bottleneck portion of the network, an up-sampling convolution with 2×2 filter, followed by a batch normalization layer, is used which is then followed by two convolution layer blocks. Features from the corresponding blocks in the encoding path are passed into the BConvLSTM after concatenation with the outputs of the corresponding block of the decoding path. In each block, outputs of BConvLSTMs are passed into two convolutional layers. At the end of the decoding path, we use a convolution layer with 1×1 filter followed by a sigmoid function as an activation function.

Bidirectional Long-Short-Term Memory Convolutions (BConvLSTM)

In the decoding path, the convolutional long-short-term memory (ConvLSTM) networks for ResBCU-Net are inspired by Azad et al. (2019) and Guo et al. (2019). The ConvLSTM to process features into two directions is used: forward and backward, known as BConvLSTM (Song et al. 2018). BConvLSTM has been implemented successively to enhance performance of neural networks (Cui et al. 1801; Guo et al. 2019). LSTMs are enhanced version of recurrent neural networks (RNNs) (Jordan 1990; Cleeremans et al. 1989; Pearlmutter 1989), which have been developed to overcome the gradient vanishing issue in long dependence of neural network in training (Fig. 7).

A single block of ConvLSTM consists of input gate, i_t; forget gate, f_t; and output gate, O_t. If $\chi_1, \chi_2, \ldots, \chi_t$ are inputs, $C_1, C_2, \ldots, C_t$ are cell-state outputs, and $h_1, h_2, \ldots, h_t$ represent hidden states; then the function of a single ConvLSTM can be represented by the following equations:

$$i_t = \sigma(\omega_{xi} * \chi_t + \omega_{hi} * h_{t-1} + \omega_{ci} o C_{t-1} + b_i)$$

$$f_t = \sigma(\omega_{xf} * \chi_t + \omega_{hf} * h_{t-1} + \omega_{cf} o C_{t-1} + b_f)$$

$$C_t = f_t o C_{t-1} + i_t o \tanh(\omega_{xc} * \chi_t + \omega_{hc} * h_{t-1} + b_c)$$

$$O_t = \sigma(\omega_{xo} * \chi_t + \omega_{ho} * h_{t-1} + \omega_{co} o C_t + b_o)$$

$$h_t = O_t o \tanh(C_t).$$

Here, $*$ and o are convolution operator and Hadamard product, respectively. The h_t is the hidden state, output, of the single ConvLSTM block. Now, in case of bidirectional ConvLSTM (BConvLSTM), the output can be represented as follows:

Fig. 7 A single block of ConvLSTM network (http://colah.github.io/posts/2015-08-Understanding-LSTMs/)

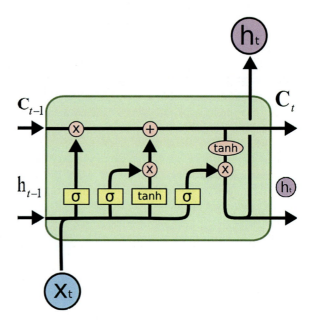

$$Y_t = \tanh(\omega_y^{\vec{h}} * \vec{h}_t + \omega_y^{\overleftarrow{h}} * \overleftarrow{h}_t + b).$$

Here, $\vec{h}_t$ and $\overleftarrow{h}_t$ are output states of forward and backward direction feature process; and the Y_t is the final output of a BConvLSTM block.

The copied features from encoding path are concatenated with corresponding outputs from decoding path and are then passed into BConvLSTM blocks. The output of these blocks then proceeds forward to the two convolution layer blocks. For training, testing, and comparison, see Badshah and Ahmad (2022) and references there in.

Conclusion

Some of the well-known active contour models for image segmentation are presented. Here both types of segmentations (global and selective) are discussed. In this chapter two-phase and multiphase segmentation models are discussed in detail. Minimization techniques for finding the optimal values and discussion about the fast numerical methods for solution of partial differential equations arising from the minimization of the models were key points of discussion in this chapter.

References

Allen, A.M., Cahn, J.W.: A microscopic theory for antiphase boundary motion and its application to antiphase domain coarsening. Acta Metall. **27**, 1085–1095 (1979)

Alvarez, L., Lions, P.-L., Morel, J.M.: Image selective smoothing and edge detection by nonlinear diffusion. SIAM J. Numer. Anal. **29**(3), 845–866 (1992)

Aubert, G., Kornprobst, P.: Mathematical Problems in Image Processing: Partial Differential Equations and the Calculus of Variations. Springer, New York (2002)

Azad, R., Asadi-Aghbolaghi, M., Fathy, M., Escalera, S.: Bi-directional convlstm U-Net with densley connected convolutions. In: Proceedings of the IEEE/CVF International Conference on Computer Vision Workshops (2019)

Badshah, N., Ahmad, A.: ResBCU-Net: deep learning approach for segmentation of skin images. Biomed. Sig. Process. Control **71**, 103137 (2022)

Badshah, N., Chen, K.: Multigrid method for the Chan-Vese model in variational segmentation. Commun. Comput. Phys. **4**(2), 294–316 (2008)

Badshah, N., Chen, K.: On two multigrid algorithms for modeling variational multiphase image segmentation. IEEE Trans. Image Process. **18**(5), 1097–1106 (2009)

Badshah, N., Chen, K.: Image selective segmentation under geometrical constraints using an active contour approach. Commun. Comput. Phys. **7**(4), 759–778 (2010)

Barash, D., Schlick, T., Israeli, M., Kimmel, R.: Multiplicative operator splittings in nonlinear diffusion: from spatial splitting to multiple timesteps. J. Math. Imaging Vis. **19**, 33–48 (2003)

Boyd, S., Parikh, N., Chu, E., Peleato, B., Eckstein, J.: Distributed optimization and statistical learning via the alternating direction method of multipliers. Found. Trends Mach. Learn. **3**, 1–122 (2010)

Brandt, A.: Multi-level adaptive solutions to boundary-value problems. Math. Comput. **31**(2), 333–390 (1977)

Bregman, L.: The relaxation method of finding the common point of convex sets and its application to the solution of problems in convex programming. USSR Comput. Math. Math. Phys. **7**(3), 200–217 (1967)

Briggs, W.L.: A Multigrid Tutorial. (1999)

Burrows, L., Guo, W., Chen, K., Torella, F.: Reproducible kernel Hilbert space based global and local image segmentation. Inverse Probl. Imaging **15**(1), 1–25 (2021)

Cai, X., Chan, R., Zeng, T.: A two-stage image segmentation method using a convex variant of the Mumford-Shah model and thresholding. SIAM J. Imaging Sci. **6**(1), 368–390 (2013)

Cai, X., Chan, R., Nikolova, M., Zeng, T.: A three-stage approach for segmenting degraded color images: smoothing, lifting and thresholding (SLaT). J. Sci. Comput. **72**, 1313–1332 (2017)

Caselles, V., Kimmel, R., Sapiro, G.: Geodesic active contours. Int. J. Comput. Vis. **22**, 61–79 (1997)

Chambolle, A., Pock, T.: A first-order primal-dual algorithm for convex problems with applications to imaging. J. Math. Imaging Vis. **40**, 120–145 (2011)

Chan, T.F., Chen, K.: An optimization based multilevel agorithm for total variation image denoising. SIAM J. Multiscale Model. Simul. (MMS) **5**(2), 615–645 (2006)

Chan, T.F., Vese, L.A.: Active Contours without Edges. IEEE Trans. Image Process. **10**(2), 266–277 (2001)

Chan, T.F., Esedoglu, S., Nikolova, M.: Algorithms for finding global minimizers of image segmentation and denoising models. SIAM J. Appl. Math. **66**(5), 1632–1648 (2006)

Chan, R., Yang, H., Zeng, T.: A two-stage image segmentation method for blurry images with poisson or multiplicative gamma noise. SIAM J. Imaging Sci. **7**(1), 98–127 (2014)

Chen, K.: Matrix Preconditioning Techniques and Applications. Cambridge University Press, Cambridge (2005)

Chen, Y., Lan, G., Ouyang, Y.: Optimal primal-dual methods for a class of saddle point problems. SIAM J. Optim. **24**(4), 1779–1814 (2014)

Cleeremans, A., Servan-Schreiber, D., McClelland, J.: Finite state automata and simple recurrent networks. Neural Comput. **1**(3), 372–381 (1989). MIT Press

Cui, Z., Ke, R., Pu, Z., Wang, Y.: Deep bidirectional and unidirectional LSTM recurrent neural network for network-wide traffic speed prediction. arXiv preprint, 1801.02143 (2018)

Deng, L.-J., Guo, W., Huang, T.-Z.: Single-image super-resolution via an iterative reproducing kernel hilbert space method. IEEE Trans. Circuits Syst. Video Technol. **26**, 2001–2014 (2016)

Geiser, J., Bartecki, K.: Additive, multiplicative and iterative splitting methods for Maxwell equations: algorithms and applications. In: International Conference of Numerical Analysis and Applied Mathematics (ICNAAM 2017)

Goldstein, T., Osher, S.: The split bregman algorithm for $l1$ regularized problems. SIAM J. Imaging Sci. **2**, 323–343 (2009)

Gout, C., Guyader, C.L., Vese, L.: Segmentation under geometrical consitions with geodesic active contour and interpolation using level set methods. Numer. Algorithms **39**, 155–173 (2005)

Guo, Y., Stein, J., Wu, G., Krishnamurthy, A.: SAU-Net: a universal deep network for cell counting. In: Proceedings of the 10th ACM International Conference on Bioinformatics, Computational Biology and Health Informatics, pp. 299–306 (2019)

Guyader, C.L., Gout, C.: Geodesic active contour under geometrical conditions theory and 3D applications. Numer. Algorithms **48**, 105–133 (2008)

He, K., Zhang, X., Ren, S., Sun, J.: Deep residual learning for image recognition. In: Proceedings of the IEEE Conference on Computer Vision and Pattern Recognition, pp. 770–778 (2016)

Ioffe, S., Szegedy, C.: Batch normalization: accelerating deep network training by reducing internal covariate shift. arXiv preprint, 1502.03167 (2015)

Jeon, M., Alexander, M., Pedrycz, W., Pizzi, N.: Unsupervised hierarchical image segmentation with level set and additive operator splitting. Pattern Recogn. Lett. **26**(10), 1461–1469 (2005)

Jordan, M.I.: Attractor dynamics and parallelism in a connectionist sequential machine. Artif. Neural Netw.: Concept Learn. 112–127 (1990)

Jumaat, A.K., Chen, K.: An optimization based multilevel algorithm for variational image segmentation models. Electron. Trans. Numer. Anal. **46**, 474–504 (2017)

Kanungo, T., Mount, D., Netanyahu, N., Piatko, C., Silverman, R., Wu, A.: An efficient k-means clustering algorithm: analysis and implementation. IEEE Trans. Pattern Anal. Mach. Intell. **24**, 881–892 (2002)

Kass, M., Witkin, A., Terzopoulos, D.: Snakes: active contour models. Int. J. Comput. Vis. **6**(4), 321–331 (1988)

Lu, T., Neittaanmaki, P., Tai, X.-C.: A parallel splitting up method for partial differential equations and its application to navier-stokes equations. RAIRO Math. Model. Numer. Anal. **26**(6), 673–708 (1992)

Mumford, D., Shah, J.: Optimal approximation by piecewise smooth functions and associated variational problems. Commun. Pure Appl. Math. **42**, 577–685 (1989)

Olaf, R., Philipp, F., Thomas, B.: U-Net: convolutional networks for biomedical image segmentation. In: International Conference on Medical Image Computing and Computer-Assisted Intervention. Springer, pp. 234–241 (2015)

Osher, S., Sethian, J.A.: Fronts propagating with curvature-dependent speed: algorithms based on Hamilton-Jacobi formulations. J. Comput. Phys. **79**(1), 12–49 (1988)

Osher, S., Burger, M., Goldfarb, D., Xu, J., Yin, W.: An iterative regularization method for total variation-based image restoration. Multiscale Model. Simul. **4**(2), 460–489 (2005)

Pearlmutter, B.: Learning state space trajectories in recurrent neural networks. Neural Comput. **1**(2), 263–269 (1989). MIT Press

Pratondo, A., Chee-Kong, C., Sim-Heng, O.: Integrating machine learning with region-based active contour models in medical image segmentation. J. Vis. Commun. Image R. **43**, 1–9 (2017)

Rada, L., Chen, K.: A new variational model with dual level set functions for selective segmentation. Commun. Comput. Phys. **12**(1), 261–283 (2012)

Rada, L., Chen, K.: Improved selective segmentation model using one level set. J. Algorithms Comput. Technol. **7**(4), 509–541 (2013)

Roberts, M., Chen, K., Li, J., Irion, K.: On an effective multigrid solver for solving a class of variational problems with application to image segmentation. Int. J. Comput. Math. **97**(10), 1–21 (2019)

Rudin, L.I., Osher, S., Fatemi, E.: Nonlinear total variation based noise removal algorithm. Physica D **60**(1–4), 259–268 (1992)

Savage, J., Chen, K.: An improved and accelerated non-linear multigrid method for total-variation denoising. Int. J. Comput. Math. **82**(8), 1001–1015 (2005)

Song, H., Wang, W., Zhao, S., Shen, J., Lam, K.: Pyramid dilated deeper ConvLSTM for video salient object detection. In: Proceedings of the European Conference on Computer Vision (ECCV), pp. 715–731 (2018)

Sussman, M., Smereka, P., Osher, S.: A level set approach for computing solutions to incompressible two-phase flow. J. Comput. Phys. **114**, 146–159 (1994)

Trottenberg, U., Schuller, A.: Multigrid. Academic, Orlando (2001)

Vese, L.A., Chan, T.F.: A multiphase level set framework for image segmentation using the Mumford and Shah model. Int. J. Comput. Vis. **50**, 271–293 (2002)

Weickert, J., Kühne, G.: Fast methods for implicit active contours models, preprint 61. Universität des Saarlandes, Saarbrücken (2002)

Weickert, J., ter Haar Romeny, B.M., Viergever, M.A.: Efficient and reliable schemes for nonlinear diffusion filtering. Scale-space theory in computer vision. Lect. Notes Comput. Sci. **1252**, 260–271 (1997)

Yang, Y., Zhao, Y., Wu, B., Wang, H.: A fast multiphase image segmentation model for gray images. Comput. Math. Appl. **67**, 1559–1581 (2014)

Yuan, Y., He, C.: Variational level set methods for image segmentation based on both L^2 and Sobolev gradients. Non Linear Anal. Real World Appl. **13**, 959–966 (2012)

Zhao, H.-K., Osher, S., Merriman, B., Kang, M.: Implicit and non parametric shape reconstruction from unorganized data using a variational level set method. Comput. Vis. Image Underst. **80**(3), 295–314 (2000)

On Variable Splitting and Augmented Lagrangian Method for Total Variation-Related Image Restoration Models

13

Zhifang Liu, Yuping Duan, Chunlin Wu, and Xue-Cheng Tai

Contents

Introduction	504
Basic Notation	507
Augmented Lagrangian Method for Total Variation-Related Image Restoration Models	508
Augmented Lagrangian Method for TV-L^2 Restoration	510
Augmented Lagrangian Method for TV-L^2 Restoration with Box Constraint	516
Augmented Lagrangian Method for TV Restoration with Non-quadratic Fidelity	519
Extension to Multichannel Image Restoration	524
The Multichannel TV Restoration Model	524
Augmented Lagrangian Method for Multichannel TV Restoration	526
Extension to High-Order Models	528
Augmented Lagrangian Method for Second-Order Total Variation Model	528
Augmented Lagrangian Method for Total Generalized Variation Model	531
Augmented Lagrangian Method for Euler Elastic-Based Model	536
Augmented Lagrangian Method for Mean Curvature-Based Model	539
Numerical Experiments	541
Conclusions	546
References	546

Z. Liu
School of Mathematical Sciences, Tianjin Normal University, Tianjin, China
e-mail: matlzhf@tjnu.edu.cn

Y. Duan
Center for Applied Mathematics, Tianjin University, Tianjin, China
e-mail: yuping.duan@tju.edu.cn

C. Wu (✉)
School of Mathematical Sciences, Nankai University, Tianjin, China
e-mail: wucl@nankai.edu.cn

X.-C. Tai
Hong Kong Center for Cerebro-cardiovascular Health Engineering (COCHE), Shatin, Hong Kong
e-mail: xtai@hkcoche.org

© Springer Nature Switzerland AG 2023
K. Chen et al. (eds.), *Handbook of Mathematical Models and Algorithms in Computer Vision and Imaging*, https://doi.org/10.1007/978-3-030-98661-2_84

Abstract

Variable splitting and augmented Lagrangian method are widely used in image processing. This chapter briefly reviews its applications for solving the total variation (TV) related image restoration problems. Due to the nonsmoothness of TV, related models and variants are nonsmooth convex or nonconvex minimization problems. Variable splitting and augmented Lagrangian method can benefit from the separable structure and efficient subsolvers, and has convergence guarantee in convex cases. We present this approach for a number of TV minimization models including TV-L^2, TV-L^1, TV with nonquadratic fidelity term, multichannel TV, high-order TV, and curvature minimization models.

Keywords

Variable splitting · Augmented lagrangian method · Total variation · Image restoration · Box constraint

Introduction

This short survey provides a brief review of the variable splitting and augmented Lagrangian method for total variation (TV)-related image restoration models. We will focus on this computational problem closely, and do not plan to touch other related topics like theoretical model analysis and algorithmic connections, which can be referred to, e.g., Aubert and Kornprobst (2010) and Glowinski et al. (2016) and references therein. Also, to keep the context as compact as possible, we would not expand all the details, although there are definitely lots of exellent works in the literature.

Total variation, which is a semi-norm of the space of functions of bounded variation, was first proposed for image denoising by Rudin, Osher, and Fatemi (ROF) in Rudin et al. (1992). In the discrete setting, it is essentially the L_1 norm of gradients and can maintain the sparse discontinuities. Therefore, it is appropriate to preserve image edges that are usually the most important features for images to recover. Owing to its edge-preserving property and convexity, total variation has been demonstrated very successful and become popular in image restoration like image denoising (Rudin et al. 1992; Le et al. 2007), image deblurring (Chan and Wong 1998; Wu and Tai 2010) and image inpainting (Bertalmio et al. 2003) and also various other types of image processing tasks including image decomposition (Vese and Osher 2003), image segmentation (Chan and Vese 2001), CT reconstruction (Persson et al. 2001), phase retrieval (Chang et al. 2016) and so on.

The total variation model has been generalized in many ways for different purposes. The original total variation regularization was proposed for gray image restoration (Rudin et al. 1992), which is the single channel case. To restore multichannel data, such as color images with RGB channels, people extended it to color TV and vectorial TV regularizations (Blomgren and Chan 1998; Sapiro and Ringach 1996). It is well-known that images recovered by total variation regularized models have the undesired staircase effect. To prevent the total variation oversharpening, there are several remarkable methods to improve the total variation regularization. These include the variable exponent TV models (Chen et al. 2006) and a wide class of high-order models, such as inf-convolution model (Chambolle and Lions 1997), second-order total variation model (Lysaker et al. 2003), bounded Hessian model (Hinterberger and Scherzer 2006), total generalized variation model (Bredies et al. 2010), and total fractional-order variation model (Zhang and Chen 2015) etc. By co-area formula, the total variation is the integral of lengths of all level curves of the intensity function. One natural extension way is thus to introduce curve curvature term for regularization. For example, Euler's elastica which contains both lengths and curvatures was proposed for image inpainting (Chan et al. 2002; Yashtini and Kang 2016), denoising (Tai et al. 2011; Duan et al. 2013), zooming (Tai et al. 2011; Duan et al. 2013), illusory contour (Kang et al. 2014), image decomposition (Liu et al. 2018), and image reconstruction (Yan and Duan 2020). Such regularity can provide strong priors for the continuity of edges. Another total variation-related geometric regularization technique we would like to mention is mean curvature minimization (Zhu and Chan 2012), which considers the image or graph in a high-dimensional space and transfers the image minimization problems to the corresponding surface minimization problems. From the viewpoint of image domain, total variation regularization was also extended to implicit surfaces, triangulated meshes and even general manifolds for image and data processing on curved spaces (Lai and Chan 2011; Wu et al. 2012) and normal vector filtering for surface denoising (Zhang et al. 2015). By exploiting the spatial interactions in images, total variation regularization was also generalized to nonlocal TV (Lou et al. 2010). By using non-convex penalty functions instead of the L_1 norm, non-convex TV regularizations got more and more attentions in recent years; see Chen et al. (2012), Hintermüller and Wu (2013), Wu et al. (2018), and Selesnick et al. (2020) and the references therein. They have been shown capable to generate good results with neat edges, as indicated by the interesting lower bound theory Nikolova (2005); Chen et al. (2012); Zeng and Wu (2018); Feng et al. (2018).

However, the non-smoothness of the total variation semi-norm gives rise to a challenge of its minimization. To overcome this problem, the common way is replacing total variation by its smoothed versions in image restoration model. Therefore, one can solve the new associated Euler-Lagrangian equation and obtain an approximate solution of the original model (Acar and Vogel 1994). For solving this Euler-Lagrangian equation, Rudin, Osher and Fatemi proposed a gradient flow method (Rudin et al. 1992). This method is slow due to strict constraints on the time step size and many methods have been proposed to improve on it. Some efficient methods are dual methods (Chambolle 2004; Chambolle and Pock 2011), the split Bregman method (Goldstein and Osher 2009) and splitting-and-penalty based methods (Wang et al. 2008), proximity algorithms (Micchelli et al. 2011), alternating direction method of multipliers (Chan et al. 2013) and augmented Lagrangian methods (Tai and Wu 2009; Wu and Tai 2010; Wu et al. 2011, 2012).

The augmented Lagrangian method was originally introduced by Hestenes and Powell for solving constrained optimization problem and further systematically studied by many researchers, such as Rockafellar (1974) and Bertsekas (1996(firstly published in 1982)). It was also widely applied to optimize unconstrained minimization problem with the aid of operator-splitting technique (Glowinski and Tallec 1989) by which one can transform the unconstrained optimization problem to its equivalent constrained versions. One of the special and very useful instance of augmented Lagrangian methods is the alternating direction method of multipliers (ADMM) (Boyd 2010; He and Yuan 2012), which is famous in optimization and statistics community and has broad applications. ADMM has been extensively studied in recent decades and has many practical variants, such as linearized ADMM (Wang and Yuan 2012), preconditioned ADMM (Deng and Yin 2016), proximal ADMM (Fazel et al. 2013), accelerated ADMM (Ouyang et al. 2015), stochastic ADMM (Chen et al. 2018) and non-convex ADMM (Li and Pong 2015; Wang et al. 2019).

Indeed, the variable splitting and augmented Lagrangian method gained great successes in solving nonlinear variational problems that arise from physics, mechanics, economics, etc. (Glowinski and Tallec 1989). The variable splitting step helps to transform a complicated problem into a constrained optimization with more variables, then an iteration based on augmented Lagrangian method is performed with several easier subproblems. Inspired by this, the method was proposed by Tai and Wu to optimize the total variation-based image restoration model in Tai and Wu (2009) and Wu and Tai (2010). As expected, augmented Lagrangian methods benefit from the periodic boundary condition which is commonly assumed for image processing problems and the L_1 norm which is included in the total variation seminorm. The augmented Lagrangian method for TV-based image restoration model has two subproblems. The periodic boundary condition allows us to solve one of the subproblems via Fourier transformation with FFT implementation in the case of deconvolution case. Meanwhile, the other subproblem with the L^1 norm has closed form solution. Despite the fact that the image processing problems are naturally in large scale, these two advantages of the augmented Lagrangian method make it efficient in minimizing the objective functionals related with the non-smooth total variation for various image processing tasks. Since Tai and Wu (2009); Wu and Tai (2010), the variable splitting and augmented Lagrangian method has been widely applied to total variation-related minimizations like the single channel case (Wu and Tai 2010; Tai and Wu 2009; Wu et al. 2011), the multichannel case (Wu and Tai 2010), high-order models (Wu and Tai 2010), TV-Stokes model (Hahn et al. 2012), Euler's elastica image restoration model (Tai et al. 2011; Duan et al. 2013; Yashtini and Kang 2016), mean curvature image denoising (Zhu et al. 2013; Myllykoski et al. 2015), total variation minimization in curved spaces for either data processing (Lai and Chan 2011; Wu et al. 2012) or normal vector-filtering based surface denoising (Zhang et al. 2015) and even more in Ramani and Fessler (2011) and Güven et al. (2016). Therein for some complicated non-convex models like Euler's elastica or mean curvature based, how to introduce the auxiliary variables is tricky and important to get stable and efficient algorithms. There are some close connections between the augmented Lagrangian method and other approaches such as split Bregman method (Goldstein and Osher 2009) and Chambolle's projection method (Chambolle 2004), and some works for improving classical augmented Lagrangian method can be found in Li et al. (2013), etc.

The content included here are organized as follows. In section "Basic Notation", we present some basic notations. In section "Augmented Lagrangian Method for Total Variation-Related Image Restoration Models", we present augmented Lagrangian methods TV restoration models with L^2 fidelity term and TV restoration models with non-quadratic fidelity. In Section "Extension to Multichannel Image Restoration", we present augmented Lagrangian methods for multichannel TV restoration. In Section "Extension to High-Order Models", we present augmented Lagrangian methods for high-order models, including second-order total variation model, total generalized variation model, Euler's elastica model, and mean curvature model. In Section "Numerical Experiments", we show some numerical experiments. We conclude this paper in Section "Conclusions".

Basic Notation

We follow Wu and Tai (2010) for most notations. As a gray image is a 2D array, we represent it by an $N \times N$ matrix, without the loss of generality. It is useful to denote the Euclidean space $\mathbb{R}^{N \times N}$ as $\mathscr{X}$ and write $\mathscr{Y} = \mathscr{X} \times \mathscr{X}$. We recall the discrete gradient operator

$$\nabla : \mathscr{X} \to \mathscr{Y}$$

$$x \to \nabla x,$$

where ∇x is given by

$$(\nabla x)_{i,j} = ((\overset{\circ}{D}_1^+ x)_{i,j}, (\overset{\circ}{D}_2^+ x)_{i,j}), i, j = 1, \ldots, N,$$

with

$$(\overset{\circ}{D}_1^+ x)_{i,j} = \begin{cases} x_{i,j+1} - x_{i,j}, & 1 \leq j \leq N-1, \\ x_{i,1} - x_{i,N}, & j = N, \end{cases}$$

$$(\overset{\circ}{D}_2^+ x)_{i,j} = \begin{cases} x_{i+1,j} - x_{i,j}, & 1 \leq i \leq N-1, \\ x_{1,j} - x_{N,j}, & i = N. \end{cases}$$

Here $\overset{\circ}{D}_1^+$ and $\overset{\circ}{D}_2^+$ are used to denote forward difference operators with periodic boundary condition for FFT algorithm implementation. We mention that other boundary conditions with corresponding implementation tricks can also be adopted.

The usual inner products and L^2 norms in the spaces $\mathscr{X}$ and $\mathscr{Y}$ are as follows. We denote

$$\langle x, z \rangle = \sum_{1 \le i,j \le N} x_{i,j} z_{i,j} \text{ and } \|x\| = \sqrt{\langle x, x \rangle},$$

for $x, z \in \mathscr{X}$; and

$$\langle w, y \rangle = \langle w^1, y^1 \rangle + \langle w^2, y^2 \rangle, \text{ and } \|y\| = \sqrt{\langle y, y \rangle},$$

for $y = (y^1, y^2) \in \mathscr{Y}$ and $w = (w^1, w^2) \in \mathscr{Y}$. At each pixel (i, j), we define

$$|y_{i,j}| = |(y^1_{i,j}, y^2_{i,j})| = \sqrt{(y^1_{i,j})^2 + (y^2_{i,j})^2}$$

as the usual Euclidean norm in $\mathbb{R}^2$. We mention that $\|x\|_{L^p}$ is used to denote the general L^p norm of $x \in \mathscr{X}$.

By using the inner products of $\mathscr{X}$ and $\mathscr{Y}$, it is clear that the discrete divergence operator, as the adjoint operator of $-\nabla$, is as follows

$$\text{div} : \mathscr{Y} \to \mathscr{X}$$

$$y = (y^1, y^2) \to \text{div } y,$$

where

$$(\text{div } y)_{i,j} = y^1_{i,j} - y^1_{i,j-1} + y^2_{i,j} - y^2_{i-1,j} = (\mathring{D}^-_1 y^1)_{i,j} + (\mathring{D}^-_2 y^2)_{i,j},$$

with backward difference operators $\mathring{D}^-_1$ and $\mathring{D}^-_2$ and periodic boundary conditions $y^1_{i,0} = y^1_{i,N}$ and $y^2_{0,j} = y^2_{N,j}$.

Augmented Lagrangian Method for Total Variation-Related Image Restoration Models

We assume $d \in \mathscr{X}$ to be an observed image. As usual, we model the degradation procedure as

$$\underline{x} \xrightarrow{\text{linear transformation}} K\underline{x} \xrightarrow{\text{noise}} d, \qquad (1)$$

where $\underline{x} \in \mathscr{X}$ is the ground truth image and $K : \mathscr{X} \to \mathscr{X}$ is a linear operator like a blur. In other cases, such as when K is a Radon transform or a subsampling, the dimensions of the observed data d and the ground truth data $\underline{x}$ may be different. However, there is no essential difficulty, and the method framework here also applies. Here the noise is not necessarily to be additive and could be Gaussian, impulsive, Poisson, or even others. The task of image restoration is to recover $\underline{x}$ from d. In this survey we only consider the case where the linear operator K is given. Even so, we usually cannot directly solve $\underline{x}$ from (1), because this is a typical inverse problem. Both the random measurement noise and the bad condition number of K bring computational difficulties. Regularization on the solution should be considered to overcome the ill-posedness.

Although the classical Tikhonov regularization has achieved great successes in lots of general inverse problems, it turns out to over smooth image edges, the most important image structure. Indeed, one of the most basic and successful image restoration models is based on total variation regularization, which reads

$$\min_{x \in \mathcal{X}} \{E(x) = F(Kx) + R(\nabla x) + B(x)\}, \tag{2}$$

where $F(Kx)$ is a fidelity term, $R(\nabla x)$ is the total variation of x (Rudin et al. 1992) defined by

$$R(\nabla x) = \mathrm{TV}(x) = \sum_{1 \leq i,j \leq N} |(\nabla x)_{i,j}|, \tag{3}$$

and $B(x)$ is an indicator function of box constraints defined as follows

$$B(x) = \begin{cases} 0, & \underline{b} \leq x_{i,j} \leq \overline{b}, \forall\, i, j, \\ +\infty, & \text{otherwise.} \end{cases}$$

Lots of researches (Le et al. 2007; Beck and Teboulle 2009; Chan et al. 2013) show that to involve this kind of constraints is useful, when the intensity range is clear. Otherwise, one can just let the box parameters $\underline{b}$ be $-\infty$ or $\overline{b}$ be $+\infty$. This model includes numerous particular cases studied in the literatures.

For further analysis and interpretation, we make the following assumptions:

- Assumption 1. $\mathrm{Null}(\nabla) \cap \mathrm{Null}(K) = \{0\}$.
- Assumption 2. $\mathrm{dom}(R \circ \nabla) \cap \mathrm{dom}(F \circ K) \cap \mathrm{dom}(B) \neq \emptyset$.
- Assumption 3. $F(z)$ is convex, proper, coercive, and lower semi-continuous.
- Assumption 4. $\mathrm{dom}(F)$ is open.

where $\mathrm{Null}(\cdot)$ is the null space of $\cdot$; $\mathrm{dom}(F) = \{z \in \mathcal{X} : F(z) < +\infty\}$ is the domain of F; and $\mathrm{dom}(R \circ \nabla), \mathrm{dom}(B), \mathrm{dom}(F \circ K)$ are similar. Here we have some comments on these assumptions, which are relatively quite natural. Since most linear operators Ks like blur kernels correspond essentially to averaging operations, Assumption 1 is reasonable. Moreover, although the fidelity terms $F(\cdot)$s are diverse by the statistics of the noise models, many of them meet all of those Assumption 3 and 4, like the following typical ones:

1. The squared L^2 fidelity (corresponding to Gaussian noise):

$$F(Kx) = \frac{\alpha}{2}\|Kx - d\|^2,$$

2. The L^1 fidelity (Nikolova 2004) (corresponding to impulsive noise):

$$F(Kx) = \alpha\|Kx - d\|_{L^1},$$

3. The Kullback-Leibler (KL) divergence fidelity (corresponding to Poisson noise, assuming $d_{i,j} > 0, \forall i, j$, as in Le et al. (2007)):

$$F(Kx) = \begin{cases} \alpha \sum_{1 \leq i,j \leq N} ((Kx)_{i,j} - d_{i,j}\log(Kx)_{i,j}), & (Kx)_{i,j} > 0, \forall\, i,\, j, \\ +\infty, & \text{otherwise,} \end{cases}$$

where $\alpha > 0$ is a parameter. Note for Poisson noise, we use the definition of the fidelity on the whole space for analysis convenience, compared to Le et al. (2007) (where $K = I$) and (Brune et al. 2009).

Under the Assumptions 1, 2, 3, and 4, it is not difficult to see that the functional $E(x)$ in (2) is convex, proper, coercive, and lower semi-continuous. Thus we have the following existence and uniqueness result, by the generalized Weierstrass theorem and Fermat's rule (Glowinski and Tallec 1989; Rockafellar and Wets 1998).

Theorem 1. *The minimization problem (2) has at least one solution x, which satisfies*

$$0 \in K^*\partial F(Kx) - \operatorname{div}\partial R(\nabla x) + \partial B(x), \tag{4}$$

with $\partial F(Kx)$ and $\partial R(\nabla x)$ being the sub-differentials (Rockafellar and Wets 1998) of F at Kx and R at ∇x, respectively. Moreover, if $F \circ K(x)$ is strictly convex, the minimizer is unique.

Next, we present to use the augmented Lagrangian method for TV regularization-based image restoration models (2) which satisfy our assumptions.

Augmented Lagrangian Method for TV-L^2 Restoration

In this section, we review the augmented Lagrangian method proposed for the TV restoration model with L^2 fidelity term (Tai and Wu 2009; Wu and Tai 2010)

$$\min_{x \in \mathscr{X}} \left\{ E_{\text{TV}}(x) = \frac{\alpha}{2}\|Kx - d\|^2 + R(\nabla x) \right\}, \tag{5}$$

where $\alpha > 0$ and $R(\nabla x)$ is defined as in (3). This model is a special case of model (2), where $F(Kx) = \frac{\alpha}{2}\|Kx - d\|^2$ and the box constraint vanishes. In the literatures, people commonly call model (5) as TV-L^2 model.

The TV-L^2 model is a fundamental model in image restoration, which is usually applied for removing Gaussian-type noise and the linear degradation like blur in image restoration problems (Rudin et al. 1992; Acar and Vogel 1994). By standard Bayesian estimation, the L^2 fidelity term is deduced from the statistical distribution of the i.i.d Gaussian noise, which guarantees that the recovered image resembles the underly truth image closely. Meanwhile, the total variation regularization preserves the sharp edges.

As we mentioned before, the total variation term is non-smooth and is a compound of the L^1 norm and the gradient operator. There is a basic idea that decouples the total variation term and treats the L^1 norm and the gradient operator separately. By combining this with variable splitting technique, the augmented Lagrangian method demonstrates this idea.

First, we introduce an auxiliary variable $y \in \mathcal{Y}$ for ∇x and convert the minimization problem (5) to an equivalent constrained optimization problem

$$\min_{x \in \mathcal{X}, y \in \mathcal{Y}} \left\{ G_{\text{TV}}(x, y) = \frac{\alpha}{2}\|Kx - d\|^2 + R(y) \right\}, \quad (6)$$
$$\text{s.t.} \quad y = \nabla x.$$

Then, we define the following augmented Lagrangian function for the constrained optimization problem (6)

$$\mathcal{L}_{\text{TV}}(x, y; \lambda) = \frac{\alpha}{2}\|Kx - d\|^2 + R(y) + \langle \lambda, y - \nabla x \rangle + \frac{\beta}{2}\|y - \nabla x\|^2, \quad (7)$$

with the Lagrange multiplier $\lambda \in \mathcal{Y}$ and a positive penalty parameter β. The augmented Lagrangian method for the problem (6) is to seek a saddle-point of the augmented Lagrangian function (7):

$$\text{Find } (x^*, y^*, \lambda^*) \in \mathcal{X} \times \mathcal{Y} \times \mathcal{Y},$$
$$\text{s.t.} \quad \mathcal{L}_{\text{TV}}(x^*, y^*; \lambda) \leq \mathcal{L}_{\text{TV}}(x^*, y^*; \lambda^*) \leq \mathcal{L}_{\text{TV}}(x, y; \lambda^*), \quad (8)$$
$$\forall (x, y, \lambda) \in \mathcal{X} \times \mathcal{Y} \times \mathcal{Y},$$

The following theorem (Glowinski and Tallec 1989; Wu and Tai 2010) reveals the relation between the solution of problem (5) and the saddle-point of problem (8).

Theorem 2. $x^* \in \mathcal{X}$ *is a solution of problem* (5) *if and only if there exist* $y^* \in \mathcal{Y}$ *and* $\lambda^* \in \mathcal{Y}$ *such that* $(x^*, y^*; \lambda^*)$ *is a saddle-point of problem* (8).

Finally, we employ an alternating direction iterative procedure in the augmented Lagrangian method to seek a saddle-point of problem (8); see Algorithm 1.

Algorithm 1 Augmented Lagrangian method for TV-L^2 model
Initialization: $x^{-1} = 0$, $y^{-1} = 0$, $\lambda^0 = 0$.
Iteration: For $k = 0, 1, \ldots$:

1. Compute (x^k, y^k) as an (approximate) minimizer of the augmented Lagrangian function (7) with the Lagrange multiplier λ^k, i.e.,

$$(x^k, y^k) \approx \arg\min_{(x,y)\in \mathscr{X}\times \mathscr{Y}} \mathscr{L}_{\mathrm{TV}}(x, y; \lambda^k), \tag{9}$$

where $\mathscr{L}_{\mathrm{TV}}(x, y; \lambda^k)$ is defined as (7).
2. Update

$$\lambda^{k+1} = \lambda^k + \beta(y^k - \nabla x^k).$$

We can see that the minimization problem (9) still can not be solved directly and exactly. Our strategy is separating the problem (9) into two subproblems with respect to x and y and minimizing them alternatively.

The Solution to Sub-problem w.r.t. x

Given y, the minimization problem (9) with respect to x is

$$\min_{x \in \mathscr{X}} \left\{ \frac{\alpha}{2} \|Kx - d\|^2 - \langle \lambda^k, \nabla x \rangle + \frac{\beta}{2} \|y - \nabla x\|^2 \right\}.$$

It is a quadratic optimization problem, whose first-order optimality condition gives a linear equation

$$(\alpha K^*K - \beta \Delta)x = \alpha K^*d - \mathrm{div}(\lambda^k + \beta y). \tag{10}$$

If K is a convolution operator like a convolution blur, the above equation under periodic boundary condition can be efficiently solved via Fourier transform with fast Fourier transform (FFT) implementation (Wang et al. 2008; Wu and Tai 2010). One can obtain its solution by

$$x = \mathscr{F}^{-1}\left(\frac{\alpha \mathscr{F}(K^*)\mathscr{F}(d) - \mathscr{F}(\mathrm{div})\mathscr{F}(\lambda^k + \beta y)}{\alpha \mathscr{F}(K^*)\mathscr{F}(K) - \beta \mathscr{F}(\Delta)} \right),$$

where $\mathscr{F}$ and $\mathscr{F}^{-1}$ denote the Fourier transform and the inverse Fourier transform. Fourier transforms of operators K^*, K, div, and Δ mean the transforms of the corresponding convolution kernels. If K is not a convolution operator, such as a Radon transform or a subsampling, we can solve the above equation (10) by other well-developed linear solvers like conjugate gradient (CG) method.

The Solution to Sub-problem w.r.t. y

Given x, the minimization problem (9) with respect to y is

$$\min_{y \in \mathscr{Y}} \left\{ R(y) + \langle \lambda^k, y \rangle + \frac{\beta}{2} \|y - \nabla x\|^2 \right\}. \tag{11}$$

According to the definition of $R(y)$, we can rewrite (11) as

$$\min_{y \in \mathscr{Y}} \left\{ \sum_{1 \leq i,j \leq N} |y_{i,j}| + \frac{\beta}{2} \sum_{1 \leq i,j \leq N} \left| y_{i,j} - \left(\nabla x - \frac{\lambda^k}{\beta} \right)_{i,j} \right|^2 \right\}, \tag{12}$$

whose solution is in closed form as follows

$$y_{i,j} = \max\left(0, 1 - \frac{1}{\beta |\eta_{i,j}|}\right) \eta_{ij}, \tag{13}$$

where $\eta = \nabla x - \lambda^k/\beta \in \mathscr{Y}$. This solution can be derived from the first-order optimality condition via the subdifferential theory (Wang et al. 2008) or the geometric explanation of the minimizer (Wu et al. 2011). We remark that the geometric method can be easily extended to higher (>2) dimensional case (Wu and Tai 2010; Wu et al. 2011) (see, e.g., multichannel image restoration and high-order models in later sections) or the case where $R(\cdot)$ is non-convex (Wu et al. 2018).

Here, we review the geometric interpretation of the formula (13) given in Wu et al. (2011). As one can see, the problem (12) is separable, and at each pixel (i, j), we can reduce it to a simple form

$$\min_{u \in \mathbb{R}^2} \left\{ |u| + \frac{\beta}{2} |u - v|^2 \right\}, \tag{14}$$

where $v \in \mathbb{R}^2$; see Fig. 1.

In fact, the minimizer of (14) locates in the same quadrant of v and inside of the solid circle with O as center and $|v|$ as radius; see Fig. 1. Without loss of generality, we consider the points inside the solid circle at the first quadrant, e.g., u. We draw a dotted circle with O as center and $|u|$ as radius, which intersects the line segment Ov at a point u^*. By the triangle inequality, we have

$$|u| + |u - v| \geq |v| = |u^*| + |u^* - v|.$$

Since $|u| = |u^*|$, we obtain

$$|u - v| \geq |u^* - v|,$$

which indicates

Fig. 1 A geometric interpretation of the formula (13)

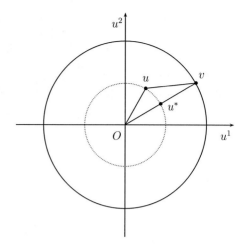

$$|u| + \frac{\beta}{2}|u - v|^2 \geq |u^*| + \frac{\beta}{2}|u^* - v|^2.$$

The above equality implies that the solution of (14) locates on the line segment Ov. Therefore, we let $u = \gamma v$ with $0 \leq \gamma \leq 1$ and simplify the problem (14) into an univariate optimization problem

$$\min_{0 \leq \gamma \leq 1} \left\{ \gamma |v| + \frac{\beta}{2}(\gamma - 1)^2 |v|^2 \right\}. \tag{15}$$

The above problem (15) can be solved exactly and has a closed form solution

$$\gamma^* = \max\left(0, 1 - \frac{1}{\beta |v|}\right).$$

According to (10) and (13), we can solve (9) by an alternating minimization procedure; see Algorithm 2.

Algorithm 2 Augmented Lagrangian method for TV-L^2 model – solve the minimization problem (9)

Initialization: $x^{k,0} = x^{k-1}$, $y^{k,0} = y^{k-1}$.
Iteration: For $l = 0, 1, \ldots, L - 1$:

- Compute $x^{k,l+1}$ by solving (10) for $y = y^{k,l}$;
- Compute $y^{k,l+1}$ from (13) for $x = x^{k,l+1}$.

Output: $x^k = x^{k,L}$, $y^k = y^{k,L}$.

Here L can be chosen using some convergence test techniques. In fact, setting $L = 1$ is sufficient to establish the convergence of the sequence (Wu and Tai 2010) generated by Algorithm 1. In this case, the augmented Lagrangian method is well-known as the alternating direction method of multipliers (Boyd 2010).

Convergence Analysis

In this section, we present some convergence results of Algorithm 1. Actually, we can verify that Algorithm 1 is convergence in two cases, i.e., when the minimization problem (9) is exactly solved in each iteration and the problem (9) is roughly solved in each iteration (Glowinski and Tallec 1989; Wu and Tai 2010). We comment that the convergence proof in Wu and Tai (2010) is based on Glowinski and Tallec (1989) but reduces the uniform convexity assumption of $R(\cdot)$. Here, we just take the main convergence results from Wu and Tai (2010) and omit the details.

In the first case, we should set $L \to \infty$ in Algorithm 2, and the inner iteration is guaranteed to converge.

Theorem 3. *The sequence $\{(x^{k,l}, y^{k,l}) : l = 0, 1, 2, \ldots\}$ generated by Algorithm 2 converges to a solution of the problem (9).*

Theorem 4. *Assume that $(x^*, y^*; \lambda^*)$ is a saddle-point of $\mathscr{L}_{\mathrm{TV}}(x, y; \lambda)$. Suppose that the minimization problem (9) is exactly solved in each iteration; i.e., $L \to \infty$ in Algorithm 2. Then the sequence $(x^k, y^k; \lambda^k)$ generated by Algorithm 1 satisfies*

$$\begin{cases} \lim_{k \to \infty} G_{\mathrm{TV}}(x^k, y^k) = G_{\mathrm{TV}}(x^*, y^*), \\ \lim_{k \to \infty} \|y^k - \nabla x^k\| = 0. \end{cases} \tag{16}$$

Since $R(y)$ is continuous, (16) indicates that x^k is a minimizing sequence of $E_{\mathrm{TV}}(\cdot)$. If we further have $\mathrm{Null}(K) = \{0\}$, then

$$\begin{cases} \lim_{k \to \infty} x^k = x^*, \\ \lim_{k \to \infty} y^k = y^*. \end{cases}$$

In the second case, we set $L = 1$ in Algorithm 2.

Theorem 5. *Assume that $(x^*, y^*; \lambda^*)$ is a saddle-point of $\mathscr{L}_{\mathrm{TV}}(x, y; \lambda)$. Suppose that the minimization problem (9) is roughly solved in each iteration, i.e., with $L = 1$ in Algorithm 2. Then the sequence $(x^k, y^k; \lambda^k)$ generated by Algorithm 1 satisfies*

$$\begin{cases} \lim_{k \to \infty} G_{\mathrm{TV}}(x^k, y^k) = G_{\mathrm{TV}}(x^*, y^*), \\ \lim_{k \to \infty} \|y^k - \nabla x^k\| = 0. \end{cases} \tag{17}$$

Since $R(y)$ is continuous, (17) indicates that x^k is a minimizing sequence of $E_{\text{TV}}(\cdot)$. If we further have $\text{Null}(K) = \{0\}$, then

$$\begin{cases} \lim_{k \to \infty} x^k = x^*, \\ \lim_{k \to \infty} y^k = y^*. \end{cases}$$

Augmented Lagrangian Method for TV-L^2 Restoration with Box Constraint

In this section, we review the augmented Lagrangian method for the TV restoration model with the L^2 fidelity term and the box constraint (Chan et al. 2013), which reads

$$\min_{x \in \mathcal{X}} \left\{ E_{\text{TVB}}(x) = \frac{\alpha}{2} \|Kx - d\|^2 + R(\nabla x) + B(x) \right\}, \tag{18}$$

where $\alpha > 0$, $R(\nabla x)$ is defined as (3), and we have $-\infty < \underline{b} \leq \bar{b} < +\infty$ in $B(x)$. This model is also a special case of model (2), where $F(Kx) = \frac{\alpha}{2}\|Kx - d\|^2$.

The box constraint is inherent in digital image processing. The nature image is stored as discrete numerical arrays in some digital media. The typical used ranges are [0, 1] and [0, 255]. It has been shown that adding the box constraint in image restoration can improve the quality of the recovered image (Beck and Teboulle 2009; Chan et al. 2013).

The original method proposed in Chan et al. (2013) is under the framework of the alternating direction method of multipliers, which is a special case of the augmented Lagrangian method. For the sake of clarity, we reformulate it in our notations and styles.

Compared with the TV-L^2 model (5), this model has one more non-differentiability term $B(x)$. Thus, we need another variable to eliminate the nondifferentiation for x. We introduce two auxiliary variables $y \in \mathcal{Y}$ and $z \in \mathcal{X}$ and rewrite the problem (18) to be the following constrained optimization problem

$$\min_{x \in \mathcal{X}, y \in \mathcal{Y}, z \in \mathcal{X}} \left\{ G_{\text{TVB}}(x, y, z) = \frac{\alpha}{2}\|Kx - d\|^2 + R(y) + B(z) \right\}$$
$$\text{s.t.} \quad \begin{pmatrix} y \\ z \end{pmatrix} = \begin{pmatrix} \nabla \\ \mathcal{I}_1 \end{pmatrix} x, \tag{19}$$

where $\mathcal{I}_1 : \mathcal{X} \to \mathcal{X}$ is the identity operator.

We define the augmented Lagrangian function for the problem (19) as follows

$$\mathcal{L}_{\text{TVB}}(x, y, z; \lambda_y, \lambda_z) = \frac{\alpha}{2}\|Kx - d\|^2 + R(y) + B(z)$$
$$+ \left\langle \begin{pmatrix} \lambda_y \\ \lambda_z \end{pmatrix}, \begin{pmatrix} y \\ z \end{pmatrix} - \begin{pmatrix} \nabla \\ \mathscr{I}_1 \end{pmatrix} x \right\rangle \quad (20)$$
$$+ \frac{1}{2}\left\| \begin{pmatrix} y \\ z \end{pmatrix} - \begin{pmatrix} \nabla \\ \mathscr{I}_1 \end{pmatrix} x \right\|^2_{\mathscr{S}},$$

where $\begin{pmatrix} \lambda_y \\ \lambda_z \end{pmatrix}$ is the Lagrangian multiplier and $\mathscr{S} = \begin{pmatrix} \beta_y \mathscr{I}_2 & \\ & \beta_z \mathscr{I}_1 \end{pmatrix}$ with the identity operator $\mathscr{I}_2 : \mathscr{Y} \to \mathscr{Y}$ and positive parameters β_y, β_z. Here $\|u\|_{\mathscr{S}}$ denotes the $\mathscr{S}$-norm, defined by $\|u\|_{\mathscr{S}} = \sqrt{\langle u, \mathscr{S}u \rangle}$.

For the augmented Lagrangian method, we consider the saddle-point problem

Find $(x^*, y^*, z^*, \lambda_y^*, \lambda_z^*) \in \mathscr{X} \times \mathscr{Y} \times \mathscr{X} \times \mathscr{Y} \times \mathscr{X}$,

s.t. $\mathcal{L}_{\text{TVB}}(x^*, y^*, z^*; \lambda_y, \lambda_z) \leq \mathcal{L}_{\text{TVB}}(x^*, y^*, z^*; \lambda_y^*, \lambda_z^*) \leq \mathcal{L}_{\text{TVB}}(x, y, z; \lambda_y^*, \lambda_z^*),$

$\forall (x, y, z, \lambda_y, \lambda_z) \in \mathscr{X} \times \mathscr{Y} \times \mathscr{X} \times \mathscr{Y} \times \mathscr{X}.$ (21)

Finally, we use an alternating direction iterative scheme in the augmented Lagrangian method to solve the saddle-point problem (21); see Algorithm 3.

Algorithm 3 Augmented Lagrangian method for TV-L^2 model with box constraint

Initialization: $x^{-1} = 0$, $\begin{pmatrix} y^{-1} \\ z^{-1} \end{pmatrix} = \begin{pmatrix} 0 \\ 0 \end{pmatrix}$, $\begin{pmatrix} \lambda_y^0 \\ \lambda_z^0 \end{pmatrix} = \begin{pmatrix} 0 \\ 0 \end{pmatrix}$.

Iteration: For $k = 0, 1, \ldots$:

1. Compute (x^k, y^k, z^k) as an (approximate) minimizer of the augmented Lagrangian functional with the Lagrange multiplier $\begin{pmatrix} \lambda_y^k \\ \lambda_z^k \end{pmatrix}$, i.e.,

$$(x^k, y^k, z^k) \approx \arg\min_{(x,y,z) \in \mathscr{X} \times \mathscr{Y} \times \mathscr{X}} \mathcal{L}_{\text{TVB}}(x, y, z; \lambda_y^k, \lambda_z^k), \quad (22)$$

where $\mathcal{L}_{\text{TVB}}(x, y, z; \lambda_y^k, \lambda_z^k)$ is as in (20).

2. Update

$$\begin{pmatrix} \lambda_y^{k+1} \\ \lambda_z^{k+1} \end{pmatrix} = \begin{pmatrix} \lambda_y^k \\ \lambda_z^k \end{pmatrix} + \begin{pmatrix} \beta_y(y^k - \nabla x^k) \\ \beta_z(z^k - x^k) \end{pmatrix}.$$

To solve the minimization problem (22), we separate it into two subproblems respect to x and $\begin{pmatrix} y \\ z \end{pmatrix}$ and employ an alternative minimization procedure.

The Solution to Sub-problem w.r.t. x

Given $\begin{pmatrix} y \\ z \end{pmatrix}$, the minimization problem (22) with respect to x reads

$$\min_{x \in \mathscr{X}} \left\{ \frac{\alpha}{2} \|Kx - d\|^2 - \left\langle \begin{pmatrix} \lambda_y^k \\ \lambda_z^k \end{pmatrix}, \begin{pmatrix} \nabla \\ \mathscr{I}_1 \end{pmatrix} x \right\rangle + \frac{1}{2} \left\| \begin{pmatrix} y \\ z \end{pmatrix} - \begin{pmatrix} \nabla \\ \mathscr{I}_1 \end{pmatrix} x \right\|_{\mathscr{S}}^2 \right\}, \quad (23)$$

whose first-order optimization condition gives a linear equation

$$(\alpha K^* K - \beta_y \Delta + \beta_z \mathscr{I}_1) x = \alpha K^* d - \mathrm{div}(\lambda_y^k + \beta_y y) + \lambda_z^k + \beta_z z. \quad (24)$$

Similar to the equation (10), the above equation can be efficiently solved by fast linear solvers such as FFT and CG.

The Solution to Sub-problem w.r.t. (y, z)

Given x, the minimization problem (22) with respect to $\begin{pmatrix} y \\ z \end{pmatrix}$ reads

$$\min_{(y,z) \in \mathscr{Y} \times \mathscr{X}} \left\{ R(y) + B(z) + \left\langle \begin{pmatrix} \lambda_y^k \\ \lambda_z^k \end{pmatrix}, \begin{pmatrix} y \\ z \end{pmatrix} \right\rangle + \frac{1}{2} \left\| \begin{pmatrix} y \\ z \end{pmatrix} - \begin{pmatrix} \nabla \\ \mathscr{I}_1 \end{pmatrix} x \right\|_{\mathscr{S}}^2 \right\}, \quad (25)$$

which can be separated into two independent minimization problems:

- y-subproblem:

$$\min_{y \in \mathscr{Y}} \left\{ R(y) + (\lambda_y^k, y) + \frac{\beta_y}{2} \|y - \nabla x\|^2 \right\}, \quad (26)$$

- z-subproblem:

$$\min_{z \in \mathscr{X}} \left\{ B(z) + (\lambda_z^k, z) + \frac{\beta_z}{2} \|z - x\|^2 \right\}. \quad (27)$$

We can obtain the minimizer of (26) from (13) and the minimizer of (27) as follows

$$z_{i,j} = \mathscr{P}_{[\underline{b}, \bar{b}]}(\xi_{i,j}), \quad \forall i, j, \quad (28)$$

where $\mathcal{P}_{[\underline{b},\bar{b}]}(\cdot)$ is the projection onto the interval $[\underline{b},\bar{b}]$ and

$$\xi = x - \frac{\lambda_z^k}{\beta_z} \in \mathcal{X}.$$

After knowing the solutions of the subproblems (23) and (25), we use the following alternative minimization procedure to solve (22); see Algorithm 4.

Algorithm 4 Augmented Lagrangian method for TV-L^2 model with box constraint – solve the minimization problem (22)

Initialization: $x^{k,0} = x^{k-1}$, $\begin{pmatrix} y^{k,0} \\ z^{k,0} \end{pmatrix} = \begin{pmatrix} y^{k-1} \\ z^{k-1} \end{pmatrix}$.

Iteration: For $l = 0, 1, 2, \ldots, L-1$:

- Compute $x^{k,l+1}$ by solving (24) for $\begin{pmatrix} y \\ z \end{pmatrix} = \begin{pmatrix} y^{k,l} \\ z^{k,l} \end{pmatrix}$;

- Compute $\begin{pmatrix} y^{k,l+1} \\ z^{k,l+1} \end{pmatrix}$ from (13) and (28) for $x = x^{k,l+1}$.

Output: $x^k = x^{k,L}$, $\begin{pmatrix} y^k \\ z^k \end{pmatrix} = \begin{pmatrix} y^{k,L} \\ z^{k,L} \end{pmatrix}$.

The convergence results of Algorithms 3 and 4 are similar to the convergence results proposed in the previous section, one can refer to Chan et al. (2013) for details.

Augmented Lagrangian Method for TV Restoration with Non-quadratic Fidelity

In this section, we review the augmented Lagrangian method proposed in Wu et al. (2011) for the TV restoration model with non-quadratic fidelity, which reads

$$\min_{x \in \mathcal{X}} \left\{ E_{\text{TVNQ}}(x) = R(\nabla x) + F(Kx) \right\}. \tag{29}$$

where $R(\nabla x)$ is defined as in (3). Here, we consider the non-quadratic fidelity $F(Kx)$ which arises for removing non-Gaussian-type noises, such as impulsive noise and Poisson noise. For impulsive noise removal, we usually use the L^1 fidelity (Nikolova 2004)

$$F(Kx) = \alpha \|Kx - d\|_{L^1}, \tag{30}$$

and for Poisson noise removal, we commonly choose the Kullback-Leibler (KL) divergence fidelity (Le et al. 2007; Brune et al. 2009)

$$F(Kx) = \begin{cases} \alpha \sum_{1 \leq i,j \leq N} ((Kx)_{i,j} - d_{i,j} \log(Kx)_{i,j}), & (Kx)_{i,j} > 0, \forall i,j, \\ +\infty, & \text{otherwise.} \end{cases} \quad (31)$$

In this section, we focus on the augmented Lagrangian method for image restoration with these two non-quadratic fidelities. For other non-quadratic fidelities, one can extend our method accordingly.

The non-quadratic fidelities (30) and (31) are non-smooth. Adopting the idea to cope with total variation term, we require one more auxiliary variable to remove the nonlinearity arising from $F(Kx)$. We first introduce two auxiliary variables y and z and reformulate (29) to an equivalent constrained optimization problem

$$\min_{x \in \mathscr{X}, y \in \mathscr{Y}, z \in \mathscr{X}} \{G_{\text{TVNQ}}(y, z) = R(y) + F(z)\}$$
$$\text{s.t.} \quad \begin{pmatrix} y \\ z \end{pmatrix} = \begin{pmatrix} \nabla \\ K \end{pmatrix} x. \quad (32)$$

We then define the augmented Lagrangian function for (32) as

$$\mathscr{L}_{\text{TVNQ}}(x, y, z; \lambda_y, \lambda_z) = R(y) + F(z)$$
$$+ \left\langle \begin{pmatrix} \lambda_y \\ \lambda_z \end{pmatrix}, \begin{pmatrix} y \\ z \end{pmatrix} - \begin{pmatrix} \nabla \\ K \end{pmatrix} x \right\rangle + \left\| \begin{pmatrix} y \\ z \end{pmatrix} - \begin{pmatrix} \nabla \\ K \end{pmatrix} x \right\|_{\mathscr{S}}^2 \quad (33)$$

with Lagrange multiplier $\begin{pmatrix} \lambda_y \\ \lambda_z \end{pmatrix}$ and $\mathscr{S} = \begin{pmatrix} \beta_y \mathscr{I}_2 & \\ & \beta_z \mathscr{I}_1 \end{pmatrix}$ and consider the saddle-point problem

$$\text{Find } (x^*, y^*, z^*, \lambda_y^*, \lambda_z^*) \in \mathscr{X} \times \mathscr{Y} \times \mathscr{X} \times \mathscr{Y} \times \mathscr{X},$$
$$\text{s.t.} \quad \mathscr{L}_{\text{TVNQ}}(x^*, y^*, z^*; \lambda_y, \lambda_z) \leq \mathscr{L}_{\text{TVNQ}}(x^*, y^*, z^*; \lambda_y^*, \lambda_z^*)$$
$$\leq \mathscr{L}_{\text{TVNQ}}(x, y, z; \lambda_y^*, \lambda_z^*),$$
$$\forall (x, y, z, \lambda_y, \lambda_z) \in \mathscr{X} \times \mathscr{Y} \times \mathscr{X} \times \mathscr{Y} \times \mathscr{X}. \quad (34)$$

Finally, we use the following iterative algorithm to solve the saddle-point problem (34); see Algorithm 5.

Algorithm 5 Augmented Lagrangian method for TV restoration with non-quadratic fidelity

Initialization: $x^{-1} = 0$, $\begin{pmatrix} y^{-1} \\ z^{-1} \end{pmatrix} = \begin{pmatrix} 0 \\ 0 \end{pmatrix}$, $\begin{pmatrix} \lambda_y^0 \\ \lambda_z^0 \end{pmatrix} = \begin{pmatrix} 0 \\ 0 \end{pmatrix}$.

Iteration: For $k = 0, 1, \ldots$:

1. Compute (x^k, y^k, z^k) as an (approximate) minimizer of the augmented Lagrangian functional with the Lagrange multipliers $\begin{pmatrix} \lambda_y^k \\ \lambda_z^k \end{pmatrix}$, i.e.,

$$(x^k, y^k, z^k) \approx \arg \min_{(x,y,z) \in \mathscr{X} \times \mathscr{Y} \times \mathscr{X}} \mathscr{L}_{\text{TVNQ}}(x, y, z; \lambda_y^k, \lambda_z^k), \tag{35}$$

where $\mathscr{L}_{\text{TVNQ}}(x, y, z; \lambda_y^k, \lambda_z^k)$ is as in (33).

2. Update

$$\begin{pmatrix} \lambda_y^{k+1} \\ \lambda_z^{k+1} \end{pmatrix} = \begin{pmatrix} \lambda_y^k \\ \lambda_z^k \end{pmatrix} + \begin{pmatrix} \beta_y(y^k - \nabla x^k) \\ \beta_z(z^k - Kx^k) \end{pmatrix}.$$

We employ an alternating minimization procedure to solve the problem (35).

The Solution to Sub-problem w.r.t. x

Given $\begin{pmatrix} y \\ z \end{pmatrix}$, we have the subproblem of x as follows

$$\min_{x \in \mathscr{X}} \left\{ -\left\langle \begin{pmatrix} \lambda_y^k \\ \lambda_z^k \end{pmatrix}, \begin{pmatrix} \nabla \\ K \end{pmatrix} x \right\rangle + \left\| \begin{pmatrix} y \\ z \end{pmatrix} - \begin{pmatrix} \nabla \\ K \end{pmatrix} x \right\|_{\mathscr{S}}^2 \right\}, \tag{36}$$

which has the optimality condition

$$(\beta_z K^* K - \beta_y \Delta) x = K^*(\lambda_z^k + \beta_z z) - \text{div}(\lambda_y^k + \beta_y y). \tag{37}$$

We can use fast linear solvers to solve the above equation, such as FFT and CG.

The Solution to Sub-problem w.r.t. (y, z)

Given x, we have the subproblem of $\begin{pmatrix} y \\ z \end{pmatrix}$ as follows

$$\min_{(y,z) \in \mathscr{Y} \times \mathscr{X}} \left\{ R(y) + F(z) + \left\langle \begin{pmatrix} \lambda_y^k \\ \lambda_z^k \end{pmatrix}, \begin{pmatrix} y \\ z \end{pmatrix} \right\rangle + \left\| \begin{pmatrix} y \\ z \end{pmatrix} - \begin{pmatrix} \nabla \\ K \end{pmatrix} x \right\|_{\mathscr{S}}^2 \right\}. \tag{38}$$

We can split it into two distinct minimization problems with respect to y and z as follows

- y-subproblem:
$$\min_{y \in \mathscr{Y}} \left\{ R(y) + (\lambda_y^k, y) + \frac{\beta_y}{2} \|y - \nabla x\|^2 \right\}; \tag{39}$$

- z-subproblem:
$$\min_{z \in \mathscr{X}} \left\{ F(z) + (\lambda_z^k, z) + \frac{\beta_z}{2} \|z - Kx\|^2 \right\}. \tag{40}$$

For the problem (39), it is the same as the problem (11) and can be solved via (13). For the problem (40), we next show its solution based on the choices of $F(\cdot)$.

For the L^1 fidelity (30), we can rewrite the z-subproblem (40) as

$$\min_{z \in \mathscr{X}} \left\{ \alpha \|z - d\|_{L^1} + \frac{\beta_z}{2} \|z - \xi\|^2 \right\}$$

where

$$\xi = Kx - \frac{\lambda_z^k}{\beta_z}.$$

It has closed form solution (Wu et al. 2011)

$$z_{i,j} = d_{i,j} + \max\left(0, 1 - \frac{\alpha}{\beta_z |\xi_{i,j} - d_{i,j}|} \right)(\xi_{i,j} - d_{i,j}), \tag{41}$$

which is a one-dimensional case of (13). In this case, the alternating minimization procedure to solve the problem (35) is described in Algorithm 6.

For the KL divergence fidelity (31), we can rewrite the z-subproblem (40) as

$$\min_{\substack{z \in \mathscr{X} \\ z_{i,j} > 0, \forall i,j}} \left\{ \alpha \sum_{1 \leq i,j \leq N} (z_{i,j} - d_{i,j} \log z_{i,j}) + \frac{\beta_z}{2} \sum_{1 \leq i,j \leq N} \left| z_{i,j} - \left(Kx - \frac{\lambda_z^k}{\beta_z} \right)_{i,j} \right|^2 \right\}.$$

It has closed form solution (Wu et al. 2011)

$$z_{i,j} = \frac{1}{2} \left(\sqrt{\left(\xi_{i,j} - \frac{\alpha}{\beta_z}\right)^2 + 4\frac{\alpha}{\beta_z} d_{i,j}} + \left(\xi_{i,j} - \frac{\alpha}{\beta_z}\right) \right), \tag{42}$$

Algorithm 6 Augmented Lagrangian method for TV restoration with the L^1 fidelity – solve the minimization problem (35)

Initialization: $x^{k,0} = x^{k-1}$, $\begin{pmatrix} y^{k,0} \\ z^{k,0} \end{pmatrix} = \begin{pmatrix} y^{k-1} \\ z^{k-1} \end{pmatrix}$.

Iteration: For $l = 0, 1, 2, \ldots, L - 1$:

- Compute $x^{k,l+1}$ by solving (37) for $\begin{pmatrix} y \\ z \end{pmatrix} = \begin{pmatrix} y^{k,l} \\ z^{k,l} \end{pmatrix}$;

- Compute $\begin{pmatrix} y^{k,l+1} \\ z^{k,l+1} \end{pmatrix}$ from (13) and (41) for $x = x^{k,l+1}$.

Output: $x^k = x^{k,L}$, $\begin{pmatrix} y^k \\ z^k \end{pmatrix} = \begin{pmatrix} y^{k,L} \\ z^{k,L} \end{pmatrix}$.

where

$$\xi = Kx - \frac{\lambda_z^k}{\beta_z}.$$

Now, the alternating minimization procedure to solve the problem (35) with the KL divergence fidelity (31) can be described in Algorithm 7.

Algorithm 7 Augmented Lagrangian method for TV restoration with the KL divergence fidelity – solve the minimization problem (35)

Initialization: $x^{k,0} = x^{k-1}$, $\begin{pmatrix} y^{k,0} \\ z^{k,0} \end{pmatrix} = \begin{pmatrix} y^{k-1} \\ z^{k-1} \end{pmatrix}$.

Iteration: For $l = 0, 1, 2, \ldots, L - 1$:

- Compute $x^{k,l+1}$ from (37) for $\begin{pmatrix} y \\ z \end{pmatrix} = \begin{pmatrix} y^{k,l} \\ z^{k,l} \end{pmatrix}$;

- Compute $\begin{pmatrix} y^{k,l+1} \\ z^{k,l+1} \end{pmatrix}$ from (13) and (42) for $x = x^{k,l+1}$.

Output: $x^k = x^{k,L}$, $\begin{pmatrix} y^k \\ z^k \end{pmatrix} = \begin{pmatrix} y^{k,L} \\ z^{k,L} \end{pmatrix}$.

The convergence results of Algorithms 5, 6 and 7 are established in Wu et al. (2011), which are similar to convergence results presented previously for Algorithms 1 and 2.

Extension to Multichannel Image Restoration

In this section, we review the augmented Lagrangian method for the multichannel TV restoration (Wu and Tai 2010). The multichannel images are widely used, such as three-channel RGB color image.

The Multichannel TV Restoration Model

We denote an M-channel image by $\boldsymbol{x} = (x_1, x_2, \ldots, x_M)$, where $x_m \in \mathscr{X}$, $\forall m = 1, 2, \ldots, M$. We mention that, at each pixel (i, j), the intensity of $\boldsymbol{x}$ is vector-valued, i.e.,

$$\boldsymbol{x}_{i,j} = ((x_1)_{i,j}, (x_2)_{i,j}, \ldots, (x_M)_{i,j}).$$

Let us define

$$\boldsymbol{\mathscr{X}} = \underbrace{\mathscr{X} \times \mathscr{X} \times \cdots \times \mathscr{X}}_{M}, \quad \boldsymbol{\mathscr{Y}} = \underbrace{\mathscr{Y} \times \mathscr{Y} \times \cdots \times \mathscr{Y}}_{M}.$$

Then we have $\boldsymbol{x} \in \boldsymbol{\mathscr{X}}$ and

$$\nabla \boldsymbol{x} = (\nabla x_1, \nabla x_2, \ldots, \nabla x_M) \in \boldsymbol{\mathscr{Y}}.$$

The usual inner products and L^2 norms in the spaces $\boldsymbol{\mathscr{X}}$ and $\boldsymbol{\mathscr{Y}}$ are as follows. We denote

$$\langle \boldsymbol{x}, \boldsymbol{z} \rangle = \sum_{1 \leq m \leq M} \langle x_m, z_m \rangle, \quad \|\boldsymbol{x}\| = \sqrt{\langle \boldsymbol{x}, \boldsymbol{x} \rangle};$$

$$\langle \boldsymbol{y}, \boldsymbol{w} \rangle = \sum_{1 \leq m \leq M} \langle y_m, w_m \rangle, \quad \|\boldsymbol{y}\| = \sqrt{\langle \boldsymbol{y}, \boldsymbol{y} \rangle}.$$

for $\boldsymbol{x}, \boldsymbol{z} \in \boldsymbol{\mathscr{X}}$ and $\boldsymbol{y}, \boldsymbol{w} \in \boldsymbol{\mathscr{Y}}$. At each pixel (i, j), we also define the following pixel-by-pixel norms

$$|\boldsymbol{x}_{i,j}| = \sqrt{\sum_{1 \leq m \leq M} (x_m)_{i,j}^2} \quad \text{and} \quad |\boldsymbol{y}_{i,j}| = \sqrt{\sum_{1 \leq m \leq M} |(y_m)_{i,j}|^2}.$$

for $\boldsymbol{x} \in \boldsymbol{\mathscr{X}}$ and $\boldsymbol{y} \in \boldsymbol{\mathscr{Y}}$.

With reference to the degradation model (1) of the gray image, here we model the multichannel image degradation procedure as

$$\underline{\boldsymbol{x}} \xrightarrow{\text{linear transformation}} K\underline{\boldsymbol{x}} \xrightarrow{\text{noise}} \boldsymbol{d},$$

where $d \in \mathscr{X}$ is an observed image and $K : \mathscr{X} \to \mathscr{X}$ is linear operator like a blur. Here the noise could be also Gaussian, impulsive, Poisson, or even others.

In this survey, we consider K as the blur operator and the noise is Gaussian type. The operator K has the form of

$$K = \begin{pmatrix} K_{11} & K_{12} & \cdots & K_{1M} \\ K_{21} & K_{22} & \cdots & K_{2M} \\ \vdots & \vdots & \ddots & \vdots \\ K_{M1} & K_{M2} & \cdots & K_{MM} \end{pmatrix},$$

where each K_{ij} is a convolution matrix. The diagonal elements of K denote within-channel blurs, while the off-diagonal elements describe cross-channel blurs. To solve $\underline{x}$, we consider the following multichannel image restoration model (Sapiro and Ringach 1996)

$$\min_{x \in \mathscr{X}} \left\{ E_{\mathrm{MTV}}(x) = \frac{\alpha}{2} \|Kx - d\|^2 + R_{\mathrm{MTV}}(\nabla x) \right\}, \tag{43}$$

where

$$R_{\mathrm{MTV}}(\nabla x) = \mathrm{TV}(x) = \sum_{1 \le i,j \le N} \sqrt{\sum_{1 \le m \le M} |(\nabla x_m)_{i,j}|^2}$$

is the vectorial TV semi-norm (Sapiro and Ringach 1996) (see Blomgren and Chan 1998 for some other choices).

Similarly as for the single channel image restoration model, here we make the following assumption:

- $\mathrm{Null}(\nabla) \cap \mathrm{Null}(K) = \{0\}$.

Under this assumption, one can verify that the functional $E_{\mathrm{MTV}}(x)$ in (43) is convex, proper, coercive, and continuous. Hence, we have the following result (Wu and Tai 2010).

Theorem 6. *The problem (43) has at least one solution x, which satisfies*

$$0 \in \alpha K^*(Kx - d) - \mathrm{div}\, \partial R_{\mathrm{MTV}}(\nabla x),$$

where $\partial R_{\mathrm{MTV}}(\nabla x)$ is the subdifferential of R_{MTV} at ∇x. Moreover, if $\mathrm{Null}(K) = \{0\}$, the minimizer is unique.

Augmented Lagrangian Method for Multichannel TV Restoration

By introducing a new variable $\mathbf{y} = (y_1, y_2, \ldots, y_M) \in \mathcal{Y}$, we first reformulate the minimization problem (43) to the following equivalent constrained optimization problem:

$$\min_{\mathbf{x} \in \mathcal{X}, \mathbf{y} \in \mathcal{Y}} \left\{ G_{\text{MTV}}(\mathbf{x}, \mathbf{y}) = \frac{\alpha}{2} \|K\mathbf{x} - \mathbf{d}\|^2 + R_{\text{MTV}}(\mathbf{y}) \right\} \quad (44)$$
$$\text{s.t.} \quad \mathbf{y} = \nabla \mathbf{x}.$$

We then define the augmented Lagrangian function as

$$\mathcal{L}_{\text{MTV}}(\mathbf{x}, \mathbf{y}; \boldsymbol{\lambda}) = \frac{\alpha}{2} \|K\mathbf{x} - \mathbf{d}\|^2 + R_{\text{MTV}}(\mathbf{y}) + \langle \boldsymbol{\lambda}, \mathbf{y} - \nabla \mathbf{x} \rangle + \frac{\beta}{2} \|\mathbf{y} - \nabla \mathbf{x}\|^2,$$

with the multiplier $\boldsymbol{\lambda} \in \mathcal{Y}$ and a positive constant β. The augmented Lagrangian method aims at seeking a saddle-point of the following problem:

$$\text{Find} \quad (\mathbf{x}^*, \mathbf{y}^*, \boldsymbol{\lambda}^*) \in \mathcal{X} \times \mathcal{Y} \times \mathcal{Y}$$
$$\text{s.t.} \quad \mathcal{L}_{\text{MTV}}(\mathbf{x}^*, \mathbf{y}^*; \boldsymbol{\lambda}) \leq \mathcal{L}_{\text{MTV}}(\mathbf{x}^*, \mathbf{y}^*; \boldsymbol{\lambda}^*) \leq \mathcal{L}_{\text{MTV}}(\mathbf{x}, \mathbf{y}; \boldsymbol{\lambda}^*)$$
$$\forall (\mathbf{x}, \mathbf{y}; \boldsymbol{\lambda}) \in \mathcal{X} \times \mathcal{Y} \times \mathcal{Y}. \quad (45)$$

Finally, an iterative procedure to solve the problem (45) is described in Algorithm 8.

Algorithm 8 Augmented Lagrangian method for the multichannel TV model

Initialization: $\mathbf{x}^{-1} = 0$, $\mathbf{y}^{-1} = 0$, $\boldsymbol{\lambda}^0 = 0$.
Iteration: For $k = 0, 1, 2, \ldots$:

1. Compute $(\mathbf{x}^k, \mathbf{y}^k)$ from

$$(\mathbf{x}^k, \mathbf{y}^k) \approx \arg \min_{(\mathbf{x}, \mathbf{y}) \in (\mathcal{X}, \mathcal{Y})} \mathcal{L}_{\text{MTV}}(\mathbf{x}, \mathbf{y}; \boldsymbol{\lambda}^k). \quad (46)$$

2. Update

$$\boldsymbol{\lambda}^{k+1} = \boldsymbol{\lambda}^k + \beta(\mathbf{y}^k - \nabla \mathbf{x}^k).$$

As for the minimization problem (46), we separate it into two subproblems with respect to $\mathbf{x}$ and $\mathbf{y}$ and minimize them alternatively.

The Solution to Sub-problem w.r.t. x

For a given y, there is the following minimization problem of variable x

$$\min_{x \in \mathcal{X}} \left\{ \frac{\alpha}{2} \|Kx - d\|^2 - \langle \lambda^k, \nabla x \rangle + \frac{\beta}{2} \|y - \nabla x\|^2 \right\}. \tag{47}$$

Applying Fourier transforms to the optimality condition of the problem (47), we obtain

$$[\alpha \mathcal{F}(K^*)\mathcal{F}(K) - \beta \mathcal{F}(\Delta)]\mathcal{F}(x) = \alpha \mathcal{F}(K^*)\mathcal{F}(d) - \mathcal{F}(\text{div})\mathcal{F}(\lambda^k + \beta y), \tag{48}$$

from which $\mathcal{F}(x)$ can be found and then x via an inverse Fourier transform (Yang et al. 2009; Wu and Tai 2010). Here applying Fourier transform to a block matrix is regarded as applying Fourier transform to each block.

The Solution to Sub-problem w.r.t. y

For a given x, there is the following minimization problem of variable y

$$\min_{y \in \mathcal{Y}} \{R_{\text{MTV}}(y) + \langle \lambda^k, y \rangle + \frac{\beta}{2} \|y - \nabla x\|^2\}.$$

It has the following closed form solution (Yang et al. 2009; Wu and Tai 2010)

$$y_{i,j} = \max\left(1 - \frac{1}{\beta |\eta_{i,j}|}, 0\right) \eta_{i,j}, \tag{49}$$

where $\eta = \nabla x - \frac{\lambda^k}{\beta} \in \mathcal{Y}$. Indeed, this solution is a high-dimensional version of (13), which can be also derived from the geometric method.

According to (48) and (49), we then have an alternating minimization procedure to (46); see Algorithm 9.

Algorithm 9 Augmented Lagrangian method for the multichannel TV model – solve the minimization problem (46)

Initialization: $x^{k,0} = x^{k-1}$, $y^{k,0} = y^{k-1}$.
Iteration: For $l = 0, 1, 2, \ldots, L - 1$:

- Compute $x^{k,l+1}$ from (48) for $y = y^{k,l}$;
- Compute $y^{k,l+1}$ from (49) for $x = x^{k,l+1}$.

Output: $x^k = x^{k,L}$, $y^k = y^{k,L}$.

We remark that the convergence results of Algorithms 3 and 4 can be directly extended for the Algorithms 8 and 9 (Wu and Tai 2010) and we omit the details.

Extension to High-Order Models

In this section, we review augmented Lagrangian methods for some high-order models, including the second-order total variation model (Lysaker et al. 2003), the total generalized variation model (Bredies et al. 2010), the Euler's elastic-based model (Chan et al. 2002; Tai et al. 2011), and the mean curvature model (Zhu and Chan 2012; Zhu et al. 2013).

Augmented Lagrangian Method for Second-Order Total Variation Model

To overcome the staircase effect, Lysaker, Lundervold, and Tai suggested regularizing the total variation of the gradient and proposed a model based on second-order derivatives (Lysaker et al. 2003). We begin with some notations to establish this second-order total variation (TV^2) model.

Let

$$\widehat{\mathcal{Y}} = \mathcal{X} \times \mathcal{X} \times \mathcal{X} \times \mathcal{X}.$$

We define the discrete Hessian operator

$$H : \mathcal{X} \to \widehat{\mathcal{Y}}$$

$$x \to Hx,$$

with

$$(Hx)_{i,j} = \begin{pmatrix} (\mathring{D}_{11}^{-+}x)_{i,j} & (\mathring{D}_{12}^{++}x)_{i,j} \\ (\mathring{D}_{21}^{++}x)_{i,j} & (\mathring{D}_{22}^{-+}x)_{i,j} \end{pmatrix},$$

where $\mathring{D}_{11}^{-+}, \mathring{D}_{12}^{++}, \mathring{D}_{21}^{++}$ and $\mathring{D}_{22}^{-+}$ are second-order difference operators and given by

$$(\mathring{D}_{11}^{-+}x)_{i,j} := (\mathring{D}_1^{-}(\mathring{D}_1^{+}x))_{i,j},$$
$$(\mathring{D}_{12}^{++}x)_{i,j} := (\mathring{D}_1^{+}(\mathring{D}_2^{+}x))_{i,j},$$
$$(\mathring{D}_{21}^{++}x)_{i,j} := (\mathring{D}_2^{+}(\mathring{D}_1^{+}x))_{i,j},$$
$$(\mathring{D}_{22}^{-+}x)_{i,j} := (\mathring{D}_2^{-}(\mathring{D}_2^{+}x))_{i,j}.$$

The usual inner product and L^2 norm in the space $\widehat{\mathcal{Y}}$ are as follows. We denote

$$\langle y, w \rangle = \langle y^1, w^1 \rangle + \langle y^2, w^2 \rangle + \langle y^3, w^3 \rangle + \langle y^4, w^4 \rangle \text{ and } \|y\| = \sqrt{\langle y, y \rangle},$$

for $y = \begin{pmatrix} y^1 & y^2 \\ y^3 & y^4 \end{pmatrix} \in \widehat{\mathcal{Y}}$ and $w = \begin{pmatrix} w^1 & w^2 \\ w^3 & w^4 \end{pmatrix} \in \widehat{\mathcal{Y}}$. At each pixel (i,j),

$$|y_{i,j}| = \sqrt{(y^1)_{i,j}^2 + (y^2)_{i,j}^2 + (y^3)_{i,j}^2 + (y^4)_{i,j}^2}$$

is the usual Euclidean norm in $\mathbb{R}^4$. By using the inner products of $\widehat{\mathcal{Y}}$ and $\mathcal{X}$ and the definitions of the finite difference operators, the adjoint operator of H is as follows

$$H^* : \widehat{\mathcal{Y}} \to \mathcal{X}$$

$$y = \begin{pmatrix} y^1 & y^2 \\ y^3 & y^4 \end{pmatrix} \to H^* y,$$

where

$$(H^* y)_{i,j} = (\mathring{D}_{11}^{+-} y^1)_{i,j} + (\mathring{D}_{21}^{--} y^1)_{i,j} + (\mathring{D}_{12}^{--} y^3)_{i,j} + (\mathring{D}_{22}^{+-} y^4)_{i,j},$$

where $\mathring{D}_{11}^{+-}$, $\mathring{D}_{12}^{--}$, $\mathring{D}_{21}^{--}$, and $\mathring{D}_{22}^{+-}$ are second-order difference operators.

By regularizing the norm of the discrete Hessian, the TV^2 model (Lysaker et al. 2003) reads

$$\min_{x \in \mathcal{X}} \left\{ E_{\text{TV}^2}(x) = \frac{\alpha}{2} \|Kx - d\|^2 + R_{\text{HO}}(Hx) \right\}, \quad (50)$$

where $\alpha > 0$, $d \in \mathcal{X}$ is the observed image, $K : \mathcal{X} \to \mathcal{X}$ is the blur operator and

$$R_{\text{HO}}(Hx) = \sum_{1 \le i,j \le N} |(Hx)_{i,j}|. \quad (51)$$

Similarly as for the total variation restoration model, we make the following assumption:

- Null$(H) \cap$ Null$(K) = \{0\}$.

Under this assumption, the functional $E_{\text{TV}^2}(x)$ in (50) is convex, proper, coercive, and continuous. Hence, we have the following result.

Theorem 7. *The problem (50) has at least one solution x, which satisfies*

$$0 \in \alpha K^*(Kx - d) + H^* \partial R_{\text{HO}}(Hx),$$

where $\partial R_{\text{HO}}(Hx)$ is the subdifferential of R_{HO} at Hx. Moreover, if Null$(K) = \{0\}$, *the minimizer is unique.*

In the following we review the augmented Lagrangian method proposed in Wu and Tai (2010) to solve (50). We first introduce a new variable $\hat{y} \in \widehat{\mathscr{Y}}$ and reformulate (50) into a constrained optimization problem

$$\min_{x \in \mathscr{X}, \hat{y} \in \widehat{\mathscr{Y}}} \left\{ G_{\mathrm{TV}^2}(x, \hat{y}) = \frac{\alpha}{2} \|Kx - d\|^2 + R_{\mathrm{HO}}(\hat{y}) \right\} \qquad (52)$$
$$\text{s.t.} \quad \hat{y} = Hx.$$

To solve (52), we define the augmented Lagrangian functional as

$$\mathscr{L}_{\mathrm{TV}^2}(x, \hat{y}; \lambda) = \frac{\alpha}{2} \|Kx - d\|^2 + R_{\mathrm{HO}}(\hat{y}) + \langle \lambda, \hat{y} - Hx \rangle + \frac{\beta}{2} \|\hat{y} - Hx\|^2, \qquad (53)$$

with the multiplier $\lambda \in \widehat{\mathscr{Y}}$ and a positive constant β, and consider the following saddle-point problem:

$$\text{Find } (x^*, \hat{y}^*, \lambda^*) \in \mathscr{X} \times \widehat{\mathscr{Y}} \times \widehat{\mathscr{Y}}$$
$$\text{s.t. } \mathscr{L}_{\mathrm{TV}^2}(x^*, \hat{y}^*; \lambda) \leq \mathscr{L}_{\mathrm{TV}^2}(x^*, \hat{y}^*; \lambda^*) \leq \mathscr{L}_{\mathrm{TV}^2}(x, \hat{y}; \lambda^*)$$
$$\forall (x, \hat{y}; \lambda) \in \mathscr{X} \times \widehat{\mathscr{Y}} \times \widehat{\mathscr{Y}}. \qquad (54)$$

We employ an iterative procedure to solve the saddle-point problem (54), which is described as Algorithm 10.

Algorithm 10 Augmented Lagrangian method for the TV^2 model

Initialization: $x^{-1} = 0, \hat{y}^{-1} = 0, \lambda^0 = 0$.
Iteration: For $k = 0, 1, 2, \ldots$:

1. Compute $(x^k, \hat{y}^k)$ from

$$(x^k, \hat{y}^k) \approx \arg \min_{(x, \hat{y}) \in (\mathscr{X}, \widehat{\mathscr{Y}})} \mathscr{L}_{\mathrm{TV}^2}(x, \hat{y}; \lambda^k). \qquad (55)$$

2. Update

$$\lambda^{k+1} = \lambda^k + \beta(\hat{y}^k - Hx^k).$$

The Solution to Sub-problem w.r.t. x

Given y, we are going to solve the following minimization problem

$$\min_{x \in \mathscr{X}} \left\{ \frac{\alpha}{2} \|Kx - d\|^2 - \langle \lambda^k, Hx \rangle + \frac{\beta}{2} \|\hat{y} - Hx\|^2 \right\}, \qquad (56)$$

the first-order optimality condition of which gives us a linear equation as follows

$$(\alpha K^*K + \beta H^*H)x = \alpha K^*d + H^*(\lambda^k + \beta \hat{y}). \tag{57}$$

This equation can solved by well-developed linear solvers such as FFT and CG.

The Solution to Sub-problem w.r.t. $\hat{y}$

Given x, we are going to solve the following minimization problem

$$\min_{\hat{y} \in \widehat{\mathscr{Y}}} \left\{ R_{\text{HO}}(\hat{y}) + (\lambda^k, \hat{y}) + \frac{\beta}{2} \|\hat{y} - Hx\|^2 \right\}, \tag{58}$$

the closed form solution of which is

$$\hat{y}_{i,j} = \max\left(0, 1 - \frac{1}{\beta |\eta_{i,j}|}\right) \eta_{ij}, \tag{59}$$

where $\eta = Hx - \frac{\lambda^k}{\beta} \in \widehat{\mathscr{Y}}$. We mention that the solution (59) is a high-dimensional version of (13), which can be also derived from the geometric method.

According to (57) and (59), we then use an iterative procedure to alternatively calculate x and $\hat{y}$; see Algorithm 11.

Algorithm 11 Augmented Lagrangian method for the TV^2 model – solve the minimization problem (55)

Initialization: $x^{k,0} = x^{k-1}$, $\hat{y}^{k,0} = \hat{y}^{k-1}$.
Iteration: For $l = 0, 1, 2, \ldots, L - 1$:

- Compute $x^{k,l+1}$ by solving (57) for $\hat{y} = \hat{y}^{k,l}$;
- Compute $\hat{y}^{k,l+1}$ from (59) for $x = x^{k,l+1}$.

Output: $x^k = x^{k,L}$, $\hat{y}^k = \hat{y}^{k,L}$.

We mention that the convergence results of the augmented Lagrangian method for the TV^2 model are straightforward as in Wu and Tai (2010) and we omit the details.

Augmented Lagrangian Method for Total Generalized Variation Model

Total generalized variation (TGV) is a very successful generalization of total variation, which involves high-order derivatives to reduce staircase effect (Bredies et al. 2010). In this section, we consider the following discrete second-order total generalized variation (Bredies et al. 2010)-based image restoration model

$$\min_{x \in \mathscr{X}, w \in \mathscr{Y}} \left\{ \frac{1}{2} \|Kx - d\|^2 + \alpha_1 R(\nabla x - w) + \alpha_0 R_{\text{HO}}(\mathscr{E}w) \right\}, \tag{60}$$

where $R(\nabla x - w)$ is defined by replacing ∇x by $\nabla x - w$ in (3), $\mathscr{E}$ denotes a distributional symmetrized gradient operator

$$\mathscr{E}: \mathscr{Y} \to \widehat{\mathscr{Y}}$$

$$w = (w^1, w^2) \to \mathscr{E}w = \frac{1}{2}(\nabla w + \nabla w^T),$$

with

$$(\mathscr{E}w)_{ij} = \frac{1}{2}(\nabla w + \nabla w^T)_{ij}$$

$$= \begin{pmatrix} (\mathring{D}_1^+ w^1)_{ij} & \frac{1}{2}((\mathring{D}_2^+ w^1)_{ij} + (\mathring{D}_1^+ w^2)_{ij}) \\ \frac{1}{2}((\mathring{D}_2^+ w^1)_{ij} + (\mathring{D}_1^+ w^2)_{ij}) & (\mathring{D}_2^+ w^2)_{ij} \end{pmatrix},$$

and $R_{\text{HO}}(\cdot)$ is defined in (51). Similarly, by using the inner products of $\widehat{\mathscr{Y}}$ and $\mathscr{Y}$ and the definitions of the finite difference operators the adjoint operator of $-\mathscr{E}$ is as follows

$$\text{div}_2 : \widehat{\mathscr{Y}} \to \mathscr{Y}$$

$$z = \begin{pmatrix} z^1 & z^3 \\ z^3 & z^2 \end{pmatrix} \to \text{div}_2 z,$$

where

$$\text{div}_2 z = \begin{pmatrix} \mathring{D}_1^- z^1 + \mathring{D}_2^- z^3 \\ \mathring{D}_1^- z^3 + \mathring{D}_2^- z^2 \end{pmatrix}$$

with

$$(\text{div}_2 z)_{ij} = \begin{pmatrix} (\mathring{D}_1^- z^1)_{ij} + (\mathring{D}_2^- z^3)_{ij} \\ (\mathring{D}_1^- z^3)_{ij} + (\mathring{D}_2^- z^2)_{ij} \end{pmatrix}.$$

Augmented Lagrangian-based methods for total generalized variation-related models can be found in Gao et al. (2018). Here, we propose the augmented Lagrangian method to solve (60). We first introduce two auxiliary variable

$y = (y^1, y^2) \in \mathscr{Y}$ and $z = \begin{pmatrix} z^1 & z^3 \\ z^3 & z^2 \end{pmatrix} \in \widehat{\mathscr{Y}}$ and transform it into an equivalent constrained optimization problem

$$\min_{x \in \mathscr{X}, w \in \mathscr{Y}, y \in \mathscr{Y}, z \in \widehat{\mathscr{Y}}} \left\{ G_{\text{TGV}}(x, y, z) = \frac{1}{2} \|Kx - d\|^2 + \alpha_1 R(y) + \alpha_0 R_{\text{HO}}(z) \right\}$$
$$\text{s.t.} \quad \begin{pmatrix} y \\ z \end{pmatrix} = \begin{pmatrix} \nabla & -\mathscr{I}_2 \\ & \mathscr{E} \end{pmatrix} \begin{pmatrix} x \\ w \end{pmatrix}. \tag{61}$$

We then define the augmented Lagrangian function as follows

$$\mathscr{L}_{\text{TGV}}(x, w, y, z; \lambda_y, \lambda_z) = \frac{1}{2} \|Kx - d\|^2 + \alpha_1 R(y) + \alpha_0 R_{\text{HO}}(z)$$
$$+ \left\langle \begin{pmatrix} \lambda_y \\ \lambda_z \end{pmatrix}, \begin{pmatrix} y \\ z \end{pmatrix} - \begin{pmatrix} \nabla & -\mathscr{I}_2 \\ & \mathscr{E} \end{pmatrix} \begin{pmatrix} x \\ w \end{pmatrix} \right\rangle \tag{62}$$
$$+ \frac{1}{2} \left\| \begin{pmatrix} y \\ z \end{pmatrix} - \begin{pmatrix} \nabla & -\mathscr{I}_2 \\ & \mathscr{E} \end{pmatrix} \begin{pmatrix} x \\ w \end{pmatrix} \right\|_{\mathscr{S}}^2,$$

where $\begin{pmatrix} \lambda_y \\ \lambda_z \end{pmatrix}$ is the Lagrange multiplier and $\mathscr{S} = \begin{pmatrix} \beta_y \mathscr{I}_2 & \\ & \beta_z \widehat{\mathscr{I}}_2 \end{pmatrix}$ with the identity operator $\widehat{\mathscr{I}}_2 : \widehat{\mathscr{Y}} \to \widehat{\mathscr{Y}}$, and consider the saddle-point problem

Find $(x^*, w^*, y^*, z^*, \lambda_y^*, \lambda_z^*) \in \mathscr{X} \times \mathscr{Y} \times \mathscr{Y} \times \widehat{\mathscr{Y}} \times \mathscr{Y} \times \widehat{\mathscr{Y}}$
s.t. $\mathscr{L}_{\text{TGV}}(x^*, w^*, y^*, z^*; \lambda_y, \lambda_z)$
$\leq \mathscr{L}_{\text{TGV}}(x^*, w^*, y^*, z^*; \lambda_y^*, \lambda_z^*)$
$\leq \mathscr{L}_{\text{TGV}}(x, w, y, z; \lambda_y^*, \lambda_z^*),$
$\forall (x, w, y, z, \lambda_y, \lambda_z) \in \mathscr{X} \times \mathscr{Y} \times \mathscr{Y} \times \widehat{\mathscr{Y}} \times \mathscr{Y} \times \widehat{\mathscr{Y}}.$ \quad (63)

Finally, the iterative algorithm for seeking a saddle point is given by Algorithm 12.

The Solution to Sub-problem w.r.t. (x, w)

Given $\begin{pmatrix} y \\ z \end{pmatrix}$, we concern with the following minimization problem

Algorithm 12 Augmented Lagrangian method for TGV model

Initialization: $\begin{pmatrix} x^{-1} \\ w^{-1} \end{pmatrix} = \begin{pmatrix} 0 \\ 0 \end{pmatrix}, \begin{pmatrix} y^{-1} \\ z^{-1} \end{pmatrix} = \begin{pmatrix} 0 \\ 0 \end{pmatrix}, \begin{pmatrix} \lambda_y^0 \\ \lambda_z^0 \end{pmatrix} = \begin{pmatrix} 0 \\ 0 \end{pmatrix}.$

Iteration: For $k = 0, 1, \ldots$:

1. Compute (x^k, w^k, y^k, z^k) from $\begin{pmatrix} \lambda_y^k \\ \lambda_z^k \end{pmatrix}$, i.e.,

$$(x^k, w^k, y^k, z^k) \approx \arg \min_{(x,w,y,z) \in \mathcal{X} \times \mathcal{Y} \times \mathcal{Y} \times \mathcal{Y}} \mathcal{L}_{\text{TGV}}(x, w, y, z; \lambda_y^k, \lambda_z^k). \tag{64}$$

2. Update

$$\begin{pmatrix} \lambda_y^{k+1} \\ \lambda_z^{k+1} \end{pmatrix} = \begin{pmatrix} \lambda_y^k \\ \lambda_z^k \end{pmatrix} + \begin{pmatrix} \beta_y(y^k - \nabla x^k + w^k) \\ \beta_z(z^k - \mathcal{E}w^k) \end{pmatrix}.$$

$$\min_{(x,w) \in \mathcal{X} \times \mathcal{Y}} \left\{ \frac{1}{2} \|Kx - d\|^2 - \left\langle \begin{pmatrix} \lambda_y^k \\ \lambda_z^k \end{pmatrix}, \begin{pmatrix} \nabla & -\mathcal{I}_2 \\ & \mathcal{E} \end{pmatrix} \begin{pmatrix} x \\ w \end{pmatrix} \right\rangle \right.$$
$$\left. + \frac{1}{2} \left\| \begin{pmatrix} y \\ z \end{pmatrix} - \begin{pmatrix} \nabla & -\mathcal{I}_2 \\ & \mathcal{E} \end{pmatrix} \begin{pmatrix} x \\ w \end{pmatrix} \right\|_{\mathcal{S}}^2 \right\}. \tag{65}$$

This problem is a quadratic optimization problem, whose optimality condition gives a linear system equations

$$\begin{pmatrix} K^*K - \beta_y \Delta & \beta_y \,\text{div} \\ -\beta_y \nabla & \beta_y - \beta_z \,\text{div}_2 \mathcal{E} \end{pmatrix} \begin{pmatrix} x \\ w \end{pmatrix} = \begin{pmatrix} K^*d - \text{div}(\lambda_y^k + \beta_y y) \\ -\lambda_y^k - \beta_y y - \text{div}_2(\lambda_z^k + \beta_z z) \end{pmatrix},$$

i.e.

$$\begin{cases} (K^*K - \beta_y \mathring{D}_1^- \mathring{D}_1^+ - \beta_y \mathring{D}_2^- \mathring{D}_2^+)x + \beta_y \mathring{D}_1^- w^1 + \beta_y \mathring{D}_2^- w^2 = g^1, \\ -\beta_y \mathring{D}_1^+ x + (\beta_y \mathscr{I} - \beta_z \mathring{D}_1^- \mathring{D}_1^+ - \frac{\beta_z}{2} \mathring{D}_2^- \mathring{D}_2^+)w^1 - \frac{\beta_z}{2} \mathring{D}_2^- \mathring{D}_1^+ w^2 = g^2, \\ -\beta_y \mathring{D}_2^+ x - \frac{\beta_z}{2} \mathring{D}_1^- \mathring{D}_2^+ w^1 + (\beta_y \mathscr{I} - \frac{\beta_z}{2} \mathring{D}_1^- \mathring{D}_1^+ - \beta_z \mathring{D}_2^- \mathring{D}_2^+)w^2 = g^3, \end{cases} \tag{66}$$

where

$$g^1 = K^*d - \mathring{D}_1^-((\lambda_y^k)^1 + \beta_y y^1) - \mathring{D}_2^-((\lambda_y^k)^2 + \beta_y y^2),$$
$$g^2 = -(\lambda_y^k)^1 - \beta_y y^1 - \mathring{D}_1^-((\lambda_z^k)^1 + \beta_z z^1) - \mathring{D}_2^-((\lambda_z^k)^3 + \beta_z z^3),$$
$$g^3 = -(\lambda_y^k)^2 - \beta_y y^2 - \mathring{D}_1^-((\lambda_z^k)^3 + \beta_z z^3) - \mathring{D}_2^-((\lambda_z^k)^2 + \beta_z z^2).$$

13 On Variable Splitting and Augmented Lagrangian Method for Total...

This linear system with periodic boundary condition can be efficiently solved by Fourier transform via FFT implementation (Yang et al. 2009). Firstly, we apply FFTs to both sides of (66) to get

$$\begin{pmatrix} a^{11} & a^{12} & a^{13} \\ a^{21} & a^{22} & a^{23} \\ a^{31} & a^{32} & a^{33} \end{pmatrix} \begin{pmatrix} \mathscr{F}(x) \\ \mathscr{F}(w^1) \\ \mathscr{F}(w^2) \end{pmatrix} = \begin{pmatrix} \mathscr{F}(g^1) \\ \mathscr{F}(g^2) \\ \mathscr{F}(g^3) \end{pmatrix}. \tag{67}$$

where a^{ij}, $(i, j = 1, \ldots 3)$ are Fourier coefficients of the operators in the left side of (66). Secondly, we solve the resulting systems by block Gaussian elimination method for $\mathscr{F}(x)$, $\mathscr{F}(w^1)$ and $\mathscr{F}(w^2)$. Finally, we apply inverse FFTs to obtain x and $w = (w^1, w^2)$.

The Solution to Sub-problem w.r.t. (y, z)

Given $\begin{pmatrix} x \\ w \end{pmatrix}$, we concern with the following minimization problem

$$\min_{(y,z) \in \mathscr{Y} \times \widehat{\mathscr{Y}}} \left\{ \alpha_1 R(y) + \alpha_0 R_{\text{HO}}(z) + \left\langle \begin{pmatrix} \lambda_y^k \\ \lambda_z^k \end{pmatrix}, \begin{pmatrix} y \\ z \end{pmatrix} \right\rangle \right.$$
$$\left. + \frac{1}{2} \left\| \begin{pmatrix} y \\ z \end{pmatrix} - \begin{pmatrix} \nabla & -\mathscr{I}_2 \\ & \mathscr{E} \end{pmatrix} \begin{pmatrix} x \\ w \end{pmatrix} \right\|_{\mathscr{S}}^2 \right\}. \tag{68}$$

It can be separated into two independent minimization problems:

- y-subproblem:

$$\min_{y \in \mathscr{Y}} \left\{ \alpha_1 R(y) + \langle \lambda_y^k, y \rangle + \frac{\beta_y}{2} \| y - \nabla x + w \|^2 \right\}; \tag{69}$$

- z-subproblem:

$$\min_{z \in \widehat{\mathscr{Y}}} \left\{ \alpha_0 R_{\text{HO}}(z) + \langle \lambda_z^k, z \rangle + \frac{\beta_z}{2} \| z - \mathscr{E}w \|^2 \right\}. \tag{70}$$

The problem (69) and (70) have the closed form solutions

$$y_{i,j} = \max\left(0, 1 - \frac{\alpha_1}{\beta_y |\eta_{i,j}|}\right) \eta_{i,j}, \text{ and } z_{i,j} = \max\left(0, 1 - \frac{\alpha_0}{\beta_z |\xi_{i,j}|}\right) \xi_{i,j}, \tag{71}$$

where

$$\eta = \nabla x - w - \frac{\lambda_y^k}{\beta_y} \in \mathcal{Y}, \text{ and } \xi = \mathcal{E}w - \frac{\lambda_z^k}{\beta_z} \in \widehat{\mathcal{Y}}.$$

After knowing the solutions of the subproblems (65) and (68), we use the following alternative minimization procedure to solve (64); see Algorithm 13.

Algorithm 13 Augmented Lagrangian method for TGV model–solve the minimization problem (64)

Initialization: $\begin{pmatrix} x^{k,0} \\ w^{k,0} \end{pmatrix} = \begin{pmatrix} x^{k-1} \\ w^{k-1} \end{pmatrix}, \begin{pmatrix} y^{k,0} \\ z^{k,0} \end{pmatrix} = \begin{pmatrix} y^{k-1} \\ z^{k-1} \end{pmatrix}.$

Iteration: For $l = 0, 1, \ldots, L - 1$:

- Compute $\begin{pmatrix} x^{k,l+1} \\ w^{k,l+1} \end{pmatrix}$ from (67) for $\begin{pmatrix} y \\ z \end{pmatrix} = \begin{pmatrix} y^{k,l} \\ z^{k,l} \end{pmatrix}$;

- Compute $\begin{pmatrix} y^{k,l+1} \\ z^{k,l+1} \end{pmatrix}$ from (71) for $\begin{pmatrix} x \\ w \end{pmatrix} = \begin{pmatrix} x^{k,l+1} \\ w^{k,l+1} \end{pmatrix}.$

Output: $\begin{pmatrix} x^k \\ w^k \end{pmatrix} = \begin{pmatrix} x^{k,L} \\ w^{k,L} \end{pmatrix}, \begin{pmatrix} y^k \\ z^k \end{pmatrix} = \begin{pmatrix} y^{k,L} \\ z^{k,L} \end{pmatrix}.$

Augmented Lagrangian Method for Euler Elastic-Based Model

As basic geometric measurements of curves, both length and curvatures are natural regularities that are widely used in various image processing problems. Euler's elastica is defined as the line energy for a smooth planar curves γ

$$E(\gamma) = \int_\gamma (a + b\kappa^2) ds, \tag{72}$$

where κ is the curvature of the curve, s is arc length, and a, b are positive constants. By summing up the Euler's elastica energies of all the level sets for an image x, it gives the following energy for image denoising task

$$\min_x \ R_{EE}(\kappa(x), \nabla x) + \frac{1}{2} \|Kx - d\|^2, \tag{73}$$

where $\kappa(x) = \text{div}(\frac{\nabla x}{|\nabla x|})$ and $R_{EE}(\kappa(x), \nabla x)$ is defined by

$$R_{EE}(\kappa(x), \nabla x) = \sum_{1 \le i,j \le N} \left(a + b\kappa^2(x_{i,j})\right) |(\nabla x)_{i,j}|.$$

Euler's elastica regularization has lots of applications in shape and image processing. However, the non-convexity, the non-smoothness, and the nonlinearity of the Euler's elastica energy make its minimization a challenging task. Chan et al. (2002) developed a computational scheme based on numerical PDEs for inpainting problem. Bae et al. (2010) presented an efficient minimization algorithm based on graph cuts for minimizing the Euler's elastica energy. Tai et al. (2011) proposed an augmented Lagrangian method based on the operator-splitting and relaxation techniques, which greatly improved the efficiency of the Euler's elastica model. Since then, operator-splitting and augmented Lagrangian method have been extensively studied for Euler's elastica (Duan et al. 2013; Yashtini and Kang 2016). Recent advances include functional lifting to get a convex, lower semi-continuous, coercive approximation of the Euler's elastica energy (Bredies et al. 2015), and a lie operator-splitting-based time discretization scheme (Deng et al. 2019). In Tai et al. (2011), Euler's elastica regularized model (73) is reformulated as the following constrained minimization problem

$$\min_{x,y,n,m} R_{EE}(\text{div } n, y) + \frac{1}{2}\|Kx - d\|^2 + I_{\mathscr{M}}(m) \quad (74)$$

$$\text{s.t.} \quad y = \nabla x, \ n = m, \ |y| = m \cdot y,$$

where $I_{\mathscr{M}}(\cdot)$ is an indicator function of the set

$$\mathscr{M} = \{m_{ij} : |m_{i,j}| \leq 1, \ \forall\, 1 \leq i, j \leq N\}.$$

Note that the variable m was introduced to relax the constraint on variable n. By requiring m to be lain in the set $\mathscr{M}$, the term $|y| - y \cdot m$ is guaranteed non-negative, which make the sub-minimization problem w.r.t. m easy to handle with. We can further define the augmented Lagrangian functional as follows

$$\begin{aligned}\mathscr{L}_{EE}(x, y, n, m; \lambda_y, \lambda_n, \lambda_m) &= R_{EE}(\text{div } n, y) + \frac{1}{2}\|Kx - d\|^2 + I_{\mathscr{M}}(m) \\ &+ \langle \lambda_y, y - \nabla x \rangle + \frac{\beta_y}{2}\|y - \nabla x\|^2 + \langle \lambda_n, n - m \rangle + \frac{\beta_n}{2}\|n - m\|^2 \\ &+ \langle \lambda_m, |y| - m \cdot y \rangle + \langle |y| - m \cdot y, \beta_m \rangle,\end{aligned} \quad (75)$$

where $\lambda_y, \lambda_n, \lambda_m$ are the Lagrange multipliers and $\beta_y, \beta_n, \beta_m$ are positive parameters. The iterative algorithm is used to find a point satisfying the first-order condition; see Algorithm 14.

Before we discuss the solution to the minimization problem (76), we define a staggered grid system in Fig. 2; see more details of the implementation in Tai et al. (2011). We separate the minimization problem (76) into subproblems to pursue the solutions in an alternative mechanism.

Algorithm 14 Augmented Lagrangian method for Euler elastic model
Initialization: $x^{-1} = 0$, $y^{-1} = 0$, $n^{-1} = 0$, $m^{-1} = 0$, $\lambda_y^0 = 0$, $\lambda_n^0 = 0$, $\lambda_m^0 = 0$.
Iteration: For $k = 0, 1, \ldots$:

1. Compute (x^k, y^k, n^k, m^k) from

$$(x^k, y^k, n^k, m^k) \approx \arg\min_{(x,y,n,m)} \mathscr{L}_{\text{EE}}(x, y, n, m; \lambda_y^k, \lambda_n^k, \lambda_m^k), \qquad (76)$$

2. Update

$$\begin{pmatrix} \lambda_y^{k+1} \\ \lambda_n^{k+1} \\ \lambda_m^{k+1} \end{pmatrix} = \begin{pmatrix} \lambda_y^k \\ \lambda_n^k \\ \lambda_m^k \end{pmatrix} + \begin{pmatrix} \beta_y(y^k - \nabla x^k) \\ \beta_n(n^k - m^k) \\ \beta_m(|y^k| - m^k \cdot y^k) \end{pmatrix}$$

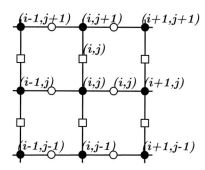

Fig. 2 The rule of indexing variables in the augmented Lagrangian functional (75): $x, z, \lambda_z, \lambda_m$ are defined on •-nodes. The first and second component of $y, n, m, \lambda_y, \lambda_n$ are defined on ○-nodes and □-node, respectively

The Solution to Sub-problem w.r.t. *x*
Given y, we solve the following minimization problem

$$\min_x \frac{1}{2}\|Kx - d\|^2 + \frac{\beta_y}{2}\|y - \nabla x\|^2 - \langle \lambda_y^k, \nabla x \rangle, \qquad (77)$$

the first-order optimal condition of which gives us

$$(K^*K - \beta_y \Delta)x = K^*d - \beta_y \operatorname{div} y - \operatorname{div} \lambda_y^k.$$

Fast numerical methods can be used to solve the above equation such as fast Fourier transform (FFT) and iterative schemes.

The Solution to Sub-problem w.r.t. *y*
Given x, n, and m, we have the subproblem of y as follows

$$\min_y \langle a + b(\operatorname{div} n)^2, |y| \rangle + \langle \lambda_y^k, y \rangle + \langle \lambda_m^k + \beta_m, |y| - m \cdot y \rangle + \frac{\beta_y}{2}\|y - \nabla x\|^2, \qquad (78)$$

which can be simplified as

$$\min_{y} \frac{\beta_y}{2} \left\| y - \left(\nabla x + (\frac{\lambda_m^k + \beta_m}{\beta_y})m - \frac{\lambda_y^k}{\beta_y} \right) \right\|^2 + \langle |y|, a + b(\operatorname{div} n)^2 + \lambda_m^k + \beta_m \rangle.$$

Such the L^1 regularized minimization problem can be efficiently solved by the closed form solution.

The Solution to Sub-problem w.r.t. m

Given n and y, the sub-minimization problem of variable m becomes

$$\min_{m} I_{\mathcal{M}}(m) - \langle \lambda_m^k, m \rangle + \frac{\beta_n}{2} \|n - m\|^2 - \langle (\lambda_m^k + \beta_m)y, m \rangle. \tag{79}$$

We can reformulate the above minimization into a quadratic problem as follows

$$\min_{m} I_{\mathcal{M}}(m) + \frac{\beta_n}{2} \left\| m - \frac{(\lambda_m^k + \beta_m)y + \lambda_m^k}{\beta_n} - n \right\|^2,$$

the optimal solution of which can be achieved by performing the one-step projection to the solution of the quadratic minimization.

The Solution to Sub-problem w.r.t. n

Given m and y, we are going to solve the following minimization problem of n

$$\min_{n} \langle b(\operatorname{div} n)^2, |y| \rangle + \langle \lambda_n^k, n \rangle + \frac{\beta_n}{2} \|n - m\|^2, \tag{80}$$

the Euler-Lagrange equation of which is

$$-2\nabla(b|y|\operatorname{div} n) + \beta_n(n - m) + \lambda_n^k = 0,$$

and can be solved by a frozen coefficient method for easier implementation (Tai et al. 2011; Yashtini and Kang 2016).

Augmented Lagrangian Method for Mean Curvature-Based Model

Mean curvature-based model (Zhu and Chan 2012) considers an image restoration problem as a surface smoothing task. A basic model is as follows

$$\min_{x} \int_{\Omega} \left| \operatorname{div} \left(\frac{\nabla x}{\sqrt{1 + |\nabla x|^2}} \right) \right| dx + \frac{\alpha}{2} \int_{\Omega} (Kx - d)^2 dx. \tag{81}$$

Originally, the smoothed mean curvature model (81) was numerically solved by the gradient descent method, which involves high-order derivatives and converges slowly in practice. Zhu et al. (2013) developed an augmented Lagrangian method for a mean curvature-based image denoising model (81), with similar ideas further studied in Myllykoski et al. (2015). Following Zhu et al. (2013), we rewrite the mean curvature-regularized model into the following constrained minimization problem

$$\min_{x,y,q,n,m} R_{\text{MC}}(q) + \frac{\alpha}{2} \|Kx - d\|^2 + I_{\mathcal{M}}(m) \tag{82}$$

$$\text{s.t. } y = \langle \nabla x, 1 \rangle, \; q = \text{div } n, \; n = m, \; |y| = y \cdot m,$$

where $R_{\text{MC}}(q)$ is defined as

$$R_{\text{MC}}(q) = \sum_{1 \leq i,j \leq N} |q_{i,j}|.$$

The corresponding augmented Lagrangian functional for the constrained minimization problem is defined as

$$\mathcal{L}_{\text{MC}}(x, y, q, m, n; \lambda_y, \lambda_q, \lambda_n, \lambda_m) = R_{\text{MC}}(q) + \frac{\alpha}{2} \|Kx - d\|^2 + I_{\mathcal{M}}(m)$$
$$+ \langle \lambda_y, \langle \nabla x, 1 \rangle \rangle + \frac{\beta_y}{2} \|y - \langle \nabla x, 1 \rangle\|^2 + \langle q - \nabla \cdot n \rangle + \frac{\beta_q}{2} \|q - \nabla \cdot n\|$$
$$+ \langle \lambda_n, n - m \rangle + \frac{\beta_n}{2} \|n - m\|^2 + \langle \lambda_m, |y| - y \cdot m \rangle + \beta_m \langle |y| - y \cdot m \rangle, \tag{83}$$

where $\lambda_y, \lambda_q, \lambda_n, \lambda_m$ are Lagrange multipliers and $\beta_y, \beta_q, \beta_n, \beta_m$ are positive parameters. The iterative algorithm is used to find a point satisfying the first-order condition; see Algorithm 15.

Algorithm 15 Augmented Lagrangian method for mean curvature-based model

Initialization: $x^{-1} = 0, y^{-1} = 0, q^{-1} = 0, n^{-1} = 0, m^{-1} = \mathbf{0}, \lambda_y^0 = 0, \lambda_q^0 = 0, \lambda_n^0 = 0, \lambda_m^0 = 0.$
Iteration: For $k = 0, 1, \ldots$:

1. Compute $(x^k, y^k, q^k, n^k, m^k)$ from

$$(x^k, y^k, q^k, n^k, m^k) \approx \arg \min_{(x,y,q,n,m)} \mathcal{L}_{\text{MC}}(x, y, q, n, m; \lambda_y^k, \lambda_q^k, \lambda_n^k, \lambda_m^k), \tag{84}$$

2. Update

$$\begin{pmatrix} \lambda_y^{k+1} \\ \lambda_q^{k+1} \\ \lambda_n^{k+1} \\ \lambda_m^{k+1} \end{pmatrix} = \begin{pmatrix} \lambda_y^k \\ \lambda_q^k \\ \lambda_n^k \\ \lambda_m^k \end{pmatrix} + \begin{pmatrix} \beta_y(y^k - \langle \nabla x^k, 1 \rangle) \\ \beta_q(q^k - \nabla \cdot n^k) \\ \beta_n(n^k - m^k) \\ \beta_m(|y^k| - y^k \cdot m^k) \end{pmatrix}$$

We can separate the minimization problem (84) into subproblems to obtain the solutions in an alternative way. Similarly as discussed for Euler's elastica model, the minimizers to the variable y, q, and m have closed form solutions, while the minimizers to the variable x and n are obtained by solving the associated Euler-Lagrange equations by either FFT or fast iterative schemes. Therefore, we omit the details here.

Numerical Experiments

In this section, we give some numerical results of augmented Lagrangian methods for solving the total variation-related image restoration models. For each model, we test only one image by considering the limit space. For more examples, please refer to literatures (Tai and Wu 2009; Wu and Tai 2010; Wu et al. 2011; Chan et al. 2013; Tai et al. 2011; Zhu et al. 2013). We perform the numerical experiments in MATLAB R2018A (Version 9.4) on a MacBook Pro with 2.3 GHz dual-core Intel Core i5 processor and 8GB memory. For each experiment, we stop the iteration until the following criterion

$$\frac{\left\|x^{k+1} - x^k\right\|}{\left\|x^k\right\|} < 1e - 3 (\text{ for multichannel case } \frac{\left\|\mathbf{x}^{k+1} - \mathbf{x}^k\right\|}{\left\|\mathbf{x}^k\right\|} < 1e - 3)$$

satisfies. We measure the quality of the restored images by the improvement of signal to noise ratio (ISNR)

$$\text{ISNR}(x^*) = 10 \log_{10} \frac{\left\|\underline{x} - x^*\right\|}{\left\|\underline{x} - d\right\|},$$

where $\underline{x}$ is the ground truth image, d is the observed image, and x^* is the recovered image. For multichannel case, we have the similar definition of ISNR. For each model, the parameter α is tuned to obtain the highest ISNR. The performances of augmented Lagrangian methods are demonstrated in Figs. 3, 4, 5, 6, 7, 8, 9, 10, and 11.

Figure 3 shows the results of augmented Lagrangian method for solving TV-L^2 model. In this experiment, we corrupt the clean image (size 512 × 512) with Gaussian blur and Gaussian noise. We set the parameters by following the recommendations in Wu and Tai (2010) and let $\beta = 10$. We report the recovered image and its ISNR in Fig. 3c. We also record the used CPU time t when the algorithm terminates. We can see that augmented Lagrangian method can solve TV-L^2 model efficiently and obtain high-quality recovered image.

Figure 4 shows the results of augmented Lagrangian method for solving TV-L^2 model with box constraint and the comparisons with TV-L^2 model. In this experiment, the degraded image (size 217 × 181) is corrupted with Gaussian blur and Gaussian noise. We set the parameters $\beta = \beta_y = 10$, and $\beta_z = 400$. We

(a) Original. Size: 512 × 512

(b) Blurry&Noisy

(c) ALM for TV-L^2 ISNR:6.83dB t=1.60s

Fig. 3 Augmented Lagrangian method (ALM) for solving TV-L^2 model. (**b**) is a corruption of (**a**) with Gaussian blur fspecial('gaussian',11,3) and Gaussian noise with variation $1e-2$; (**c**) is the recovered result

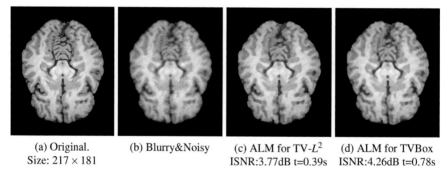

(a) Original. Size: 217 × 181

(b) Blurry&Noisy

(c) ALM for TV-L^2 ISNR:3.77dB t=0.39s

(d) ALM for TVBox ISNR:4.26dB t=0.78s

Fig. 4 Augmented Lagrangian method for solving TV-L^2 model with box constraint (TVBox). (**b**) is a corruption of (**a**) with Gaussian blur fspecial('gaussian',5,1.5) and Gaussian noise with variation $1e-3$; (**c**) and (**d**) are the recovered results

(a) Original. Size: 512 × 512

(b) Blurry&Noisy 50% salt & pepper

(c) ALM for TV-L^1 ISNR:20.34dB t=4.7s

Fig. 5 Augmented Lagrangian method for solving TV-L^1 model. (**b**) is a corruption of (**a**) with Gaussian blur fspecial('gaussian',11,3) and 50% salt and pepper noise; (**c**) is the recovered result

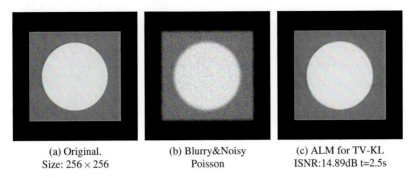

(a) Original.　　　　(b) Blurry&Noisy　　　(c) ALM for TV-KL
Size: 256 × 256　　　　Poisson　　　　ISNR:14.89dB t=2.5s

Fig. 6 Augmented Lagrangian method for solving TV-KL model. (**b**) is a corruption of (**a**) with Gaussian blur `fspecial('gaussian',11,3)` and Poisson noise; (**c**) is the recovered result

(a) Lena in color.　　　(b) Blurry&Noisy　　　(c) ALM for MTV
Size: 512 × 512 × 3　　　　　　　　　　　ISNR:2.99dB t=6.21s

Fig. 7 Augmented Lagrangian method for multichannel TV (MTV) restoration (**b**) is a corruption of (**a**) with within-channel Gaussian blur `fspecial('gaussian',21,5)`, and Gaussian noise with variation $1e-3$; (**c**) is the recovered result

report the recovered images and their ISNRs in Fig. 4c, d. We also record the used CPU times t when the algorithms terminate. We can see that augmented Lagrangian method can solve TV-L^2 model with box constraint efficiently and obtain high-quality recovered image. The TV-L^2 model with box constraint gains higher ISNR than the TV-L^2 model.

Figures 5 and 6 show the results of augmented Lagrangian methods for TV-L^1 model and TV-KL model. In the experiment for TV-L^1 model, the observed image (size 512 × 512) is degraded with Gaussian blur and 50% salt and pepper noise. We set $\beta_y = 20$ and $\beta_z = 100$. In the experiment for TV-KL model, the observed image (size 256 × 256) is corrupted with Gaussian kernel and Poisson noise. We let $\beta_y = 20$ and $\beta_z = 20$. We can see that augmented Lagrangian methods can recover high-quality images in these two experiments and the CPU costs are low.

Figure 7 shows the results of augmented Lagrangian method for multichannel TV restoration. In this experiment, the degraded image is generated by first blurring the ground truth image (size 512 × 512 × 3) with within-channel Gaussian blur and then adding Gaussian noise to the blurred image. We set $\beta = 100$. We also can see that

Fig. 8 Augmented Lagrangian method for solving TV^2 model. (**b**) is a corruption of (**a**) with Gaussian blur `fspecial('gaussian', 11,3)` and Gaussian noise with variation $1e-2$; (**c**) and (**d**) are the recovered results

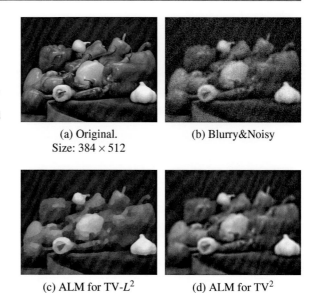

(a) Original. Size: 384×512

(b) Blurry&Noisy

(c) ALM for TV-L^2 ISNR:8.87dB t=1.08s

(d) ALM for TV^2 ISNR:8.99dB t=2.26s

Fig. 9 Augmented Lagrangian method for solving TGV model. (**b**) is a corruption of (**a**) with Gaussian blur `fspecial('gaussian', 5,1.5)` and Gaussian noise with variation $1e-2$; (**c**) and (**d**) are the recovered results

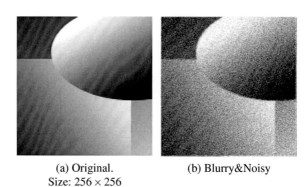

(a) Original. Size: 256×256

(b) Blurry&Noisy

(c) ALM for TV-L^2 ISNR:6.68dB t=0.40s

(d) ALM for TGV ISNR:6.76dB t=1.84s

13 On Variable Splitting and Augmented Lagrangian Method for Total...

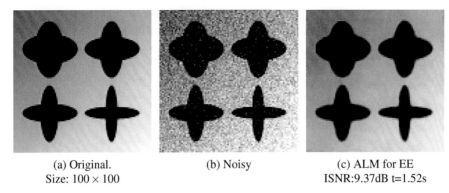

(a) Original.
Size: 100×100

(b) Noisy

(c) ALM for EE
ISNR:9.37dB t=1.52s

Fig. 10 Augmented Lagrangian method for solving Euler's elastica (EE) based image denoising model. (**b**) is a corruption of (**a**) with Gaussian noise with variation $1e - 2$; (**c**) is the recovered result

(a) Original.
Size: 256×256

(b) Noisy

(c) ALM for MC
ISNR:8.10dB t=4.17s

Fig. 11 Augmented Lagrangian method for solving mean curvature (MC)-based image denoising model. (**b**) is a corruption of (**a**) with Gaussian noise with variation $1e - 2$; (**c**) is the recovered result

augmented Lagrangian method can restore high-quality multichannel image with a low CPU cost.

Figures 8 and 9 show the results of augmented Lagrangian methods for solving TV^2 model and TGV model and the comparisons with TV-L^2 model. In the experiment for TV^2 model, the degraded image (size 384×512) is generated with Gaussian blur and Gaussian noise. We set $\beta = 10$. In the experiment for TGV model, the degraded image (size 256×256) is also generated with Gaussian blur and Gaussian noise. We let $(\alpha_0, \alpha_1) = (1.0, 0.1)$, $\beta_y = 10$ and $\beta_z = 20$. We report the recovered images and their ISNRs in Figs. 8c, d and 9c, d. We also record the used CPU times t when the algorithms terminate. We can see that augmented Lagrangian method can solve TV^2 model and TGV model efficiently and obtain high-quality recovered images. The TV^2 model and TGV model, which use high-order regularization, can suppress the staircase effect well.

Figures 10 and 11 show the results of augmented Lagrangian methods for solving Euler's elastica-based image denoising model and mean curvature-based image denoising model. Both these two models include curvature term in the regularization and are non-convex and highly nonlinear. We generate the degraded images Figs. 10b, 11b by adding Gaussian noise to the clean images Figs. 10a and 11a, respectively. In the experiment for Euler's elastica based model, we use $\beta_y = 200$, $\beta_n = 500$ and $\beta_m = 1$. In the experiment for mean curvature-based model, we use $\beta_y = 40$, $\beta_q = 1e5$, $\beta_n = 1e5$ and $\beta_m = 40$. We report the recovered images and their ISNRs and show the CPU costs in Figs. 10c and 11c. We can see that augmented Lagrangian methods can solve non-convex curvature-based models efficiently and obtain high-quality recovered images.

Conclusions

In this survey, we have reviewed variable splitting and augmented Lagrangian methods for total variation-related image restoration models. Due to the closed form solutions of subproblems and fast linear solvers like the FFT implementations, these methods are efficient for both total variation-related convex models and non-convex Euler's elastica and mean curvature-based models.

Acknowledgments Tai is supported by NSFC/RGC Joint Research Scheme (N_HKBU214/19), Initiation Grant for Faculty Niche Research Areas(RC-FNRA-IG/19-20/SCI/01) and CRF (C1013-21GF).

References

Acar, R., Vogel, C.R.: Analysis of bounded variation penalty methods for ill-posed problems. Inverse Prob. **10**(6), 1217–1229 (1994)

Aubert, G., Kornprobst, P.: Mathematical Problems in Image Processing: Partial Differential Equations and the Calculus of Variations, 2nd edn. Springer, New York (2010)

Bae, E., Shi. J., Tai, X.C.: Graph cuts for curvature based image denoising. IEEE Trans. Image Process **20**(5), 1199–1210 (2010)

Beck, A., Teboulle, M.: Fast gradient-based algorithms for constrained total variation image denoising and deblurring problems. IEEE Trans. Image Process. **18**(11), 2419–2434 (2009)

Bertalmio, M., Vese, L., Sapiro, G., Osher, S.: Simultaneous structure and texture image inpainting. IEEE Trans. Image Process. **12**(8), 882–889 (2003)

Bertsekas, D.P.: Constrained Optimization and Lagrange Multiplier Methods. Optimization and Neural Computation Series, Athena Scientific, Belmont, Mass (1996(firstly published in 1982))

Blomgren, P., Chan, T.F.: Color TV: Total variation methods for restoration of vector-valued images. IEEE Trans. Image Process. **7**(3), 304–309 (1998)

Boyd, S.: Distributed optimization and statistical learning via the alternating direction method of multipliers. Found. Trends Mach. Learn. **3**(1), 1–122 (2010)

Bredies, K., Kunisch, K., Pock, T.: Total generalized variation. SIAM J. Imaging Sci. **3**(3), 492–526 (2010)

Bredies, K., Pock, T., Wirth, B.: A convex, lower semicontinuous approximation of Euler's elastica energy. SIAM J. Math. Anal. **47**(1), 566–613 (2015)

Brune, C., Sawatzky, A., Burger, M.: Bregman-em-tv methods with application to optical nanoscopy. In: Tai, X.C., Mørken, K., Lysaker, M., Lie, K.A. (eds.) Scale Space and Variational Methods in Computer Vision. Springer, Berlin/Heidelberg, pp. 235–246 (2009)

Chambolle, A.: An algorithm for total variation minimization and applications. J. Math. Imaging Vis. **20**(1/2), 89–97 (2004)

Chambolle, A., Lions, P.L.: Image recovery via total variation minimization and related problems. Numer. Math. **76**(2), 167–188 (1997)

Chambolle, A., Pock, T.: A first-order primal-dual algorithm for convex problems with applications to imaging. J. Math. Imaging Vis. **40**(1), 120–145 (2011)

Chan, R.H., Tao, M., Yuan, X.: Constrained total variation deblurring models and fast algorithms based on alternating direction method of multipliers. SIAM J. Imaging Sci. **6**(1), 680–697 (2013)

Chan, T., Wong, C.K.: Total variation blind deconvolution. IEEE Trans. Image Process. **7**(3), 370–375 (1998)

Chan, T.F., Vese, L.A.: Active contours without edges. IEEE Trans. Image Process. **10**(2), 266–277 (2001)

Chan, T.F., Kang, S.H., Shen, J.: Euler's elastica and curvature-based inpainting. SIAM J. Appl. Math. **63**(2), 564–592 (2002)

Chang, H., Lou, Y., Ng, M., Zeng, T.: Phase retrieval from incomplete magnitude information via total variation regularization. SIAM J. Sci. Comput. **38**(6), A3672–A3695 (2016)

Chen, C., Chen, Y., Ouyang, Y., Pasiliao, E.: Stochastic accelerated alternating direction method of multipliers with importance sampling. J. Optim. Theory Appl. **179**(2), 676–695 (2018)

Chen, X., Ng, M.K., Zhang, C.: Non-Lipschitz ℓ_p-regularization and box constrained model for image restoration. IEEE Trans. Image Process. **21**(12), 4709–4721 (2012)

Chen, Y., Levine, S., Rao, M.: Variable exponent, linear growth functionals in image restoration. SIAM J. Appl. Math. **66**(4), 1383–1406 (2006)

Deng, L.J., Glowinski, R., Tai, X.C.: A new operator splitting method for the Euler elastica model for image smoothing. SIAM J. Imaging Sci. 12(2):1190–1230 (2019)

Deng, W., Yin, W.: On the global and linear convergence of the generalized alternating direction method of multipliers. J. Sci. Comput. **66**(3), 889–916 (2016)

Duan, Y., Wang, Y., Hahn, J.: A fast augmented Lagrangian method for Euler's elastica models. Numer. Math. Theory Methods Appl. **006**(001), 47–71 (2013)

Fazel, M., Pong, T.K., Sun, D., Tseng, P.: Hankel matrix rank minimization with applications to system identification and realization. SIAM J. Matrix Anal. Appl. **34**(3), 946–977 (2013)

Feng, X., Wu, C., Zeng, C.: On the local and global minimizers of ℓ_0 gradient regularized model with box constraints for image restoration. Inverse Prob. **34**(9), 095,007 (2018)

Gao, Y., Liu, F., Yang, X.: Total generalized variation restoration with non-quadratic fidelity. Multidim. Syst. Sign. Process. **29**(4), 1459–1484 (2018)

Glowinski, R., Tallec, P.L.: Augmented Lagrangians and Operator-Splitting Methods in Nonlinear Mechanics. SIAM, Philadelphia (1989)

Glowinski, R., Osher, S.J., Yin, W. (eds.): (2016) Splitting Methods in Communication, Imaging, Science, and Engineering. Springer, Cham

Goldstein, T., Osher, S.: The split Bregman method for L1-regularized problems. SIAM J. Imaging Sci. **2**(2), 323–343 (2009)

Güven, H.E., Güngör. A., Çetin, M.: An augmented Lagrangian method for complex-valued compressed SAR imaging. IEEE Trans. Comput. Imag. **2**(3), 235–250 (2016)

Hahn, J., Wu, C., Tai, X.C.: Augmented Lagrangian method for generalized TV-Stokes model. J. Sci. Comput. **50**(2), 235–264 (2012)

He, B., Yuan, X.: On the $o(1/n)$ convergence rate of the Douglas-Rachford alternating direction method. SIAM J. Numer. Anal. **50**(2), 700–709 (2012)

Hinterberger, W., Scherzer, O.: Variational methods on the space of functions of bounded Hessian for convexification and denoising. Computing **76**(1–2), 109–133 (2006)

Hintermüller, M., Wu, T.: Nonconvex TV^q-models in image restoration: Analysis and a trust-region regularization–based superlinearly convergent solver. SIAM J. Imaging Sci. **6**(3), 1385–1415 (2013)

Kang, S.H., Zhu, W., Jianhong, J.: Illusory shapes via corner fusion. SIAM J. Imaging Sci. **7**(4), 1907–1936 (2014)

Lai, R., Chan, T.F.: A framework for intrinsic image processing on surfaces. Comput. Vis. Image Und **115**(12), 1647–1661 (2011)

Le, T., Chartrand, R., Asaki, T.J.: A variational approach to reconstructing images corrupted by Poisson noise. J. Math. Imaging Vis. **27**, 257–263 (2007)

Li, C., Yin, W., Jiang, H., Zhang, Y.: An efficient augmented Lagrangian method with applications to total variation minimization. Comput. Optim. Appl. **56**(3), 507–530 (2013)

Li, G., Pong, T.K.: Global convergence of splitting methods for nonconvex composite optimization. SIAM J. Optim. **25**(4), 2434–2460 (2015)

Liu, Z., Wali, S., Duan, Y., Chang, H., Wu, C., Tai, X.C.: Proximal ADMM for Euler's elastica based image decomposition model. Numer. Math. Theory Methods Appl. **12**(2), 370–402 (2018)

Lou, Y., Zhang, X., Osher, S., Bertozzi, A.L.: Image recovery via nonlocal operators. J. Sci. Comput. **42**(2), 185–197 (2010)

Lysaker, M., Lundervold, A., Tai, X.: Noise removal using fourth-order partial differential equation with applications to medical magnetic resonance images in space and time. IEEE Trans. Image Process. **12**(12), 1579–1590 (2003)

Micchelli, C.A., Shen, L., Xu, Y.: Proximity algorithms for image models: denoising. Inverse Prob. **27**(4), 045,009 (2011)

Myllykoski, M., Glowinski, R., Karkkainen, T., Rossi, T.: A new augmented Lagrangian approach for L^1-mean curvature image denoising. SIAM J. Imaging Sci. **8**(1), 95–125 (2015)

Nikolova, M.: A variational approach to remove outliers and impulse noise. J. Math. Imaging Vis. **20**(1–2), 99–120 (2004)

Nikolova, M.: Analysis of the recovery of edges in images and signals by minimizing nonconvex regularized least-squares. Multiscale Model. Simul. **4**(3), 960–991 (2005)

Ouyang, Y., Chen, Y., Lan, G., Pasiliao, E.: An accelerated linearized alternating direction method of multipliers. SIAM J. Imaging Sci. **8**(1), 644–681 (2015)

Persson, M., Bone, D., Elmqvist, H.: Total variation norm for three-dimensional iterative reconstruction in limited view angle tomography. Phys. Med. Biol. **46**(3), 853–866 (2001)

Ramani, S., Fessler, J.A.: Parallel MR image reconstruction using augmented Lagrangian methods. IEEE Trans. Med. Imaging **30**(3), 694–706 (2011)

Rockafellar, R.T.: Augmented Lagrange multiplier functions and duality in nonconvex programming. SIAM J. Control **12**(2), 268–285 (1974)

Rockafellar, R.T., Wets, R.J.B.: Variational Analysis. Springer, Berlin/Heidelberg (1998)

Rudin, L., Osher, S., Fatemi, E.: Nonlinear total variation based noise removal algorithms. Physica D **60**, 259–268 (1992)

Sapiro, G., Ringach, D.: Anisotropic diffusion of multivalued images with applications to color filtering. IEEE Trans. Image Process. **5**, 1582–1586 (1996)

Selesnick, I., Lanza, A., Morigi, S., Sgallari, F.: Non-convex total variation regularization for convex denoising of signals. J. Math. Imaging Vis. **62**(6), 825–841 (2020)

Tai, X.C., Wu, C.: Augmented Lagrangian method, dual methods and split Bregman iteration for ROF model. In: Scale Space and Variational Methods in Computer Vision, Second International Conference, SSVM 2009, Voss, 1–5 June 2009. Proceedings, pp 502–513 (2009)

Tai, X.C., Hahn, J., Chung, G.J.: A fast algorithm for Euler's elastica model using augmented Lagrangian method. SIAM J. Imaging Sci. **4**(1), 313–344 (2011)

Vese, L.A., Osher, S.J.: Modeling textures with total variation minimization and oscillating patterns in image processing. J. Sci. Comput. **19**(1/3), 553–572 (2003)

Wang, Y., Yuan, X.: The linearized alternating direction method of multipliers for dantzig selector. SIAM J. Sci. Comput. **34**(5), A2792–A2811 (2012)

Wang, Y., Yang, J., Yin, W., Zhang, Y.: A new alternating minimization algorithm for total variation image reconstruction. SIAM J. Imaging Sci. **1**(3), 248–272 (2008)

Wang, Y., Yin, W., Zeng, J.: Global convergence of ADMM in nonconvex nonsmooth optimization. J. Sci. Comput. **78**(1), 29–63 (2019)

Wu, C., Tai, X.C.: Augmented Lagrangian method, dual methods, and split Bregman iteration for ROF, vectorial TV, and high order models. SIAM J. Imaging Sci. **3**(3), 300–339 (2010)

Wu, C., Zhang, J., Tai, X.C.: Augmented Lagrangian method for total variation restoration with non-quadratic fidelity. Inverse Probl. Imaging **5**(1), 237–261 (2011)

Wu, C., Zhang, J., Duan, Y., Tai, X.C.: Augmented lagrangian method for total variation based image restoration and segmentation over triangulated surfaces. J. Sci. Comput. **50**(1), 145–166 (2012)

Wu, C., Liu, Z., Wen, S.: A general truncated regularization framework for contrast-preserving variational signal and image restoration: Motivation and implementation. Sci. China Math. **61**(9), 1711–1732 (2018)

Yan, M., Duan, Y.: Nonlocal elastica model for sparse reconstruction. J. Math. Imaging Vis. **62**, 532–548 (2020)

Yang, J., Yin, W., Zhang, Y., Wang, Y.: A fast algorithm for edge-preserving variational multichannel image restoration. SIAM J. Imaging Sci. **2**(2), 569–592 (2009)

Yashtini, M., Kang, S.H.: A fast relaxed normal two split method and an effective weighted TV approach for Euler's elastica image inpainting. SIAM J. Imaging Sci. **9**(4), 1552–1581 (2016)

Zeng, C., Wu, C.: On the edge recovery property of noncovex nonsmooth regularization in image restoration. SIAM J. Numer. Anal. **56**(2), 1168–1182 (2018)

Zeng, C., Wu, C.: On the discontinuity of images recovered by noncovex nonsmooth regularized isotropic models with box constraints. Adv. Comput. Math. **45**(2), 589–610 (2019)

Zhang, H., Wu, C., Zhang, J., Deng, J.: Variational mesh denoising using total variation and piecewise constant function space. IEEE Trans. Vis. Comput. Graphics **21**(7), 873–886 (2015)

Zhang, J., Chen, K.: A total fractional-order variation model for image restoration with nonhomogeneous boundary conditions and its numerical solution. SIAM J. Imaging Sci. **8**(4), 2487–2518 (2015)

Zhu, W., Chan, T.: Image denoising using mean curvature of image surface. SIAM J. Imaging Sci. **5**(1), 1–32 (2012)

Zhu, W., Tai, X.C., Chan, T.: Augmented Lagrangian method for a mean curvature based image denoising model. Inverse Prob. Imaging **7**(4), 1409–1432 (2013)

Sparse Regularized CT Reconstruction: An Optimization Perspective

14

Elena Morotti and Elena Loli Piccolomini

Contents

Introduction	552
Tomographic Imaging	554
Mathematics of Sparse Tomography	557
Lambert Beer's Law	557
The Radon Transform and Its Discretization	559
The Filtered Back Projection Algorithm	559
Model-Based Approaches for Sparse-View CT	561
From Lambert-Beer's Law to a Linear System	561
Implementation of the Forward Operator M	562
The Optimization Framework	564
Iterative Algorithms for Optimization	566
Regularization: Little or Too Much?	566
Toward the Convergence of the Iterative Method	568
New Frontiers of CT Reconstruction with Deep Learning	569
Case Study: Reconstruction of Digital Breast Tomosynthesis Images	570
DBT 3D Imaging	571
Model and Analysis	572
Reconstructions of the Accreditation Phantom	574
Reconstructions of a Human Dataset	576
Distance-Driven Approach for 3D CT Imaging	578
Code Parallelization	579
Conclusion	581
References	581

E. Morotti
Department of Political and Social Sciences, University of Bologna, Bologna, Italy
e-mail: elena.morotti4@unibo.it

E. L. Piccolomini (✉)
Department of Computer Science and Engineering, University of Bologna, Bologna, Italy
e-mail: elena.loli@unibo.it

© Springer Nature Switzerland AG 2023
K. Chen et al. (eds.), *Handbook of Mathematical Models and Algorithms in Computer Vision and Imaging*, https://doi.org/10.1007/978-3-030-98661-2_123

Abstract

In Computed Tomography (CT), decreasing the X-rays dose is essential to reduce the negative effects of radiation exposure on the human health. One possible way to accomplish it is to reduce the number of projections acquired, hence the name of *sparse* CT. Traditional methods for image reconstruction cannot recover reliable images in this case: the lack of information due to the missed projections produces strong artifacts. Alternatively, optimization frameworks are flexible models where incorporated regularization functions impose regularity constraints on the solution, thus avoiding unwanted artifacts and contrasting noise propagation. Since the iterative methods solving the optimization problem calculate more accurate solutions as iterations (and computational time) increase, it is possible to choose a better reconstructed image at the expense of execution time, or viceversa. Parallel implementations of the iterative solvers significantly reduce the computational time, allowing for a large number of iterations in a prefixed short time.

Here, the effectiveness of the optimization approach is shown on the case study of 3D reconstruction of breast images from tomosynthesis with tests on real projection data.

Keywords

Sparse-view CT · Tomographic image reconstruction · Model-based iterative methods · Breast tomosynthesis

Introduction

X-ray computed tomography (CT) is an imaging technique which has first been experimented in the medical area, as the evolution of the projection radiography. In particular, medical imaging was born not long after Wilhelm Röntgen discovered X-rays in 1895, as soon as scientists realized X-ray capability of crossing objects: for decades 2D planar images (projection radiographies) have been used to investigate the inner parts of human bodies. However, these images represent a mean of the information of the 3D scanned object which is squeezed on a 2D plane. In the 1930s, a new mathematical theory by Johann Radon published in 1917, the studies by the physician Grossmann, together with the desire to overcome the averaging process of the conventional X-ray radiography, led to the definition of tomography as a new tool for object inspection. Since the advent of computers in the 1970s, CT has raised and revolutionized the non-intrusive diagnostic imaging by allowing the three-dimensional orientation of anatomy to be reconstructed in transverse (cross-sectional) sections.

To achieve it, the CT imaging device acquires several projections of the same slice of the object under exam, from angled views in a round trajectory. Then, a software reconstructs the digital image from the acquired projection data. Hence, tomographic image reconstruction mathematically represents an *inverse problem*.

14 Sparse Regularized CT Reconstruction: An Optimization Perspective

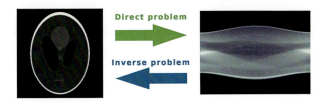

Fig. 1 In CT imaging, the direct problem (from the object to the data) is represented by the acquisition of the sinogram, whereas the inverse problem (from the data to the object) is the reconstruction of the image

In an inverse problem, in fact, only the measurements of an effect are known, as given data, whereas the cause represents the unknown that must be retrieved as solution of the problem. The cause-effect pair in CT is represented in Fig. 1, which shows the well-known Shepp-Logan brain phantom as scanned object and its projection dataset, acquired by the CT system in an entire 2D scan and organized as a *sinogram* image (as introduced in the following). The inverse problem aims at recovering the phantom image as accurately as possible from the sinogram.

Mathematically, inverse problems are generally *ill-posed* in the sense of Hadamard (1902), i.e., one of the following conditions is not satisfied:

1. At least one solution of the problem exists.
2. The solution of the problem is unique.
3. The solution continuously depends on the data.

Traditional methods for CT cannot face the ill-posedness and compute images with unwanted artifacts and noise. To face this, a more recent approach models the CT imaging process as an optimization problem where the inverse problem is solved by inverting the discrete model, represented by a linear system, constrained by means of regularization functions. Imposing regularization allows to choose a good solution among the infinite possible ones.

In particular, the optimization problem is solved by iterative algorithms (called *model-based iterative algorithms*). They converge to the problem solution in many iterations, but they should possibly compute a good solutions far before convergence. In fact, a slow convergence would make model-based iterative algorithms not usable on real systems, where very fast executions are required for clinical needs. However, acceleration techniques make iterative algorithms produce good solutions in few iterations, and efficient parallel executions on low-cost GPU boards greatly reduce the execution time; hence the optimization approach is effective in real applications.

The aim of this chapter is:

- to derive the optimization framework from the mathematical model of CT;
- to highlight the flexibility of the optimization framework, where different regularization terms can be easily incorporated and different iterative algorithms can be used for solving the minimization problem;

- to show that the solution of the optimization problem computed by an iterative method converges toward an accurate reconstruction;
- to present a 3D real case of limited angle tomography with an example of parallel execution on GPU boards, demonstrating that it is possible to achieve short execution times compatible to the speed and quality standards in clinical settings.

The chapter is organized as follows. The next section contains a brief survey both on the CT scan geometries (with particular attention to few-view protocols) and on the mathematics of CT imaging. Then, the regularized optimization framework is presented for the CT image reconstruction; examples of iterative reconstructions as solution of the optimization problem from a 2D phantom prove the effectiveness of the approach. Finally, a case study on 3D breast tomosynthesis is analyzed with results from a parallel implementation on GPUs.

Tomographic Imaging

From the primordial systems to the most modern gantries currently used in medicine and industrial applications, many studies have been led by different research groups, collecting engineers, physicists, mathematicians, and computer scientists, with the aim of improving both the technologies and the reconstruction software. For each prefixed angled position of the X-ray source, first-generation CT devices performed long-lasting projections where parallel rays allowed simple reconstruction algorithms (top-left image in Fig. 2). Among the numerous developments, the shift from parallel- to fan-beam X-ray projections has been the most significant. Fan-beam geometries are preferred today, since they enable to acquire all the single-view measurements in one fan simultaneously (top-right image in Fig. 2). However, computation speedups are required when recovering objects from fan-beam projections in real scenarios (Averbuch et al. 2011).

Historically, a further step forward has been the blooming of 3D CT imaging systems. The first developments led to helical CT, where the X-ray source walked on a narrow helical trajectory scanning a volume with fan beams, slice by slice. As depicted in Fig. 2, another approach exploits cone-beam projections to run over a volume in just one scan. In this case, the X-ray source rotates on a circular planar trajectory.

In the last years, many tomographic devices have been designed to fit different medical needs, and, on the other hand, interesting technical, anthropomorphic, forensic, and archeological as well as paleontological applications of CT have been developed too (Hughes 2011; De Chiffre et al. 2014). As a consequence, the CT technique is evolving into new inquiring forms. In particular, motivated by an increasing focus on the potentially harmful effects of X-ray ionizing radiation, a recent trend in CT research is to develop safer protocols to reduce the radiation dose per patient. This allows to apply CT techniques to a wider class of medical examinations, including vascular, dental, orthopedic, musculoskeletal, chest, and mammographic imaging. Safer protocols are of interest not only for medicine but

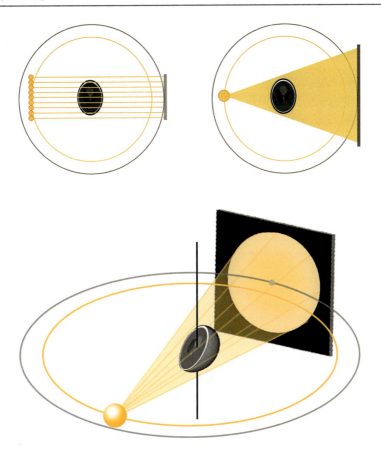

Fig. 2 Sketches of tomographic devices, from the primordial technology with parallel X-ray scans (top left) to the most modern solution exploiting fan beams for 2D (top right) and cone beams (bottom) for 3D CT

also for material science and cultural heritage, to prevent damage to the subject under study, due to excessive radiations.

Specifically, there are two main techniques allowing for a significant reduction of the total radiation exposure per patient. The first one, usually named *low-dose CT*, consists in reducing the X-ray tube current at each scan. In this case, the geometry traditionally used in CT, where up to one thousand projections are taken along the circular trajectory, does not change, but the measured data presents higher quantum noise. The second practical way to lower the radiation consists in reducing the number of X-ray projections. The resulting protocols are labeled as *sparse tomography* (or sparse-view, few-view tomography), and it leads to incomplete tomographic data, but very fast examinations (Kubo et al. 2008; Yu et al. 2009). Figure 3 shows a graphical draft of the reconstruction process. In the first row, the classical full-dose CT case is represented; in the second row, a *sparse-view*

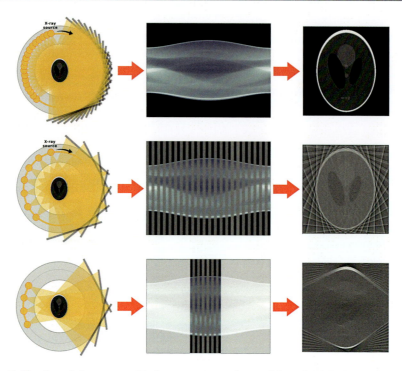

Fig. 3 Sketches of the tomographic image reconstruction workflow, for full-view, sparse-view full-angle, and limited-angle protocols (from top to bottom, respectively). From the different geometries on the left, the acquired projections and the reconstructed image of the Shepp-Logan phantom. The missing portions of sinogram in the sparse-view and limited-angle protocols are depicted in light gray

full-angle tomography is considered where a reduced number of views are taken in the whole circular orbit. A different sparse-view geometry using few projections is called *limited-angle tomography* (see the third row of Fig. 3). Here, a further reduction of X-ray scans is made by limiting the source trajectory to a C-shaped path, i.e., by restricting the 360-degree angular scanning interval to a range smaller than 180 degrees. In some tomographic applications, the human anatomy does not allow a complete circular motion to the X-ray source; thus, the use of a reduced range is mandatory and the resulting technique is called *tomosynthesis*. An example is breast imaging, where the breast is in a stationary position between the detector surface and the compression plate (Wu et al. 2004; Zhang et al. 2006; Reiser et al. 2009; Barca et al. 2021). The source moves through a quite limited arc (at most 80 degrees) over the breast. Another possible reason for using limited angles is the impossibility of probing through a ball in the center of the target, such as in nondestructive testing (Quinto 1993).

A low radiation dose and high in-plane resolution make sparse tomography an attractive alternative to full-view computed tomography. However, the incompleteness of the projection data results in image artifacts that may disable diagnostic interpretation. As depicted in Fig. 3, the sets of projection data are severely subsampled in case of sparse-view and limited-angle acquisitions with respect to the full-dose case. The resulting lack of information causes well-studied artifacts on the images reconstructed with the algorithms traditionally used for full-view protocols. However, thanks to the efficiency of new reconstruction approaches, some low-dose and sparse-view protocols have already been approved for screening tests: safer tomographic exams can indeed be led without compromising the reliability of their diagnosis (Mueller and Siltanen 2012; He et al. 2018).

Mathematics of Sparse Tomography

What is there behind the X-ray imaging techniques? From a physical point of view, the projection data reflects the absorption of the photons constituting the X-rays, and the image of the scanned object is a picture of the attenuation coefficient map in pseudo-colors. The physical model describing photons absorption in terms of attenuation coefficients is described in the Lambert Beer's law.

Lambert Beer's Law

All the physical mechanisms leading to the attenuation of radiation intensity (i.e., reduction of photons) measured by a detector behind a homogeneous object are usually described by a single attenuation coefficient $\mu = \mu(w) \geq 0$ depending on the crossed point w. The total attenuation of a monochromatic X-ray beam passing through a dense object of thickness Δw can be calculated in the following way (Buzug 2011):

$$m(w + \Delta w) = m(w) - \mu(w)m(w)\Delta w \tag{1}$$

where $m(w)$ is the intensity of the incoming beam. Reordering (1) and computing the limit, it holds:

$$\lim_{\Delta w \to 0} \frac{m(w + \Delta w) - m(w)}{\Delta w} = \frac{dm}{dw} = -\mu(w)m(w). \tag{2}$$

By assuming the object to be homogeneous (i.e., $\mu(w) = \mu$ along the entire path Δw), the solution of the differential equation (2) computed by separation of variables and integration is:

$$\ln|m(w)| = -\mu w + C. \tag{3}$$

Imposing the initial condition $m(0) = m_0$ (where m_0 is the known emitted photon count, as in Fig. 4) and considering that all the measured intensities are positive quantities, the previous equation can be written as:

Fig. 4 Scheme of the X-ray absorption by an infinitesimal object. The number m_0 of input photons is reduced to $m < m_0$ at output after crossing a thickness $d\omega$

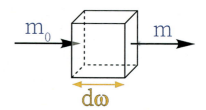

$$m(w) = m_0 e^{-\mu w}. \tag{4}$$

Equation (4) is known as the *Lambert Beer's law* of attenuation.

In practice, the attenuation coefficient $\mu(w)$ is not constant along the ray path. In this case the solution for the intensity measured after a running length W is given by:

$$m = m_0 e^{-\int_0^W \mu(w)dw} \tag{5}$$

and the *projection integral* of μ along a segment of length W is computed as:

$$\mathscr{P}_W \mu = -\ln\left(\frac{m}{m_0}\right) = \int_0^W \mu(w)dw. \tag{6}$$

By setting the plane coordinates as $w = (x, y)$ (the attenuation coefficient is a continuous function $\mu(w) = \mu(x, y)$ over the spatial domain of the slice) and naming L the integration path, the following relation holds:

$$-\ln\left(\frac{m}{m_0}\right) = +\int_L \mu(x, y)dw \tag{7}$$

by assuming the air coefficient outside the object $\mu(x, y) = 0$.

Suppose to rotate the xy-plane of an angle Φ and to set a change of variables from (x, y) to (t, s) as in Fig. 5. By considering the X-ray parallel to the direction of the vector θ with $t = \bar{t}$, the projection of the attenuation coefficient μ along L becomes:

$$\int_L \mu(x, y)dw = \int_{-\infty}^{+\infty} \mu(\bar{t}\theta^\perp + s\theta)ds. \tag{8}$$

Since the direction of the vector θ is uniquely determined by the rotation angle Φ, it is convenient to denote with θ also the rotation angle. Now, by considering the X-ray parallel beam emitted from the θ-angled position, the projection of the whole object described by μ is the map $\mathscr{P}_\theta \mu : \mathbb{R} \to \mathbb{R}$ such that:

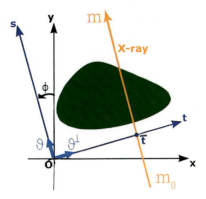

Fig. 5 Scheme of the X-ray absorption by an object, to illustrate the rotation of the coordinate system used in (8)

$$\mathcal{P}_\theta \mu(t) = \int_{-\infty}^{+\infty} \mu(t\theta^\perp + s\theta)ds, \quad \forall\, t \in \mathbb{R}. \tag{9}$$

The Radon Transform and Its Discretization

In 1917, a well-known paper by the Austrian mathematician Johann Radon provided the mathematical foundation for tomographic imaging reconstruction. The *Radon transform of* μ is defined as the map $\mathcal{R} : [0, 2\pi] \times \mathbb{R} \to \mathbb{R}$ such that:

$$(\mathcal{R}\mu)(\theta, t) = \mathcal{P}_\theta \mu(t), \quad \forall \theta \in [0, 2\pi], \forall t \in \mathbb{R} \tag{10}$$

In other words, the Radon transform $\mathcal{R}$ of an object slice described by μ is the set of projections acquired along the full-angle circular trajectory, in a continuous model.

The process defining the full-dose tomography represents a discrete realization of the (continuous) Radon transform. The graphical representation of all the measured data, in the bidimensional case, is called *sinogram*, and it is represented in Fig. 3 for full-view, sparse-view, and limited-angle geometries. As it is clearly visible, in case of sparse-view and limited-angle protocols, the incomplete projections provide only a portion of the entire sinogram, making the corresponding inverse problems trickier and the reconstruction process more complicated than in the full-view case.

The Filtered Back Projection Algorithm

Historically, the first technique implemented to reconstruct CT images from projections is the *filtered back projection* (FBP) (Feldkamp et al. 1984). To recover the attenuation coefficient function, the basic idea of FBP is to project backward every data onto the original ray path causing such absorption (see Kak and Slaney (2001) for more details).

The FBP algorithm is still implemented in many commercial systems, since it computes the output image in a very short time, which is a fundamental request in medical setting. However, it is well known that in the case of few views the FBP algorithm produces images corrupted by artifacts and noise (Natterer 2001).

Figure 6 shows some FBP reconstructions of the well-known Shepp-Logan digital phantom obtained at different sparse geometries. The sparsity is boosted by decreasing the angular range (from top to bottom) and the number of views (from left to right). The FBP image quality deteriorates: the large angular step characterizing sparse-view projections leads to streaking artifacts on the image, whereas a limited-angle acquisition produces a swiped band corresponding to the lost projecting directions. In the last row, where the scan is limited to a 60-degree arc, the object inside the brain is deformed and not distinguishable, regardless of the number of projection numbers.

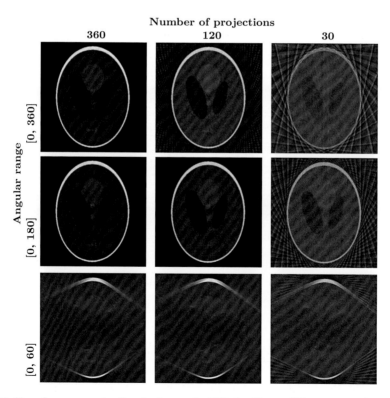

Fig. 6 Shepp Logan reconstructions by the popular FBP algorithm, at different geometric settings

14 Sparse Regularized CT Reconstruction: An Optimization Perspective

Model-Based Approaches for Sparse-View CT

A valid alternative to FBP for sparse-view CT image reconstruction is represented by model-based iterative methods, which derive from the discretization of Lambert-Beer's law (4).

From Lambert-Beer's Law to a Linear System

In the real discrete setting, both the scanned object and the system detector are discrete. The attenuation coefficient function $\mu(x, y)$ is discretized into an image of $N = N_x \times N_y$ picture elements (pixels), with values $f_{i,j}, \forall i \in 1, \ldots, N_x, \ j \in 1, \ldots, N_y$, which can be re-ordered in a vector f.

The detector is made of n_p recording units of length δ_x μm; hence, at each X-ray shot, n_p is the number of measured data. Figure 7 depicts a graphical example of the discrete CT configuration where $N_x = N_y = 4$ and $n_p = 7$. The whole scan is constituted by N_θ projections acquired at equally spaced θ_k angles, $\forall k = 1, \ldots, N_\theta$, and performed in the angular range $[-\Theta, +\Theta]$. Let $N_d = N_\theta \cdot n_p$ be the total number of data: in classical CT $N_d \gg N$, while $N_d < N$ in case of sparse tomography.

Fixing the k-th projection (acquired from the θ_k-th angled position) and calling m_i the photon counting measured at the i-th recording unit (with $i \in 1, \ldots, n_p$), from equation (7), it is possible to define:

$$g_i = -\ln\left(\frac{m_i}{m_0}\right) \quad \forall i \in 1, \ldots, n_p. \tag{11}$$

The line integral of equation (7) can be discretized into a sum over all the pixels; hence:

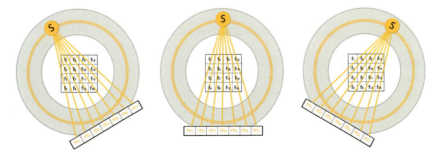

Fig. 7 Scheme of the scanning process for three different angled projections. The sources rotate around the 2D object along a circular trajectory. The slice of interest is discretized into $N = 16$ pixels and the detector has $n_p = 7$ recording units

$$g_i = \sum_{j=0}^{N} M_{i,j}^{\theta_k} f_j \quad \forall i \in 1, \ldots, n_p. \tag{12}$$

In matrix-vector notation, equation (12) becomes:

$$g^{\theta_k} = M^{\theta_k} f \tag{13}$$

where M^{θ_k} is a matrix of size $n_p \times N$ and $g^{\theta_k} = \{g_i\}_{i=1,\ldots,n_p}$ is a vector collecting the projections obtained from the angle θ_k (hence g^{θ_k} is the discretization of (9)).

Collecting together all the equations (13) for $k = 1, \ldots N_\theta$, the following large size linear system is obtained:

$$\begin{bmatrix} \underline{\quad M^{\theta_1} \quad} \\ \underline{\quad M^{\theta_2} \quad} \\ \underline{\quad M^{\theta_3} \quad} \\ \underline{\quad M^{\theta_4} \quad} \\ \ldots \\ \ldots \\ \ldots \\ \ldots \\ \underline{\quad M^{\theta_{N_\theta}} \quad} \end{bmatrix} \cdot \begin{bmatrix} f_1 \\ f_2 \\ f_3 \\ f_4 \\ f_5 \\ \ldots \\ \ldots \\ \ldots \\ \ldots \\ f_N \end{bmatrix} = \begin{bmatrix} g^{\theta_1} \\ g^{\theta_2} \\ g^{\theta_3} \\ g^{\theta_4} \\ \ldots \\ \ldots \\ \ldots \\ g^{\theta_{N_\theta}} \end{bmatrix} \tag{14}$$

Equation (14) represents the discretization of the Radon transform (10). Using a more compact notation, the CT process is described by the linear system:

$$Mf = g \tag{15}$$

where $M \in \mathbb{R}^{N_d} \times \mathbb{R}^N$, $f \in \mathbb{R}^N$, and $g \in \mathbb{R}^{N_d}$.

Implementation of the Forward Operator M

The most crucial issue in the discrete formulation concerns the computation of the matrix coefficients $M_{i,j}$: although very simple in principle, elaborate computer algorithms and a significant amount of computer time are required to determine its entries. Really, the matrix M is the mathematical description of the physical process of CT data acquisition; hence, it must mirror the forward projection of a slice onto the detector units, for all the scanning views. We recall that M is obtained by collecting the matrices M^{θ_k} corresponding to the single projections at angle θ_k, $k = 1, \ldots N_\theta$ as in (14). Different algorithms have been proposed in literature

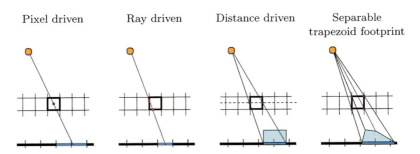

Fig. 8 2D schemes of different approaches to compute the projection matrix M

to efficiently compute the value $M_{i,j}^{\theta_k}$ as the contribution of the object element f_j onto the detector unit g_i. The most common are *pixel-driven, ray-driven, distance-driven*, and *separable trapezoid footprints*. Figure 8 schematically draws the idea behind each approach.

Historically, the first proposed approach has been the *pixel-driven* (Peters 1981) one: according to the geometry of the device, the f_j pixel is projected from its center onto the element g_i of the detector; its contribution is split among the adjacent measuring units with a linear (or more complex) interpolation routine (Harauz and Ottensmeyer 1983; Fessler 1997). When the spatial resolution of the reconstruction is much bigger than the detector cell size, too few rays are taken into account, and it may happen that some detector cells do not receive any values at all (which is, of course, unrealistic).

In the *ray-driven* (or ray-casting) approach (Lacroute and Levoy 1994; Matej et al. 2004; O'Connor and Fessler 2006), only a straight line is considered reaching the center of each detector unit g_i from the source, and then for each element f_j crossed by the line, $M_{i,j}^{\theta_k}$ is proportional to the length of segment intersecting f_j.

In the *distance-driven* approach, proposed by De Mann in 2002, the idea is to project onto the detector, for each element f_j, not only a point but the element in-plane expansion. This provides a linear shadow, enlarged for the height of f_j, creating a rectangular footprint over one or more detector elements. For each element g_i, the value of $M_{i,j}^{\theta_k}$ is proportional to the area of the portion of rectangle built on it. An extension of the distance-driven algorithm to the 3D case is presented for the case study on limited-angle tomography in a following section.

The *separable trapezoid footprint* algorithm was introduced in 2010 by Long and Fessler. In this method, all the vertices of the element f_j are projected onto the detector, and the element footprint is approximated by a trapeze, to shape a more accurate footprint than in the distance driven case.

The last two methods better model the physical nature of X-ray beams; hence, they compute more accurate projection matrices at the expense of a higher computational cost. All these approaches are conceptually straightforward to be generalized to the 3D case.

Some final considerations about the matrix M implementing the forward operator:

- M is a very sparse matrix, because few pixels are effective for a single value of a projection; hence, each row has mostly zero elements;
- M is under-determined in case of few views; hence, no unique solution exists for the linear system (15);
- M cannot be stored because of its huge dimensions, neither in sparse form, for most of the real CT imaging: whenever we need a matrix product, M must be recalculated element by element and this represents a noticeable computational effort.

The Optimization Framework

In case of sparse-view CT, the linear system (15) is under-determined ($N \gg N_d$); hence, it has infinite possible solutions. Moreover, due to the ill-posedness of the inverse problem and to the lack of data, unwanted artifacts corrupt the solutions.

The model-based approach is introduced to overcome these numerical controversies, by adding some a priori information. The resulting formulation can be stated as a minimization problem involving a data-fitting function $\mathcal{F}$ and a prior operator $\mathcal{R}$ (acting here as a regularizer). The optimization framework is flexible and can be stated as an unconstrained minimization on the objective function $\mathcal{G}$ as:

$$\arg\min_{x} \mathcal{G}(f) = \mathcal{F}(f) + \lambda \mathcal{R}(f) \qquad (16)$$

where $\lambda \geq 0$ is a regularization parameter, or as a constrained minimization:

$$\arg\min_{f} \mathcal{R}(f) \quad s.t. \quad \mathcal{F}(f) \leq \epsilon^2. \qquad (17)$$

or

$$\arg\min_{f} \mathcal{F}(f) \quad s.t. \quad \mathcal{R}(f) \leq \sigma^2, \qquad (18)$$

where $\epsilon \geq 0$ and $\sigma \geq 0$ are estimates of the noise and of the value of $\mathcal{R}(f)$ in the object, respectively.

A meaningful physical constraint to impose is the non-negativity of the solution which reflects the non-negativity property of the linear attenuation coefficient μ; hence, model (16) could be reinforced as:

$$\arg\min_{f \geq 0} \mathcal{G}(f) = \mathcal{F}(f) + \lambda \mathcal{R}(f). \qquad (19)$$

A detailed overview of model-based methods can be found in Graff and Sidky (2015).

Common choices for $\mathcal{F}(f)$ are the least squares (LS) function:

$$LS(f) = \|Mf - g\|_2^2 \qquad (20)$$

or the weighted least squares (WLS) function (Thibault et al. 2007):

$$WLS(f) = \|\sum_{i=1}^{N_d} W_i(Mf - g)_i\|_2^2 \qquad (21)$$

where W_i are positive weights.

Focusing on the regularization $\mathcal{R}(f)$, different functions have been proposed in literature. The most widely used convex regularizer in sparse-view CT is the total variation (TV) defined as (Vogel 2002):

$$TV(f) = \sum_{j=1}^{N} \|\nabla f_j\|_2. \qquad (22)$$

or in its smoothed differentiable form:

$$TV_\beta(f) = \sum_{j=1}^{N} \sqrt{\|\nabla f_j\|_2^2 + \beta^2} \qquad (23)$$

where β is a small positive parameter (Vogel 2002). The TV function is chosen by many authors because of its excellent shape recovering and denoising properties, even if it is known that it can produce staircasing effects when the regularization parameter is too high (Sidky et al. 2009; Choi et al. 2010; Ritschl et al. 2011; Hashemi et al. 2013; Graff and Sidky 2015; Luo et al. 2017). Alternative choices preserving convexity are the total generalized variation (TGV) (Niu et al. 2014), the weighted TV (Yu and Zeng 2014), the normal-dose induced non-local means filter (Huang et al. 2013), and the tight frame (Jia et al. 2011) regularizers. Recently, an l1/l2 regularizer has been proposed in Wang et al. (2021).

To reduce the TV oversmoothing, also the non-convex and non-differentiable TpV regularization function:

$$TpV(f) = \|\nabla f\|_p^p, \quad 0 \leq p \leq 1 \qquad (24)$$

has been proposed (Sidky et al. 2014).

Iterative Algorithms for Optimization

For the solution of the minimization problem expressed in one of the formulations (16), (17), (18), and (19), a suitable optimization algorithm is used. For clinical applications, not only an accurate reconstruction but also a low computational time is required. Hence, the optimization algorithm should meet the following demands:

- have a fast error decreasing in the initial iterations;
- have a low computational cost per iteration, to efficiently run the solver in a short time;
- have a limited request of memory, to solve real-size problems on commercially affordable hardware. For this reason first-order descent methods are generally preferred to methods exploiting second-order information, which require further storage space.

Various iterative methods have been considered and efficiently used in CT reconstruction, such as the scaled gradient projection (Loli Piccolomini and Morotti 2016; Loli Piccolomini et al. 2018) and alternate directions of multipliers method (ADMM) (Wang et al. 2021) for the solution of a convex problem or the proximal dual hybrid gradient (PDHG, also known as Chambolle-Pock) in the non-convex case (Sidky et al. 2014). A new method accelerating both ADMM and PDHG has been recently proposed in Liu et al. (2021).

Regularization: Little or Too Much?

Both the unconstrained (16) and constrained (17), (18), and (19) minimization formulations depend on a parameter: λ, ϵ, or σ. The amount of regularization on the solution depends on the choice of this parameter.

To investigate the effects of regularization on the reconstructed image, a dataset freely downloadable from the web page of the Finnish Inverse Problems Society www.fips.fi/dataset.php is considered (the relative documentation can be found in Bubba et al. 2016). The object in exam is a lotus root (see Fig. 9) which has been filled with several objects of different shapes, sizes, and attenuation coefficients.

The scanning process consists in 120 fan-beam projections, performed from a circular trajectory with angular step size $\Delta_\theta = 3$ degrees; each real projection array has been downsampled into 429 recorded values; hence, the sinogram is a data matrix of size 429×120 and it is shown in Fig. 9. The dataset also provides the forward projector, as a sparse matrix of size $51{,}480 \times 65{,}536$; hence, the reconstruction will be an image of 256×256 pixels.

The reconstructions in Fig. 10 are obtained with the minimization model (19) setting $\mathcal{F}(f)$ as the LS function (20) and $\mathcal{R}(f)$ as the TV_β function defined in (23) (with $\beta_{TV} = 10^{-3}$). The images are computed with different values of the

14 Sparse Regularized CT Reconstruction: An Optimization Perspective

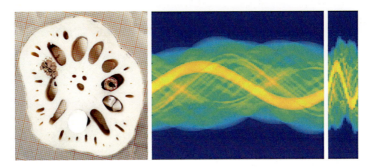

Fig. 9 On the left: a picture of the lotus root, filled with different materials. At the center: the sinogram of the lotus dataset with 120 sparse projections. On the right: the sinogram with 20 highly sparse views

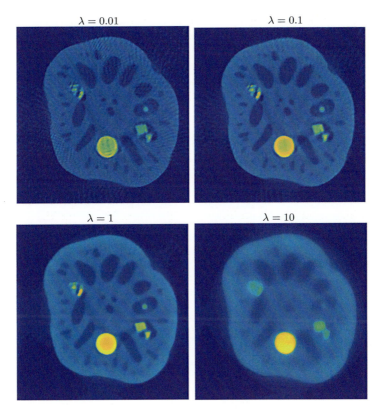

Fig. 10 Results achieved at convergence, for increasing values of the regularization parameter $\lambda = 0.01, 0.1, 1, 10$

regularization parameter λ, getting the increasing values $\lambda = 0.01, 0.1, 1, 10$. The artifacts visible in the reconstruction with lowest value of λ disappear when the regularization parameter increases. However, a too large value of λ blurs the image as shown in the bottom row of Fig. 10.

Toward the Convergence of the Iterative Method

Figure 11 reports the lotus images reconstructed at 10, 50, and 100 iterations and at convergence (about 1000 iterations) using the sinogram with 120 projections over 360 degrees. The regularization parameter λ is set to 1 in all the tests. From the zoomed crops aside each reconstructed image, it is visible how the objects of interest are better enhanced and detected with increasing iterations. It is also evident that after very few iterations, the contours of the objects are defined, whereas more iterations are necessary to obtain a good contrast. Moreover, Fig. 11 confirms that the chosen model well approximates the desired image and that the iterative method is converging toward the problem solution. In practice, the more iterations are executed, the better will be the reconstructed image.

Finally, some considerations about the model when applied to a sparser geometry can be deduced from Fig. 12, where the images are reconstructed from only 20

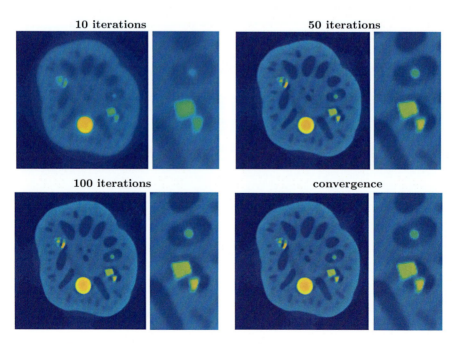

Fig. 11 Results obtained with $\lambda = 1$ at 10, 50, and 100 iterations and at convergence, from 120 projections

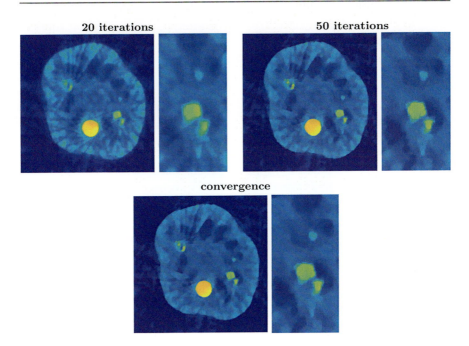

Fig. 12 Results obtained with $\lambda = 1$ at 20 and 50 iterations and at convergence, from 20 projections

projections over 360 degrees (with angular step of 18 degrees) in 20 and 50 iterations and at convergence (145 iterations).

In this case of very sparse-view full-angle CT, some artifacts are present in all the reconstructions, and more iterations must be performed to achieve reasonable results, compared to the previous geometry with many more projections. In 20 iterations, not all the objects are detectable and they have low contrast with the background. However, increasing the iterations enhances the images better, and the results obtained at the algorithm convergence are very promising.

By the way, these tests show the importance of running the reconstructing solvers for a longer time, when the CT problem is characterized by a severe subsampling, and it mirrors the difficulty to back-project the dataset and fit it, in case of few tomographic projections.

New Frontiers of CT Reconstruction with Deep Learning

Since few years ago, deep learning (DL)-based methods have emerged over fully conventional or variational approaches for sparse-view tomographic reconstruction (Wang et al. 2018). In the first experiments, neural networks have been mainly used as a postprocessing tool to remove artifacts and noise from fast reconstructions (typically obtained with analytical solver, as FBP). Such approach is usually called learnt postprocessing (LPP). Here the network learns from a set of *ground truth* images reconstructed from full-dose acquisitions (see, e.g., Han and Ye (2018), Pelt et al. (2018), Zhang et al. (2019), Schnurr et al. (2019), Urase et al. (2020), Morotti et al. (2021) and the references therein). However, in their inspiring work (Sidky et al. 2020), Sidky et al. have claimed that the popular LPP schemes lack of mathematical characterization and a new framework has been recently proposed in Evangelista et al. (2022) to face this drawback.

Neural networks have been also introduced into model-based schemes to improve their efficiency. In the so-called *unrolling* (or unfolding) strategies, each iteration is executed by a layer of the neural network which learns, in the training phase, some parameters of the optimization algorithm (Monga et al. 2021). The proposals differ for the considered iterative scheme and for the block-per-iteration learned by the neural network. For instance, in 2017, Adler and Öktem have developed a partially learned gradient descent algorithm, whereas they have worked on the Chambolle-Pock scheme in Adler and Öktem (2018). In Gupta et al. (2018) a convolutional neural network is trained to act like a projector in a gradient descent algorithm, whereas in Xiang et al. (2021) both the proximal operator and gradient operator of an unrolled FISTA scheme are learned. In Zhang et al. (2020) the neural network learns the initial iterate of the inner conjugate gradient solver in a splitting scheme for optimization. A different approach is constituted by the plug-and-play scheme. In this case, the minimization problem is solved by a splitting optimization method, such as ADMM, and the neural network is plugged in the denoising substep of the method at each iteration (Venkatakrishnan et al. 2013; He et al. 2018).

Case Study: Reconstruction of Digital Breast Tomosynthesis Images

Digital breast tomosynthesis (DBT) is a quite recent development of the mammographic imaging system for breast tumor detection. DBT, in fact, provides a volumetric breast reconstruction as a stack of 2D images, each representing a cross-sectional slice of the breast itself (Cavicchioli et al. 2020). The detection of breast cancer by mammography suffers from the obscuring effect of overlapping breast tissue, due to the projection onto a flat image of all the breast volume: the cancer can be masked by surrounding overlapping structures, especially in woman with radiographically dense breasts. On the contrary, DBT has the advantage of separating the anatomical tissues, and this generally reduces false-negative diagnosis (Fig. 13). At the same time, DBT provides a low radiation dose (comparable to the radiation dose used in one standard mammography), since the X-ray source emits only few projecting cone beams from few angled points along a narrow C-shaped trajectory. In 2011, the Food and Drug Administration (the federal agency of the US Department of Health and Human Services) recommended the DBT technique over mammography as breast cancer screening in the USA, due to its established

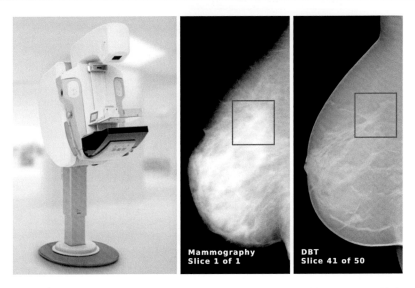

Fig. 13 A modern DBT device on the left and a comparison between a 2D mammographic image and a DBT image slice, showing the same spiculated mass

higher accuracy in the most important breast diagnostic imaging tasks, i.e., finding microcalcifications and suspected masses (Andersson et al. 2008; Das et al. 2010).

DBT 3D Imaging

DBT puts into practice a limited-angle sparse tomographic protocol for three-dimensional imaging; hence, its image reconstruction is not trivial technically. As schematically reported in Fig. 14 where a Cartesian axis system is introduced for clarity, in a modern DBT machinery, the X-ray source moves on the $YZ-$plane, drawing an arc which spans 11 to 60 degrees typically (hence $\Theta \approx 5$ to 30 degrees, according to the notation previously introduced). From equally spaced angled points on such trajectory, $N_\theta = 9 - 25$ projection images are acquired by the detector. The detector is flat, built as a $n_x \times n_y$ grid of recording units with a uniform sensitive area of $\delta_x \times \delta_y$ μm^2. Typically, δ_x and δ_y are 85–160 μm. Moreover, the detector is fixed on a $XY-$plane and stationary during the whole scanning process.

The breast volume is numerically discretized into $N_v = N_x \times N_y \times N_z$ volumetric elements (called *voxels*) of size $\Delta_x \times \Delta_y \times \Delta_z$ μm^3. Due to the high resolution of the projection images, DBT allows for very high in-plane resolution (i.e., the resolution on the reconstructed slices which are parallel to the detector plane): Δ_x and Δ_y are smaller than 0.1 mm. On the contrary, because of the severe narrowness of the scanning range $[-\Theta, +\Theta]$, DBT is unfeasible to reconstruct thin slices as classical CT and its Z-axis resolution Δ_z is 1 to 1.5 mm typically.

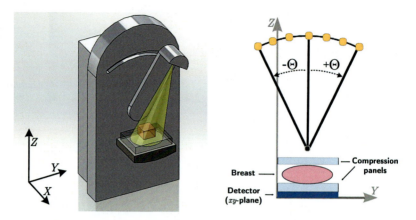

Fig. 14 On the left, a sketch of a modern DBT device where the Cartesian axis system is added for clarity. On the right, a view of the DBT geometry, projected onto the YZ-plane

In contrast to classical medical CT, DBT also makes use of *soft X-rays* with few tens of electron volts: this choice helps to reduce the provided radiations and it is further motivated by the anatomical structure of the breast. In breast imaging, there are no bones nor metallic objects, but adipose and fibro-glandular tissues that have very low attenuating properties: breast materials would not capture many photons from high-radiation X-rays. Since much more photon scattering occurs, this choice provides noisier data; nevertheless, it also allows to detect the breast objects in a more distinguishable way.

A further relevant feature of DBT imaging is due to its actual use in hospitals and clinics, where the high frequency of DBT screening tests makes long executions too expensive for a variety of reasons. As a consequence, an iterative solver can perform few iterations and it is stopped far before its convergence, typically. Such disadvantage is partially alleviated by parallel implementations (Jia et al. 2010; Matenine et al. 2015; Cavicchioli et al. 2020), but as the allowed computational time is shorter than 1 min, the huge amount of data and the complexity of the matrix computation make only four or five iterations feasible.

Model and Analysis

TV-Based Framework

All the following reconstructions are computed as solutions of the non-negative constrained and differentiable optimization problem:

$$\arg\min_{f \geq 0} \mathcal{G}(f) = LS(f) + \lambda TV_\beta(f). \qquad (25)$$

As solver, the scaled gradient projection (SGP) method, which is a gradient descent-like algorithm, is used (Loli Piccolomini et al. 2018). It is a first-order accelerated method, already proposed in Loli Piccolomini and Morotti (2021) for real 3D subsampled tomography. Essentially, the method follows a gradient projection approach accelerated by choosing the step lengths with Barzilai-Borwein techniques and by introducing a suitable scaling matrix improving the matrix conditioning. Its convergence to the unique minimum of (25) is proved in Bonettini and Prato (2015), under feasible assumptions. Numerically, the SGP solver runs until the following stopping condition on the objective function g is satisfied by an iterate $f^{(k)}$:

$$\left| \frac{f(x^{(k)}) - f(x^{(k-1)})}{f(x^{(k)})} \right| < 10^{-6}. \qquad (26)$$

A comparison among different solvers is out of the scope of this paper. However, results on the same data with different iterative methods can be found in Loli Piccolomini and Morotti (2021).

Measure and Graphics of Merits for 3D Tomography

To quantitatively evaluate the digitally reconstructed objects of interest, two widely used measure of merits are used in literature: the contrast-to-noise ratio (CNR) and the full width at half maximum (FWHM).

The CNR measure on a mass is calculated as:

$$CNR_{MS} = \frac{\mu_{MS} - \mu_{BG}}{\sigma_{MS} - \sigma_{BG}} \qquad (27)$$

where μ and σ are the mean and standard deviation computed on the reconstructed volume, in small regions located inside the mass (MS) or in the background (BG). Similarly, the CNR measure on a microcalcification is defined as:

$$CNR_{MC} = \frac{m_{MC} - \mu_{BG}}{\sigma_{BG}} \qquad (28)$$

where m_{MC} is the maximum intensity inside the considered microcalcification (MC). Higher values of the CNR indices reflect a better detection of an object from the background.

To compute the FWHM parameter, the transverse slice (parallel to the XY-plane) where the microcalcification lies must be considered, and then it is required to extract the plane profile (PP) of the MC, along the Y direction. The FWMH index is thus computed as:

$$FWHM = 2\sqrt{2\ln(2)}d \qquad (29)$$

where d is the standard deviation of the Gaussian curve fitting the PP. In particular,

$$w = FWHM \cdot \Delta_y \tag{30}$$

approximates the width of the examined microcalcification. The plane profiles are also useful tools to evaluate the reconstruction accuracy on the transverse plane.

To estimate the solver effectiveness along the Z direction, which is the most challenging purpose in DBT imaging, it is convenient to extract the artifact spread function (ASF) vector from the digital reconstruction. The ASF components are computed on a microcalcification as:

$$ASF(z) = \frac{|\mu_{MC}(z) - \mu_{BG}(z)|}{|\mu_{MC}(\bar{z}) - \mu_{BG}(\bar{z})|}, \quad \forall z = 1, \ldots, N_z \tag{31}$$

where $\mu(z)$ is the mean of the reconstructed values inside a circular region of three pixels diameter inside the considered MC and in the background, $\bar{z}$ corresponds to the slice where the object is on focus, and N_z is the total number of discrete slices. Similarly, we compute the ASF for the masses.

Reconstructions of the Accreditation Phantom

The tests here reported are performed on the *Giotto Class* digital system by the Italian I.M.S. Giotto Spa company in Bologna (IMS Giotto Class). To get the considered data, the source executes $N_\theta = 11$ scans from equally spaced angles in an approximately 30-degree range. The detector has squared pixel pitch of 85 μm, whereas the reconstructed voxel dimensions along the three Cartesian axes are $\Delta_x = \Delta_y = 90\,\mu\text{m}$ and $\Delta_z = 1\,\text{mm}$, respectively.

The scanned object is a breast imaging phantom, the model 020 of BR3D, produced by CIRS Tissue Simulation and Phantom company (Computerized Imaging Reference Systems). It is characterized by a heterogeneous background, where adipose-like and gland-like tissues are mixed in about 50:50 ratio. Inside, objects of interest for breast cancer detection are inserted at the same depth: they are acrylic spheres simulating breast masses (MSs), acrylic short fibers, and clusters of calcium carbonate specks simulating microcalcifications (MCs), of different dimensions and thickness.

Running the gradient descent solver, the convergence criterion (26) stops the execution after 44 iterations. The fast decreasing behavior of the $\mathcal{G}$ function along the iterative reconstruction process is remarkable, as visible in Fig. 15. The objective function exhibits a very fast reduction in the first five iterations, whereas it has a very flat trend from ten iterations on, as confirmed by the red-labeled values. Indeed, the reconstructed images are visually almost indistinguishable after 30 iterations.

Figure 16 presents the reconstructions of a 4.7 mm mass and of a cluster with the 165-μm-thick MCs, obtained in 5, 15, and 30 iterations. In Fig. 17 the corresponding PP and ASF plots are reported.

Simulated and anatomical masses are larger than microcalcifications, but their lower photon absorption capability makes their detection difficult. In fact, even if

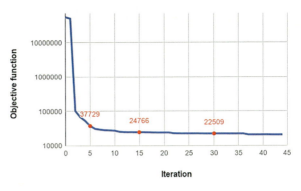

Fig. 15 Objective function values vs. iteration number for the iterative reconstruction of the phantom test. The red labels outline the function values at 5, 15, and 30 iterations

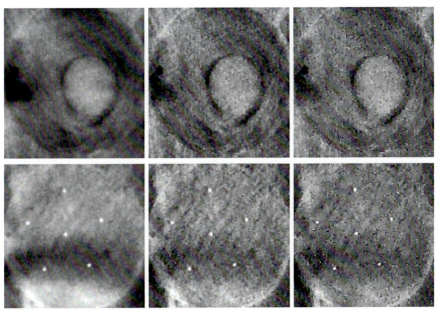

Fig. 16 Crops of a reconstructed slice on BR3D phantom, obtained in 5, 15, and 30 iterations (from left to right). First row: zooms in of a mass. Last row: zooms in of a MC cluster

visible in only five iterations, the mass tends to present smooth edges, and more iterations are required to enhance the mass contrast to the background (see the first row of Fig. 16 and the corresponding plane profile in Fig. 17). The perfect location of the mass at its correct depth still remains critical, since it tends to be out of focus and blurred along the Z direction.

In spite of their smallness, microcalcifications are immediately visible on the earliest model-based reconstructions, as high absorbing structures of a breast. In fact, all the six MCs of the reported cluster are clearly detected in only five iterations, but again the effect of the TV regularization needs longer executions to make them less blurry and more contrasted from the background. It is remarked by the PP

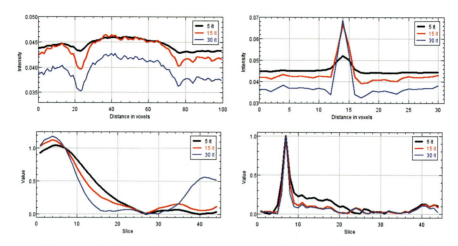

Fig. 17 Plane profiles (first row) and ASF plots (second row) of the mass and one microcalcification from the BR3D phantom reconstructions shown in Fig. 16. In all the plots: black line corresponds to five iterations, red line to 15 iterations, and blue line to 30 iterations

Table 1 FWHM index (29) and w measures (30) computed on images reconstructed in 5, 15, and 30 SGP iterations. In the first column are the actual diameters of the microcalcification spheres of the BR3D phantom

Diameters (μm) of the MC	FWHM			w (μm)		
	5 it.	15 it.	30 it.	5 it.	15 it.	30 it.
230	4.77	3.32	2.70	430	299	243
165	3.52	2.65	2.32	317	238	209
130	–	2.05	1.52	–	185	137

plots of Fig. 17. Even the object detection along the Z axis improves with ongoing iterations, as deductible from the depth-oriented inspection by the ASF plot. The FWHM values and the corresponding MCs width w (reported in Table 1) denote that the regularized iterative approach is indeed effective in recovering very small microcalcifications: MCs of 130 μm width, which should approximately fill inside only two voxels, are not discernible from the background in only five iterations (the FWHM is not measurable here), but they can be well recovered after more iterations with a good approximation of their real size.

At last, the increasing values of CNR in Table 2 denote the strong effect of the regularized model in denoising the objects of interest.

Reconstructions of a Human Dataset

The performances of an iterative model-based reconstruction are further confirmed when it is used on real screening DBT datasets. For example, the considered breast contains here a microcalcification and a mass, on the same reconstructed slice, and

14 Sparse Regularized CT Reconstruction: An Optimization Perspective

Table 2 CNR measure for microcalcifications as in (28) and for masses as in (27) computed on images reconstructed in 5, 15, and 30 SGP iterations. In the first column are the actual diameters of the considered objects of the BR3D phantom

	Diameters (μm)	5 it.	15 it.	30 it.
MS	4700	0.82	1.07	1.66
MS	3100	0.87	1.00	1.33
MC	230	24.21	33.34	38.00
MC	165	10.03	19.00	28.00
MC	130	7.27	11.02	17.00

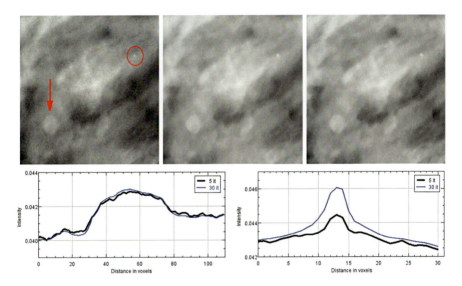

Fig. 18 Results obtained after 5, 15, and 30 SGP iterations on a human breast dataset. First row: zooms in of a 440 × 400 pixels region presenting both a spherical mass (pointed by the arrow) and a microcalcification (identified by the circle). Last row: plane profiles on the mass and on the microcalcification. In the plots: black line corresponds to 5 iterations and blue line to 30 iterations

the images in Fig. 18 zoom over such objects of interest on the reconstructions computed in 5, 15, and 30 iterations. Figure 18 also shows the plot profiles of the mass and the microcalcification. In this case, the mass detection is already effective in the earliest reconstruction and its gray level intensity does not change remarkably, but the denoising effect of the TV prior in the last iterations is evident on the PP. Also the microcalcification is detected in few iterations, even if a more time-consuming SGP execution enhances the contrast of the object with respect to the background and the corresponding FWHM values (reported in Table 3) confirm its getting more and more defined, from 5 to 30 iterations.

Table 3 FWHM measures on the microcalcification visible in Fig. 18

	FWHM		
	5 it.	15 it.	30 it.
MC	8.57	7.81	7.29

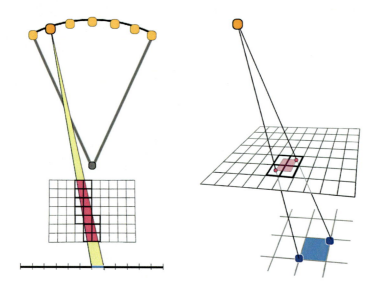

Fig. 19 The distance-driven approach for the forward projection on a DBT-like device. On the left, the process is seen on the YZ-plane (hence the detector is reduced to a 1D array and the volume to a grid of voxels); on the right, one volume slice is considered over the detector. In all the images, one detector unit is considered and remarked in blue, whereas its backward projection cone is highlighted in pink, defining the voxels that are indeed involved in the forward projections

Distance-Driven Approach for 3D CT Imaging

In DBT, the projection matrix M can be efficiently computed with a distance-driven (DD) approach. The standard DD extension to 3D imaging is presented in De Man and Basu (2004) for a general CT process, but due to the presence of a flat and stationary detector, it is necessary to specifically tune the algorithm for DBT devices.

Recalling the notation used in this chapter, Δ_x, Δ_y, and Δ_z are the spatial resolution of the discretization into voxels of the volume, respectively, along the Cartesian axis, whereas δ_x, δ_y are the dimensions of each detector unit mounted on the DBT machinery. For prefixed scanning angle θ_k and element g_i of the k-th projection (where $k \in \{1, \ldots, N_\theta\}, i \in \{1, \ldots, n_p\}$), the i-th row of the forward projection operator M^{θ_k} models the X-ray cone having as a basis the i-th detector pixel itself and vertex on the X-ray source (see Fig. 19 as reference); then the DD algorithm determines the backward footprints of the detector unit onto each object slice, at its middle height $\frac{\Delta_z}{2}$ (as indicated on the left image in Fig. 20). Fixing one object slice and the j-th voxel on it (as the green one in Fig. 20), let A_i be the area of

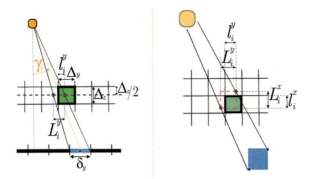

Fig. 20 The distance-driven approach for the forward projection on a DBT-like device. On the left, the process is seen on the YZ-plane for one slice; on the right, it is projected onto the XY-plane. In all the images, one detector unit is considered and remarked in blue, whereas its backward projection area is highlighted in pink, and the considered voxel is green

the backward footprints onto the slice (dashed pink in Fig. 20) of the i-th pixel and $a_{i,j}$ be the area of the intersection between the pink and the gray squares, as denoted in Fig. 20. The matrix element is computed as:

$$M_{i,j}^{\theta_k} = \frac{\Delta_z}{\alpha_i \gamma_i} \frac{a_{i,j}}{A_i} \tag{32}$$

for all the voxels on that slice.

In this equation, α_i and γ_i are the in- and out-of-plane angles, respectively, i.e., the two angles describing the X-ray linking the source to the center of the i-th detector pixel. Drawing the perpendicular from the X-ray source to the detector and tracing the X-ray reaching the middle point of the i-th pixel, γ_i is the angle between these two elements on the YZ−plane, as shown in Fig. 20, while α_i is that angle on the XY−plane. Moreover, the factor $\frac{\Delta_z}{\alpha_i \gamma_i}$ can be interpreted as a normalization by the popular $\frac{1}{r^2}$ term, which is known as the *inverse-square* physical law, stating that a specific physical quantity (like the photon intensity in our case) is inversely proportional to the square of the distance from the source of that physical quantity.

Code Parallelization

The required accuracy on the breast digital volume and the resolution of the detector make the DBT problem of very high dimensions. The magnitude of the involved numerical objects prevents the storage of the system matrix M on the hardware; hence, its entries must be computed at each invocation of the matrix itself. This causes an extremely long execution of the optimization solver (which also impacts on the number of iterations allowed in a real clinical setting). In fact, by profiling a serial execution of an iterative solver, two main kernels can be identified as heavy computational tasks in each iteration, and they are the forward and the backward applications of the matrix operator, i.e., the steps with the matrix-vector products involving M and M^T, respectively.

To set a realistic example, consider a volume with $N = 1.5 \cdot 10^8$ voxels to be recovered from $N_\theta = 11$ views of 3000×1500 pixel projection images (resulting in $N_d \approx 5 \cdot 10^7$ data). Table 4 reports the output of the profiling analysis of the scaled gradient projection algorithm, compiled on an i7 high-end computer with 32 GB of RAM and 1 TB of solid state disk (Cavicchioli et al. 2020). In such a configuration, almost 90% of the computational time is spent for forward and backward projections in a gradient descent solver, where both the kernels occur only once per iteration. A third task addressing all the computations for the TV function covers 5% of the execution time per iteration, whereas only the 8% is spent for all the remaining SGP steps.

By parallelizing the C code on NVIDIA GPU by means of the CUDA SDK, the execution times drastically go down: GPU implementation exploits the massive parallel architecture of graphical boards and distributes work to hundreds of small cores. However, if the algorithm cannot store all the necessary variables in the GPU memory entirely, many data transfers between the CPU and the GPU are required during each iteration of the solver (see Fig. 21): as visible from the second row of

Table 4 Results of the profiling of the iterative solver, according to its different implementations on a CPU (Intel i7 7700K CPU at 4.3 GHz, 32 GB of RAM, and 1 TB of solid state disk) and on the Titan V board by NVIDIA (12 GB of RAM and 5120 CUDA cores). In each row: the computing time of the four considered kernels, the whole iteration time, the number of feasible iterations in 50 s, and the resulting speedup (with respect to the serial implementation). All the times are relative to a single iteration of a gradient descent-like solver and are expressed in milliseconds

	Forward (ms)	Backward (ms)	TV (ms)	Other (ms)	1 iter. (ms)	Iters. in 50 s	Speedup
Serial	235,368	237,556	23,841	39,735	536,500	–	–
Parallel on CPU	270	263	1229	7613	9375	5	**57×**
Parallel on GPU	116	110	372	548	1146	50	**468×**

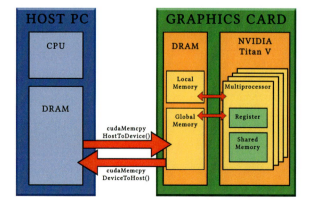

Fig. 21 Logical view of a system composed by host and accelerator. The data stored in the host memory (DRAM) must be transferred to the graphics card memory (global memory) to execute the parallel computation, and then the results must be transferred back to be saved in DRAM

Table 4, the resulting parallel execution achieves a 57× speedup with respect to the serial one, allowing for only five iterations in less than 1 min. On the contrary, if the GPU has a larger global memory, a higher level of parallelism can be exploited to completely run the SGP solver on the GPU so that one iteration requires about only 1 s (reflecting an impressive speedup of almost 470). This means to achieve a close-to-convergence reconstruction in less than 1 min.

Conclusion

Nowadays the medical world aims at enlarging the class of CT exams with new, safe, and fast X-ray protocols, which can be defined by reducing the number of projection views. Model-based iterative methods are efficient methods for sparse-view CT image reconstruction, since they solve an optimization problem where a priori information are embedded by means of a regularization function. When approaching convergence, iterative solvers achieve very accurate images where low-contrast objects and very small structures are well detected and shaped. On a case study on real projections of 3D breast tomosynthesis, model-based approaches reconstruct in very few iterations images where the objects of interest, such as masses and microcalcifications, are clearly distinguishable. Moreover, a parallel reconstruction of breast imaging on a GPU board can be obtained from real data in less than 1 min, a time compatible with clinical requests.

Indeed, if the main drawback of iterative solver lays in their high computational costs and slow executions, the ongoing development of GPU boards (which are more and more powerful and affordable) paves the way to almost real-time reconstructions, making this approach feasible for real-life applications.

Finally, the flexibility of the optimization framework also allows to incorporate external information by means of neural networks to improve the quality of the reconstructed image.

References

Computerized Imaging Reference Systems: https://www.cirsinc.com/products/a11/51/br3d-breast-imaging-phantom/. BR3D Breast Imaging Phantom, Model 020

IMS Giotto Class: http://www.imsgiotto.com/

Adler, J., Öktem, O.: Solving ill-posed inverse problems using iterative deep neural networks. Inverse Probl. **33**(12), 124007 (2017)

Adler, J., Öktem, O.: Learned primal-dual reconstruction. IEEE Trans. Med. Imaging **37**(6), 1322–1332 (2018)

Andersson, I., Ikeda, D.M., Zackrisson, S., Ruschin, M., Svahn, T., Timberg, P., Tingberg, A.: Breast tomosynthesis and digital mammography: a comparison of breast cancer visibility and birads classification in a population of cancers with subtle mammographic findings. Eur. Radiol. **18**(12), 2817–2825 (2008)

Averbuch, A., Sedelnikov, I., Shkolnisky, Y.: CT reconstruction from parallel and fan-beam projections by a 2-d discrete radon transform. IEEE Trans. Image Process. **21**(2), 733–741 (2011)

Barca, P., Lamastra, R., Tucciariello, R., Traino, A., Marini, C., Aringhieri, G., Caramella, D., Fantacci, M.: Technical evaluation of image quality in synthetic mammograms obtained from 15° and 40° digital breast tomosynthesis in a commercial system: a quantitative comparison. Phys. Eng. Sci. Med. **44**(1), 23–35 (2021). cited By 0

Bonettini, S., Prato, M.: New convergence results for the scaled gradient projection method. Inv. Probl. **31**(9), 1196–1211 (2015)

Bubba, T.A., Hauptmann, A., Huotari, S., Rimpeläinen, J., Siltanen, S.: Tomographic x-ray data of a lotus root filled with attenuating objects. arXiv preprint arXiv:1609.07299 (2016)

Buzug, T.M.: Computed tomography. In: Springer Handbook of Medical Technology, pp. 311–342. Springer, Muller and Siltanen, Philadelphia(2011)

Cavicchioli, R., Hu, J., Loli Piccolomini, E., Morotti, E., Zanni, L.: A first-order primal-dual algorithm for convex problems with applications to imaging. GPU acceleration of a model-based iterative method for digital breast tomosynthesis. Sci. Rep. **10**(1), 120–145 (2020)

Choi, K., Wang, J., Zhu, L., Suh, T.-S., Boyd, S.P., Xing, L.: Compressed sensing based cone-beam computed tomography reconstruction with a first-order method. Med. Phys. **37**(9), 5113–5125 (2010)

Das, M., Gifford, H.C., O'Connor, J.M., Glick, S.J. Penalized maximum likelihood reconstruction for improved microcalcification detection in breast tomosynthesis. IEEE Trans. Med. Imaging **30**(4), 904–914 (2010)

De Chiffre, L., Carmignato, S., Kruth, J.-P., Schmitt, R., Weckenmann, A.: Industrial applications of computed tomography. CIRP Ann. **63**(2), 655–677 (2014)

De Man, B., Basu, S.: Distance-driven projection and backprojection. In: 2002 IEEE Nuclear Science Symposium Conference Record, vol. 3, pp. 1477–1480. IEEE (2002)

De Man, B., Basu, S.: Distance-driven projection and backprojection in three dimensions. Phys. Med. Biol. **49**(11), 2463 (2004)

Evangelista, D., Morotti, E., Piccolomini, E.L.: Rising a new framework for few-view tomographic image reconstruction with deep learning. arXiv preprint arXiv:2201.09777 (2022)

Feldkamp, L.A., Davis, L.C., Kress, J.W.: Practical cone-beam algorithm. J. Opt. Soc. Am. A **1**(6), 612–619 (1984)

Fessler, J.A.: Equivalence of pixel-driven and rotation-based backprojectors for tomographic image reconstruction (1997)

Graff, C., Sidky, E.: Compressive sensing in medical imaging. Appl. Opt. **54**(8), C23–C44 (2015)

Gupta, H., Jin, K.H., Nguyen, H.Q., McCann, M.T., Unser, M.: CNN-based projected gradient descent for consistent CT image reconstruction. IEEE Trans. Med. Imaging **37**(6), 1440–1453 (2018)

Hadamard, J.: Sur les problèmes aux dérivées partielles et leur signification physique, pp. 49–52. Princeton University Bulletin, Natterer, Stuttgart (1902)

Han, Y., Ye, J.C.: Framing U-NET via deep convolutional framelets: application to sparse-view CT. IEEE Trans. Med. Imaging **37**(6), 1418–1429 (2018)

Harauz, G., Ottensmeyer, F.: Interpolation in computing forward projections in direct three-dimensional reconstruction. Phys. Med. Biol. **28**(12), 1419 (1983)

Hashemi, S., Beheshti, S., Gill, P.R., Paul, N.S., Cobbold, R.S.: Fast fan/parallel beam CS-based low-dose CT reconstruction. In: 2013 IEEE International Conference on Acoustics, Speech and Signal Processing, pp. 1099–1103. IEEE (2013)

He, J., Yang, Y., Wang, Y., Zeng, D., Bian, Z., Zhang, H., Sun, J., Xu, Z., Ma, J.: Optimizing a parameterized plug-and-play ADMM for iterative low-dose CT reconstruction. IEEE Trans. Med. Imaging **38**(2), 371–382 (2018)

He, Y., Luo, S., Wu, X., Yang, H., Zhang, B.B., Bleyer, M., Chen, G.: Computed tomography angiography with 3d reconstruction in diagnosis of hydronephrosis cause by aberrant renal vessel: a case report and mini review. J. X-Ray Sci. Technol. **26**(1), 125–131 (2018)

Huang, J., Zhang, Y., Ma, J., Zeng, D., Bian, Z., Niu, S., Feng, Q., Liang, Z., Chen, W.: Iterative image reconstruction for sparse-view CT using normal-dose image induced total variation prior. PloS One **8**(11), e79709 (2013)

Hughes, S.: CT scanning in archaeology. In: Saba, L. (ed.) Computed Tomography-Special Applications, pp. 57–70. InTech Europe, Buzug, Berlin (2011)

Jia, X., Dong, B., Lou, Y., Jiang, S.B.: GPU-based iterative cone-beam CT reconstruction using tight frame regularization. Phys. Med. Biol. **56**(13), 3787 (2011)

Jia, X., Lou, Y., Li, R., Song, W.Y., Jiang, S.B.: GPU-based fast cone beam CT reconstruction from undersampled and noisy projection data via total variation. Med. Phys. **37**(4), 1757–1760 (2010)

Kak, A.C., Slaney, M.: Principles of Computerized Tomographic Imaging. SIAM, Kak, Philadelphia (2001)

Kubo, T., Lin, P.-J.P., Stiller, W., Takahashi, M., Kauczor, H.-U., Ohno, Y., Hatabu, H.: Radiation dose reduction in chest CT: a review. Am. J. Roentgenol. **190**(2), 335–343 (2008)

Lacroute, P., Levoy, M.: Fast volume rendering using a shear-warp factorization of the viewing transformation. In: Proceedings of the 21st Annual Conference on Computer Graphics and Interactive Techniques, pp. 451–458 (1994)

Liu, Y., Xu, Y., Yin, W.: Acceleration of primal–dual methods by preconditioning and simple subproblem procedures. J. Sci. Comput. **86**(2), 1–34 (2021)

Loli Piccolomini, E., Coli, V., Morotti, E., Zanni, L.: Reconstruction of 3D X-ray CT images from reduced sampling by a scaled gradient projection algorithm. Comput. Optim. Appl. **71**, 171–191 (2018)

Loli Piccolomini, E., Morotti, E.: A fast TV-based iterative algorithm for digital breast tomosynthesis image reconstruction. J. Algorithms Comput. Technol. **10**(4), 277–289 (2016)

Loli Piccolomini, E., Morotti, E.: A model-based optimization framework for iterative digital breast tomosynthesis image reconstruction. J. Imaging **7**(2), 36 (2021)

Long, Y., Fessler, J.A., Balter, J.: 3d forward and back-projection for x-ray CT using separable footprints with trapezoid functions. In: Proceedings of First International Conference on Image Formation in X-Ray Computed Tomography, pp. 216–219 (2010)

Luo, X., Yu, W., Wang, C.: An image reconstruction method based on total variation and wavelet tight frame for limited-angle CT. IEEE Access **6**, 1–1 (2017)

Matej, S., Fessler, J.A., Kazantsev, I.G.: Iterative tomographic image reconstruction using fourier-based forward and back-projectors. IEEE Trans. Med. Imaging **23**(4), 401–412 (2004)

Matenine, D., Goussard, Y., Després, P.: GPU-accelerated regularized iterative reconstruction for few-view cone beam CT. Med. Phys. **42**(4), 1505–1517 (2015)

Monga, V., Li, Y., Eldar, Y.C.: Algorithm unrolling: interpretable, efficient deep learning for signal and image processing. IEEE Sig. Process. Mag. **38**(2), 18–44 (2021)

Morotti, E., Evangelista, D., Loli Piccolomini, E.: A green prospective for learned post-processing in sparse-view tomographic reconstruction. J. Imaging **7**(8), 139 (2021)

Mueller, J.L., Siltanen, S.: Linear and Nonlinear Inverse Problems with Practical Applications. SIAM, Huges, Croatia (2012)

Natterer, F.: The Mathematics of Computerized Tomography. SIAM, Hadamard, Princeton (2001)

Niu, S., Gao, Y., Bian, Z., Huang, J., Chen, W., Yu, G., Liang, Z., Ma, J.: Sparse-view x-ray CT reconstruction via total generalized variation regularization. Phys. Med. Biol. **59**(12), 2997 (2014)

O'Connor, Y., Fessler, J.A.: Fourier-based forward and back-projectors in iterative fan-beam tomographic image reconstruction. IEEE Trans. Med. Imaging **25**(5), 582–589 (2006)

Pelt, D.M., Batenburg, K.J., Sethian, J.A.: Improving tomographic reconstruction from limited data using mixed-scale dense convolutional neural networks. J. Imaging **4**(11), 128 (2018)

Peters, T.: Algorithms for fast back-and re-projection in computed tomography. IEEE Trans. Nucl. Sci. **28**(4), 3641–3647 (1981)

Quinto, E.T.: Singularities of the x-ray transform and limited data tomography in R^2 and R^3. SIAM J. Math. Anal. **24**(5), 1215–1225 (1993)

Reiser, I., Bian, J., Nishikawa, R.M., Sidky, E.Y., Pan, X.: Comparison of reconstruction algorithms for digital breast tomosynthesis. arXiv:0908.2610 (2009)

Ritschl, L., Bergner, F., Fleischmann, C., Kachelrieß, M.: Improved total variation-based CT image reconstruction applied to clinical data. Phys. Med. Biol. **56**(6), 1545–1561 (2011)

Schnurr, A.-K., Chung, K., Russ, T., Schad, L.R., Zöllner, F.G.: Simulation-based deep artifact correction with convolutional neural networks for limited angle artifacts. Zeitschrift für Medizinische Physik **29**(2), 150–161 (2019)

Sidky, E., Chartrand, R., Boone, J., Pan, X.: Constrained TpV-minimization for enhanced exploitation of gradient sparsity: application to CT image reconstruction. IEEE J. Transl. Eng. Health Med. **2**, 1800418 (2014)

Sidky, E.Y., Kao, C.M., Pan, X.: Accurate image reconstruction from few-views and limited-angle data in divergent-beam CT. J. Xray Sci. Technol. **14**(2), 119–139 (2009)

Sidky, E.Y., Lorente, I., Brankov, J.G., Pan, X.: Do cnns solve the CT inverse problem? IEEE Trans. Biomed. Eng. **68**(6), 1799–1810 (2020)

Thibault, J.-B., Sauer, K.D., Bouman, C.A., Hsieh, J.: A three-dimensional statistical approach to improved image quality for multislice helical CT. Med. Phys. **34**(11), 4526–4544 (2007)

Urase, Y., Nishio, M., Ueno, Y., Kono, A.K., Sofue, K., Kanda, T., Maeda, T., Nogami, M., Hori, M., Murakami, T.: Simulation study of low-dose sparse-sampling CT with deep learning-based reconstruction: usefulness for evaluation of ovarian cancer metastasis. Appl. Sci. **10**(13), 4446 (2020)

Venkatakrishnan, S.V., Bouman, C.A., Wohlberg, B.: Plug-and-play priors for model based reconstruction. In: 2013 IEEE Global Conference on Signal and Information Processing, pp. 945–948. IEEE (2013)

Vogel, C.R.: Computational Methods for Inverse Problems. SIAM, Philadelphia (2002)

Wang, C., Tao, M., Nagy, J.G., Lou, Y.: Limited-angle CT reconstruction via the l_1/l_2 minimization. SIAM J. Imaging Sci. **14**(2), 749–777 (2021)

Wang, G., Ye, J.C., Mueller, K., Fessler, J.A.: Image reconstruction is a new frontier of machine learning. IEEE Trans. Med. Imaging **37**(6), 1289–1296 (2018)

Wu, T., Moore, R.H., Rafferty, E.A., Kopans, D.B.: A comparison of reconstruction algorithms for breast tomosynthesis. Med. Phys. **31**(9), 2636 (2004)

Xiang, J., Dong, Y., Yang, Y.: Fista-net: learning a fast iterative shrinkage thresholding network for inverse problems in imaging. IEEE Trans. Med. Imaging **40**(5), 1329–1339 (2021)

Yu, L., Liu, X., Leng, S., Kofler, J.M., Ramirez-Giraldo, J.C., Qu, M., Christner, J., Fletcher, J.G., McCollough, C.H.: Radiation dose reduction in computed tomography: techniques and future perspective. Imaging Med. **1**(1), 65 (2009)

Yu, W., Zeng, L.: A novel weighted total difference based image reconstruction algorithm for few-view computed tomography. PloS One **9**(10), e109345 (2014)

Zhang, H., Liu, B., Yu, H., Dong, B.: Metainv-net: meta inversion network for sparse view CT image reconstruction. IEEE Trans. Med. Imaging **40**(2), 621–634 (2020)

Zhang, T., Gao, H., Xing, Y., Chen, Z., Zhang, L.: Dualres-UNET: limited angle artifact reduction for computed tomography. In: 2019 IEEE Nuclear Science Symposium and Medical Imaging Conference (NSS/MIC), pp. 1–3. IEEE (2019)

Zhang, Y., Chan, H.H.-P., Sahiner, B., Wei, J., Goodsitt, M., Hadjiiski, L.M.L., Ge, J., Zhou, C.: A comparative study of limited-angle cone-beam reconstruction methods for breast tomosynthesis. Med. Phys. **33**(10), 3781 (2006)

Recent Approaches for Image Colorization

15

Fabien Pierre and Jean-François Aujol

Contents

Context and Modeling	586
Challenge	586
Mathematical Modeling of Colorization	587
Range of Chrominance	588
Color Diffusion	589
State-of-the-Art of Color Diffusion	590
Coupled Total Variation for Image Colorization	592
Constrained TV-L2 Debiasing Algorithm	594
Exemplar-Based Colorization	599
Morphing-Based Approach	600
Segmentation-Based Techniques	601
Patch-Based Methods	602
A Variational Model for Image Colorization with Channel Coupling	605
Colorization from Dataset	607
Coupled Approaches	608
Coupling Manual Approach with Exemplar-Based Colorization	609
Coupling CNN with a Variational Approach	611
Conclusion and Future Works	618
References	619

F. Pierre (✉)
LORIA, UMR CNRS 7503, Université de Lorraine, INRIA projet Tangram, Nancy, France
e-mail: fabien.pierre@loria.fr

J.-F. Aujol (✉)
Univ. Bordeaux, Bordeaux INP, CNRS, IMB, UMR 5251, Talence, France
e-mail: jean-francois.aujol@math.u-bordeaux.fr

© Springer Nature Switzerland AG 2023
K. Chen et al. (eds.), *Handbook of Mathematical Models and Algorithms in Computer Vision and Imaging*, https://doi.org/10.1007/978-3-030-98661-2_55

Abstract

In the last years, image and video colorization has been considered from many points of view. The technique consists of the addition of a color component to a grayscale image. This operation needs additional priors which can be given by manual intervention of the user from an example image or be extracted from a large dataset of color images. A very large variety of approaches has been used to solve this problem, like PDE models, non-local methods, variational frameworks, learning approaches, etc. In this chapter, we aim at providing a general overview of state-of-the-art approaches with a focus on few representative methods. Moreover, some recent techniques from the different types of priors (manual, exemplar-based, dataset-based) are explained and compared. The organization of the chapter aims at describing the evolution of the techniques in relation to each other. A focus on some efficient strategies is proposed for each kind of methodology.

Keywords

Image colorization · Variational approaches · Deep learning · Patch-based methods

Context and Modeling

Challenge

Image colorization consists of the transformation of a grayscale image into a color one. The reverse transformation, i.e., turning a color image into a grayscale one, is based on visual assumptions and it is also an active research topic (Kuhn et al. 2008; Cui et al. 2010; Song et al. 2013). Image colorization is useful for the entertainment industry to make old film productions attractive to young people, for instance. In France, in 2014, *Apocalypse*, a historical documentary by I. Clarke and D. Costelle, was made from archives colorized by F. Montpellier of the ImaginColor company. The broadcast gathered over 18.5% of viewers over the age bracket 11–14 during the first 2 episodes (Lannaud 2009). The colorization for movies is mostly performed manually, which is a very tedious work. As an example, the colorization of about 4 hours of video sequences for the *Apocalypse* documentary required 47 weeks by F. Montpellier and his team. Image colorization can also be used to help a user to analyze an image, for example, for sensor fusion in Zheng and Essock (2008). For instance, to assist in airport security screening, color is added to the X-ray scanner result based on the density of the objects, so that the operator can know their composition and quickly interpret the result (Abidi et al. 2006). Image colorization can also be used to restore artistic heritage, for example, Fornasier (2006) or Wolfgang Baatz Massimo Fornasier and Schönlieb (2008). This old subject started with the ability of screens to display color. A first approach, very basic, consists of matching each grayscale to a color (Gonzalez and Woods 2008). However, it is impossible to recover every color without additional information (there are 256 gray levels and about 16 million colors displayable on standard screens). In existing approaches, this information can be added by three ways: the first one directly adds color to the image by the user (see, e.g., the approach of Levin et al. 2004), the second one provides an example image (also called source image, see, e.g., the method of Welsh et al. 2002), and the third one uses a deep learning approach based on a large database (see for instance the method of Zhang et al. 2016).

In this chapter, we propose a general overview of colorization methods which have been described in the literature with a focus on few representative approaches. This review is not based on the application point of view but it has been done from a methodological perspective. The term "automatic" has been widely used, but it means in fact that the algorithms are able to assist the user. For manual methods, the diffusion of the colors put by the user is automatic, and for exemplar-based approaches, the diffusion of colors from a given reference image to the target one is automatic but actually it requires the choice of the source image. For dataset-based colorization, the colorization is automatic after training on a large dataset given by the user. In this chapter, an overview of the three different approaches to colorize images (manual, exemplar-based, and dataset-based) is proposed. In particular, a highlight on a variational model is used as a thread along the chapter because this model enables some coupling of different approaches such as manual with exemplar-based. More generally, we focus on the different strategies available among state-of-the-art methods for each kind of methodology. Moreover, a final section proposes an overview of coupled strategies.

In this chapter, the mathematical modeling of the colorization problem is reviewed in section "Mathematical Modeling of Colorization". Next, in section "Range of Chrominance", we recall the definition of the range of the solution, and we present an algorithm to compute an orthogonal projection onto this set. The three next sections deal with, respectively, the manual, the exemplar-based, and the dataset-based colorization. Finally, in section "Coupled Approaches", we propose an overview about the coupling of some techniques within a variational formulation.

Mathematical Modeling of Colorization

In order to model the colorization problem, let us consider the luminance-chrominance color spaces. The results of this section are based on the papers (Pierre et al. 2015c, 2017b) that can be considered as the state of the art for luminance specification. In all state-of-the-art approaches, the grayscale image is considered as the luminance channel of a color image. The luminance can be defined as a weighted average of the RGB channels:

$$Y = 0.299R + 0.587G + 0.114B. \qquad (1)$$

Some other definitions are also sometimes used. For instance, the L channel of the CIE Lab color space can be used. In order to preserve its content, colorization methods must always require that the luminance channel of the image of interest is equal to the target image. Most methods compute only the two chrominance channels, complementary to the luminance, which is enough to provide a displayable color image.

Some different spaces have been introduced, such as YUV, YCbCr, YIQ, etc. The transformation from RGB to YUV is linear and defined with the following matrices:

$$R, G, B, Y \in [0, 255], U \in [-111.18, 111.18], V \in [-156.825, 156.825]. \quad (2)$$

$$\begin{pmatrix} Y \\ U \\ V \end{pmatrix} = \begin{pmatrix} 0,299 & 0,587 & 0,114 \\ -0,14713 & -0,28886 & 0,436 \\ 0,615 & -0,51498 & -0,10001 \end{pmatrix} \begin{pmatrix} R \\ G \\ B \end{pmatrix}. \quad (3)$$

Let us notice that the main problem raised by these color spaces is that all the luminance-chrominance values cannot be converted into a RGB color between 0 and 255. Thus, some additional techniques have to be employed to recover the RGB color image (Pierre et al. 2015c). These techniques are out of the scope of this chapter, but the reader has to keep in mind that they are essential to compute the final result. The next section recalls the basis of gamut problem in the case of the YUV color space.

Range of Chrominance

The natural problem arising when editing a color while keeping its luminance or intensity constant is the preservation of the RGB standard range of the produced image. Most of the methods of the literature work directly in the RGB space (Nikolova and Steidl 2014; Fitschen et al. 2015; Pierre et al. 2015c), since it is easier to maintain the standard range. Nevertheless, working in the RGB space needs to process three channels, while two chrominance channels are enough to edit a color image while keeping the luminance.

Description of the Range

In this section, we recall the geometric description of the set of chrominance values which correspond to a particular luminance level and which are contained in the RGB standard range. Let us denote by $T(y, u, v)$ the invertible linear operator mapping YUV colors onto the RGB ones. This operator corresponds to the inverse of the operation described in Equation (3).

Proposition 1. *Let y be a value of luminance between 0 and 255. The set of chrominance values (u, v) that satisfy $T(y, u, v) \in [0, 255]^3$ is a convex polygon.*

Remark 1. For a given luminance, the chrominance values out of this polygon can be transformed into the RGB space, but they are out of the bounds of the RGB cube. A truncation of the coordinates is usually done, but it generally changes both the luminance and the hue of the result.

15 Recent Approaches for Image Colorization

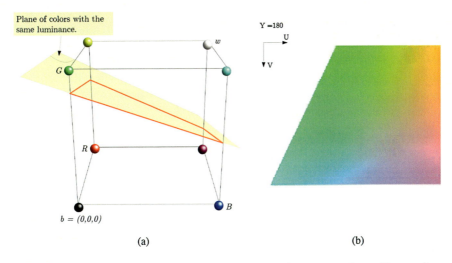

Fig. 1 The set of the RGB colors with a particular luminance is a convex polygon. The map from RGB to YUV being affine, the set of the corresponding chrominances is also a convex polygon. (**a**) Set of the RGB colors with a fixed luminance. (**b**) Corresponding colors in the YUV space

Proof. [of Proposition 1] The intuition of the proof is given in Fig. 1. The set of the colors in the RGB cube whose luminance is equal to a particular value y is a convex polygon (see, e.g., Pierre et al. 2015c). Indeed, the set of colors with a particular luminance is an affine plane in $\mathbb{R}^3$ and the intersection of the RGB cube with it is a polygon. The transformation of the RGB values into the YUV space being affine, the set of corresponding colors is thus also a convex polygon included in the set $Y = y$. □

Orthogonal Projection onto the Convex Range

Pixel-wise, the valid chrominances are contained in a convex polygon that has, at most, six edges. The numerical computation of the vertex coordinates has been detailed in Pierre et al. (2017b). When the vertices are computed, and denoted by P1, P2, etc., the orthogonal projection onto the polygon is computed as follows.

The algorithm first checks if the corresponding RGB value is between 0 and 255. If so, the point is its own orthogonal projection. If not, the orthogonal projection is onto one of the edges and can be computed for each of them. Finally, the closest result is retained as the solution. The algorithm is summarized in Algorithm 1 and illustrated in Fig. 2.

Color Diffusion

In this section, we first summarize the state-of-the-art methods. We then present a strategy for image colorization based on the total variation minimization. This framework uses some recent state-of-the-art approaches in order to diffuse color

Algorithm 1 Algorithm computing projection $P_\mathcal{R}$

Require: X: chrominance vector; Y luminance value.
1: **if** $RGB(Y, X) \notin [0, 255]^3$ **then**
2:   **for** $i = 1 : n$ **do**
3:     $j \leftarrow i + 1 \mod n$
4:     $\alpha \leftarrow \left\langle \overrightarrow{P_i P_j} | \overrightarrow{P_i X} \right\rangle / \left(\| \overrightarrow{P_i P_j} \|_2 \right)$
5:     **if** $\alpha > 1$ **then**
6:       $X_{i,j} \leftarrow P_j$
7:     **else if** $\alpha < 0$ **then**
8:       $X_{i,j} \leftarrow P_i$
9:     **else**
10:       $X_{i,j} \leftarrow P_i + \alpha \overrightarrow{P_i P_j}$
11:     **end if**
12:   **end for**
13:   $X \leftarrow \arg\min_{X_{i,j}} \| X - X_{i,j} \|_2$
14: **end if**

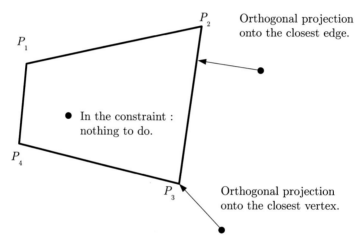

Fig. 2 To compute the orthogonal projection, different cases can appear. If the YUV color respects the constraint, the projection is the identity. Otherwise, the orthogonal projection onto the closest edge or vertex should be done

strokes on grayscale images. We review some work addressing a coupled total variation with a L2 data-fidelity term. Since this estimator is biased, we then review a debiasing strategy that can be applied on this last model.

State-of-the-Art of Color Diffusion

Some papers of the literature aim at helping the user to perform manual colorization. This is done by a diffusion of the colors over the grayscale image by various techniques. The diffusion approaches can also take inspiration from manual colorization to improve the results of other colorization approaches. In this section, we will describe the diffusion techniques proposed in the literature. This chapter is based on the papers (Pierre et al. 2014b, 2015a, 2017b) that are competitive methods of the literature. Let us remark that there does not exist a perfect diffusion method, all the state-of-the-art approaches having their advantages and drawbacks.

In order to perform manual colorization, a user manually adds color strokes. These are called *scribbles*, and they consist of a set of pixels for which the chrominance channels are defined. Many methods using this process have been proposed. For example, the method of Levin et al. (2004) solves an optimization problem for diffusing scribbles on the target image, assuming that the chrominance must have small variations when the luminance does vary a lot. Specifically, the following functional is minimized:

$$H(U) = \sum_r \left(U(r) - \sum_{r \sim s} w_{rs} U(s) \right)^2, \qquad (4)$$

where $r \sim s$ means that pixels r and s are neighbors and U is a chrominance channel (the same functional is minimized for the channel V). w_{rs} denotes the weights which can be either:

$$w_{rs} \propto e^{(Y(r)-Y(s))^2/2\sigma^2},$$

or:

$$w_{rs} \propto 1 + \frac{1}{\sigma_r^2}(Y(r) - \mu_r)(Y(s) - \mu_r),$$

where μ_r and σ_r denote the mean and the variance of the neighborhood of the pixel r. The two types of weights are more or less sensitive to the variation of contrast. The authors of Luan et al. (2007) include texture similarity in the model of Levin et al. (2004) to improve the diffusion process.

The authors of Yatziv and Sapiro (2006) have proposed a simple and fast method using geodesic distance to weight for each pixel the melting of the colors given by the scribbles. For each pixel of the grayscale image, the geodesic distance from the scribble is computed with respect to the gradient of the image. Next, a weighted average of the chrominances given by the scribbles is computed. The weights are computed from a function depending on the geodesic distance. This method enables a diffusion of the chrominance on constant parts of the image with respect to a function having similar properties as the inverse function:

- $\lim_{r \to 0} w(r) = \infty$;
- $\lim_{r \to \infty} w(r) = 0$;
- $\lim_{r \to \infty} w(r + r_0)/w(r) = 1$.

Yatziv et al. have proposed experimental results with the function $\frac{1}{r^b}$ with $1 \leq b \leq 6$.

The authors of Kawulok et al. (2012) have extended this method to textured images by introducing texture descriptors in the diffusion potential.

Some methods are designed as a propagation of the colors from neighbors to neighbors. Some colors are given by strokes drawn by the user. In this way, some of the image pixels are colored. The algorithm then propagates the color to their neighbors with a rule based on the values of the grayscale image. To this aim, the authors of Heu et al. (2009) give an explicit formula for melting the neighbor colors, whereas the ones of Lagodzinski and Smolka (2008) provide a modeling based on probabilistic distance transform, and the authors of Kim et al. (2010) use random walks.

It was also proposed to use diffusion through the regularization of non-local graphs. The method proposed by Lézoray et al. (2008) is based on the regularity of the image. This is modeled as a graph, each pixel being represented by a vertex and each neighborhood relationship by an edge. A local graph is considered, where each edge represents a relationship of eight neighborhoods. The weight of an edge being inversely proportional to the difference between gray levels, the minimization of an energy depending on these weights (see, e.g., Lézoray et al. 2007a) enables to diffuse the chrominances on the constant parts of the image. If a non-local graph is designed with a weight which depends on the distance between patches, a set of pixels is considered constant if the patches are similar. Thus, the color of the scribbles is diffused between pixels close in the graph, therefore belonging to similar textures.

Inspired by the PDE diffusion scheme (Perona and Malik 1990), some chrominance diffusion including a guidance with Di Zenzo tensor structure computed from grayscale image was proposed independently by Peter et al. (2017) and by Drew and Finlayson (2011).

The authors of Quang et al. (2010) have proposed a variational approach in chromaticity-brightness color space (see, e.g., Chan et al. 2001) to interpolate the missing colors. The *reproducing kernel Hilbert spaces* (RKHS) are used to compute a link between the chromaticity and brightness channels. Jin et al. (2016) introduced a variational model with the coupling of contour directions. Based on Mumford-Shah-type functional, the authors of Jung and Kang (2016) introduced a novel variational image colorization model. In the following, we present a recent state-of-the-art method based on total variation minimization. This approach enables to combine various strategies of the literature.

Coupled Total Variation for Image Colorization

In the following we focus on a variational model to denoise the chrominance channels of an image while keeping the luminance unchanged. Similarly to the colorization model of Pierre et al. (2015a), we want to find the minimizer $\hat{u}(c)$ of the denoising functional:

$$\hat{u}(c) = \arg\min\nolimits_{u=(U,V)} \mathrm{TV}_{\mathcal{C}}(u) + \lambda \int_{\Omega} \|u(x) - c(x)\|^2 \, dx + \chi_{\mathcal{R}}(u), \quad (5)$$

with

$$\mathrm{TV}_{\mathcal{C}}(u) = \int_{\Omega} \sqrt{\gamma \|\nabla Y(x)\|^2 + \|\nabla U(x)\|^2 + \|\nabla V(x)\|^2} \, dx, \quad (6)$$

where Y, U, and V are the luminance and chrominance channels. This term is a coupled total variation which enforces the chrominance channels to have a contour at the same location as the luminance ones. γ is a parameter which enforces the coupling of the channels. Some other total variation formulations have been proposed to couple the channels; see for instance Kang and March (2007) or Caselles et al. (2009).

The fidelity-data term is a classical L^2 norm between chrominance channels of the unknown u and the data c. For each pixel, the chrominance values live onto the convex polygon denoted by $\mathcal{R}$ and described in section "Range of Chrominance". This last assumption ensures that the final solution lies onto the RGB cube, avoiding the final truncation that leads to modification of the luminance channel. Model (5) is convex and it can be turned into a saddle-point problem of the form:

$$\min_{u \in \mathbb{R}^2} \max_{z \in \mathbb{R}^6} \frac{\lambda}{2} \|u - c\|^2 + \langle \nabla u | z_{1,\ldots,4} \rangle + \langle \gamma \nabla Y | z_{5,\ldots,6} \rangle - \chi_{B(0,1)}(z) + \chi_{\mathcal{R}}(u). \quad (7)$$

The primal-dual algorithm (Chambolle and Pock 2011) used to compute such saddle-point problem is recalled in Algorithm 2, where $P_{\mathcal{R}}$ is the orthogonal projection described in Algorithm 1 and $P_{\mathcal{B}}$ is defined as follows for one pixel:

$$P_{\mathcal{B}}(z) = \frac{(z_{1,\ldots,4}, z_{5,6} - \sigma \nabla Y)}{\max\left(1, \|z_{1,\ldots,4}, z_{5,6} - \sigma \nabla Y\|_2\right)}. \quad (8)$$

Algorithm 2 Minimization of (7)

1: $u^0 = c$
2: $z^0 \leftarrow \nabla u$
3: **for** $n \geq 0$ **do**
4: $\quad z^{n+1} \leftarrow P_{\mathcal{B}}\left(z^n + \sigma \nabla \bar{u}^n\right)$
5: $\quad u^{n+1} \leftarrow P_{\mathcal{R}}\left(\dfrac{u^n + \tau\left(\mathrm{div}(z^{n+1}) + \lambda c\right)}{1 + \tau\lambda}\right)$
6: $\quad \bar{u}^{n+1} \leftarrow 2u^{n+1} - u^n$
7: **end for**
8: set $\hat{u}(c) = u^{n+1}$ and $\hat{z} = z^{n+1}$.

The results produced by Algorithm 2 are promising, but with a low data parameter λ, they are drab (see, e.g., Pierre et al. 2017b).

Constrained TV-L2 Debiasing Algorithm

In this section we present a debiased algorithm for correcting the loss of colorfulness of the solution given by the optimum of (5).

The CLEAR Method (Deledalle et al. 2017)

The CLEAR method (Deledalle et al. 2017) can be applied for debiasing estimators $\hat{u}(c)$ obtained as:

$$\hat{u}(c) \in \arg\min_{u \in \mathbb{R}^p} F(u, c) + G(u), \tag{9}$$

where F is a convex data-fidelity term with respect to the data c and G is a convex regularizer. For G being the total variation regularization, the estimator $\hat{u}(c)$ is generally computed by an iterative algorithm, and it presents a loss of contrast with respect to the data c. In order to debias this estimator, the CLEAR method refits the data c with respect to some structural information contained in the biased estimator $\hat{u}$. This information is encoded by the Jacobian of the biased estimator with respect to the data c:

$$J_{\hat{u}}(c)d = \lim_{\varepsilon \to 0} \frac{\hat{u}(c + \varepsilon d) - \hat{u}(c)}{\varepsilon}. \tag{10}$$

For instance, when G is the anisotropic TV regularization, the Jacobian contains the information concerning the support of the solution $\hat{u}$, on which a projection of the data can be computed.

In general case, the CLEAR method relies on the *refitting estimator* $\mathcal{R}_{\hat{u}}(c)$ of the data c from the biased estimation $\hat{u}(c)$:

$$\mathcal{R}_{\hat{u}}(c) \in \arg\min_{h \in \mathcal{H}} \|h(c) - c\|_2^2 \tag{11}$$

where $\mathcal{H}$ is defined as the set of maps $h : \mathbb{R}^n \to \mathbb{R}^p$ satisfying $\forall c \in \mathbb{R}^n$:

$$h(c) = \hat{u}(c) + \rho J_{\hat{u}(c)}(c - \hat{u}(c)), \text{ with } \rho \in \mathbb{R}. \tag{12}$$

A closed formula for ρ can be given:

$$\rho = \begin{cases} \dfrac{\langle J_{\hat{u}(c)}(\delta) | \delta \rangle}{\|J_{\hat{u}(c)}(\delta)\|_2^2} & \text{if } J_{\hat{u}(c)}(\delta) \neq 0 \\ 1 & \text{otherwise.} \end{cases}, \tag{13}$$

where $\delta = c - \hat{u}(c)$. In practice, the global value ρ allows to recover most of the bias in the whole image domain.

An algorithm is then proposed in Deledalle et al. (2017) to compute the numerical value of $J_{\hat{u}(c)}(c - \hat{u}(c))$. The process is based on the differentiation of the algorithm providing $\hat{u}(c)$.

It is important to notice that the CLEAR method applies well for estimators obtained from the resolution of unconstrained minimization problems of the form (9). Nevertheless, it is not adapted to the denoising problem (5) that contains an additional constraint $\chi_\mathcal{R}(u)$ as CLEAR may violate the constraint.

Direct Extension of CLEAR to Constrained Problems

Extending the CLEAR method to the constrained model (5) requires to take the constraint into account in the axioms of the refitting model (11). The main difference with the original model is the addition of the constraint $\chi_\mathcal{R}(u)$. We can first notice that the refitting axioms $h(c) = Ac + b$ for some $A \in \mathbb{R}^{p \times n}, b \in \mathbb{R}^p$ and $J_h(c) = \rho J_{\hat{u}}(c)$ for some $\rho \in \mathbb{R}$ are in line with the introduction of the constraint. In particular, the definition of the Jacobian $J_{\hat{u}}$ in Equation (10) remains valid with the constraint, since $\hat{u}(c)$ and $\hat{u}(c+\varepsilon d)$ are still in the closed convex $\mathcal{R}$. The computation of the ρ parameter in Equation (13) may nevertheless produce, from Equation (12), an estimation out of the constraint that has to be post-processed. This points out the main difference between the constrained and the unconstrained debiased estimator.

In Deledalle et al. (2017), the value of ρ is computed from the minimization of a map from $\mathbb{R}$ to $\mathbb{R}$ defined as follows:

$$\rho \mapsto \left\| \left(I_d - \rho J_{\hat{u}(c)} \right) \left(\hat{u}(c) - c \right) \right\|_2^2. \tag{14}$$

In the case of the constrained problem, the function to be minimized is written as:

$$\rho \mapsto \| \hat{u}(c) + \rho J_{\hat{u}(c)} \left(c - \hat{u}(c) \right) - c \|_2^2 + \chi_\mathcal{R}(\hat{u}(c) + \rho J_{\hat{u}(c)} \left(c - \hat{u}(c) \right)). \tag{15}$$

Let us denote by ρ the value defined in Equation (13). In the case when the constraint is fulfilled, i.e., when $\hat{u}(c) + \rho J_{\hat{u}(c)} \left(c - \hat{u}(c) \right) \in \mathcal{R}$, then the minimum of (15) is reached with ρ.

If not, since function (15) is convex, it is possible to compute explicitly the minimizer. The value $\rho = 0$ is in the domain of the functional because $\hat{u}(c) \in \mathcal{R}$. The idea is to find the maximum value of ρ such that $\hat{u}(c) + \rho J_{\hat{u}(c)} \delta \in \mathcal{R}$. In this case, since $\mathcal{R}$ is a convex polygon, this computation can be done with a ray-tracing algorithm (Williams et al. 2005). To this aim, we can parametrize the segment $[\hat{u}(c), \hat{u}(c) + \rho J_{\hat{u}(c)} \delta]$:

$$\tilde{\rho} = \max_{t \in [0,1]} t\rho \text{ such that } \hat{u}(c) + t\rho J_{\hat{u}(c)} \left(c - \hat{u}(c) \right) \in \mathcal{R}. \tag{16}$$

Equation (16) can thus be directly solved by the maximum value t such that $\hat{u}(c) + t\rho J_{\hat{u}(c)} \left(c - \hat{u}(c) \right)$ intersects the border of $\mathcal{R}$.

Direct Debiasing Process

Let us summarize the refitting algorithm designed for model (5). The first step consists of computing a solution of (5) with Algorithm 2. This iterative algorithm provides at convergence a first biased solution $\hat{u}(c)$ and its dual variable $\hat{z}$. Once this solution has been computed, the differentiated algorithm, presented in Algorithm 3, is applied in the direction $\delta = c - \hat{u}(c)$. This algorithm requires the definition of the operator $\Pi_{\hat{z}}(\tilde{z})$ which is the linearized version of the projection $P_{\mathcal{B}}$ around $\hat{z}$ and which reads (Deledalle et al. 2017):

$$\Pi_{\hat{z}}(\tilde{z}) = \begin{cases} \tilde{z} & \text{if } \|\hat{z}\| < 1 \\ \dfrac{1}{\|\hat{z}\|}\left(\tilde{z} - \dfrac{\langle \hat{z}|\tilde{z}\rangle}{\|\hat{z}\|^2}\hat{z}\right) & \text{otherwise.} \end{cases} \tag{17}$$

Finally, the ray-tracing is applied to obtain $\tilde{\rho}$ and get the debiased solution as $\hat{u}(c) + \tilde{\rho} J_{\hat{u}(c)}(c - \hat{u}(c))$.

Algorithm 3 Differentiation of Algorithm 2 for computing $J_{\hat{u}(c)}\delta$ from $(\hat{u}(c), \hat{z})$

1: $\tilde{u}^0 = \delta, \overline{\tilde{u}}^0 = \delta$
2: $\tilde{z}^0 \leftarrow \nabla \tilde{u}$
3: **for** $n \geq 0$ **do**
4: $\quad \tilde{z}^{n+1} \leftarrow \Pi_{\hat{z}}\left(\tilde{z}^n + \sigma \nabla \overline{\tilde{u}}^n\right)$
5: $\quad \tilde{u}^{n+1} \leftarrow \dfrac{\tilde{u}^n + \tau \left(\text{div}(\tilde{z}^{n+1}) + \lambda\delta\right)}{1 + \tau\lambda}$
6: $\quad \overline{\tilde{u}}^{n+1} \leftarrow 2\tilde{u}^{n+1} - \tilde{u}^n$
7: **end for**
8: $J_{\hat{u}(c)}\delta = \tilde{u}^{n+1}$.

Unfortunately, this direct approach does not lead to valuable results on general cases. Indeed, if for one particular pixel the solution $\hat{u}(c)$ is saturated, and if the debiased solution is out of $\mathcal{R}$, then $\tilde{\rho} = 0$ is the unique global ρ satisfying $\hat{u}(c) + \rho J_{\hat{u}(c)}(c - \hat{u}(c)) \in \mathcal{R}$. Thus, the debiased solution is equal to the biased one, and the debiasing algorithm has no action.

In the next section, we propose a model with an adaptive ρ parameter, depending on the pixel, to tackle this saturated value issue.

Adaptive Debiasing Model for Constrained Problems

For a pixel ω such that $\hat{u}(c)_\omega + \rho J_{\hat{u}(c),\omega}(c_\omega - \hat{u}(c)_\omega)$ fulfills the constraint, ρ is the best value to refit the model according to the hypothesis of model (11). Here, $J_{\hat{u}(c),\omega}$ denotes the value of $J_{\hat{u}(c)}$ in pixel ω.

On the other hand, if for a pixel ω, the values of $\hat{u}(c)_\omega$ and $J_{\hat{u}(c),\omega}(c_\omega - \hat{u}(c)_\omega)$ are such that $\hat{u}(c)_\omega + \rho J_{\hat{u}(c),\omega}(c_\omega - \hat{u}(c)_\omega) \notin \mathcal{R}$, the ρ value has to be adapted. Thus, let us define for a pixel ω the adapted $\tilde{\rho}_\omega$ as follows:

$$\tilde{\rho}_\omega = \max_{t_\omega \in [0,1]} t_\omega \rho \text{ such that } \hat{u}(c)_\omega + t_\omega \rho J_{\hat{u}(c),\omega}\left(c_\omega - \hat{u}(c)_\omega\right) \in \mathcal{R}. \tag{18}$$

The constrained refitting model is then defined pixel-wise as:

$$R_{\hat{u}}^{\mathcal{R}}(c) = \hat{u}(c)_\omega + \tilde{\rho}_\omega J_{\hat{u}(c),\omega}\left(c_\omega - \hat{u}(c)_\omega\right) \tag{19}$$

This definition ensures that the debiased estimation fulfills the constraint. Moreover, if the debiasing method of Deledalle et al. (2017) produces an estimation that fulfills the constraint, this solution is retained. Notice however that the CLEAR hypothesis $J_h(c) = \rho J_{\hat{u}}(c)$ for some $\rho \in \mathbb{R}$ in model (11) is not fulfilled anymore. In numerical experiments, for most pixels, the values of $\tilde{\rho}_\omega$ computed with this method are the same as with Model (11).

As illustrated by Fig. 3, such a local debiasing strategy realizes an oblique projection onto $\mathcal{R}$ (Figs. 4 and 5).

Computation of the Oblique Projection

In Pierre et al. (2017b), an algorithm used to compute the oblique projection when the constraint is the chrominance set for a particular value of luminance (see, e.g., section "Range of Chrominance") is proposed. To simplify the notation, the problem is considered for a single pixel ω and one set $u := \hat{u}(c)_\omega$, $c := J_{\hat{u}(c),\omega}\left(c_\omega - \hat{u}(c)_\omega\right)$ and $\rho \in \mathbb{R}$ computed by the algorithm of Deledalle et al. (2017).

For $u + \rho c \notin \mathcal{R}$, the maximum value of $t \in [0, 1]$ such that $u + t\rho c \in \mathcal{R}$ is computed. Since $u \in \mathcal{R}$, thus if $u + \rho c \notin \mathcal{R}$, the segment $[u, u + \rho c]$ crosses one edge of the polygon.

One considers this problem by testing it into the RGB space. Indeed, the edges in the chrominance space correspond to edges in the RGB one, and the intersections between them correspond to intersections in the RGB space. In RGB, the problem of finding the intersection between an edge and the polygon is reduced to computing the intersection between the edge and the cube faces because the edges of the polygon are included in the cube by construction (see, e.g., Fig. 1a).

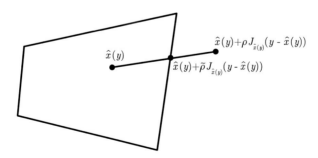

Fig. 3 The refitting of the method of Deledalle et al. (2017) may be out of the constraint. An oblique projection onto this constraint is able to respect most of hypotheses of Model (11) while fulfilling the constraint

The transformation of the chrominance values $u = (U, V)$ to the RGB space with the luminance value Y is denoted by $T_Y(u)$. From the expression of the standard transformation from RGB to YUV, we have $T_Y(u) = Y(1, 1, 1)^t + L(U, V)$ with L a linear function. The following equalities come:

$$\begin{aligned} T_Y(u + \rho c) &= Y(1, 1, 1)^t + L(u + \rho c) \\ &= Y(1, 1, 1)^t + L(u) + \rho L(c) \\ &= T_Y(u) + \rho T_Y(c) - \rho Y(1, 1, 1)^t. \end{aligned} \qquad (20)$$

It is required to compute $\tilde{\rho}$ such that $T_Y(u + \tilde{\rho} c)$ is at the boundary of the RGB cube. To this aim, the 6 different values $\tilde{\rho}_c^v$ with $c \in \{R, G, B\}$ and $v \in \{0, 255\}$ corresponding to the cases where the 3 coordinates of $T_Y(u + \tilde{\rho} c)$ are equal to 0 or 255 are computed. For instance, if the first coordinate R of $T_Y(u + \tilde{\rho} c)$ is equal to 255, we have:

$$T_Y(u + \tilde{\rho}_R^{255} c)_R = 255 \qquad (21)$$

$$T_Y(u)_1 + \tilde{\rho}_R^{255} T_Y(c)_R - \tilde{\rho}_R^{255} Y = 255. \qquad (22)$$

so that

$$\tilde{\rho}_R^{255} = \frac{255 - T_Y(u)_R}{T_Y(c)_R - Y}. \qquad (23)$$

For each of the six values $\tilde{\rho}_c^v$ computed as in Equation (23), one can compute $t_c^v = \frac{\tilde{\rho}_c^v}{\rho}$. The values t_c^v that are between 0 and 1 correspond to an intersection of the segment $[u, u + \rho c]$ with the boundaries of $\mathcal{R}$. One finally takes $t^* = \min_{t_c^v \in [0;1]} t_c^v$ and the result of Equation (18) is given by $t^* \rho$.

Figure 4 and 5 show some numerical results to compare Models (5) and (19). One can remark that Model (5) fits well the contours of images in comparison to the standard TVL2 model on chrominance channels. Moreover, the debiasing approach improves the colorfulness of the results in comparison with Model (5) and it has the advantage of well fitted contours.

To summarize, to design a suitable variational model for image colorization, the three main ingredients are the coupled total variation, the orthogonal projection onto the range of the problem, and the debiasing algorithm. This variational model is a basis for image colorization in many paradigms. In the next sections, some concrete cases of application of this model are presented in the case of exemplar-based approaches or coupled with manual techniques or CNN-based framework.

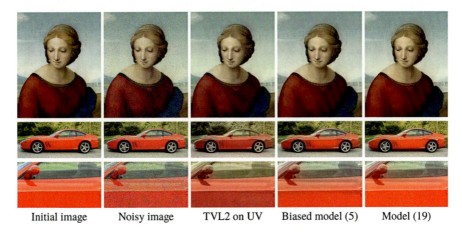

Fig. 4 Results of chrominance channels with a TV-L2 model on chrominance, with the biased method, and with the unbiased method. The debiasing algorithm produces more colorful results

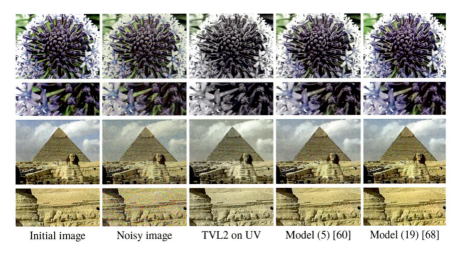

Fig. 5 The advantage of the coupled total variation (5) on the TV-L2 model has been shown in Pierre et al. (2015a). In Pierre et al. (2017b), it is refined in a better colorfulness-preserving model

Exemplar-Based Colorization

The manual methods enable the user to choose the color in each pixel of the image. Nevertheless, their main drawback is the tedious work needed for complex scenes, for instance with textures. In exemplar-based image colorization methods, the color information is provided by a color image called *source image*. The grayscale image to colorize is called *target image*. This color image can be chosen by the user or automatically provided from a database with an indexation algorithm.

The results available in this chapter are based on Pierre et al. (2014b,c, 2015a) which are among the most recent methods in patch-based colorization and on Persch et al. (2017) which is the current most competitive method for exemplar-based colorization of face images.

In order to transfer the colors from the source image to the target one, three concepts have been proposed in the literature. One of them is based on geometry, the two others are based on texture similarities. The first one is specifically well adapted to face colorization. In the first part of this section, we will review the work of Persch et al. (2017) which is the current most competitive method for exemplar-based colorization of face images. Next, we will present an overview of segmentation-based approaches which use the texture similarities on the segmented parts of the images to transfer colors. Finally, we present patch-based technique which avoids the requirement of an efficient segmentation method and which can be coupled with a variational model.

Morphing-Based Approach

In this section, we describe the model of Persch et al. (2017). The authors compute the morphing map between the two grayscale images I_{temp} and I_{tar} with a model inspired by Berkels et al. (2015). This results in the deformation sequence φ which produces the resulting map Φ from the template image to the target one. Due to the discretization of the images, the map Φ is defined, for images of size $n \times m$, on the discrete grid $\mathcal{G} := \{1 \ldots n\} \times \{1 \ldots m\}$:

$$\Phi : \mathcal{G} \to [1, n] \times [1, m], \quad x \mapsto \Phi(x), \tag{24}$$

where $\Phi(x)$ is the position in the source image which corresponds to the pixel $x \in \mathcal{G}$ in the target image. Now we colorize the target image by computing its chrominance channels, denoted by $(U_{\text{tar}}(x), V_{\text{tar}}(x))$ at position x as

$$\bigl(U_{\text{tar}}(x), V_{\text{tar}}(x)\bigr) := \bigl(U(\Phi(x)), V(\Phi(x))\bigr). \tag{25}$$

The chrominance channels of the target image are defined on the image grid $\mathcal{G}$, but usually $\Phi(x) \notin \mathcal{G}$. Therefore, the values of the chrominance channels at $\Phi(x)$ have to be computed by interpolation. In the algorithm, bilinear interpolation is simply used, which is defined for $\Phi(x) = (p, q)$ with $(p, q) \in [i, i+1] \times [j, j+1]$, $(i, j) \in \{1, \ldots, m-1\} \times \{1, \ldots, n-1\}$ by

$$U(\Phi(x)) = U(p, q)$$

$$:= (i+1-p, p-i) \begin{pmatrix} U(i, j) & U(i, j+1) \\ U(i+1, j) & U(i+1, j+1) \end{pmatrix} \begin{pmatrix} j+1-q \\ q-j \end{pmatrix}. \tag{26}$$

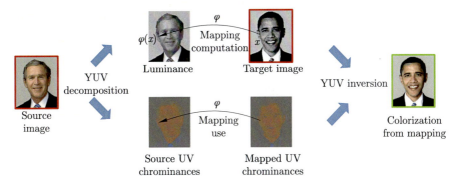

Fig. 6 Overview of the color transfer. The mapping φ is computed from a model inspired by Berkels et al. (2015) between the luminance channel of the source image and the target one. From this map, the chrominances of the source image are mapped. Finally, from these chrominances and the target image the colorization result is computed

Finally, a colorized RGB image is computed from its luminance $I_{\text{tar}} = Y_{\text{tar}}$ and the chrominance channels.

Figure 6 summarizes the color transfer method.

The technique proposed in Persch et al. (2017) is adapted to faces. To address the problem of colorization of textured images, geometric similarities are not reliable. Texture similarities have to be obviously compared. Such approaches are reviewed in the next sections.

Segmentation-Based Techniques

In order to transfer the colors from the source image to the target one, a lot of approaches are based on an image segmentation technique in order to compare the statistical attributes of the textures. For instance, the authors of Irony et al. (2005) proposed to compute the best correspondence between the target image and some segmented parts of the source image. From these correspondences, some *micro-scribbles* are drawn of the target image from the source image and the color strokes are then propagated by the diffusion technique in Levin et al. (2004). In Sỳkora et al. (2004), the author used a segmentation approach to colorize images of old cartoons. The method of Gupta et al. (2012) extracts various descriptors from superpixel segmentation (see, e.g., Ren and Malik 2003; Achanta et al. 2012) from target image and matches them with the ones of the target image with these various descriptors (SURF, mean, standard deviation, Gabor filters, etc.). The method hence draws one scribble for each superpixel from this matching. The final color is computed from the optimization of a criterion which favors a spatial consistency of the colors as done in Levin et al. (2004). A similar approach has been proposed in Kuzovkin et al. (2015).

The efficiency of these methods depends on the preliminary segmentation of the images. In the next section, we propose an overview of patch-based techniques which avoid this preliminary step.

Patch-Based Methods

The first patch-based method for image colorization is the one proposed by Welsh et al. (2002), which is widely inspired by the texture synthesis algorithm introduced by the authors of Efros and Leung (1999). It is based on the patch similarities in the colorization process.

First, a luminance remapping (see, e.g., Hertzmann et al. 2001) is done as a first step: in order to make the luminance values more comparable between the source image and the target one, an affine mapping is used on the luminance of the source image in order to better match the histogram of the luminance channel. Indeed, the range of the luminance channels could be different and the comparison of these channels could be senseless.

Next, for each pixel of the target image, the algorithm compare the patch centered in this pixel with a set of patches extracted from the luminance channel of the source image. Once the closest patch is computed, the chrominance values of the pixel at the center of the patch of the source image are extracted and provided to the considered pixel in the target image (see, e.g., Fig. 7). In combination with the luminance of the target image and the chrominance values extracted from the source image, a RGB color is given.

The set of reference patches extracted from the source image is a subset of patches randomly chosen in this way: the image is divided within a regular grid and one pixel is chosen randomly on each part of this grid (see, e.g., Fig. 7b).

The authors of Di Blasi and Reforgiato (2003) proposed an improvement which speeds up the patch research with a tree-clustering algorithm inspired from Wei and Levoy (2000). Next, the authors of Chen and Ye (2011) proposed an improvement based on a Bayesian approach.

The patch-based approaches suffer from two drawbacks, which are the difficulty choosing a reliable metric to compare the patches and the spatial coherency in the border of two areas with different textures. We will see in the following how to overcome these limitations.

The patch-based approaches need some metrics in order to compare the patches. Unfortunately, there does not exist any perfect metric, each of them having its advantages and drawbacks. In most computer vision problems, the algorithms have to distinguish objects or textures with the same accuracy and the same sensitivity as human visual system. Metrics for texture comparison are based on numerical data. The link between this data and the human visual system is done by features that are vectors which describe the local statistic of the image.

The most simple metrics are based on the mean or the standard deviation of the patches, whereas some others use histograms, Fourier transform, SURF features (Bay et al. 2006), structure tensors, co-variance matrices, Gabor features, etc.

15 Recent Approaches for Image Colorization

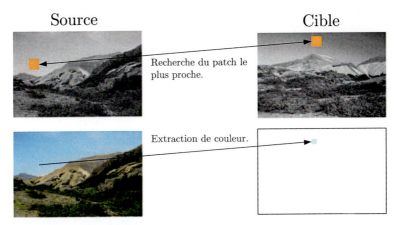

(a) Search of the candidates.

(b) Sub-sampling on a regular grid.

Fig. 7 For each pixel of the target image, the method compares the patch centered on the pixel with the ones of the gray-scale version of the source. Next, the method retains the color of the central pixel of the closest patch (see (**a**)). To speed up the algorithm, the search is not performed among all pixels, but only on a sub-sampling (see (**b**))

Based on various patch metrics, it is thus possible to get many exemplar-based colorization results. In the following, we focus on the fusion of such results to obtain only one final result.

Experimentally, the authors of Bugeau and Ta (2012) have used the following descriptors:

- The standard deviation on 5×5 and 3×3 patches
- The spectrum amplitude (FFT) on 7×7, 9×9, and 11×11 patches
- Difference in L^1 norm of the cumulative histograms on 7×7, 9×9, and 11×11 patches

Fig. 8 Some methods of the literature begin with the search of C candidates per pixel (here $C = 8$)

These descriptors are used by the authors of Bugeau and Ta (2012) to extract eight color candidates for each pixel in the same way as done in Welsh et al. (2002). For each metric, the method retains the pixel of the source image corresponding to the closest patch with respect to this metric. After this step, for each pixel of the target image, eight pixels of the source image can match. To summarize, each pixel having its luminance and eight chrominance values coming from the matched pixels (see, e.g., Fig. 8), eight colors are available, called *color candidates*. In the work of Bugeau and Ta (2012), the colors are used directly, whereas in Pierre et al. (2014a) an oblique projection in the RGB color space is proposed in order to avoid some artificial modification of the hue due to gamut problems.

Some other metrics could be used. For instance, whereas the method of Charpiat et al. (2008) is not based on patch decomposition, it uses a local representation with SURF descriptors to predict color in each pixel. Let us mention that this method also requires numerous and complex steps.

With multiple color candidates coming from various descriptors, a choice has to be done among them. In the following we will consider a generic number of color candidates denoted by C. The aim of the methods described hereinafter consists of the selection of one of the color candidates. Let us notice that the choice of an ideal metric based on metric learning has been proposed in Pierre et al. (2015b) but with rather worst results than the state of the art due to a lack of spatial regularization of the results. In order to retain only one color per pixel, the authors of Bugeau and Ta (2012) proposed to compute a median of the candidates based on an order between them computed with a standard PCA of the set of colors. This PCA is required because there is no natural order in the RGB space of colors. The method of Lézoray et al. (2005, 2007b) provides an order in the set of colors, but it requires some neighborhood information which is not available here.

Let us remark that the method of Bugeau and Ta (2012) does not use the spatial regularization or spatial coherency of the color to choose a color candidate. The authors of Jin et al. (2019) proposed an extension to exemplar-based colorization of Jung and Kang (2016) with color inference based on patch descriptors (DFT and variance of patches). A variational method similar to Pierre et al. (2015a) is proposed to regularize the final results.

A Variational Model for Image Colorization with Channel Coupling

In Pierre et al. (2015a), the authors have proposed a functional that selects a color among candidates extracted from a patch-based method, inspired by the method of Bugeau et al. (2014), in order to tackle some issues (numerical cost of numerical scheme, halo effects, etc.). Assume that C candidates are available in each pixel of a domain Ω and assume that two chrominance channels are available for each candidate. Let us denote for each pixel at position x the i-th candidate by $c_i(x)$, $u(x) = (U(x), V(x))$ stands for chrominances to compute, and $w(x) = \{w_i(x)\}$ with $i = 1, \ldots, C$ for the candidate weights. Let us minimize the following functional with respect to (u, w):

$$F(u, w) := TV_{\mathfrak{C}}(u) + \frac{\lambda}{2} \int_\Omega \sum_{i=1}^{C} w_i \|u(x) - c_i(x)\|_2^2 \, dx + \chi_{\mathcal{R}}(u(x)) + \chi_{\Delta}(w(x)). \tag{27}$$

The central part of this model is based on the term

$$\int_\Omega \sum_{i=1}^{C} w_i(x) \|u(x) - c_i(x)\|_2^2 \, dx. \tag{28}$$

This term is a weighted average of some L^2 norms with respect to the candidates c_i. The weights w_i can be seen as a probability distribution of the c_i. For instance, if $w_1 = 1$ and $w_i = 0$ for $2 \le i \le C$, the minimum of F with respect to u is equal to the minimization of

$$TV_{\mathfrak{E}}(u) + \frac{\lambda}{2} \int_{\Omega} \|u(x) - c_1(x)\|_2^2 \, dx + \chi_{\mathcal{R}}(u(x)). \tag{29}$$

To simplify the notations, the dependence of each value to the position x of the current pixel will be removed in the following. For instance, the second term of (27) will be denoted by $\int_{\Omega} \sum_{i=1}^{C} w_i \|u - c_i\|_2^2 \, dx$.

This model is a classical one with a fidelity-data term $\int_{\Omega} \sum_{i=1}^{C} w_i \|u - c_i\|_2^2$ and a regularization term $TV_{\mathfrak{E}}(u)$ defined in Equation (6). Since the first step of the method extracts many candidates, we propose averaging the fidelity-data term issued from each candidate. This average is weighted by w_i. Thus, the term

$$\int_{\Omega} \sum_{i=1}^{C} w_i \|u - c_i\|_2^2 \tag{30}$$

connects the candidate color c_i to the color u that will be retained. The minimum of this term with respect to u is reached when u is equal to the weighted average of candidates c_i.

Since the average is weighted by w_i, these weights are constrained to be onto the probability simplex. This constraint is formalized by $\chi_{\Delta}(w)$ whose value is 0 if $w \in \Delta$ and $+\infty$ otherwise, with Δ defined as:

$$\Delta := \left\{ (w_1, \cdots, w_C) \text{ s.t. } 0 \leq w_i \leq 1 \text{ and } \sum_{i=1}^{C} w_i = 1 \right\}. \tag{31}$$

In order to compute a suitable solution for the problem in (27), the authors of Pierre et al. (2015a) propose a primal-dual algorithm with alternating minimization of the terms depending of w. They also proposed numerical experiments showing the convergence of their algorithm. Let us note that this recent reference shows that the convergence of such numerical schemes can be demonstrated after smoothing of the total variation term. Among all the numerical schemes proposed in the references (Pierre et al. 2015a; Tan et al. 2019), we choose the methodology having the best convergence rate as well as a convergence proof. This scheme is given in Algorithm 2 in Tan et al. (2019). This algorithm is a block-coordinate forward-backward algorithm. To increase the speed-up of the convergence, Algorithm 2 of Tan et al. (2019) is initialized with the result of 500 iterations of the primal-dual algorithm of Pierre et al. (2015a). Whereas this algorithm has no guaranty of convergence, the authors of Tan et al. (2019) have experimentally observed that it numerically converges faster.

Unfortunately, the functional (27) is highly non-convex and it contains many critical points. More precisely, the functional is convex with respect to u with fixed w, and reversely, it is convex with respect to w for fixed u. Nevertheless, the functional is not convex with respect to the joint variables (u, w). Thus, even if the numerical scheme would converge to a local minimum, the solution of the problem depends on the initialization.

The dependence to the initialization implies an influence of the source image for exemplar-based colorization, and it does not enable a fully automatic image colorization within this paradigm. In the next section, we will show how the colorization from datasets can be used to tackle this last limitation.

Colorization from Dataset

The third colorization approach uses some large image databases (Zhang et al. 2016). Neural networks (convolutional neural networks, generative adversarial networks, autoencoder, recursive neural networks) have also been used successfully leading to a significant number of recent contributions. The survey proposed in this section is based on the paper (Mouzon et al. 2019). This literature can be divided into two categories of methods. The first evaluates the statistical distribution of colors for each pixel (Zhang et al. 2016; Royer et al. 2017; Chen et al. 2018). The network computes, for each pixel of the grayscale images, the probability distribution of the possible colors. The second takes a grayscale image as input and provides a color image as output, mostly in the form of chrominance channels (Iizuka et al. 2016; Larsson et al. 2016; Cao et al. 2017; Isola et al. 2017; Deshpande et al. 2017; Guadarrama et al. 2017; He et al. 2018; Su et al. 2018). Some methods use a mixture of both (e.g., Zhang et al. 2017).

Both techniques require image resizing that is either done by deconvolution layers or performed a posteriori with standard interpolation techniques.

In the case of Zhang et al. (2016), the network computes a probability distribution of the color on a down-sampled version of the original image. The choice of a color in each pixel at high resolution is made by linear interpolation without taking into account the grayscale image. Hence, the contours of chrominance and luminance may be not aligned, producing halo effects. Figure 9 shows some gray halo effects at the bottom of the cat that are visible on the red part, near the tail. On the other hand, in comparison to the other approaches of the state of the art, the method of Zhang et al. (2016) produces images which are shinier.

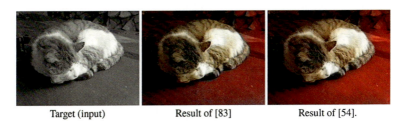

Fig. 9 Example of halo effects produced by the method of Zhang et al. (2016). Based on a variational model, the approach of Mouzon et al. (2019) is able to remove such artifacts

Below, the CNN described in Zhang et al. (2016) is presented in detail. The method of Zhang et al. (2016) is based on a discretization of the CIE Lab color space into C = 313 colors. This number of reference colors comes from the intersection gamut of the RGB color space and the discretization of the Lab space. The authors designed a CNN based on a VGG network (Simonyan and Zisserman 2015) in order to compute a statistical distribution of the C colors in each pixel. The input of the network is the L lightness channel of the Lab transform of an image of size 256 × 256. The output is a distribution of probability over a set of 313 couples of a, b chrominance values for each pixel of a 64 × 64 size image. The quantification of the color space in 313 colors is computed from two assumptions. First, the colors are regularly spaced onto the CIE Lab color space. On this color space two colors are close with respect to the Euclidean norm when the human visual system feels them close. The second assumption that rules the set of colors is the respect of the RGB gamut. The colors have to be displayable onto a standard screen.

To train this CNN, the database ImageNet (Deng et al. 2009) is used without the grayscale images. The images are resized at size 256×256 and then transformed into the CIE Lab color space. The images are then resized at size 64 × 64 to compute the a and b channels. The loss function used is the cross-entropy between the luminance (a, b) of the training image and the distribution over the 313 original colors. Let us denote by Δ the probability simplex in C = 313 dimensions.

Denoting by $(\hat{w}_i(x))_{i=1..C} \in \Delta^N$ the probability distribution of dimension C in the N pixels of the 64 × 64 image (over a domain Ω), and denoting by $(w_i(x))$ the ground truth distribution computed with a soft-encoding scheme (see Zhang et al. 2016 for details), the loss function is given by:

$$L(\hat{w}, w) = -\sum_{x \in \Omega} \sum_{i=1}^{C} w_i(x) \log(\hat{w}_i(x)). \tag{32}$$

The forward propagation in the network provides a probability distribution over the C colors. In order to compute a colorization result, a choice among all these colors has to be performed. Basically, the authors of Zhang et al. (2016) proposed an annealed mean in each pixel, independently. After that, a resizing of the (a, b) channels at original size is done and recombined with brightness channel to obtain the color image.

Nevertheless, this recombination is done without taking into account any spatial consideration. In the next section, we will describe some approaches that couple some previously described algorithms.

Coupled Approaches

Neither the exemplar-based methods, nor the manual techniques, nor the deep learning approaches are able to colorize images without some defects. All of them having drawbacks or advantages, we propose to describe some coupling approaches that rely on different types of methods in the literature. First, a framework to couple exemplar-based approach and manual colorization is described. A coupling of variational method with deep learning is then recalled.

Coupling Manual Approach with Exemplar-Based Colorization

A method can be considered interactive when the user can influence the result of the colorization process. Nevertheless, the interactivity can be difficult to reach. Indeed, if a method computes a result with a too long delay, the user cannot stand to an intermediary result in order to see the influence of his intervention. The results and the survey proposed in this section are based on the papers (Pierre et al. 2014b, 2015a) which have led to a software (Pierre et al. 2016).

Some of the exemplar-based methods enable some interaction with the user, for instance, the *swatches* approach of Welsh et al. (2002) in which the user distinguishes some parts of the image by drawing some rectangles on the source and target images where the textures are similar. The method then colorizes some parts of the target image with the specified parts of the source image. Finally, the method computes a solution for all the remaining uncolored pixels of the image based on the already colorized parts. The advantage of this framework is that the user can easily distinguish or associate the textures of the different images, which is difficult with an automatic method. At the opposite, the exemplar-based method is reliable to well colorize an image from its own parts, because the textures are more similar. With this method, a contextual information is added.

The framework of Chia et al. (2011) exploits the huge quantity of data available on the Internet. Nevertheless, the user has to manually segment and label the objects of the target image. Next, for each labeled object, the images with the same label are found on the Internet and used as source images. The image research is based on superpixel extraction (Comaniciu and Meer 2002) as well as graph-based optimization.

In the work of Ding et al. (2012), the scribbles are automatically generated and the user is invited to associate a color to each scribble. Then, the phases of the wavelet in the quaternion space are computed in order to propagate the colors along the lines of equal phase. Indeed, the wavelets in quaternion space are a measure of contours.

The method proposed in Pierre et al. (2014b) consists of a combination of the method of Bugeau and Ta (2012) and the one of Yatziv and Sapiro (2006). The approach uses a GPU implementation to compute a solution of model (27) that enables to colorize an image of size 370×600 in approximately 1 s. This computation time enables an extension of the exemplar-based approach of Pierre et al. (2014c) by including an interaction with the user, which leads to a software for colorization (Pierre et al. 2016).

The scribbles can be given in advance or added step by step by the user. When a source image is added, the first step consists of the extraction of C candidates as in section "Patch-Based Methods" and the corresponding weights are initialed with the value $w = 1/C$.

The information given by the scribbles influences the weights and the candidate number. More precisely, for each pixel of the image, a new candidate is added for each scribble. When a candidate is introduced, its weight is initialized for the minimization process with a value depending on the geodesic distance in a similar way as Yatziv and Sapiro (2006).

The geodesic distance, denoted by D, is computed with the fast marching algorithm (Sethian 1999) with a potential equal to $\left(0.001 + \|\nabla u\|_2^2\right)^{-4}$ given by Chan and Vese (2001). D is normalized to get values between 0 and 1. The implementation of Peyré (2008) can be used to compute it.

The pixels having a low geodesic distance from a scribble get its color, whereas those having a high geodesic distance are not influenced by the user intervention. The w variable is composed of concatenation of uniform weights for the color candidates coming from the source image with the patch extraction and the weight coming from the geodesic distance. The values are then projected onto the probability simplex Δ with the algorithm of Chen and Ye (2011). The u variable is initialized with $\sum_i w_i c_i$ and the functional (27) is minimized using this initialization.

In Fig. 10, we show a first example of colorization using both manual and exemplar-based approaches. Figure 10a and b shows the source and target images. Figure 10c corresponds to exemplar-based colorization done without manual scribble. In this first result, the sky is not suitably colorized since it appears brown instead of blue, as well as the door in ruins. Moreover, some blue blotches appear on the floor. Figure 10d shows the corrections done by the user by adding three scribbles on the exemplar-based result (Fig. 10c). Figure 10e illustrates the advantage of the combination of the methods. Indeed, the work provided by the user is of lower quality than the full manual colorization. It also shows that Model (27) is able to enhance contours.

Figure 11 shows the additional results and illustrate the advantage of using the joint model instead of using only the source image (fourth column) or only scribbles (fifth column). Colorization results in the last column in Fig. 11 are visually better than the ones computed from only one information source. This experiment shows

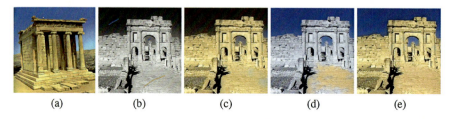

Fig. 10 Colorization using manual and exemplar-based approach. (**a**) Source image. (**b**) Target image with three scribbles. (**c**) Exemplar-based colorization. (**d**) Manual colorization. (**e**) Both

15 Recent Approaches for Image Colorization

Source.　　　Scribbles.　　Exemplar-based.　　Manual.　　　Joint.

Fig. 11 Advantage of the joint approach, compared to manual and exemplar-based colorization. From left to right: source, target with scribbles added by the user, exemplar-based result, scribble-based results, and finally the joint approach

also that old photographies and faces are difficult to colorize with exemplar-based approaches since they require more scribbles. This statement has been done in Chen et al. (2004). Indeed, old pictures contain a lot of noise and textures. Face image contains smooth parts, for instance skin or background, with no textures. This kind of images is hard to colorize with assumption of texture similarities. Nevertheless, it is possible to compute a suitable result with the joint method, as well as morphing-based approach presented in section "Morphing-Based Approach". Let us remark that the scribbles given by the user have naturally a local influence, but this influence can be also considered as global. For instance, on the last row in Fig. 11, the blue scribble in the arch also improves the color of the sky in the left-hand part of the image.

Coupling CNN with a Variational Approach

In the following, we recall the results given in the paper Mouzon et al. (2019) which consists of a coupling between a variational approach and the output of the CNN of Zhang et al. (2016). Next, we perform numerical comparisons with the original CNN approach of Zhang et al. (2016).

Coupling the CNN with a Variational Method

In image colorization, convolutional neural networks can be used to compute in each pixel a set of possible colors and their associated probabilities (Zhang et al. 2016). However, since the final choice is made without taking into account the regularity of the image, this leads to halo effects. To improve this, we first propose to adapt the functional of Pierre et al. (2015a) to the regularization of such results within the framework of colorization. The method of Pierre et al. (2015a) being able to choose between several color candidates in each pixel, it will be quite easy to use the color distribution provided by the CNN described in Zhang et al. (2016). In addition, the numerical results of Pierre et al. (2015a) demonstrate the ability to remove halos, which is relevant to the limitations of Zhang et al. (2016). This functional will have to face two main problems: on the one hand, the transition from a low to a high resolution and, on the other hand, the maintenance of a higher saturation than the current methods.

In this section, a method to couple the prediction power of CNN with the precision of variational methods is described. To this aim, let us remark that the variable w of the functional (27) represents the ratio of each color candidate which is represented in the final result. This comes from the fact that, for a given vector $w \in \mathbb{R}^C$, the minimum of

$$\sum_{i=1}^{C} w_i \|u - c_i\| \qquad (33)$$

with respect to u is given by

$$\sum_{i=1}^{C} w_i c_i. \qquad (34)$$

Thus, it can be seen as a probability distribution of the colors in the desired color image, which has exactly the same purpose as the one of the CNN in Zhang et al. (2016).

Figure 12 shows an overview of Mouzon et al. (2019). First, the grayscale image, considered as the luminance L, is given as an input to the CNN. The output of the CNN is a probability distribution over 313 possible chromaticity at low resolution (64 × 64). In order to initialize the minimization algorithm, the output weights of the CNN can be used. The CNN provides a coarse-scale output that needs an up-sampling before producing a suitable output at original definition. Two ways can be considered. For the first one, the variational method can be used at coarse scale (low definition), and then an interpolation can be performed to recover a result at fine scale (high definition). For the second one, the probability distributions can be interpolated to get a high-definition array. In the following, the second approach will be preferred. Indeed, the interpolation of a color image produces a decrease of the saturation that makes the images drabber. By interpolating the probability distributions instead of the color images, the variational method will be able to compute a color for each pixel based on a coupling of the channels at high resolution. The given probability distribution is then used as the initialization value for the numerical scheme. As it was still proposed in Pierre et al. (2015a), the variable u is initialized with $\sum_{i=1}^{C} w_i c_i$. After the iterations of the functional, the result, denoted by (u^*, w^*), provides some binary weights (see, e.g., Pierre et al. (2015a), Section 2.3.2) and a regularized result u^* that gives two chromaticity

15 Recent Approaches for Image Colorization

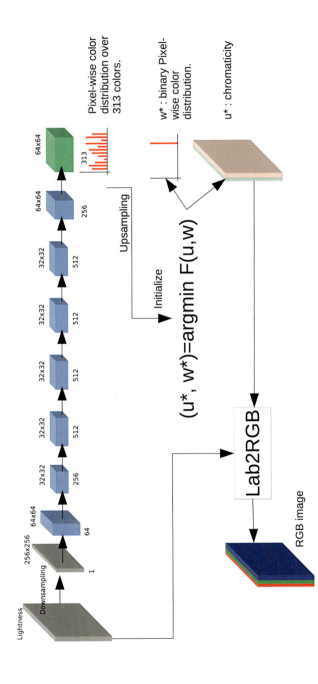

Fig. 12 Overview of method (Mouzon et al. 2019). A CNN computes color distribution on each pixel. A variational method selects then a color for each pixel based on a regularity hypothesis

channels, *a* and *b*, at initial definition. Recombined with the luminance L and transformed into the RGB space, that produces a color image.

Let us remark that the authors of Zhang et al. (2016) proposed to first produce the color image and then to resize it with bi-cubic interpolation. Unfortunately, upsampling or down-sampling images with bi-linear or bi-cubic interpolations reduces the saturation of the colors and makes them drabber than the original. To avoid that, we propose here the opposite approach: we first up-sample the color distribution, and then we compute a color image at full definition by using it. Since the numerical scheme is used at full definition, the required memory of the algorithm for all the weights and the colors is a limitation to process high-resolution images on a standard PC. To tackle this issue, we propose to select some of the 313 colors. This selection is done with respect to the probability distribution of the colors, by choosing the ten highest modes.

This choice of ten modes has been done experimentally. For most images, eight or nine candidates are enough and taking more of them does not improve the result, but it increases the computational time. On the other hand, taking less candidates decreases the quality of the result on a significant number of images. Finally, the number of ten is a fair trade-off.

The training step of the CNN is done as in Zhang et al. (2016). The variational step is not taken into account during the training process. Indeed, the relation between the initialization of the weights and the result is not analytically described and the gradient back-propagation algorithms are not suitable for this problem. Thus, the training is done by feeding the CNN with a grayscale image as input and a color distribution as output. The variational step remains independent of the full framework during the training step. Its integration will be the purpose of future works.

In the next section, numerical results are presented.

Numerical Results

In this section we show a qualitative comparison between Zhang et al. (2016) and the framework of Mouzon et al. (2019). A lot of results provided by Zhang et al. (2016) are accurate and reliable. We show on these examples that the method of Mouzon et al. (2019) does not reduce the quality of the images. We then propose some comparisons with erroneous results of Zhang et al. (2016), which shows that the method of Mouzon et al. (2019) is reliable to fully automatically colorize images without artifacts and halo effects. A time comparison between the CNN inference computation and the variational step will be proposed to show that the regularization of the result is not a burden on the CNN approach. Finally, to show the limitation of CNN in image colorization, we show some results where neither the approach of Zhang et al. (2016) nor the framework of Mouzon et al. (2019) is able to produce some reliable results.

Figure 13 shows the colorization results of the method of Zhang et al. (2016). Whereas it is hard to see that the method of Mouzon et al. (2019) produces a

15 Recent Approaches for Image Colorization

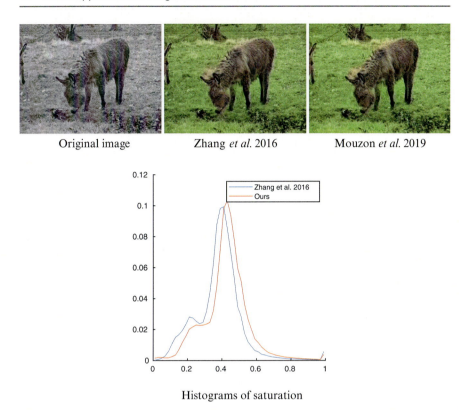

Fig. 13 Results of Zhang et al. (2016) compared with Mouzon et al. (2019). The histogram of the saturation shows the second result is shinier than the first one. Indeed, the average value of the saturation is higher for the model of Mouzon et al. (2019) (0.4228) than the one of Zhang et al. (2016) (0.3802). (**a**) Original image. (**b**) Zhang et al. 2016. (**c**) Mouzon et al. 2019. (**d**) Histograms of saturation

shinier result than the result of Zhang et al. (2016) unless being a calibration expert, the histogram of the saturation is able to show the improvement. Indeed, since the histogram is right-shifted, it means that globally, the saturation is higher on the result of Mouzon et al. (2019). Quantitatively, the average of the saturation is equal to 0.4228 for the method of Mouzon et al. (2019), while it is equal to 0.3802 for the method of Zhang et al. (2016). This improvement comes from the fact that the method of Mouzon et al. (2019) selects one color among the ones given by the results of the CNN, whereas the method of Zhang et al. (2016) computes the annealed mean of them. The mean of the chrominances of the colors produces a decrease of the saturation and makes the colors drabber. By using a selection algorithm based on the image regularization, the method of Mouzon et al. (2019) is able to avoid this drawback.

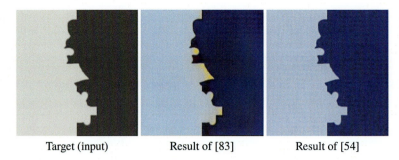

Fig. 14 Comparison of Mouzon et al. (2019) with Zhang et al. (2016). This example provides a proof of concept. The method of Mouzon et al. (2019) is able to remove the halo effects on the colorization result of Zhang et al. (2016)

The result in Fig. 14 is a proof of concept for the proposed framework. We can see a toy example which is automatically colorized by the method of Zhang et al. (2016). The result given by the method of Zhang et al. (2016) produces some halo effect near the only contour of the image, which is unnatural. The regularization of the result is able to remove this halo effect and to recover an image looking less artificial. This toy example contains only two constant parts. The aim of the variational method is to couple the contours of the chrominance channels and the ones of the luminance. The result produced with the method of Mouzon et al. (2019) contains no halo effect, showing the benefits of their framework.

In Fig. 15, we review some results and we compare them to the method of Zhang et al. (2016). For the lion, first line, a misalignment of the colors with the grayscale image is visible (a part of the lion is colorized in blue and a part of the sky is brown beige). This is a typical case of halo effect where the framework of Mouzon et al. (2019) is able to remove the artifacts. For the image of mountaineer, on the result of Zhang et al. (2016) some pink stains appear. With the method of Mouzon et al. (2019), the minimization of the total variation ensures the regularity of the image, and thus it removes these strains.

Figure 16 shows additional results. The first line is an old port card. Its colorization is reliable with the CNN, and, in addition, the variational approach makes it a little bit shinier. This example shows the ability of the approach of Mouzon et al. (2019) to colorize historical images. In the second example, most of the image is well colorized by the original method of Zhang et al. (2016). Nevertheless, the lighthouse and the right-side building contain some orange halos that are not reliable. With the variational method, the colors are convincing. Additional results are available in http://www.fabienpierre.fr/colorization.

The computational time of the CNN forward pass is about 1.5 s in GPU, whereas the minimization of the variational model (27) is about 15 s in Matlab on CPU.

15 Recent Approaches for Image Colorization

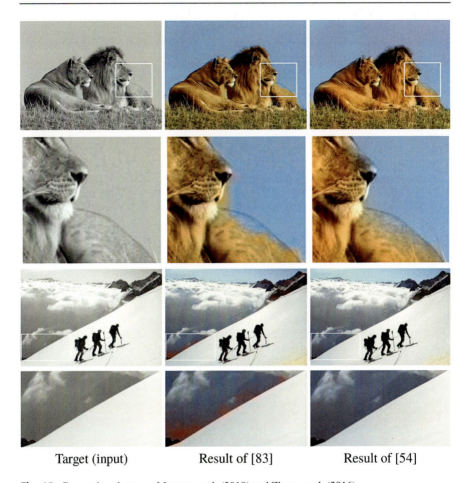

Fig. 15 Comparison between Mouzon et al. (2019) and Zhang et al. (2016)

In Pierre et al. (2017a), the authors provide a computation time almost equal to 1 s with unoptimized GPU implementation. Since the minimization scheme of Tan et al. (2019) is approximately the same, the computational time would be almost equal. Thus, the computational time of the approach of Mouzon et al. (2019) is not a burden in comparison with the method of Zhang et al. (2016).

In Fig. 17, a failure case is shown. In this case, since the minimization of the variational model strongly depends on its initialization, the method of Mouzon et al. (2019) is not able to recover its realistic colors. Actually, fully automatic colorization remains an open problem.

Fig. 16 Additional comparisons of Mouzon et al. (2019) with Zhang et al. (2016)

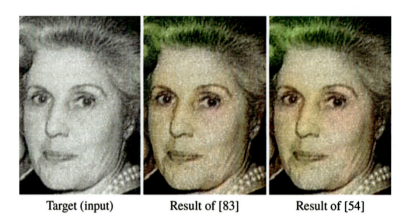

Fig. 17 Fail case. The prediction of the CNN is not able to recover a reliable color

Conclusion and Future Works

In this chapter, we have shown that image colorization has known a huge progress during the last 10 years by introducing a wide number of methods and approaches. Some extensions of these techniques have been proposed for video colorization but with limited number of frames. Future works could consider this application with more success. In this work, we have shown some limitations to colorization which let the topic open for active research. Joint approaches have shown their efficiency, and a combination of deep leaning with manual approaches could enhance the human system interface for image and video colorization.

Acknowledgments This study has been carried out with financial support from the French Research Agency through the PostProdLEAP project (ANR-19-CE23-0027-01).

References

Abidi, B.R., Zheng, Y., Gribok, A.V., Abidi, M.A.: Improving weapon detection in single energy x-ray images through pseudocoloring. IEEE Trans. Syst. Man Cybern. Part C **36**(6), 784–796 (2006)

Achanta, R., Shaji, A., Smith, K., Lucchi, A., Fua, P., Süsstrunk, S.: Slic superpixels compared to state-of-the-art superpixel methods. IEEE Trans. Pattern Anal. Mach. Intell. **34**(11), 2274–2282 (2012)

Bay, H., Tuytelaars, T., Van Gool, L.: Surf: speeded up robust features. In: European Conference on Computer Vision, pp. 404–417. Springer (2006)

Berkels, B., Effland, A., Rumpf, M.: Time discrete geodesic paths in the space of images. SIAM J. Imaging Sci. **8**(3), 1457–1488 (2015)

Bugeau, A., Ta, V.T.: Patch-based image colorization. In: IEEE International Conference on Pattern Recognition, pp. 3058–3061 (2012)

Bugeau, A., Ta, V.T., Papadakis, N.: Variational exemplar-based image colorization. IEEE Trans. Image Proces. **23**(1), 298–307 (2014)

Cao, Y., Zhou, Z., Zhang, W., Yu, Y.: Unsupervised diverse colorization via generative adversarial networks. In: Joint European Conference on Machine Learning and Knowledge Discovery in Databases, pp. 151–166. Springer (2017)

Caselles, V., Facciolo, G., Meinhardt, E.: Anisotropic cheeger sets and applications. SIAM J. Imaging Sci. **2**(4), 1211–1254 (2009)

Chambolle, A., Pock, T.: A first-order primal-dual algorithm for convex problems with applications to imaging. J. Math. Imaging Vis. **40**(1), 120–145 (2011)

Chan, T.F., Vese, L.A.: Active contours without edges. IEEE Trans. Image Proces. **10**(2), 266–277 (2001)

Chan, T.F., Kang, S.H., Shen, J.: Total variation denoising and enhancement of color images based on the cb and hsv color models. J. Vis. Commun. Image Represent. **12**(4), 422–435 (2001)

Charpiat, G., Hofmann, M., Schölkopf, B.: Automatic image colorization via multimodal predictions. In: European Conference on Computer Vision, pp. 126–139. Springer (2008)

Chen, Y., Ye, X.: Projection onto a simplex. arXiv preprint arXiv:1101.6081 (2011)

Chen, T., Wang, Y., Schillings, V., Meinel, C.: Grayscale image matting and colorization. In: Asian Conference on Computer Vision, pp. 1164–1169 (2004)

Chen, Y., Luo, Y., Ding, Y., Yu, B.: Automatic colorization of images from chinese black and white films based on cnn. In: 2018 IEEE International Conference on Audio, Language and Image Processing, pp. 97–102 (2018)

Chia, A.Y.S., Zhuo, S., Kumar, R.G., Tai, Y.W., Cho, S.Y., Tan, P., Lin, S.: Semantic colorization with internet images. In: ACM SIGGRAPH ASIA (2011)

Comaniciu, D., Meer, P.: Mean shift: a robust approach toward feature space analysis. IEEE Trans. Pattern Anal. Mach. Intell. **24**(5), 603–619 (2002)

Cui, M., Hu, J., Razdan, A., Wonka, P.: Color-to-gray conversion using isomap. Vis. Comput. **26**(11), 1349–1360 (2010)

Deledalle, C.A., Papadakis, N., Salmon, J., Vaiter, S.: Clear: covariant least-square re-fitting with applications to image restoration. SIAM J. Imaging Sci. **10**(1), 243–284 (2017)

Deng, J., Dong, W., Socher, R., Li, L.J., Li, K., Fei-Fei, L.: Imagenet: a large-scale hierarchical image database. In: IEEE Conference on Computer Vision and Pattern Recognition, pp. 248–255 (2009)

Deshpande, A., Lu, J., Yeh, M.C., Chong, M.J., Forsyth, D.A.: Learning diverse image colorization. In: IEEE Conference on Computer Vision and Pattern Recognition, pp. 2877–2885 (2017)

Di Blasi, G., Reforgiato, D.: Fast colorization of gray images. Eurographics Italian (2003). http://citeseerx.ist.psu.edu/viewdoc/download?doi=10.1.1.99.6839&rep=rep1&type=pdf

Ding, X., Xu, Y., Deng, L., Yang, X.: Colorization using quaternion algebra with automatic scribble generation. In: Advances in Multimedia Modeling (2012)

Drew, M.S., Finlayson, G.D.: Improvement of colorization realism via the structure tensor. Int. J. Image Graph. **11**(04), 589–609 (2011)

Efros, A.A., Leung, T.K.: Texture synthesis by non-parametric sampling. In: IEEE International Conference on Computer Vision, vol. 2, pp. 1033–1038 (1999)

Fitschen, J.H., Nikolova, M., Pierre, F., Steidl, G.: A variational model for color assignment. In: Scale Space and Variational Methods in Computer Vision, pp. 437–448 (2015)

Fornasier, M.: Nonlinear projection recovery in digital inpainting for color image restoration. J. Math. Imaging Vis. **24**(3), 359–373 (2006)

Gonzalez, R.C., Woods, R.E.: Digital Image Processing, 3rd edn. Upper Saddle River, Pearson (2008)

Guadarrama, S., Dahl, R., Bieber, D., Shlens, J., Norouzi, M., Murphy, K.: Pixcolor: pixel recursive colorization. In: British Machine Vision Conference (2017)

Gupta, R.K., Chia, A.Y.S., Rajan, D., Ng, E.S., Zhiyong, H.: Image colorization using similar images. In: ACM International Conference on Multimedia, pp. 369–378 (2012)

He, M., Chen, D., Liao, J., Sander, P.V., Yuan, L.: Deep exemplar-based colorization. ACM Trans. Graph. **37**(4), 47:1–47:16 (2018)

Hertzmann, A., Jacobs, C.E., Oliver, N., Curless, B., Salesin, D.H.: Image analogies. In: ACM Computer Graphics and Interactive Techniques, pp. 327–340 (2001)

Heu, J.H., Hyun, D.Y., Kim, C.S., Lee, S.U.: Image and video colorization based on prioritized source propagation. In: IEEE International Conference on Image Processing, pp. 465–468 (2009)

Iizuka, S., Simo-Serra, E., Ishikawa, H.: Let there be color!: joint end-to-end learning of global and local image priors for automatic image colorization with simultaneous classification. ACM Trans. Graph. **35**(4), 1–11 (2016)

Irony, R., Cohen-Or, D., Lischinski, D.: Colorization by example. In: Eurographics Symposium on Rendering, vol. 2. Citeseer (2005)

Isola, P., Zhu, J.Y., Zhou, T., Efros, A.A.: Image-to-image translation with conditional adversarial networks. In: IEEE Conference on Computer Vision and Pattern Recognition (2017)

Jin, Z., Zhou, C., Ng, M.K.: A coupled total variation model with curvature driven for image colorization. Inverse Prob. Imaging **10**(1930–8337), 1037 (2016). https://doi.org/10.3934/ipi.2016031

Jin, Z., Min, L., Ng, M.K., Zheng, M.: Image colorization by fusion of color transfers based on DFT and variance features. Comput. Math. Appl. **77**, 2553–2567 (2019)

Jung, M., Kang, M.: Variational image colorization models using higher-order mumford–shah regularizers. J. Sci. Comput **68**(2), 864–888 (2016). https://doi.org/10.1007/s10915-015-0162-9

Kang, S.H., March, R.: Variational models for image colorization via chromaticity and brightness decomposition. IEEE Trans. Image Proces. **16**(9), 2251–2261 (2007)

Kawulok, M., Kawulok, J., Smolka, B.: Discriminative textural features for image and video colorization. IEICE Trans. Inf. Syst. **95-D**(7), 1722–1730 (2012)

Kim, T.H., Lee, K.M., Lee, S.U.: Edge-preserving colorization using data-driven random walks with restart. In: IEEE International Conference on Image Processing, pp. 1661–1664 (2010)

Kuhn, G.R., Oliveira, M.M., Fernandes, L.A.: An improved contrast enhancing approach for color-to-grayscale mappings. Vis. Comput. **24**(7–9), 505–514 (2008)

Kuzovkin, D., Chamaret, C., Pouli, T.: Descriptor-based image colorization and regularization. In: Computational Color Imaging, pp. 59–68. Springer, Cham (2015)

Lagodzinski, P., Smolka, B.: Digital image colorization based on probabilistic distance transformation. In: 50th International Symposium ELMAR, vol. 2, pp. 495–498 (2008)

Lannaud, C.: Fallait-il coloriser la guerre? L'express (2009). Disponible en ligne sur http://www.lexpress.fr/culture/tele/fallait-il-coloriser-la-guerre_789380.html

Larsson, G., Maire, M., Shakhnarovich, G.: Learning representations for automatic colorization. In: European Conference on Computer Vision, pp. 1–16. Springer (2016)

Levin, A., Lischinski, D., Weiss, Y.: Colorization using optimization. In: ACM Transactions on Graphics, vol. 23–3, pp. 689–694 (2004)

Lézoray, O., Meurie, C., Elmoataz, A.: A graph approach to color mathematical morphology. In: IEEE International Symposium on Signal Processing and Information Technology, pp. 856–861 (2005)

Lézoray, O., Elmoataz, A., Bougleux, S.: Graph regularization for color image processing. Comput. Vis. Image Underst. **107**(1), 38–55 (2007a)

Lézoray, O., Elmoataz, A., Meurie, C.: Mathematical morphology in any color space. In: IAPR/IEEE International Conference on Image Analysis and Processing, Computational Color Imaging Workshop (2007b)

Lézoray, O., Ta, V.T., Elmoataz, A.: Nonlocal graph regularization for image colorization. In: IEEE International Conference on Pattern Recognition, pp. 1–4 (2008)

Luan, Q., Wen, F., Cohen-Or, D., Liang, L., Xu, Y.Q., Shum, H.Y.: Natural image colorization. In: Proceedings of the 18th Eurographics Conference on Rendering Techniques, EGSR'07, pp. 309–320. Eurographics Association, Aire-la-Ville (2007). https://doi.org/10.2312/EGWR/EGSR07/309-320

Mouzon, T., Pierre, F., Berger, M.O.: Joint CNN and variational model for fully-automatic image colorization. In: SSVM 2019 – Seventh International Conference on Scale Space and Variational Methods in Computer Vision, Hofgeismar (2019). https://hal.archives-ouvertes.fr/hal-02059820

Nikolova, M., Steidl, G.: Fast hue and range preserving histogram specification: theory and new algorithms for color image enhancement. IEEE Trans. Image Proces. **23**(9), 4087–4100 (2014)

Perona, P., Malik, J.: Scale-space and edge detection using anisotropic diffusion. IEEE Trans. Pattern Anal. Mach. Intell. **12**(7), 629–639 (1990)

Persch, J., Pierre, F., Steidl, G.: Exemplar-based face colorization using image morphing. J. Imaging **3**(4), 48 (2017)

Peter, P., Kaufhold, L., Weickert, J.: Turning diffusion-based image colorization into efficient color compression. IEEE Trans. Image Proces. **26**(2), 860–869 (2017)

Peyré, G.: Toolbox fast marching – a toolbox for fast marching and level sets computations (2008). http://www.mathworks.com/matlabcentral/fileexchange/loadFile.do?objectId=6110&objectType=FILE

Pierre, F., Aujol, J.F., Bugeau, A., Ta, V.T.: Hue constrained image colorization in the RGB space. Preprint (2014a). Disponible en ligne sur https://hal.archives-ouvertes.fr/hal-00995724/document

Pierre, F., Aujol, J.F., Bugeau, A., Ta, V.T.: A unified model for image colorization. In: Color and Photometry in Computer Vision (ECCV Workshop), pp. 1–12 (2014b)

Pierre, F., Aujol, J.F., Bugeau, A., Ta, V.T., Papadakis, N.: Exemplar-based colorization in RGB color space. In: IEEE International Conference on Image Processing, pp. 1–5 (2014c)

Pierre, F., Aujol, J.F., Bugeau, A., Papadakis, N., Ta, V.T.: Luminance-chrominance model for image colorization. SIAM J. Imaging Sci. **8**(1), 536–563 (2015a)

Pierre, F., Aujol, J.F., Bugeau, A., Ta, V.T.: Combinaison linéaire optimale de métriques pour la colorisation d'images. In: XXVème colloque GRETSI, pp. 1–4 (2015b)

Pierre, F., Aujol, J.F., Bugeau, A., Ta, V.T.: Luminance-hue specification in the RGB space. In: Scale Space and Variational Methods in Computer Vision, pp. 413–424 (2015c)

Pierre, F., Aujol, J.F., Bugeau, A., Ta, V.T.: Colociel. Dépôt Agence de Protection des Programmes No IDDN.FR.001.080021.000.S.P.2016.000.2100 (2016). Disponible en ligne sur http://www.labri.fr/perso/fpierre/colociel_v1.zip

Pierre, F., Aujol, J.F., Bugeau, A., Ta, V.T.: Interactive video colorization within a variational framework. SIAM J. Imaging Sci. **10**(4), 2293–2325 (2017a) a

Pierre, F., Aujol, J.F., Deledalle, C.A., Papadakis, N.: Luminance-guided chrominance denoising with debiased coupled total variation. In: International Workshop on Energy Minimization Methods in Computer Vision and Pattern Recognition, pp. 235–248. Springer (2017b)

Quang, M.H., Kang, S.H., Le, T.M.: Image and video colorization using vector-valued reproducing kernel hilbert spaces. J. Math. Imaging Vis. **37**(1), 49–65 (2010)

Ren, X., Malik, J.: Learning a classification model for segmentation. In: IEEE International Conference on Computer Vision, pp. 10–17 (2003)

Royer, A., Kolesnikov, A., Lampert, C.H.: Probabilistic image colorization. In: British Machine Vision Conference (2017)

Sethian, J.A.: Level Set Methods and Fast Marching Methods: Evolving Interfaces in Computational Geometry, Fluid Mechanics, Computer Vision, and Materials Science, vol. 3. Cambridge University Press, Cambridge (1999)

Simonyan, K., Zisserman, A.: Very deep convolutional networks for large-scale image recognition. In: International Conference on Learning Representations (2015)

Song, M., Tao, D., Chen, C., Bu, J., Yang, Y.: Color-to-gray based on chance of happening preservation. Neurocomputing **119**, 222–231 (2013)

Su, Z., Liang, X., Guo, J., Gao, C., Luo, X.: An edge-refined vectorized deep colorization model for grayscale-to-color images. Neurocomputing **311**, 305–315 (2018)

Sỳkora, D., Buriánek, J., Žára, J.: Unsupervised colorization of black-and-white cartoons. In: Proceedings of the 3rd International Symposium on Non-photorealistic Animation and Rendering, pp. 121–127. ACM (2004)

Tan, P., Pierre, F., Nikolova, M.: Inertial alternating generalized forward–backward splitting for image colorization. J. Math. Imaging Vis. **61**(5), 672–690 (2019)

Wei, L.Y., Levoy, M.: Fast texture synthesis using tree-structured vector quantization. In: ACM Computer Graphics and Interactive Techniques, pp. 479–488. Press/Addison-Wesley Publishing Co. (2000)

Welsh, T., Ashikhmin, M., Mueller, K.: Transferring color to greyscale images. In: ACM Transactions on Graphics, vol. 21–3, pp. 277–280. ACM (2002)

Williams, A., Barrus, S., Morley, R.K., Shirley, P.: An efficient and robust ray-box intersection algorithm. In: ACM SIGGRAPH 2005 Courses, p. 9 (2005)

Wolfgang Baatz Massimo Fornasier, P.A.M., Schönlieb, C.B.: Inpainting of ancient austrian frescoes. In: Sarhangi, R., Séquin, C.H. (eds.) Bridges Leeuwarden: Mathematics, Music, Art, Architecture, Culture, pp. 163–170. Tarquin Publications, London (2008). Disponible en ligne sur http://archive.bridgesmathart.org/2008/bridges2008-163.html

Yatziv, L., Sapiro, G.: Fast image and video colorization using chrominance blending. IEEE Trans. Image Proces. **15**(5), 1120–1129 (2006)

Zhang, R., Isola, P., Efros, A.A.: Colorful image colorization. In: European Conference on Computer Vision, pp. 1–16. Springer (2016)

Zhang, R., Zhu, J.Y., Isola, P., Geng, X., Lin, A.S., Yu, T., Efros, A.A.: Real-time user-guided image colorization with learned deep priors. ACM Trans. Graph. **9**(4), 119:1–119:11 (2017)

Zheng, Y., Essock, E.A.: A local-coloring method for night-vision colorization utilizing image analysis and fusion. Inf. Fusion **9**(2), 186–199 (2008)

Numerical Solution for Sparse PDE Constrained Optimization

16

Xiaoliang Song and Bo Yu

Contents

Introduction	624
Finite Element Approximation and Error Estimates	632
An Inexact Heterogeneous ADMM Algorithm	642
An Inexact Heterogeneous ADMM Algorithm	642
Convergence Results of ihADMM	645
An Inexact Majorized Accelerated Block Coordinate Descent Method for (D_h)	652
An Inexact Block Symmetric Gauss-Seidel Iteration	653
Inexact Majorized Accelerate Block Coordinate Descent (imABCD) Method	656
A sGS-imABCD Algorithm for (D_h)	659
Numerical Results	663
Algorithmic Details	663
Examples	664
Conclusion	673
References	673

Abstract

In this chapter, elliptic PDE-constrained optimal control problems with L^1-control cost (L^1-EOCP) are considered. Motivated by the success of the first-order methods, we give an overview on two efficient first-order methods to solve L^1-EOCP: inexact heterogeneous alternating direction method of multipliers (ihADMM) and an inexact symmetric Gauss-Seidel (sGS)-based 2-block majorized accelerated block coordinate descent (ABCD) method (sGS-imABCD). Different from the classical ADMM, the ihADMM adopts two

---

X. L. Song · B. Yu (✉)
School of Mathematical Sciences, Dalian University of Technology, Dalian, Liaoning, China
e-mail: songxiaoliang@dlut.edu.cn; yubo@dlut.edu.cn

© Springer Nature Switzerland AG 2023
K. Chen et al. (eds.), *Handbook of Mathematical Models and Algorithms in Computer Vision and Imaging*, https://doi.org/10.1007/978-3-030-98661-2_129

different weighted inner products to define the augmented Lagrangian function in two subproblems, respectively. Benefiting from such different weighted techniques, two subproblems of ihADMM can be efficiently implemented. Furthermore, theoretical results on the global convergence as well as the iteration complexity results $o(1/k)$ for ihADMM are given. A common approach to solve the L^1-EOCP is directly solving the primal problem. Based on the dual problem of L^1-EOCP, which can be reformulated as a multi-block unconstrained convex composite minimization problem, an efficient inexact ABCD method is introduced for solving L^1-EOCP. The design of this method combines an inexact 2-block majorized ABCD and the recent advances in the inexact sGS technique for solving a multi-block convex composite quadratic programming whose objective contains a nonsmooth term involving only the first block.

Keywords

PDE-constrained optimization · Sparsity · Finite element · ADMM · Symmetric Gauss-Seidel accelerated block coordinate descent

Introduction

We study the following linear-quadratic elliptic PDE-constrained optimal control problem with L^1-control cost and piecewise box constraints on the control:

$$\begin{cases} \min_{(y,u) \in Y \times U} J(y,u) = \frac{1}{2}\|y - y_d\|^2_{L^2(\Omega)} + \frac{\alpha}{2}\|u\|^2_{L^2(\Omega)} + \beta\|u\|_{L^1(\Omega)} \\ \text{s.t.} \quad Ly = u + y_r \text{ in } \Omega, \\ \qquad y = 0 \quad \text{on } \Gamma, \\ \qquad u \in U_{ad} = \{v(x) | a \leq v(x) \leq b, \text{ a.e. on } \Omega\} \subseteq U, \end{cases} \tag{P}$$

where $Y := H^1_0(\Omega)$, $U := L^2(\Omega)$, $\Omega \subseteq \mathbb{R}^n$ ($n = 2$ or 3) is a convex, open, and bounded domain with $C^{1,1}$- or polygonal boundary Γ; the desired state $y_d \in L^2(\Omega)$ and the source term $y_r \in L^2(\Omega)$ are given; and $a \leq 0 \leq b$ and $\alpha, \beta > 0$. Moreover, the operator L is a second-order linear elliptic differential operator. It is well-known that L^1-norm could lead to sparse optimal control, i.e., the optimal control with small support. Such an optimal control problem (P) plays an important role for the placement of control devices (Stadler 2009). In some cases, it is difficult or undesirable to place control devices all over the control domain and one hopes to localize controllers in small and effective regions, and the L^1-solution gives information about the optimal location of the control devices.

Through this chapter, let us suppose the elliptic PDEs involved in (P) which are of the form

16 Numerical Solution for Sparse PDE Constrained Optimization

$$Ly = u + y_r \quad \text{in } \Omega,$$
$$y = 0 \quad \text{on } \partial\Omega, \tag{1}$$

satisfy the following assumption:

Assumption 1. *The linear second-order differential operator L is defined by*

$$(Ly)(x) := -\sum_{i,j=1}^{n} \partial_{x_j}(a_{ij}(x)y_{x_i}) + c_0(x)y(x), \tag{2}$$

where functions $a_{ij}(x), c_0(x) \in L^\infty(\Omega)$, $c_0 \geq 0$, and it is uniformly elliptic, i.e., $a_{ij}(x) = a_{ji}(x)$ and there is a constant $\theta > 0$ such that

$$\sum_{i,j=1}^{n} a_{ij}(x)\xi_i\xi_j \geq \theta\|\xi\|^2 \quad \text{for a.a. } x \in \Omega \text{ and } \forall \xi \in \mathbb{R}^n. \tag{3}$$

The weak formulation of (1) is given by

$$\text{Find } y \in H_0^1(\Omega) : a(y, v) = (u + y_r, v)_{L^2(\Omega)} \quad \forall v \in H_0^1(\Omega), \tag{4}$$

with the bilinear form

$$a(y, v) = \int_\Omega (\sum_{i,j=1}^{n} a_{ji} y_{x_i} v_{x_j} + c_0 y v) dx, \tag{5}$$

or in short $Ay = B(u + y_r)$, where $A \in \mathcal{L}(Y, Y^*)$ is the operator induced by the bilinear form a, i.e., $Ay = a(y, \cdot)$ and $B \in \mathcal{L}(U, Y^*)$ is defined by $Bu = (u, \cdot)_{L^2(\Omega)}$. Since the bilinear form $a(\cdot, \cdot)$ is symmetric and U, Y are Hilbert spaces, we have $A^* \in \mathcal{L}(Y, Y^*) = A$ and $B^* \in \mathcal{L}(Y, U)$ with $B^*v = v$ for any $v \in Y$.

Remark 1. Although we assume that the Dirichlet boundary condition $y = 0$ holds, it should be noted that the assumption is not a restriction and our considerations can also carry over to the more general boundary conditions of Robin type:

$$\frac{\partial y}{\partial \nu} + \gamma y = g \quad \text{on } \partial\Omega,$$

where $g \in L^2(\partial\Omega)$ is given and $\gamma \in L^\infty(\partial\Omega)$ is nonnegative coefficient. Furthermore, it is assumed that the control satisfies $a \leq u \leq b$, where a and b have opposite signs. First, we should emphasize that this condition is required in practice, e.g., the placement of control devices. In addition, please also note that this condition is not a restriction from the point of view of the algorithm. If one has, e.g., $a > 0$ on Ω, the L^1-norm in U_{ad} is in fact a linear function, and thus the problem can also be handled by our method.

Optimal control problems with $\alpha > 0$, $\beta = 0$ and their numerical realization have been studied intensively in recent papers; see, e.g., Hinze (2005), Falk (1973), Geveci (1979), Rösch (2006), Casas and Tröltzsch (2003), Meyer and Rösch (2004) and the references cited there. Let us first comment on known results on error estimates of control-constrained optimal control problems. Basic a priori error estimates were derived by Falk (1973) and Geveci (1979) where Falk considered distributed controls, while Geveci concentrated on the Neumann boundary controls. Both the authors proved optimal L^2-error estimates $O(h)$ for piecewise constant approximations of the control variables. Convergence results for the approximations of the controls by piecewise linear, globally continuous elements can be found in Casas and Tröltzsch (2003), where Casas and Tröltzsch proved order $O(h)$ in the case of linear-quadratic control problems. Later Casas (2007) proved order $o(h)$ for the control problems governed by semilinear elliptic equations and quite general cost functions. In Rösch (2006) for the first time proved that the error order is $O(h^{\frac{3}{2}})$ under some special assumptions on the continuous solutions. However, his proof was just done in one dimension. All previous papers were devoted to the full discretization. Recently, a variational discretization concept is introduced by Hinze (2005). More precisely, the state variable and the state equation are discretized, but there is no discretization of the control. He showed that the control error is of order $O(h^2)$. In certain situations, the same convergence order can also be achieved by a special postprocessing procedure; see Meyer and Rösch (2004).

For the study of optimal control problems with sparsity promoting terms, as far as we know, the first paper devoted to this study is published by Stadler (2009), in which structural properties of the control variables were analyzed in the case of the linear-quadratic elliptic optimal control problem. In 2011, a priori and a posteriori error estimates were first given by Wachsmuth and Wachsmuth in (2011) for piecewise linear control discretizations, in which the convergence rate is obtained to be of order $O(h)$ under the L^2 norm. However, from the point of view of the algorithm, the resulting discretized L^1-norm

$$\|u_h\|_{L^1(\Omega_h)} := \int_{\Omega_h} \left|\sum_{i=1}^{N_h} u_i \phi_i(x)\right| dx, \qquad (6)$$

does not have a decoupled form with respect to the coefficients $\{u_i\}$, where $\{\phi_i(x)\}$ are the piecewise linear nodal basis functions. Hence, the authors introduced an alternative discretization of the L^1-norm which relies on a nodal quadrature formula:

$$\|u_h\|_{L^1_h(\Omega)} := \sum_{i=1}^{N_h} |u_i| \int_{\Omega_h} \phi_i(x) dx. \qquad (7)$$

Obviously, this quadrature incurs an additional error, although the authors proved that this approximation does not change the order of error estimates. In a sequence of papers (Casas et al. 2012a,b), for the non-convex case governed by a semilinear elliptic equation, Casas et al. proved second-order necessary and sufficient optimality conditions. Using the second-order sufficient optimality conditions, the authors provide error estimates of order h w.r.t. the L^∞ norm for three different choices of the control discretization. It should be pointed out that, for the piecewise linear control discretization case, a similar approximation technique to the one introduced in Wachsmuth and Wachsmuth (2011) is also used for the discretizations of the L^2 norm and L^1 norm of the control.

Apart from using L^1-norm to induce sparsity, Clason and Kunisch in (2011) investigated elliptic control problems with measure-valued controls to promote the sparsity of the control. They discussed the existence and uniqueness of the corresponding dual problems. Subsequently, in 2012, Casas et al. in (2012) studied the optimality conditions and provided a priori finite element error estimates for the case of linear-quadratic elliptic control problems with a measure-valued control, in which the control measure was approximated by a linear combination of Dirac measures.

To numerically solve the problem (P), there are two possible ways. One is called *First discretize, then optimize*, and another approach is called *First optimize, then discretize* (Collis and Heinkenschloss 2002). Independently of where discretization is located, the resulting finite dimensional equations are quite large. Thus, both of these cases require us to consider proposing an efficient algorithm. In this chapter, we focus on the *First discretize, then optimize* approach to solve (P) and employ the piecewise linear finite elements to discretize (P).

Next, let us mention some existing numerical methods for solving problem (P). Since problem (P) is nonsmooth, thus applying semismooth Newton (SSN) methods is used to be a priority in consideration of their locally superlinear convergence. A special semismooth Newton method with the active set strategy, called the primal-dual active set (PDAS) method, is introduced in Bergounioux et al. (1999) for control-constrained elliptic optimal control problems. It is proved to have the locally superlinear convergence (see Ulbrich (2002), Ulbrich (2003), Hinze et al. (2009) for more details). Mesh-independence results for the SSN method were established in Hintermüller and Ulbrich (2004). Additionally, the authors in Porcelli et al. (2017) showed that a saddle point system with 2×2 block structure should be solved by employing some Krylov subspace methods with a good preconditioner at each iteration step of the SSN method. However, the 2×2 block linear system is obtained by reducing a 3×3 block linear system with bringing additional computation for linear system involving the mass matrix. Furthermore, the coefficient matrix of the Newton equation would change with every iteration due to the change of the active set. In this case, it is clear that forming a uniform preconditioner, which is used to precondition the Krylov subspace methods for solving the Newton equations, is difficult. For a survey of how to precondition saddle point problems, we refer to Herzog and Ekkehard (2010).

Undeniably, employing the SSN method can derive the solution with high precision. However, it should be mentioned that in general solving Newton equations is expensive.

Recently, for the finite dimensional large-scale optimization problem, some efficient first-order algorithms, such as iterative shrinkage/soft thresholding algorithms (ISTA) (Blumensath and Davies 2008), accelerated proximal gradient (APG)-based method (Beck and Teboulle 2009), ADMM (Fazel et al. 2013; Chen and Toh 2017; Li et al. 2015, 2016), etc., have become the state-of-the-art algorithms. Thanks to the iteration complexity $O(1/k^2)$, a fast inexact proximal (FIP) method in function space, which is actually the APG method, was proposed to solve the problem (P) in Schindele and Borzì (2016). As we know, the efficiency of the FIP method depends on how close the step-length is to the Lipschitz constant. However, in general, choosing an appropriate step-length is difficult since the Lipschitz constant is usually not available analytically. Thus, this disadvantage largely limits the efficiency of APG method.

In this chapter, we will focus first on the ADMM algorithm. The classical ADMM was originally proposed by Glowinski and Marroco (1975) and Gabay and Mercier (1976), and it has found lots of efficient applications in a broad spectrum of areas. In particular, we refer to Boyd et al. (2011) for a review of the applications of ADMM in the areas of distributed optimization and statistical learning. We give a brief sketch of ADMM for the following finite dimensional linearly constrained convex programming problem with two blocks of functions and variables:

$$\begin{cases} \min f(u) + g(z) \\ \text{s.t. } A_1 u + A_2 z = c, \\ u \in U, \ z \in Z, \end{cases} \tag{8}$$

where $f(u) : \mathbb{R}^n \to \mathbb{R}$ and $g(z) : \mathbb{R}^m \to \mathbb{R}$ are both closed, proper, and convex functions (but not necessary smooth); $A_1 \in \mathbb{R}^{p \times n}$, $A_2 \in \mathbb{R}^{p \times m}$ and $c \in \mathbb{R}^p$; $U \subset \mathbb{R}^n$ and $Z \subset \mathbb{R}^m$ are given closed, convex, and non-empty sets. Let

$$\mathcal{L}_\sigma(u, z, \lambda; \sigma) = f(u) + g(z) + \langle \lambda, A_1 u + A_2 z - c \rangle + \frac{\sigma}{2} \| A_1 u + A_2 z - c \|^2 \tag{9}$$

be the augmented Lagrangian function of (8) with the Lagrange multiplier $\lambda \in \mathbb{R}^p$ and the penalty parameter $\sigma > 0$. For a given $\tau \in \left(0, \frac{\sqrt{5}+1}{2}\right)$, the classical ADMM is described as follows:

$$\begin{cases} u^{k+1} = \arg\min_{u \in U} \mathcal{L}_\sigma(u, z^k, \lambda^k; \sigma), \\ z^{k+1} = \arg\min_{z \in Z} \mathcal{L}_\sigma(u^{k+1}, z, \lambda^k; \sigma), \\ \lambda^{k+1} = \lambda^k + \tau \rho (A_1 u^{k+1} + A_2 z^{k+1} - c). \end{cases} \tag{10}$$

Thanks to the separable structure of the objective function, each subproblem in (10) involves only one block of $f(u)$ and $g(z)$ and could be solved easily. Under some trivial assumptions, the classical ADMM for solving (8) has global convergence and sublinear convergence rate at least.

Motivated by the success of the finite dimensional ADMM algorithm, it is reasonable to consider extending the ADMM to infinite dimensional optimal control problems, as well as the corresponding discretized problems. In 2016, the authors Elvetun and Nielsen (2014) adapted the split Bregman method (equivalent to the classical ADMM) to handle PDE-constrained optimization problems with total variation regularization. However, for the discretized problem, the authors did not take advantage of the inherent structure of problem and still used the classical ADMM to solve it. In this chapter, making full use of inherent structure of problem, we aim to design an appropriate ADMM-type algorithm to solve problem (P). In order to employ the ADMM algorithm and obtain a separable by adding an artificial variable z, we can separate the smooth and nonsmooth terms and equivalently reformulate problem (P) as

$$\begin{cases} \min_{(y,u,z) \in Y \times U \times U} J(y, u, z) = \frac{1}{2} \|y - y_d\|_{L^2(\Omega)}^2 + \frac{\alpha}{4} \|u\|_{L^2(\Omega)}^2 + \frac{\alpha}{4} \|z\|_{L^2(\Omega)}^2 \\ \qquad\qquad + \beta \|z\|_{L^1(\Omega)} \\ \text{s.t.} \qquad Ay = u + y_r \quad \text{in } \Omega, \\ \qquad\qquad y = 0 \quad \text{on } \partial \Omega, \\ \qquad\qquad u = z, \\ \qquad\qquad z \in U_{ad} = \{v(x) | a \leq v(x) \leq b, \text{a.e on } \Omega\} \subseteq U. \end{cases} \quad (\tilde{P})$$

However, when the classical ADMM is directly used to solve $(\overline{\text{DP}}_h)$, i.e., the discrete version of $(\tilde{P})$, there is no well-structure as in continuous case and the corresponding subproblems cannot be efficiently solved. Thus, making use of the inherent structure of $(\overline{\text{DP}}_h)$, a heterogeneous ADMM is proposed. Meanwhile, sometimes it is unnecessary to exactly compute the solution of each subproblem even if it is doable, especially at the early stage of the whole process. For example, if a subproblem is equivalent to solving a large-scale or ill-condition linear system, it is a natural idea to use the iterative methods such as some Krylov-based methods. Hence, taking the inexactness of the solutions of associated subproblems into account, a more practical inexact heterogeneous ADMM (ihADMM) is proposed. Different from the classical ADMM, we utilize two different weighted inner products to define the augmented Lagrangian function for two subproblems, respectively. Specifically, based on the M_h-weighted inner product, the augmented Lagrangian function with respect to the u-subproblem in k-th iteration is defined as

$$\mathcal{L}_\sigma(u, z^k; \lambda^k) = f(u) + g(z^k) + \langle \lambda, M_h(u - z^k) \rangle + \frac{\sigma}{2} \|u - z^k\|_{M_h}^2,$$

where M_h is the mass matrix. On the other hand, for the z-subproblem, based on the W_h-weighted inner product, the augmented Lagrangian function in k-th iteration is defined as

$$\mathcal{L}_\sigma(u^{k+1}, z; \lambda^k) = f(u^{k+1}) + g(z) + \langle \lambda, M_h(u^{k+1} - z) \rangle + \frac{\sigma}{2}\|u^{k+1} - z\|^2_{W_h},$$

where the lumped mass matrix W_h is diagonal.

Benefiting from different weighted techniques, each subproblem of ihADMM for $(\overline{DP_h})$ can be efficiently solved. Specifically, the u-subproblem of ihADMM, which results in a large-scale linear system, is the main computation cost in whole algorithm. W_h-weighted technique makes z-subproblem have a decoupled form and admit a closed-form solution given by the soft thresholding operator and the projection operator onto the box constraint $[a, b]$. Moreover, global convergence and the iteration complexity result $o(1/k)$ in non-ergodic sense for our ihADMM will be proved. Taking the precision of discretized error into account, we should mention that using our ihADMM algorithm to solve problem $(\overline{DP_h})$ is highly enough and efficient in obtaining an approximate solution with moderate accuracy.

As far as we know, most of the aforementioned papers are devoted to solving the primal problem. Based on the special structure of the dual problem, we will also consider using the duality-based approach for (P). The dual of problem (P) can be written, in its equivalent minimization form, as

$$\min \Phi(\lambda, \mu, p) := \frac{1}{2}\|A^*p - y_d\|^2_{L^2(\Omega)} + \frac{1}{2\alpha}\| - p + \lambda + \mu\|^2_{L^2(\Omega)} \qquad \text{(D)}$$
$$+ \langle p, y_r \rangle_{L^2(\Omega)} + \delta_{\beta B_\infty(0)}(\lambda) + \delta^*_{U_{ad}}(\mu) - \frac{1}{2}\|y_d\|^2_{L^2(\Omega)},$$

where $p \in H_0^1(\Omega)$, $\lambda, \mu \in L^2(\Omega)$, $B_\infty(0) := \{\lambda \in L^2(\Omega) : \|\lambda\|_{L^\infty(\Omega)} \leq 1\}$, and for any given non-empty, closed convex subset C of $L^2(\Omega)$, $\delta_C(\cdot)$ is the indicator function of C. Based on the L^2-inner product, we define the conjugate of $\delta_C(\cdot)$ as follows:

$$\delta^*_C(w^*) = \sup_{w \in C} \langle w^*, w \rangle_{L^2(\Omega)}.$$

Although the duality-based approach has been introduced in Clason and Kunisch (2011) for elliptic control problems without control constraints in nonreflexive Banach spaces, the authors did not take advantage of the structure of the dual problem and still used semismooth Newton methods to solve the Moreau-Yosida regularization of the dual problem. In the chapter, in terms of the structure of problem (D), we aim to design an algorithm which could efficiently and fast solve the dual problem (D).

By setting $x = (\mu, \lambda, p)$, $x_0 = \mu$, and $x_1 = \lambda$, it is quite clear that our dual problem (D) belongs to a general class of multi-block convex optimization problems of the form

$$\min F(x_0, x) := \varphi(x_0) + \psi(x_1) + \phi(x_0, x), \qquad (11)$$

where $x_0 \in X_0$, $x = (x_1, \ldots, x_s) \in X := X_1 \times \ldots \times X_s$ and each X_i is a finite dimensional real Euclidean space. The functions φ, ψ, and ϕ are three closed proper convex functions. Thanks to the structure of (11), in 2015, Chambolle and Dossa (2015) proposed the accelerated alternative descent (AAD) algorithm to solve the problem (11) in which the joint objective function ϕ was quadratic. But the disadvantage is that the AAD method does not take the inexactness of the solutions of the associated subproblems into account. As we know, in some case, it is either impossible or extremely expensive to exactly compute the solution of each subproblem even if it is doable, especially at the early stage of the whole process. For example, if a subproblem is equivalent to solving a large-scale or ill-condition linear system, it is a natural idea to use the iterative methods such as some Krylov-based methods. Hence, it is not suitable for the practical application. Subsequently, when ϕ is a general closed proper convex function and $\arg\min_{x_0} \varphi(x_0) + \phi(x_0, x)$ could be computed exactly, Sun et al. (2016) proposed an inexact accelerated block coordinate descent (iABCD) method to solve least squares semidefinite programming (LSSDP) via its dual. The basic idea of the iABCD method is firstly applying the Danskin-type theorem to reduce the two block nonsmooth terms into only one block and then using APG method to solve the reduced problem. More importantly, the powerful inexact symmetric Gauss-Seidel (sGS) decomposition technique developed in Li et al. (2015) is the key for designing the iABCD method. Additionally, the authors proved that the iABCD method has the $O(1/k^2)$ iteration complexity when the subproblems are solved approximately subject to certain inexactness criteria.

However, for the situation the subproblem with respect to block x_0 could not be solved exactly, one could not no longer use Danskin-type theorem to achieve the goal of reducing it into one block nonsmooth term. To overcome the above bottlenecks, in her PhD thesis (Cui 2016, Chapter 3), Cui proposed an inexact majorized accelerated block coordinate descent (imABCD) method for solving the following unconstrained convex optimization problems with coupled objective functions:

$$\min_{v,w} f(v) + g(w) + \phi(v, w). \tag{12}$$

Under suitable assumptions and certain inexactness criteria, the author can prove that the inexact mABCD method also enjoys the impressive $O(1/k^2)$ iteration complexity.

In this chapter, which is inspired by the success of the iABCD and imABCD methods, we combine their virtues and propose an inexact sGS-based majorized ABCD method (called sGS-imABCD) to solve problem (D). The design of this method combines an inexact 2-block majorized ABCD and the recent advances in the inexact sGS technique. Owing to the convergence results of imABCD method which are given in Cui (2016, Chapter 3), our proposed algorithm could be proven having the $O(1/k^2)$ iteration complexity as well. Moreover, some truly implementable inexactness criteria controlling the accuracy of the generated imABCD subproblems are analyzed. Specifically, because of two nonsmooth subproblems having the closed-form solutions, it is easy to see that the main computation of our sGS-imABCD algorithm is in solving p-subproblems, which is equivalent to solving the 2×2 block saddle point linear system twice at each iteration. It should be pointed out that the coefficient matrix of the saddle point linear system is fixed. To efficiently solve the linear system, a preconditioned GMRES method is used which leads to the rapid convergence and the robustness with respect to the mesh size h. More importantly, at first glance, it appears that we would need to solve the linear system twice. In practice, in order to avoid this situation and improve the efficiency of our sGS-imABCD algorithm, we design a strategy to approximate the solution for the second linear system. Thus, when a residual error condition is satisfied, the linear system need only to be solved once instead of twice. We should emphasize that such a saving can be significant, especially in the middle and later stages of the whole algorithm. Thus, in terms of the amount of calculation and the discretized error, our sGS-imABCD algorithm is superior to the semismooth Newton method.

Finite Element Approximation and Error Estimates

The goal of this section is to study the approximation of problems (P) and ($\widetilde{P}$) by finite elements.

To achieve our aim, we first consider a family of regular and quasi-uniform triangulations $\{\mathcal{T}_h\}_{h>0}$ of $\bar{\Omega}$. For each cell $T \in \mathcal{T}_h$, let us define the diameter of the set T by $\rho_T := \operatorname{diam} T$ and define σ_T to be the diameter of the largest ball contained in T. The mesh size of the grid is defined by $h = \max_{T \in \mathcal{T}_h} \rho_T$. We suppose that the following regularity assumptions on the triangulation are satisfied which are standard in the context of error estimates.

Assumption 2. *There exist two positive constants κ and τ such that*

$$\frac{\rho_T}{\sigma_T} \leq \kappa \quad \text{and} \quad \frac{h}{\rho_T} \leq \tau,$$

hold for all $T \in \mathcal{T}_h$ and all $h > 0$. Let us define $\bar{\Omega}_h = \bigcup_{T \in \mathcal{T}_h} T$, and let $\Omega_h \subset \Omega$ and Γ_h denote its interior and its boundary, respectively. In the case that Ω is a convex polyhedral domain, we have $\Omega = \Omega_h$. In the case that Ω has a $C^{1,1}$-boundary Γ, we assumed that $\bar{\Omega}_h$ is convex and that all boundary vertices of $\bar{\Omega}_h$ are contained in Γ, such that $|\Omega \backslash \Omega_h| \leq \hat{c} h^2$, where $|\cdot|$ denotes the measure of the set and $\hat{c} > 0$ is a constant.

On account of the homogeneous boundary condition of the state equation, we use

$$Y_h = \left\{ y_h \in C(\bar{\Omega}) \mid y_{h|T} \in \mathcal{P}_1 \text{ for all } T \in \mathcal{T}_h \text{ and } y_h = 0 \text{ in } \bar{\Omega} \backslash \Omega_h \right\}$$

as the discretized state space, where $\mathcal{P}_1$ denotes the space of polynomials of degree less than or equal to 1. For a given source term y_r and right-hand side $u \in L^2(\Omega)$, we denote by $y_h(u)$ the approximated state associated with u, which is the unique solution for the following discretized weak formulation:

$$\int_{\Omega_h} \left(\sum_{i,j=1}^n a_{ij} y_{h x_i} v_{h x_j} + c_0 y_h v_h \right) dx = \int_{\Omega_h} (u + y_r) v_h dx \quad \forall v_h \in Y_h. \quad (13)$$

Moreover, $y_h(u)$ can also be expressed by $y_h(u) = \mathcal{S}_h(u + y_r)$, in which $\mathcal{S}_h$ is a discretized version of $\mathcal{S}$ and an injective, self-adjoint operator. The following error estimates are well-known.

Lemma 1 (Ciarlet 1978, Theorem 4.4.6). *For a given* $u \in L^2(\Omega)$, *let* y *and* $y_h(u)$ *be the unique solution of (4) and (13), respectively. Then there exists a constant* $c_1 > 0$ *independent of* h, u, *and* y_r *such that*

$$\|y - y_h(u)\|_{L^2(\Omega)} + h\|\nabla y - \nabla y_h(u)\|_{L^2(\Omega)} \leq c_1 h^2 (\|u\|_{L^2(\Omega)} + \|y_r\|_{L^2(\Omega)}). \quad (14)$$

In particular, this implies $\|\mathcal{S} - \mathcal{S}_h\|_{L^2 \to L^2} \leq c_1 h^2$ *and* $\|\mathcal{S} - \mathcal{S}_h\|_{L^2 \to H^1} \leq c_1 h$.

Considering the homogeneous boundary condition of the adjoint state equation (1), we use

$$U_h = \left\{ u_h \in C(\bar{\Omega}) \mid u_{h|T} \in \mathcal{P}_1 \text{ for all } T \in \mathcal{T}_h \text{ and } u_h = 0 \text{ in } \bar{\Omega} \setminus \Omega_h \right\},$$

as the discretized space of the control u and artificial variable z.

For a given regular and quasi-uniform triangulation $\mathcal{T}_h$ with nodes $\{x_i\}_{i=1}^{N_h}$, let $\{\phi_i(x)\}_{i=1}^{N_h}$ be a set of nodal basis functions associated with nodes $\{x_i\}_{i=1}^{N_h}$, where the basis functions satisfy the following properties:

$$\phi_i(x) \geq 0, \quad \|\phi_i(x)\|_\infty = 1 \quad \forall i = 1, 2, \ldots, N_h, \quad \sum_{i=1}^{N_h} \phi_i(x) = 1. \quad (15)$$

The elements $z_h \in U_h$, $u_h \in U_h$, and $y_h \in Y_h$ can be represented in the following forms, respectively:

$$u_h = \sum_{i=1}^{N_h} u_i \phi_i(x), \quad z_h = \sum_{i=1}^{N_h} z_i \phi_i(x), \quad y_h = \sum_{i=1}^{N_h} y_i \phi_i(x),$$

and $u_h(x_i) = u_i$, $z_h(x_i) = z_i$, and $y_h(x_i) = y_i$ hold.

Let $U_{ad,h}$ denote the discretized feasible set, which is defined by

$$U_{ad,h} := U_h \cap U_{ad} = \left\{ z_h = \sum_{i=1}^{N_h} z_i \phi_i(x) \mid a \leq z_i \leq b, \forall i = 1, \ldots, N_h \right\} \subset U_{ad}.$$

Following the approach of Carstensen (1999), for the error analysis further below, let us introduce a quasi-interpolation operator $\Pi_h : L^1(\Omega_h) \to U_h$ which provides interpolation estimates. For an arbitrary $w \in L^1(\Omega)$, the operator Π_h is constructed as follows:

$$\Pi_h w = \sum_{i=1}^{N_h} \pi_i(w) \phi_i(x), \quad \pi_i(w) = \frac{\int_{\Omega_h} w(x) \phi_i(x) \mathrm{d}x}{\int_{\Omega_h} \phi_i(x) \mathrm{d}x}. \tag{16}$$

And we know that

$$w \in U_{ad} \Rightarrow \Pi_h w \in U_{ad,h}, \quad \text{for all } w \in L^1(\Omega). \tag{17}$$

Based on the assumption on the mesh and the control discretization, we extend $\Pi_h w$ to Ω by taking $\Pi_h w = w$ for every $x \in \Omega \setminus \Omega_h$ and have the following estimates of the interpolation error. For the detailed proofs, we refer to Carstensen (1999) and de Los Reyes et al. (2008).

Lemma 2. *There is a constant c_2 independent of h such that*

$$h \|z - \Pi_h z\|_{L^2(\Omega)} + \|z - \Pi_h z\|_{H^{-1}(\Omega)} \leq c_2 h^2 \|z\|_{H^1(\Omega)},$$

holds for all $z \in H^1(\Omega)$.

Now, we can consider a discretized version of problem $(\widetilde{P})$ as

$$\begin{cases} \min J_h(y_h, u_h, z_h) = \frac{1}{2} \|y_h - y_d\|_{L^2(\Omega_h)}^2 + \frac{\alpha}{4} \|u_h\|_{L^2(\Omega_h)}^2 + \frac{\alpha}{4} \|z_h\|_{L^2(\Omega_h)}^2 \\ \qquad\qquad\qquad + \beta \|z_h\|_{L^1(\Omega_h)} \\ \text{s.t.} \quad y_h = S_h(u_h + y_r), \\ \qquad u_h = z_h, \\ \qquad z_h \in U_{ad,h}, \end{cases} \tag{$\widetilde{P}_h$}$$

where

$$\|z_h\|_{L^2(\Omega_h)}^2 = \int_{\Omega_h} \left(\sum_{i=1}^{N_h} z_i \phi_i(x) \right)^2 \mathrm{d}x, \tag{18}$$

$$\|z_h\|_{L^1(\Omega_h)} = \int_{\Omega_h} |\sum_{i=1}^{N_h} z_i \phi_i(x)| dx. \tag{19}$$

This implies, for problem (P), we have the following discretized version:

$$\begin{cases} \min_{(y_h,u_h,z_h)\in Y_h\times U_h\times U_h} J_h(y_h,u_h,z_h) = \frac{1}{2}\|y_h - y_d\|^2_{L^2(\Omega_h)} + \frac{\alpha}{2}\|u_h\|^2_{L^2(\Omega_h)} \\ \qquad\qquad\qquad\qquad\qquad + \beta\|u_h\|_{L^1(\Omega_h)} \\ \text{s.t.} \qquad y_h = \mathcal{S}_h(u_h + y_r), \\ \qquad\qquad u_h \in U_{ad,h}. \end{cases} \tag{P_h}$$

For problem (P_h), in Wachsmuth and Wachsmuth (2011), the authors gave the following error estimates results.

Theorem 1 (Wachsmuth and Wachsmuth 2011, Proposition 4.3). *Let (y,u) be the optimal solution of problem (P), and (y_h, u_h) be the optimal solution of problem (P_h). For every $h_0 > 0$, $\alpha_0 > 0$, there is a constant $C > 0$ such that for all $0 < \alpha \leq \alpha_0$, $0 < h \leq h_0$ it holds*

$$\|u - u_h\|_{L^2(\Omega)} \leq C(\alpha^{-1}h + \alpha^{-\frac{3}{2}}h^2), \tag{20}$$

where C is a constant independent of h and α.

However, the resulting discretized problem ($\widetilde{P}_h$) is not in a decoupled form as the finite dimensional l^1-regularization optimization problem usually does, since (18) and (19) do not have a decoupled form. Thus, if we directly apply ADMM algorithm to solve the discretized problem, then the z-subproblem cannot have a closed-form solution. Thus, directly solving ($\widetilde{P}_h$), it cannot make full use of the advantages of ADMM. In order to overcome this bottleneck, we introduce the nodal quadrature formulas to approximately discretized the L^2-norm and L^1-norm. Let

$$\|z_h\|_{L^2_h(\Omega_h)} := \left(\sum_{i=1}^{N_h} (z_i)^2 \int_{\Omega_h} \phi_i(x) dx\right)^{\frac{1}{2}}, \tag{21}$$

$$\|z_h\|_{L^1_h(\Omega_h)} := \sum_{i=1}^{N_h} |z_i| \int_{\Omega_h} \phi_i(x) dx, \tag{22}$$

and call them L^2_h- and L^1_h-norm, respectively.

It is obvious that the L_h^2-norm and the L_h^1-norm can be considered as a weighted l^2-norm and a weighted l^1-norm of the coefficient of z_h, respectively. Both of them are norms on U_h. In addition, the L_h^2-norm is a norm induced by the following inner product:

$$\langle z_h, v_h \rangle_{L_h^2(\Omega_h)} = \sum_{i=1}^{N_h}(z_i v_i)\int_{\Omega_h}\phi_i(x)dx \quad \text{for } z_h, v_h \in U_h. \tag{23}$$

More importantly, the following properties hold.

Proposition 1 (Wathen 1987, Table 1). $\forall\ z_h \in U_h$, *the following inequalities hold:*

$$\|z_h\|_{L^2(\Omega_h)}^2 \leq \|z_h\|_{L_h^2(\Omega_h)}^2 \leq c\|z_h\|_{L^2(\Omega_h)}^2, \ \text{where } c = \begin{cases} 4 & \text{if } n=2, \\ 5 & \text{if } n=3. \end{cases} \tag{24}$$

$$\int_{\Omega_h} |\sum_{i=1}^n z_i\phi_i(x)|\ dx \leq \|z_h\|_{L_h^1(\Omega_h)}. \tag{25}$$

Thus, based on (22) and (21), we derive a new discretized optimal control problems

$$\begin{cases} \min J_h(y_h, u_h, z_h) = \dfrac{1}{2}\|y_h - y_d\|_{L^2(\Omega_h)}^2 + \dfrac{\alpha}{4}\|u_h\|_{L^2(\Omega_h)}^2 \\ \qquad\qquad\qquad\quad + \dfrac{\alpha}{4}\|z_h\|_{L_h^2(\Omega_h)}^2 + \beta\|z_h\|_{L_h^1(\Omega_h)} \\ \text{s.t.} \qquad\quad y_h = S_h u_h, \\ \qquad\qquad\quad u_h = z_h, \\ \qquad\qquad\quad z_h \in U_{ad,h}. \end{cases} \tag{$\widetilde{\text{DP}}_h$}$$

It should be mentioned that the approximate L_h^1 was already used in Wachsmuth and Wachsmuth (2011, Section 4.4). However, different from their discretization schemes, in this chapter, in order to keep the separability of the discrete L^2-norm with respect to z, we use (21) to approximately discretize it. In addition, although these nodal quadrature formulas incur additional discrete errors, it will be proven that these approximation steps will not change the order of error estimates as shown in (20); see Theorem 1.

To give the error estimates, we first introduce the Karush-Kuhn-Tucker (KKT) conditions. It is clear that problem $(\widetilde{\text{P}})$ is continuous and strongly convex. Therefore, the existence and uniqueness of solution of $(\widetilde{\text{P}})$ are obvious.

16 Numerical Solution for Sparse PDE Constrained Optimization

Theorem 2 (First-Order Optimality Condition). *Under Assumption 1, (y^*, u^*, z^*) is the optimal solution of $(\widetilde{P})$, if and only if there exists adjoint state $p^* \in H_0^1(\Omega)$ and Lagrange multiplier $\lambda^* \in L^2(\Omega)$, such that the following conditions hold in the weak sense:*

$$y^* = S(u^* + y_r), \tag{26a}$$

$$p^* = S^*(y^* - y_d), \tag{26b}$$

$$\frac{\alpha}{2}u^* + p^* + \lambda^* = 0, \tag{26c}$$

$$u^* = z^*, \tag{26d}$$

$$z^* \in U_{ad}, \tag{26e}$$

$$\left\langle \frac{\alpha}{2}z^* - \lambda^*, \tilde{z} - z^* \right\rangle_{L^2(\Omega)} + \beta(\|\tilde{z}\|_{L^1(\Omega)} - \|z^*\|_{L^1(\Omega)}) \geq 0, \quad \forall \tilde{z} \in U_{ad}. \tag{26f}$$

Moreover, we have

$$u^* = P_{U_{ad}}\left(\frac{1}{\alpha}\operatorname{soft}(-p^*, \beta)\right), \tag{27}$$

where the projection operator $P_{U_{ad}}(\cdot)$ and the soft thresholding operator $\operatorname{soft}(\cdot, \cdot)$ are defined as follows, respectively:

$$P_{U_{ad}}(v(x)) := \max\{a, \min\{v(x), b\}\},$$

$$\operatorname{soft}(v(x), \beta) := \operatorname{sgn}(v(x)) \circ \max(|v(x)| - \beta, 0). \tag{28}$$

In addition, the optimal control u has the regularity $u \in H^1(\Omega)$.

Analogous to the continuous problem $(\widetilde{P})$, the discretized problem $(\widetilde{DP}_h)$ is also a strictly convex problem, which is uniquely solvable. We derive the following first-order optimality conditions, which are necessary and sufficient for the optimal solution of $(\widetilde{DP}_h)$.

Theorem 3 (Discrete First-order Optimality Condition). (u_h, z_h, y_h) *is the optimal solution of $(\widetilde{DP}_h)$, if and only if there exist an adjoint state p_h and a Lagrange multiplier λ_h, such that the following conditions are satisfied:*

$$y_h = S_h(u_h + y_r), \tag{29a}$$

$$p_h = S_h^*(y_h - y_d), \tag{29b}$$

$$\frac{\alpha}{2}u_h + p_h + \lambda_h = 0, \tag{29c}$$

$$u_h = z_h, \tag{29d}$$

$$z_h \in U_{ad,h}, \tag{29e}$$

$$\left\langle \frac{\alpha}{2} z_h, \tilde{z}_h - z_h \right\rangle_{L_h^2(\Omega_h)} - (\lambda_h, \tilde{z}_h - z_h)_{L^2(\Omega_h)}$$

$$+ \beta \left(\|\tilde{z}_h\|_{L_h^1(\Omega_h)} - \|z\|_{L_h^1(\Omega_h)} \right) \geq 0, \tag{29f}$$

$$\forall \tilde{z}_h \in U_{ad,h}.$$

Now, let us start to do error estimation. Let (y, u, z) be the optimal solution of problem $(\widetilde{P})$, and (y_h, u_h, z_h) be the optimal solution of problem $(\widetilde{DP}_h)$. We have the following results.

Theorem 4. *Let (y, u, z) be the optimal solution of problem $(\widetilde{P})$, and (y_h, u_h, z_h) be the optimal solution of problem $(\widetilde{DP}_h)$. For any $h > 0$ small enough and $\alpha_0 > 0$, there is a constant C such that for all $0 < \alpha \leq \alpha_0$,*

$$\frac{\alpha}{2} \|u - u_h\|_{L^2(\Omega)}^2 + \frac{1}{2} \|y - y_h\|_{L^2(\Omega)}^2 \leq C(h^2 + \alpha h^2 + \alpha^{-1} h^2 + h^3 + \alpha^{-1} h^4 + \alpha^{-2} h^4),$$

where C is a constant independent of h and α.

Proof. Due to the optimality of z and z_h, z and z_h satisfy (26f) and (29f), respectively. Let us use the test function $z_h \in U_{ad,h} \subset U_{ad}$ in (26f) and the test function $\tilde{z}_h := \Pi_h z \in U_{ad,h}$ in (29f); thus, we have

$$\left\langle \frac{\alpha}{2} z - \lambda, z_h - z \right\rangle_{L^2(\Omega)} + \beta \left(\|z_h\|_{L^1(\Omega)} - \|z\|_{L^1(\Omega)} \right) \geq 0, \tag{30}$$

$$\left\langle \frac{\alpha}{2} z_h, \tilde{z}_h - z_h \right\rangle_{L_h^2(\Omega_h)} - (\lambda_h, \tilde{z}_h - z_h)_{L^2(\Omega_h)} + \beta \left(\|\tilde{z}_h\|_{L_h^1(\Omega_h)} - \|z_h\|_{L_h^1(\Omega_h)} \right) \geq 0. \tag{31}$$

Because $z_h = 0$ on $\bar{\Omega} \setminus \Omega_h$, the integrals over Ω can be replaced by integrals over Ω_h in (30), and it can be rewritten as

$$\left\langle \frac{\alpha}{2} z - \lambda, z - z_h \right\rangle_{L^2(\Omega_h)} + \beta \left(\|z\|_{L^1(\Omega_h)} - \|z_h\|_{L^1(\Omega_h)} \right) \leq \left\langle \lambda - \frac{\alpha}{2} z, z \right\rangle_{L^2(\Omega \setminus \Omega_h)}$$

$$- \beta \|z\|_{L^1(\Omega \setminus \Omega_h)} \leq \langle \lambda, z \rangle_{L^2(\Omega \setminus \Omega_h)} \leq ch^2, \tag{32}$$

where the last inequality follows from the boundedness of λ and z and the assumption $|\Omega \setminus \Omega_h| \leq \hat{c} h^2$.

By the definition of the quasi-interpolation operator in (16) and (24) in Proposition 1, we have

$$\langle z_h, \tilde{z}_h - z_h \rangle_{L_h^2(\Omega_h)} = \langle z_h, \tilde{z}_h \rangle_{L_h^2(\Omega_h)} - \|z_h\|_{L_h^2(\Omega_h)}^2 \leq \langle z_h, z - z_h \rangle_{L^2(\Omega_h)}. \tag{33}$$

Thus, (31) can be rewritten as

$$\left\langle -\frac{\alpha}{2} z_h + \lambda_h, z - z_h \right\rangle_{L^2(\Omega_h)} + \langle \lambda_h, \tilde{z}_h - z \rangle_{L^2(\Omega_h)}$$
$$- \beta \left(\|\tilde{z}_h\|_{L_h^1(\Omega_h)} - \|z_h\|_{L_h^1(\Omega_h)} \right) \leq 0. \tag{34}$$

Adding up and rearranging (32) and (34), we obtain

$$\frac{\alpha}{2} \|z - z_h\|_{L^2(\Omega_h)}^2 \leq \langle \lambda - \lambda_h, z - z_h \rangle_{L^2(\Omega_h)} - \langle \lambda_h, \tilde{z}_h - z \rangle_{L^2(\Omega_h)}$$
$$+ \beta \left(\|z_h\|_{L^1(\Omega_h)} - \|z\|_{L^1(\Omega_h)} + \|\tilde{z}_h\|_{L_h^1(\Omega_h)} - \|z_h\|_{L_h^1(\Omega_h)} \right) + ch^2$$
$$= \underbrace{\left\langle \frac{\alpha}{2}(u_h - u) + p_h - p, z - z_h \right\rangle_{L^2(\Omega_h)}}_{I_1} + \underbrace{\left\langle \frac{\alpha}{2} u_h + p_h, \tilde{z}_h - z \right\rangle_{L^2(\Omega_h)}}_{I_2} \tag{35}$$
$$+ \underbrace{\beta \left(\|z_h\|_{L^1(\Omega_h)} - \|z\|_{L^1(\Omega_h)} + \|\tilde{z}_h\|_{L_h^1(\Omega_h)} - \|z_h\|_{L_h^1(\Omega_h)} \right)}_{I_3} + ch^2,$$

where the second inequality follows from (26c) and (29c).

Next, we first estimate the third term I_3. By (25) in Proposition 1, we have $\|z_h\|_{L^1(\Omega_h)} \leq \|z_h\|_{L_h^1(\Omega_h)}$. And following from the definition of $\tilde{z}_h = \Pi_h(z)$ and the non-negativity and partition of unity of the nodal basis functions, we get

$$\|\tilde{z}_h\|_{L_h^1(\Omega_h)} = \|\Pi_h(z)\|_{L_h^1(\Omega_h)} = \sum_{i=1}^{N_h} \left| \frac{\int_{\Omega_h} z(x)\phi_i dx}{\int_{\Omega_h} \phi_i dx} \right| \int_{\Omega_h} \phi_i dx \leq \|z\|_{L^1(\Omega_h)}. \tag{36}$$

Thus, we have $I_3 \leq 0$.

For the terms I_1 and I_2, from $u = z$, $u_h = z_h$, we get

$$I_1 + I_2 = -\frac{\alpha}{2} \|u - u_h\|_{L^2(\Omega_h)}^2 + \langle p_h - p, \tilde{z}_h - z_h \rangle_{L^2(\Omega_h)}$$
$$+ \left\langle \frac{\alpha}{2} u + p, \tilde{z}_h - z \right\rangle_{L^2(\Omega_h)} + \frac{\alpha}{2} \langle u_h - u, \tilde{z}_h - z \rangle_{L^2(\Omega_h)}.$$

Then (35) can be rewritten as

$$\frac{\alpha}{2}\|z - z_h\|^2_{L^2(\Omega_h)} + \frac{\alpha}{2}\|u - u_h\|^2_{L^2(\Omega_h)} \leq \underbrace{\langle p_h - p, \tilde{z}_h - z_h\rangle_{L^2(\Omega_h)}}_{I_4} + \underbrace{\left\langle \frac{\alpha}{2}u + p, \tilde{z}_h - z\right\rangle_{L^2(\Omega_h)}}_{I_5}$$

$$+ \underbrace{\frac{\alpha}{2}\langle u_h - u, \tilde{z}_h - z\rangle_{L^2(\Omega_h)}}_{I_6} + ch^2.$$
(37)

For the term I_4, let $\tilde{p}_h = S_h^*(y - y_d)$, and we have

$$I_4 = \langle p_h - \tilde{p}_h + \tilde{p}_h - p, \tilde{z}_h - z_h\rangle_{L^2(\Omega_h)}$$

$$= -\|y - y_h\|^2_{L^2(\Omega_h)} + \underbrace{\langle y_h - y, (S_h - S)(\tilde{z}_h + y_r) - S(z - \tilde{z}_h))\rangle_{L^2(\Omega_h)}}_{I_7}$$

$$+ \underbrace{(y - y_d, (S_h - S)(\tilde{z}_h - z_h))_{L^2(\Omega_h)}}_{I_8}.$$

Consequently,

$$\frac{\alpha}{2}\|z - z_h\|^2_{L^2(\Omega_h)} + \frac{\alpha}{2}\|u - u_h\|^2_{L^2(\Omega_h)} + \|y - y_h\|^2_{L^2(\Omega_h)} \leq I_5 + I_6 + I_7 + I_8 + ch^2.$$
(38)

In order to further estimate (38), we will discuss each of these items from I_5 to I_8 in turn. Firstly, from the regularity of the optimal control u, i.e., $u \in H^1(\Omega)$, and (27), we know that

$$\|u\|_{H^1(\Omega)} \leq \frac{1}{\alpha}\|p\|_{H^1(\Omega)} + \left(\frac{\beta}{\alpha} + |a| + b\right)M(\Omega),$$
(39)

where $M(\Omega)$ denotes the measure of the Ω. Then we have

$$\left\|\frac{\alpha}{2}u + p\right\|_{H^1(\Omega)} \leq \frac{3}{2}\|p\|_{H^1(\Omega)} + \frac{1}{2}(\beta + \alpha|a| + \alpha b)M(\Omega).$$

Moreover, due to the boundedness of the optimal control u, the state y, the adjoint state p, and the operator S, we can choose a large enough constant $L > 0$ independent of α, h and a constant α_0, such that for all $0 < \alpha \leq \alpha_0$ and $h > 0$, the following inequation holds:

$$\frac{3}{2}\|p\|_{H^1(\Omega)} + (\beta + \alpha a + \alpha b)M(\Omega) + \|y - y_d\|_{L^2(\Omega)} + \|y_r\|_{L^2(\Omega)}$$
$$+ \|S\|_{\mathcal{L}(H^{-1}, L^2)} + \sup_{u_h \in U_{ad,h}} \|u_h\| \leq L. \quad (40)$$

From (40) and $u = z$, we have $\|z\|_{H^1(\Omega)} \leq \alpha^{-1} L$. Thus, for the term I_5, utilizing Lemma 2, we have

$$I_5 \leq \|\frac{\alpha}{2}u + p\|_{H^1(\Omega_h)} \|\tilde{z}_h - z\|_{H^{-1}(\Omega_h)} \leq c_2 L \|z\|_{H^1(\Omega_h)} h^2 \leq c_2 L^2 \alpha^{-1} h^2. \tag{41}$$

For terms I_6 and I_7, using Hölder's inequality, Lemma 1, and Lemma 2, we have

$$I_6 \leq \frac{\alpha}{4} \|u_h - u\|_{L^2(\Omega_h)}^2 + \frac{\alpha}{4} \|\tilde{z}_h - z\|_{L^2(\Omega_h)}^2 \leq \frac{\alpha}{4} \|u_h - u\|_{L^2(\Omega_h)}^2 + \frac{c_2^2 L^2 \alpha^{-1}}{4} h^2, \tag{42}$$

and

$$I_7 \leq \frac{1}{2} \|y - y_h\|_{L^2(\Omega_h)}^2 + 2 \|S_h - S\|_{\mathcal{L}(L^2,L^2)}^2 (\|\tilde{z}_h\|_{L^2(\Omega_h)}^2 + \|y_r\|_{L^2(\Omega_h)}^2)$$
$$+ \|S\|_{\mathcal{L}(H^{-1},L^2)} \|z - \tilde{z}_h\|_{H^{-1}(\Omega_h)}^2 \tag{43}$$
$$\leq \frac{1}{2} \|y - y_h\|_{L^2(\Omega_h)}^2 + 2 c_1^2 L^2 h^4 + c_2^2 L^3 \alpha^{-2} h^4.$$

Finally, about the term I_8, we have

$$I_8 \leq \|y - y_d\|_{L^2(\Omega_h)} \|S_h - S\|_{\mathcal{L}(L^2,L^2)} (\|\tilde{z}_h - z\|_{L^2(\Omega_h)} + \|z - z_h\|_{L^2(\Omega_h)})$$
$$\leq c_1 L h^2 (c_2 L \alpha^{-1} h + \|z - z_h\|_{L^2(\Omega_h)})$$
$$\leq \frac{\alpha}{4} \|z - z_h\|_{L^2(\Omega_h)}^2 + c_1 c_2 \alpha^{-1} L^2 h^3 + 4 c_1^2 L^2 \alpha^{-1} h^4. \tag{44}$$

Substituting (41), (42), (43), and (44) into (38) and rearranging, we get

$$\frac{\alpha}{2} \|u - u_h\|_{L^2(\Omega_h)}^2 + \frac{1}{2} \|y - y_h\|_{L^2(\Omega_h)}^2 \leq C(h^2 + \alpha^{-1} h^2 + \alpha^{-1} h^3 + \alpha^{-1} h^4 + \alpha^{-2} h^4),$$

where $C > 0$ is a properly chosen constant. Using again the assumption $|\Omega \setminus \Omega_h| \leq ch^2$, we can get

$$\frac{\alpha}{2} \|u - u_h\|_{L^2(\Omega)}^2 + \frac{1}{2} \|y - y_h\|_{L^2(\Omega)}^2 \leq C(h^2 + \alpha h^2 + \alpha^{-1} h^2 + h^3 + \alpha^{-1} h^4 + \alpha^{-2} h^4).$$

Corollary 1. *Let (y, u, z) be the optimal solution of problem $(\widetilde{P})$, and (y_h, u_h, z_h) be the optimal solution of problem $(\widetilde{DP}_h)$. For every $h_0 > 0$, $\alpha_0 > 0$, there is a constant $C > 0$ such that for all $0 < \alpha \leq \alpha_0$, $0 < h \leq h_0$ it holds*

$$\|u - u_h\|_{L^2(\Omega)} \leq C(\alpha^{-1} h + \alpha^{-\frac{3}{2}} h^2),$$

where C is a constant independent of h and α. □

An Inexact Heterogeneous ADMM Algorithm

In this section, we will introduce the ihADMM algorithm with the aim of solving $(\widetilde{\mathrm{DP}_h})$ to moderate accuracy. Firstly, let us define following stiffness and mass matrices:

$$K_h = \left(a_h(\phi_i, \phi_j)\right)_{i,j=1}^{N_h}, \quad M_h = \left(\int_{\Omega_h} \phi_i \phi_j \mathrm{d}x\right)_{i,j=1}^{N_h},$$

where the bilinear form $a_h(\cdot, \cdot)$ is defined as

$$a_h(y, v) = \int_{\Omega_h} \left(\sum_{i,j=1}^{n} a_{ji} y_{x_i} v_{x_i} + c_0 y v\right) \mathrm{d}x.$$

Due to the quadrature formulas (21) and (22), a lumped mass matrix $W_h = \mathrm{diag}\left(\int_{\Omega_h} \phi_i(x) \mathrm{d}x\right)_{i=1}^{N_h}$ is introduced. Moreover, by (24) in Proposition 1, we have the following results about the mass matrix M_h and the lump mass matrix W_h.

Proposition 2. $\forall\, z \in \mathbb{R}^{N_h}$, the following inequalities hold:

$$\|z\|_{M_h}^2 \le \|z\|_{W_h}^2 \le c \|z\|_{M_h}^2, \quad \text{where} \quad c = \begin{cases} 4 & \text{if } n = 2, \\ 5 & \text{if } n = 3. \end{cases}$$

An Inexact Heterogeneous ADMM Algorithm

Denoting by $y_{d,h} := \sum_{i=1}^{N_h} y_d^i \phi_i(x)$ and $y_{c,h} := \sum_{i=1}^{N_h} y_r^i \phi_i(x)$ the L^2-projection of y_d and y_r onto Y_h, respectively, and identifying discretized functions with their coefficient vectors, we can rewrite the problem $(\widetilde{\mathrm{DP}_h})$ as a matrix-vector form:

$$\begin{cases} \min_{(y,u,z) \in \mathbb{R}^{3N_h}} \dfrac{1}{2}\|y - y_d\|_{M_h}^2 + \dfrac{\alpha}{4}\|u\|_{M_h}^2 + \dfrac{\alpha}{4}\|z\|_{W_h}^2 + \|W_h z\|_1 \\ \text{s.t.} \quad K_h y = M_h(u + y_r), \\ \quad\quad u = z, \\ \quad\quad z \in [a, b]^{N_h}. \end{cases} \quad (\overline{\mathrm{DP}_h})$$

By Assumption 1, we have the stiffness matrix K_h is a symmetric positive definite matrix. Then problem $(\overline{\mathrm{DP}_h})$ can be rewritten as the following reduced form:

$$\begin{cases} \min_{(u,z)\in\mathbb{R}^{2N_h}} f(u) + g(z) \\ \text{s.t.} \quad u = z. \end{cases} \qquad (\overline{\text{RDP}}_h)$$

where

$$f(u) = \frac{1}{2}\|K_h^{-1}M_h(u + y_r) - y_d\|_{M_h}^2 + \frac{\alpha}{4}\|u\|_{M_h}^2,$$
$$g(z) = \frac{\alpha}{4}\|z\|_{W_h}^2 + \beta\|W_h z\|_1 + \delta_{[a,b]^{N_h}}(z). \qquad (45)$$

To solve $(\overline{\text{RDP}}_h)$ by using ADMM-type algorithm, we first introduce the augmented Lagrangian function for $(\overline{\text{RDP}}_h)$. According to three possible choices of norms ($\mathbb{R}^{N_h}$ norm, W_h-weighted norm, and M_h-weighted norm), for the augmented Lagrangian function, there are three versions as follows: for given $\sigma > 0$,

$$\mathcal{L}_\sigma^1(u, z; \lambda) := f(u) + g(z) + \langle \lambda, u - z \rangle + \frac{\sigma}{2}\|u - z\|^2, \qquad (46)$$

$$\mathcal{L}_\sigma^2(u, z; \lambda) := f(u) + g(z) + \langle \lambda, M_h(u - z) \rangle + \frac{\sigma}{2}\|u - z\|_{W_h}^2, \qquad (47)$$

$$\mathcal{L}_\sigma^3(u, z; \lambda) := f(u) + g(z) + \langle \lambda, M_h(u - z) \rangle + \frac{\sigma}{2}\|u - z\|_{M_h}^2. \qquad (48)$$

Then based on these three versions of augmented Lagrangian function, we give the following four versions of ADMM-type algorithm for $(\overline{\text{RDP}}_h)$ at k-th ineration: for given $\tau > 0$ and $\sigma > 0$,

$$\begin{cases} u^{k+1} = \arg\min_u f(u) + \langle \lambda^k, u - z^k \rangle + \sigma/2\|u - z^k\|^2, \\ z^{k+1} = \arg\min_z g(z) + \langle \lambda^k, u^{k+1} - z \rangle + \sigma/2\|u^{k+1} - z\|^2, \\ \lambda^{k+1} = \lambda^k + \tau\sigma(u^{k+1} - z^{k+1}). \end{cases} \qquad (\text{ADMM1})$$

$$\begin{cases} u^{k+1} = \arg\min_u f(u) + \langle \lambda^k, M_h(u - z^k) \rangle + \sigma/2\|u - z^k\|_{W_h}^2, \\ z^{k+1} = \arg\min_z g(z) + \langle \lambda^k, M_h(u^{k+1} - z) \rangle + \sigma/2\|u^{k+1} - z\|_{W_h}^2, \\ \lambda^{k+1} = \lambda^k + \tau\sigma(u^{k+1} - z^{k+1}). \end{cases}$$
$$(\text{ADMM2})$$

$$\begin{cases} u^{k+1} = \arg\min_u \ f(u) + \langle \lambda^k, M_h(u - z^k) \rangle + \sigma/2 \|u - z^k\|_{M_h}^2, \\ z^{k+1} = \arg\min_z \ g(z) + \langle \lambda^k, M_h(u^{k+1} - z) \rangle + \sigma/2 \|u^{k+1} - z\|_{M_h}^2, \\ \lambda^{k+1} = \lambda^k + \tau\sigma(u^{k+1} - z^{k+1}). \end{cases}$$
(ADMM3)

$$\begin{cases} u^{k+1} = \arg\min_u \ f(u) + \langle \lambda^k, M_h(u - z^k) \rangle + \sigma/2 \|u - z^k\|_{M_h}^2, \\ z^{k+1} = \arg\min_z \ g(z) + \langle \lambda^k, M_h(u^{k+1} - z) \rangle + \sigma/2 \|u^{k+1} - z\|_{W_h}^2, \\ \lambda^{k+1} = \lambda^k + \tau\sigma(u^{k+1} - z^{k+1}). \end{cases}$$
(ADMM4)

As one may know, (ADMM1) is actually the classical ADMM for $(\overline{\text{RDP}_h})$. The remaining three ADMM-type algorithms are proposed based on the structure of $(\overline{\text{RDP}_h})$. Now, let us start to analyze and compare the advantages and disadvantages of the four algorithms. Firstly, we focus on the z-subproblem in each algorithm. Since both identity matrix I and lumped mass matrix W_h are diagonal, it is clear that all the z-subproblems in (ADMM1), (ADMM2), and (ADMM4) have a closed form solution, except for the z-subproblem in (ADMM3). Specifically, for z-subproblem in (ADMM1), the closed-form solution could be given by

$$z^k = P_{U_{ad}} \left((\frac{\alpha}{2} W_h + \sigma I)^{-1} W_h \text{soft}(W_h^{-1}(\sigma u^{k+1} + \lambda^k), \beta) \right). \tag{49}$$

Similarly, for z-subproblems in (ADMM2) and (ADMM4), the closed-form solutions could be given by

$$z^{k+1} = P_{U_{ad}} \left(\frac{1}{\sigma + 0.5\alpha} \text{soft}\left(\sigma u^{k+1} + W_h^{-1} M_h \lambda^k, \beta\right) \right), \tag{50}$$

Next, let us analyze the structure of u-subproblem in each algorithm. For (ADMM1), the first subproblem at k-th iteration is equivalent to solving the following linear system:

$$\begin{bmatrix} M_h & 0 & K_h \\ 0 & \frac{\alpha}{2} M_h + \sigma I & -M_h \\ K_h & -M_h & 0 \end{bmatrix} \begin{bmatrix} y^{k+1} \\ u^{k+1} \\ p^{k+1} \end{bmatrix} = \begin{bmatrix} M_h y_d \\ \sigma z^k - \lambda^k \\ M_h y_r \end{bmatrix}. \tag{51}$$

Similarly, the u-subproblem in (ADMM2) can be converted into the following linear system:

$$\begin{bmatrix} M_h & 0 & K_h \\ 0 & \frac{\alpha}{2}M_h + \sigma W_h & -M_h \\ K_h & -M_h & 0 \end{bmatrix} \begin{bmatrix} y^{k+1} \\ u^{k+1} \\ p^{k+1} \end{bmatrix} = \begin{bmatrix} M_h y_d \\ \sigma W_h z^k - M_h \lambda^k \\ M_h y_r \end{bmatrix}. \qquad (52)$$

However, the u-subproblem in both (ADMM3) and (ADMM4) can be rewritten as

$$\begin{bmatrix} M_h & 0 & K_h \\ 0 & (0.5\alpha + \sigma)M_h & -M_h \\ K_h & -M_h & 0 \end{bmatrix} \begin{bmatrix} y^{k+1} \\ u^{k+1} \\ p^{k+1} \end{bmatrix} = \begin{bmatrix} M_h y_d \\ M_h (\sigma z^k - \lambda^k) \\ M_h y_r \end{bmatrix}. \qquad (53)$$

In (53), since $p^{k+1} = (0.5\alpha + \sigma)u^{k+1} - \sigma z^k + \lambda^k$, it is obvious that (53) can be reduced into the following system by eliminating the variable p without any computational cost:

$$\begin{bmatrix} \frac{1}{0.5\alpha+\sigma} M_h & K_h \\ -K_h & M_h \end{bmatrix} \begin{bmatrix} y^{k+1} \\ u^{k+1} \end{bmatrix} = \begin{bmatrix} \frac{1}{0.5\alpha+\sigma}(K_h(\sigma z^k - \lambda^k) + M_h y_d) \\ -M_h y_r \end{bmatrix}, \qquad (54)$$

while, reduced forms of (51) and (52), both involve the inversion of M_h.

For abovementioned reasons, we prefer to use (ADMM4), which is called the heterogeneous ADMM (hADMM). However, in general, it is expensive and unnecessary to exactly compute the solution of saddle point system (54) even if it is doable, especially at the early stage of the whole process. Based on the structure of (54), it is a natural idea to use the iterative methods such as some Krylov-based methods. Hence, taking the inexactness of the solution of u-subproblem into account, a more practical inexact heterogeneous ADMM (ihADMM) algorithm is proposed.

Due to the inexactness of the proposed algorithm, we first introduce an error tolerance. Throughout this chapter, let $\{\epsilon_k\}$ be a summable sequence of nonnegative numbers, and define

$$C_1 := \sum_{k=0}^{\infty} \epsilon_k \leq \infty, \quad C_2 := \sum_{k=0}^{\infty} \epsilon_k^2 \leq \infty. \qquad (55)$$

The details of our ihADMM algorithm is shown in Algorithm 1 to solve $(\overline{DP_h})$.

Convergence Results of ihADMM

For the ihADMM (Algorithm 1), in this section, we establish the global convergence and the iteration complexity results in non-ergodic sense for the sequence generated by Algorithm 1.

Algorithm 1 Inexact heterogeneous ADMM algorithm for $(\overline{DP}_h)$

Input : $(z^0, u^0, \lambda^0) \in \text{dom}(\delta_{[a,b]}(\cdot)) \times \mathbb{R}^n \times \mathbb{R}^n$ and parameters $\sigma > 0, \tau > 0$. let $\{\epsilon_k\}$ be a summable sequence of nonnegative numbers, and define

$$C_1 := \sum_{k=0}^{\infty} \epsilon_k \leq \infty, \quad C_2 := \sum_{k=0}^{\infty} \epsilon_k^2 \leq \infty.$$

Set $k = 1$.
Output : u^k, z^k, λ^k
Step 1 Find an minimizer (inexact)

$$u^{k+1} = \arg\min f(u) + (M_h \lambda^k, u - z^k) + \frac{\sigma}{2} \|u - z^k\|_{M_h}^2 - \langle \delta^k, u \rangle,$$

where the error vector δ^k satisfies $\|\delta^k\|_2 \leq \epsilon_k$
Step 2 Compute z^k as follows:

$$z^{k+1} = \arg\min g(z) + (M_h \lambda^k, u^{k+1} - z) + \frac{\sigma}{2} \|u^{k+1} - z\|_{W_h}^2$$

Step 3 Compute

$$\lambda^{k+1} = \lambda^k + \tau\sigma(u^{k+1} - z^{k+1}).$$

Step 4 If a termination criterion is not met, set $k := k + 1$ and go to Step 1

Before giving the proof of Theorem 5, we first provide a lemma, which is useful for analyzing the non-ergodic iteration complexity of ihADMM and introduced in Chen and Toh (2017).

Lemma 3. *If a sequence* $\{a_i\} \in \mathbb{R}$ *satisfies the following conditions:*

$$a_i \geq 0 \text{ for any } i \geq 0 \quad \text{and} \quad \sum_{i=0}^{\infty} a_i = \bar{a} < \infty.$$

Then we have $\min_{i=1,\ldots,k} \{a_i\} \leq \frac{\bar{a}}{k}$, *and* $\lim_{k \to \infty} \{k \cdot \min_{i=1,\ldots,k} \{a_i\}\} = 0.$ □

For the convenience of the iteration complexity analysis below, we define the function $R_h : (u, z, \lambda) \to [0, \infty)$ by

$$R_h(u, z, \lambda) = \|M_h \lambda + \nabla f(u)\|^2 + \text{dist}^2(0, -M_h \lambda + \partial g(z)) + \|u - z\|^2. \quad (56)$$

By the definitions of $f(u)$ and $g(z)$ in (45), it is obvious that $f(u)$ and $g(z)$ are both closed, proper, and convex functions. Since M_h and K_h are symmetric positive definite matrices, we know the gradient operator ∇f is strongly monotone, and we have

$$\langle \nabla f(u_1) - \nabla f(u_2), u_1 - u_2 \rangle = \|u_1 - u_2\|_{\Sigma_f}^2, \tag{57}$$

where $\Sigma_f = \frac{\alpha}{2} M_h + M_h K_h^{-1} M_h K_h^{-1} M_h$ is symmetric positive definite. Moreover, the subdifferential operator ∂g is a maximal monotone operators, e.g.,

$$\langle \varphi_1 - \varphi_2, z_1 - z_2 \rangle \geq \frac{\alpha}{2} \|z_1 - z_2\|_{W_h}^2 \quad \forall \varphi_1 \in \partial g(z_1), \varphi_2 \in \partial g(z_2). \tag{58}$$

For the subsequent convergence analysis, we denote

$$\bar{u}^{k+1} := \arg\min f(u) + \langle M_h \lambda^k, u - z^k \rangle + \frac{\sigma}{2} \|u - z^k\|_{M_h}^2, \tag{59}$$

$$\bar{z}^{k+1} := P_{U_{ad}} \left(\frac{1}{\sigma + 0.5\alpha} \text{soft} \left(\sigma \bar{u}^{k+1} + W_h^{-1} M_h \lambda^k, \beta \right) \right), \tag{60}$$

which are the exact solutions at the $(k+1)$-th iteration in Algorithm 1. The following results show the gap between (u^{k+1}, z^{k+1}) and $(\bar{u}^{k+1}, \bar{z}^{k+1})$ in terms of the given error tolerance $\|\delta^k\|_2 \leq \epsilon_k$.

Lemma 4. *Let $\{(u^{k+1}, z^{k+1})\}$ be the sequence generated by Algorithm 1, and $\{\bar{u}^{k+1}\}, \{\bar{z}^{k+1}\}$ be defined in (59) and (60). Then for any $k \geq 0$, we have*

$$\|u^{k+1} - \bar{u}^{k+1}\| = \|(\sigma M_h + \Sigma_f)^{-1} \delta^k\| \leq \rho \epsilon_k, \tag{61}$$

$$\|z^{k+1} - \bar{z}^{k+1}\| \leq \frac{\sigma}{\sigma + 0.5\alpha} \|u^{k+1} - \bar{u}^{k+1}\| \leq \frac{\rho \sigma}{\sigma + 0.5\alpha} \epsilon_k, \tag{62}$$

where $\rho := \|(\sigma M_h + \Sigma_f)^{-1}\|$. □

Proof. By the optimality conditions at point (u^{k+1}, z^{k+1}) and $(\bar{u}^{k+1}, \bar{z}^{k+1})$, we have

$$\Sigma_f u^{k+1} - M_h K_h^{-1} M_h y_d + M_h \lambda^k + \sigma M_h (u^{k+1} - z^k) - \delta^k = 0,$$

$$\Sigma_f \bar{u}^{k+1} - M_h K_h^{-1} M_h y_d + M_h \lambda^k + \sigma M_h (\bar{u}^{k+1} - z^k) = 0;$$

thus,

$$u^{k+1} - \bar{u}^{k+1} = (\sigma M_h + \Sigma_f)^{-1} \delta^k$$

which implies (61). From (50) and (60), and the fact that the projection operator $\Pi_{[a,b]}(\cdot)$ and soft thresholding operator $\text{soft}(\cdot, \cdot)$ are nonexpansive, we get

$$\|z^{k+1} - \bar{z}^{k+1}\| \leq \frac{\sigma}{\sigma + 0.5\alpha} \|u^{k+1} - \bar{u}^{k+1}\|,$$

which implies (62). The proof is completed.

Next, for $k \geq 0$, we define

$$r^k = u^k - z^k, \quad \bar{r}^k = \bar{u}^k - \bar{z}^k$$
$$\tilde{\lambda}^{k+1} = \lambda^k + \sigma r^{k+1}, \quad \bar{\lambda}^{k+1} = \lambda^k + \tau\sigma \bar{r}^{k+1}, \quad \hat{\lambda}^{k+1} = \lambda^k + \sigma \bar{r}^{k+1},$$

and give two inequalities which is essential for establishing both the global convergence and the iteration complexity of our ihADMM. For the details of the proof, one can see in Appendix.

Proposition 3. *Let* $\{(u^k, z^k, \lambda^k)\}$ *be the sequence generated by Algorithm 1 and* (u^*, z^*, λ^*) *be the KKT point of problem* $(\overline{\text{RDP}}_h)$. *Then for* $k \geq 0$, *we have*

$$\langle \delta^k, u^{k+1} - u^* \rangle + \frac{1}{2\tau\sigma} \|\lambda^k - \lambda^*\|^2_{M_h} + \frac{\sigma}{2} \|z^k - z^*\|^2_{M_h}$$
$$- \frac{1}{2\tau\sigma} \|\lambda^{k+1} - \lambda^*\|^2_{M_h} - \frac{\sigma}{2} \|z^{k+1} - z^*\|^2_{M_h} \geq \|u^{k+1} - u^*\|^2_T$$
$$+ \frac{\sigma}{2} \|z^{k+1} - z^*\|^2_{2W_h - M_h} + \frac{\sigma}{2} \|r^{k+1}\|^2_{W_h - \tau M_h} + \frac{\sigma}{2} \|u^{k+1} - z^k\|^2_{M_h}, \quad (63)$$

where $T := \Sigma_f - \frac{\sigma}{2}(W_h - M_h)$. □

Proposition 4. *Let* $\{(u^k, z^k, \lambda^k)\}$ *be the sequence generated by Algorithm 1,* (u^*, z^*, λ^*) *be the KKT point of the problem* $(\overline{\text{RDP}}_h)$ *and* $\{\bar{u}^k\}$ *and* $\{\bar{z}^k\}$ *be two sequences defined in (59) and (60), respectively. Then for* $k \geq 0$, *we have*

$$\frac{1}{2\tau\sigma} \|\lambda^k - \lambda^*\|^2_{M_h} + \frac{\sigma}{2} \|z^k - z^*\|^2_{M_h} - \frac{1}{2\tau\sigma} \|\bar{\lambda}^{k+1} - \lambda^*\|^2_{M_h} - \frac{\sigma}{2} \|\bar{z}^{k+1} - z^*\|^2_{M_h}$$
$$\geq \|\bar{u}^{k+1} - u^*\|^2_T + \frac{\sigma}{2} \|\bar{z}^{k+1} - z^*\|^2_{2W_h - M_h} + \frac{\sigma}{2} \|\bar{r}^{k+1}\|^2_{W_h - \tau M_h} + \frac{\sigma}{2} \|\bar{u}^{k+1} - z^k\|^2_{M_h}, \quad (64)$$

where $T := \Sigma_f - \frac{\sigma}{2}(W_h - M_h)$. □

Then based on former results, we have the following convergence results.

Theorem 5. *Let* $(y^*, u^*, z^*, p^*, \lambda^*)$ *be the KKT point of* $(\overline{\text{DP}}_h)$, *then the sequence* $\{(u^k, z^k, \lambda^k)\}$ *is generated by Algorithm 1 with the associated state* $\{y^k\}$ *and adjoint state* $\{p^k\}$, *and then for any* $\tau \in (0, 1]$ *and* $\sigma \in (0, \frac{1}{4}\alpha]$, *we have*

$$\lim_{k \to \infty} \{\|u^k - u^*\| + \|z^k - z^*\| + \|\lambda^k - \lambda^*\|\} = 0 \quad (65)$$

16 Numerical Solution for Sparse PDE Constrained Optimization

$$\lim_{k \to \infty} \{\|y^k - y^*\| + \|p^k - p^*\|\} = 0 \tag{66}$$

Moreover, there exists a constant C only depending on the initial point (u^0, z^0, λ^0) and the optimal solution (u^*, z^*, λ^*) such that for $k \geq 1$,

$$\min_{1 \leq i \leq k} \{R_h(u^i, z^i, \lambda^i)\} \leq \frac{C}{k}, \quad \lim_{k \to \infty} \left(k \times \min_{1 \leq i \leq k} \{R_h(u^i, z^i, \lambda^i)\} \right) = 0. \tag{67}$$

where $R_h(\cdot)$ is defined as in (56).

Proof. It is easy to see that (u^*, z^*) is the unique optimal solution of discrete problem $(\overline{\text{RDP}_h})$ if and only if there exists a Lagrangian multiplier λ^* such that the following Karush-Kuhn-Tucker (KKT) conditions hold:

$$-M_h \lambda^* = \nabla f(u^*), \tag{68a}$$

$$M_h \lambda^* \in \partial g(z^*), \tag{68b}$$

$$u^* = z^*. \tag{68c}$$

In the inexact heterogeneous ADMM iteration scheme, the optimality conditions for (u^{k+1}, z^{k+1}) are

$$\delta^k - (M_h \lambda^k + \sigma M_h(u^{k+1} - z^k)) = \nabla f(u^{k+1}), \tag{69a}$$

$$M_h \lambda^k + \sigma W_h(u^{k+1} - z^{k+1}) \in \partial g(z^{k+1}). \tag{69b}$$

Next, let us first prove the **global convergence of iteration sequences**, e.g., establish the proof of (65) and (66).

The first step is to show that $\{(u^k, z^k, \lambda^k)\}$ is bounded. We define the following sequence θ^k and $\bar{\theta}^k$ with:

$$\begin{aligned}
\theta^k &= \left(\frac{1}{\sqrt{2\tau\sigma}} M_h^{\frac{1}{2}}(\lambda^k - \lambda^*), \sqrt{\frac{\sigma}{2}} M_h^{\frac{1}{2}}(z^k - z^*) \right), \\
\bar{\theta}^k &= \left(\frac{1}{\sqrt{2\tau\sigma}} M_h^{\frac{1}{2}}(\bar{\lambda}^k - \lambda^*), \sqrt{\frac{\sigma}{2}} M_h^{\frac{1}{2}}(\bar{z}^k - z^*) \right).
\end{aligned} \tag{70}$$

According to Proposition 1, for any $\tau \in (0, 1]$ and $\sigma \in (0, \frac{1}{4}\alpha]$ for, we have $\Sigma_f - \frac{\sigma}{2}(W_h - M_h) \succ 0$, and $W_h - \tau M_h \succ 0$. Then, by Proposition 4, we get $\|\bar{\theta}^{k+1}\|^2 \leq \|\theta^k\|^2$. As a result, we have

$$\|\theta^{k+1}\| \leq \|\bar{\theta}^{k+1}\| + \|\bar{\theta}^{k+1} - \theta^{k+1}\| = \|\theta^k\| + \|\bar{\theta}^{k+1} - \theta^{k+1}\|. \tag{71}$$

Employing Lemma 4, we get

$$\|\bar{\theta}^{k+1} - \theta^{k+1}\|^2 = \frac{1}{2\tau\sigma}\|\bar{\lambda}^{k+1} - \lambda^{k+1}\|_{M_h}^2 + \frac{\sigma}{2}\|\bar{z}^{k+1} - z^{k+1}\|_{M_h}^2 \qquad (72)$$
$$\leq (2\tau + 1/2)\sigma\|M_h\|\rho^2\epsilon_k^2 \leq 5/2\sigma\|M_h\|\rho^2\epsilon_k^2,$$

which implies $\|\bar{\theta}^{k+1} - \theta^{k+1}\| \leq \sqrt{5/2\sigma\|M_h\|}\rho\epsilon_k$. Hence, for any $k \geq 0$, we have

$$\|\theta^{k+1}\| \leq \|\theta^k\| + \sqrt{5/2\sigma\|M_h\|}\rho\epsilon_k \leq \|\theta^0\| + \sqrt{5/2\sigma\|M_h\|}\rho\sum_{k=0}^{\infty}\epsilon_k$$
$$= \|\theta^0\| + \sqrt{5/2\sigma\|M_h\|}\rho C_1 \equiv \bar{\rho}. \qquad (73)$$

From $\|\bar{\theta}^{k+1}\| \leq \|\theta^k\|$, for any $k \geq 0$, we also have $\|\bar{\theta}^{k+1}\| \leq \bar{\rho}$. Therefore, the sequences $\{\theta^k\}$ and $\{\bar{\theta}^k\}$ are bounded. From the definition of $\{\theta^k\}$ and the fact that $M_h \succ 0$, we can see that the sequences $\{\lambda^k\}$ and $\{z^k\}$ are bounded. Moreover, from updating technique of λ^k, we know $\{u^k\}$ is also bounded. Thus, due to the boundedness of the sequence $\{(u^k, z^k, \lambda^k)\}$, we know the sequence has a subsequence $\{(u^{k_i}, z^{k_i}, \lambda^{k_i})\}$ which converges to an accumulation point $(\bar{u}, \bar{z}, \bar{\lambda})$. Next we should show that $(\bar{u}, \bar{z}, \bar{\lambda})$ is a KKT point and equal to (u^*, z^*, λ^*).

Again employing Proposition 4, we can derive

$$\sum_{k=0}^{\infty}\left(\|\bar{u}^{k+1} - u^*\|_T^2 + \frac{\sigma}{2}\|\bar{z}^{k+1} - z^*\|_{2W_h - M_h}^2 + \frac{\sigma}{2}\|\bar{r}^{k+1}\|_{W_h - \tau M_h}^2 + \frac{\sigma}{2}\|\bar{u}^{k+1} - z^k\|_{M_h}^2\right)$$
$$\leq \sum_{k=0}^{\infty}(\|\theta^k\|^2 - \|\theta^{k+1}\|^2 + \|\theta^{k+1}\|^2 - \|\bar{\theta}^{k+1}\|^2) \leq \|\theta^0\|^2 + 2\bar{\rho}\sqrt{5/2\sigma\|M_h\|}\rho C_1 < \infty. \qquad (74)$$

Note that $T \succ 0$, $W_h - M_h \succ 0$, $W_h - \tau M_h \succ 0$ and $M_h \succ 0$, then we have

$$\lim_{k\to\infty}\|\bar{u}^{k+1} - u^*\| = 0, \quad \lim_{k\to\infty}\|\bar{z}^{k+1} - z^*\| = 0,$$
$$\lim_{k\to\infty}\|\bar{r}^{k+1}\| = 0, \quad \lim_{k\to\infty}\|\bar{u}^{k+1} - z^k\| = 0. \qquad (75)$$

From the Lemma 4, we can get

$$\|u^{k+1} - u^*\| \leq \|\bar{u}^{k+1} - u^*\| + \|u^{k+1} - \bar{u}^{k+1}\| \leq \|\bar{u}^{k+1} - u^*\| + \rho\epsilon_k,$$
$$\|z^{k+1} - z^*\| \leq \|\bar{z}^{k+1} - z^*\| + \|z^{k+1} - \bar{z}^{k+1}\| \leq \|\bar{z}^{k+1} - z^*\| + \rho\epsilon_k. \qquad (76)$$

From the fact that $\lim_{k\to\infty}\epsilon_k = 0$ and (75), by taking the limit of both sides of (76), we have

$$\lim_{k\to\infty} \|u^{k+1} - u^*\| = 0, \quad \lim_{k\to\infty} \|z^{k+1} - z^*\| = 0,$$
$$\lim_{k\to\infty} \|r^{k+1}\| = 0, \quad \lim_{k\to\infty} \|u^{k+1} - z^k\| = 0. \tag{77}$$

Now taking limits for $k_i \to \infty$ on both sides of (69a), we have

$$\lim_{k_i\to\infty} (\delta^{k_i} - (M_h \lambda^{k_i} + \sigma M_h(u^{k_i+1} - z^{k_i}))) = \lim_{k_i\to\infty} \nabla f(u^{k_i+1}),$$

which results in $-M_h\bar{\lambda} = \nabla f(u^*)$. Then from (68a), we know $\bar{\lambda} = \lambda^*$. At last, to complete the proof, we need to show that λ^* is the limit of the sequence of $\{\lambda^k\}$. From (73), we have for any $k > k_i$, $\|\theta^{k+1}\| \le \|\theta^{k_i}\| + \sqrt{5/2\sigma\|M_h\|}\bar{\rho}\sum_{j=k_i}^{k}\epsilon_j$. Since $\lim_{k_i\to\infty} \|\theta^{k_i}\| = 0$ and $\sum_{k=0}^{\infty}\epsilon_k < \infty$, we have that $\lim_{k\to\infty}\|\theta^k\| = 0$, which implies $\lim_{k\to\infty}\|\lambda^{k+1} - \lambda^*\| = 0$. Hence, we have proved the convergence of the sequence $\{(u^{k+1}, z^{k+1}, \lambda^{k+1})\}$, which completes the proof of (65). For the proof of (66), it is easy to show by the definition of the sequence $\{(y^k, p^k)\}$; here we omit it.

At last, we establish the proof of (67), e.g., **the iteration complexity results in non-ergodic sense for the sequence generated by the ihADMM.**

Firstly, by the optimality condition (69a) and (69b) for (u^{k+1}, z^{k+1}), we have

$$\delta^k + (\tau - 1)\sigma M_h r^{k+1} - \sigma M_h(z^{k+1} - z^k) = M_h\lambda^{k+1} + \nabla f(u^{k+1}), \tag{78a}$$
$$\sigma(W_h - \tau M_h)r^{k+1} \in -M_h\lambda^{k+1} + \partial g(z^{k+1}). \tag{78b}$$

By the definition of R_h and denoting $w^{k+1} := (u^{k+1}, z^{k+1}, \lambda^{k+1})$, we derive

$$R_h(w^{k+1}) = \|M_h\lambda^{k+1} + \nabla f(u^{k+1})\|^2 + \text{dist}^2(0, -M_h\lambda^{k+1} + \partial g(z^{k+1}))$$
$$+ \|u^{k+1} - z^{k+1}\|^2 \le 2\|\delta^k\|^2 + \eta\|r^{k+1}\|^2 + 4\sigma^2\|M_h\|\|u^{k+1} - z^k\|_{M_h}^2, \tag{79}$$

where $\eta := 2(\tau - 1)^2\sigma^2\|M_h\|^2 + 2\sigma^2\|M_h\|^2 + \sigma^2\|W_h - \tau M_h\|^2 + 1$.

In order to get a upper bound for $R_h(w^{k+1})$, we will use (63) in Proposition 3. First, by the definition of θ^k and (73), for any $k \ge 0$ we can easily have

$$\|\lambda^k - \lambda^*\| \le \bar{\rho}\sqrt{\frac{2\tau\sigma}{\|M_h^{-1}\|}}, \quad \|z^k - z^*\| \le \bar{\rho}\sqrt{\frac{2}{\sigma\|M_h^{-1}\|}}. \qquad \square$$

Next, we should give a upper bound for $\langle \delta^k, u^{k+1} - u^* \rangle$:

$$\langle \delta^k, u^{k+1} - u^* \rangle \le \|\delta^k\|(\|u^{k+1} - z^{k+1}\| + \|z^{k+1} - z^*\|)$$

$$\le \left(\left(1 + \frac{2}{\sqrt{\tau}}\right)\frac{2\sqrt{2}\bar{\rho}}{\sqrt{\tau\sigma\|M_h^{-1}\|}}\right)\|\delta^k\| \equiv \bar{\eta}\|\delta^k\|. \tag{80}$$

Then by (63) in Proposition 3, we have

$$\sum_{k=0}^{\infty}\left(\frac{\sigma}{2}\|r^{k+1}\|_{W_h-\tau M_h}^2 + \frac{\sigma}{2}\|u^{k+1} - z^k\|_{M_h}^2\right) \le \sum_{k=0}^{\infty}(\|\theta^k\| - \|\theta^{k+1}\|) + \sum_{k=0}^{\infty}\langle \delta^k, u^{k+1} - u^*\rangle$$

$$\le \|\theta^0\| + \bar{\eta}\sum_{k=0}^{\infty}\|\delta^k\| \le \|\theta^0\| + \bar{\eta}\sum_{k=0}^{\infty}\epsilon^k = \|\theta^0\| + \bar{\eta}C_1. \tag{81}$$

Hence,

$$\sum_{k=0}^{\infty}\|r^{k+1}\|^2 \le \frac{2(\|\theta^0\| + \bar{\eta}C_1)}{\sigma\|(W_h - \tau M_h)^{-1}\|}, \quad \sum_{k=0}^{\infty}\|u^{k+1} - z^k\|_{M_h}^2 \le \frac{2(\|\theta^0\| + \bar{\eta}C_1)}{\sigma}. \tag{82}$$

By substituting (82) to (79), we have

$$\sum_{k=0}^{\infty} R_h(w^{k+1}) \le 2\sum_{k=0}^{\infty}\|\delta^k\|^2 + \eta\sum_{k=0}^{\infty}\|r^{k+1}\|^2 + 4\sigma^2\|M_h\|\sum_{k=0}^{\infty}\|u^{k+1} - z^k\|_{M_h}^2$$

$$\le C := 2C_2 + \eta\frac{2(\|\theta^0\| + \bar{\eta}C_1)}{\sigma\|(W_h - \tau M_h)^{-1}\|} + 4\sigma^2\|M_h\|\frac{2(\|\theta^0\| + \bar{\eta}C_1)}{\sigma} \tag{83}$$

Thus, by Lemma 3, we know (67) holds. Therefore, combining the obtained global convergence results, we complete the whole proof of the Theorem 5. □

An Inexact Majorized Accelerated Block Coordinate Descent Method for (D_h)

In this section, we consider solving problem (P) by a duality-based approach. Thus, for the purpose of numerical implementation, we first give the finite element discretizations of (D) as follows:

$$\min_{\mu,\lambda,p \in \mathbb{R}^{N_h}} \Phi_h(\mu, \lambda, p) := \frac{1}{2}\|K_h p - M_h y_d\|_{M_h^{-1}}^2 + \frac{1}{2\alpha}\|\lambda + \mu - p\|_{M_h}^2 + \langle M_h y_r, p\rangle$$

$$+ \delta_{[-\beta,\beta]}(\lambda) + \delta_{[a,b]}^*(M_h\mu) - \frac{1}{2}\|y_d\|_{M_h}^2. \tag{D_h}$$

It is clear that problem (D$_h$) is a convex composite minimization problem whose objective is the sum of a coupled quadratic function involving three blocks of variables and two separable nonsmooth functions involving only the first and second block, respectively. In the following sections, benefiting from the structure of (D$_h$), we aim to propose an efficient and fast algorithm to solve it.

An Inexact Block Symmetric Gauss-Seidel Iteration

We first introduce the symmetric Gauss-Seidel (sGS) technique proposed recently by Li, Sun, and Toh (Li et al. 2016). It is a powerful tool to solve a convex minimization problem whose objective is the sum of a multi-block quadratic function and a nonsmooth function involving only the first block, which plays an important role in our subsequent algorithms designs for solving the PDE-constraints optimization problems.

Let $s \geq 2$ be a given integer and $\mathcal{X} := \mathcal{X}_1 \times \mathcal{X}_2 \times \ldots \times \mathcal{X}_s$ where each $\mathcal{X}_i$ is a real finite dimensional Euclidean space. The sGS technique aims to solve the following unconstrained nonsmooth convex optimization problem approximately:

$$\min \phi(x_1) + \frac{1}{2}\langle x, \mathcal{H}x\rangle - \langle r, x\rangle, \tag{84}$$

where $x \equiv (x_1, \ldots, x_s) \in \mathcal{X}$ with $x_i \in \mathcal{X}_i, i = 1, \ldots, s, \phi : \mathcal{X}_1 \to (-\infty, +\infty]$ is a closed proper convex function, $\mathcal{H} : \mathcal{X} \to \mathcal{X}$ is a given self-adjoint positive semidefinite linear operator, and $r \equiv (r_1, \ldots, r_s) \in \mathcal{X}$ is a given vector.

For notational convenience, we denote the quadratic function in (84) as

$$h(x) := \frac{1}{2}\langle x, \mathcal{H}x\rangle - \langle r, x\rangle, \tag{85}$$

and the block decomposition of the operator $\mathcal{H}$ as

$$\mathcal{H}x := \begin{pmatrix} \mathcal{H}_{11} & \mathcal{H}_{12} & \cdots & \mathcal{H}_{1s} \\ \mathcal{H}_{12}^* & \mathcal{H}_{22} & \cdots & \mathcal{H}_{2s} \\ \vdots & \vdots & \ddots & \vdots \\ \mathcal{H}_{1s}^* & \mathcal{H}_{2s}^* & \cdots & \mathcal{H}_{ss} \end{pmatrix} \begin{pmatrix} x_1 \\ x_2 \\ \vdots \\ x_s \end{pmatrix}, \tag{86}$$

where $\mathcal{H}_{ii} : \mathcal{X}_i \to \mathcal{X}_i, i = 1, \ldots, s$ are self-adjoint positive semidefinite linear operators and $\mathcal{H}_{ij} : \mathcal{X}_j \to \mathcal{X}_i, i = 1, \ldots, s-1, j > i$ are linear maps whose adjoints are given by $\mathcal{H}_{ij}^*$. Here, we assume that $\mathcal{H}_{ii} \succ 0, \forall i = 1, \ldots, s$. Then, we consider a splitting of $\mathcal{H}$:

$$\mathcal{H} = \mathcal{D} + \mathcal{U} + \mathcal{U}^*, \tag{87}$$

where

$$\mathcal{U} := \begin{pmatrix} 0 & \mathcal{H}_{12} & \cdots & \mathcal{H}_{1s} \\ & \ddots & \cdots & \mathcal{H}_{2s} \\ & & \ddots & \mathcal{H}_{(s-1)s} \\ & & & 0 \end{pmatrix}, \qquad (88)$$

denotes the strict upper triangular part of $\mathcal{H}$ and $\mathcal{D} := \mathrm{Diag}(\mathcal{H}_{11}, \ldots, \mathcal{H}_{ss}) \succ 0$ is the diagonal of $\mathcal{H}$. For later discussions, we also define the following self-adjoint positive semidefinite linear operator:

$$\mathrm{sGS}(\mathcal{H}) := \mathcal{T} = \mathcal{U}\mathcal{D}^{-1}\mathcal{U}^*. \qquad (89)$$

For any $x \in \mathcal{X}$, we define

$$x_{\leq i} := (x_1, x_2, \ldots, x_i), \quad x_{\geq i} := (x_i, x_{i+1}, \ldots, x_s), \quad i = 1, \ldots, s,$$

with the convention $x_{\leq 0} = x_{\geq 0} = \emptyset$. Moreover, in order to solve problem (84) inexactly, we introduce the following two error tolerance vectors:

$$\delta' := (\delta'_1, \ldots, \delta'_s), \quad \delta := (\delta_1, \ldots, \delta_s),$$

with $\delta'_1 = \delta_1$. Define

$$\Delta(\delta', \delta) = \delta + \mathcal{U}\mathcal{D}^{-1}(\delta - \delta'). \qquad (90)$$

Given $\bar{x} \in \mathcal{X}$, we consider solving the following problem:

$$x^+ := \arg\min_x \left\{ \phi(x_1) + h(x) + \frac{1}{2}\|x - \bar{x}\|_{\mathcal{T}}^2 - \langle x, \Delta(\delta', \delta) \rangle \right\}, \qquad (91)$$

where $\Delta(\delta', \delta)$ could be regarded as the error term. Then, the following sGS decomposition theorem, which is established by Li et al. in (2015), shows that computing x^+ in (91) is equivalent to computing in an inexact block symmetric Gauss-Seidel-type sequential updating of the variables $x_1, \ldots, x_s$.

Theorem 6 (Li et al. 2015, Theorem 2.1). *Assume that the self-adjoint linear operators $\mathcal{H}_{ii}$ are positive definite for all $i = 1, \ldots, s$. Then, it holds that*

$$\mathcal{H} + \mathcal{T} = (\mathcal{D} + \mathcal{U})\mathcal{D}^{-1}(\mathcal{D} + \mathcal{U}^*) \succ 0. \qquad (92)$$

Furthermore, given $\bar{x} \in \mathcal{X}$, for $i = s, \ldots, 2$, suppose we have computed $x'_i \in \mathcal{X}_i$ defined as follows:

$$x_i' := \arg\min_{x_i \in \mathcal{X}_i} \phi(\bar{x}_1) + h(\bar{x}_{\leq i-1}, x_i, x'_{\geq i+1}) - \langle \delta_i', x_i \rangle$$

$$= \mathcal{H}_{ii}^{-1} \left(r_i + \delta_i' - \sum_{j=1}^{i-1} \mathcal{H}_{ji}^* \bar{x}_j - \sum_{j=i+1}^{s} \mathcal{H}_{ij} x_j' \right), \tag{93}$$

then the optimal solution x^+ defined by (91) can be obtained exactly via

$$\begin{cases} x_1^+ = \arg\min_{x_1 \in \mathcal{X}_1} \phi(x_1) + h(x_1, x'_{\geq 2}) - \langle \delta_1, x_1 \rangle, \\ x_i^+ = \arg\min_{x_i \in \mathcal{X}_i} \phi(x_1^+) + h(x_{\leq i-1}^+, x_i, x'_{\geq i+1}) - \langle \delta_i, x_i \rangle \\ \quad = \mathcal{H}_{ii}^{-1} \left(r_i + \delta_i - \sum_{j=1}^{i-1} \mathcal{H}_{ji}^* x_j^+ - \sum_{j=i+1}^{s} \mathcal{H}_{ij} x_j' \right), \quad i = 2, \ldots, s. \end{cases} \tag{94}$$

Remark 2. (a). In *(93)* and *(94)*, x_i' and x_i^+ should be regarded as inexact solutions to the corresponding minimization problems without the linear error terms $\langle \delta_i', x_i \rangle$ and $\langle \delta_i, x_i \rangle$. Once these approximate solutions have been computed, they would generate the error vectors δ_i' and δ_i as follows:

$$\delta_i' = \mathcal{H}_{ii} x_i' - \left(r_i - \sum_{j=1}^{i-1} \mathcal{H}_{ji}^* \bar{x}_j - \sum_{j=i+1}^{s} \mathcal{H}_{ij} x_j' \right), \quad i = s, \ldots, 2,$$

$$\delta_1 \in \partial \phi(x_1^+) + \mathcal{H}_{11} x_1^+ - \left(r_1 - \sum_{j=2}^{s} \mathcal{H}_{1j} x_j' \right),$$

$$\delta_i = \mathcal{H}_{ii} x_i^+ - \left(r_i - \sum_{j=1}^{i-1} \mathcal{H}_{ji}^* x_j^+ - \sum_{j=i+1}^{s} \mathcal{H}_{ij} x_j' \right), \quad i = 2, \ldots, s.$$

With the above known error vectors, we have that x_i' and x_i^+ are the exact solutions to the minimization problems in *(93)* and *(94)*, respectively.

(b). In actual implementations, assuming that for $i = s, \ldots, 2$, we have computed x_i' in the backward GS sweep for solving *(93)*, then when solving the subproblems in the forward GS sweep in *(94)* for $i = 2, \ldots, s$, we may try to estimate x_i^+ by using x_i', and in this case the corresponding error vector δ_i would be given by

$$\delta_i = \delta_i' + \sum_{j=1}^{i-1} \mathcal{H}_{ji}^* (x_j' - \bar{x}_j).$$

In practice, we may accept such an approximate solution $x_i^+ = x_i'$ for $i = 2, \ldots, s$, if the corresponding error vector satisfies an admissible condition such as $\|\delta_i\| \leq c\|\delta_i'\|$ for some constant $c > 1$, say $c = 10$. □

In order to estimate the error term $\Delta(\delta', \delta)$ in (90), we have the following proposition.

Proposition 5 (Li et al. 2015, Proposition 2.1). *Suppose that $\widehat{\mathcal{H}} = \mathcal{H} + \mathcal{T}$ is positive definite. Let $\xi = \|\widehat{\mathcal{H}}^{-1/2} \Delta(\delta', \delta)\|$. It holds that*

$$\xi \leq \|\mathcal{D}^{-1/2}(\delta - \delta')\| + \|\widehat{\mathcal{H}}^{-1/2} \delta'\|. \tag{95}$$

Obviously, by choosing $v = \mu$ and $w = (\lambda, p)$ and taking

$$f(v) = \delta_{[a,b]}^*(M_h \mu), \tag{96}$$

$$g(w) = \delta_{[-\beta,\beta]}(\lambda), \tag{97}$$

$$\phi(v,w) = \frac{1}{2}\|K_h p - M_h y_d\|_{M_h^{-1}}^2 + \frac{1}{2\alpha}\|\lambda + \mu - p\|_{M_h}^2 + \langle M_h y_r, p \rangle - \frac{1}{2}\|y_d\|_{M_h}^2, \tag{98}$$

(D_h) belongs to a general class of unconstrained, multi-block convex optimization problems with coupled objective function, that is,

$$\min_{v,w} \theta(v,w) := f(v) + g(w) + \phi(v,w), \tag{99}$$

where $f : \mathcal{V} \to (-\infty, +\infty]$ and $g : \mathcal{W} \to (-\infty, +\infty]$ are two convex functions (possibly nonsmooth), $\phi : \mathcal{V} \times \mathcal{W} \to (-\infty, +\infty]$ is a smooth convex function, and $\mathcal{V}, \mathcal{W}$ are real finite dimensional Hilbert spaces.

Inexact Majorized Accelerate Block Coordinate Descent (imABCD) Method

It is well-known that taking the inexactness of the solutions of associated subproblems into account is important for the numerical implementation. Thus, let us give a brief sketch of the inexact majorized accelerate block coordinate descent (imABCD) method which is proposed by Cui in (2016, Chapter 3) for the case ϕ being a general smooth function. To deal with the general model (99), we need some more conditions and assumptions on ϕ.

Assumption 3. *The convex function $\phi : \mathcal{V} \times \mathcal{W} \to (-\infty, +\infty]$ is continuously differentiable with Lipschitz continuous gradient.* □

16 Numerical Solution for Sparse PDE Constrained Optimization

Let us denote $z := (v, w) \in \mathcal{V} \times \mathcal{W}$. In Hiriart-Urruty et al. 1984, Theorem 2.3, Hiriart-Urruty and Nguyen provide a second-order mean value theorem for ϕ, which states that for any z' and z in $\mathcal{V} \times \mathcal{W}$, there exists $z'' \in [z', z]$ and a self-adjoint positive semidefinite operator $\mathcal{G} \in \partial^2 \phi(z'')$ such that

$$\phi(z) = \phi(z') + \langle \nabla \phi(z'), z - z' \rangle + \frac{1}{2} \|z' - z\|_{\mathcal{G}}^2,$$

where $\partial^2 \phi(z'')$ denotes the Clarke's generalized Hessian at given z'' and $[z', z]$ denotes the line segment connecting z' and z. Under Assumption 3, it is obvious that there exist two self-adjoint positive semidefinite linear operators $\mathcal{Q}$ and $\widehat{\mathcal{Q}}$: $\mathcal{V} \times \mathcal{W} \to \mathcal{V} \times \mathcal{W}$ such that for any $z \in \mathcal{V} \times \mathcal{W}$,

$$\mathcal{Q} \preceq \mathcal{G} \preceq \widehat{\mathcal{Q}}, \quad \forall \mathcal{G} \in \partial^2 \phi(z).$$

Thus, for any $z, z' \in \mathcal{V} \times \mathcal{W}$, it holds

$$\phi(z) \geq \phi(z') + \langle \nabla \phi(z'), z - z' \rangle + \frac{1}{2} \|z' - z\|_{\mathcal{Q}}^2,$$

and

$$\phi(z) \leq \hat{\phi}(z; z') := \phi(z') + \langle \nabla \phi(z'), z - z' \rangle + \frac{1}{2} \|z' - z\|_{\widehat{\mathcal{Q}}}^2.$$

Furthermore, we decompose the operators $\mathcal{Q}$ and $\widehat{\mathcal{Q}}$ into the following block structures:

$$\mathcal{Q}z := \begin{pmatrix} \mathcal{Q}_{11} & \mathcal{Q}_{12} \\ \mathcal{Q}_{12}^* & \mathcal{Q}_{22} \end{pmatrix} \begin{pmatrix} v \\ w \end{pmatrix}, \quad \widehat{\mathcal{Q}}z := \begin{pmatrix} \widehat{\mathcal{Q}}_{11} & \widehat{\mathcal{Q}}_{12} \\ \widehat{\mathcal{Q}}_{12}^* & \widehat{\mathcal{Q}}_{22} \end{pmatrix} \begin{pmatrix} v \\ w \end{pmatrix}, \quad \forall z = (v, w) \in \mathcal{U} \times \mathcal{V},$$

and assume $\mathcal{Q}$ and $\widehat{\mathcal{Q}}$ satisfy the following conditions.

Assumption 4 (Cui 2016, Assumption 3.1). *There exist two self-adjoint positive semidefinite linear operators $\mathcal{D}_1 : \mathcal{U} \to \mathcal{U}$ and $\mathcal{D}_2 : \mathcal{V} \to \mathcal{V}$ such that*

$$\widehat{\mathcal{Q}} := \mathcal{Q} + \mathrm{Diag}(\mathcal{D}_1, \mathcal{D}_2).$$

Furthermore, $\widehat{\mathcal{Q}}$ satisfies that $\widehat{\mathcal{Q}}_{11} \succ 0$ and $\widehat{\mathcal{Q}}_{22} \succ 0$. □

Remark 3. It is important to note that Assumption 4 is a realistic assumption in practice. For example, when ϕ is a quadratic function, we could choose $\mathcal{Q} = \mathcal{G} = \nabla^2 \phi$. If we have $\mathcal{Q}_{11} \succ 0$ and $\mathcal{Q}_{22} \succ 0$, then Assumption 4 holds automatically. We should point out that ϕ is a quadratic function for many problems in the practical application, such as the SDP relaxation of a binary integer nonconvex quadratic

(BIQ) programming, the SDP relaxation for computing lower bounds for quadratic assignment problems (QAPs), and so on, and one can refer to Sun et al. (2016). Fortunately, it should be noted that the function ϕ defined in (98) for our problem (D_h) is quadratic and thus we can choose $Q = \nabla^2 \phi$.

We can now present the inexact majorized ABCD algorithm for the general problem (99) as follows.

Algorithm 2 (An inexact majorized ABCD algorithm for (99))

Input: $(v^1, w^1) = (\tilde{v}^0, \tilde{w}^0) \in \text{dom}(f) \times \text{dom}(g)$. Let $\{\epsilon_k\}$ be a summable sequence of nonnegative numbers, and set $t_1 = 1, k = 1$.
Output: $(\tilde{v}^k, \tilde{w}^k)$
Iterate until convergence:
Step 1 Choose error tolerance $\delta_v^k \in \mathcal{U}$, $\delta_w^k \in \mathcal{V}$ such that

$$\max\{\delta_v^k, \delta_w^k\} \leq \epsilon_k.$$

Compute

$$\begin{cases} \tilde{v}^k = \arg\min_{v \in \mathcal{V}}\{f(v) + \hat{\phi}(v, w^k; v^k, w^k) - \langle \delta_v^k, v \rangle\}, \\ \tilde{w}^k = \arg\min_{w \in \mathcal{W}}\{g(w) + \hat{\phi}(\tilde{v}^k, w; v^k, w^k) - \langle \delta_w^k, w \rangle\}. \end{cases}$$

Step 2 Set $t_{k+1} = \frac{1 + \sqrt{1 + 4t_k^2}}{2}$ and $\beta_k = \frac{t_k - 1}{t_{k+1}}$, compute

$$v^{k+1} = \tilde{v}^k + \beta_k(\tilde{v}^k - \tilde{v}^{k-1}), \quad w^{k+1} = \tilde{w}^k + \beta_k(\tilde{w}^k - \tilde{w}^{k-1}).$$

Here we state the convergence result without proving. For the detailed proof, one could see Cui (2016, Chapter 3). This theorem builds a solid foundation for our subsequent proposed algorithm.

Theorem 7 (Cui 2016, Theorem 3.2). *Suppose that Assumption 4 holds and the solution set Ω of the problem (99) is non-empty. Let $z^* = (v^*, w^*) \in \Omega$. Assume that $\sum_{k=1}^{\infty} k\epsilon_k < \infty$. Then the sequence $\{\tilde{z}^k\} := \{(\tilde{v}^k, \tilde{w}^k)\}$ generated by the Algorithm 2 satisfies that*

$$\theta(\tilde{z}^k) - \theta(z^*) \leq \frac{2\|\tilde{z}^0 - z^*\|_S^2 + c_0}{(k+1)^2}, \quad \forall k \geq 1,$$

where c_0 is a constant number and $\mathcal{S} := \text{Diag}(\mathcal{D}_1, \mathcal{D}_2 + \mathcal{Q}_{22})$. □

A sGS-imABCD Algorithm for (D$_h$)

Now, we can apply Algorithm 2 to our problem (D$_h$), where μ is taken as one block, and (λ, p) are taken as the other one. Let us denote $z = (\mu, \lambda, p)$. Since ϕ defined in (98) for (D$_h$) is quadratic, we can take

$$Q := \frac{1}{\alpha} \begin{pmatrix} M_h & M_h & -M_h \\ M_h & M_h & -M_h \\ -M_h & -M_h & M_h + \alpha K_h M_h^{-1} K_h \end{pmatrix}, \quad (100)$$

where

$$Q_{11} := \frac{1}{\alpha} M_h, \quad Q_{22} := \frac{1}{\alpha} \begin{pmatrix} M_h & -M_h \\ -M_h & M_h + \alpha K_h M_h^{-1} K_h \end{pmatrix}.$$

Additionally, we assume that there exist two self-adjoint positive semidefinite operators $\mathcal{D}_1$ and $\mathcal{D}_2$, such that Assumption 4 holds. It implies that we should majorize $\phi(\mu, \lambda, p)$ at $z' = (\mu', \lambda', p')$ as

$$\phi(z) \leq \hat{\phi}(z; z') = \phi(z) + \frac{1}{2}\|\mu - \mu'\|_{\mathcal{D}_1}^2 + \frac{1}{2}\left\| \begin{pmatrix} \lambda \\ p \end{pmatrix} - \begin{pmatrix} \lambda' \\ p' \end{pmatrix} \right\|_{\mathcal{D}_2}^2. \quad (101)$$

Thus, the framework of imABCD for (D$_h$) is given below.

Algorithm 3 (imABCD algorithm for (D$_h$))

Input: $(\mu^1, \lambda^1, p^1) = (\tilde{\mu}^0, \tilde{\lambda}^0, \tilde{p}^0) \in \mathrm{dom}(\delta^*_{[a,b]}) \times [-\beta, \beta] \times \mathbb{R}^{N_h}$. Set $k = 1, t_1 = 1$.
Output: $(\tilde{\mu}^k, \tilde{\lambda}^k, \tilde{p}^k)$
Iterate until convergence
Step 1 Compute

$$\tilde{\mu}^k = \arg\min \delta^*_{[a,b]}(M_h \mu) + \phi(\mu, \lambda^k, p^k) + \frac{1}{2}\|\mu - \mu^k\|_{\mathcal{D}_1}^2 - \langle \delta_\mu^k, \mu \rangle,$$

$$(\tilde{\lambda}^k, \tilde{p}^k) = \arg\min \delta_{[-\beta,\beta]}(\lambda) + \phi(\tilde{\mu}^k, \lambda, p) + \frac{1}{2}\left\| \begin{pmatrix} \lambda \\ p \end{pmatrix} - \begin{pmatrix} \lambda^k \\ p^k \end{pmatrix} \right\|_{\mathcal{D}_2}^2 - \langle \delta_\lambda^k, \lambda \rangle - \langle \delta_p^k, p \rangle.$$

Step 2 Set $t_{k+1} = \frac{1+\sqrt{1+4t_k^2}}{2}$ and $\beta_k = \frac{t_k-1}{t_{k+1}}$, compute

$$\mu^{k+1} = \tilde{\mu}^k + \beta_k(\tilde{\mu}^k - \tilde{\mu}^{k-1}), \quad p^{k+1} = \tilde{p}^k + \beta_k(\tilde{p}^k - \tilde{p}^{k-1}), \quad \lambda^{k+1} = \tilde{\lambda}^k + \beta_k(\tilde{\lambda}^k - \tilde{\lambda}^{k-1}).$$

Next, another key issue that should be considered is how to choose the operators $\mathcal{D}_1$ and $\mathcal{D}_2$. As we know, choosing the appropriate and effective operators $\mathcal{D}_1$ and $\mathcal{D}_2$ is an important thing from the perspective of both theory analysis and numerical implementation. Note that for numerical efficiency, the general principle is that both $\mathcal{D}_1$ and $\mathcal{D}_2$ should be chosen as small as possible such that $\tilde{\mu}^k$ and $(\tilde{\lambda}^k, \tilde{p}^k)$ could take larger step-lengths while the corresponding subproblems still could be solved relatively easily.

First, for the proximal term $\frac{1}{2}\|\mu - \mu^k\|_{\mathcal{D}_1}^2$, in order to make the subproblem of the block μ having an analytical solution, and from Proposition (1), we choose

$$\mathcal{D}_1 := \frac{1}{\alpha} c_n M_h W_h^{-1} M_h - \frac{1}{\alpha} M_h, \quad \text{where } c_n = \begin{cases} 4 & \text{if } n = 2, \\ 5 & \text{if } n = 3. \end{cases}$$

Next, we will focus on how to choose the operator $\mathcal{D}_2$. If we ignore the proximal term $\frac{1}{2}\left\|\begin{pmatrix}\lambda\\p\end{pmatrix} - \begin{pmatrix}\lambda^k\\p^k\end{pmatrix}\right\|_{\mathcal{D}_2}^2$ and the error terms, it is obvious that the subproblem of the block (λ, p) belongs to the form (84), which can be rewritten as

$$\min \delta_{[-\beta,\beta]}(\lambda) + \frac{1}{2}\left\langle \begin{pmatrix}\lambda\\p\end{pmatrix}, \mathcal{H}\begin{pmatrix}\lambda\\p\end{pmatrix}\right\rangle - \left\langle r, \begin{pmatrix}\lambda\\p\end{pmatrix}\right\rangle, \tag{102}$$

where $\mathcal{H} = Q_{22} = \frac{1}{\alpha}\begin{pmatrix} M_h & -M_h \\ -M_h & M_h + \alpha K_h M_h^{-1} K_h \end{pmatrix}$ and $r = \begin{pmatrix} \frac{1}{\alpha} M_h \tilde{\mu}^k \\ M_h y_r - K_h y_d - \frac{1}{\alpha} M_h \tilde{\mu}^k \end{pmatrix}$. Since the objective function of (102) is the sum of a two-block quadratic function and a nonsmooth function involving only the first block, thus the inexact sGS technique, which is introduced in Section , can be used to solve (102). To achieve our goal, we choose

$$\tilde{\mathcal{D}}_2 = \text{sGS}(Q_{22}) = \frac{1}{\alpha}\begin{pmatrix} M_h(M_h + \alpha K_h M_h^{-1} K_h)^{-1} M_h & 0 \\ 0 & 0 \end{pmatrix}.$$

Then according to Theorem 6, we can solve the (λ, p)-subproblem by the following procedure:

16 Numerical Solution for Sparse PDE Constrained Optimization

$$\begin{cases} \hat{p}^k = \arg\min \frac{1}{2}\|K_h p - M_h y_d\|^2_{M_h^{-1}} + \frac{1}{2\alpha}\|p - \lambda^k - \tilde{\mu}^k + \alpha y_r\|^2_{M_h} - \langle \hat{\delta}_p^k, p \rangle, \\ \tilde{\lambda}^k = \arg\min \frac{1}{2\alpha}\|\lambda - (\hat{p}^k - \tilde{\mu}^k)\|^2_{M_h} + \delta_{[-\beta,\beta]}(\lambda), \\ \tilde{p}^k = \arg\min \frac{1}{2}\|K_h p - M_h y_d\|^2_{M_h^{-1}} + \frac{1}{2\alpha}\|p - \tilde{\lambda}^k - \tilde{\mu}^k + \alpha y_r\|^2_{M_h} - \langle \hat{\delta}_p^k, p \rangle. \end{cases} \quad (103)$$

However, it is easy to see that the λ-subproblem is coupled about the variable λ since the mass matrix M_h is not diagonal; thus, there is no closed-form solution for λ. To overcome this difficulty, we can take advantage of the relationship between the mass matrix M_h and the lumped mass matrix W_h and add a proximal term $\frac{1}{2\alpha}\|\lambda - \lambda^k\|^2_{W_h - M_h}$ to the λ-subproblem. Fortunately, we have

$$\text{sGS}(Q_{22}) = \text{sGS}\left(Q_{22} + \frac{1}{\alpha}\begin{bmatrix} W_h - M_h & 0 \\ 0 & 0 \end{bmatrix}\right),$$

which implies that the proximal term $\frac{1}{2\alpha}\|\lambda - \lambda^k\|^2_{W_h - M_h}$ has no influence on the sGS technique. Thus, we can choose $\mathcal{D}_2$ as follows:

$$\mathcal{D}_2 = \text{sGS}(Q_{22}) + \frac{1}{\alpha}\begin{pmatrix} W_h - M_h & 0 \\ 0 & 0 \end{pmatrix}.$$

Based on the choice of $\mathcal{D}_1$ and $\mathcal{D}_2$, we get the majorized Hessian matrix $\widehat{Q}$ as follows:

$$\widehat{Q} = Q + \frac{1}{\alpha}\begin{pmatrix} c_n M_h W_h^{-1} M_h - M_h & 0 & 0 \\ 0 & M_h(M_h + \alpha K_h M_h^{-1} K_h)^{-1} M_h + W_h - M_h & 0 \\ 0 & 0 & 0 \end{pmatrix}. \quad (104)$$

Then, according to the choice of $\mathcal{D}_1$ and $\mathcal{D}_2$, we give the detailed framework of our inexact sGS based majorized ABCD method (called sGS-imABCD) for (D$_h$) as follows.

Based on Theorem 7, we can show our Algorithm 4 (sGS-imABCD) also has the following $O(1/k^2)$ iteration complexity.

Theorem 8. *Assume that $\sum_{i=k}^{\infty} k\epsilon_k < \infty$. Let $\{\tilde{z}^k\} := \{(\tilde{\mu}^k, \tilde{\lambda}^k, \tilde{p}^k)\}$ be the sequence generated by Algorithm 4. Then we have*

$$\Phi_h(\tilde{z}^k) - \Phi_h(z^*) \le \frac{2\|\tilde{z}^0 - z^*\|_S^2 + c_0}{(k+1)^2}, \quad \forall k \ge 1,$$

where c_0 is a constant number, $S := \mathrm{Diag}(\mathcal{D}_1, \mathcal{D}_2 + \mathcal{Q}_{22})$, and $\Phi_h(\cdot)$ is the objective function of the dual problem (D_h). □

Algorithm 4 (sGS-imABCD algorithm for (D_h))

Input: $(\mu^1, \lambda^1, p^1) = (\tilde{\mu}^0, \tilde{\lambda}^0, \tilde{p}^0) \in \mathrm{dom}(\delta^*_{[a,b]}) \times [-\beta, \beta] \times \mathbb{R}^{N_h}$. Let $\{\epsilon_k\}$ be a nonincreasing sequence of nonnegative numbers such that $\sum_{k=1}^{\infty} k\epsilon_k < \infty$. Set $k = 1, t_1 = 1$.

Output: $(\tilde{\mu}^k, \tilde{\lambda}^k, \tilde{p}^k)$

Iterate until convergence

Step 1 Choose error tolerance $\delta^k_\mu, \hat{\delta}^k_p, \delta^k_p$ such that

$$\max\{\|\delta^k_\mu\|, \|\hat{\delta}^k_p\|, \|\delta^k_p\|\} \le \epsilon_k.$$

Compute

$$\tilde{\mu}^k = \arg\min \frac{1}{2\alpha}\|\mu - (p^k - \lambda^k)\|^2_{M_h} + \delta^*_{[a,b]}(M_h\mu) + \frac{1}{2}\|\mu - \mu^k\|^2_{\mathcal{D}_1} - \langle \delta^k_\mu, \mu \rangle,$$

$$\hat{p}^k = \arg\min \frac{1}{2}\|K_h p - M_h y_d\|^2_{M_h^{-1}} + \frac{1}{2\alpha}\|p - \lambda^k - \tilde{\mu}^k + \alpha y_r\|^2_{M_h} - \langle \hat{\delta}^k_p, p \rangle,$$

$$\tilde{\lambda}^k = \arg\min \frac{1}{2\alpha}\|\lambda - (\hat{p}^k - \tilde{\mu}^k)\|^2_{M_h} + \delta_{[-\beta,\beta]}(\lambda) + \frac{1}{2\alpha}\|\lambda - \lambda^k\|^2_{W_h - M_h},$$

$$\tilde{p}^k = \arg\min \frac{1}{2}\|K_h p - M_h y_d\|^2_{M_h^{-1}} + \frac{1}{2\alpha}\|p - \tilde{\lambda}^k - \tilde{\mu}^k + \alpha y_r\|^2_{M_h} - \langle \delta^k_p, p \rangle.$$

Step 2 Set $t_{k+1} = \frac{1+\sqrt{1+4t_k^2}}{2}$ and $\beta_k = \frac{t_k - 1}{t_{k+1}}$, compute

$$\mu^{k+1} = \tilde{\mu}^k + \beta_k(\tilde{\mu}^k - \tilde{\mu}^{k-1}), \quad p^{k+1} = \tilde{p}^k + \beta_k(\tilde{p}^k - \tilde{p}^{k-1}), \quad \lambda^{k+1} = \tilde{\lambda}^k + \beta_k(\tilde{\lambda}^k - \tilde{\lambda}^{k-1}).$$

Proof. By Proposition 1, we know that $c_n M_h W_h^{-1} M_h - M_h \succ 0$, $M_h(M_h + \alpha K_h M_h^{-1} K_h)^{-1} M_h \succ 0$, $W_h - M_h \succ 0$. Moreover, since stiffness and mass matrices are symmetric positive definite matrices, it is noticed that Assumption 4 is valid for our $\widehat{Q}$ which is defined in (104). Thus, according to Theorem 7, we can establish the convergence of Algorithm 4. □

Remark 4. Let $\tau_h = 2\|\tilde{z}^0 - z^*\|_S^2 + c_0$. It is obvious that τ_h is independent of the parameter β, whereas it depends on the parameter α and will increase with the decrease of α.

Numerical Results

In this section, we will first use Example 1 and Example 2 to evaluate the numerical behavior of the ihADMM and use Example 3 and Example 4 to evaluate the numerical behavior of the sGS-imABCD.

Algorithmic Details

We begin by describing the algorithmic details which are common to all examples.

Discretization. The discretization was carried out by using piecewise linear and continuous finite elements. The assembly of mass and the stiffness matrices, as well as the lump mass matrix, was left to the iFEM software package. To present the finite element error estimate results, it is convenient to introduce the experimental order of convergence (EOC), which for some positive error functional $E(h)$ with $h > 0$ is defined as follows: Given two grid sizes $h_1 \neq h_2$, let

$$\text{EOC} := \frac{\log E(h_1) - \log E(h_2)}{\log h_1 - \log h_2}. \tag{105}$$

It follows from this definition that if $E(h) = O(h^\gamma)$, then EOC $\approx \gamma$. The error functional $E(\cdot)$ investigated in the present section is given by $E_2(h) := \|u - u_h\|_{L^2(\Omega)}$.

Initialization. For all numerical examples, we choose $u = 0$ as initialization u^0 for all algorithms.

In Example 1 and Example 2, for comparison with ihADMM, we will also show the numerical results obtained by the classical ADMM and the APG algorithm, and the PDAS with line search. For the classical ADMM and our ihADMM, the penalty parameter σ was chosen as $\sigma = 0.1\alpha$. About the step-length τ, we choose $\tau = 1.618$ for the classical ADMM, and $\tau = 1$ for our ihADMM. For the PDAS method, the parameter in the active set strategy was chosen as $c = 1$. For the APG method, we estimate an approximation for the Lipschitz constant L with a backtracking method. In the numerical experiments, we measure the accuracy of an approximate optimal solution by using the corresponding K-K-T residual error for each algorithm. For the purpose of showing the efficiency of our ihADMM, we report the numerical results obtained by running the classical ADMM and the APG method to compare with the results obtained by our ihADMM. In this case, we terminate all the algorithms when $\eta < 10^{-6}$ with the maximum number of iterations set at 500.

In Example 3 and Example 4, for comparison with sGS-imABCD, we will also show the numerical results obtained by the ihADMM and APG methods for $(\overline{\text{DP}_h})$. For the ihADMM method, the step-length τ for Lagrangian multipliers λ was chosen as $\tau = 1$, and the penalty parameter σ was chosen as $\sigma = 0.1\alpha$. For the APG method, we estimate an approximation to the Lipschitz constant L with a backtracking method with $\eta = 1.4$ and $L^0 = 10^{-8}$. In the numerical experiments, we terminate all the algorithms when the corresponding relative residual $\eta < 10^{-7}$.

Examples

Example 1.

$$\begin{cases} \min_{(y,u) \in H_0^1(\Omega) \times L^2(\Omega)} J(y, u) = \frac{1}{2}\|y - y_d\|_{L^2(\Omega)}^2 + \frac{\alpha}{2}\|u\|_{L^2(\Omega)}^2 + \beta\|u\|_{L^1(\Omega)} \\ \text{s.t.} \quad -\Delta y = u + y_c \quad \text{in } \Omega, \\ \qquad y = 0 \quad \text{on } \partial\Omega, \\ \qquad u \in U_{ad} = \{v(x) | a \leq v(x) \leq b, \text{a.e on } \Omega\}. \end{cases}$$

Here, we consider the problem with control $u \in L^2(\Omega)$ on the unit square $\Omega = (0, 1)^2$ with $\alpha = 0.5$, $\beta = 0.5$, $a = -0.5$, and $b = 0.5$. It is a constructed problem; thus, we set $y^* = \sin(\pi x_1)\sin(\pi x_2)$ and $p^* = 2\beta \sin(2\pi x_1)\exp(0.5x_1)\sin(4\pi x_2)$. Then through $u^* = \Pi_{U_{ad}}\left(\frac{1}{\alpha}\text{soft}(-p^*, \beta)\right)$, $y_c = y^* - \mathcal{S}u^*$, and $y_d = \mathcal{S}^{-*}p^* + y^*$, we can construct the example for which we know the exact solution.

The error of the control u w.r.t the L^2-norm and the EOC for control are presented in Table 1. They also confirm that indeed the convergence rate is of order $O(h)$. Numerical results for the accuracy of solution, number of iterations, and CPU time obtained by our ihADMM, classical ADMM, and APG methods are shown in Table 1. As a result from Table 1, we can see that our proposed ihADMM method is an efficient algorithm to solve problem $(\overline{DP_h})$ to medium accuracy. Moreover, it is obvious that our ihADMM outperforms the classical ADMM and the APG method in terms of CPU time, especially when the discretization is in a fine level. It is worth noting that although the APG method requires less number of iterations when the termination condition is satisfied, the APG method spends much time on backtracking step with the aim of finding an appropriate approximation for the Lipschitz constant. This is the reason that our ihADMM has better performance than the APG method in actual numerical implementation. Furthermore, the numerical results in terms of iteration numbers illustrate the mesh-independent performance of the ihADMM and the APG method, except for the classical ADMM.

16 Numerical Solution for Sparse PDE Constrained Optimization

Table 1 Example 1: The convergence behavior of our ihADMM, classical ADMM, and APG for $(\overline{DP_h})$. In the table, #dofs stands for the number of degrees of freedom for the control variable on each grid level

h	#dofs	E_2	EOC	Index	ihADMM	Classical ADMM	APG
2^{-3}	49	0.2925	–	iter	27	32	13
				residual η	7.15e-07	7.55e-07	6.88e-07
				CPU time/s	0.19	0.23	0.18
2^{-4}	225	0.1127	1.3759	iter	31	44	13
				residual η	9.77e-07	9.91e-07	8.23e-07
				CPU times/s	0.37	0.66	0.32
2^{-5}	961	0.0457	1.3390	iter	31	58	12
				residual η	7.41e-07	8.11e-07	7.58e-07
				CPU time/s	1.02	2.32	1.00
2^{-6}	3969	0.0161	1.3944	iter	32	76	14
				residual η	7.26e-07	8.10e-07	7.88e-07
				CPU time/s	4.18	9.12	4.25
2^{-7}	16129	0.0058	1.4132	iter	31	94	14
				residual η	5.33e-07	7.85e-07	4.45e-07
				CPU time/s	17.72	65.82	26.25
2^{-8}	65025	0.0019	1.4503	iter	32	127	13
				residual η	6.88e-07	8.93e-07	7.47e-07
				CPU time/s	70.45	312.65	80.81
2^{-9}	261121	0.0007	1.4542	iter	31	255	13
				residual η	7.43e-07	7.96e-07	6.33e-07
				CPU time/s	525.28	4845.31	620.55

Example 2.

$$\begin{cases} \min_{(y,u) \in Y \times U} J(y, u) = \frac{1}{2}\|y - y_d\|^2_{L^2(\Omega)} + \frac{\alpha}{2}\|u\|^2_{L^2(\Omega)} + \beta\|u\|_{L^1(\Omega)} \\ \text{s.t.} \quad -\Delta y = u, \quad \text{in } \Omega = (0, 1) \times (0, 1) \\ \qquad y = 0, \quad \text{on } \partial\Omega \\ \qquad u \in U_{ad} = \{v(x) | a \leq v(x) \leq b, \text{ a.e on } \Omega\}, \end{cases}$$

where the desired state $y_d = \frac{1}{6}\sin(2\pi x)\exp(2x)\sin(2\pi y)$ and the parameters $\alpha = 10^{-5}$, $\beta = 10^{-3}$, $a = -30$, and $b = 30$. In addition, the exact solutions of the problem are unknown. Instead, we use the numerical solutions computed on a grid with $h^* = 2^{-10}$ as reference solutions.

The error of the control u w.r.t the L^2 norm with respect to the solution on the finest grid ($h^* = 2^{-10}$) and the experimental order of convergence (EOC) for control are presented in Table 2. They confirm the linear rate of convergence w.r.t. h.

Numerical results for the accuracy of solution, number of iterations, and CPU time obtained by our ihADMM, classical ADMM, and APG methods are also shown in Table 2. Experiment results show that the ADMM has evident advantage over the classical ADMM and the APG method in computing time. Furthermore, the numerical results in terms of iteration numbers also illustrate the mesh-independent performance of our ihADMM. These results demonstrate that our ihADMM is highly efficient in obtaining an approximate solution with moderate accuracy.

Table 2 Example 2: The convergence behavior of ihADMM, classical ADMM, and APG for $(\overline{\mathrm{DP}_h})$

h	#dofs	E_2	EOC	Index	ihADMM	Classical ADMM	APG
2^{-3}	49	6.6122	–	iter	40	48	18
				residual η	8.22e-07	8.65e-07	7.96e-07
				CPU time/s	0.30	0.51	0.24
2^{-4}	225	2.6314	1.3293	iter	41	56	18
				residual η	7.22e-07	8.01e-07	7.58e-07
				CPU times/s	0.45	0.71	0.44
2^{-5}	961	1.2825	1.1831	iter	40	69	19
				residual η	8.12e-07	8.01e-07	7.90e-07
				CPU time/s	1.60	3.05	1.58
2^{-6}	3969	0.7514	1.0458	iter	42	85	18
				residual η	6.11e-07	7.80e-07	6.45e-07
				CPU time/s	7.25	14.62	7.45
2^{-7}	16129	0.2930	1.1240	iter	40	108	18
				residual η	6.35e-07	7.11e-07	5.62e-07
				CPU time/s	33.85	101.36	34.39
2^{-8}	65025	0.1357	1.1213	iter	41	132	19
				residual η	7.55e-07	7.83e-07	7.57e-07
				CPU time/s	158.62	508.65	165.75
2^{-9}	261121	0.0958	1.0181	iter	42	278	18
				residual η	5.25e-07	5.56e-07	4.85e-07
				CPU time/s	1781.98	11788.52	1860.11
2^{-10}	1046529	–	–	iter	41	500	19
				residual η	8.78e-07	Error	8.47e-07
				CPU time/s	42033.79	Error	44131.27

16 Numerical Solution for Sparse PDE Constrained Optimization

Example 3.

$$\begin{cases} \min_{(y,u) \in H_0^1(\Omega) \times L^2(\Omega)} & J(y,u) = \frac{1}{2}\|y - y_d\|_{L^2(\Omega)}^2 + \frac{\alpha}{2}\|u\|_{L^2(\Omega)}^2 + \beta\|u\|_{L^1(\Omega)} \\ \text{s.t.} & -\Delta y = u + y_r \quad \text{in } \Omega, \\ & y = 0 \quad \text{on } \partial\Omega, \\ & u \in U_{ad} = \{v(x) | a \le v(x) \le b, \text{a.e on } \Omega\}. \end{cases}$$

Here, we consider the problem with control $u \in L^2(\Omega)$ on the unit square $\Omega = (0,1)^2$ with $\alpha = 0.5$, $\beta = 0.5$, $a = -0.5$, and $b = 0.5$. It is a constructed problem; thus, we set $y^* = \sin(2\pi x_1)\exp(0.5x_1)\sin(4\pi x_2)$ and $p^* = 2\beta\sin(2\pi x_1)\exp(0.5x_1)\sin(4\pi x_2)$.

The error of the control u w.r.t the L^2 norm and the experimental order of convergence (EOC) for control are presented in Tables 3 and 5. They also confirm that indeed the convergence rate is of order $O(h)$. Comparing the error results from Tables 3 and 5, it is obvious to see that solving the dual problem (D$_h$) could get better error results than that from solving ($\overline{\text{DP}_h}$).

Numerical results for the accuracy of solution, number of iterations, and CPU time obtained by our proposed sGS-imABCD method for (D$_h$) are also shown in Table 3. As a result we obtain from Table 3, one can see that our proposed sGS-imABCD method is an efficient algorithm to solve problem (D$_h$) to high accuracy. It should be pointed out that iter.$\tilde{p}$-block denotes the iterations of $\tilde{p}$ in Table 3. It is clear that p-subproblem almost always not be computed twice, which demonstrates the efficiency of our strategy to predict the solution of $\tilde{p}$-subproblem. Furthermore, the numerical results in terms of iteration numbers illustrate the mesh-independent performance of our proposed sGS-imABCD method. Additionally, in Table 4, we list the numbers of iteration steps and the relative residual errors of PMHSS-preconditioned GMRES method for the $\hat{p}$-subproblem on mesh $h = 2^{-7}$

Table 3 Example 3: The performance of sGS-imABCD for (D$_h$). In the table, #dofs stands for the number of degrees of freedom for the control variable on each grid level

h	#dofs	iter.sGS-imABCD	iter.$\tilde{p}$-block	residual η	CPU time/s	E_2	EOC
2^{-3}	49	13	4	6.60e-08	0.14	0.1784	-
2^{-4}	225	13	4	6.32e-08	0.20	0.0967	0.8834
2^{-5}	961	12	3	7.38e-08	0.33	0.0399	1.0803
2^{-6}	3969	13	3	9.78e-08	2.04	0.0155	1.1749
2^{-7}	16129	12	3	6.66e-08	8.25	0.0052	1.2754
2^{-8}	65025	10	3	7.05e-08	52.15	0.0017	1.3388
2^{-9}	261121	9	2	5.19e-08	312.82	0.0006	1.3617

Table 4 Example 3: The convergence behavior of GMRES for $\hat{p}$-block subproblem

h	iter.sGS-imABCD	iter.GMRES of $\hat{p}$-block	Relative residual error of GMRES
2^{-7}	1	8	1.30e-07
	2	4	1.07e-07
	3	4	5.26e-08
	4	4	1.56e-08
	5	4	2.05e-09
	6	4	1.58e-09
	7	4	1.23e-09
	8	4	1.29e-10
	9	2	1.16e-10
	10	2	1.07e-10
	11	2	5.98e-11
	12	2	1.30e-11
2^{-8}	1	8	6.31e-08
	2	4	2.18e-08
	3	4	8.43e-09
	4	4	3.18e-09
	5	4	1.07e-09
	6	4	5.53e-10
	7	4	5.25e-11
	8	4	5.90e-12
	9	2	4.86e-12
	10	2	4.18e-12

and $h = 2^{-8}$. From Table 4, we can see that the number of iteration steps of the PMHSS-preconditioned GMRES method is roughly independent of the mesh size h.

As a comparison, numerical results obtained by the our proposed sGS-imABCD method for (D$_h$) and the iwADMM and APG methods for ($\overline{\text{DP}_h}$) are shown in Table 5. As a result from Table 5, it can be observed that our sGS-imABCD is faster and more efficient than the iwADMM and APG methods in terms of the iterations and CPU times.

At last, in order to show the robustness of our proposed sGS-imABCD method with respect to the parameters α and β, we also test the same problem with different values of α and β on mesh $h = 2^{-8}$. The results are presented in Table 6. From Table 6, it is obvious to see that our method could solve problem (D$_h$) to high accuracy for all tested values of α and β within 50 iterations. More importantly, from the results, we can see that when α is fixed, the number of iteration steps of the sGS-imABCD method remains nearly constant for β ranging from 0.005 to 1. However, for a fixed β, as α increases from 0.005 to 0.5, the number of iteration steps of the sGS-imABCD method changes drastically. These observations indicate that the sGS-imABCD method shows the β-independent convergence property, whereas it does not have the same convergence property with respect to the parameter α.

Table 5 Example 3: The convergence behavior of sGS-imABCD for (D_h), ihADMM, and APG for $(\overline{DP_h})$. In the table, #dofs stands for the number of degrees of freedom for the control variable on each grid level. $E_2 = \min\{E_2(sGS-imABCD), E_2(ihADMM), E_2(APG)\}$

h	#dofs	E_2	EOC	Index of performance	sGS-imABCD	ihADMM	APG
2^{-3}	49	0.2925	–	iter	13	32	16
				residual η	6.25e-08	6.33e-08	3.51e-08
				CPU time/s	0.16	0.23	0.22
2^{-4}	225	0.1127	1.3759	iter	12	36	18
				residual η	6.34e-08	8.91e-08	7.23e-08
				CPU times/s	0.24	0.44	0.45
2^{-5}	961	0.0457	1.3390	iter	13	40	16
				residual η	7.10e-08	7.42e-08	8.88e-08
				CPU time/s	0.47	1.17	2.98
2^{-6}	3969	0.0161	1.3944	iter	14	44	16
				residual η	4.05e-08	9.10e-08	6.60e-08
				CPU time/s	2.62	6.04	4.86
2^{-7}	16129	0.0058	1.4132	iter	12	50	16
				residual η	6.43e-08	9.80e-08	8.45e-08
				CPU time/s	10.22	29.53	30.63
2^{-8}	65025	0.0019	1.4503	iter	10	53	17
				residual η	7.05e-08	8.93e-08	8.88e-08
				CPU time/s	60.45	160.24	92.60
2^{-9}	261121	0.0007	1.4542	iter	10	54	18
				residual η	5.21e-08	7.96e-08	3.24e-08
				CPU time/s	395.78	915.71	859.22

It should be pointed out that the numerical results are also consistent with the theoretical conclusion based on Theorem 8.

Example 4.

$$\begin{cases} \min_{(y,u)\in Y\times U} J(y,u) = \frac{1}{2}\|y-y_d\|^2_{L^2(\Omega)} + \frac{\alpha}{2}\|u\|^2_{L^2(\Omega)} + \beta\|u\|_{L^1(\Omega)} \\ \text{s.t.} \quad -\Delta y = u \quad \text{in } \Omega = (0,1)\times(0,1), \\ \qquad y = 0 \quad \text{on } \partial\Omega, \\ \qquad u \in U_{ad} = \{v(x)|a \leq v(x) \leq b, \text{a.e on } \Omega\}, \end{cases}$$

where the desired state $y_d = \frac{1}{6}\sin(2\pi x)\exp(2x)\sin(2\pi y)$ and the parameters $\alpha = 10^{-5}$, $\beta = 10^{-3}$, $a = -30$, and $b = 30$. In addition, the exact solution of

Table 6 Example 3: The performance of sGS-imABCD for (D$_h$) with different values of α and β

h	α	β	iter.sGS-imABCD	residual error η about K-K-T
2^{-8}	0.005	0.005	49	7.59e-08
		0.05	48	8.86e-08
		0.5	46	6.76e-08
		1	48	5.49e-08
	0.05	0.005	23	8.74e-08
		0.05	25	7.26e-08
		0.5	22	5.77e-08
		1	23	7.63e-08
	0.5	0.005	12	6.51e-08
		0.05	11	8.80e-08
		0.5	10	7.05e-08
		1	12	8.53e-08

Table 7 Example 4: The performance of sGS-imABCD for (D$_h$). In the table, #dofs stands for the number of degrees of freedom for the control variable on each grid level

h	#dofs	iter. sGS-imABCD	No. $\tilde{p}$-block	residual η	CPU time/s	E_2	EOC
2^{-3}	49	37	12	8.67e-08	0.64	5.5408	–
2^{-4}	225	30	10	7.32e-08	0.65	2.4426	1.1817
2^{-5}	961	22	8	8.38e-08	0.73	1.1504	1.1340
2^{-6}	3969	22	7	6.83e-08	4.65	0.4380	1.2203
2^{-7}	16129	16	5	6.46e-08	16.60	0.1774	1.2413
2^{-8}	65025	15	3	6.36e-08	105.70	0.1309	1.0807
2^{-9}	261121	15	3	5.65e-08	1158.62	0.0406	1.1821
2^{-10}	1046529	16	3	4.50e-08	24008.07	–	–

the problem is unknown. In this case, using a numerical solution as the reference solution is a common method. For more details, one can see Hinze et al. (2009). In our practice implementation, we use the numerical solution computed on a grid with $h^* = 2^{-10}$ as the reference solution. It should be emphasized that choosing the solution that computed on mesh $h^* = 2^{-10}$ is reliable. As shown below, when $h^* = 2^{-10}$, the scale of data is 1046529.

In Table 7, we report the numerical results obtained by our proposed sGS-imABCD method for solving (D$_h$). As a result, one can see that our proposed sGS-imABCD method is an efficient algorithm to solve problem (D$_h$) to high accuracy. In addition, the errors of the control u with respect to the solution on the finest grid ($h^* = 2^{-10}$) and the results of EOC for control are also presented in Table 7, which confirm the error estimate result as shown in Theorem 1. For the sake of comparison, in Table 9, we report the numerical results obtained by

Table 8 Example 4: The convergence behavior of GMRES for $\hat{p}$-block subproblem

h	iter.sGS-imABCD	iter.GMRES of $\hat{p}$-block	Relative residual error of GMRES
2^{-7}	1	7	1.54e-04
	2	7	1.12e-05
	3	8	7.25e-06
	4	8	3.95e-06
	5	8	3.85e-06
	6	8	2.66e-06
	7	8	3.33e-06
	8	8	2.60e-06
	9	8	1.86e-06
	10	8	1.15e-06
	11	8	1.28e-06
	12	7	8.68e-07
	13	7	9.26e-07
	14	7	5.17e-07
	15	7	7.76e-07
	16	7	7.39e-07
2^{-8}	1	7	1.50e-04
	2	7	1.11e-05
	3	8	7.23e-06
	4	8	9.61e-06
	5	9	5.56e-06
	6	10	7.37e-07
	7	8	3.98e-06
	8	8	2.34e-06
	9	8	1.96e-06
	10	8	1.15e-06
	11	8	1.27e-06
	12	7	8.36e-07
	13	7	8.16e-07
	14	7	4.38e-07
	15	7	7.61e-07

sGS-imABCD method for solving (D$_h$) and iwADMM and APG methods for ($\overline{\text{DP}}_h$). Comparing the error results from Tables 7 and 9, we can see that directly solving (D$_h$) can get better error results than that from solving (D$_h$) and ($\overline{\text{DP}}_h$). Obviously, this conclusion shows the efficiency of our dual-based approach which can avoid the additional error caused by the approximation of L^1-norm. Furthermore, from Table 7, the numerical results in terms of iteration numbers illustrate the mesh-independent performance of our proposed sGS-imABCD method.

In addition, in Table 8, numbers of iteration steps and the relative residual errors of PMHSS-preconditioned GMRES method for the $\hat{p}$-subproblem on mesh $h = 2^{-7}$

Table 9 Example 4: The convergence behavior of sGS-imABCD, ihADMM, and APG for ($\overline{\text{DP}}_h$)

h	#dofs	E_2	EOC	Index of performance	sGS-imABCD	ihADMM	APG
2^{-3}	49	6.6122	–	iter	40	56	44
				residual η	6.06e-08	8.36e-08	9.92e-08
				CPU time/s	0.72	0.42	0.60
2^{-4}	225	2.6314	1.3293	iter	16	55	39
				residual η	9.94e-08	9.14e-08	9.74e-08
				CPU times/s	0.48	0.62	1.03
2^{-5}	961	1.2825	1.1831	iter	21	51	29
				residual η	5.36e-08	8.59e-08	8.31e-06
				CPU time/s	0.99	1.707	3.84
2^{-6}	3969	0.7514	1.0458	iter	22	46	29
				residual η	9.91e-08	6.83e-08	9.38e-08
				CPU time/s	4.95	8.34	11.94
2^{-7}	16129	0.29304	1.1240	iter	20	46	24
				residual η	9.89e-08	5.85e-08	9.36e-08
				CPU time/s	20.83	38.93	45.85
2^{-8}	65025	0.1357	1.1213	iter	20	48	20
				residual η	4.99e-08	8.39e-08	9.05e-08
				CPU time/s	143.88	219.27	181.11
2^{-9}	261121	0.0958	1.0181	iter	18	50	20
				residual η	9.05e-08	7.04e-08	8.84e-08
				CPU time/s	1272.25	2227.48	1959.11

and $h = 2^{-8}$ are presented, which shows that the PMHSS-preconditioned GMRES method is roughly independent of the mesh size h.

As a result from Table 9, it can be also observed that our sGS-imABCD is faster and more efficient than the iwADMM and APG methods in terms of the iteration numbers and CPU times. The numerical performance of our proposed sGS-imABCD method clearly demonstrates the importance of our method.

Finally, to show the influence of the parameters α and β on our proposed sGS-imABCD method, we also test Example 4 with different values of α and β on mesh $h = 2^{-8}$. The results are presented in Table 10. From Table 10, it is obvious to see that our proposed sGS-imABCD method is independent of the parameter β. However, its convergence rate depends on α. It also confirms the convergence results of Theorem 8.

Table 10 Example 4: The performance of sGS-imABCD for (D$_h$) with different values of α and β

h	α	β	iter.sGS-imABCD	residual error η about K-K-T
2^{-8}	10^{-6}	0.0005	26	8.37e-08
		0.001	27	8.40e-08
		0.005	26	9.77e-08
		0.008	28	2.47e-08
	10^{-5}	0.0005	13	5.44e-08
		0.001	15	6.36e-08
		0.005	14	8.60e-08
		0.008	13	8.17e-08
	10^{-4}	0.0005	5	9.84e-08
		0.001	4	3.71e-08
		0.005	5	9.23e-08
		0.008	5	5.22e-08

Conclusion

In this chapter, elliptic PDE-constrained optimal control problems with L^1-control cost (L^1-EOCP) are considered. By taking advantage of inherent structures of the problem, we introduce an inexact heterogeneous ADMM (ihADMM) to solve discretized problems. Furthermore, theoretical results on the global convergence as well as the iteration complexity results $o(1/k)$ for ihADMM were given. Instead of solving the primal problem, we introduce a duality-based approach. By taking advantage of the structure of dual problem, and combining the inexact majorized ABCD (imABCD) method and the recent advances in the inexact symmetric Gauss-Seidel (sGS) technique, we introduce the sGS-imABCD method to solve the dual problem.

References

Beck, A., Teboulle, M.: A fast iterative shrinkage-thresholding algorithm for linear inverse problems. SIAM J. Imaging Sci. **2**, 183–202 (2009)

Bergounioux, M., Ito, K., Kunisch, K.: Primal-dual strategy for constrained optimal control problems, SIAM J. Control Optim. **37**, 1176–1194 (1999)

Blumensath, T., Davies, M.E.: Iterative Thresholding for Sparse Approximations. J. Fourier Anal. Appl. **14**, 629–654 (2008)

Boyd, S., Parikh, N., Chu, E., Peleato, B., Eckstein, J.: Distributed optimization and statistical learning via the alternating direction method of multipliers, Found. Trends® Mach. Learn. **3**, 1–122 (2011)

Carstensen, C.: Quasi-interpolation and a posteriori error analysis in finite element methods. ESAIM: Math. Model. Numer. Anal. **33**, 1187–1202 (1999)

Casas, E.: Using piecewise linear functions in the numerical approximation of semilinear elliptic control problems. Adv. Comput. Math. **26**, 137–153 (2007)

Casas, E., Tröltzsch, F.: Error estimates for linear-quadratic elliptic control problems. Analysis and optimization of differential systems, pp. 89–100. Springer (2003)

Casas, E., Clason,C., Kunisch, K.: Approximation of elliptic control problems in measure spaces with sparse solutions. SIAM J. Control Optim. **50**, 1735–1752 (2012)

Casas, E., Herzog, R., Wachsmuth, G.: Approximation of sparse controls in semilinear equations by piecewise linear functions. Numer. Math. **122**, 645–669 (2012a)

Casas, E., Herzog, R., Wachsmuth, G.: Optimality conditions and error analysis of semilinear elliptic control problems with L^1 cost functional. SIAM J. Optim. **22**, 795–820 (2012b)

Chambolle, A., Dossa, C.: A remark on accelerated block coordinate descent for computing the proximity operators of a sum of convex functions (2015). https://hal.archives-ouvertes.fr/hal-01099182

Chen, L. Sun, D.F., Toh, K.C.: An efficient inexact symmetric Gauss-Seidel based majorized ADMM for high-dimensional convex composite conic programming. Math. Program. **161**(1), 237–270 (2017)

Ciarlet, P.G.: The finite element method for elliptic problems. Math. Comput. **36**, xxviii+530 (1978)

Clason, C., Kunisch, K.: A duality-based approach to elliptic control problems in non-reflexive Banach spaces. ESAIM Control Optim. Calc. Var. **17**, 243–266 (2011)

Collis, S.S., Heinkenschloss, M.: Analysis of the streamline upwind/Petrov Galerkin method applied to the solution of optimal control problems. CAAM TR02–01 (2002)

Cui, Y.: Large scale composite optimization problems with coupled objective functions: theory, algorithms and applications. PhD thesis, National University of Singapore (2016)

de Los Reyes, J.C., Meyer, C., Vexler, B.: Finite element error analysis for state-constrained optimal control of the Stokes equations. Control. Cybern. **37**, 251–284 (2008)

Elvetun, O.L., Nielsen, B.F.: The split bregman algorithm applied to PDE-constrained optimization problems with total variation regularization. Comput. Optim. Appl. **64**, 1–26 (2014)

Falk, R.S.: Approximation of a class of optimal control problems with order of convergence estimates. J. Math. Anal. Appl. **44**, 28–47 (1973)

Fazel, M., Pong, T.K., Sun, D.F., Tseng, P.: Hankel matrix rank minimization with applications to system identification and realization. SIAM J. Matrix Anal. Appl. **34**, 946–977 (2013)

Gabay, D., Mercier, B.: A dual algorithm for the solution of nonlinear variational problems via finite element approximation. Comput. Math. Appl. **2**, 17–40 (1976)

Geveci, T.: On the approximation of the solution of an optimal control problem problem governed by an elliptic equation. RAIRO-Analyse numérique. **13**, 313–328 (1979)

Glowinski, R., Marroco, A.: Sur l'approximation, par éléments finis d'ordre un, et la résolution, par pénalisation-dualité d'une classe de problèmes de dirichlet non linéaires, Revue française d'automatique, informatique, recherche opérationnelle. Analyse numérique **9**, 41–76 (1975)

Hintermüller, M., Ulbrich, M.: A mesh-independence result for semismooth Newton methods. Math. Program. **101**, 151–184 (2004)

Hiriart-Urruty, J.-B., Strodiot, J.-J., Nguyen, V.H.: Generalized Hessian matrix and second-order optimality conditions for problems with $C^{1,1}$ data. Appl. Math. Optim. **11**, 43–56 (1984)

Herzog, R., Ekkehard S.: Preconditioned conjugate gradient method for optimal control problems with control and state constraints. SIAM J. Matrix Anal. Appl. **31**, 2291–2317 (2010)

Hinze, M.: A variational discretization concept in control constrained optimization: the linear-quadratic case. Comput. Optim. Appl. **30**, 45–61 (2005)

Hinze, M., Pinnau, R., Ulbrich, M., Ulbrich, S.: Optimization with PDE Constraints, Mathematical Modelling: Theory and Applications, p. 23. Springer, New York (2009)

Li, X.D., Sun, D.F., Toh, K.C.: QSDPNAL: A two-phase Newton-CG proximal augmented Lagrangian method for convex quadratic semidefinite programming problems (2015). *arXiv*:1512.08872

Li, X.D., Sun, D.F., Toh, K.C.: A Schur complement based semi-proximal ADMM for convex quadratic conic programming and extensions. Math. Program. **155**, 333–373 (2016)

Meyer, C., Rösch, A.: Superconvergence properties of optimal control problems. SIAM J. Control Optim. **43**, 970–985 (2004)

Porcelli, M., Simoncini, V., Stoll, M.: Preconditioning PDE-constrained optimization with L^1-sparsity and control constraints. Comput. Math. Appl. **74**, 1059–1075 (2017)

Rösch, A.: Error estimates for linear-quadratic control problems with control constraints. Optim. Methods Softw. **21**, 121–134 (2006)

Schindele, A., Borzì, A.: Proximal methods for elliptic optimal control problems with sparsity cost functional. Appl. Math. **7**, 967–992 (2016)

Sun, D.F., Toh, K.C., Yang, L.Q.: An Efficient Inexact ABCD Method for Least Squares Semidefinite Programming. SIAM J. Optim. **26**, 1072–1100 (2016)

Stadler, G.: Elliptic optimal control problems with L^1-control cost and applications for the placement of control devices. Comp. Optim. Appls. **44**, 159–181 (2009)

Ulbrich, M.: Nonsmooth Newton-like methods for variational inequalities and constrained optimization problems in function spaces. Habilitation thesis, Fakultät für Mathematik, Technische Universität München (2002)

Ulbrich, M.: Semismooth Newton methods for operator equations in function spaces. SIAM J. Optim. **13**, 805–842 (2003)

Wachsmuth, G., Wachsmuth D.: Convergence and regularisation results for optimal control problems with sparsity functional. ESAIM Control Optim. Calc. Var. **17**, 858–886 (2011)

Wathen, A.J.: Realistic eigenvalue bounds for the Galerkin mass matrix. IMA J. Numer. Anal. **7**, 449–457 (1987)

Game Theory and Its Applications in Imaging and Vision

17

Anis Theljani, Abderrahmane Habbal, Moez Kallel, and Ke Chen

Contents

Introduction to Game Theory and Paradigm	678
Applications of Game Theory in Image Restoration and Segmentation	681
Applications of Game Theory in Image Registration	683
Introduction to Image Registration	683
Application of Game Theory to a Simple Registration Model	685
Application of Game Theory to Registering Images Requiring Bias Correction	688
Game Models in Deep Learning	696
Generative Adversarial Networks (GANs)	696
GANs for Image Generation: A Two-Player Game	698
GANs for Image Segmentation: A Two-Player Game	699
Conclusion	703
References	703

A. Theljani (✉) · K. Chen (✉)
Department of Mathematical Sciences, University of Liverpool Mathematical Sciences Building, Liverpool, UK
e-mail: theljani@liverpool.ac.uk; k.chen@liverpool.ac.uk

A. Habbal (✉)
Université Côte d'Azur, Inria, Sophia Antipolis, France

Modeling and Data Science, Mohammed VI Polytechnic University, Benguerir, Morocco
e-mail: Abderrahmane.Habbal@univ-cotedazur.fr

M. Kallel (✉)
Laboratory for Mathematical and Numerical Modeling in Engineering Science (LAMSIN), University of Tunis El Manar, National Engineering School of Tunis, Tunis-Belvédère, Tunisia
e-mail: moez.kallel@enit.utm.tn

© Crown 2023
K. Chen et al. (eds.), *Handbook of Mathematical Models and Algorithms in Computer Vision and Imaging*, https://doi.org/10.1007/978-3-030-98661-2_102

Abstract

It is very common to see many terms in a variational model from Imaging and Vision, each aiming to optimize some desirable measure. This is naturally so because we desire several objectives in an objective functional. Among these is data fidelity which in itself is not unique and often one hopes to have both L_1 and L_2 norms to be small for instance, or even two differing fidelities: one for geometric fitting and the other for statistical closeness. Regularity is another demanding quantity to be settled on. Apart from combination models where one wants both minimizations to be achieved (e.g., total generalized variation or infimal convolution) in some balanced way through an internal parameter, quite often, we demand both gradient and curvature based terms to be minimized; such demand can be conflicted. A conflict is resolved by a suitable choice of parameters which can be a daunting task. Overall, it is fair to state that many variational models for Imaging and Vision try to make multiple decisions through one complicated functional.

Game theory deals with situations involving multiple decision makers, each making its optimal strategies. When assigning a decision (objective) by a variational model to a player by associating it with a game framework, many complicated functionals from Imaging and Vision modeling may be simplified and studied by game theory. The decoupling effect resulting from game theory reformulation is often evident when dealing with the choice of competing parameters. However, the existence of solutions and equivalence to the original formulations are emerging issues to be tackled.

This chapter first presents a brief review of how game theory works and then focuses on a few typical Imaging and Vision problems, where game theory has been found useful for solving joint problems effectively.

Keywords

Noncooperative game theory · Nash equilibria · Joint restoration and segmentation · Image registration · Deep learning

Introduction to Game Theory and Paradigm

Game theory deals with situations involving multiple decision makers. Each decision maker owns the control on some variable known as his action. All actions are collected in an overall variable known as a strategy. Each of the decision makers owns a specific cost function, to be minimized, which depends on the overall strategy variable. Decision makers are also termed by players or agents, and cost functions could also be replaced by payoffs, to be maximized instead. For readers who are familiar with, let us rephrase the classical optimization problems as follows: optimization deals with situations where a single decision maker owns control over one single overall strategy (all optimization variables), and optimizes a single cost/payoff function, possibly subject to constraints.

To start with some comprehensive and easy-reading reference, the book (Gibbons 1992) introduces, most if not all, the must-have material, including the earliest models of Cournot and Bertrand, those of Stackelberg and actually illustrates with many examples how the game theory first emerged from the need to model economic behavior.

We focus in this introduction on noncooperative games, which means that the players do not share the same cost function, or they do not aggregate their costs into a single one (e.g., a weighted sum). We do not consider as well finite or discrete games, where the set of strategies is either finite (e.g., prisoner's dilemma) or discrete (e.g., games on graphs).

Noncooperative games may be static or dynamic. Roughly speaking, in a dynamic game, players sequentially observe actions of other players and then choose their optimal responses. In a static game, players choose their best responses to the others without exchange (or communication) of information. Remark that the notion of time involved in games is not necessarily the physical time involved in, for example, state equations. As well, a static game could be played by players whose cost functions are constrained by, for example, unsteady fluid mechanics. Games may also be with complete information, meaning that all players know each other's strategy spaces and cost functionals (including their own ones). The failure of this assumption is termed as a game with incomplete information, see Gibbons (1992) for details.

Noncooperative games may also be differential and/or stochastic.

Differential games involve state equations governed by system of differential equations. They model a huge variety of competitive interactions, in social behavior, economics, biology among many others, predator-prey, pursuit-evasion games, and so on (Isaacs 1999). Stochastic games theory, starting from the seminal paper by Shapley (1953), occupies nowadays most of the game theory publications, and a vast literature is dedicated to stochastic differential games (Friedman 1972), robust games (Nishimura et al. 2009), games on random graphs, or agents learning games (Hu and Wellman 2003), among many other branches, and it is definitely out of the scope of the introductory section to review all aspects of the field. See also the introductory book (Neyman and Sorin 2003) to the basic concepts of the stochastic games theory.

Solutions to noncooperative games are called equilibria. Contrarily to classical optimization, the definition of an equilibrium depends on the game setting (game rules). Within the static with complete information setting, a relevant one is the so-called Nash equilibrium (**NE**).

We consider primarily the standard static, under complete information, Nash equilibrium problem (**NEP**) (Gibbons 1992).

Definition 1. An **NEP** consists of $p \geq 2$ decision makers (i.e., players), where each player $i \in \{1, \ldots, p\}$ tries to solve his optimization problem:

$$(\mathcal{P}_i) \quad \min_{\mathbf{x}_i \in \mathbb{X}_i} y_i(\mathbf{x}), \tag{1}$$

where $\mathbf{y}(\mathbf{x}) = \left[y_1(\mathbf{x}), \ldots, y_p(\mathbf{x})\right] : \mathbb{X} \subset \mathbb{R}^n \to \mathbb{R}^p$ (with $n \geq p$) denotes a vector of cost functions (a.k.a. pay-off or utility functions), y_i denotes the specific cost function of player i, and the strategy variable $\mathbf{x}$ consists of block components $\mathbf{x}_1, \ldots, \mathbf{x}_p$ $(\mathbf{x} = (\mathbf{x}_j)_{1 \leq j \leq p})$.

Each block $\mathbf{x}_i$ denotes the action variable of player i and $\mathbb{X}_i$ its corresponding action space and $\mathbb{X} = \prod_i \mathbb{X}_i$. We shall use the convention $y_i(\mathbf{x}) = y_i(\mathbf{x}_i, \mathbf{x}_{-i})$ when we need to emphasize the role of $\mathbf{x}_i$.

Definition 2. A Nash equilibrium (**NE**) $\mathbf{x}^* \in \mathbb{X}$ is a strategy such that:

$$(\mathbf{NE}) \quad \forall i, \ 1 \leq i \leq p, \quad \mathbf{x}_i^* = \arg\min_{\mathbf{x}_i \in \mathbb{X}_i} y_i(\mathbf{x}_i, \mathbf{x}_{-i}^*). \tag{2}$$

In other words, when all players have chosen to play an (**NE**), then no single player has incentive to move from his $\mathbf{x}_i^*$. Let us however mention by now that, generically, Nash equilibria are not efficient, that is, do not belong to the underlying set of best compromise solutions, called Pareto front, of the objective vector $(y_i(\mathbf{x}))_{\mathbf{x} \in \mathbb{X}}$.

An important class of games are the so-called potential games. As introduced in the survey paper (David and Hernández-Lerma Onésimo 2016), in the static case, a noncooperative game is said to be a potential game if there is a real-valued function, called a potential function, such that a strategy profile that optimizes the potential function is a Nash equilibrium for the game. This is precisely one of the key properties of potential games; namely, in a potential game one can find Nash equilibria by optimizing a single function rather than using a fixed-point argument as is typically done for noncooperative games.

From application side, few papers are dedicated to engineering applications involving partial differential state equations where distributed parameters are seen as Nash strategies. In Habbal et al. (2004), a Nash game is set up between two physical processes, heat transfer and structural mechanics, using cooling and structural material densities (like as in topology optimization) as Nash strategies. Nash games could also be used to model biological processes, as introduced in Habbal (2005), where tumoral angiogenesis is modeled as a Nash game between pro- and anti-angiogenic factors and involves porous media and elasticity state equations. In Roy et al. (2017), Nash strategies are used to model the cognitive process of pedestrian avoidance, with Fokker-Planck state equations.

Engineering applications involving multidisciplinary optimization may also benefit from reframing within a Nash game framework, see Desideri et al. (2014) for an overview and Benki et al. (2015) for an original application in nonlinear mechanics. Finally, and in close connection to image processing, ill-posed inverse problems may find a strikingly efficient benefit in being reformulated as Nash games. See Habbal and Kallel (2013) for a novel approach in solving data recovery problems, and Habbal et al. (2019); Chamekh et al. (2019) in devising new algorithms to solve the coupled data recovery and parameter or shape identification problems.

Applications of Game Theory in Image Restoration and Segmentation

There are two classical problems associated with image processing: the image denoising (restoration) and contour identification (segmentation). To address these issues, there are various approaches, such as the stochastic modeling, the wavelet approach and the variational approach leading to the partial differential equations. Image restoration is an inverse problem which consists of finding the original image from another observed, often linked by the equation, $I_0 = \mathcal{T}I + v$, where $\mathcal{T}$ is a linear operator modeling the blur, I a (mathematical) image defined by the intensity (or gray level), and v represents the noise (Gaussian for example). Image segmentation is the process of extracting objects from an image, and can be formulated as finding a finite collection $\{\Omega_i\}_{i=1}^{K}$ of disjoint open subsets of Ω, where Ω is an open and bounded subset of $\mathbb{R}^2$ and represents the image domain. The restoration and segmentation of the image can be performed simultaneously. In this case, one has to solve a minimization problem of a sum of two energies (see, e.g., Mumford-Shah functional (Mumford 1989)). One favors image regularization and the other detects and enhances the contours presented in the image. If the regularization term of the energy is favored over the segmentation term, then the contours are smoothed and hence destroyed. On the other hand, if the segmentation contribution to the energy is made stronger than the regularization contribution, then we might obtain an oversegmented image.

A game-theoretic approach was proposed in Kallel et al. (2014) to simultaneously restore and segment noisy images. The method is based on iterative negotiation between the two antagonistic processes, segmentation and restoration, where acceptable solutions arise then as stationary (noncooperative) decisions. In this work, the game theory concepts are used and define two players: one is interested in the regularization of the image and the other is concerned with its segmentation. Each of two players will try to increase his profit by making an adequate decision until a "Nash equilibrium" is reached. More specifically, the restoration player's goal is to minimize the functional

$$\mathcal{J}_1(I, C) = \int_{\Omega} (I - I_0)^2 \, dx + \mu \int_{\Omega \setminus C} |\nabla I|^2 \, dx, \tag{3}$$

and the segmentation player's objective is to minimize the functional

$$\mathcal{J}_2(I, C) = \sum_{i=1}^{K} \int_{\Omega_i} (I_0 - I_i)^2 \, dx + \nu |C|, \quad \text{where } I_i = \frac{1}{|\Omega_i|} \int_{\Omega_i} I(x) \, dx. \tag{4}$$

The functional (4) is inspired from the Mumford-Shah one and it is obtained by replacing the restriction of I in each connected component Ω_i of Ω with its mean over Ω_i. To summarize this approach, the authors consider a two-player static of complete information game where the first player is restoration, and the second is segmentation. Restoration minimizes the cost $\mathcal{J}_1(I, C)$ with action on the intensity field I, while segmentation minimizes the cost $\mathcal{J}_2(I, C)$ with action on the discontinuity set C. In this case, solving the game amounts to finding a Nash equilibrium (**NE**), defined as a pair of strategies (I^*, C^*), such that

$$\begin{cases} I^* = \operatorname{argmin}_I \mathcal{J}_1(I, C^*), \\ C^* = \operatorname{argmin}_C \mathcal{J}_2(I^*, C). \end{cases} \qquad (5)$$

The minimizer I^* is sought in the Sobolev space $H^1(\Omega \setminus C^*)$ and C^* is sought in the set of the union of curves made of a finite set of $C^{1,1}$-arcs.

To compute this equilibrium, they use the classical iterative method with relaxation (Uryas'ev 1994) as described in Algorithm 1. The main advantage of using this algorithm is that $\overline{I}^{(k)}$ and $\overline{C}^{(k)}$ can be numerically computed, separately and parallelly, using descent algorithms.

Algorithm 1 Nash equilibrium algorithm

1: Initial guess: $S^{(0)} = (I^{(0)}, C^{(0)})$. Set $k = 0$.
2: **repeat**
3: $\quad \overline{I}^{(k)} = \operatorname{argmin}_I \mathcal{J}_1(I, C^{(k)})$
4: $\quad \overline{C}^{(k)} = \operatorname{argmin}_C \mathcal{J}_2(I^{(k)}, C)$
5: $\quad S^{(k+1)} = (I^{(k+1)}, C^{(k+1)}) = \tau S^{(k)} + (1-\tau)(\overline{I}^{(k)}, \overline{C}^{(k)})$ {for τ fixed, $0 < \tau < 1$}
6: $\quad k = k + 1$
7: **until** $S^{(k)}$ converges

Finally, the authors use a level-set approach to get rid of the tricky control dependence of functional spaces. After, a numerical study is carried on some real images in order to evaluate the effectiveness of the proposed algorithm. In particular, they show that by decoupling the Mumford-Shah functional using the game algorithm, the dependence on the regularization parameters μ and ν is uncorrelated and the choice of their values becomes more flexible and natural. On the other hand, the dependence of the functional $\mathcal{J}_2$ only on the mean of I in each connected component has a significant effect on the speed of convergence. In Fig. 1, a numerical result using only one level-set function is represented. The top row displays the evolution of curves over the corresponding images $I^{(k)}$, $k \in \{0, 10, 50\}$. The bottom row displays the final segmentation result (second image) and denoised image (third image) with $PSNR = 31.98$. For this case, the algorithm converges after 135 iterations.

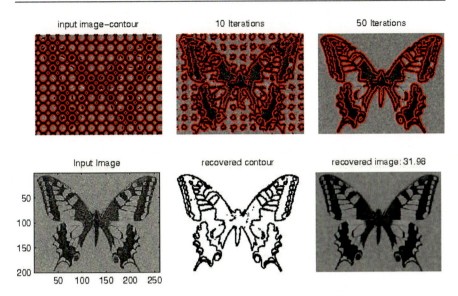

Fig. 1 Top row: noisy image with Gaussian noise (variance = 0.2) and initial contour, evolution by iterations. Bottom row: segmentation and restoration of image by the proposed algorithm with ($\nu = 0.2, \mu = 0.01$), for $k = 135$. CPU time = 117 sec

Applications of Game Theory in Image Registration

There exist many image registration models: each is designed for one class of problems. It is challenging to find an universally robust model that can deal with all registration problems, due to the inherent difficulties of image registration. The previous section discussed how game theory can be used to enhance a model for image restoration and segmentation. Here we shall see that game theory is also a natural tool to reformulate an image registration model in achieving better performance and robustness.

In this section, we review recent works on using game theory to design and reformulate the traditional variational models for deformable image registration. The advantages gained will be in reduction of the burden of tuning many parameters; hence a more robust model is obtained. The ideas are generally applicable to almost other variational models.

Introduction to Image Registration

Image registration (Chen et al. 2019) aims to align two given images through mapping one (the template image T) to the other (the reference image R) so that the aligned (or registered) image $T(\phi)$ may be used to give us complementary information from T to R, or highlight the differences between T and R. Here $\phi(x) = x + \mathbf{u}(x)$ where $\mathbf{u}(x) = (u_1(x), u_2, \ldots, u_d(x))$ is the unknown map concerned if $x \in \Omega \subset \mathbb{R}^d$. Practically $d = 2, 3$ are more common.

To find ϕ, a typical variational model takes the form ; Chen et al. (2019)

$$\min_{\mathbf{u}} \mathcal{J}(\mathbf{u}) = F(\mathbf{u}) + \alpha S(\mathbf{u}) + \beta C(\mathbf{u}) \tag{6}$$

where $F(\mathbf{u})$, $S(\mathbf{u})$, $C(\mathbf{u})$ are, respectively, the fitting terms to align T, R, the regularization term to overcome the ill-posedness of minimizing the fitting term alone and the control term to ensure the underlying map ϕ does not have folding (e.g., by making ϕ diffeomorphic).

Flexibilities exist for specifying each of the three terms in (6) differently, though none of these flexibilities is sufficient to construct a robust model for a wider class of problems than with a fixed choice of terms.

First, since the fitting term F is supposed to measure the dissimilarity of T, R, it has many possible choices especially for multi-modality pairs of T, R (e.g., T is from MRI and R is from ultrasound).

For single modal images (e.g., when both T, R are CT images), a popular choice for F is the SSD (sum of squared differences)

$$F(\mathbf{u}) = \int_\Omega |T(x+\mathbf{u}) - R(x)|^2 dx.$$

For multimodal image pairs, one may take the popular choice of mutual information (Maes et al. 1997). This statistical measure has also been improved a few times since 1997. One alternative is the normalized gradient differences (NGD)

$$F(\mathbf{u}) = \int_\Omega |\nabla_n T(x+\mathbf{u}) - \nabla_n R(x)|^2 dx$$

where $\nabla_n T = \nabla T / |\nabla T|$; however, we remark that this fitting term is not very robust, a better variant is proposed in Theljani and Chen (2019a).

Second, as for designing the regularizer S, one way is to regularize all deformation directions individually:

$$S(u) = S(u_1) + \ldots + S(u_d) \tag{7}$$

but one may introduce some coupling between these individual terms.

Finally, the control term C is designed to ensure $\det(\nabla \phi) > 0$. If it makes sense to achieve volume or area preservation in features of T, R, that is, $\det(\nabla \phi) = 1$, a simple method is to define

$$C(\mathbf{u}) = \int_\Omega (\det(\nabla \phi) - 1)^2 dx.$$

However, if this is not appropriate for other applications, a robust method seems to define

$$C(\mathbf{u}) = \int_\Omega \Phi(\mu(\phi))^2 dx,$$

where Φ is some smooth function (Zhang and Chen 2018) and μ is the Beltrami coefficient for the same mapping ϕ projected to a complex plane with $\phi = \phi_1(x) + i\phi_2(x)$ with $d = 2$. The central idea is the equivalence relationship $|\mu| < 1 \Leftrightarrow \det(\nabla\phi) > 0$, which facilitates the design of an unconstrained optimization problem (Lam and Lui 2014).

Of course, it is entirely appropriate to propose a minimization problem like (6) without its third term, and to add the constraint $\det(\nabla\phi) > 0$ as done in Zhang et al. (2016) and Thompson and Chen (2019). However, nonlinear constraints are not easy to deal with in numerical implementations.

One drawback of the Beltrami coefficient is that such a quantity μ does not exist when $d \geq 3$, though there are some recent attempts to generalize it to high dimensions. The recent work by Zhang and Chen (2020) designed a 3D Beltrami coefficient-like quantity that possesses the same property as 2D, and hence extended the classical work.

Another method to replace the third term in (6) is the so-called inverse consistent formulation where the folding is avoided by simultaneously registering T to R by ϕ and also R to T by ψ. The central idea is $\phi(\psi) = I$ or $\psi(\phi) = I$ so that the map is inversely consistent and does not fold. See Christensen et al. (2007), Thompson and Chen (2019), Theljani and Chen (2019c) and Chen and Ye (2010).

Application of Game Theory to a Simple Registration Model

To illustrate the idea of using the game theory, let us first consider the diffusion registration model for simple modal images before we elaborate on more robust models in later subsections.

Let us start with the simple diffusion model (Fischer and Modersitzki 2002) which takes the following form:

$$\min_{\mathbf{u} \in W^{1,2}(\Omega)} \mathcal{J}(\mathbf{u}) = \int_\Omega |\nabla \mathbf{u}|^2 dx + \alpha M(\mathbf{u}) \qquad (8)$$

where $M(\cdot)$ is a similarity measure. One application using game theory for this model is to consider two different similarity measures. For the simple case of monomodal images, using the sum of squared differences is used because of the grey value constancy assumption. However, in some scenarios, the SSD has a big drawback: it is quite susceptible to slight changes in brightness, which often appear in natural scenes. Therefore, it is useful to allow some small variations in the grey value and help to determine the displacement vector by a criterion that is invariant under grey value changes. Thus, to have a model which is less sensitive to illumination variations, it is interesting to combine SSD with another measure, which can capture more information, such as gradients, and fulfill the gradient constancy assumption.

Coupled Measures: Nongame Approach

The combination of the two measures can be done as a classical variational formulation where one has to optimize one single energy which couples both measures. In case where the SSD is combined with the NGD for monomodal image registration, the natural vibrational approach consists of solving

$$\min_{\mathbf{u} \in W^{1,2}(\Omega)} \mathcal{J}(\mathbf{u}) = \int_\Omega |\nabla \mathbf{u}|^2 dx + \lambda_1 \int_\Omega |T(x+\mathbf{u}) - R(x)|^2 dx + \lambda_2$$
$$\times \int_\Omega |\nabla_n T(x+\mathbf{u}) - \nabla_n R(x)|^2 dx \qquad (9)$$

This approach may lead to a solution which is sensible to choice of the weighting parameters λ_1 and λ_2 between the two measures. In fact, if more weights are put on the SSD term, it seems that the model does not work because the SSD will not handle well the regions in the images that distorted by varying illumination. Only few regions are well registered where there is no big difference in the intensity variation between the two images. Reversely, if the NGD contribution to the model is too much strong by taking large value of λ_2, then the solution seems to be well registered in regions of varying intensity whereas the registration quality is poorer than the SSD model in clean regions, that is, nor varying intensities.

Coupled Measures: Game Approach

In game formulation, the combination of the two measures can be done differently from the classical approach. We can design a game where the two measures are incorporated in different models that have some communications through a coupling term. As an example, we could consider the following game model: Find a Nash equilibrium (**NE**) $(\mathbf{u}^*, \mathbf{v}^*)$ such that

$$\begin{cases} \mathbf{u}^* = \mathrm{argmin}_{\mathbf{u} \in W^{1,2}(\Omega)} \mathcal{J}_1(\mathbf{u}, \mathbf{v}^*), \\ \mathbf{v}^* = \mathrm{argmin}_{\mathbf{v} \in W^{1,2}(\Omega)} \mathcal{J}_2(\mathbf{u}^*, \mathbf{v}), \end{cases} \qquad (10)$$

where

$$\mathcal{J}_1(\mathbf{u}, \mathbf{v}) = \int_\Omega |\nabla \mathbf{u}|^2 dx + \int_\Omega |T(x+\mathbf{u}) - R(x)|^2 dx + \lambda \int_\Omega (\mathbf{u}-\mathbf{v})^2 dx, \qquad (11)$$

$$\mathcal{J}_2(\mathbf{u}, \mathbf{v}) = \int_\Omega |\nabla \mathbf{v}|^2 dx + \int_\Omega |\nabla_n T(x+\mathbf{v}) - \nabla_n R(x)|^2 dx + \lambda \int_\Omega (\mathbf{u}-\mathbf{v})^2 dx \qquad (12)$$

The first energy uses the sum of squared difference as similarity measure, whereas the second energy uses the normalized gradient difference term (NGD). The third part is a coupling term which serves for the communication between the two players u and v. The first player tries to minimize his one cost $\mathcal{J}_1(\cdot)$ taking into account the information about the gradient consistency coming from the second player v through the coupling term, and vice versa.

Examples

Figure 2 shows an example of using game model for a pair of MRI images registration. We assess the registration quality by measuring the normalized cross correlation coefficient (NCC) between the registered image $T(\mathbf{u})$ and R (closer NCC to 1 means better registration). Mainly, the example illustrates how two players in a game model can cooperate to achieve better registration quality. However, by considering two separate models, that is, no communication for $\lambda = 0$, the first model in (11) is unable to achieve an acceptable result.

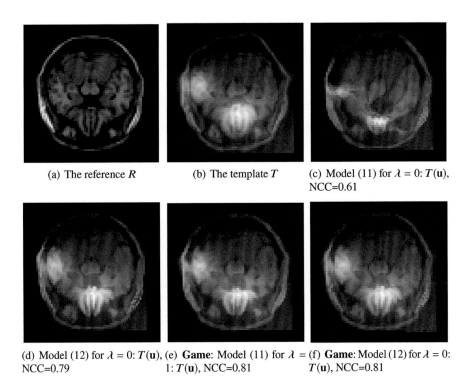

(a) The reference R (b) The template T (c) Model (11) for $\lambda = 0$: $T(\mathbf{u})$, NCC=0.61

(d) Model (12) for $\lambda = 0$: $T(\mathbf{u})$, NCC=0.79 (e) **Game**: Model (11) for $\lambda = 1$: $T(\mathbf{u})$, NCC=0.81 (f) **Game**: Model (12) for $\lambda = 0$: $T(\mathbf{u})$, NCC=0.81

Fig. 2 Example 1: the game approach for registering a pair of MRI images. The template image T contains some undesirable artifact. Clearly, the game approach is able to cope with this case because of the use of two different measures. (**a**) The reference R (**b**) The template T (**c**) Model (11) for $\lambda = 0$: $T(\mathbf{u})$, NCC=0.61 (**d**) Model (12) for $\lambda = 0$: $T(\mathbf{u})$, NCC=0.79 (**e**) **Game**: Model (11) for $\lambda = 1$: $T(\mathbf{u})$, NCC=0.81 (**f**) **Game**: Model (12) for $\lambda = 0$: $T(\mathbf{u})$, NCC=0.81

Application of Game Theory to Registering Images Requiring Bias Correction

In many real-life applications, even a pair of monomodality images acquired from the same source can differ from each other, leading to inaccurate registration results. The difference is often presented as an undesirable artifact either caused by the device itself (spatially homogeneous signal response, bias field, and shading in MRI images) or caused by the imaging modality itself such as perfusion CT which creates some high contrasted regions in the image. In order to obtain accurate registration results and to cope with these problems, many models have been developed for intensity correction (Aghajani et al. 2016; Ebrahimi and Martel 2009; Ghaffari and Fatemizadeh 2018; Li et al. 2009; Rak et al. 2017; Kim and Tagare 2014). It is important to note that, without intensity correction, both monomodality and multimodality models may fail to register the images correctly because bias introduces incorrect intensity values or false edges.

The artifacts can be of either additive or multiplicative type (Modersitzki and Wirtz 2006; Chumchob and Chen 2012; Ghaffari and Fatemizadeh 2018). It has been generally accepted that the image T with bias field, generally presented as a mixed type, relates to the "true" unbiased image T^* via the following affine like intensity relationship: $T = mT^* + s$, where $m(\mathbf{x})$ and $s(\mathbf{x})$ are responsible for the intensity-correction. Rigorously speaking, the word "affine" is misleading because both m, s are never constants so the model is highly nontrivial. Once m, s are found or estimated, the registration task is to find the deformation field $\mathbf{u}$ such that $T^*(\mathbf{u}) \approx R$. Denote by $T_c(\mathbf{u}) = T^*(\mathbf{u})$ the corrected and registered image of T. Hence the equivalent statement to the model $T = mT^* + s$ is

$$R_1 = mR + s, \quad T(\mathbf{u}) \approx R_1, \quad \text{with} \quad T_c(\mathbf{u}) = \frac{T(\mathbf{u}) - s}{m} \approx R, \tag{13}$$

where $T(\mathbf{u})$ is the uncorrected and registered image, carrying the bias field features from T and aligned with R, that is, one may minimize one of these fidelity terms for $m, s, \mathbf{u}$ in some norm:

$$\|mR + s - T(\mathbf{u})\|, \quad \left\|\frac{T(\mathbf{u}) - s}{m} - R\right\|.$$

Any model building on minimization of the above quantities may be much simplified if one of the unknowns is dropped (i.e., $m \equiv 1$ or $s \equiv 0$); however, a full model including both m and s always gives better results in solution quality. In fact, in many cases, intensity correction by either multiplicative or additive model is not always enough (Wang and Pan 2014; Vovk et al. 2007; Park et al. 2019) thus a combined model is necessary.

Non-game Approach

A classical variational approach for joint full bias correction and image registration consists in solving the multivariate optimization problem

$$\text{JM} \qquad \mathcal{J}(\mathbf{u}, m, s) := \lambda \int_\Omega |mR + s - T(\mathbf{u})|^2 d\mathbf{x} + \mathcal{R}(\mathbf{u}, s, m), \qquad (14)$$

where $\mathcal{R}(\mathbf{u}, s, m)$ will be chosen to be the same as comparable models shortly. Since m is not a constant function, the first term in (14) is not convenient for numerical implementation for solving the subproblems. The authors in Theljani and Chen (2019b) proposed a variant to this term. They transformed the multiplicative term into an additive one since the latter is more convenient (a simple filtering problem). This transformation was obtained by applying a splitting method to the bias model (13). The splitting leads the additive problems

$$K_l = m_l + R_l, \quad T(\mathbf{u}) = e^{K_l} + s, \qquad (15)$$

which is easier to handle, assuming $m, R > 0$. Here $R_l = \ln(R)$ is known since R is given, $m_l = \log(m)$, and K_l is the intermediate quantity as a spitting variable. The application of a logarithmic transform in the context of intensity transformations increases the contrast between certain intensity values (Duan et al. 2015; Chang et al. 2017; Bansal et al. 2004; Van Leemput et al. 1999). After applying the penalty method to incorporate the constraints (15), the new variational model takes the following form

$$\text{CV} \qquad \min_{\mathbf{u}, s, m_l, K_l} \{\mathcal{J}(\mathbf{u}, s, m_l, K_l) =$$

$$\mathcal{R}(\mathbf{u}, s, m_l, K_l) + \lambda_1 \int_\Omega |T(\mathbf{u}) - e^{K_l} - s|^2 d\mathbf{x} + \lambda_2 \int_\Omega |m_l + R_l - K_l|^2 d\mathbf{x}\}$$

$$(16)$$

where $\mathbf{u}$ is the main deformation field variable, $\mathcal{R}(\cdot)$ contains regularization terms associated to all four unknowns (to be specified), and the rest of the energy are two fidelity terms. Clearly there are no multiplicative terms in (16) as designed. One would normally specify $\mathcal{R}(\cdot)$ and try to solve the joint optimization problem by some techniques, for example, the alternating direction method of multipliers (ADMM) (Boyd et al. 2011) or Augmented Lagrangian (Bonnans et al. 2006; Boyd et al. 2011). The problem (16) is split into 4 subproblems for each of the main variables: $\mathbf{u}, s, m_l, K_l$. There are two challenges: i) choosing the 5 parameters (assuming there are 3 new parameters from $\mathcal{R}(\cdot)$) suitably is a highly nontrivial task; ii) one cannot avoid coupling all 4 variables in any subproblem. This challenge can be solved using a game theory formulation as described in the sequel.

Game Model

It was shown in Theljani and Chen (2019b) that it is more convenient to reformulate (16) to another form using the Nash game idea where both of these two challenges are overcome: first, each subproblem will have one parameter which can be tuned for that subproblem in an easier way; second, it is possible to modify the above subproblems to reduce couplings and hence improve convergence. The authors demonstrated that the game model offers a better solution for two main aspects: choice of underlying parameters and proof of solution existence. In fact, the K_l subproblem in model (16) has three terms and involves two penalty parameters λ_1 and λ_2, which are pretended to be large enough. The solution will be sensitive to these two parameters and the optimal choice is nontrivial. We shall reformulate this problem to yield only one parameter (instead of two) by considering a game approach that has a separable structure and makes the model less sensitive to these parameters. The joint model (16) was reformulated as a game where the solution is a **Nash equilibrium** defined by $(A_1, A_2, A_3, A_4) = (\mathbf{u}, s, m_l, K_l)$ in the space $\mathcal{X} = \mathcal{W} \times W^{1,2}(\Omega) \times W^{1,2}(\Omega) \times W^{1,2}(\Omega)$ where $\mathcal{W} = W^{2,2}(\Omega, \mathbb{R}^2) \cap W_0^{1,2}(\Omega, \mathbb{R}^2)$. The space $\mathcal{X}$ is endowed with the following norm:

$$\|z\|_\mathcal{X} = \left(\|\mathbf{u}\|_\mathcal{W}^2 + \|\nabla s\|_{W^{1,2}(\Omega)}^2 + \|\nabla m_l\|_{W^{1,2}(\Omega)}^2 + \|\nabla K_l\|_{W^{1,2}(\Omega)}^2 \right)^{1/2},$$

where $\|\mathbf{u}\|_\mathcal{W} = \left(\|\nabla \mathbf{u}\|_2^2 + \|\nabla^2 \mathbf{u}\|_2^2 \right)^{1/2}$. The game formulation allows many choices of energies $\mathcal{R}_i(\cdot)$ and $\mathcal{G}_i(\cdot)$ whose terms may not be part of each other. The choice of the different energies leads to either potential or non-potential games (Monderer and Shapley 1996).

The Potential Game

The potential game structure is very important because it makes easy to prove the existence of Nash equilibrium (**NE**) (Nash 1950, 1951). One example is to make the particular choice of the following energies $\mathcal{J}_i(\cdot) = \mathcal{R}_i(\cdot) + \mathcal{G}_i(\cdot)$ with

$$\begin{cases} \mathcal{R}_1(\mathbf{u}) = \|\mathbf{u}\|_\mathcal{W}^2, & \mathcal{G}_1(\mathbf{u}, s, m_l, K_l) = \lambda_1 \int_\Omega |T(\mathbf{u}) - e^{K_l} - s|^2 d\mathbf{x}, \\ \mathcal{R}_2(s) = \int_\Omega |\nabla s|^2 d\mathbf{x}, & \mathcal{G}_2(\mathbf{u}, s, m_l, K_l) = \lambda_2 \int_\Omega |T(\mathbf{u}) - e^{K_l} - s|^2 d\mathbf{x}, \\ \mathcal{R}_3(m_l) = \int_\Omega |\nabla m_l|^2 d\mathbf{x}, & \mathcal{G}_3(\mathbf{u}, s, m_l, K_l) = \lambda_3 \int_\Omega |m_l + R_l - K_l|^2 d\mathbf{x}, \\ \mathcal{R}_4(K_l) = \int_\Omega |\nabla K_l|^2 d\mathbf{x}, & \mathcal{G}_4(\mathbf{u}, s, m_l, K_l) = \lambda_4 \int_\Omega |m_l + R_l - K_l|^2 d\mathbf{x} \\ & + \lambda_5 \int_\Omega |T(\mathbf{u}) - e^{K_l} - s|^2 d\mathbf{x}, \end{cases}$$

(17)

where $\mathcal{R}_i(\cdot)$ is the regularization term in energy i. There are many possible choices of regularization leading to different solution spaces. For the deformation $\mathbf{u}$, the authors in Theljani and Chen (2019b) used regularizers based on combined first and second-order derivatives. Using only the first-order derivatives, that is, H^1 semi-norm, is sensitive to affine preregistration. We avoid this problem by combining it with the second-order derivative term which are not sensitive to (affine) preregistration as it has the affine transformations in its kernel. Moreover, this choice penalizes oscillations and also allows smooth transformations in order to get visually pleasing registration results. The variables K_l, m_l, and s are chosen in the space $W^{1,2}(\Omega)$ and we could consider different spaces such as $W^{2,2}(\Omega)$ or the space of bounded variation functions $BV(\Omega)$. The formulation in (17) is special cases of game formulation known as a potential game (**PG**) (Monderer and Shapley 1996) which amounts to find a minimizer of an energy $\mathcal{L}(\cdot) = \sum_i^4 \mathcal{J}_i(\mathbf{u}, s, m_l, K_l)$ in (16) – then the game model reduces to an ADMM algorithm if alternating iterations are used or a Nash equilibrium of (16) is a minimizer of $\sum_i^4 \mathcal{J}_i(\mathbf{u}, s, m_l, K_l)$. We refer the reader to Monderer and Shapley (1996), Attouch and Soueycat (2008) and Attouch et al. (2008) for more details about potential game in PDEs.

The Non-potential Game

Instead of (17), it is possible to modify $\mathcal{J}_3$, $\mathcal{J}_4$ to get new subproblems which lead to a better model than (17); the new energies to be minimized are still denoted by $\mathcal{J}_i = \mathcal{R}_i + \mathcal{G}_i$, for $i = 1, 2, 3, 4$, with all terms defined in (17) except these three new terms, that is,

$$\begin{cases} \mathcal{R}_1(\mathbf{u}) = \|\mathbf{u}\|_{\mathcal{W}}^2, & \mathcal{G}_1(\mathbf{u}, s, m_l, K_l) = \lambda_1 \int_\Omega |T(\mathbf{u}) - e^{K_l} - s|^2 d\mathbf{x}, \\ \mathcal{R}_2(s) = \int_\Omega |\nabla s|^2 d\mathbf{x}, & \mathcal{G}_2(\mathbf{u}, s, m_l, K_l) = \lambda_2 \int_\Omega |T(\mathbf{u}) - e^{K_l} - s|^2 d\mathbf{x}, \\ \mathcal{R}_3(m_l) = \int_\Omega |\nabla m_l|^2 d\mathbf{x}, & \mathcal{G}_3(\mathbf{u}, s, m_l, K_l) = \lambda_3 \int_\Omega |m_l + R_l - \ln(T(\mathbf{u}) - s)|^2 d\mathbf{x}, \\ \mathcal{R}_4(K_l) = \int_\Omega |\nabla K_l|^2 d\mathbf{x} + \iota_\Lambda(K_l), & \mathcal{G}_4(\mathbf{u}, s, m_l, K_l) = \lambda_4 \int_\Omega |m_l + R_l - K_l|^2 d\mathbf{x}, \end{cases} \quad (18)$$

where $\Lambda = \{K_l \in L^2(\Omega); K_{\min} \leq K_l \leq K_{\max}\}$ is a closed and convex set; and $\iota_\Lambda(\cdot)$ is a projection into Λ. The variables K_l are bounded for theoretical reasons in order to prove the existence of a Nash equilibrium (**NE**). In this case, an **NE** is not a minimizer of $\sum_i^4 \mathcal{J}_i(\mathbf{u}, s, m_l, K_l)$, which makes the proof of the existence difficult. Formally this Nash game problem is called a non-potential game (denoted by **NPG**). Clearly the essential simplification is in $\mathcal{G}_4$ and there are other possible alternative formulations, for example, using L_1 semi-norm. These changes simplify the K_l-problem in (17), equivalently in (16), where the K_l-energy has three terms and which necessitates two regularization parameters λ_4 and λ_5. Whereas, in the game approach (18), the same problem consists only of regularization and one fidelity term, that is, has only one parameter λ_4. Moreover, to discuss any theory for (18), the non-convexity should be addressed, e.g. the energy $\mathcal{G}_1(\cdot)$ is non-convex w.r.t $\mathbf{u}$. Non-convexity means that we cannot apply the Nash theorem (Nash 1951) to show the existence of an **NE**.

Iterative Algorithm

To compute the (NE), the authors in Theljani and Chen (2019b) used alternating Forward-Backward algorithm (ADMM-like), by means of the following iterative process:

Algorithm 2 Forward-Backward algorithm for computing a Nash equilibrium

- Set $k = 0$ and choose an initial guess $\mathbf{z}^{(0)} = (\mathbf{u}^{(0)}, s^{(0)}, m_l^{(0)}, K_l^{(0)})$.
- Step 1: Compute (in parallel) $(\mathbf{u}^{(k+1)}, s^{(k+1)}, m_l^{(k+1)}, K_l^{(k+1)})$ solution of

$$\overline{\mathbf{u}}^{(k)} = \mathbf{u}^k - \gamma \nabla \mathcal{G}_{\mathbf{u}}(\mathbf{u}^k, s^k, m_l^k, K_l^k), \qquad \mathbf{u}^{(k+1)} = \mathbf{prox}_{\gamma \mathcal{R}_1}(\overline{\mathbf{u}}^{(k)}) \qquad (19)$$

$$\overline{s}^{(k)} = s^k - \gamma \nabla \mathcal{G}_s(\mathbf{u}^k, s^k, m_l^k, K_l^k), \qquad s^{(k+1)} = \mathbf{prox}_{\gamma \mathcal{R}_2}(\overline{s}^{(k)}) \qquad (20)$$

$$\overline{m_l}^{(k)} = m_l^k - \gamma \nabla \mathcal{G}_{m_l}(\mathbf{u}^k, s^k, m_l^k, K_l^k), \qquad m_l^{(k+1)} = \mathbf{prox}_{\gamma \mathcal{R}_3}(\overline{m_l}^{(k)}) \qquad (21)$$

$$\overline{K_l}^{(k)} = K_l^k - \gamma \nabla \mathcal{G}_{K_l}(\mathbf{u}^k, s^k, m_l^k, K_l^k), \qquad K_l^{(k+1)} = \mathbf{prox}_{\gamma \mathcal{R}_4}(\overline{K_l}^{(k)}) \qquad (22)$$

- If $\frac{\|\mathbf{z}^{(k+1)} - \mathbf{z}^{(k)}\|_2}{\|\mathbf{z}^{(k)}\|_2} \leq \epsilon$, stop. Otherwise $k = k + 1$, go to Step 1.

Examples

The experiments show that the game approach can have significant robustness in presence of bias noise and varying illumination. In all examples, the weighting parameters were fixed as $\lambda_1 = 200$ for the **u**-subproblem, $\lambda_2 = 20$ for the s-subproblem, $\lambda_3 = 1$ for the m_l-subproblem, and $\lambda_4 = 5$ for the K_l-subproblem. A multi-resolution technique was used to initialize the displacement **u** in order to avoid local minima and to speed up registration. The game model, denoted by "**Game**," is compared with joint models (14) "**JM**" and the classical variational model (16) denoted by "**CV**." The last models are the more natural choices for the class of joint problems. The authors also compared with the Mutual Information based multi-modality model where they minimize an energy which uses the same regulrizer $\mathcal{R}_1(\cdot)$ and the Mutual Information as similarity measure (denoted by "**MI**" below). Numerical experiments on "**MI**" are performed using the publicly available image registration toolbox – Flexible algorithms for image registration (FAIR) (http://www.siam.org/books/fa06/), where the implementation is based on the Gauss-Newton method.

In the examples, they show the registered images $T(\mathbf{u})$ and the corrected images $T_c(\mathbf{u})$. The latter are defined by the formula $T_c = (T(\mathbf{u}) - s)/e^{m_l}$ for '**Game** and '**CV** , and $T_c = (T(\mathbf{u}) - s)/m$ for **JM**. In contrast, the final registered image for **MI** is just $T(\mathbf{u})$. The normalized correlation coefficient (NCC) between T_c and R and between $T(\mathbf{u})$ and R was used as evaluation metric to quantify the performance of the models and the comparison (the closer the NCC is to 1, the better is the alignment).

Example 1: MRI Images

Figure 3 shows an example of registering two MRI images using the completive models. The moving image T (synthetically enhanced) contains some bias field and varying illumination. The results of all models are displayed and they show that except '**MI**, all models perform well in most parts of the image. However,

17 Game Theory and Its Applications in Imaging and Vision

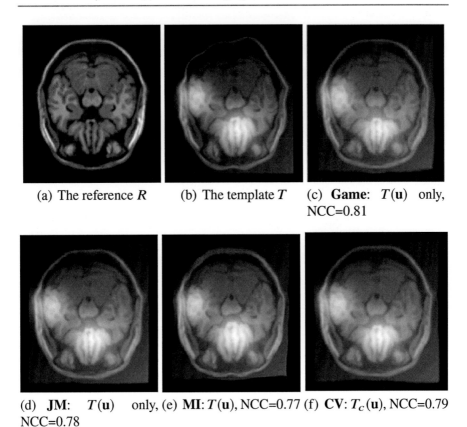

(a) The reference R (b) The template T (c) **Game**: $T(\mathbf{u})$ only, NCC=0.81

(d) **JM**: $T(\mathbf{u})$ only, NCC=0.78 (e) **MI**: $T(\mathbf{u})$, NCC=0.77 (f) **CV**: $T_c(\mathbf{u})$, NCC=0.79

Fig. 3 Example 2: Comparison of 3 different models to register MRI T-1 and T-2 images. From this figure and Fig. 4, we see that **Game** model gives the best registration result. (**a**) The reference R (**b**) The template T (**c**) **Game**: $T(\mathbf{u})$ only, NCC=0.81 (**d**) **JM**: $T(\mathbf{u})$ only, NCC=0.78 (**e**) **MI**: $T(\mathbf{u})$, NCC=0.77 (**f**) **CV**: $T_c(\mathbf{u})$, NCC=0.79

in the middle of the images, the game model is the most advantageous and this can be observed in the zoomed details in Fig. 4. For the parameters tuning, the authors tested different values and they are tabulated in Table 1 which indicates the registration results for different parameters λ_i ($i = 1, ..., 4$). The table shows that the game approach is stable.

Example 2: Application to Perfusion CT Registration

In Fig. 5, pair of CT and Perfusion CT lung images are registered. In the middle of the images T and R, there is a big difference because of the high contrast in T which makes inefficient the use of classical monomodal measures. The registered images using "**Game**, "**CV**, "**JM**, and "**MI** models are shown. We easily see that **Game** model gives a satisfactory result and the corrected part of the moving image is very similar to the middle part of the reference whereas the registration is not good.

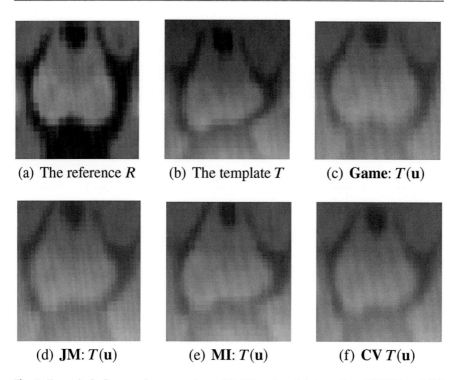

Fig. 4 Example 2: Compared zoom regions of 5 different models to register MRI T-1 and T-2 images. Again **Game** model is the best in solving the registration and the intensity correction jointly, whereas **JM** model cannot solve both problem jointly, only the image correction task is successful. (**a**) The reference R (**b**) The template T (**c**) **Game**: $T(\mathbf{u})$ (**d**) **JM**: $T(\mathbf{u})$ (**e**) **MI**: $T(\mathbf{u})$ (**f**) **CV** $T(\mathbf{u})$

Table 1 Parameters tuning for the pair of MRI images in Fig. 3 using **Game**. In the first column, we fix the parameters λ_3 and λ_4 and we vary the parameters λ_1 and λ_2. In the third column, we vary λ_1 and λ_3 where λ_2 and λ_4 are fixed, whereas, in the last column, we vary λ_1 and λ_4 for fixed λ_2 and λ_3. The NCC errors for the different values of parameters are comparable

		Parameters	
λ_1	λ_2 | NCC	λ_3 | NCC	λ_4 | NCC
100	05 |NCC=0.77	0.5 |NCC=0.78	01 |NCC=0.78
150	15 | NCC=0.79	01 |NCC=0.80	05 |NCC=0.80
200	20 |NCC=0.80	05 |NCC=0.80	20 |NCC=0.79
250	40 | NCC=0.79	10 |NCC=0.77	50 |NCC=0.78
	$\lambda_3 = 1$ and $\lambda_4 = 5$	$\lambda_2 = 20$ and $\lambda_4 = 5$	$\lambda_2 = 20$ and $\lambda_3 = 1$

17 Game Theory and Its Applications in Imaging and Vision 695

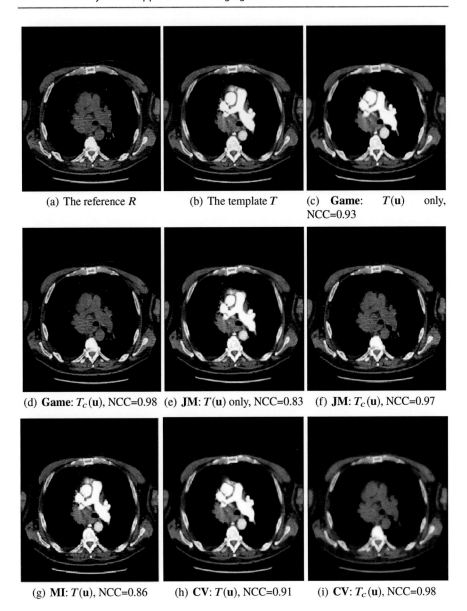

Fig. 5 Example 4: Comparison of 5 different models in registering CT and perfusion CT images. **Game** model performs the best. (**a**) The reference R (**b**) The template T (**c**) **Game**: $T(\mathbf{u})$ only, NCC=0.93 (**d**) **Game**: $T_c(\mathbf{u})$, NCC=0.98 (**e**) **JM**: $T(\mathbf{u})$ only, NCC=0.83 (**f**) **JM**: $T_c(\mathbf{u})$, NCC=0.97 (**g**) **MI**: $T(\mathbf{u})$, NCC=0.86 (**h**) **CV**: $T(\mathbf{u})$, NCC=0.91 (**i**) **CV**: $T_c(\mathbf{u})$, NCC=0.98

The result of both registration and correction is satisfactory and this underlines the performance of this model in solving both problems jointly and efficiently which is not the case for **CV**, **JM**, and **MI** as they only handle the correction task correctly and fail in registration. For this particular example, $T(\mathbf{u})$ is very useful as clinicians like to where the contrasts from perfusion CT ("artifacts") would be located on the CT.

Game Models in Deep Learning

Game theory is a crucial element in building artificial intelligence (**AI**) models today for solving a multitasking models. In fact, a model which is designed to have multitasking property is the natural setting for Nash game formulation where the problem can effectively solved by considering different networks and different losses, one for each task. Good model would involve interaction between game theory and deep learning, that is, deep learning games. It is a very recent and interesting technique in artificial intelligence which uses neural networks and game strategies. Game environments and models are increasingly becoming popular training mechanisms for machine learning such as generative adversarial networks (Goodfellow et al. 2014), which have become one of the most successful frameworks for unsupervised generative modeling. Game theory is also recently used in reinforcing learning (Sutton et al. 2018) where various agents in the model compete against each other to improve the overall behavior. These both approaches represent the most recent powerful game models in artificial intelligence and have been used in different challenges and applications. In the sequel, we discuss the generative adversarial networks approach and its application in some image processing problems.

Generative Adversarial Networks (GANs)

Generative Adversarial modeling is a particular case of deep learning models which is based on the competition between two networks, pitting one against the other (thus the "adversarial"). Originality, it was developed for the image generation task from random samples (Goodfellow et al. 2014). It has progressed remarkably with the advent of convolution neural networks (CNNs) and is widely used for various imaging problems, mainly in the unsupervised learning context.

Generative vs Discriminative Algorithms
The generative adversarial models are based on the competition of two neural blocks, the discriminator and the generator. The generator is a convolutional neural network designed to create new instances of an object. The discriminator, on the other hand, is a "deconvolutional" neural network that determines the authenticity of the object and whether or not it is part of a set of true data. In terms of optimization, a backpropagation is used to make sure that the parameters in both networks are optimized by minimizing and/or maximizing a specific losses between

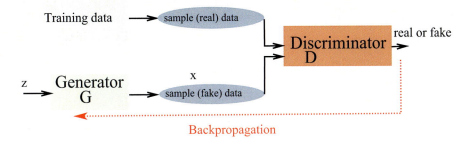

Fig. 6 The architecture of GAN models

true and generated data. They are trained in an adversarial and iterative manner until convergence is achieved when both are satisfied, that is, equilibrium situation. The illustration in Fig. 6 gives a rough idea on the work-flow of the generator and discriminator in the Generative Adversarial Networks.

Theory and Numerics

GANs models are an infinite zeros-sum minmax game where Nash equilibrium (**NE**) is considered as a saddle point and the existence result is not straightforward. The existence of saddle points, equivalently a Nash equilibria, in infinite action games requires some "strong" properties like convexity and concavity of the loss functions, which is not always true as these losses are mostly nonconvex w.r.t networks are the weights of the network.

In various studies, existence was considered only for local Nash equilibrium or for Mixed Nash equilibrium (**MNE**), that is, with respect to probability distributions.

In practice, the training of GANs is considered as a tricky matter. In fact, reaching a Nash equilibrium for GANs through an optimization algorithm can be difficult to prove theoretically. Empirically, it has been observed that common algorithms, such as Stochastic Gradient Descent (SGD), lead to unstable training. Some studies on the convergence behaviors of gradient based training have been evolving throughout the years. The local convergence behavior has been studied in Nagarajan and Kolter (2017) and Heusel et al. (2017). The gradient-based optimization is proved to converge assuming that the discriminator and the generator is convex over the network parameters (Nowozin et al. 2016). However, even though research has been focused on understanding the training dynamics of GANs (Balduzzi et al. 2018; Gemp and Mahadevan 2018; Gidel et al. 2018a,b), a provably convergent algorithm for general GANs, even under reasonably strong assumptions, is still lacking.

GANs have been used for various image processing tasks with satisfactory results: images generation, image deblurring (Kupyn et al. 2018), image registration (Mahapatra et al. 2018), image classification, etc. In the sequel, we describe the GANs framework for the image generation problem which is a particular case of two-player game. We also give an example of using GANs for solving the image segmentation problem.

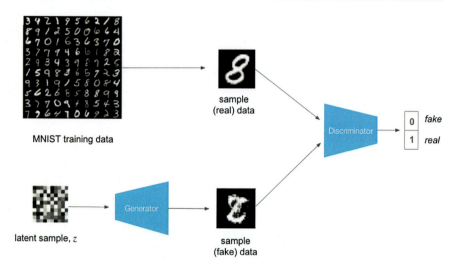

Fig. 7 The model architecture of GANs model for the image generation problem

GANs for Image Generation: A Two-Player Game

We consider the example of handwritten digits generation using generative adversarial network (Goodfellow et al. 2014) trained on the MNIST dataset (http://yann.lecun.com/exdb/mnist/). The aim is to be able to generate new digits from a random vector x of size 784. As mentioned, the GANs model is composed by the two networks, Generator **G** and the discriminator **D**. The generator takes the input random vector z (noise) and tries to generate a 28×28 image which is intended to be very close to the original images of MNIST dataset. Whereas the discriminator **D** takes generated images by **G** and tries to discriminate between them and real data. It is a binary classification network which turns the probability that the generated image by **G** belongs to real dataset, that is, a class 1 means that it is real and 0 means fake (Fig. 7). Theoretically, GANs is a game model which is designed to compete the two networks **G** and **D** by solving the following min-max problem

$$\min_{G} \max_{D} \mathcal{J}(\mathbf{D}, \mathbf{G}) = \mathbf{E}_{x \sim p_{\text{data}}(x)} \log[\mathbf{D}(x)] + \mathbf{E}_{z \sim p_{\text{data}}(z)} \log[(1 - \mathbf{D}(\mathbf{G}(z)))] \quad (23)$$

where $\mathbf{E}_{x \sim p_{\text{data}}(x)}$ is the expected value over all real data instances. It is easy to prove the existence of Nash equilibrium for this model as it is two-player zero-sum minimax game. However, the main challenge in GANs is the training as finding a Nash equilibrium is not straightforward. The model is trained in alternating way; the **D**-problem consists of solving the maximization problem

$$\min_{G} \mathcal{J}(\mathbf{D}, \mathbf{G}) = \mathbf{E}_{x \sim p_{\text{data}}(x)} \log[\mathbf{D}(x)] + \mathbf{E}_{z \sim p_{\text{data}}(z)} \log[(1 - \mathbf{D}(\mathbf{G}(z)))], \quad (24)$$

where the first allows to recognize real images, whereas the second helps to recognize fake ones. The **G**-problem consists in solving the minimization problem

$$\min_G \mathcal{J}(\mathbf{D}, \mathbf{G}) = \mathbf{E}_{x \sim p_{\text{data}}(x)} \log[(1 - \mathbf{D}(\mathbf{G}(x))] \qquad (25)$$

The GANs training algorithm involves training both the discriminator and the generator nets in parallel. The algorithm used in the original 2014 paper by Goodfellow (Goodfellow et al. 2014) is summarized in the figure below:

Algorithm 3 Mini batch stochastic gradient descent training of generative adversarial nets

    for number of training iterations **do**
        for k steps **do**
- Sample mini batch of m noise samples $\{z^{(1)}, \cdots, z^{(m)}\}$ from noise prior $p_g(z)$.
- Sample mini batch of m examples $\{x^{(1)}, \cdots, x^{(m)}\}$ from data generating distribution $p_{\text{data}}(x)$.
- Update the discriminator by ascending its stochastic gradient:

$$\nabla_{\theta_d} \frac{1}{m} \sum_{i=1}^{m} \left[\log \mathbf{D}\left(x^{(i)}\right) + \log\left(1 - \mathbf{D}\left(\mathbf{G}\left(z^{(i)}\right)\right)\right) \right].$$

    end for
- Sample mini batch of m noise samples $\{x^{(1)}, \cdots, x^{(m)}\}$ from noise prior $p_g(z)$.
- Update the generator by descending its stochastic gradient:

$$\nabla_{\theta_g} \frac{1}{m} \sum_{i=1}^{m} \log\left(1 - \mathbf{D}\left(\mathbf{G}\left(z^{(i)}\right)\right)\right).$$

endfor

With iteration, Generator **G** gets stronger and stronger at generating the real images and the discriminator **D** also gets stronger and stronger at identifying which one is real, which one is fake.

Examples
Few examples of images created by GANs for MNIST dataset are given in Fig. 8.

GANs for Image Segmentation: A Two-Player Game

Several approaches for the image segmentation problem based on the GANs framework were proposed in Luc et al. (2016), Mahapatra et al. (2018), and Tanner et al. (2018). We describe here the proposed GANs model in Luc et al. (2016) for the particular case of semantic segmentation.

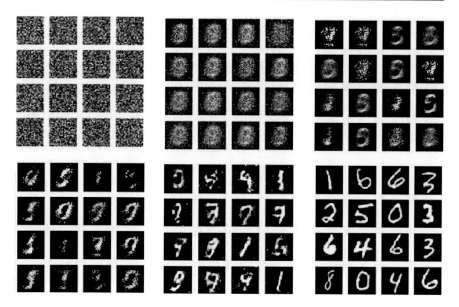

Fig. 8 Starting from random noise images, the generator gradually learns with iterations to emulate the features of the training dataset; it produces like-handwritten digits

The idea consists of using a generative adversarial networks (GANs) for RGB images segmentation where the trained network takes an RGB image x of size $H \times W \times 3$ as inputs and outputs the segmented image which is represented as a class label at each pixel location independently.

Generator and Discriminator

The generator is a segmentation CNN model which predicts a segmentation class from the input x by minimizing a segmentation loss. Its goal is to produce segmentation maps that are hard to distinguish from ground-truth ones for the adversarial model. The discriminator **D** uses the generated maps by **G** and compares it to the ground truth data in order to discriminate segmentation maps coming either from the ground truth or from the segmentation network. The model is summarized in Fig. 9.

Model Loss

The generator and the discriminator are trained together to optimize global loss function which is a weighted sum two terms. Given a data set of N training color images x_n of size $H \times W \times C$ and a corresponding label maps y_n, the authors defined a global loss as

17 Game Theory and Its Applications in Imaging and Vision

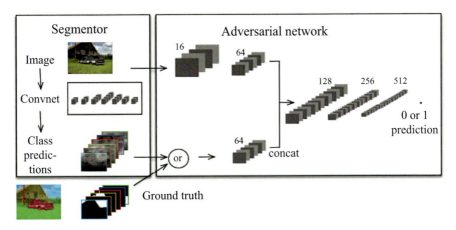

Fig. 9 Figure taken from Luc et al. (2016). Overview of the proposed approach. Left: segmentation net takes RGB image as input, and produces per-pixel class predictions. Right: Adversarial net takes label map as input and produces class label (1=ground truth, or 0=synthetic). Adversarial optionally also takes RGB image as input

$$\mathcal{J}(\mathbf{G}, \mathbf{D}) = \sum_{n=1}^{N} \ell_{mce}(\mathbf{G}(x_n), y_n) - \lambda \left[\ell_{bce}(\mathbf{D}(x_n, y_n), 1) + \ell_{bce}(\mathbf{D}(x_n, \mathbf{G}(x_n)), 0) \right]$$
(26)

where $\lambda = 10$ controls the contribution of the two terms, that is, the multi-class cross-entropy loss

$$\ell_{mce}(\mathbf{G}(x_n), y_n) = - \sum_{i=1}^{H \times W} \sum_{c=1}^{C} y_{ni} \log(\mathbf{G}(x_n)_c),$$

and the binary cross-entropy loss

$$\ell_{bce}(z_1, z_2) = - \left[z_2 \log z_1 + (1 - z_2) \log(1 - z_1) \right]$$

The term $\ell_{mce}(\mathbf{G}(x_n), y_n)$ denotes the multi-class cross-entropy loss for predictions $\mathbf{G}(x_n)$ and is a standard loss for semantic segmentation models. It encourages the segmentation model to predict the right class label at each pixel location independently. The Discriminator output $\mathbf{D}(x_n, y_n) \in [0, 1]$ represents the scalar probability of y_n being the ground truth label map of x_n, or being a fake map produced by Generator $\mathbf{G}$. The second part of the loss is for adversarial convolutional network and is large if the adversarial network can discriminate the generated segmentation map by Generator $\mathbf{G}$ from ground-truth label maps.

Similar to all GANs models, this is a min-max game model where the full loss is minimized with respect to Generator $\mathbf{G}$ of the segmentation, and maximized with

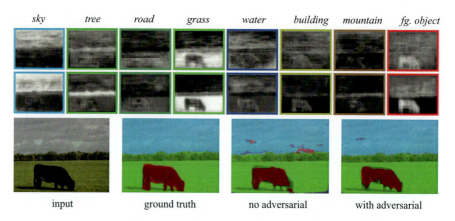

Fig. 10 Figure taken from Luc et al. (2016). Segmentations on Stanford Background. Class probabilities without (first row) and with (second row) adversarial training. In the last row the class labels are superimposed on the image

respect the adversarial model **D** model.

$$\min_{\mathbf{G}} \max_{\mathbf{D}} \mathcal{J}(\mathbf{G}, \mathbf{D}). \tag{27}$$

Training
The model is trained in alternating way; the **D**-problem consists in solving the minimization problem

$$\min_{\mathbf{D}} \ell_{bce}(\mathbf{D}(x_n, y_n), 1) + \ell_{bce}(\mathbf{D}(x_n, \mathbf{G}(x_n)), 0), \tag{28}$$

where the first allows to recognize real labels, whereas the second helps to recognize fake ones. The **G**-problem consists in solving the minimization problem

$$\min_{\mathbf{G}} \sum \ell_{mce}(\mathbf{G}(x_n), y_n) - \lambda \ell_{bce}(\mathbf{D}(x_n, \mathbf{G}(x_n)), 0) \tag{29}$$

The GANs training algorithm involves training both the discriminator and the generator nets in parallel.

Example
The numerical example in Fig. 10 illustrate a comparison between the segmentation results using adversarial (GANs) and non-adversarial approaches. The results state that GANs approach clearly enhances the segmentation better than a classical deep learning approach, that is, non-adversarial.

Conclusion

Mathematical modeling of Vision and Imaging problems do naturally lead to formulations where antagonistic optimal decisions are aimed at. To this end, the recourse to a non-cooperative game paradigm seems to be very promising. The present chapter has addressed major imaging problematics, namely restoration versus segmentation and registration. Game theory can also be applied in various aspects of artificial intelligence, in particular for Adversarial Machine learning. Generative adversarial deep learning models have taken advantages from the game theory to reinvent generative models for different image processing problems. The authors have provided different illustrations of the strikingly efficient ability of game theory to address difficult concurrent optimization problems arising from these problematics.

References

Aghajani, K., Manzuri, M.T., Yousefpour, R.: A robust image registration method based on total variation regularization under complex illumination changes. Comput. Meth. Prog. Biomed. **134**, 89–107 (2016)

Attouch, H., Bolte, J., Redont, P., Soubeyran, A.: Alternating proximal algorithms for weakly coupled convex minimization problems. applications to dynamical games and pde's. J. Convex Anal. **15**(3), 485 (2008)

Attouch, H., Soueycatt, M.: Augmented lagrangian and proximal alternating direction methods of multipliers in hilbert spaces. applications to games, pde's and control. Pac. J. Optim. **5**(1), 17–37 (2008)

Balduzzi, D., Racaniere, S., Martens, J., Foerster, J., Tuyls, K., Graepel, T.: The mechanics of n-player differentiable games. arXiv preprint arXiv:1802.05642 (2018)

Bansal, R., Staib, L.H., Peterson, B.S.: Correcting nonuniformities in MRI intensities using entropy minimization based on an elastic model. In: International Conference on Medical Image Computing and Computer-Assisted Intervention, pp. 78–86. Springer (2004)

Benki, A., Habbal, A., Mathis, G., Beigneux, O.: Multicriteria shape design of an aerosol can. J. Comput. Design Eng. 11 (2015). https://doi.org/10.1016/j.jcde.2015.03.003. https://hal.inria.fr/hal-01144269

Bonnans, J.F., Gilbert, J.C., Lemaréchal, C., Sagastizábal, C.A.: Numerical Optimization: Theoretical and Practical Aspects. Springer Science & Business Media (2006)

Boyd, S., Parikh, N., Chu, E., Peleato, B., Eckstein, J., et al.: Distributed optimization and statistical learning via the alternating direction method of multipliers. Found. Trends® Mach. Learn. **3**(1), 1–122 (2011)

Chamekh, R., Habbal, A., Kallel, M., Zemzemi, N.: A nash game algorithm for the solution of coupled conductivity identification and data completion in cardiac electrophysiology. Math. Modell. Nat. Phenom. **14**(2), 15 (2019). https://doi.org/10.1051/mmnp/2018059. https://hal.archives-ouvertes.fr/hal-01923819

Chang, H., Huang, W., Wu, C., Huang, S., Guan, C., Sekar, S., Bhakoo, K.K., Duan, Y.: A new variational method for bias correction and its applications to rodent brain extraction. IEEE Trans. Med. Imaging **36**(3), 721–733 (2017)

Chen, K., Lui, L.M., Modersitzki, J.: Image and surface registration. In: Handbook of Numerical Analysis – Processing, Analyzing and Learning of Images, Shapes, and Forms, vol. 20. Elsevier (2019)

Chen, Y., Ye, X.: Inverse consistent deformable image registration. In: The Legacy of Alladi Ramakrishnan in the Mathematical Sciences, pp. 419–440. Springer (2010)

Christensen, G.E., Song, J.H., Lu, W., ElNaqa, I., Low, D.A.: Tracking lung tissue motion and expansion/compression with inverse consistent image registration and spirometry. Med. Phys. **34**, 2155–2163 (2007)

Chumchob, N., Chen, K.: Improved variational image registration model and a fast algorithm for its numerical approximation. Numer. Meth. Partial Differen. Equations **28**(6), 1966–1995 (2012)

Mumford, D.J.S.: Optimal approximations by piecewise smooth functions and variational problems. Commun. Pure Appl. Math. **42**, 577–685 (1989)

Desideri, J.A., Duvigneau, R., Habbal, A.: Multiobjective design optimization using nash Games. In: M. Vasile, V.M. Becerra (eds.) Computational Intelligence in the Aerospace Sciences, Progress in Astronautics and Aeronautics. American Institute of Aeronautics and Astronautics (AIAA) (2014). https://hal.inria.fr/hal-00923584

Duan, Y., Chang, H., Huang, W., Zhou, J., Lu, Z., Wu, C.: The $l\_\{0\}$ regularized mumford–shah model for bias correction and segmentation of medical images. IEEE Trans. Image Process. **24**(11), 3927–3938 (2015)

Ebrahimi, M., Martel, A.L.: A general pde-framework for registration of contrast enhanced images. In: International Conference on Medical Image Computing and Computer-Assisted Intervention, pp. 811–819. Springer (2009)

Fischer, B., Modersitzki, J.: Fast diffusion registration. Contemp. Math. **313**, 117–12 (2002)

Friedman, A.: Stochastic differential games. J. Differen. Equ. **11**(1), 79–108 (1972)

Gemp, I., Mahadevan, S.: Global convergence to the equilibrium of gans using variational inequalities. arXiv preprint arXiv:1808.01531 (2018)

Ghaffari, A., Fatemizadeh, E.: Image registration based on low rank matrix: Rank-regularized ssd. IEEE Trans. Med. Imaging **37**(1), 138–150 (2018)

Gibbons, R.S.: Game Theory for Applied Economists. Princeton University Press (1992)

Gidel, G., Berard, H., Vignoud, G., Vincent, P., Lacoste-Julien, S.: A variational inequality perspective on generative adversarial networks. arXiv preprint arXiv:1802.10551 (2018a)

Gidel, G., Hemmat, R.A., Pezeshki, M., Lepriol, R., Huang, G., Lacoste-Julien, S., Mitliagkas, I.: Negative momentum for improved game dynamics. arXiv preprint arXiv:1807.04740 (2018b)

Goodfellow, I., Pouget-Abadie, J., Mirza, M., Xu, B., Warde-Farley, D., Ozair, S., Courville, A., Bengio, Y.: Generative adversarial nets. In: Advances in Neural Information Processing Systems, pp. 2672–2680 (2014)

Habbal, A.: A topology Nash game for tumoral antiangiogenesis. Struct. Multidiscip. Optim. **30**(5), 404–412 (2005)

Habbal, A., Kallel, M.: Neumann-Dirichlet nash strategies for the solution of elliptic cauchy problems. SIAM J. Control. Optim. **51**(5), 4066–4083 (2013). https://hal.inria.fr/hal-00923574

Habbal, A., Kallel, M., Ouni, M.: Nash strategies for the inverse inclusion Cauchy-Stokes problem. Inverse Prob. Imag. **13**(4), 36 (2019). https://doi.org/10.3934/ipi.2019038. https://hal.inria.fr/hal-01945094

Habbal, A., Petersson, J., Thellner, M.: Multidisciplinary topology optimization solved as a Nash game. Int. J. Numer. Meth. Engng **61**, 949–963 (2004)

Heusel, M., Ramsauer, H., Unterthiner, T., Nessler, B., Hochreiter, S.: Gans trained by a two time-scale update rule converge to a local nash equilibrium. In: Advances in Neural Information Processing Systems, pp. 6626–6637 (2017)

Hu, J., Wellman, M.P.: Nash q-learning for general-sum stochastic games. J. Mach. Learn. Res. **4**(Nov), 1039–1069 (2003)

Isaacs, R.: Differential Games: A Mathematical Theory with Applications to Warfare and Pursuit, Control and Optimization. Courier Corporation (1999)

Kallel, M., Aboulaich, R., Habbal, A., Moakher, M.: A nash-game approach to joint image restoration and segmentation. Appl. Math. Model. **38**(11-12), 3038–3053 (2014)

Kim, Y., Tagare, H.D.: Intensity nonuniformity correction for brain mr images with known voxel classes. SIAM J. Imag. Sci. **7**(1), 528–557 (2014)

Kupyn, O., Budzan, V., Mykhailych, M., Mishkin, D., Matas, J.: Deblurgan: Blind motion deblurring using conditional adversarial networks. In: Proceedings of the IEEE Conference on Computer Vision and Pattern Recognition, pp. 8183–8192 (2018)

Lam, K.C., Lui, L.M.: Landmark- and intensity-based registration with large deformations via quasi-conformal maps. SIAM J. Imag. Sci. **7**(4), 2364–2392 (2014)

Li, C., Gatenby, C., Wang, L., Gore, J.C.: A robust parametric method for bias field estimation and segmentation of mr images. In: IEEE Conference on Computer Vision and Pattern Recognition, 2009. CVPR 2009, pp. 218–223. IEEE (2009)

Luc, P., Couprie, C., Chintala, S., Verbeek, J.: Semantic segmentation using adversarial networks. arXiv preprint arXiv:1611.08408 (2016)

Maes, F., Collignon, A., Vandermeulen, D., Marchal, G., Suetens, P.: Multimodality image registration by maximization of mutual information. IEEE Trans. Med. Imaging **16**(2), 187–198 (1997)

Mahapatra, D., Antony, B., Sedai, S., Garnavi, R.: Deformable medical image registration using generative adversarial networks. In: 2018 IEEE 15th International Symposium on Biomedical Imaging (ISBI 2018), pp. 1449–1453. IEEE (2018)

Modersitzki, J.: FAIR: Flexible Algorithms for Image Registration. SIAM publications (2009)

Modersitzki, J., Wirtz, S.: Combining homogenization and registration. In: International Workshop on Biomedical Image Registration, pp. 257–263. Springer (2006)

Monderer, D., Shapley, L.S.: Potential games. Games Econom. Behav. **14**(1), 124–143 (1996)

Nagarajan, V., Kolter, J.Z.: Gradient descent gan optimization is locally stable. In: Advances in Neural Information Processing Systems, pp. 5585–5595 (2017)

Nash, J.: Equilibrium points in n-person games. Proc. Natl. Acad. Sci. USA **36**(1), 48–49 (1950)

Nash, J.: Non-cooperative games. Ann. Math. 286–295 (1951)

Neyman, A., Sorin, S.: Stochastic Games and Applications, vol. 570. Springer Science & Business Media (2003)

Nishimura, R., Hayashi, S., Fukushima, M.: Robust nash equilibria in n-person non-cooperative games: Uniqueness and reformulation. Pac. J. Optim. **5**(2), 237–259 (2009)

Nowozin, S., Cseke, B., Tomioka, R.: f-gan: Training generative neural samplers using variational divergence minimization. In: Advances in Neural Information Processing Systems, pp. 271–279 (2016)

Park, C.R., Kim, K., Lee, Y.: Development of a bias field-based uniformity correction in magnetic resonance imaging with various standard pulse sequences. Optik **178**, 161–166 (2019)

Rak, M., König, T., Tönnies, K.D., Walke, M., Ricke, J., Wybranski, C.: Joint deformable liver registration and bias field correction for mr-guided hdr brachytherapy. Int. J. Comput. Assist. Radiol. Surg. **12**(12), 2169–2180 (2017)

Roy, S., Borzì, A., Habbal, A.: Pedestrian motion modeled by FP-constrained Nash games. R. Soc. Open Sci. (2017). https://doi.org/10.1098/rsos.170648. https://hal.inria.fr/hal-01586678

Uryas'ev, S., Rubinstein, R.Y.: On relaxation algorithms in computation of noncooperative equilibria. IEEE Trans. Autom. Control **39**, 1263–1267 (1994)

David, S., Hernández-Lerma Onésimo, G.: A survey of static and dynamic potential games. Sci. China Math. **59**(11), 2075–2102 (2016)

Shapley, L.S.: Stochastic games. Proc. Natl. Acad. Sci. **39**(10), 1095–1100 (1953)

Sutton, R.S., Barto, A.G., et al.: Introduction to Reinforcement Learning, 2nd edn. MIT Press Cambridge (2018)

Tanner, C., Ozdemir, F., Profanter, R., Vishnevsky, V., Konukoglu, E., Goksel, O.: Generative adversarial networks for mr-ct deformable image registration. arXiv preprint arXiv:1807.07349 (2018)

Theljani, A., Chen, K.: An augmented lagrangian method for solving a new variational model based on gradients similarity measures and high order regularization for multimodality registration. Inv. Prob. Imag. **13**, 309–335 (2019a)

Theljani, A., Chen, K.: A nash game based variational model for joint image intensity correction and registration to deal with varying illumination. Inv. Prob. **36**, 034002 (2019b)

Theljani, A., Chen, K.: A variational model for diffeomorphic multi-modal image registration using a new correlation like measure. submitted (2019c)

Thompson, T., Chen, K.: An effective diffeomorphic model and its fast multigrid algorithm for registration of lung ct images improved optimization methods for image registration problems. J. Comput. Meth. Appl. Math. (2019)

Thompson, T., Chen, K.: A more robust multigrid algorithm for diffusion type registration models. J. Comput. Appl. Math. **361**, 502–527 (2019)

Van Leemput, K., Maes, F., Vandermeulen, D., Suetens, P.: Automated model-based bias field correction of mr images of the brain. IEEE Trans. Med. Imaging **18**(10), 885–896 (1999)

Vovk, U., Pernus, F., Likar, B.: A review of methods for correction of intensity inhomogeneity in MRI. IEEE Trans. Med. Imaging **26**(3), 405–421 (2007)

Wang, L., Pan, C.: Nonrigid medical image registration with locally linear reconstruction. Neurocomputing **145**, 303–315 (2014)

Zhang, D., Chen, K.: A novel diffeomorphic model for image registration and its algorithm. J. Math. Imaging Vision **60**, 1261–1283 (2018)

Zhang, D., Chen, K.: 3D orientation-preserving variational models for accurate image registration. SIAM J. Imaging Sci. **13**, 1653–1691 (2020)

Zhang, J., Chen, K., Yu, B.: A novel high-order functional based image registration model with inequality constraint. Comput. Math. Appl. **72**, 2887–2899 (2016)

First-Order Primal–Dual Methods for Nonsmooth Non-convex Optimization

18

Tuomo Valkonen

Contents

Introduction	708
Sample Problems	709
Outline	710
Bregman Divergences	711
Primal–Dual Proximal Splitting	713
Optimality Conditions and Proximal Points	714
Algorithm Formulation	715
Block Adaptation	716
Convergence Theory	718
A Fundamental Estimate	718
Ellipticity of the Bregman Divergences	720
Ellipticity for Block-Adapted Methods	723
Nonsmooth Second-Order Conditions	724
Second-Order Growth Conditions for Block-Adapted Methods	727
Convergence of Iterates	729
Convergence of Gaps in the Convex-Concave Setting	732
Inertial Terms	733
A Generalization of the Fundamental Theorem	733
Inertia (Almost) as Usually Understood	735
Improvements to the Basic Method Without Dual Affinity	737
Further Directions	741
Acceleration	741
Stochastic Methods	741
Alternative Bregman Divergences	741
Alternative Approaches	742
Functions on Manifolds and Hadamard Spaces	743
References	745

T. Valkonen (✉)
Center for Mathematical Modeling, Escuela Politécnica Nacional, Quito, Ecuador
Department of Mathematics and Statistics, University of Helsinki, Helsinki, Finland
e-mail: tuomo.valkonen@iki.fi

© Springer Nature Switzerland AG 2023
K. Chen et al. (eds.), *Handbook of Mathematical Models and Algorithms in Computer Vision and Imaging*, https://doi.org/10.1007/978-3-030-98661-2_93

Abstract

We provide an overview of primal–dual algorithms for nonsmooth and nonconvex-concave saddle-point problems. This flows around a new analysis of such methods, using Bregman divergences to formulate simplified conditions for convergence.

Keywords

Primal-dual · Nonsmooth · Nonconvex · Optimization · Inverse problems

Introduction

Interesting imaging problems can often be written in the general form

$$\min_{x \in X} \max_{y \in Y} \ F(x) + K(x, y) - G_*(y), \tag{S}$$

where X and Y are Banach spaces, $K \in C^1(X, Y)$, and $F : X \to \overline{\mathbb{R}}$ and $G_* : Y \to \overline{\mathbb{R}}$ are convex, proper, lower semicontinuous functions with G_* the preconjugate of some $G : Y^* \to \overline{\mathbb{R}}$, meaning $G = (G_*)^*$. The functions F and G_* may be nonsmooth. In this chapter, we provide an overview of proximal-type primal–dual algorithms for this class of problems together with a simplified analysis, based on Bregman divergences.

▷ Notation, Conventions, and Basic Convex Analysis

As is standard in optimization, all vector/Banach/Hilbert spaces in this chapter are over the real field without it being explicitly mentioned. For basic definitions of convex analysis, such as the (pre)conjugate and the subdifferential, see the glossary at the end of the chapter or textbooks such as Hiriart-Urruty and Lemaréchal (2004), Rockafellar (1972), Clason and Valkonen (2020), and Ekeland and Temam (1999).

A common instance of (S) is when $K(x, y) = \langle Ax | y \rangle$ for a linear operator $A \in \mathbb{L}(X; Y^*)$ with $\langle \cdot | \cdot \rangle : Y^* \times Y \to \mathbb{R}$ denoting the dual product. Then (S) arises from writing G in terms of its (pre)conjugate G_* in

$$\min_{x \in X} \ F(x) + G(Ax). \tag{1}$$

We now discuss sample imaging and inverse problems of the types (S) and (1) and then outline our approach to solving them in the rest of the chapter.

Sample Problems

Optimization problems of the type (1) can effectively model linear inverse problems; typically one would attempt to minimize the sum of a data term and a regularizer

$$\min_{x \in X} \Phi(z - Tx) + G(Ax), \qquad (2)$$

where

- $T :\in \mathbb{L}(X; \mathbb{R}^n)$ is a forward operator, mapping our unknown x into a finite number of measurements.
- Φ models noise ν in the data $z \in \mathbb{R}^n$; for normal-distributed noise, $\Phi(z) = \frac{1}{2}\|z\|^2$.
- $G \circ A$ is a typically nonsmooth regularization term that models our prior assumptions on what a good solution to the ill-posed problem $z = Tx + \nu$ should be; in imaging, what "looks good."

For conventional total variation regularization on a domain $\Omega \subset \mathbb{R}^m$, one would take $G(y^*) = \alpha \|y^*\|_{\mathcal{M}(\Omega;\mathbb{R}^m)}$ the Radon norm of the measure $y^* \in \mathcal{M}(\Omega;\mathbb{R}^m)$ weighted by the regularization parameter $\alpha > 0$ and $A = D \in \mathbb{L}(BV(\Omega); \mathcal{M}(\Omega;\mathbb{R}^m))$ the distributional derivative (Ambrosio et al. 2000). Simple examples of a *linear* forward operator T include:

- the identity for denoising (Rudin et al. 1992)
- a convolution operation for deblurring or deconvolution (Vogel and Oman 1998)
- a subsampling operator for inpainting (Shen and Chan 2002)
- the Fourier transform for magnetic resonance imaging (MRI) (Nishimura 1996; Lustig et al. 2007)
- the Radon transform for computational (CT) or positron emission tomography (PET) (Ollinger and Fessler 1997)

The last two examples would frequently be combined with subsampling for reconstruction from limited data.

In many important problems, T is, however, nonlinear:

- a pointwise application of $(r, \phi) \mapsto re^{-i\phi}$ for phase and amplitude reconstruction for velocity-encoded magnetic resonance imaging (Valkonen 2014)
- a pointwise application of $u \mapsto s_0 - se^{-\langle u,b \rangle}$ to model the Stejskal–Tanner equation in diffusion tensor imaging (Valkonen 2014; Kingsley 2006)
- the solution operator of nonlinear partial differential equation (PDE) for several forms of tomography from magnetic and electric to acoustic and optical (Nishimura 1996; Ollinger and Fessler 1997; Arridge et al. 2011; Kuchment and Kunyansky 2011; Hunt 2014; Trucu et al. 2009; Uhlmann 2009; Lipponen et al. 2011)

In the last example, the PDE governs the physics of measurement, typically relating boundary measurements and excitations to interior data. The methods we study in this chapter are applied to electrical impedance tomography in Jauhiainen et al. (2020) and Mazurenko et al. (2020).

How to fit a nonlinear forward operator T into the framework (S) that requires both F and G_* to be convex? If the noise model $\Phi : \mathbb{R}^n \to \overline{\mathbb{R}}$ is convex, proper, and lower semicontinuous, we can write (2) using the Fenchel conjugate Φ^* and $K_{TA}(x, (y_1, y_2)) := \langle z - T(x) | y_1 \rangle + \langle Ax | y_2 \rangle$ as

$$\min_{x \in X} \max_{(y_1, y_2) \in \mathbb{R}^n \times Y} K_{TA}(x, (y_1, y_2)) - \Phi^*(y_1) - G_*(y_2). \tag{3}$$

This is of the form (S) for the functions $\tilde{F} \equiv 0$ and $\tilde{G}_*(y_1, y_2) := \Phi^*(y_1) - G_*(y_2)$. Even for linear T, although (2) is readily of the form (1) and hence (S), this reformulation may allow expressing (2) in the form (S) with both $\tilde{F}$ and $\tilde{G}_*$ "prox-simple." We will make this concept, important for the effective realization of algorithms, more precise in section "Primal–Dual Proximal Splitting."

Finally, fully general K in (S) was shown in Clason et al. (2020) to be useful for highly nonsmooth and non-convex problems, such as the Geman and Geman (1984). Indeed, the "0-function"

$$|t|_0 := \begin{cases} 0, & t = 0, \\ 1, & t \neq 0, \end{cases}$$

can be written as

$$|t|_0 = \sup_{s \in \mathbb{R}} \rho(st) \quad \text{for} \quad \rho(t) = 2t - t^2.$$

For the (anisotropic) Potts model, this is applied pixelwise on a discretized image gradient computed for an $n_1 \times n_2$ image by $\nabla_h : \mathbb{R}^{n_1 n_2} \to \mathbb{R}^{2 \times n_1 n_2}$ (Clason et al. 2020):

$$\min_{x \in \mathbb{R}^{n_1 n_2}} \max_{y \in \mathbb{R}^{2 \times n_1 n_2}} \frac{1}{2} \|b - x\|_2^2 + \sum_{i=1}^{n_1} \sum_{j=1}^{n_2} \rho(\langle [\nabla_h x]_{ij}, y_{ij} \rangle), \tag{4}$$

where $b \in \mathbb{R}^{n_1 n_2}$ is the image to be segmented.

Outline

We introduce in section "Primal–Dual Proximal Splitting" methods for (S) inspired by the primal–dual proximal splitting (PDPS) of Chambolle and Pock (2011) and Pock et al. (2009) for bilinear K, commonly known as the Chambolle–Pock method. We work in Banach spaces, as was done in Hohage and Homann (2014). To be able to define proximal-type methods in Banach spaces, in section "Bregman Divergences," we introduce and recall the crucial properties of the so-called Bregman divergences.

Our main reason for working with Bregman divergences is, however, not the generality of Banach spaces. Rather, they provide a powerful proof tool to deal with the general K in (S). This approach allows us in section "Convergence Theory" to significantly simplify and better explain the original convergence proofs and conditions of Chambolle and Pock (2011), Valkonen (2014), Clason et al. (2019), Clason et al. (2020), and Mazurenko et al. (2020). Without additional effort, they also allow us to present block-adapted methods like those in Valkonen and Pock (2017), Valkonen (2019), and Mazurenko et al. (2020).

Our overall approach and the internal organization of section "Convergence Theory" centers around the following three main ingredients of the convergence proof:

(i) A **three-point identity**, satisfied by all Bregman divergences (shown in section "Bregman Divergences" and employed in section "A Fundamental Estimate")
(ii) **(Semi-)ellipticity** of the algorithm-defining Bregman divergences (concept defined in section "Bregman Divergences," specific Bregman divergence in section "Primal–Dual Proximal Splitting," and its ellipticity verified in sections "Ellipticity of the Bregman Divergences," and "Ellipticity for Block--Adapted Methods" through several examples)
(iii) A **nonsmooth second-order growth** condition around a solution of (S) (treated in sections "Nonsmooth Second-Order Conditions" and "Second-Order Growth Conditions for Block-Adapted Methods")

With these basic ingredients, we then prove convergence in sections "Convergence of Iterates" and "Convergence of Gaps in the Convex-Concave Setting." In the present overview, with focus on key concepts and aiming to avoid technical complications, we only cover, weak, strong, and linear convergence of iterates, and the convergence of gap functionals when K is convex-concave.

In section "Inertial Terms" we improve the basic method by adding dependencies to earlier iterates, a form of inertia. This is needed to develop an effective algorithm for K not affine in y, including the aforementioned formulation of the Potts segmentation model. We finish in section "Further Directions" with pointers to alternative methods and further extensions.

Bregman Divergences

The norm and inner product in a (real) Hilbert space X satisfy the three-point identity:

$$\langle x-y, x-z\rangle_X = \frac{1}{2}\|x-y\|_X^2 - \frac{1}{2}\|y-z\|_X^2 + \frac{1}{2}\|x-z\|_X^2 \quad (x, y, z \in X). \tag{5}$$

This is crucial for convergence proofs of optimization methods (Valkonen 2020), so we would like to have something similar in Banach spaces—or other more general spaces. Towards this end, we let $J : X \to \mathbb{R}$ be a Gâteaux-differentiable function[1]. Then one can define the asymmetric Bregman divergence:

$$B_J(z, x) := J(z) - J(x) - \langle DJ(x)|z - x\rangle_X \quad (x, z \in X). \tag{6}$$

This function is non-negative *if and only if*[2] the generating function J is convex; it is not in general a true distance, as it can happen that $B_J(x, z) = 0$ although $x = z$.

Writing D_1 for the Gâteaux derivative with respect to the first parameter, we have

$$D_1 B_J(x, z) = DJ(z) - DJ(x). \tag{7}$$

Moreover, the Bregman divergence satisfies for any $\bar{x} \in X$ the three-point identity

$$\begin{aligned}\langle D_1 B_J(x, z)|x - \bar{x}\rangle_X &= \langle DJ(x) - DJ(z)|x - \bar{x}\rangle_X \\ &= B_J(\bar{x}, x) - B_J(\bar{x}, z) + B_J(x, z).\end{aligned} \tag{8}$$

Indeed, writing the right-hand side out, we have

$$\begin{aligned}B_J(\bar{x}, x) - B_J(\bar{x}, z) + B_J(x, z) &= [J(\bar{x}) - J(x) - \langle DJ(x)|\bar{x} - x\rangle_X] \\ &\quad - [J(\hat{x}) - J(z) - \langle DJ(z)|\hat{x} - z\rangle_X] \\ &\quad + [J(x) - J(z) - \langle DJ(z)|x - z\rangle_X],\end{aligned}$$

which immediately gives the three-point identity.

Example 1. In a Hilbert space X, the standard generating function $J = N_X := \frac{1}{2}\|\cdot\|_X^2$ yields $B_J(z, x) = \frac{1}{2}\|z - x\|_X^2$, so (8) recovers (5).

We will frequently require B_J to be non-negative or semi-elliptic ($\gamma = 0$) or elliptic ($\gamma > 0$) within some $\Omega \subset X$. These notions mean that

$$B_J(z, x) \geq \frac{\gamma}{2}\|z - x\|_X^2 \quad (x, z \in \Omega). \tag{9}$$

---

[1] The differentiability assumption is for notational and presentational simplicity; otherwise we would need to write the Bregman divergence as $B_J^p(z, x) := J(z) - J(x) - \langle p|z - x\rangle_X$ for some subdifferential p of J and define explicit updates of this subdifferential in algorithms.

[2] For the entirely algebraic proof of the "only if," see Hiriart-Urruty and Lemaréchal 2004, Theorem 4.1.1.

Equivalently, this defines J to be (γ-strongly) subdifferentiable within Ω. When $\Omega = X$, we simply call B_J (semi-)elliptic and J (γ-strongly) subdifferentiable[3].

We will in section "Inertial Terms" also need a Cauchy inequality for Bregman divergences. We base this on strong subdifferentiability and the smoothness property (10) in the next lemma. The latter holding with $\Omega = X$ implies that DJ is L-Lipschitz and in Hilbert spaces is equivalent to this property; see Bauschke and Combettes (2017, Theorem 18.15) or Valkonen (2020, Appendix C).

Lemma 1. *Suppose $J : X \to \mathbb{R}$ is Gâteaux-differentiable and γ-strongly subdifferentiable within Ω and satisfies for some $L > 0$ the subdifferential smoothness*

$$\frac{1}{2L}\|DJ(x) - DJ(y)\|_{X^*}^2 \leq J(x) - J(y) - \langle DJ(y)|x - y\rangle \quad (x, y \in \Omega). \tag{10}$$

Then, for any $\alpha > 0$,

$$|\langle D_1 B_J(x, y)|z - x\rangle| \leq \frac{L}{\alpha} B_J(x, y) + \frac{\alpha}{\gamma} B_J(z, x) \quad (x, y, z \in \Omega).$$

Proof. By Cauchy's inequality and (7),

$$|\langle D_1 B_J(x, y)|z - x\rangle| \leq \frac{1}{2\alpha}\|DJ(x) - DJ(y)\|_{X^*}^2 + \frac{\alpha}{2}\|z - x\|_X^2.$$

By the strong convexity, $\frac{\gamma}{2}\|z - x\|_X^2 \leq B_J(z, x)$, and by the smoothness property (10), $\frac{1}{2L}\|DJ(x) - DJ(y)\|_{X^*}^2 \leq B_J(x, y)$. Together these estimates yield the claim. □

Primal–Dual Proximal Splitting

We now formulate a basic version of our primal–dual method. Later in section "Inertial Terms" we improve the algorithm to be more effective when K is not affine in y.

> Notation

Throughout the manuscript, we combine the primal and dual variables x and y into variables involving the letter u:

$$u = (x, y), \quad u^k = (x^k, y^k), \quad \hat{u} = (\hat{x}, \hat{y}), \quad \text{etc.}$$

---

[3] In Banach spaces strong subdifferentiability is implied by strong convexity, as defined without subdifferentials. In Hilbert spaces the two properties are equivalent.

Optimality Conditions and Proximal Points

We define the Lagrangian as

$$\mathcal{L}(x, y) := F(x) + K(x, y) - G_*(y).$$

A saddle point $\hat{u} = (\hat{x}, \hat{y})$ of the problem (S) satisfies, by definition

$$\mathcal{L}(\hat{x}, y) \leq \mathcal{L}(\hat{x}, \hat{y}) \leq \mathcal{L}(x, \hat{y}) \quad \text{for all } u = (x, y) \in X \times Y.$$

Writing $D_x K$ and $D_y K$ for the Gâteaux derivatives of K with respect to the two variables, if K is convex-concave, basic results in convex analysis (Ekeland and Temam 1999; Bauschke and Combettes 2017) show that

$$- D_x K(\hat{x}, \hat{y}) \in \partial F(\hat{x}) \quad \text{and} \quad D_y K(\hat{x}, \hat{y}) \in \partial G_*(\hat{y}) \tag{11}$$

is necessary and sufficient for $\hat{u}$ to be saddle point. If K is C^1, the theory of generalized subdifferentials of Clarke (1990) still indicates[4] the necessity of (11).

We can alternatively write (11) as

$$0 \in H(\hat{u}) := \begin{pmatrix} \partial F(\hat{x}) + D_x K(\hat{x}, \hat{y}) \\ \partial G_*(\hat{y}) - D_y K(\hat{x}, \hat{y}) \end{pmatrix}. \tag{12}$$

If X and Y were Hilbert spaces, we could in principle use the classical proximal point method (Minty 1961; Rockafellar 1976) to solve (12): given step length parameters $\tau_k > 0$, iteratively solve u^{k+1} from

$$0 \in H(u^{k+1}) + \tau_k^{-1}(u^{k+1} - u^k). \tag{13}$$

If K were bilinear, H would be a so-called monotone operator and convergence of iterates would follow from Rockafellar (1976). In practice the steps of the method are too expensive to realize as the primal and dual iterates x^{k+1} and y^{k+1} are coupled: generally, one cannot solve one before the other.

Fortunately, the iterates can be decoupled by introducing a preconditioner that switches $D_x K(x^{k+1}, y^{k+1})$ on the first line of $H(u^{k+1})$ to $D_x K(x^k, y^k)$. This gives rise to the primal–dual proximal splitting (PDPS), introduced in Chambolle and Pock (2011) and Pock et al. (2009) for bilinear $K(x, y) = \langle Ax|y \rangle$. That the PDPS is actually a preconditioned proximal point method was first observed in He and Yuan (2012). In the following, we describe its extension from Valkonen (2014) and Clason et al. (2019, 2020) to general K and the general problem (S). To simplify the proofs and concepts in them, we work with Bregman divergences, at no cost in Banach spaces.

---

[4]The Fermat rule $0 \in \partial_C[F + K(\cdot, \hat{y})](\hat{x})$ holds. Since F is convex and $K(\cdot, \hat{y})$ is C^1, $\hat{x}$ is a regular point of both, so also the subdifferential sum rule holds. We argue $G_* + K(\hat{y}, \cdot)$ similarly.

Algorithm Formulation

Given Gâteaux-differentiable functions $J_X : X \to \overline{\mathbb{R}}$ and $J_Y : Y \to \overline{\mathbb{R}}$ with the corresponding Bregman divergences $B_X := B_{J_X}$ and $B_Y := B_{J_Y}$, we define

$$J^0(x, y) := J_X(x) + J_Y(y) - K(x, y). \tag{14}$$

Introducing the short-hand notation $B^0 := B_{J^0}$, we propose to solve (12) through the iterative solution of

$$0 \in H(u^{k+1}) + D_1 B^0(u^{k+1}, u^k) \tag{15}$$

for u^{k+1}. Inserting (12) and (7) for $J = J^0$ as defined in (14), we expand and rearrange this implicitly defined method as:

Primal–dual Bregman-proximal splitting (PDBS)

Iteratively over $k \in \mathbb{N}$, solve for x^{k+1} and y^{k+1}:

$$DJ_X(x^k) - D_x K(x^k, y^k) \in DJ_X(x^{k+1}) + \partial F(x^{k+1}) \quad \text{and}$$
$$DJ_Y(y^k) - D_y K(x^k, y^k) \in DJ_Y(y^{k+1}) + \partial G_*(y^{k+1}) - 2D_y K(x^{k+1}, y^{k+1}). \tag{16}$$

We readily obtain x^{k+1} if the inverse of $DJ_X + \tau \partial F$ has an analytical closed-form expression. In this case we say that F is prox-simple with respect to J_X. For y^{k+1}, the same is true if K is affine in y and G_* is prox-simple with respect to J_Y. If, however, K is not affine in y, it is practically unlikely that $\partial G_* - 2D_y K(x^{k+1}, \cdot)$ would be prox-simple. We will therefore improve the method for general K in section "Inertial Terms," after first studying fundamental ideas behind convergence proofs in the following section "Convergence Theory."

If X and Y are Hilbert spaces with $J_X = \tau^{-1} N_X$ and $J_Y = \sigma^{-1} N_Y$, the standard generating functions divided by some step length parameters $\tau, \sigma > 0$, (16) becomes

Primal–dual proximal splitting (PDPS)

Iterate over $k \in \mathbb{N}$:

$$x^{k+1} := \mathrm{prox}_{\tau F}(x^k - \tau \nabla_x K(x^k, y^k)),$$
$$y^{k+1} := \mathrm{prox}_{\sigma[G_* - 2K(x^{k+1}, \cdot)]}(y^k - \sigma \nabla_y K(x^k, y^k)). \tag{17}$$

The proximal map is defined as

$$\operatorname{prox}_{\tau F}(x) := (I + \tau \partial F)^{-1}(x) = \arg\min_{\tilde{x} \in X} \left(\tau F(\tilde{x}) + \frac{1}{2}\|\tilde{x} - x\|_X^2 \right).$$

When this map has an analytical closed-form expression, we say that F is prox-simple (without reference to J_X). In finite dimensions, several worked out proximal maps may be found online (Chierchia et al. 2019) or in the book (Beck 2017). Some extend directly to Hilbert spaces or by superposition to L^2.

Remark 1. For K affine in y, i.e., $K(x, y) = \langle A(x) | y \rangle$ for some differentiable $A : X \to Y^*$, the dual update of (17) reduces to

$$\begin{aligned} y^{k+1} &= \operatorname{prox}_{\sigma G_*}(y^k + \sigma[2\nabla_y K(x^{k+1}, y^k) - \nabla_y K(x^k, y^k)]) \\ &= \operatorname{prox}_{\sigma G_*}(y^k + \sigma[2\nabla A(x^{k+1}) - \nabla A(x^k)]). \end{aligned}$$

This corresponds to the "linearized" variant of the NL-PDPS of Valkonen (2014). The "exact" variant, studied in further detail in Clason et al. (2019), updates

$$y^{k+1} := \operatorname{prox}_{\sigma G_*}(y^k + \sigma \nabla_y K(2x^{k+1} - x^k, y^k)).$$

If K is bilinear, the two variants are the exactly same PDPS of Chambolle and Pock (2011). For K not affine in y, the method is neither the generalized PDPS of Clason et al. (2020) nor the version for convex-concave K from Hamedani and Aybat (2018).

Block Adaptation

We now derive a version of the PDBS (16) adapted to the structure of

$$F(x) = \sum_{j=1}^{m} F_j(x_j) \quad \text{and} \quad G_*(y) = \sum_{\ell=1}^{n} G_{\ell*}(y_\ell),$$

where $x = (x_1, \ldots, x_m)$ and $y = (y_1, \ldots, y_n)$ in the (for simplicity) Hilbert spaces $X = \prod_{j=1}^{m} X_j$ and $Y = \prod_{\ell=1}^{n} Y_k$, and $F_j : X_j \to \overline{\mathbb{R}}$ and $G_{\ell*} : Y_\ell \to \overline{\mathbb{R}}$ are convex, proper, and lower semicontinuous.

For some "blockwise" step length parameters $\tau_j, \sigma_\ell > 0$, we take

$$J_X(x) = \sum_{j=1}^{m} \tau_j^{-1} N_{X_j}(x_j) \quad \text{and} \quad J_Y(y) = \sum_{\ell=1}^{n} \sigma_\ell^{-1} N_{Y_\ell}(y_\ell)$$

If K is now affine in y, observing Remark 1, (16) readily transforms into:

Block-adapted PDPS for K affine in y

Iteratively over $k \in \mathbb{N}$, for all $j = 1, \ldots, m$ and $\ell = 1, \ldots, n$, update:

$$x_j^{k+1} := \mathrm{prox}_{\tau_j F_j}(x_j^k - \tau_j \nabla_{x_j} K(x^k, y^k)),$$
$$y_\ell^{k+1} := \mathrm{prox}_{\sigma_\ell G_{\ell*}}(y_\ell^k + \sigma_\ell[2\nabla_{y_\ell} K(x^{k+1}, y^k) - \nabla_{y_\ell} K(x^k, y^k)]). \tag{18}$$

The idea is that the blockwise step length parameters adapt the algorithm to the structure of the problem. We will return their choices in the examples of section "Ellipticity for Block-Adapted Methods."

Performance gains

Correct adaptation of the blockwise step length parameters to the specific problem structure can yield significant performance gains compared to not exploiting the block structure (Pock and Chambolle 2011; Jauhiainen et al. 2020; Mazurenko et al. 2020).

Remark 2. For bilinear K, (18) is the "diagonally preconditioned" method of Pock and Chambolle (2011), or an unaccelerated non-stochastic variant of the methods in Valkonen (2019). For K affine in y, (18) differs from the methods in Mazurenko et al. (2020) by placing the over-relaxation in the dual step outside K, compare Remark 1.

Recall the saddle-point formulation (3) for inverse problems with nonlinear forward operators. We can now adapt step lengths to the constituent dual blocks.

Example 2. Let $A_1 \in C^1(X; Y_1^*)$ and $A_2 \in \mathbb{L}(X; Y_2^*)$, and suppose the convex functions $G_1 : Y_1^* \to \overline{\mathbb{R}}$ and $G_2 : Y_2^* \to \overline{\mathbb{R}}$ have the preconjugates G_{1*} and G_{2*}. Then we can write the problem

$$\min_{x \in X} G_1(A_1(x)) + G_2(A_2 x) + F(x).$$

in the form (S) with $G_*(y_1, y_2) = G_{1*}(y_1) + G_{2*}(y_2)$ and $K(x, y) = \langle A_1(x) | y_1 \rangle + \langle A_2 x | y_2 \rangle$. The algorithm (18) specializes as

$$x^{k+1} := \mathrm{prox}_{\tau F}(x^k - \tau[\nabla A_1(x^k)^* y_1 + A_2^* y_2]),$$
$$y_1^{k+1} := \mathrm{prox}_{\sigma_1 G_{1*}}(y_1^k + \sigma_1[2A_1(x^{k+1}) - A_1(x^k)]),$$
$$y_2^{k+1} := \mathrm{prox}_{\sigma_2 G_{2*}}(y_2^k + \sigma_2[A_2(2x^{k+1} - x^k)])$$

for some step length parameters $\tau, \sigma_1, \sigma_2 > 0$. We return to their choices and the local neighborhood of convergence in Examples 8 and 17 after developing the necessary convergence theory.

Convergence Theory

We now seek to understand when the basic version (15) of the PDBS convergences. The organization of this section centers around the three main ingredients of the convergence proof, as discussed in the Introduction:

(i) the three-point identity (8) employed in the general-purpose estimate of section "A Fundamental Estimate"
(ii) the (semi-)ellipticity of the algorithm-generating Bregman divergences B_{J_0} for J^0 as in (14), verified for several examples in sections "Ellipticity of the Bregman Divergences" and "Ellipticity for Block-Adapted Methods"
(iii) a second-order growth condition on (S), verified for several examples in sections "Nonsmooth Second-Order Conditions" and "Second-Order Growth Conditions for Block-Adapted Methods"

With these basic ingredients, we then prove various convergence results in sections "Convergence of Iterates" and "Convergence of Gaps in the Convex-Concave Setting." The usefulness of both (ii) and (iii) will become apparent from the fundamental estimates and examples of the next section "A Fundamental Estimate."

A Fundamental Estimate

We start with a simple estimate applicable to general methods of the form

$$0 \in H(u^{k+1}) + D_1 B(u^{k+1}, u^k) \tag{BP}$$

for some set-valued $H : U \rightrightarrows U^*$ and a Bregman divergence $B := B_J$ generated by some Gâteaux-differentiable $J : U \to \mathbb{R}$. We analyze (BP) following the "testing" ideas introduced in Valkonen (2020), extending them to the Bregman–Banach space setting, however in a simplified constant-metric setting that cannot model accelerated methods. The generic gap functional $\mathcal{G}(u^{k+1}, \bar{u})$ in the next result models any function value differences available from H. Its non-negativity will provide the basis for the aforementioned second-order growth conditions of sections "Nonsmooth Second-Order Conditions" and "Second-Order Growth Conditions for Block-Adapted Methods." We provide an example and interpretation after the theorem.

Theorem 1. On a Banach space U, let $H : U \rightrightarrows U^*$, and let $B := B_J$ be generated by a Gâteaux-differentiable $J : U \to \mathbb{R}$. Suppose (BP) is solvable for $\{u^{k+1}\}_{k \in \mathbb{N}}$ given an initial iterate $u^0 \in U$. Let $N \geq 1$. If for all $k = 0, \ldots, N-1$, for some $\bar{u} \in U$ and $\mathcal{G}(u^{k+1}, \bar{u}) \in \mathbb{R}$, the fundamental condition

$$\langle h^{k+1} | u^{k+1} - \bar{u} \rangle \geq \mathcal{G}(u^{k+1}, \bar{u}) \quad (h^{k+1} \in H(u^{k+1})) \tag{C}$$

holds, then so do the quantitative Δ-Féjer monotonicity

$$B(\bar{u}, u^{k+1}) + B(u^{k+1}, u^k) + \mathcal{G}(u^{k+1}, \bar{u}) \leq B(\bar{u}, u^k) \tag{F}$$

and the descent inequality

$$B(\bar{u}, u^N) + \sum_{k=0}^{N-1} B(u^{k+1}, u^k) + \sum_{k=0}^{N-1} \mathcal{G}(u^{k+1}, \bar{u}) \leq B(\bar{u}, u^0). \tag{D}$$

Proof. We can write (BP) as

$$0 = h^{k+1} + D_1 B(u^{k+1}, u^k) \quad \text{for some} \quad h^{k+1} \in H(u^{k+1}). \tag{19}$$

Testing (19) by applying $\langle \cdot | u^{k+1} - \bar{u} \rangle$, we obtain

$$0 = \langle h^{k+1} + D_1 B(u^{k+1}, u^k) | u^{k+1} - \bar{u} \rangle.$$

We use the three-point identity (8) to transform this into

$$B(\bar{u}, u^k) = \langle h^{k+1} | u^{k+1} - \bar{u} \rangle + B(\bar{u}, u^{k+1}) + B(u^{k+1}, u^k).$$

Inserting (C), we obtain (F). Summing the latter over $k = 0, \ldots, N-1$ yields (D). □

Example 3. If $H = \partial F$ for a convex function F, then by the definition of the convex subdifferential, (C) holds with the gap functional

$$\mathcal{G}(u, \bar{u}) = F(u) - F(\bar{u}).$$

If we take $\bar{u}$ is a minimizer of F, then the gap functional is non-negative and indeed positive if u is also not minimizer. This is why it is called a gap functional.

Consider then for some step length parameter $\tau > 0$ the proximal point method (13) in a Hilbert space X, that is, taking $B = \tau^{-1} N_X$

$$u^{k+1} := \operatorname{prox}_{\tau F}(x^k), \quad \text{equivalently} \quad 0 \in \partial F(u^{k+1}) + \tau(u^{k+1} - u^k).$$

Then (D) reads

$$\frac{1}{2\tau}\|u^N - \bar{u}\|_X^2 + \sum_{k=0}^{N-1} \frac{1}{2}\|u^{k+1} - u^k\|_X^2 + \sum_{k=0}^{N-1} \tau(F(u^{k+1}) - F(\bar{u})) \leq \frac{1}{2}\|\bar{u} - u^0\|_X^2. \tag{20}$$

With $\bar{u}$ a minimizer, this clearly forces $F(u^N) \searrow F(\bar{u})$ as $N \nearrow \infty$, suggesting why we call (D) the "descent inequality."

If our problem is non-convex, then we can try to locally ensure second-order growth by imposing $\mathcal{G}(u^{k+1}, \bar{u}) \geq 0$. Verifying this for the PDBS will be the topic of sections "Nonsmooth Second-Order Conditions" and "Second-Order Growth Conditions for Block-Adapted Methods." If B is not given by the standard generating function N_X on a Hilbert spaces X, then to get from (D) an estimate like (20) on norms, we can assume the ellipticity or at least semi-ellipticity of the overall Bregman divergence B. Verifying this for $B = B_{J^0}$ with J^0 given in (14) is our next topic.

Ellipticity of the Bregman Divergences

As just discussed, for Theorem 1 to provide estimates that we can use to prove the convergence of the PDBS, we need at least the semi-ellipticity of B^0 generated by J^0 given in (14). Deriving simple conditions that ensure such semi-ellipticity or ellipticity is the topic of the present subsection. To do this, we need the "basic" Bregman divergences B_X and B_Y on both spaces X and Y to be elliptic:

> **Standing assumption**
>
> In this subsection, we assume that B_X is τ^{-1}-elliptic and B_Y is σ^{-1}-elliptic for some $\tau, \sigma > 0$. This is true for the Hilbert-space PDPS (17) where τ and σ are the primal and dual step length parameters.

The examples that follow the next general lemma will provide improved estimates.

Lemma 2. *Suppose $K \in C^1(X \times Y)$ is Lipschitz-continuously differentiable with the factor L_{DK} in a convex subdomain $\Omega \subset X \times Y$. Then for $u, u' \in \Omega$*

$$B_K(u', u) \leq \frac{L_{DK}}{2}\|u' - u\|_{X \times Y}^2. \tag{21}$$

Consequently, if B_X is τ^{-1}-elliptic and B_Y is σ^{-1}-elliptic and $1 \geq \max\{\tau, \sigma\}L_{DK}$, then B^0 is semi-elliptic (elliptic if the inequality is strict) within Ω.

Proof. By definition, $B_K(u', u) = K(u') - K(u) - \langle DK(u)|u' - u\rangle$. Using the mean value equality in $\mathbb{R}$ with the chain rule and the Cauchy–Schwarz inequality, we get

$$B_K(u', u) = \int_0^1 \langle DK(u+t(u'-u)) - DK(u) | u' - u \rangle \, dt \leq \int_0^1 t L_{DK} \|u' - u\|_{X \times Y}^2 \, dt.$$

Calculating the last integral yields (21).

For the (semi-)ellipticity, we need $B^0(u, u') \geq \frac{\epsilon}{2} \|u - u'\|_{X \times Y}^2$ for some $\epsilon > 0$ ($\epsilon = 0$) and all $u, u' \in \Omega$. Since B_X and B_Y are τ^{-1}- and σ^{-1}-elliptic, we have

$$B^0(u', u) = B_X(x', x) + B_Y(y', y) - B_K(u', u) \tag{22}$$

$$\geq \frac{1}{2\tau} \|x' - x\|_X^2 + \frac{1}{2\sigma} \|y' - y\|_Y^2 - B_K(u'u).$$

Using (21), therefore $B^0(u', u) \geq \frac{\tau^{-1} - L_{DK}}{2} \|x' - x\|_X^2 + \frac{\sigma^{-1} - L_{DK}}{2} \|y' - y\|_Y^2$. Thus B^0 is ϵ-elliptic when $\tau^{-1}, \sigma^{-1} \geq L_{DK} + \epsilon$. This gives the claim. □

We now provide several examples of ellipticity. In practice, to guarantee ellipticity, we would choose $\tau, \sigma > 0$ to satisfy the stated conditions.

Example 4. Suppose $K(x, y) = E(x)$ with DE L_{DE}-Lipschitz in $\Omega = X \times Y$. Then $L_{DK} = L_{DE}$, so we recover the standard-for-gradient-descent step length bound $1 \geq \tau L_{DE}$ for B^0 to be semi-elliptic in Ω (elliptic if the inequality is strict).

Example 5. If $K(x, y) = \langle Ax | y \rangle$ for $A \in \mathbb{L}(X; Y^*)$, then B^0 is elliptic under the standard-for-PDPS (Chambolle and Pock 2011) step length condition

$$1 > \tau \sigma \|A\|^2.$$

Indeed

$$\langle DK(u + t(u' - u)) - DK(u) | u' - u \rangle = 2t \langle A(x - x') | y - y' \rangle.$$

Therefore, taking any $w > 1$, we easily improve (21) to

$$B_K(u', u) \leq \|A\| \|x' - x\|_X \|y' - y\|_Y \tag{23}$$

$$\leq \frac{w \|A\|}{2} \|x' - x\|_X^2 + \frac{w^{-1} \|A\|}{2} \|y' - y\|_Y^2 \quad (u, u' \in X \times Y).$$

By (22), B^0 is therefore ϵ-elliptic if $\tau^{-1} \geq w \|A\| + \epsilon$ and $\sigma^{-1} \geq w^{-1} \|A\| + \epsilon$. Taking $w = \sigma \|A\|/(1 - \sigma \epsilon)$, this holds if $1 \geq \tau \sigma \|A\|^2/(1 - \sigma \epsilon) + \tau \epsilon$. Since $\epsilon > 0$ was arbitrary, the claimed step length condition follows.

Example 6. Suppose $K(x, y) = \langle A(x)|y\rangle$ with A and DA Lipschitz with the respective factors $L_A, L_{DA} \geq 0$. Then B^0 is elliptic within $\Omega = X \times B(0, \rho_y)$ if

$$1 > \tau\sigma L_A^2 + \tau \frac{L_{DA}\rho_y}{2}.$$

Indeed, for any $w > 1$, using the mean value equality as in the proof of Lemma 2, we deduce

$$\begin{aligned}B_K(u', u) &= \langle A(x') - A(x)|y'\rangle - \langle DA(x)(x' - x)|y\rangle \\ &= \langle A(x') - A(x)|y' - y\rangle + \langle A(x') - A(x) - DA(x)(x' - x)|y\rangle \\ &\leq L_A\|x' - x\|_X\|y' - y\|_Y + \frac{L_{DA}\|y'\|}{2}\|x' - x\|_X^2 \\ &\leq \frac{wL_A + L_{DA}\|y\|}{2}\|x' - x\|_X^2 + \frac{w^{-1}L_A}{2}\|y' - y\|_Y^2.\end{aligned} \quad (24)$$

If $\rho_y > 0$ is such that $\|y\| \leq \rho_y$, taking $w = \sigma L_A/(1-\sigma\epsilon)$, similarly to Example 5, we deduce the claimed bound.

We can combine the examples above:

Example 7. As in Example 2, take $K(x, (y_1, y_2)) = \langle A_1(x)|y_1\rangle + \langle A_2x|y_2\rangle$ with $A_1 \in C^1(X; Y_1^*)$ and $A_2 \in \mathbb{L}(X; Y_2^*)$. Then B^0 is elliptic within $\Omega = X \times B(0, \rho_y)$ if

$$1 > \tau\sigma(L_{A_1}^2 + \|A_2\|^2) + \tau \frac{L_{DA_1}\rho_{y_1}}{2}.$$

Indeed, we bound B_K by summing (23) for A_1 and (24) for A_2. This yields for any $w_1, w_2 > 0$ the estimate

$$\begin{aligned}B_K(u', u) &\leq \frac{w_1 L_{A_1} + L_{DA_1}\|y_1\|}{2}\|x - x'\|_X^2 + \frac{w_1^{-1}L_{A_1}}{2}\|y_1' - y_1\|_Y^2 \\ &+ \frac{w_2\|A_2\|}{2}\|x' - x\|_X^2 + \frac{w_2^{-1}\|A_2\|}{2}\|y_2' - y_2\|_{Y_2}^2.\end{aligned} \quad (25)$$

Taking $w_1 = \sigma L_{A_1}/(1-\sigma\epsilon)$ and $w_2 = \sigma\|A_2\|/(1-\sigma\epsilon)$ and using (22), we deduce the claimed ellipticity for small enough $\epsilon > 0$.

Remark 3. In Examples 6 and 7, we needed a bound on the dual variable y. In the latter, as an improvement, this was only needed on the subspace Y_1 of nonbilinearity. An ad hoc solution is to introduce the bound into the problem. In the

Hilbert case, Clason et al. (2019, 2020) secure such bounds by taking the primal step length τ small enough and arguing as in Theorem 1 individually on the primal and dual iterates.

Ellipticity for Block-Adapted Methods

We now study ellipticity for block-adapted methods. The goal is to obtain faster convergence by adapting the blockwise step length parameters to the problem structure (connections between blocks) and the local (blockwise) properties of the problem.

> **Standing assumption**

In this subsection, we assume F, G_*, J_X, and J_Y to have the form of section "Block Adaptation." In particular, X and Y are (products of) Hilbert spaces, and

$$B^0(u', u) = \sum_{j=1}^{m} \frac{1}{2\tau_j} \|x_j' - x_j\|_{X_j}^2 + \sum_{\ell=1}^{n} \frac{1}{2\sigma_\ell} \|y_\ell' - y_\ell\|_{Y_\ell}^2 - B_K(u', u). \qquad (26)$$

We start by refining the two-block Example 7 to be adapted to the blocks:

Example 8. Let $K(x, (y_1, y_2)) = \langle A_1(x)|y_1\rangle + \langle A_2 x|y_2\rangle$ with $A_1 \in C^1(X; Y_1^*)$ and $A_2 \in \mathbb{L}(X; Y_2^*)$ as in Examples 2 and 7. Write $\tau = \tau_1$. Using (25) in (26) for $m = 1$ and $n = 2$ with (25), we see B^0 to be ϵ-elliptic within $\Omega = X \times B(0, \rho_{y_1}) \times Y_2$ if $\tau^{-1} \geq w_1 L_{A_1} + L_{DA_1}\rho_{y_1} + w_2\|A_2\| + \epsilon$ and $\sigma_1^{-1} \geq w_1^{-1} L_{A_1}$ as well as $\sigma_2^{-1} \geq w_2^{-1}\|A_2\| + \epsilon$. Taking $w_1 = \sigma_1 L_{A_1}/(1 - \sigma_1\epsilon)$ and $w_2 = \sigma_2\|A_2\|/(1 - \sigma_2\epsilon)$, B^0 is therefore elliptic (some $\epsilon > 0$) within Ω if $1 > \tau(\sigma_1 L_{A_1}^2 + \sigma_2\|A_2\|^2) + \tau \frac{L_{DA_1}\rho_{y_1}}{2}$.

Example 9. In Example 8, if both $A_1 \in \mathbb{L}(X; Y_1^*)$ and $A_2 \in \mathbb{L}(X; Y_2^*)$, then B^0 is elliptic within $\Omega = X \times Y_1 \times Y_2$ if $1 > \tau(\sigma_1\|A_1\|^2 + \sigma_2\|A_2\|^2)$.

Example 10. Suppose we can write $K(x, y) = \sum_{j=1}^{m}\sum_{\ell=1}^{n} K_{j\ell}(x_j, y_\ell)$ with each $K_{j\ell}$ Lipschitz-continuously differentiable with the factor $L_{j\ell}$. Following Lemma 2

$$B_K(u', u) \leq \sum_{j=1}^{m}\sum_{\ell=1}^{n} \frac{L_{j\ell}}{2}(\|x_j' - x_j\|^2 + \|y_\ell' + y_\ell\|^2). \qquad (27)$$

Consequently, using (26), we see that B^0 is ϵ-elliptic if $1 \geq \tau_j(\sum_{\ell=1}^{n} L_{j\ell} + \epsilon)$ and $1 \geq \sigma_\ell(\sum_{j=1}^{m} L_{j\ell} + \epsilon)$ for all $j = 1, \ldots, m$ and $\ell = 1, \ldots, n$.

Example 11. If $K(x, y) = \sum_{j=1}^{m} \sum_{\ell=1}^{m} \langle A_{j\ell} x_j | y_\ell \rangle$ for some $A_{j\ell} \in \mathbb{L}(X_j; Y_\ell^*)$, then following Example 5, for arbitrary $w_{j\ell} > 0$

$$B_K(u', u) \leq \sum_{j=1}^{m} \sum_{\ell=1}^{m} \|A_{j\ell}\| \|x_j' - x_j\| \|y_j' - y_j\|$$

$$\leq \sum_{j=1}^{m} \sum_{\ell=1}^{n} \left(\frac{w_{j\ell} \|A_{j\ell}\|}{2} \|x_j' - x_j\|^2 + \frac{w_{j\ell}^{-1} \|A_{j\ell}\|}{2} \|y_\ell' - x_\ell\|^2 \right).$$

Using (26), B^0 is thus ϵ-elliptic if $1 \geq \tau_j(\epsilon + \sum_{\ell=1}^{n} w_{j\ell} \|A_{j\ell}\|)$ and $1 \geq \sigma_\ell(\epsilon + \sum_{j=1}^{m} w_{j\ell}^{-1} \|A_{j\ell}\|)$ for all $j = 1, \ldots, m$ and $\ell = 1, \ldots, n$. We can use the factors $w_{j\ell}$ to adapt the algorithm to the different blocks for potentially better convergence.

Nonsmooth Second-Order Conditions

We now study conditions for (C) to hold with $\mathcal{G}(\cdot, \bar{u}) \geq 0$. We start by writing out the condition for the PDBS.

Lemma 3. *Let $\bar{u} = (\bar{x}, \bar{y}) \in X \times Y$, and suppose for some $\mathcal{G}(u, \bar{u}) \in \mathbb{R}$ and a neighborhood $\Omega_{\bar{u}} \subset X \times Y$ that for all $u = (x, y) \in \Omega_{\bar{u}}$, $x^* \in \partial F(x)$ and $y^* \in \partial G_*(y)$*

$$\langle x^* + D_x K(x, y) | x - \bar{x} \rangle + \langle y^* - D_y K(x, y) | y - \bar{y} \rangle \geq \mathcal{G}(u, \bar{u}). \tag{C^2}$$

Let $\{u^{k+1}\}_{k \in \mathbb{N}}$ be generated by the PDBS (16) for some $u^0 \in X \times Y$, and suppose $\{u^k\}_{k \in \mathbb{N}} \subset \Omega_{\bar{u}}$. Then with $B = B^0$ the fundamental condition (C) and the quantitative Δ-Féjer monotonicity (F) hold for all $k \in \mathbb{N}$, and the descent inequality (D) holds for all $N \geq 1$.

Proof. Theorem 1 proves (F) and (D) if we show (C^2). For H in (12), we have

$$h^{k+1} = \begin{pmatrix} x_{k+1}^* + D_x K(x^{k+1}, y^{k+1}) \\ y_{k+1}^* - D_y K(x^{k+1}, y^{k+1}) \end{pmatrix} \in H(u^{k+1}) \quad \text{with} \quad \begin{cases} x_{k+1}^* \in \partial F(x^{k+1}), \\ y_{k+1}^* \in \partial G_*(y^{k+1}). \end{cases}$$

Thus (C) expands as (C^2) for $u = u^{k+1}$ and $(x^*, y^*) = (x_{k+1}^*, y_{k+1}^*)$. □

In section "Convergence of Gaps in the Convex-Concave Setting" on the convergence of gap functionals, we will consider general $\bar{u}$ in (C^2). For the moment, we however fix a root $\bar{u} = \hat{u} \in H^{-1}(0)$. Then

… 18 First-Order Primal–Dual Methods for Nonsmooth Non-convex Optimization

$$0 = \begin{pmatrix} \hat{x}^* + D_x K(\hat{x}, \hat{y}) \\ \hat{y}^* - D_y K(\hat{x}, \hat{y}) \end{pmatrix} \in H(\hat{u}) \quad \text{with} \quad \begin{cases} \hat{x}^* \in \partial F(\hat{x}), \\ \hat{y}^* \in \partial G_*(\hat{y}). \end{cases} \quad (28)$$

Since we assume F and G_* to be convex; their subdifferentials are monotone. When K is not convex-concave and to obtain strong convergence of iterates even when it is, we will need some strong monotonicity of the subdifferentials, but only *at a solution*. Specifically, for $\gamma > 0$, we say that $T : X \rightrightarrows X^*$ is γ-strongly monotone at $\hat{x}$ for $\hat{x}^* \in T(\hat{x})$ if

$$\langle x^* - \hat{x}^* | x - \hat{x} \rangle \geq \gamma \|x - \hat{x}\|_X^2 \quad (x \in X, \; x^* \in T(x)). \quad (29)$$

If $\gamma = 0$, we drop the word "strong." For $T = \partial F$, (29) follows from the γ-strong subdifferentiability of F.

> **Standing assumption**

Throughout the rest of this subsection, we assume (28) to hold and that ∂F is (γ_F-strongly) monotone at $\hat{x}$ for $\hat{x}^*$ and ∂G_* is (γ_{G_*}-strongly) monotone at $\hat{y}$ for $\hat{y}^*$.

Lemma 4. *The nonsmooth second-order growth condition* (C²) *holds provided*

$$\gamma_F \|x - \hat{x}\|^2 + \gamma_{G_*} \|y - \hat{y}\|^2 \geq B_K(\hat{u}, u) + B_K(u, \hat{u}) + \mathcal{G}(u, \hat{u}) \quad (u \in \Omega_{\bar{u}}), \quad (30)$$

equivalently

$$\gamma_F \|x - \hat{x}\|^2 + \gamma_{G_*} \|y - \hat{y}\|^2 \geq a_K(\hat{u}, u) + a_K(u, \hat{u}) + \mathcal{G}(u, \hat{u}) \quad (u \in \Omega_{\bar{u}}) \quad (30')$$

for

$$a_K(u, \bar{u}) := K(x, y) - K(\bar{x}, \bar{y}) + \langle D_x K(x, y) | \bar{x} - x \rangle + \langle D_y K(\bar{x}, \bar{y}) | \bar{y} - y \rangle. \quad (31)$$

Note that (30) involves the symmetrized Bregman divergence $B_K^S(u, u') := B_K(u, u') + B_K(u', u)$ generated by K.

Proof. Inserting the zero of (28) in (C²), we rewrite the latter as

$$\langle x^* - \hat{x}^* | x - \hat{x} \rangle + \langle y^* - \hat{y}^* | y - \hat{y} \rangle \geq \langle D_x K(x, y) - D_x K(\hat{x}, \hat{y}) | \hat{x} - x \rangle$$
$$+ \langle D_y K(x, y) - D_y K(\hat{x}, \hat{y}) | y - \hat{y} \rangle + \mathcal{G}(u^{k+1}, \hat{u}).$$

Using the assumed strong monotonicities, and the definitions of B_K and a_K, this is immediately seen to hold when (30) or (30′) does. □

Example 12. If K is convex-concave, the next Lemma 5 and Lemma 4 prove (C²) for

$$\mathcal{G}(u, \hat{u}) = \gamma_F \|x - \hat{x}\|^2 + \gamma_{G_*} \|y - \hat{y}\|^2 \geq 0 \quad \text{within} \quad \Omega_{\hat{u}} = X \times Y.$$

This is in particular true for $K(x, y) = \langle Ax | y \rangle + E(x)$ with $A \in \mathbb{L}(X; Y^*)$ and $E \in C^1(X)$ convex.

Lemma 5. *Suppose $K : X \times Y \to \mathbb{R}$ is Gâteaux-differentiable and convex-concave. Then $a_K(u, \bar{u}) \leq 0$ and $B_K^S(u, \bar{u}) \leq 0$ for all $u, \bar{u} \in X \times Y$.*

Proof. The convexity of $K(\cdot, y)$ and the concavity of $K(\bar{x}, \cdot)$ show

$$K(x, y) - K(\bar{x}, y) + \langle D_x K(x, y) | \bar{x} - x \rangle \leq 0 \quad \text{and}$$
$$K(\bar{x}, y) - K(\bar{x}, \bar{y}) + \langle D_y K(\bar{x}, \bar{y}) | \bar{y} - y \rangle \leq 0.$$

Summing these two estimates proves $a_K(u, \bar{u}) \leq 0$, consequently $B_K^S(u, \bar{u}) = a_K(u, \bar{u}) + a_K(\bar{u}, u) \leq 0$. □

Example 13. Suppose K has L_{DK}-Lipschitz derivative within $\Omega \subset X \times Y$. If $\hat{u} \in \Omega$, then by Lemma 2, $B_K(u, \hat{u}), B_K(\hat{u}, u) \leq \frac{L_{DK}}{2} \|u - \hat{u}\|_{X \times Y}^2$ for $u \in \Omega$. Thus (C²) holds by Lemma 4 with $\Omega_{\hat{u}} = \Omega$ and

$$\mathcal{G}(u, \hat{u}) = (\gamma_F - L_{DK}) \|x - \hat{x}\|^2 + (\gamma_{G_*} - L_{DK}) \|y - \hat{y}\|^2.$$

This is non-negative if $\gamma_F, \gamma_{G_*} \geq L_{DK}$.

Example 14. Let $K(x, y) = \langle A(x) | y \rangle$ for some $A \in \mathbb{L}(X; Y^*)$ such that DA is Lipschitz with the factor $L_{DA} \geq 0$. For some $\tilde{\gamma}_F, \tilde{\gamma}_{G_*} \geq 0$ and $\rho_y, \hat{\rho}_x, \alpha > 0$, let either

(a) $\tilde{\gamma}_F \geq \frac{L_{DA}}{2}(\rho_y + \|\hat{y}\|_Y)$, $\tilde{\gamma}_{G_*} \geq 0$, and $\Omega_{\hat{u}} = X \times B(0, \rho_y)$; or
(b) $\tilde{\gamma}_F > L_{DA}\left(\|\hat{y}\|_Y + \frac{\alpha}{2}\right)$, $\tilde{\gamma}_{G_*} \geq \frac{L_{DA}}{2\alpha} \hat{\rho}_x^2$, and $\Omega_{\hat{u}} = B(\hat{x}, \hat{\rho}_x) \times Y$.

Then Lemma 4 proves (C²) with

$$\mathcal{G}(u, \hat{u}) = (\gamma_F - \tilde{\gamma}_F) \|x - \hat{x}\|^2 + (\gamma_{G_*} - \tilde{\gamma}_{G_*}) \|y - \hat{y}\|^2.$$

To see this, we need to prove (30′). Now

$$a_K(u, \hat{u}) := \langle A(x) - A(\hat{x}) + DA(x)(\hat{x} - x) | y \rangle \quad (u, \hat{u} \in X \times Y). \tag{32}$$

Arguing with the mean value equality and the Lipschitz assumption as in Lemma 2, we get $a_K(\hat{u}, u) + a_K(u, \hat{u}) \leq \frac{L_{DA}}{2}(\|y\|_Y + \|\hat{y}\|_Y)\|x - \hat{x}\|^2$. Thus (a) implies (30′). By (32), the mean-value equality, and the Lipschitz assumption, also

$$a_K(u, \hat{u}) + a_K(\hat{u}, u) = \langle [DA(x) - DA(\hat{x})](\hat{x} - x) | \hat{y} \rangle$$
$$+ \langle A(x) - A(\hat{x}) + DA(x)(\hat{x} - x) | y - \hat{y} \rangle$$
$$\leq L_{DA}\|x - \hat{x}\|_X^2 (\|\hat{y}\|_Y + \tfrac{1}{2}\|y - \hat{y}\|_Y).$$

Using Cauchy's inequality and (b) we deduce (30′).

Remark 4. In the last two examples, we need to bound some of the iterates and to initialize close enough to a solution. Showing that the iterates stay in a local neighborhood is a large part of the work in Clason et al. (2019, 2020), as discussed in Remark 3.

Second-Order Growth Conditions for Block-Adapted Methods

We now study second-order growth for problems with a block structure as in section "Block Adaptation":

> **Standing assumption**

In this subsection, F and G_* are as in section "Block Adaptation," each component subdifferential ∂F_j now (γ_{F_j}-strongly) monotone at $\hat{x}_j$ for $\hat{x}_j^*$ and each $\partial G_{\ell*}$ ($\gamma_{G_{\ell*}}$-strongly) monotone at $\hat{y}_\ell$ for $\hat{y}_\ell^*$. Here $\hat{x}_j, \hat{x}_j^*, \hat{y}_\ell$, and $\hat{y}_\ell^*$ are the components of $\hat{x}$, $\hat{x}^*, \hat{y}$, and $\hat{y}^*$ in the corresponding subspace, assumed to satisfy the critical point condition (28).

As only some of the component functions may have $\gamma_{F_j}, \gamma_{G_{\ell*}} > 0$, through detailed analysis of the block structure, we hope to obtain (strong) convergence on some subspaces even if the entire primal or dual variables might not converge.

Similarly to Lemma 4 we prove:

Lemma 6. *Suppose for some neighborhood $\Omega_{\hat{u}} \subset X \times Y$ that*

$$\Delta_{k+1} := \sum_{j=1}^m \tilde{\gamma}_{F_j} \|x_j - \hat{x}_j\|_{X_j}^2 + \sum_{\ell=1}^n \tilde{\gamma}_{G_{\ell*}} \|y_\ell - \hat{y}_\ell\|_{Y_\ell}^2 \geq a_K(\hat{u}, u) + a_K(u, \hat{u})$$

for some $\tilde{\gamma}_{F_j}, \tilde{\gamma}_{G_{\ell*}} \geq 0$ for all $u \in \Omega_{\hat{u}}$. Then (C²) holds with

$$\mathcal{G}(u, \hat{u}) = \sum_{j=1}^{m}(\gamma_{F_j} - \tilde{\gamma}_{F_j})\|x_j - \hat{x}_j\|_{X_j}^2 + \sum_{\ell=1}^{n}(\gamma_{G_{\ell*}} - \tilde{\gamma}_{G_{\ell*}})\|y_\ell - \hat{y}_\ell\|_{Y_\ell}^2. \quad (33)$$

In the convex-concave case, we can transfer all strong monotonicity into $\mathcal{G}$:

Example 15. If K is convex-concave, then by Lemmas 5 and 6, (C²) holds with $\Omega_{\hat{u}} = X \times Y$ and $\mathcal{G}$ as in (33) for $\tilde{\gamma}_{F_j} = 0$ and $\tilde{\gamma}_{G_{\ell*}} = 0$. We have $\mathcal{G}(\,\cdot\,, \hat{u}) \geq 0$ always.

Example 16. As in Example 10, suppose we can write $K(x, y) = \sum_{j=1}^{m}\sum_{\ell=1}^{n} K_{j\ell}(x_j, y_\ell)$ with each $K_{j\ell}$ Lipschitz-continuously differentiable with the factor $L_{j\ell}$ in Ω. Then using (27) and Lemma 6, we see (C²) to hold with $\Omega_{\hat{u}} = \Omega$ and $\mathcal{G}$ as in (33) with

$$\tilde{\gamma}_{F_j} = \sum_{\ell=1}^{n} L_{j\ell} \quad (j=1,\ldots,m) \quad \text{and} \quad \tilde{\gamma}_{G_{\ell*}} = \sum_{j=1}^{m} L_{j\ell} \quad (\ell=1,\ldots,n).$$

Thus $\mathcal{G}(\,\cdot\,, \hat{u}) \geq 0$ if $\gamma_{F_j} \geq \sum_{\ell=1}^{n} L_{j\ell}$ and $\gamma_{G_{\ell*}} \geq \sum_{j=1}^{m} L_{j\ell}$ for all ℓ and j.

The special case of Example 10 with each $K_{j\ell}$ bilinear, corresponding to Example 11 for ellipticity, is covered by Example 15.

We consider in detail the two dual block setup of Examples 2 and 8:

Example 17. As in Example 2, let $K(x, y) = \langle A_1(x)|y_1\rangle + \langle A_2x|y_2\rangle$ for $A_1 \in C^1(X; Y_1^*)$ and $A_2 \in \mathbb{L}(X; Y_2^*)$. Then, as in (32),

$$a_K(u, \bar{u}) = \langle A_1(x) - A_1(\bar{x}) + DA_1(x)(\bar{x} - x)|y_1\rangle,$$

which does not depend on A_2. For any $\alpha, \rho_y, \hat{\rho}_x > 0$, let either

(a) $\tilde{\gamma}_F \geq \frac{L_{DA_1}}{2}(\rho_{y_1} + \|\hat{y}_1\|_{Y_1}), \tilde{\gamma}_{G_{1*}} \geq 0$, and $\Omega_{\hat{u}} = X \times B(0, \rho_{y_1})$; or
(b) $\tilde{\gamma}_F > L_{DA_1}\left(\|\hat{y}_1\|_{Y_1} + \frac{\alpha}{2}\right), \tilde{\gamma}_{G_{1*}} \geq \frac{L_{DA_1}}{2\alpha}\hat{\rho}_x^2$, and $\Omega_{\hat{u}} = B(\hat{x}, \hat{\rho}_x) \times Y$.

Arguing as in Example 14 and using Lemma 6, we then see (C²) to hold with $\mathcal{G}$ as in (33) and $\tilde{\gamma}_{G_{2*}} = 0$. In this case $\mathcal{G}(\,\cdot\,, \hat{u})$ is non-negative if $\gamma_F \geq \tilde{\gamma}_F$ and $\gamma_{G_{1*}} \geq \tilde{\gamma}_{G_{1*}}$.

Convergence of Iterates

We are now ready to prove the convergence of the iterates. We start with weak convergence and proceed to strong and linear convergence. For weak convergence in infinite dimensions, we need some further technical assumptions. We recall that a set-valued map $T : X \rightrightarrows X^*$ is weak-to-strong (weak-*-to-strong) outer semicontinuous if $x_k^* \in T(x^k)$ and $x^k \rightharpoonup x$ ($x^k \stackrel{*}{\rightharpoonup} x$) and $x_k^* \to x^*$ imply $x^* \in T(x)$. The nonreflexive case of the next assumption covers spaces of functions of bounded variation (Ambrosio et al. 2000, Remark 3.12), important for total variation based imaging.

Assumption 1. Each of the spaces X and Y is, individually, either a reflexive Banach space or the dual of separable space. The operator $H : X \times Y \rightrightarrows X^* \times Y^*$ is weak(-*)-to-strong outer semicontinuous, where we mean by "weak(-*)" that we take the weak topology if the space is reflexive and weak-* otherwise, individually on X and Y.

Subdifferentials of lower semicontinuous convex functions are weak(-*)-to-strong outer semicontinuous[5], so the outer semicontinuity of H depends mainly on K.

Example 18. If X and Y are finite-dimensional, Assumption 1 holds if $K \in C^1(X; Y)$.

Example 19. More generally, Assumption 1 holds if $K \in C^1(X \times Y)$ and DK is continuous from the weak(-*) topology to the strong topology.

Example 20. If $K = \langle Ax | y \rangle + E(x)$ for $A \in \mathbb{L}(X; Y^*)$ and $E \in C^1(X)$ convex, then H satisfies Assumption 1. Indeed, it can be shown that H is maximal monotone, hence weak(-*) outer semicontinuous similarly to convex subdifferentials.

> **Verification of the conditions**

To verify the nonsmooth second-order growth condition (C^2) for each of the following Theorems 2, 3, and 4, we point to sections "Nonsmooth Second-Order Conditions" and "Second-Order Growth Conditions for Block-Adapted Methods." For the verification of the (semi-)ellipticity of B^0, we point to sections "Ellipticity of the Bregman Divergences" and "Ellipticity for Block-Adapted Methods." As special cases of the PDBS (16), the theorems apply to the Hilbert-space PDPS (17) and its block adaptation (18). Then J_X and J_Y are continuously differentiable and convex.

---

[5]This result seems difficult to find in the literature for Banach spaces but follows easily from the definition of the subdifferential: If $F(x) \geq F(x^k) + \langle x_k^* | x - x^k \rangle$ and $x_k^* \to \hat{x}^*$ as well as $x^k \rightharpoonup$ (or $\stackrel{*}{\rightharpoonup}$) $\hat{x}$, then, using the fact that $\{\|x^k - \hat{x}\|\}_{k \in \mathbb{N}}$ is bounded, in the limit $F(x) \geq F(\hat{x}) + \langle \hat{x}^* | x - \hat{x} \rangle$.

Theorem 2 (Weak convergence). *Let F and G_* be convex, proper, and lower semicontinuous; $K \in C^1(X \times Y)$; and both $J_X \in C^1(X)$ and $J_Y \in C^1(Y)$ convex. Suppose Assumption 1 holds and for some $\hat{u} \in H^{-1}(0)$ that*

(i) (C^2) *holds with* $\mathcal{G}(\,\cdot\,, \hat{u}) \geq 0$ *within* $\Omega_{\hat{u}} \subset X \times Y$.
(ii) B^0 *is elliptic within* $\Omega \ni \hat{u}$.

Let $\{u^{k+1}\}_{k \in \mathbb{N}}$ be generated by the PDBS (16) for any initial u^0, and suppose $\{u^k\}_{k \in \mathbb{N}} \subset \Omega \cap \Omega_{\hat{u}}$. Then there exists at least one cluster point of $\{u^k\}_{k \in \mathbb{N}}$, and all weak(-) cluster points belong to $H^{-1}(0)$.*

Proof. Lemma 3 establishes (D) for $B = B^0$ and all $N \geq 1$. With $\epsilon > 0$ the factor of ellipticity of B^0, it follows

$$\frac{\epsilon}{2} \|u^N - \hat{u}\|_{X \times Y}^2 + \frac{\epsilon}{2} \sum_{k=0}^{N-1} \|u^{k+1} - u^k\|_{X \times Y}^2 \leq B^0(\hat{u}, u^0) \quad (N \geq 1).$$

Clearly $\|u^{k+1} - u^k\| \to 0$ while $\{\|u^k - \hat{u}\|\}_{k \in \mathbb{N}}$ is bounded. Using the Eberlein–Šmulyan theorem in a reflexive X or Y and the Banach–Alaoglu theorem otherwise (X or Y the dual of a separable space), we may therefore find a subsequence of $\{u^k\}_{k \in \mathbb{N}}$ converging weakly(-*) to some $\bar{x}$. Since $J^0 \in C^1(X \times Y)$, we deduce $D_1 B^0(u^{k+1}, u^k) \to 0$. Consequently (15) implies that $0 \in \limsup_{k \to \infty} H(u^{k+1})$, where we write "lim sup" for the Painlevé–Kuratowski outer limit of a sequence of sets in the strong topology. Since H is weak(-*)-to-strong outer semicontinuous by Assumption 1, it follows that $0 \in H(\hat{u})$. Therefore, there exists at least one cluster point of $\{u^k\}_{k \in \mathbb{N}}$ belonging to $H^{-1}(0)$. Repeating the argument on any weak(-*) convergent subsequence, we deduce that all cluster points belong to $H^{-1}(0)$. □

Remark 5. For a unique weak limit, we may in Hilbert spaces use the quantitative Féjer monotonicity (F) with Opial's lemma (Opial 1967; Browder 1967). For bilinear K the result is relatively immediate, as B^0 is a squared matrix-weighted norm; see Valkonen (2020). Otherwise a variable-metric Opial's lemma (Clason et al. 2019) and additional work based on the Brezis–Crandall–Pazy lemma (Brezis et al. 1970, Corollary 20.59 (iii)) are required; see Clason et al. (2019) for $K(x, y) = \langle A(x) | y \rangle$ and Clason et al. (2020) for general K.

Theorem 3 (Strong convergence). *Let F and G_* be convex, proper, and lower semicontinuous; $K \in C^1(X \times Y)$; and both $J_X \in C(X)$ and $J_Y \in C(Y)$ convex and Gâteaux-differentiable. Suppose for some $\hat{u} \in H^{-1}(0)$ that*

(i) (C^2) *holds with* $\mathcal{G}(\,\cdot\,, \hat{u}) \geq 0$ *within* $\Omega_{\hat{u}} \subset X \times Y$.
(ii) B^0 *is semi-elliptic within* $\Omega \ni \hat{u}$.

Let $\{u^{k+1}\}_{k \in \mathbb{N}}$ be generated by the PDBS (16) for any initial u^0, and suppose $\{u^k\}_{k \in \mathbb{N}} \subset \Omega \cap \Omega_{\hat{u}}$. Then $\mathcal{G}(u^{k+1}, \hat{u}) \to 0$ as $N \to \infty$.

In particular, if $\mathcal{G}(u, \hat{u}) \geq \|P(u - \hat{u})\|_Z^2$ for some $P \in \mathbb{L}(X; Z)$, then $Px^N \to P\hat{x}$ strongly in Z and the ergodic sequence $\tilde{x}_P^N := \frac{1}{N}\sum_{k=0}^{N-1} Px^{k+1} \to P\hat{x}$ at rate $O(1/N)$.

Proof. Lemma 3 establishes (D). By the semi-ellipticity of B^0, then $\sum_{k=0}^{N-1}\mathcal{G}(u^{k+1}, \hat{u}) \leq B^0(\hat{u}, u^0)$, $(N \in \mathbb{N})$. Since $\mathcal{G}(u^{k+1}, \hat{u}) \geq 0$, this shows that $\mathcal{G}(u^N, \hat{u}) \to 0$. The strong convergence of the primal variable for quadratically minorized $\mathcal{G}$ is then immediate whereas following by Jensen's inequality gives the ergodic convergence claim. □

Example 21. In section "Nonsmooth Second-Order Conditions," we can take $Pu = \sqrt{\gamma_F - \tilde{\gamma}_F}x$ if $\gamma_F > \tilde{\gamma}_F$ or $Pu = \sqrt{\gamma_{G_*} - \tilde{\gamma}_{G_*}}y$ if $\gamma_{G_*} > \tilde{\gamma}_{G_*}$. The examples of section "Second-Order Growth Conditions for Block-Adapted Methods" for $x = (x_1, \ldots, x_m)$ and $y = (y_1, \ldots, y_n)$ may allow $Pu = \sqrt{\gamma_{F_j} - \tilde{\gamma}_{F_j}}x_j$ or $Pu = \sqrt{\gamma_{G_{\ell*}} - \tilde{\gamma}_{G_{\ell*}}}y_\ell$.

Remark 6. Under similar conditions as Theorem 3, it is possible to obtain $O(1/N^2)$ convergence rates; see Chambolle and Pock (2011) and Valkonen (2020) for the convex-concave case and Clason et al. (2019, 2020) in general.

Theorem 4 (Linear convergence). *Let F and G_* be convex, proper, and lower semicontinuous; $K \in C^1(X \times Y)$; and both $J_X \in C(X)$ and $J_Y \in C(Y)$ convex and Gâteaux-differentiable. Suppose for some $\gamma > 0$ and $\hat{u} \in H^{-1}(0)$ that*

(i) (C²) holds with $\mathcal{G}(u, \hat{u}) \geq \gamma B^0(\hat{u}, u)$ within $\Omega_{\hat{u}} \subset X \times Y$.
(ii) B^0 is elliptic within $\Omega \supset \hat{u}$.

Let $\{u^{k+1}\}_{k \in \mathbb{N}}$ be generated by the PDBS (16) for any initial u^0, and suppose $\{u^k\}_{k \in \mathbb{N}} \subset \Omega \cap \Omega_{\hat{u}}$. Then $B^0(\hat{u}, u^N) \to 0$ and $u^N \to \hat{u}$ at a linear rate.

In particular, if $\mathcal{G}(u, \hat{u}) \geq \gamma \|u - \hat{u}\|^2$, $(k \in \mathbb{N})$, for some $\gamma > 0$, and J^0 is Lipschitz-continuously differentiable, then $u^N \to \hat{u}$ at a linear rate.

Proof. Lemma 3 establishes the quantitative Δ-Féjer monotonicity (F). Using (i), this yields $(1 + \gamma)B^0(\hat{u}, u^{k+1}) \leq B^0(\hat{u}, u^k)$. By the semi-ellipticity of B^0, the claimed linear convergence of $B^0(\hat{u}, u^N) \to 0$ follows. Since B^0 is assumed elliptic, also $u^N \to \hat{u}$ linearly. If J^0 is Lipschitz-continuously differentiable, then, similarly to Lemma 2, $B^0(\hat{u}, u^{k+1}) \leq L_{DJ}\|u^{k+1} - \hat{u}\|^2$ for some $L_{DJ} > 0$. Thus $\mathcal{G}(u^{k+1}, \hat{u}) \geq \gamma H_{DJ}^{-1} B^0(\hat{u}, u^{k+1})$, so the main claim establishes the particular claim. □

Example 22. J^0 is Lipschitz-continuously differentiable if X and Y are Hilbert spaces with $J_X = \tau^{-1}N_X$ and $J_Y = \sigma^{-1}N_Y$, and K is Lipschitz-continuously differentiable.

Convergence of Gaps in the Convex-Concave Setting

We finish this section by studying the convergence of gap functionals in the convex-concave setting.

Lemma 7. *Suppose F and G_* are convex, proper, and lower semicontinuous and $K \in C^1(X \times Y)$ is convex-concave on $\operatorname{dom} F \times \operatorname{dom} G_*$. Then (C²) holds for all $\bar{u} \in X \times Y$ with $\Omega_{\bar{u}} = X \times Y$ and $\mathcal{G} = \mathcal{G}^{\mathcal{L}}$ the Lagrangian gap*

$$\mathcal{G}^{\mathcal{L}}(u, \bar{u}) := \mathcal{L}(x, \bar{y}) - \mathcal{L}(\bar{x}, y)$$
$$= [F(x) + K(x, \bar{y}) - G_*(\bar{y})] - [F(\bar{x}) + K(\hat{x}, y) - G_*(y)].$$

This functional is non-negative if $\bar{u} \in H^{-1}(0)$.
  Moreover, if $\sum_{k=0}^{N-1} \mathcal{G}^{\mathcal{L}}(u^{k+1}, \bar{u}) \leq M(\bar{u})$ for some $M(\bar{u}) \geq 0$, for all $\bar{u} \in X \times Y$ and all $N \in \mathbb{N}$, and we define the ergodic sequence $\tilde{u}^N := \frac{1}{N} \sum_{k=0}^{N-1} u^{k+1}$, then

(i) $0 \leq \frac{1}{N} \sum_{k=0}^{N-1} \mathcal{G}^{\mathcal{L}}(u^{k+1}, \hat{u}) \to 0$ *at the rate $O(1/N)$ for $\hat{u} \in H^{-1}(0)$.*
(ii) $0 \leq \mathcal{G}^{\mathcal{L}}(\tilde{u}^N, \hat{u}) \to 0$ *at the rate $O(1/N)$ for $\hat{u} \in H^{-1}(0)$.*
(iii) *If $M \in C(X \times Y)$ and $\Omega \subset X \times Y$ is bounded with $\Omega \cap H^{-1}(0) \neq \emptyset$, then $0 \leq \mathcal{G}_\Omega(\tilde{u}^N) \to 0$ at the rate $O(1/N)$ for the partial gap $\mathcal{G}_\Omega(u) := \sup_{\bar{u} \in \Omega} \mathcal{G}^{\mathcal{L}}(u, \bar{u})$.*

The convergence results in Lemma 7 are ergodic because they apply to sequences of running averages. To understand the partial gap, we recall that with $K(x, y) = \langle Ax|y \rangle$ bilinear Fenchel–Rockafellar's theorem shows that the duality gap $\mathcal{G}^D(u) := [F(x) + G_*(Ax)] + [F_*(-A^*y) + G^*_*(y)] \geq 0$ and is zero *if and only if* $u \in H^{-1}(0)$. The duality gap can be written $\mathcal{G}^D(u) = \mathcal{G}_{X \times Y}(u)$.

Proof. By the convex-concavity of K and the definition of the subdifferential

$$\langle D_x K(x, y) | x - \bar{x} \rangle - \langle D_y K(x, y) | y - \bar{y} \rangle$$
$$\geq [K(x, y) - K(\bar{x}, y)] - [K(x, y) - K(x, \bar{y})] = K(x, \bar{y}) - K(\bar{x}, y).$$

for all $(x, y) \in X \times Y$. Also using $x^* \in \partial F(x^{k+1})$ and $y^* \in \partial G_*(y^{k+1})$ with the definition of the convex subdifferential, we see that $\mathcal{G} = \mathcal{G}^{\mathcal{L}}$ satisfies (C²). The non-negativity of $\mathcal{G}(\,\cdot\,, \hat{u})$ follows by similar reasoning, first using that

$$K(x, \hat{y}) - K(\hat{x}, y) \geq \langle D_x K(\hat{x}, \hat{y}) | x - \hat{x} \rangle - \langle D_y K(\hat{x}, \hat{y}) | y - \hat{y} \rangle \qquad (34)$$

for all $(x, y) \in X \times Y$ and following by the definition of the subdifferential applied to $-D_x K(\hat{x}, \hat{y}) \in \partial F(\hat{x})$ and $D_y K(\hat{x}, \hat{y}) \in \partial G_*(\hat{y})$.
  For (i)–(iii), we first observe that the semi-ellipticity of B^0 and (C²) imply $\sum_{k=0}^{N-1} \mathcal{G}^{\mathcal{L}}(u^{k+1}, \bar{u}) \leq M(\bar{u})$. Dividing by N and using that $\mathcal{G}^{\mathcal{L}}(u^{k+1}, \hat{u}) \geq 0$ for $\bar{u} \in H^{-1}(0)$, we obtain (i). Jensen's inequality then gives $\mathcal{G}^{\mathcal{L}}(\tilde{u}^{k+1}, \bar{u}) \leq M(\bar{u})/N$,

hence (ii) for $\bar{u} \in H^{-1}(0)$. Finally, taking the supremum over $\bar{u} \in \Omega$ gives (iii) because M is bounded on bounded sets. □

In the following theorem, we may in particular take $K(x, y) = \langle Ax|y\rangle$ bilinear or $K(x, y) = \langle Ax|y\rangle + E(x)$ with E convex. Lemma 2 and Examples 4 and 5 provide step length conditions that ensure the semi-ellipticity required of B^0 in Theorem 5.

Theorem 5 (Gap convergence). *Let $F : X \to \overline{\mathbb{R}}$ and $G_* : Y \to \overline{\mathbb{R}}$ be convex, proper, and lower semicontinuous. Also let $K \in C^1(X \times Y)$ be convex-concave within $\mathrm{dom}\, F \times \mathrm{dom}\, G_*$. Finally, let $J_X \in C^1(X)$ and $J_Y \in C^1(Y)$ convex. If B^0 is semi-elliptic, then the iterates $\{u^{k+1}\}_{k \in \mathbb{N}}$ generated by the PDBS (16) for any initial $u^0 \in X \times Y$ satisfies Lemma 7 (i)–(iii).*

Proof. By Lemma 7, holds with $\mathcal{G} = \mathcal{G}^{\mathcal{L}}$. Hence by Lemma 3, (D) holds. Since B^0 is semi-elliptic, this implies that $\sum_{k=0}^{N-1} \mathcal{G}(u^{k+1}, \bar{u}) \leq M(\bar{u}) := B^0(\bar{u}, u^0)$ for all $N \in \mathbb{N}$. Since J_X, J_Y, and K are continuously differentiable, $M \in C^1(X \times Y)$. The rest follows from the second part of Lemma 7. □

Inertial Terms

We now generalize (BP), making the involved Bregman divergences dependent on the iteration k and earlier iterates

$$0 \in H(u^{k+1}) + D_1 B_{k+1}(u^{k+1}, u^k) + D_1 B^-_{k+1}(u^k, u^{k-1}), \tag{IPP}$$

for $B_{k+1} := B_{J_{k+1}}$ and $B^-_{k+1} := B_{J^-_{k+1}}$ generated by $J_{k+1}, J^-_{k+1} : U \to \mathbb{R}$. We take $u^{-1} := u^0$ for this to be meaningful for $k = 0$. Our main reason for introducing the dependence on u^{k-1} is to improve (16) and (17) to be explicit in K when K is not affine in y: Otherwise the dual step of those methods is in general not practical to compute unlike the affine case of Remark 1. Along the way we also construct a more conventional inertial method.

A Generalization of the Fundamental Theorem

We realign indices to get a simple fundamental condition to verify on each iteration.

Theorem 6. *On a Banach space U, let $H : U \rightrightarrows U^*$, and let $J_k, J^-_k : U \to \overline{\mathbb{R}}$ be Gâteaux-differentiable with the corresponding Bregman divergences $B_k := B_{J_k}$ and $B^-_k := B_{J^-_k}$ for all $k = 1, \ldots, N$. Suppose (IPP) is solvable for $\{u^{k+1}\}_{k \in \mathbb{N}}$ given an initial iterate $u^0 \in U$. If for all $k = 0, \ldots, N-1$, for some $\bar{u} \in U$ and $\mathcal{G}(u^{k+1}, \bar{u}) \in \mathbb{R}$, for all $h^{k+1} \in H(u^{k+1})$ the modified fundamental condition*

$$\langle h^{k+1}|u^{k+1} - \bar{u}\rangle \geq [(B_{k+2} + B^-_{k+3}) - (B_{k+1} + B^-_{k+2})](\bar{u}, u^{k+1}) + \mathcal{G}(u^{k+1}, \bar{u}) \tag{IC}$$

holds, and B_{k+1}^- satisfies the general Cauchy inequality

$$\langle D_1 B_{k+1}^-(u^k, u) | u^k - u' \rangle \le B_{k+1}'(u^k, u) + B_{k+1}''(u', u^k) \quad (u, u' \in X) \quad (35)$$

for some $B_{k+1}', B_{k+1}'' : U \times U \to \mathbb{R}$, then we have the modified descent inequality

$$[B_{N+1} + B_{N+2}^- - B_{N+1}''](\bar u, u^N) + \sum_{k=0}^{N-1} [B_{k+1} + B_{k+2}^- - B_{k+1}'' - B_{k+2}'](u^{k+1}, u^k)$$

$$+ \sum_{k=0}^{N-1} \mathcal{G}(u^{k+1}, \bar u) \le [B_1 + B_2^-](\bar u, u^0). \quad \text{(ID)}$$

Proof. We can write (IPP) as

$$0 = h^{k+1} + D_1 B_{k+1}(u^{k+1}, u^k) + D_1 B_{k+1}^-(u^k, u^{k-1}) \text{ for some } h^{k+1} \in H(u^{k+1}). \quad (36)$$

Testing (IPP) by applying $\langle \cdot | u^{k+1} - \bar u \rangle$, we obtain

$$0 = \langle h^{k+1} + D_1 B_{k+1}(u^{k+1}, u^k) + D_1 B_{k+1}^-(u^k, u^{k-1}) | u^{k+1} - \bar u \rangle.$$

Summing over $k = 0, \ldots, N-1$ and using $u^{-1} = u^0$ to eliminate $B_1^-(u^0, u^{-1}) = 0$, we rearrange

$$0 = S_N + \sum_{k=0}^{N-1} \langle h^{k+1} + D_1 [B_{k+1} + B_{k+2}^-](u^{k+1}, u^k) | u^{k+1} - \bar u \rangle \quad (37)$$

for

$$S_N := \langle D_1 B_{J_{N+1}^-}(u^N, u^{N-1}) | \bar u - u^N \rangle + \sum_{k=0}^{N-1} \langle D_1 B_{J_{k+1}^-}(u^k, u^{k-1}) | u^{k+1} - u^k \rangle.$$

Abbreviating $\bar B_{k+1} := B_{k+1} + B_{k+2}^-$ and using (IC) and the three-point identity (8) in (37), we obtain

$$0 \ge S_N + \sum_{k=0}^{N-1} \left(\bar B_{k+2}(\bar u, u^{k+1}) - \bar B_{k+1}(\bar u, u^k) + \bar B_{k+1}(u^{k+1}, u^k) + \mathcal{G}(u^{k+1}, \bar u) \right).$$

Using the generalized Cauchy inequality (35) and, again, that $u^{-1} = u^0$, we get

$$S_N \geq -B'_{N+1}(u^N, u^{N-1}) - B''_{N+1}(\bar{u}, u^N) - \sum_{k=0}^{N-1} \left(B'_{k+1}(u^k, u^{k-1}) + B''_{k+1}(u^{k+1}, u^k) \right)$$

$$= -B''_{N+1}(\bar{u}, u^N) - \sum_{k=0}^{N-1} [B''_{k+1} + B'_{k+2}](u^{k+1}, u^k).$$

These two inequalities yield (ID). □

Inertia (Almost) as Usually Understood

We take $J_{k+1} = J^0$ and $J^-_{k+1} = -\lambda_k J^0$ for some $\lambda_k \in \mathbb{R}$. We then expand (IPP) as

Inertial PDBS

Iteratively over $k \in \mathbb{N}$, solve for x^{k+1} and y^{k+1}:

$$(1 + \lambda_k)[DJ_X(x^k) - D_x K(x^k, y^k)] - \lambda_k[DJ_X(x^{k-1}) - D_x K(x^{k-1}, y^{k-1})]$$
$$\in DJ_X(x^{k+1}) + \partial F(x^{k+1}),$$
$$(1 + \lambda_k)[DJ_Y(y^k) - D_y K(x^k, y^k)] - \lambda_k[DJ_Y(y^{k-1}) - D_y K(x^{k-1}, y^{k-1})]$$
$$\in DJ_Y(y^{k+1}) + \partial G_*(y^{k+1}) - 2D_y K(x^{k+1}, y^{k+1})$$
(38)

If X and Y are Hilbert spaces with $J_X = \tau^{-1} N_X$ and $J_Y = \sigma^{-1} N_Y$, the standard generating functions divided by some step length parameters $\tau, \sigma > 0$, and $K(x, y) = \langle Ax|y \rangle$ for $A \in \mathbb{L}(X; Y)$, (38) reduces to the inertial method of Chambolle and Pock (2015):

Inertial PDPS for bilinear K

With initial $\tilde{x}^0 = x^0$ and $\tilde{y}^0 = y^0$, iterate over $k \in \mathbb{N}$:

$$\begin{aligned}
x^{k+1} &:= \text{prox}_{\tau F}(\tilde{x}^k - \tau A^* \tilde{y}^k), \\
y^{k+1} &:= \text{prox}_{\sigma G_*}(\tilde{y}^k + \sigma A(2x^{k+1} - \tilde{x}^k)), \\
\tilde{x}^{k+1} &:= (1 + \lambda_{k+1})x^{k+1} - \lambda_{k+1} x^k, \\
\tilde{y}^{k+1} &:= (1 + \lambda_{k+1})y^{k+1} - \lambda_{k+1} y^k.
\end{aligned}$$
(39)

More generally, however, (38) does not directly apply inertia to the iterates. It applies inertia to K.

The general Cauchy inequality (35) automatically holds by the three-point identity (8) with $J''_{k+1} = J'_{k+1} = J^-_{k+1}$ if $B^-_{k+1} \geq 0$, which is to say that J^-_{k+1} is convex. This is the case if $\lambda_k \leq 0$. For usual inertia we, however, want $\lambda_k > 0$. We will therefore use Lemma 1, requiring:

Assumption 2. For some $\beta > 0$, in a domain $\Omega \subset X \times Y$

$$|\langle D_1 B^0(u^k, u) | u^k - u \rangle| \leq B^0(u^k, u) + \beta B^0(u', u^k) \quad (u, u', u^k \in \Omega). \tag{40}$$

Moreover, the parameters $\{\lambda_k\}_{k \in \mathbb{N}}$ are non-increasing and for some $\epsilon > 0$

$$0 \leq \lambda_{k+1} \leq \frac{1 - \epsilon - \lambda_k \beta}{2} \quad (k \in \mathbb{N}). \tag{41}$$

Example 23. Suppose the generating function J^0 is γ-strongly subdifferentiable (i.e., B^0 is γ-elliptic, see sections "Ellipticity of the Bregman Divergences" and "Ellipticity for Block-Adapted Methods") within $\Omega \subset X \times Y$ and satisfies the subdifferential smoothness property (10) with the factor $L > 0$. Then by Lemma 1, (40) holds with $\beta = L\gamma^{-1}$ in some domain $\Omega \subset X \times Y$.

As a particular case, let X and Y be Hilbert spaces with the standard generating functions $J_X = \tau^{-1} N_X$ and $J_Y = \sigma^{-1} N_Y$. Also let DK be L_{DK}-Lipschitz within Ω. Then J^0 is Lipschitz with factor $L = \max\{\sigma^{-1}, \tau^{-1}\} + L_{DK}$. Consequently the required subdifferential smoothness property (10) holds with the same factor L; see Bauschke and Combettes (2017, Theorem 18.15) or Valkonen (2020, Appendix C).

We computed L_{DK} for some specific K in section "Ellipticity of the Bregman Divergences."

Example 24. If $K(x, y) = \langle Ax | y \rangle$ with $A \in \mathbb{L}(X; Y^*)$, and if $J_X = \tau^{-1} N_X$, $J_Y = \sigma^{-1} N_Y$, in Hilbert spaces X and Y, then $B^0(u', u) = \frac{1}{2\tau}\|x - x'\|^2 + \frac{1}{2\sigma}\|y - y'\|^2 + \langle A(x - x') | y - y' \rangle$. By standard Cauchy inequality, (40) holds for $\beta = 1$ in $\Omega = X \times Y$. Consequently the next example recovers the upper bound for λ in Chambolle and Pock (2015):

Example 25. The bound (41) holds for some $\epsilon > 0$ if $\lambda_k \equiv \lambda$ for $0 \leq \lambda < 1/(2+\beta)$.

Lemma 8. *Suppose Assumption 2 holds and that* (C2) *holds within* $\Omega_{\bar{u}}$ *for some* $\bar{u} \in \Omega$ *and* $\mathcal{G}(u, \bar{u})$. *Given* $u^0 \in \Omega$, *suppose the iterates generated by the inertial PDBS* (38) *satisfy* $\{u^k\}_{k=0}^N \subset \Omega_{\bar{u}} \cap \Omega$. *Then*

$$\epsilon B^0(\bar{u}, u^N) + \epsilon \sum_{k=0}^{N-1} B^0(u^{k+1}, u^k) + \sum_{k=0}^{N-1} \mathcal{G}(u^{k+1}, \bar{u}) \leq (1 - \lambda_1) B^0(\bar{u}, u^0). \tag{42}$$

Proof. Since $B_{k+1} = B^0$ and $B^-_{k+1} = -\lambda_k B^0$ for all $k \in \mathbb{N}$,

$$(B_{k+2} + B^-_{k+3}) - (B_{k+1} + B^-_{k+2}) = (\lambda_{k+1} - \lambda_{k+2})B^0.$$

Since λ_k is decreasing and B^0 is semi-elliptic within $\Omega \supset \{u^k, \bar{u}\}$, we deduce that $(\lambda_{k+1} - \lambda_{k+2})B^0(\bar{u}, u^k) \geq 0$. Consequently (IC) holds if (C) does. By the proof of Lemma 3, (IC) then holds if (C²) does. Using (40), (35) holds with $B'_{k+1} = \lambda_k B_0$ and $B''_{k+1} = \lambda_k \beta B_0$. Referring to Theorem 6, we now obtain (ID). We expand

$$[B_{N+1} + B^-_{N+2} - B''_{N+1}](\bar{u}, u^N) = (1 - \lambda_{k+1} - \lambda_k \beta) B^0(\bar{u}, u^N) \quad \text{and}$$
$$[B_{k+1} + B^-_{k+2} - B''_{k+1} - B'_{k+2}](u^{k+1}, u^k) = (1 - \lambda_{k+1} - \lambda_k\beta - \lambda_{k+1})B^0(u^{k+1}, u^k).$$

Since $\bar{u}, u^k \in \Omega$ for all $k = 0, \ldots, N$, using the ellipticity of B^0 within Ω as well as (41), we now estimate the first from below by $\epsilon B^0(\bar{u}, u^N)$ and the second by $\epsilon B^0(u^{k+1}, u^k)$. Thus (ID) produces (42). □

We may now proceed as in sections "Convergence of Gaps in the Convex–Concave Setting" and "Convergence of Iterates" to prove convergence. For the verification of Assumption 2, we can use Examples 23, 24, and 25.

Theorem 7 (Convergence, inertial method). *Theorems 2, 3, and 5 apply to the iterates $\{u^{k+1}\}_{k \in \mathbb{N}}$ generated by the inertial PDBS (38) if we replace the assumptions of (semi-)ellipticity of B^0 with Assumption 2.*

Proof. We replace Lemma 3 and (D) by Lemma 8 and (42) in the proofs of Theorems 2, 3, and 5. Observe that Assumption 2 implies that B^0 is (semi-)elliptic. □

Remark 7. The inertial PDPS is improved in Valkonen (2020) to yield *non-ergodic* convergence of the Lagrangian gap. To do the "inertial unrolling" that leads to such estimates, one, however, needs to correct for the anti-symmetry introduced by K into H.

Remark 8. Since Theorem 6 does not provide the quantitative Δ-Féjer monotonicity used in Theorem 4, we cannot prove linear convergence using our present simplified "testing" approach lacking the "testing parameters" of Valkonen (2020).

Improvements to the Basic Method Without Dual Affinity

We now have the tools to improve the basic PDBS (16) to enjoy prox-simple steps for general K not affine in y. Compared to (14) we amend $J_{k+1} = J^0$ by taking

$$J_{k+1}(x, y) := J_X(x) + J_Y(y) - K(x, y) + 2K(x^{k+1}, y) \qquad (43)$$
$$= J^0(x, y) + 2K(x^{k+1}, y).$$

This would be enough for K to be explicit in the algorithm; however, proofs of convergence would practically require G_* to be strongly convex even in the convex-concave case. To fix this, we introduce the inertial term generated by

$$J_{k+1}^-(u) := [J^0 - J_k](u) = -2K(x^k, y). \qquad (44)$$

As always, we write B_{k+1}, B^0, and B_{k+1}^- for the Bregman divergences generated by J_{k+1}, J^0, and J_{k+1}^-.
Since

$$D_1[B_{k+1} - B^0](u^k, u^{k-1}) + D_1 B_{k+1}^-(u^k, u^{k-1}) = (0, \tilde{y}_{k+1}^*)$$

for

$$\tilde{y}_{k+1}^* = 2[D_y K(x^{k+1}, y^{k+1}) - D_y K(x^{k+1}, y^k) - D_y K(x^k, y^k) + D_y K(x^k, y^{k-1})],$$

the algorithm (IPP) expands similarly to (16) as the

Modified PDBS

Iteratively over $k \in \mathbb{N}$, solve for x^{k+1} and y^{k+1}:

$$DJ_X(x^k) - D_x K(x^k, y^k) \in DJ_X(x^{k+1}) + \partial F(x^{k+1}) \quad \text{and}$$
$$DJ_Y(y^k) + [2D_y K(x^{k+1}, y^k) + D_y K(x^k, y^k) - 2D_y(x^k, y^{k-1})] \qquad (45)$$
$$\in DJ_Y(y^{k+1}) + \partial G_*(y^{k+1}).$$

The method reduces to the basic PDBS (16) when K is affine in y. In Hilbert spaces X and Y with $J_X = \tau^{-1} N_X$ and $J_Y = \sigma^{-1} N_Y$, we can rearrange (45) as

Modified PDPS

Iterate over $k \in \mathbb{N}$:

$$x^{k+1} := \operatorname{prox}_{\tau F}(x^k - \tau \nabla_x K(x^k, y^k)),$$
$$y^{k+1} := \operatorname{prox}_{\sigma G_*}(y^k + \sigma[2\nabla_y K(x^{k+1}, y^k) + \nabla_y K(x^k, y^k) - 2\nabla_y K(x^k, y^{k-1})]).$$
$$(46)$$

18 First-Order Primal–Dual Methods for Nonsmooth Non-convex Optimization

Remark 9. The modified PDPS (46) is slightly more complicated than the method in Clason et al. (2020), which would update

$$y^{k+1} := \text{prox}_{\sigma G_*}(y^k + \sigma \nabla_y K(2x^{k+1} - x^k, y^k)).$$

Likewise, (45) is different from the algorithm presented in Hamedani and Aybat (2018) for convex-concave K. It would, for the standard generating functions, update[6]

$$y^{k+1} := \text{prox}_{\sigma G_*}(y^k + \sigma[2\nabla_y K(x^{k+1}, y^k) - \nabla_y K(x^k, y^{k-1})]).$$

We could produce this method by taking $J_{k+1}^-(u) = -K(x^k, y)$. However, the convergence proofs would require some additional steps.

The main difference to the overall analysis of section "Convergence Theory" is in bounding from below the Bregman divergences in (ID). We now have

$$B_{N+1} + B_{N+2}^- - B_{N+1}'' = B^0 - B_{N+1}'' \quad \text{and} \tag{47a}$$

$$B_{k+1} + B_{k+2}^- - B_{k+1}'' - B_{k+2}' = B^0 - B_{k+1}'' - B_{k+2}'. \tag{47b}$$

If $D_y K(x^k, \cdot)$ is $L_{DK,y}$-Lipschitz

$$\langle D_1 B_{k+1}^-(u^k, u) | u^k - u' \rangle = 2\langle D_y K(x^k, y^k) - D_y K(x^k, y) | y^k - y' \rangle \tag{48}$$

$$\leq \sqrt{L_{DK,y}} \|y - y^k\|^2 + \sqrt{L_{DK,y}} \|y' - y^k\|^2$$

$$=: B_{k+1}'(u^k, u) + B_{k+1}''(u', u^k).$$

Therefore, for the modified descent inequality (ID) to be meaningful, we require:

Assumption 3. We assume that $\|D_y K(x, y) - D_y K(x, y')\| \leq L_{DK,y} \|y - y'\|$ when $(x, y), (x, y') \in \Omega$ for some domain $\Omega \subset X \times Y$. Moreover, for some $\epsilon \geq 0$, we have

$$B^0(u, u') \geq \frac{\epsilon}{2} \|u - u'\|_{X \times Y}^2 + 2\sqrt{L_{DK,y}} \|y - y'\|_Y^2 \quad (u, u' \in \Omega). \tag{49}$$

We say that the present assumption holds *strongly* if $\epsilon > 0$.

---

[6]Note that Hamedani and Aybat (2018) uses the historical ordering of the primal and dual updates from Chambolle and Pock (2011), prior to the proof-simplifying discovery of the proximal point formulation in He and Yuan (2012). Hence our y^k is their y^{k+1}.

Example 26. If K is affine in y, $L_{DK,y} = 0$. Therefore, Assumption 3 reduces to the (semi-)ellipticity of B^0, which can be verified as in sections "Ellipticity of the Bregman Divergences" and "Ellipticity for Block-Adapted Methods."

Example 27. Generally, it is easy to see that if one of the results of section "Ellipticity of the Bregman Divergences" holds with $\tilde{\sigma} = 1/(\sigma^{-1} - 4\sqrt{L_{DK,y}}) > 0$ in place of σ, then (49) holds. In particular, if K has L_{DK}-Lipschitz derivative within Ω, then Lemma 2 gives the condition $1 \geq L_{DK} \max\{\tau, \sigma/(1 - 4\sigma\sqrt{L_{DK,y}})\}$ and $1 > 4\sigma\sqrt{L_{DK,y}}$ for (49) to hold with $\epsilon = 0$. The assumption holds strongly if the first inequality is strict.

Similarly to Lemma 8, we now have the following replacement for Lemma 3:

Lemma 9. *Suppose Assumption 3 holds and* (C²) *holds within* $\Omega_{\bar{u}}$ *for some* $\bar{u} \in X \times Y$ *and* $\mathcal{G}(u, \bar{u})$. *Given* $u^0 \in X \times Y$, *suppose the iterates generated by the modified PDBS* (45) *satisfy* $\{u^k\}_{k=0}^N \subset \Omega_{\bar{u}}$. *Then*

$$\epsilon B^0(\bar{u}, u^N) + \epsilon \sum_{k=0}^{N-1} B^0(u^{k+1}, u^k) + \sum_{k=0}^{N-1} \mathcal{G}(u^{k+1}, \bar{u}) \leq [B_1 + B_2^-](\bar{u}, u^0). \quad (50)$$

Proof. Inserting (43) and (44), (IC) reduces to (C), which follows from (C²) as in Lemma 3. We verify (35) via (48) and Assumption 3. Thus Theorem 6 proves (ID). Inserting (47) and (49) with B'_{k+1} and B''_{k+1} from (48) into (ID) proves (50). □

We may now proceed as in sections "Convergence of Gaps in the Convex–Concave Setting" and "Convergence of Iterates" to prove convergence. For the verification of Assumption 3, we can use Examples 26 and 27.

Theorem 8 (Convergence, modified method). *Theorems 2, 3, and 5 apply to the iterates* $\{u^{k+1}\}_{k \in \mathbb{N}}$ *generated by the modified PDBS* (45) *if we replace the assumptions of semi-ellipticity (resp. ellipticity) of* B^0 *with Assumption 3 holding (strongly).*

Proof. We replace Lemma 3 and (D) by Lemma 9 and (50) in Theorems 2, 3, and 5. Observe that (strong) Assumption 3 implies the (semi-)ellipticity of B^0. □

Now we have a locally convergent method (46) with easily implementable steps to tackle problems such as Potts segmentation (4) (Clason et al. 2020).

Further Directions

We close by briefly reviewing some things not covered, other possible extensions, and alternative algorithms.

Acceleration

To avoid technical detail, we did not cover $O(1/N^2)$ acceleration. The fundamental ingredients of proof are, however, exactly the same as we have used: sufficient second-order growth and ellipticity of the Bregman divergences B_k^0, which are now iteration-dependent. Additionally, a portion of the second-order growth must be used to make the metrics B_k^0 grow as $k \to \infty$. For bilinear K in Hilbert spaces, such an argument can be found in Valkonen (2020); for $K(x, y) = \langle A(x) | y \rangle$ in Clason et al. (2019); and for general K in Clason et al. (2020). As mentioned in Remarks 1 and 9, the algorithms in the latter two differ slightly from the ones presented here.

Stochastic Methods

It is possible to refine the block-adapted (18) and its accelerated version into stochastic methods. The idea is to take on each step subsets of primal-blocks $S(i) \subset \{1, \ldots, m\}$ and dual blocks $V(i+1) \subset \{1, \ldots, n\}$ and to only update the corresponding x_j^{k+1} and y_ℓ^{k+1}. Full discussion of such technical algorithms is outside the scope of our present overview. We refer to Valkonen (2019) for an approach covering block-adapted acceleration and both primal and dual randomization in the case of bilinear K, but see also Chambolle et al. (2018) for a more basic version. For more general K affine in y, see Mazurenko et al. (2020).

Alternative Bregman Divergences

We have used Bregman divergences as a proof tool, in the end opting for the standard quadratic generating functions on Hilbert spaces. Nevertheless, our theory works for arbitrary Bregman divergences. The practical question is whether F and G_* remain prox-simple with respect to such a divergence. This can be the case for the "entropic distance" generated on $L^1(\Omega; [0, \infty))$ by

$$J(x) := \begin{cases} \int_\Omega x(t) \ln x(t) \, dt, & x \geq 0 \text{ a.e. on } \Omega, \\ \infty, & \text{otherwise} \end{cases}$$

See, for example, Burger et al. (2019) for a Landweber method (gradient descent on regularized least squares) based on such a distance.

Alternative Approaches

The derivative $D_1 B^0$ in (15) can be seen as a preconditioner, replacing $\tau(u - u')$ in the proximal point method (13). Our choice of B^0 is not the only option.

Consider the problem

$$\min_{x \in X} F(x) + E(x). \tag{51}$$

Provided E is differentiable and F prox-simple, i.e., the proximal map of F has a closed-form expression, (1) can be solved by forward–backward splitting methods as first introduced in Lions and Mercier (1979). In a Hilbert space X, this can be written

$$x^{k+1} := \mathrm{prox}_{\tau F}(x^k - \tau \nabla E(x^k)). \tag{52}$$

Variants based on Bregman divergences were introduced in Nemirovski and Yudin (1983) under the name "mirror prox" or "mirror descent"; see also the review Chambolle and Pock (2016). The method and convergence proofs for it can be derived from our primal–dual approach. Indeed, if we take $G_* \equiv \delta_{\{0\}}$ as the indicator function of zero, and $K(x, y) = E(x)$ for some $E \in C^1(X)$, then (S) is equivalent to (51). Now the dual step of (17) is $y^{k+1} := 0$, and the primal step is (52).

Forward–backward splitting is especially popular under the name iterative soft thresholding (ISTA) in the context of sparse reconstruction (i.e., regularization of linear inverse problems with ℓ^1 penalties), see, e.g., Chambolle et al. (1998), Daubechies et al. (2004), and Beck and Teboulle (2009). However, forward–backward splitting has limited applicability in imaging and inverse problems due to the joint prox-simplicity and smoothness requirements. Sometimes these can be circumvented by considering so-called dual problems (Beck and Teboulle 2009).

Let then E be Gâteaux-differentiable and $F = G \circ A$ for a nonsmooth function F and a linear operator A in (51), i.e., consider the problem

$$\min_{x \in X} E(x) + G(Ax),$$

Forward–backward splitting is impractical as $G \circ A$ is in general not prox-simple. Assuming G to have the preconjugate G_*, we can write this problem as an instance of (S) with $F = 0$ and $K(x, y) = E(x) + \langle Ax|y \rangle$. Therefore the methods we have presented are applicable. However, in this instance, also $J^0(u) := \frac{1}{2}\|u\|^2_{X \times Y} + \frac{1}{2}\|A^* y\|^2_{X^*}$ would produce an algorithm with realizable steps. In analogy to the PDPS, it might be called the primal–dual explicit spitting (PDES). The method was introduced in Loris and Verhoeven (2011) for $E(z) = \frac{1}{2}\|b - z\|^2$

as the "generalized iterative soft thresholding" (GIST), but has also been called the primal–dual fixed point method (PDFP, Chen et al. 2013) and the proximal alternating predictor corrector (PAPC, Drori et al. 2015).

The classical Augmented Lagrangian method solves the saddle point problem

$$\min_x \max_y \ F(x) + \frac{\tau}{2}\|E(x)\|^2 + \langle E(x)|y\rangle, \qquad (53)$$

alternatingly for x and y. The alternating directions method of multipliers (ADMM) of Gabay (1983) and Arrow et al. (1958) takes $E(x) = Ax_1 + Bx_2 - c$ and $F(x) = F_1(x_1) + F_2(x_2)$ for $x = (x_1, x_2)$ and alternates between solving (53) for x_1, x_2, and y, using the most recent iterate for the other variables. The method cannot be expressed in our Bregman divergence framework, as the preconditioner $D_1 B_{k+1}(\cdot, x^k)$ would need to be nonsymmetric. The steps of the method are potentially expensive, each itself being an optimization problem. Hence the *preconditioned ADMM* of Zhang et al. (2011), which is equivalent to the PDPS, and the classical Douglas–Rachford splitting (DRS, Douglas and Rachford 1956) are applied to appropriate problems (Chambolle and Pock 2011; Clason and Valkonen 2020). The preconditioned ADMM was extended to nonlinear E in Benning et al. (2016).

Based on derivations avoiding the Lipschitz gradient assumption (cocoercivity) in forward–backward splitting, Malitsky and Tam (2018) moves the over-relaxation step $\bar{x}^{k+1} := 2x^{k+1} - x^k$ of the PDPS outside the proximal operators. This amounts to taking $J^-_{k+1} = \lambda_k K$ in section "Inertia (Almost) as Usually Understood" instead of $J^-_{k+1}(x, y) = \lambda_k J^0 = \lambda_k[\tau^{-1} J_X(x) + \sigma^{-1} J_Y(y) - K(x, y)]$, so is "partial inertia"; compare the "corrected inertia" of Valkonen (2020).

An over-relaxed variant of the same idea may be found in Bredies and Sun (2015). We have not discussed over-relaxation of entire algorithms. To briefly relate it to the basic inertia of (39), the latter "rebases" the algorithm at the inertial iterate $\tilde{u}^k$ constructed from u^k and u^{k-1}, whereas over-relaxation would construct $\tilde{u}^k$ from u^k and $\tilde{u}^{k-1}$. The derivation in Bredies and Sun (2015) is based on applying Douglas–Rachford splitting on a lifted problem. The basic over-relaxation of the PDPS is known as the Condat–Vũ method (Condat 2013; Vũ 2013).

Functions on Manifolds and Hadamard Spaces

The PDPS has been extended in Begmann et al. (2019) to functions on Riemannian manifolds: the problem $\min_{x \in \mathcal{M}} F(x) + G(Ex)$, where $E : \mathcal{M} \to \mathcal{N}$ with $\mathcal{M}$ and $\mathcal{N}$ Riemannian manifolds. In general, between manifolds, there are no linear maps, so E is nonlinear. Indeed, besides introducing a theory of conjugacy for functions on manifolds, the algorithm presented in Begmann et al. (2019) is based on the NL-PDPS of Valkonen (2014); Clason et al. (2019).

Convergence could only be proved on Hadamard manifolds, which are special: a type of three-point inequality holds (do Carmo 2013, Lemma 12.3.1). Indeed, in even more general Hadamard spaces with the metric d, for any three points $x^{k+1}, x^k, \bar{x}$, we have (Bačák 2014, Corollary 1.2.5)

$$\frac{1}{2}d(x^k, x^{k+1})^2 + \frac{1}{2}d(x^{k+1}, \bar{x})^2 - \frac{1}{2}d(x^k, \bar{x})^2 \leq d(x^k, x^{k+1})d(\bar{x}, x^{k+1}). \quad (54)$$

Therefore, given a function f on such a space, to derive a simple proximal point algorithm, having constructed the iterate x^k, we might try to find x^{k+1} such that

$$f(x^{k+1}) + d(x^k, x^{k+1}) \leq f(x^k).$$

Multiplying this inequality by $d(\bar{x}, x^{k+1})$ and using the three-point inequality (54)

$$\frac{1}{2}d(x^k, x^{k+1})^2 + \frac{1}{2}d(x^{k+1}, \bar{x})^2 + [f(x^{k+1}) - f(x^k)]d(\bar{x}, x^{k+1}) \leq \frac{1}{2}d(x^k, \bar{x})^2.$$

If the space is bounded, $d(\bar{x}, x^{k+1}) \leq C$, so since $f(x^k) \geq f(x^{k+1})$, we may telescope and proceed as before to obtain convergence.

The Hadamard assumption is restrictive: if a Banach space is Hadamard, it is Hilbert, while a Riemannian manifold is Hadamard if it is simply connected with a non-positive sectional curvature (Bačák 2014, Section 1.2).

Glossary

The extended reals	We define $\overline{\mathbb{R}} := [-\infty, \infty]$.
A convex function	A function $F : X \to \overline{\mathbb{R}}$ is convex if for all $x, x' \in X$ and $\lambda \in (0, 1)$, we have $$F(\lambda x + (1-\lambda)x') \leq F(\lambda x) + F((1-\lambda)x').$$
A concave function	A function $F : X \to \overline{\mathbb{R}}$ is concave if $-f$ is convex.
A convex-concave function	A function $K : X \times Y \to \overline{\mathbb{R}}$ is convex-concave if $K(\cdot, y)$ is convex for all $y \in Y$, and $K(x, \cdot)$ is concave for all $x \in X$.
The dual space	We write X^* for the dual space of a topological vector (Banach, Hilbert) space X.
Set-valued map	We write $A : X \rightrightarrows Y$ if A is a set-valued map between the spaces X and Y.
Derivative	We write $DF : X \to X^*$ for the derivative of a Gâteaux-differentiable function $F : X \to \mathbb{R}$.
Convex subdifferential	This is the map $\partial F : X \rightrightarrows X^*$ for a convex $F : X \to \overline{\mathbb{R}}$. By definition $x^* \in \partial F(x)$ at $x \in X$ if and only if $$F(x') - F(x) \geq \langle x^* | x' - x \rangle \quad (x' \in X).$$

Fenchel conjugate	This is the function $f^*: X^* \to \overline{\mathbb{R}}$ defined for $F: X \to \overline{\mathbb{R}}$ by
$$f^*(x^*) := \sup_{x \in X} \langle x^*	x \rangle - F(x) \quad (x^* \in X^*).$$
Fenchel preconjugate	If $X = (X_*)^*$ is the dual space of some space X_* and $F: X \to \overline{\mathbb{R}}$, then $f_*: X_* \to \overline{\mathbb{R}}$ is the preconjugate of f if $f = (f_*)^*$.
Proximal map	For a function $F: X \to \overline{\mathbb{R}}$, this can be defined as
$$\operatorname{prox}_F(x) := \arg\min_{\tilde{x} \in X} \left(F(\tilde{x}) + \frac{1}{2}\|\tilde{x} - x\|_X^2 \right).$$
Distributional derivative	It arises from integration by parts: If $u: \mathbb{R}^n \supset \Omega \to \mathbb{R}$ is differentiable and $\phi \in C_c^\infty(\Omega; \mathbb{R}^n)$, then
$$\int_\Omega \langle \nabla u, \varphi \rangle \, dx = -\int_\Omega u \operatorname{div} \varphi \, dx.$$
If now u is not differentiable, we *define* the distribution $D \in C_c^\infty(\Omega; \mathbb{R}^n)^*$ by
$$Du(\varphi) := -\int_\Omega u \operatorname{div} \varphi \, dx.$$
If Du is bounded (as a linear operator), it can be presented as a vector Radon measure (Federer 1969), the space denoted $\mathcal{M}(\Omega; \mathbb{R}^n)$.
Indicator function	For a set A, we define
$$\delta_A(x) := \begin{cases} 0, & x \in A, \\ \infty, & x \notin A. \end{cases}$$

References

Ambrosio, L., Fusco, N., Pallara, D.: Functions of Bounded Variation and Free Discontinuity Problems. Oxford University Press (2000)

Arridge, S.R., Kaipio, J.P., Kolehmainen, V., Tarvainen, T.: Optical imaging. In: Scherzer, O. (ed.) Handbook of Mathematical Methods in Imaging, pp. 735–780. Springer, New York (2011). https://doi.org/10.1007/978-0-387-92920-0_17

Arrow, K.J., Hurwicz, L., Uzawa, H.: Studies in Linear and Non-linear Programming. Stanford University Press (1958)

Bačák, M.: Convex Analysis and Optimization in Hadamard Spaces, Nonlinear Analysis and Applications. De Gruyter (2014)

Bauschke, H.H., Combettes, P.L.: Convex Analysis and Monotone Operator Theory in Hilbert Spaces. CMS Books in Mathematics, 2 edition. Springer (2017). https://doi.org/10.1007/978-3-319-48311-5

Beck, A.: First-Order Methods in Optimization. SIAM (2017). https://doi.org/10.1137/1.9781611974997

Beck, A., Teboulle, M.: Fast gradient-based algorithms for constrained total variation image denoising and deblurring problems. IEEE Trans. Image Process. **18**, 2419–2434 (2009). https://doi.org/10.1109/tip.2009.2028250

Beck, A., Teboulle, M.: A fast iterative shrinkage-thresholding algorithm for linear inverse problems. SIAM J. Imaging Sci. **2**, 183–202 (2009). https://doi.org/10.1137/080716542

Begmann, R., Herzog, R., Tenbrick, D., Vidal-Núñez, J.: Fenchel duality for convex optimization and a primal dual algorithm on Riemannian manifolds (2019). arXiv:1908.02022

Benning, M., Knoll, F., Schönlieb, C.B., Valkonen, T.: Preconditioned ADMM with nonlinear operator constraint. In: System Modeling and Optimization: 27th IFIP TC 7 Conference, CSMO 2015, Sophia Antipolis, 29 June–3 July 2015, Revised Selected Papers, pp. 117–126. Springer (2016). https://doi.org/10.1007/978-3-319-55795-3_10. arXiv:1511.00425

Bredies, K., Sun, H.: Preconditioned Douglas–Rachford splitting methods for convex-concave saddle-point problems. SIAM J. Numer. Anal. **53**, 421–444 (2015). https://doi.org/10.1137/140965028

Brezis, H., Crandall, M.G., Pazy, A.: Perturbations of nonlinear maximal monotone sets in Banach space. Commun. Pure Appl. Math. **23**, 123–144 (1970). https://doi.org/10.1002/cpa.3160230107

Browder, F.E.: Convergence theorems for sequences of nonlinear operators in Banach spaces. Mathematische Zeitschrift **100**, 201–225 (1967). https://doi.org/10.1007/bf01109805

Burger, M., Resmerita, E., Benning, M.: An entropic Landweber method for linear ill-posed problems (2019) arXiv:1906.10032

Chambolle, A., DeVore, R.A., Lee, N.Y., Lucier, B.J.: Nonlinear wavelet image processing: variational problems, compression, and noise removal through wavelet shrinkage. IEEE Trans. Image Process. **7**, 319–335 (1998). https://doi.org/10.1109/83.661182

Chambolle, A., Ehrhardt, M., Richtárik, P., Schönlieb, C.: Stochastic primal-dual hybrid gradient algorithm with arbitrary sampling and imaging applications. SIAM J. Optim. **28**, 2783–2808 (2018). https://doi.org/10.1137/17m1134834

Chambolle, A., Pock, T.: A first-order primal-dual algorithm for convex problems with applications to imaging. J. Math. Imaging Vis. **40**, 120–145 (2011). https://doi.org/10.1007/s10851-010-0251-1

Chambolle, A., Pock, T.: On the ergodic convergence rates of a first-order primal–dual algorithm. Math. Program. 1–35 (2015). https://doi.org/10.1007/s10107-015-0957-3

Chambolle, A., Pock, T.: An introduction to continuous optimization for imaging. Acta Numer. **25**, 161–319 (2016). https://doi.org/10.1017/s096249291600009x

Chen, P., Huang, J., Zhang, X.: A primal-dual fixed point algorithm for convex separable minimization with applications to image restoration. Inverse Probl. **29**, 025011 (2013). https://doi.org/10.1088/0266-5611/29/2/025011

Chierchia, G., Chouzenoux, E., Combettes, P.L., Pesquet, J.C.: The Proximity Operator Repository (2019). http://proximity-operator.net. Online resource

Clarke, F.: Optimization and Nonsmooth Analysis. Society for Industrial and Applied Mathematics (1990). https://doi.org/10.1137/1.9781611971309

Clason, C., Mazurenko, S., Valkonen, T.: Acceleration and global convergence of a first-order primal-dual method for nonconvex problems. SIAM J. Optim. **29**, 933–963 (2019). https://doi.org/10.1137/18m1170194. arXiv:1802.03347

Clason, C., Mazurenko, S., Valkonen, T.: Primal-dual proximal splitting and generalized conjugation in nonsmooth nonconvex optimization. Appl. Math. Optim. (2020). https://doi.org/10.1007/s00245-020-09676-1. arXiv:1901.02746

Clason, C., Valkonen, T.: Introduction to Nonsmooth Analysis and Optimization (2020). arXiv:2001.00216. Work in progress

Condat, L.: A primal–dual splitting method for convex optimization involving lipschitzian, proximable and linear composite terms. J. Optim. Theory Appl. **158**, 460–479 (2013). https://doi.org/10.1007/s10957-012-0245-9

Daubechies, I., Defrise, M., De Mol, C.: An iterative thresholding algorithm for linear inverse problems with a sparsity constraint. Commun. Pure Appl. Math. **57**, 1413–1457 (2004). https://doi.org/10.1002/cpa.20042

do Carmo, M.P.: Riemannian Geometry. Mathematics: Theory & Applications. Birkhäuser (2013)

Douglas Jim, J., Rachford, H.H.: On the numerical solution of heat conduction problems in two and three space variables. Trans. Am. Math. Soc. **82**, 421–439 (1956). https://doi.org/10.2307/1993056

Drori, Y., Sabach, S., Teboulle, M.: A simple algorithm for a class of nonsmooth convex-concave saddle-point problems. Oper. Res. Lett. **43**, 209–214 (2015). https://doi.org/10.1016/j.orl.2015.02.001

Ekeland, I., Temam, R.: Convex Analysis and Variational Problems. SIAM (1999)

Federer, H.: Geometric Measure Theory. Springer (1969)

Gabay, D.: Applications of the method of multipliers to variational inequalities. In: Fortin, M., Glowinski, R. (eds.) Augmented Lagrangian Methods: Applications to the Numerical Solution of Boundary-Value Problems. Studies in Mathematics and Its Applications, vol. 15, pp. 299–331. North-Holland (1983)

Geman, S., Geman, D.: Stochastic relaxation, Gibbs distributions, and the Bayesian restoration of images. IEEE Trans. Pattern Anal. Mach. Intell. **6**, 721–741 (1984). https://doi.org/10.1109/tpami.1984.4767596

Hamedani, E.Y., Aybat, N.S.: A primal-dual algorithm for general convex-concave saddle point problems (2018). arXiv:1803.01401

He, B., Yuan, X.: Convergence analysis of primal-dual algorithms for a saddle-point problem: from contraction perspective. SIAM J. Imaging Sci. **5**, 119–149 (2012). https://doi.org/10.1137/100814494

Hiriart-Urruty, J.B., Lemaréchal, C.: Fundamentals of Convex Analysis. Grundlehren Text Editions. Springer (2004)

Hohage, T., Homann, C.: A Generalization of the Chambolle-Pock Algorithm to Banach Spaces with Applications to Inverse Problems (2014). arXiv:1412.0126

Hunt, A.: Weighing without touching: applying electrical capacitance tomography to mass flowrate measurement in multiphase flows. Meas. Control **47**, 19–25 (2014). https://doi.org/10.1177/0020294013517445

Jauhiainen, J., Kuusela, P., Seppänen, A., Valkonen, T.: Relaxed Gauss–Newton methods with applications to electrical impedance tomography. SIAM J. Imaging Sci. **13**, 1415–1445 (2020). https://doi.org/10.1137/20m1321711. arXiv:2002.08044

Kingsley, P.: Introduction to diffusion tensor imaging mathematics: Parts I–III. Concepts Magn. Reson. Part A **28**, 101–179 (2006). https://doi.org/10.1002/cmr.a.20048

Kuchment, P., Kunyansky, L.: Mathematics of photoacoustic and thermoacoustic tomography. In: Scherzer, O. (ed.) Handbook of Mathematical Methods in Imaging, pp. 817–865. Springer, New York (2011). https://doi.org/10.1007/978-0-387-92920-0_19

Lions, P., Mercier, B.: Splitting algorithms for the sum of two nonlinear operators. SIAM J. Numer. Anal. **16**, 964–979 (1979). https://doi.org/10.1137/0716071

Lipponen, A., Seppänen, A., Kaipio, J.P.: Nonstationary approximation error approach to imaging of three-dimensional pipe flow: experimental evaluation. Meas. Sci. Technol. **22**, 104013 (2011). https://doi.org/10.1088/0957-0233/22/10/104013

Loris, I., Verhoeven, C.: On a generalization of the iterative soft thresholding algorithm for the case of non-separable penalty. Inverse Probl. **27**, 125007 (2011). https://doi.org/10.1088/0266-5611/27/12/125007

Lustig, M., Donoho, D., Pauly, J.M.: Sparse MRI: the application of compressed sensing for rapid MR imaging. Magn. Reson. Med. **58**, 1182–1195 (2007). https://doi.org/10.1002/mrm.21391

Malitsky, Y., Tam, M.K.: A forward-backward splitting method for monotone inclusions without cocoercivity (2018). arXiv:1808.04162

Mazurenko, S., Jauhiainen, J., Valkonen, T.: Primal-dual block-proximal splitting for a class of non-convex problems, Electron. Trans. Numer. Anal. **52**, 509–552 (2020). https://doi.org/10.1553/etna_vol52s509. arXiv:1911.06284

Minty, G.J.: On the maximal domain of a "monotone" function. Mich. Math. J. **8**, 135–137 (1961)

Nemirovski, A.S., Yudin, D.: Problem Complexity and Method Efficiency in Optimization (Translated from Russian). Wiley Interscience Series in Discrete Mathematics. Wiley (1983)

Nishimura, D.: Principles of Magnetic Resonance Imaging. Stanford University (1996)

Ollinger, J.M., Fessler, J.A.: Positron-emission tomography. IEEE Signal Process. Mag. **14**, 43–55 (1997). https://doi.org/10.1109/79.560323

Opial, Z.: Weak convergence of the sequence of successive approximations for nonexpansive mappings. Bull. Am. Math. Soc. **73**, 591–597 (1967). https://doi.org/10.1090/s0002-9904-1967-11761-0

Pock, T., Chambolle, A.: Diagonal preconditioning for first order primal-dual algorithms in convex optimization. In: 2011 IEEE International Conference on Computer Vision (ICCV), pp. 1762–1769. IEEE (2011). https://doi.org/10.1109/iccv.2011.6126441

Pock, T., Cremers, D., Bischof, H., Chambolle, A.: An algorithm for minimizing the Mumford-Shah functional. In: 12th IEEE Conference on Computer Vision, pp. 1133–1140. IEEE (2009). https://doi.org/10.1109/iccv.2009.5459348

Rockafellar, R.T.: Convex Analysis. Princeton University Press (1972)

Rockafellar, R.T.: Monotone operators and the proximal point algorithm. SIAM J. Optim. **14**, 877–898 (1976). https://doi.org/10.1137/0314056

Rudin, L., Osher, S., Fatemi, E.: Nonlinear total variation based noise removal algorithms. Physica D **60**, 259–268 (1992)

Shen, J., Chan, T.F.: Mathematical models for local nontexture inpaintings. SIAM J. Appl. Math. **62**, 1019–1043 (2002). https://doi.org/10.1137/s0036139900368844

Trucu, D., Ingham, D.B., Lesnic, D.: An inverse coefficient identification problem for the bio-heat equation. Inverse Probl. Sci. Eng. **17**, 65–83 (2009). https://doi.org/10.1080/17415970802082880

Uhlmann, G.: Electrical impedance tomography and Calderón's problem. Inverse Probl. **25**, 123011 (2009). https://doi.org/10.1088/0266-5611/25/12/123011

Valkonen, T.: A primal-dual hybrid gradient method for non-linear operators with applications to MRI. Inverse Probl. **30**, 055012 (2014). https://doi.org/10.1088/0266-5611/30/5/055012. arXiv:1309.5032

Valkonen, T.: Block-proximal methods with spatially adapted acceleration. Electron. Trans. Numer. Anal. **51**, 15–49 (2019). https://doi.org/10.1553/etna_vol51s15. arXiv:1609.07373

Valkonen, T.: Inertial, corrected, primal-dual proximal splitting. SIAM J. Optim. **30**, 1391–1420 (2020). https://doi.org/10.1137/18m1182851. arXiv:1804.08736

Valkonen, T.: Testing and non-linear preconditioning of the proximal point method. Appl. Math. Optim. **82** (2020). https://doi.org/10.1007/s00245-018-9541-6. arXiv:1703.05705

Valkonen, T., Pock, T.: Acceleration of the PDHGM on partially strongly convex functions. J. Math. Imaging Vis. **59**, 394–414 (2017) https://doi.org/10.1007/s10851-016-0692-2. arXiv:1511.06566

Vogel, C.R., Oman, M.E.: Fast, robust total variation-based reconstruction of noisy, blurred images. IEEE Trans. Image Process. **7**, 813–824 (1998). https://doi.org/10.1109/83.679423

Vũ, B.C.: A splitting algorithm for dual monotone inclusions involving cocoercive operators. Adv. Comput. Math. **38**, 667–681 (2013). https://doi.org/10.1007/s10444-011-9254-8

Zhang, X., Burger, M., Osher, S.: A unified primal-dual algorithm framework based on Bregman iteration. J. Sci. Comput. **46**, 20–46 (2011). https://doi.org/10.1007/s10915-010-9408-8